Lecture Notes in Computer Science 10405

Commenced Publication in 1973
Founding and Former Series Editors:
Gerhard Goos, Juris Hartmanis, and Jan van Leeuwen

More information about this series at http://www.springer.com/series/7407

Lecture Notes in Computer Science 10405

Commenced Publication in 1973
Founding and Former Series Editors:
Gerhard Goos, Juris Hartmanis, and Jan van Leeuwen

More information about this series at http://www.springer.com/series/7407

Osvaldo Gervasi · Beniamino Murgante
Sanjay Misra · Giuseppe Borruso
Carmelo M. Torre · Ana Maria A.C. Rocha
David Taniar · Bernady O. Apduhan
Elena Stankova · Alfredo Cuzzocrea (Eds.)

Computational Science and Its Applications – ICCSA 2017

17th International Conference
Trieste, Italy, July 3–6, 2017
Proceedings, Part II

 Springer

Editors
Osvaldo Gervasi ⓘ
University of Perugia
Perugia
Italy

Beniamino Murgante ⓘ
University of Basilicata
Potenza
Italy

Sanjay Misra ⓘ
Covenant University
Ota
Nigeria

Giuseppe Borruso ⓘ
University of Trieste
Trieste
Italy

Carmelo M. Torre ⓘ
Polytechnic University of Bari
Bari
Italy

Ana Maria A.C. Rocha ⓘ
University of Minho
Braga
Portugal

David Taniar ⓘ
Monash University
Clayton, VIC
Australia

Bernady O. Apduhan
Kyushu Sangyo University
Fukuoka
Japan

Elena Stankova ⓘ
Saint Petersburg State University
Saint Petersburg
Russia

Alfredo Cuzzocrea ⓘ
University of Trieste
Trieste
Italy

ISSN 0302-9743   ISSN 1611-3349 (electronic)
Lecture Notes in Computer Science
ISBN 978-3-319-62394-8   ISBN 978-3-319-62395-5 (eBook)
DOI 10.1007/978-3-319-62395-5

Library of Congress Control Number: 2017945283

LNCS Sublibrary: SL1 – Theoretical Computer Science and General Issues

Printed on acid-free paper

This Springer imprint is published by Springer Nature
The registered company is Springer International Publishing AG
The registered company address is: Gewerbestrasse 11, 6330 Cham, Switzerland

Preface

These multiple volumes (LNCS volumes 10404, 10405, 10406, 10407, 10408, and 10409) consist of the peer-reviewed papers from the 2017 International Conference on Computational Science and Its Applications (ICCSA 2017) held in Trieste, Italy, during July 3–6, 2017.

ICCSA 2017 was a successful event in the ICCSA conference series, previously held in Beijing, China (2016), Banff, Canada (2015), Guimarães, Portugal (2014), Ho Chi Minh City, Vietnam (2013), Salvador, Brazil (2012), Santander, Spain (2011), Fukuoka, Japan (2010), Suwon, South Korea (2009), Perugia, Italy (2008), Kuala Lumpur, Malaysia (2007), Glasgow, UK (2006), Singapore (2005), Assisi, Italy (2004), Montreal, Canada (2003), (as ICCS) Amsterdam, The Netherlands (2002), and San Francisco, USA (2001).

Computational science is a main pillar of most present research as well as industrial and commercial activities and plays a unique role in exploiting ICT innovative technologies. The ICCSA conference series have been providing a venue to researchers and industry practitioners to discuss new ideas, to share complex problems and their solutions, and to shape new trends in computational science.

Apart from the general tracks, ICCSA 2017 also include 43 international workshops, in various areas of computational sciences, ranging from computational science technologies to specific areas of computational sciences, such as computer graphics and virtual reality. Furthermore, this year ICCSA 2017 hosted the XIV International Workshop on Quantum Reactive Scattering. The program also features three keynote speeches and four tutorials.

The success of the ICCSA conference series in general, and ICCSA 2017 in particular, is due to the support of many people: authors, presenters, participants, keynote speakers, session chairs, Organizing Committee members, student volunteers, Program Committee members, international Advisory Committee members, international liaison chairs, and various people in other roles. We would like to thank them all.

We would also like to thank Springer for their continuous support in publishing the ICCSA conference proceedings.

July 2017
Giuseppe Borruso
Osvaldo Gervasi
Bernady O. Apduhan

Welcome to Trieste

We were honored and happy to have organized this extraordinary edition of the conference, with so many interesting contributions and participants coming from more than 46 countries around the world!

Trieste is a medium-size Italian city lying on the north-eastern border between Italy and Slovenia. It has a population of nearly 200,000 inhabitants and faces the Adriatic Sea, surrounded by the Karst plateau.

It is quite an atypical Italian city, with its history being very much influenced by belonging for several centuries to the Austro-Hungarian empire and having been through several foreign occupations in history: by French, Venetians, and the Allied Forces after the Second World War. Such events left several footprints on the structure of the city, on its buildings, as well as on culture and society!

During its history, Trieste hosted people coming from different countries and regions, making it a cosmopolitan and open city. This was also helped by the presence of a commercial port that made it an important trade center from the 18th century on. Trieste is known today as a 'City of Science' or, more proudly, presenting itself as the 'City of Knowledge', thanks to the presence of several universities and research centers, all of them working at an international level, as well as of cultural institutions and traditions. The city has a high presence of researchers, more than 35 per 1,000 employed people, much higher than the European average of 6 employed researchers per 1,000 people.

The University of Trieste, the origin of such a system of scientific institutions, dates back to 1924, although its roots go back to the end of the 19th century under the Austro-Hungarian Empire. The university today employs nearly 1,500 teaching, research, technical, and administrative staff with a population of more than 16,000 students.

The university currently has 10 departments: Economics, Business, Mathematical, and Statistical Sciences; Engineering and Architecture; Humanities; Legal, Language, Interpreting, and Translation Studies; Mathematics and Geosciences; Medicine, Surgery, and Health Sciences; Life Sciences; Pharmaceutical and Chemical Sciences; Physics; Political and Social Sciences.

We trust the participants enjoyed the cultural and scientific offerings of Trieste and will keep a special memory of the event.

Giuseppe Borruso

Organization

ICCSA 2017 was organized by the University of Trieste (Italy), University of Perugia (Italy), Monash University (Australia), Kyushu Sangyo University (Japan), University of Basilicata (Italy), and University of Minho, (Portugal).

Honorary General Chairs

Antonio Laganà	University of Perugia, Italy
Norio Shiratori	Tohoku University, Japan
Kenneth C.J. Tan	Sardina Systems, Estonia

General Chairs

Giuseppe Borruso	University of Trieste, Italy
Osvaldo Gervasi	University of Perugia, Italy
Bernady O. Apduhan	Kyushu Sangyo University, Japan

Program Committee Chairs

Alfredo Cuzzocrea	University of Trieste, Italy
Beniamino Murgante	University of Basilicata, Italy
Ana Maria A.C. Rocha	University of Minho, Portugal
David Taniar	Monash University, Australia

International Advisory Committee

Jemal Abawajy	Deakin University, Australia
Dharma P. Agrawal	University of Cincinnati, USA
Marina L. Gavrilova	University of Calgary, Canada
Claudia Bauzer Medeiros	University of Campinas, Brazil
Manfred M. Fisher	Vienna University of Economics and Business, Austria
Yee Leung	Chinese University of Hong Kong, SAR China

International Liaison Chairs

Ana Carla P. Bitencourt	Universidade Federal do Reconcavo da Bahia, Brazil
Maria Irene Falcão	University of Minho, Portugal
Robert C.H. Hsu	Chung Hua University, Taiwan
Tai-Hoon Kim	Hannam University, Korea
Sanjay Misra	University of Minna, Nigeria
Takashi Naka	Kyushu Sangyo University, Japan

Rafael D.C. Santos   National Institute for Space Research, Brazil
Maribel Yasmina Santos   University of Minho, Portugal

Workshop and Session Organizing Chairs

Beniamino Murgante   University of Basilicata, Italy
Sanjay Misra   Covenant University, Nigeria
Jorge Gustavo Rocha   University of Minho, Portugal

Award Chair

Wenny Rahayu   La Trobe University, Australia

Publicity Committee Chair

Stefano Cozzini   Democritos Center, National Research Council, Italy
Elmer Dadios   De La Salle University, Philippines
Hong Quang Nguyen   International University (VNU-HCM), Vietnam
Daisuke Takahashi   Tsukuba University, Japan
Shangwang Wang   Beijing University of Posts and Telecommunications,
                                  China

Workshop Organizers

Agricultural and Environmental Big Data Analytics (AEDBA 2017)

Sandro Bimonte   IRSTEA, France
André Miralles   IRSTEA, France

Advances in Data Mining for Applications (AMDMA 2017)

Carlo Cattani   University of Tuscia, Italy
Majaz Moonis   University of Massachusettes Medical School, USA
Yeliz Karaca   IEEE, Computer Society Association

Advances Smart Mobility and Transportation (ASMAT 2017)

Mauro Mazzei   CNR, Italian National Research Council, Italy

Advances in Information Systems and Technologies for Emergency Preparedness and Risk Assessment and Mitigation (ASTER 2017)

Maurizio Pollino   ENEA, Italy
Marco Vona   University of Basilicata, Italy
Beniamino Murgante   University of Basilicata, Italy

Advances in Web-Based Learning (AWBL 2017)

Mustafa Murat Inceoglu   Ege University, Turkey
Birol Ciloglugil   Ege University, Turkey

Big Data Warehousing and Analytics (BIGGS 2017)

Maribel Yasmina Santos   University of Minho, Portugal
Monica Wachowicz   University of New Brunswick, Canada
Joao Moura Pires   NOVA de Lisboa University, Portugal
Rafael Santos   National Institute for Space Research, Brazil

Bio-inspired Computing and Applications (BIONCA 2017)

Nadia Nedjah   State University of Rio de Janeiro, Brazil
Luiza de Macedo Mourell   State University of Rio de Janeiro, Brazil

Computational and Applied Mathematics (CAM 2017)

M. Irene Falcao   University of Minho, Portugal
Fernando Miranda   University of Minho, Portugal

Computer-Aided Modeling, Simulation, and Analysis (CAMSA 2017)

Jie Shen   University of Michigan, USA and Jilin University, China
Hao Chenina   Shanghai University of Engineering Science, China
Chaochun Yuan   Jiangsu University, China

Computational and Applied Statistics (CAS 2017)

Ana Cristina Braga   University of Minho, Portugal

Computational Geometry and Security Applications (CGSA 2017)

Marina L. Gavrilova   University of Calgary, Canada

Central Italy 2016 Earthquake: Computational Tools and Data Analysis for Emergency Response, Community Support, and Reconstruction Planning (CIEQ 2017)

Alessandro Rasulo   Università degli Studi di Cassino e del Lazio
                            Meridionale, Italy
Davide Lavorato   Università degli Studi di Roma Tre, Italy

Computational Methods for Business Analytics (CMBA 2017)

Telmo Pinto   University of Minho, Portugal
Claudio Alves   University of Minho, Portugal

Chemistry and Materials Sciences and Technologies (CMST 2017)

Antonio Laganà   University of Perugia, Italy
Noelia Faginas Lago   University of Perugia, Italy

Computational Optimization and Applications (COA 2017)

Ana Maria Rocha   University of Minho, Portugal
Humberto Rocha   University of Coimbra, Portugal

Cities, Technologies, and Planning (CTP 2017)

Giuseppe Borruso   University of Trieste, Italy
Beniamino Murgante   University of Basilicata, Italy

Data-Driven Modelling for Sustainability Assessment (DAMOST 2017)

Antonino Marvuglia   Luxembourg Institute of Science and Technology, LIST,
                                 Luxembourg
Mikhail Kanevski   University of Lausanne, Switzerland
Beniamino Murgante   University of Basilicata, Italy
Janusz Starczewski   Częstochowa University of Technology, Poland

Databases and Computerized Information Retrieval Systems (DCIRS 2017)

Sultan Alamri   College of Computing and Informatics, SEU, Saudi
                                 Arabia
Adil Fahad   Albaha University, Saudi Arabia
Abdullah Alamri   Jeddah University, Saudi Arabia

Data Science for Intelligent Decision Support (DS4IDS 2016)

Filipe Portela   University of Minho, Portugal
Manuel Filipe Santos   University of Minho, Portugal

Deep Cities: Intelligence and Interoperability (DEEP_CITY 2017)

Maurizio Pollino   ENEA, Italian National Agency for New Technologies,
    Energy and Sustainable Economic Development, Italy
Grazia Fattoruso   ENEA, Italian National Agency for New Technologies,
    Energy and Sustainable Economic Development, Italy

Emotion Recognition (EMORE 2017)

Valentina Franzoni   University of Rome La Sapienza, Italy
Alfredo Milani   University of Perugia, Italy

Future Computing Systems, Technologies, and Applications (FISTA 2017)

Bernady O. Apduhan   Kyushu Sangyo University, Japan
Rafael Santos   National Institute for Space Research, Brazil

Geographical Analysis, Urban Modeling, Spatial Statistics (Geo-and-Mod 2017)

Giuseppe Borruso   University of Trieste, Italy
Beniamino Murgante   University of Basilicata, Italy
Hartmut Asche   University of Potsdam, Germany

Geomatics and Remote Sensing Techniques for Resource Monitoring and Control (GRS-RMC 2017)

Eufemia Tarantino   Polytechnic of Bari, Italy
Rosa Lasaponara   Italian Research Council, IMAA-CNR, Italy
Antonio Novelli   Polytechnic of Bari, Italy

Interactively Presenting High-Quality Graphics in Cooperation with Various Computing Tools (IPHQG 2017)

Masataka Kaneko   Toho University, Japan
Setsuo Takato   Toho University, Japan
Satoshi Yamashita   Kisarazu National College of Technology, Italy

Web-Based Collective Evolutionary Systems: Models, Measures, Applications (IWCES 2017)

Alfredo Milani   University of Perugia, Italy
Rajdeep Nyogi   Institute of Technology, Roorkee, India
Valentina Franzoni   University of Rome La Sapienza, Italy

Computational Mathematics, and Statistics for Data Management and Software Engineering (IWCMSDMSE 2017)

M. Filomena Teodoro   Lisbon University and Portuguese Naval Academy,
                              Portugal
Anacleto Correia   Portuguese Naval Academy, Portugal

Land Use Monitoring for Soil Consumption Reduction (LUMS 2017)

Carmelo M. Torre   Polytechnic of Bari, Italy
Beniamino Murgante   University of Basilicata, Italy
Alessandro Bonifazi   Polytechnic of Bari, Italy
Massimiliano Bencardino   University of Salerno, Italy

Mobile Communications (MC 2017)

Hyunseung Choo   Sungkyunkwan University, Korea

Mobile-Computing, Sensing, and Actuation - Fog Networking (MSA4FOG 2017)

Saad Qaisar   NUST School of Electrical Engineering and Computer
                              Science, Pakistan
Moonseong Kim   Korean Intellectual Property Office, South Korea

Physiological and Affective Computing: Methods and Applications (PACMA 2017)

Robertas Damasevicius   Kaunas University of Technology, Lithuania
Christian Napoli   University of Catania, Italy
Marcin Wozniak   Silesian University of Technology, Poland

Quantum Mechanics: Computational Strategies and Applications (QMCSA 2017)

Mirco Ragni   Universidad Federal de Bahia, Brazil
Ana Carla Peixoto   Universidade Estadual de Feira de Santana, Brazil
  Bitencourt
Vincenzo Aquilanti   University of Perugia, Italy

Advances in Remote Sensing for Cultural Heritage (RS 2017)

Rosa Lasaponara   IRMMA, CNR, Italy
Nicola Masini   IBAM, CNR, Italy Zhengzhou Base, International
                         Center on Space Technologies for Natural and
                         Cultural Heritage, China

Scientific Computing Infrastructure (SCI 2017)

Elena Stankova   Saint Petersburg State University, Russia
Alexander Bodganov   Saint Petersburg State University, Russia
Vladimir Korkhov   Saint Petersburg State University, Russia

Software Engineering Processes and Applications (SEPA 2017)

Sanjay Misra   Covenant University, Nigeria

Sustainability Performance Assessment: Models, Approaches and Applications Toward Interdisciplinarity and Integrated Solutions (SPA 2017)

Francesco Scorza   University of Basilicata, Italy
Valentin Grecu   Lucia Blaga University on Sibiu, Romania
Jolanta Dvarioniene   Kaunas University, Lithuania
Sabrina Lai   Cagliari University, Italy

Software Quality (SQ 2017)

Sanjay Misra   Covenant University, Nigeria

Advances in Spatio-Temporal Analytics (ST-Analytics 2017)

Rafael Santos   Brazilian Space Research Agency, Brazil
Karine Reis Ferreira   Brazilian Space Research Agency, Brazil
Maribel Yasmina Santos   University of Minho, Portugal
Joao Moura Pires   New University of Lisbon, Portugal

Tools and Techniques in Software Development Processes (TTSDP 2017)

Sanjay Misra   Covenant University, Nigeria

Challenges, Trends, and Innovations in VGI (VGI 2017)

Claudia Ceppi	University of Basilicata, Italy
Beniamino Murgante	University of Basilicata, Italy
Lucia Tilio	University of Basilicata, Italy
Francesco Mancini	University of Modena and Reggio Emilia, Italy
Rodrigo Tapia-McClung	Centro de Investigación en Geografía y Geomática "Ing Jorge L. Tamayo", Mexico
Jorge Gustavo Rocha	University of Minho, Portugal

Virtual Reality and Applications (VRA 2017)

Osvaldo Gervasi	University of Perugia, Italy

Industrial Computational Applications (WICA 2017)

Eric Medvet	University of Trieste, Italy
Gianfranco Fenu	University of Trieste, Italy
Riccardo Ferrari	Delft University of Technology, The Netherlands

XIV International Workshop on Quantum Reactive Scattering (QRS 2017)

Niyazi Bulut	Fırat University, Turkey
Noelia Faginas Lago	University of Perugia, Italy
Andrea Lombardi	University of Perugia, Italy
Federico Palazzetti	University of Perugia, Italy

Program Committee

Jemal Abawajy	Deakin University, Australia
Kenny Adamson	University of Ulster, UK
Filipe Alvelos	University of Minho, Portugal
Paula Amaral	Universidade Nova de Lisboa, Portugal
Hartmut Asche	University of Potsdam, Germany
Md. Abul Kalam Azad	University of Minho, Portugal
Michela Bertolotto	University College Dublin, Ireland
Sandro Bimonte	CEMAGREF, TSCF, France
Rod Blais	University of Calgary, Canada
Ivan Blečić	University of Sassari, Italy
Giuseppe Borruso	University of Trieste, Italy
Yves Caniou	Lyon University, France
José A. Cardoso e Cunha	Universidade Nova de Lisboa, Portugal
Rui Cardoso	University of Beira Interior, Portugal
Leocadio G. Casado	University of Almeria, Spain
Carlo Cattani	University of Salerno, Italy

Mete Celik	Erciyes University, Turkey
Alexander Chemeris	National Technical University of Ukraine KPI, Ukraine
Min Young Chung	Sungkyunkwan University, Korea
Gilberto Corso Pereira	Federal University of Bahia, Brazil
M. Fernanda Costa	University of Minho, Portugal
Gaspar Cunha	University of Minho, Portugal
Alfredo Cuzzocrea	ICAR-CNR and University of Calabria, Italy
Carla Dal Sasso Freitas	Universidade Federal do Rio Grande do Sul, Brazil
Pradesh Debba	The Council for Scientific and Industrial Research (CSIR), South Africa
Hendrik Decker	Instituto Tecnológico de Informática, Spain
Frank Devai	London South Bank University, UK
Rodolphe Devillers	Memorial University of Newfoundland, Canada
Prabu Dorairaj	NetApp, India/USA
M. Irene Falcao	University of Minho, Portugal
Cherry Liu Fang	U.S. DOE Ames Laboratory, USA
Edite M.G.P. Fernandes	University of Minho, Portugal
Jose-Jesús Fernandez	National Centre for Biotechnology, CSIS, Spain
María Antonia Forjaz	University of Minho, Portugal
María Celia Furtado Rocha	PRODEB-Pós Cultura/UFBA, Brazil
Akemi Galvez	University of Cantabria, Spain
Paulino Jose Garcia Nieto	University of Oviedo, Spain
Marina Gavrilova	University of Calgary, Canada
Jerome Gensel	LSR-IMAG, France
María Giaoutzi	National Technical University, Athens, Greece
Andrzej M. Goscinski	Deakin University, Australia
Alex Hagen-Zanker	University of Cambridge, UK
Malgorzata Hanzl	Technical University of Lodz, Poland
Shanmugasundaram Hariharan	B.S. Abdur Rahman University, India
Eligius M.T. Hendrix	University of Malaga/Wageningen University, Spain/The Netherlands
Tutut Herawan	Universitas Teknologi Yogyakarta, Indonesia
Hisamoto Hiyoshi	Gunma University, Japan
Fermin Huarte	University of Barcelona, Spain
Andrés Iglesias	University of Cantabria, Spain
Mustafa Inceoglu	EGE University, Turkey
Peter Jimack	University of Leeds, UK
Qun Jin	Waseda University, Japan
Farid Karimipour	Vienna University of Technology, Austria
Baris Kazar	Oracle Corp., USA
Maulana Adhinugraha Kiki	Telkom University, Indonesia
DongSeong Kim	University of Canterbury, New Zealand
Taihoon Kim	Hannam University, Korea
Ivana Kolingerova	University of West Bohemia, Czech Republic

Additional Reviewers

A. Alwan Al-Juboori Ali	School of Computer Science and Technology, China
Aceto Lidia	University of Pisa, Italy
Acharjee Shukla	Dibrugarh University, India
Afreixo Vera	University of Aveiro, Portugal
Agra Agostinho	University of Aveiro, Portugal
Aguilar Antonio	University of Barcelona, Spain
Aguilar José Alfonso	Universidad Autónoma de Sinaloa, Mexico
Aicardi Irene	Politecnico di Torino, Italy
Alberti Margarita	University of Barcelona, Spain
Alberto Rui	University of Lisbon, Portugal
Ali Salman	University of Magna Graecia, Italy
Alvanides Seraphim	University at Newcastle, UK
Alvelos Filipe	Universidade do Minho, Portugal
Amato Alba	Seconda Università degli Studi di Napoli, Italy
Amorim Paulo	Instituto de Matemática da UFRJ (IM-UFRJ), Brazil
Anderson Roger	University of California Santa Cruz, USA
Andrianov Serge	Saint Petersburg State University, Russia
Andrienko Gennady	Fraunhofer-Institut für Intelligente Analyse- und Informationssysteme, Germany
Apduhan Bernady	Kyushu Sangyo University, Japan
Aquilanti Vincenzo	University of Perugia, Italy
Asche Hartmut	Potsdam University, Germany
Azam Samiul	United International University, Bangladesh
Azevedo Ana	Athabasca University, USA
Bae Ihn-Han	Catholic University of Daegu, South Korea
Balacco Gabriella	Polytechnic of Bari, Italy
Balena Pasquale	Polytechnic of Bari, Italy
Barroca Filho Itamir	Universidade Federal do Rio Grande do Norte, Brazil
Behera Ranjan Kumar	Indian Institute of Technology Patna, India
Belpassi Leonardo	National Research Council, Italy
Bentayeb Fadila	Université Lyon, France
Bernardino Raquel	Universidade da Beira Interiore, Portugal
Bertolotto Michela	University Collegue Dublin, UK
Bhatta Bijaya	Utkal University, India
Bimonte Sandro	IRSTEA, France
Blecic Ivan	University of Cagliari, Italy
Bo Carles	ICIQ, Spain
Bogdanov Alexander	Saint Petersburg State University, Russia
Bollini Letizia	University of Milano-Bicocca, Italy
Bonifazi Alessandro	Polytechnic of Bari, Italy
Bonnet Claude-Laurent	Université de Bordeaux, France
Borgogno Mondino Enrico Corrado	University of Turin, Italy
Borruso Giuseppe	University of Trieste, Italy

Bostenaru Maria	Ion Mincu University of Architecture and Urbanism, Romania
Boussaid Omar	Université Lyon 2, France
Braga Ana Cristina	University of Minho, Portugal
Braga Nuno	University of Minho, Portugal
Brasil Luciana	Instituto Federal Sao Paolo, Brazil
Cabral Pedro	Universidade NOVA de Lisboa, Portugal
Cacao Isabel	University of Aveiro, Portugal
Caiaffa Emanuela	Enea, Italy
Campagna Michele	University of Cagliari, Italy
Caniato Renhe Marcelo	Universidade Federal de Juiz de Fora, Brazil
Canora Filomena	University of Basilicata, Italy
Caradonna Grazia	Polytechnic of Bari, Italy
Cardoso Rui	Beira Interior University, Portugal
Caroti Gabriella	University of Pisa, Italy
Carravilla Maria Antonia	Universidade do Porto, Portugal
Cattani Carlo	University of Salerno, Italy
Cefalo Raffaela	University of Trieste, Italy
Ceppi Claudia	Polytechnic of Bari, Italy
Cerreta Maria	University Federico II of Naples, Italy
Chanet Jean-Pierre	UR TSCF Irstea, France
Chaturvedi Krishna Kumar	University of Delhi, India
Chiancone Andrea	University of Perugia, Italy
Choo Hyunseung	Sungkyunkwan University, South Korea
Ciabo Serena	University of l'Aquila, Italy
Coletti Cecilia	University of Chieti, Italy
Correia Aldina	Porto Polytechnic, Portugal
Correia Anacleto	CINAV, Portugal
Correia Elisete	University of Trás-Os-Montes e Alto Douro, Portugal
Correia Florbela Maria da Cruz Domingues	Instituto Politécnico de Viana do Castelo, Portugal
Cosido Oscar	University of Cantabria, Spain
Costa e Silva Eliana	University of Minho, Portugal
Costa Graça	Instituto Politécnico de Setúbal, Portugal
Costantini Alessandro	INFN, Italy
Crispim José	University of Minho, Portugal
Cuzzocrea Alfredo	University of Trieste, Italy
Danese Maria	IBAM, CNR, Italy
Daneshpajouh Shervin	University of Western Ontario, USA
De Fazio Dario	IMIP-CNR, Italy
De Runz Cyril	University of Reims Champagne-Ardenne, France
Deffuant Guillaume	Institut national de recherche en sciences et technologies pour l'environnement et l'agriculture, France
Degtyarev Alexander	Saint Petersburg State University, Russia
Devai Frank	London South Bank University, UK
Di Leo Margherita	JRC, European Commission, Belgium

Dias Joana	University of Coimbra, Portugal
Dilo Arta	University of Twente, The Netherlands
Dvarioniene Jolanta	Kaunas University of Technology, Lithuania
El-Zawawy Mohamed A.	Cairo University, Egypt
Escalona Maria-Jose	University of Seville, Spain
Faginas-Lago, Noelia	University of Perugia, Italy
Falcinelli Stefano	University of Perugia, Italy
Falcão M. Irene	University of Minho, Portugal
Faria Susana	University of Minho, Portugal
Fattoruso Grazia	ENEA, Italy
Fenu Gianfranco	University of Trieste, Italy
Fernandes Edite	University of Minho, Portugal
Fernandes Florbela	Escola Superior de Tecnologia e Gest ão de Bragancca, Portugal
Fernandes Rosario	USP/ESALQ, Brazil
Ferrari Riccardo	Delft University of Technology, The Netherlands
Figueiredo Manuel Carlos	University of Minho, Portugal
Florence Le Ber	ENGEES, France
Flouvat Frederic	University of New Caledonia, France
Fontes Dalila	Universidade do Porto, Portugal
Franzoni Valentina	University of Perugia, Italy
Freitas Adelaide de Fátima Baptista Valente	University of Aveiro, Portugal
Fusco Giovanni	Università di Bari, Italy
Gabrani Goldie	Tecpro Syst. Ltd., India
Gaido Luciano	INFN, Italy
Gallo Crescenzio	University of Foggia, Italy
Garaba Shungu	University of Connecticut, USA
Garau Chiara	University of Cagliari, Italy
Garcia Ernesto	University of the Basque Country, Spain
Gargano Ricardo	Universidade Brasilia, Brazil
Gavrilova Marina	University of Calgary, Canada
Gensel Jerome	IMAG, France
Gervasi Osvaldo	University of Perugia, Italy
Gioia Andrea	Polytechnic University of Bari, Italy
Giovinazzi Sonia	University of Canterbury, New Zealand
Gizzi Fabrizio	National Research Council, Italy
Gomes dos Anjos Eudisley	Universidade Federal da Paraíba, Brazil
Gonzaga de Oliveira Sanderson Lincohn	Universidade Federal de Lavras, Brazil
Gonçalves Arminda Manuela	University of Minho, Braga, Portugal
Gorbachev Yuriy	Geolink Technologies, Russia
Grecu Valentin	University of Sibiu, Romania
Gupta Brij	Cancer Biology Research Center, USA
Hagen-Zanker Alex	University of Surrey, UK

Hamaguchi Naoki	Tokyo Kyoiku University, Japan
Hanazumi Simone	University of Sao Paulo, Brazil
Hanzl Malgorzata	University of Lodz, Poland
Hayashi Masaki	University of Calgary, Canada
Hendrix Eligius M.T.	Operations Research and Logistics Group, The Netherlands
Henriques Carla	Inst. Politécnico de Viseu, Portugal
Herawan Tutut	State Polytechnic of Malang, Indonesia
Hsu Hui-Huang	National Chiao Tung University, Taiwan
Ienco Dino	La Maison de la télédétection de Montpellier, France
Iglesias Andres	Universidad de Cantabria, Spain
Imran Rabeea	NUST Islamabad, Pakistan
Inoue Kentaro	National Technical University of Athens, Greece
Josselin Didier	Université d'Avignon et des Pays de Vaucluse, France
Kaneko Masataka	Kisarazu National College of Technology, Japan
Kang Myoung-Ah	Blaise Pascal University, France
Karampiperis Pythagoras	National Center of Scientific Research, Athens, Greece
Kavouras Marinos	University of Athens, Greece
Kolingerova Ivana	University of West Bohemia, Czech Republic
Korkhov Vladimir	Saint Petersburg State University, Russia
Kotzinos Dimitrios	University of Cergy Pontoise, France
Kulabukhova Nataliia	Saint Petersburg State University, Russia
Kumar Dileep	SR Engineering College, India
Kumar Lov	National Institute of Technology, Rourkela, India
Kumar Pawan	Institute for Advanced Study, Princeton, USA
Laganà Antonio	University of Perugia, Italy
Lai Sabrina	Università di Cagliari, Italy
Lanza Viviana	Lombardy Regional Institute for Research, Italy
Lasala Piermichele	Università di Foggia, Italy
Laurent Anne	Laboratoire d'Informatique, de Robotique et de Microélectronique de Montpellier, France
Lavorato Davide	University of Rome, Italy
Le Duc Tai	Sungkyunkwan University, South Korea
Legatiuk Dmitrii	Bauhaus University, Germany
Li Ming	University of Waterloo, Canada
Lima Ana	University of São Paulo (UNIFESP), Brazil
Liu Xin	École polytechnique fédérale de Lausanne, Switzerland
Lombardi Andrea	University of Perugia, Italy
Lopes Cristina	Instituto Superior de Contabilidade e Administracao do Porto, Portugal
Lopes Maria João	Instituto Universitário de Lisboa, Portugal
Lourenço Vanda Marisa	Universidade NOVA de Lisboa, Portugal
Machado Jose	University of Minho, Portugal
Maeda Yoichi	Tokai University, Japan
Majcen Nineta	Euchems, Belgium
Malonek Helmuth	Universidade de Aveiro, Portugal

Mancini Francesco	University of Modena and Reggio Emilia, Italy
Mandanici Emanuele	Università di Bologna, Italy
Manganelli Benedetto	Università degli studi della Basilicata, Italy
Manso Callejo Miguel Angel	Universidad Politécnica de Madrid, Spain
Margalef Tomas	Autonomous University of Barcelona, Spain
Marques Jorge	University of Coimbra, Portugal
Martins Bruno	Universidade de Lisboa, Portugal
Marvuglia Antonino	Public Research Centre Henri Tudor, Luxembourg
Mateos Cristian	Universidad Nacional del Centro, Argentina
Mauro Giovanni	University of Trieste, Italy
McGuire Michael	Towson University, USA
Medvet Eric	University of Trieste, Italy
Milani Alfredo	University of Perugia, Italy
Millham Richard	Durban University of Technoloy, South Africa
Minghini Marco	Polytechnic University of Milan, Italy
Minhas Umar	University of Waterloo, Ontario, Canada
Miralles André	La Maison de la télédétection de Montpellier, France
Miranda Fernando	Universidade do Minho, Portugal
Misra Sanjay	Covenant University, Nigeria
Modica Giuseppe	Università Mediterranea di Reggio Calabria, Italy
Molaei Qelichi Mohamad	University of Tehran, Iran
Monteiro Ana Margarida	University of Coimbra, Portugal
Morano Pierluigi	Polytechnic University of Bari, Italy
Moura Ana	Universidade de Aveiro, Portugal
Moura Pires João	Universidade NOVA de Lisboa, Portugal
Mourão Maria	ESTG-IPVC, Portugal
Murgante Beniamino	University of Basilicata, Italy
Nagy Csaba	University of Szeged, Hungary
Nakamura Yasuyuki	Nagoya University, Japan
Natário Isabel Cristina Maciel	University Nova de Lisboa, Portugal
Nemmaoui Abderrahim	Universidad de Almeria (UAL), Spain
Nguyen Tien Dzung	Sungkyunkwan University, South Korea
Niyogi Rajdeep	Indian Institute of Technology Roorkee, India
Novelli Antonio	University of Bari, Italy
Oliveira Irene	University of Trás-Os-Montes e Alto Douro, Portugal
Oliveira José A.	Universidade do Minho, Portugal
Ottomanelli Michele	University of Bari, Italy
Ouchi Shunji	Shimonoseki City University, Japan
Ozturk Savas	Scientific and Technological Research Council of Turkey, Turkey
P. Costa M. Fernanda	Universidade do Minho, Portugal
Painho Marco	NOVA Information Management School, Portugal
Panetta J.B.	Tecnologia Geofísica Petróleo Brasileiro SA, PETROBRAS, Brazil

Pantazis Dimos	Otenet, Greece
Papa Enrica	University of Amsterdam, The Netherlands
Pardede Eric	La Trobe University, Australia
Parente Claudio	Università degli Studi di Napoli Parthenope, Italy
Pathan Al-Sakib Khan	Islamic University of Technology, Bangladesh
Paul Prantosh K.	EIILM University, Jorethang, Sikkim, India
Pengő Edit	University of Szeged, Hungary
Pereira Ana	IPB, Portugal
Pereira José Luís	Universidade do Minho, Portugal
Peschechera Giuseppe	Università di Bologna, Italy
Pham Quoc Trung	HCMC University of Technology, Vietnam
Piemonte Andreaa	University of Pisa, Italy
Pimentel Carina	Universidade de Aveiro, Portugal
Pinet Francois	IRSTEA, France
Pinto Livio	Polytechnic University of Milan, Italy
Pinto Telmo	Universidade do Minho, Portugal
Pinet Francois	IRSTEA, France
Poli Giuliano	Université Pierre et Marie Curie, France
Pollino Maurizio	ENEA, Italy
Portela Carlos Filipe	Universidade do Minho, Portugal
Prata Paula	Universidade Federal de Sergipe, Brazil
Previl Carlo	University of Quebec in Abitibi-Témiscamingue (UQAT), Canada
Prezioso Giuseppina	Università degli Studi di Napoli Parthenope, Italy
Pusatli Tolga	Cankaya University, Turkey
Quan Tho	Ho Chi Minh, University of Technology, Vietnam
Ragni Mirco	Universidade Estadual de Feira de Santana, Brazil
Rahman Nazreena	Biotechnology Research Centre, Malaysia
Rahman Wasiur	Technical University Darmstadt, Germany
Rashid Sidra	National University of Sciences and Technology (NUST) Islamabad, Pakistan
Rasulo Alessandro	Università degli studi di Cassino e del Lazio Meridionale, Italy
Raza Syed Muhammad	Sungkyunkwan University, South Korea
Reis Ferreira Gomes Karine	Instituto Nacional de Pesquisas Espaciais, Brazil
Requejo Cristina	Universidade de Aveiro, Portugal
Rocha Ana Maria	University of Minho, Portugal
Rocha Humberto	University of Coimbra, Portugal
Rocha Jorge	University of Minho, Portugal
Rodriguez Daniel	University of Berkeley, USA
Saeki Koichi	Graduate University for Advanced Studies, Japan
Samela Caterina	University of Basilicata, Italy
Sannicandro Valentina	Polytechnic of Bari, Italy
Santiago Júnior Valdivino	Instituto Nacional de Pesquisas Espaciais, Brazil
Sarafian Haiduke	Pennsylvania State University, USA

Santos Daniel	Universidade Federal de Minas Gerais, Portugal
Santos Dorabella	Instituto de Telecomunicações, Portugal
Santos Eulália	SAPO, Portugal
Santos Maribel Yasmina	Universidade de Minho, Portugal
Santos Rafael	University of Toronto, Canada
Santucci Valentinoi	University of Perugia, Italy
Sautot Lucil	MR TETIS, AgroParisTech, France
Scaioni Marco	Polytechnic University of Milan, Italy
Schernthanner Harald	University of Potsdam, Germany
Schneider Michel	ISIMA, France
Schoier Gabriella	University of Trieste, Italy
Scorza Francesco	University of Basilicata, Italy
Sebillo Monica	University of Salerno, Italy
Severino Ricardo Jose	Universidade de Minho, Portugal
Shakhov Vladimir	Russian Academy of Sciences (Siberian Branch), Russia
Sheeren David	Toulouse Institute of Technology, France
Shen Jie	University of Michigan, USA
Silva Elsa	INESC Tec, Porto, Portugal
Sipos Gergely	MTA SZTAKI Computer and Automation Research Institute, Hungary
Skarga-Bandurova Inna	Technological Institute of East Ukrainian National University, Ukraine
Skoković Dražen	University of Valencia, Spain
Skouteris Dimitrios	SNS, Italy
Soares Inês Soares Maria Joana	Universidade de Minho, Portugal
Soares Michel	Federal University of Sergipe, Brazil
Sokolovski Dmitri	Ikerbasque, Basque Foundation for Science, Spain
Sousa Lisete	Research, FCUL, CEAUL, Lisboa, Portugal
Stener Mauro	Università di Trieste, Italy
Sumida Yasuaki	Center for Digestive and Liver Diseases, Nara City Hospital, Japan
Suri Bharti	Guru Gobind Singh Indraprastha University, India
Sørensen Claus Aage Grøn	University of Aarhus, Denmark
Tajani Francesco	University of Rome, Italy
Takato Setsuo	Kisarazu National College of Technology, Japan
Tanaka Kazuaki	Hasanuddin University, Indonesia
Taniar David	Monash University, Australia
Tapia-McClung Rodrigo	The Center for Research in Geography and Geomatics, Mexico
Tarantino Eufemia	Polytechnic of Bari, Italy
Teixeira Ana Paula	Federal University of Ceará, Fortaleza, Brazil
Teixeira Senhorinha	Universidade do Minho, Portugal
Teodoro M. Filomena	Instituto Politécnico de Setúbal, Portugal
Thill Jean-Claude	University at Buffalo, USA
Thorat Pankaj	Sungkyunkwan University, South Korea

Sponsoring Organizations

ICCSA 2017 would not have been possible without the tremendous support of many organizations and institutions, for which all organizers and participants of ICCSA 2017 express their sincere gratitude:

University of Trieste, Trieste, Italy
(http://www.units.it/)

University of Perugia, Italy
(http://www.unipg.it)

University of Basilicata, Italy
(http://www.unibas.it)

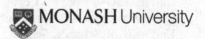

Monash University, Australia
(http://monash.edu)

Kyushu Sangyo University, Japan
(www.kyusan-u.ac.jp)

Universidade do Minho
Escola de Engenharia

Universidade do Minho, Portugal
(http://www.uminho.pt)

Contents – Part II

Workshop on Bio-inspired Computing and Applications (BIONCA 2017)

Workshop on Computational and Applied Mathematics (CAM 2017)

Workshop on Computational and Applied Statistics (CAS 2017)

Workshop on Computational Methods for Business Analytics (CMBA 2017)

Workshop on Agricultural and Environmental Big Data Analytics (AEDBA 2017)

Lack of Data: Is It Enough Estimating the Coffee Rust with Meteorological Time Series?

David Camilo Corrales[1,2(✉)], German Gutierrez[2], Jhonn Pablo Rodriguez[1(✉)], Agapito Ledezma[2], and Juan Carlos Corrales[1(✉)]

[1] Telematics Engineering Group, University of Cauca, Popayán, Colombia
{dcorrales,jhonnpablo,jcorral}@unicauca.edu.co
[2] Computer Science Department, Carlos III University of Madrid, Leganes, Spain
davidcamilo.corrales@alumnos.uc3m.es
{ggutierr,ledezma}@inf.uc3m.es
http://www.unicauca.edu.co,http://www.uc3m.es

Abstract. Rust is the most economically important coffee disease in the world. Coffee rust epidemics have affected a number of countries including: Colombia, Brazil and Central America. Researchers try to predict the Incidence Rate of Rust (IRR) through supervised learning models, nevertheless the available IRR measurements are few, then the data set does not represent a sample trustworthy of the population. In this paper we use Cubic Spline Interpolation algorithm to increase the measurements of Incidence Rate of Rust and subsequently we construct different subsets of meteorological time series: (i) *Daily meteorology*, (ii) *Meteorological variation*, and (iii) *Previous meteorology* using M5 Regression Tree, Support Vector Regression and Multi-Layer Perceptron. *Previous meteorology* with Multi-Layer Perceptron have shown better results in measures as Pearson Coefficient Correlation of 0.81 and Mean Absolute Error $= 7.41\%$.

Keywords: Coffee rust · Incidence rate of rust · Regression models · Time series · Interpolation

1 Introduction

Coffee rust is the main disease in coffee crops in the world [5]. The Rust disease has reduced considerably the coffee production in Colombia (by 31% on average during the epidemic years compared with 2007), Brazil (where climate conditions favour the disease, losses can reach about 35%, and sometimes even more than 50%) [5] and Central America (by 16% in 2013 compared with 2011–2012 and by 10% in 2013–2014 compared with 2012–2013). These reductions have had direct impacts on the livelihoods of thousands of smallholders and harvesters. For these populations, particularly in Central America, coffee is often the only source of income used to buy food and supplies for the cultivation of basic grains. As a result, the coffee rust epidemic has had indirect impacts on food security [2].

© Springer International Publishing AG 2017
O. Gervasi et al. (Eds.): ICCSA 2017, Part II, LNCS 10405, pp. 3–16, 2017.
DOI: 10.1007/978-3-319-62395-5_1

Several supervised learning models to predict some features of coffee rust have been proposed. Brazilian researchers monitoring the coffee rust incidence through Fuzzy Decision Trees, Bayesian Networks, Support Vector Machine at the Experimental Farm of the Procafé Foundation, in Varginha, Minas Gerais, Brazil, during 8 years (October, 1998–October, 2006), which includes 182 examples and 23 attributes that involves weather conditions and physical properties of crop.

In [17] the authors use SVM regression; rather than predicting the change in incidence, they consider actual incidence. They trained a SVM that provided a correlation of about 0.94 between predicted and actual incidences. However, when they try to devise an alarm system for predicting values above a given threshold, the number of false negatives turned to be too high.

Support Vector Regression was adapted interpreting intervals as classes in [16]. This approach presents a framework where the costs of false negatives are higher than that of false positives, and both are higher than the cost of warning predictions.

In [5] the authors compare two decision tree methods for the warning of the coffee rust disease: a fuzzy and a classic model. They built six datasets based on two different infection rate of rust levels: 5 and 10 % points. The fuzzy model, namely FUZZYDT, is based on the classical C4.5 decision tree, but incorporates all the interesting characteristics of fuzzy logic related to interpretability and handling of continuous attributes.

The approach represented in [20] constructs two Bayesian Networks (ignoring the weather variables) using the CaMML (Causal Minimum Message Length). They run CaMML with different structural priors, and tries to learn the best causal structure to take into account the data, i.e., to describe the information about relationships between variables.

Similarly, Colombian researchers use Support Vector Machine, Decision Trees, Artificial Neural Networks, Bayesian Networks, and Ensemble Methods for estimating the incidence rate of rust. The coffee rust records were collected trimonthly for 18 plots, closest to weather station at the Experimental Farm of the Supracafé Enterprise, in Cajibio, Cauca, Colombia, during the period of 2011–2013. The dataset includes 147 instances and 21 attributes that involves weather conditions, physical properties of crop, and crop management [6].

The work [9] builds three dataset based on different subset of attributes. The dataset were evaluated with three classifiers: Support Vector Regression, Backpropagation Neural Network, and M5 Regression Tree. These classifiers provide a correlation of about 0.47, 0.4549, and 0.4532 to the dataset with 13 attributes.

In [10] it is proposed theoretically an early warning system for coffee rust based on Error Correcting Output Codes with binary Support Vector Machines. The authors suggest to build a dataset with features as: plant density, shadow level, soil acidity, previous month rainfall, previous nighttime rainfall intensity, and relative humidity in the previous days.

The authors [15] propose a rule extraction approach to detect coffee rust from decision tree induction and expert knowledge using graph-based representation. The patterns found were modeled according to meteorological variables related to coffee rust dataset.

The works [7,8] develop an empirical multi-classifier based on Cascade Generalization method. The multi-classifier is composed by: backpropagation neural network to select among two experts classifiers: Regression tree (M5) to detect the incidence rate of rust greater than 7.18%; or support vector regression to detect the incidence rate of rust less than 7.18%.

However, the main drawback of the related works mentioned above is the low number of instances to try to predict a continuous value: Incidence Rate of Rust (values are 0%–100%); if the available examples are few, the data set does not represent a sample trustworthy of the population, then the classifiers will be not inaccurate [7]. This paper proposes three data set for estimating Incidence Rate of Rust through meteorological time series. We used cubic spline interpolation to increase the measurements of Incidence Rate of Rust. The remainder of this paper is organized as follows: Sect. 2 describes the coffee rust context and the regression models used; Sect. 3 the three data set proposed for estimating Incidence Rate of Rust through meteorological time series; Sect. 4 presents experimental results and Sect. 5 conclusions and future works.

2   Background

In this section we explained the coffee rust context, the dataset employed, cubic spline interpolation algorithm, concepts of times series, and the regression models used.

2.1   Coffee Rust

Coffee rust is caused by the fungus *Hemileia vastatrix*, a parasite that affects the leaves of the genus *Coffea*. Among the cultivated species, *Coffea arabica* is the most severely attacked. The disease causes defoliation that, when acute, can lead to the death of branches and heavy crop losses. The first symptoms are small yellowish spots that appear on the underside of leaves, where the fungus has penetrated through the stomata. These spots then grow, coalesce and produce uredospores with their distinctive orange colour. Chlorotic spots can be observed on the upper surface of the leaves. During the last stage of the disease, the spots become necrotic [2]. The progression of coffee rust depends on four factors that appears simultaneously [23]:

- **The host:** variety of coffee plants which are susceptible and resistant to rust. Varieties as *Típica*, *Borbón* and *Caturra* suffer severe rust attacks, while *Castillo* multiline variety is highly resistant to rust.
- **Pathogenic organism:** *Hemileia vastatrix* lifecycle begins with the germination of uredospores in 2–4 h under optimum conditions. Within 24–48 h, infection is completed. Once the infection is completed, the underside of the leaf is colonized and sporulation will occur through the stomata [19].

– **Weather conditions:** constant precipitations mainly in the afternoon and night with cloudy sky, high humidity in the plants and low temperatures are relevant factors for germination of rust. Disease spread and development is usually limited to the rainy season, while in dry periods the rust incidence is very low.
– **Agronomic practices:** is referred to properties of crop sowing (plant spacing, percentage of the shade, etc.), application of fungicides and fertilizers on coffee crops with aim to avoid several rust attacks.

In Colombia the rust progress is evaluated through the methodology developed by Centro Nacional de Investigaciones de Café (Cenicafé) [23], which is explained as follows:

Measurement Process of Rust in Colombian Coffee Crops: The Incidence Rate of Rust (IRR) is calculated for a plot with an area lower or equal to one hectare. The methodology is composed of three steps: (i) Farmer must be standing in the middle of the first furrow and he has to choose one coffee tree and pick out the branch with greater foliage for each level (high, medium, low); the leaves of the selected branches are counted as well as the infected ones with rust. (ii) Farmer must repeat step (i) for every tree in the plot until 60 trees are selected. The farmer must take in consideration that the same number of trees must be selected in every furrow (e.g. if plot has 30 furrows, the farmer selects two coffee trees for each furrow). (iii) Finalized step (i) and (ii), the leaves of the coffee trees selected (LCT) are added as well as the infected leaves with rust (ILR). Later it must be computed the Incidence Rate of Rust (IRR) using the following formula:

$$IRR = \frac{ILR}{LCT} \, 100 \qquad (1)$$

For the purposes of this research, IRR was used as the dependent variable, collected by Cenicafé in the next Colombian region:

Jazmín Village: The data used in this paper was obtained from Jazmín Village which is a coffee growing area sowing with *Caturra* variety in 45 farms approximately, monitored by Cenicafé and located in Santa Rosa de Cabal, Colombia (4°55'00"N, 75°38'0"W). The initial data set for this study records 43 measurements of Incidence Rate of Rust (average of 30 coffee trees per measurement) and 1024 samples for six daily meteorological attributes around 26/02/1986 and 15/12/1988. The input features available were: minimum, average and maximum temperature (*MinTem, AvgTem, MaxTem*), sun hours (*SunHours*), accumulated rain during day and night (*RainDay, RainNight*).

2.2   Time Series

A time series is just a sequence of time values which are usually regularly sampled in time: $(x_1, x_2, ..., x_{t-1}, x_t)$. There are univariate time series and multivariate time series. Univariate time series refers to a sequence of one scalar measure (e.g. Temperature at main cities taken hourly or twice a day). Additionally, multivariate time series show several different measures (or time series variables), and the aim of the studies related with them is *"to model and explain the interactions and co-movements among a group of time series variables"* [27]. Time series are used in regression models to try to give an answer for the relationship between its past and future values (forecast). Below we present the regression models used.

2.3 Regression Models

In this section we briefly describe the different regression models applied to forecast the IRR. In this work we have applied three techniques to achieve a regression model, one of them a linear model, M5 regression tree, and two non-linear models Support Vector Regression and Multilayer Perceptron, which are also universal approximators [14].

M5 Regression Tree, developed by Quinlan [21], and its result is a piecewise linear model based on trees. Some of the characteristic that make M5 more suitable than other algorithms based on regression trees (CART, classification and regression trees [3]; MARS, Multivariate Adaptive Regression Splines [11]) are the following: the computational requirements of M5 grows in a way that can tackle problems with high dimensionality, and the trees generated with M5 are smaller than other methods as CART and MARS.

Support Vector Regression (SVR): A Support Vector Machine tries to get a hyperplane as a boundary for decision so that the separation between the patterns at each side (one side for each class) of the hyperplane is maximum. A hyperplane is a subspace which number of dimensions is just one less that surrounds the hyperplane (e.g. a line is an hyperplane within a surface, e.g. a plane is a hyperplane in a 3-Dimensional space. In fact a hyperplane split its ambient space in just two parts. The approach that SVM follows to define the hyperplanes is based on statistical learning theory (see [26] for a better understanding of the topic), so that SVM is an approximate implementation of Structural Risk Minimisation (SRM) principle, instead of ANN that employs Empirical Risk Minimisation (ERM) principle ("SRM minimises an upper bound on the expected risk, as opposed to ERM that minimises the error on the training data" [12]). SVM can be also applied to regression tasks [25], in this case a distance measure is included into the loss function [24].

Multi-Layer Perceptron: Is a feed-forward artificial neural network [14], which is a universal approximator. The MLP is made up by nodes that are organised in layers: input, hidden and output layers. In this MLP there are no short-cuts, so each node within a layer is connected to every node in the following layer.

The topology of the MLP (i.e. number of hidden layers, number of nodes in each hidden layer, the activation function of each neuron, etc.) is fixed, and established by an expert or follows default parameters given by a software platform. Nonetheless, the memory and the nonlinear model itself is given by the weights of each connection, which are established throughout a learning process (back-propagation learning algorithm).

3   Data Pre-processing

To increase the measurements of the Incidence Rate of Rust, we use an interpolation algorithm from 26/02/1986 and 15/12/1988. Subsequently we construct three data set to estimate the IRR based on meteorological time series, these are presented:

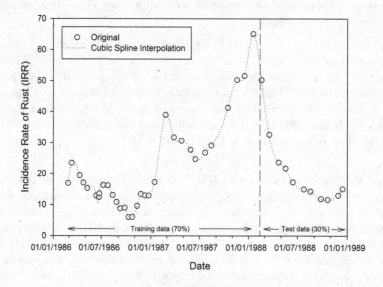

Fig. 1. Interpolation of IRR

Cubic Spline Interpolation: Having an account of the nature of problem, we decide to use cubic spline interpolation because the imputed data are inside the natural range of values, except in particular cases where quite extreme values are obtained [4].

The concept of spline is originated from conscripting technique of using a flexible strip known as spline to draw the smooth curve through a set of points. Interpolation task is just to approximate a function to a set (or sequence) of points with a certain precision which depends on the candidate function and on one or more parameters [18]. There are some issues about interpolation, but in this work we used spline cubic interpolation, which approximation candidate is a piecewise third-order polynomial function (so first and second derivatives of the candidate function are continuous at the nodes, i.e. the known points to be interpolated).

Figure 1 shows 43 original measurements of IRR represented by white dots. The remaining 981 black dots are interpolated data obtained by the Cubic Spline algorithm (executed with the function `splinefun` from the `stats` package of CRAN R [22]).

Once interpolated the IRR, we proposed three data set to estimate the IRR in Jazmín Village based on meteorological time series, we show the scheme below:

Daily Meteorology: Values of meteorological attributes and IRR dependent variable are organized by day:

$$IRR_t = \Big\{ MinTem_t, AvgTem_t, MaxTem_t, SunHours_t,$$
$$RainDay_t, RainNight_t \Big\} \tag{2}$$

In (2) *IRR*, *MinTem*, *AvgTem*, *MaxTem*, *SunHours*, *RainDay*, and *RainNight* were measured at t moment, where t correspond to one day of 26/02/1986 and 15/12/1988.

Meteorological Variation (MV): Values of meteorological attributes is a variation Δt for IRR in t:

$$IRR_t = \Big\{ MinTem_{\Delta t}, AvgTem_{\Delta t}, MaxTem_{\Delta t}, SunHours_{\Delta t},$$
$$RainDay_{\Delta t}, RainNight_{\Delta t} \Big\} \tag{3}$$

where Δt is the absolute value of the difference of a meteorological attribute between $t-7$ and $t-1$. For instance:

$$MinTem_{\Delta t} = |MinTem_{t-7} - MinTem_{t-1}| \tag{4}$$

and Δt is applied for all meteorological attributes.

Previous Meteorology (PM): represents the last $t-7$ measurements of the meteorological attributes for current IRR:

$$IRR_t = \Big\{ \big(MinTem_k\big)_{k=t-1}^{t-7}, \big(AvgTem_k\big)_{k=t-1}^{t-7}, \big(MaxTem_k\big)_{k=t-1}^{t-7},$$
$$\big(SunHours_k\big)_{k=t-1}^{t-7}, \big(RainDay_k\big)_{k=t-1}^{t-7}, \big(RainNight_k\big)_{k=t-1}^{t-7} \Big\} \tag{5}$$

where each meteorological attribute is separated in seven sub-attributes corresponding to last $t-7$ measurements. For example:

$$\big(AvgTem_k\big)_{k=t-1}^{t-7} = \Big\{ AvgTem_{t-1}, AvgTem_{t-2}, AvgTem_{t-3},$$
$$AvgTem_{t-4}, AvgTem_{t-5}, AvgTem_{t-6}, AvgTem_{t-7} \Big\} \tag{6}$$

We take the last $t - 7$ meteorological measurements to observe the evolution of disease considering that infection is completed at 24–48 h once the uredospores germinates.

4   Experimental Results

This section reports a number of experiments carried out to estimate the IRR in Jazmín Village presented above. The Software tool to carry out all the systems indicated in Subsect. 2.3 (M5, SVR and MLP) is the data mining software "workbench" called WEKA (Waikato Environment for Knowledge Analysis) [13], specifically version 3.6.12. In order to perform M5, SVR and MLP models and their algorithms for learning process, we have set the parameters suggested by Weka[1, 2, 3]. For instance, for MLP some of the default parameters are: 1 hidden layer with hn hidden nodes where $hn = (nInputs + nClasses)/2$, $nInputs = 6 \times 7$ (for the 6 meteorological measures for the previous 7 days) and $nClasses = 1$ as the MLP afford a regression problem.

We use Leave-p-out cross-validation with $p = 30\%$. Figures 2, 3 and 4 show the results for $p = 30\%$ of M5 Regression Tree, Support Vector Regression and Multi-Layer Perceptron for time series proposed: (i) *Daily meteorology*, (ii) *Meteorological variation*, and (iii) *Previous meteorology*.

To estimate the IRR through *Daily meteorology* and *Meteorological variation* (Figs. 2 and 3 respectively), the predicted IRR are vastly different from real IRR values. Whereas the regression models of *Previous meteorology* (Fig. 4) present a similar behavior of IRR measurements of the training set but remains different of real IRR values of the test set. Multi-Layer Perceptron of *Previous meteorology* is the best approximation although some IRR predicted are outliers with values higher than 100% and less than 0%.

Table 1 shows Pearson's Correlation Coefficient (PCC) and Mean Absolute Error (MAE) to test M5 Regression Tree, Support Vector Regression and Multi-Layer Perceptron for mentioned time series. Again, the best values of the evaluation measures for all regression models were obtained for *Previous meteorology (PM)*. Multi-Layer Perceptron has the highest value of correlation (PCC = 0.81) and the least difference among predicted and real IRR measurement (MAE = 7.41%), overcoming M5 and SVR.

As we could see in Fig. 1 that IRR measurements of training set present periodic fluctuations among 26/02/1986–09/05/1986, 23/06/1986–09/10/1986, 23/10/1986–15/06/1987 and 16/06/1987–11/02/1988, unlike the IRR measures of test set without trend among dates 12/02/1988–15/12/1988. Hence, the regression models cannot estimate IRR values that they have not learned.

To improve the results of regression models we need more information about properties of coffee plots to analyze the seasonality behavior of IRR. We expect

---

[1] http://weka.sourceforge.net/doc.dev/weka/classifiers/functions/
MultilayerPerceptron.html.

[2] http://weka.sourceforge.net/doc.dev/weka/classifiers/functions/SMO.html.

[3] http://weka.sourceforge.net/doc.dev/weka/classifiers/trees/m5/M5Base.html.

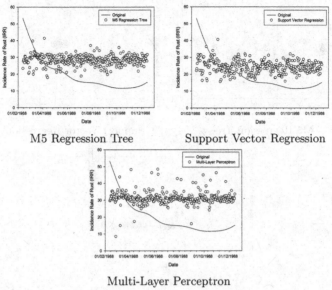

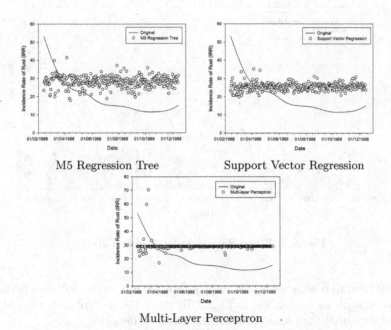

Fig. 2. Daily meteorology

Fig. 3. Meteorological variation

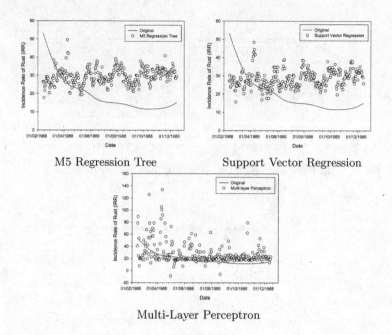

M5 Regression Tree   Support Vector Regression

Multi-Layer Perceptron

Fig. 4. Previous meteorology

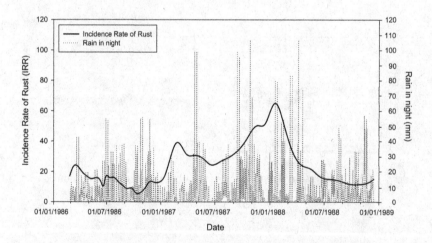

Fig. 5. IRR and Rain night for Jazmín Village

that IRR tend to peak after higher rainy seasons and then decline after performing rust control, as we can see in Fig. 5. To prove the control rust affirmation,
we need data about application of fungicides on coffee plots.

In Fig. 5 we can observe higher accumulated rain in the night during seasons 17/06/1986–07/12/1986 and 20/04/1987–2/11/1987, before the two highest

Table 1. PCC and Mean Absolute Error measures

Regression models	Daily meteorology		Meteorological variation		Previous meteorology	
	r	MAE	r	MAE	r	MAE
M5	0.31	11.43%	0.12	12.04%	0.47	10.53%
SVR	0.25	11.38%	0.12	11.93%	0.48	10.02%
MLP	0.28	12.36%	0.14	12.20%	0.81	7.41%

peaks of IRR in 01/03/1987 and 18/01/1988 as stated in [23]: constant precipitations in the night is a relevant factor for germination of rust (Subsect. 2.1).

5 Conclusions and Future Works

This paper describes three data set for estimating the Incidence Rate of Rust based on meteorological time series: (i) *Daily meteorology*, (ii) *Meteorological variation*, and (iii) *Previous meteorology*. Cubic spline algorithm was used to interpolate IRR measurements during 26/02/1986 and 15/12/1988. The results show that Multi-Layer Perceptron of *Previous meteorology* is the best approximation with PCC = 0.81 and MAE = 7.41%. This approach is helpful as first approximation to test in Central American Countries where there was a crisis of coffee rust during the years 2008–2013 [2].

The hypothesis of Experimental Results presented above is that IRR tend to peak for rainy seasons and decline when it is performed rust control, nevertheless for prove it, we need to know the reliability of IRR measurement, obtaining more information about how was the Incidence Rate of Rust collected: plots, farms. Therefore it is necessary to take into account:

– *Distance between meteorological station and coffee trees:* if a weather station is away from a coffee plot, the weather measurements are inaccurate, because coffee plot can have micro-climate influenced by: coffee plot orography and properties of crop sowing such as: plant spacing, shade on coffee trees, etc. Unfortunately the weather stations are very expensive to have one per coffee plot.
– *Information about application of fungicides on coffee plots:* if fungicides are applied on coffee plots before germination of rust, the weather conditions can not be relevant factors to increase the rust incidence. We consider necessary this information to build a correct regression model based on meteorological variables.
– *Consider a margin of error in IRR measurements:* the insufficient data due to the expensive collection process that requires large expenditures of money and time [7]. The farmers must select 3 branches for each 60 coffee trees (minimum) per plot [23]. Usually one plot have 10000, 5000, or 2500 coffee trees [1], given that the maximum number of IRR measures that we can obtain for one plot are: 0.6%, 1.2%, or 2.4% respectively. Besides a coffee farm has over one coffee plot.

We propose as future work, the use Multi-Layer Perceptron of *Previous mete-orology* to estimate the Incidence Rate of Rust, addressing the considerations mentioned below:

- Obtain meteorological information nearby coffee plots through forecast weather models such as meteoblue[4]. This weather models are based on NMM (Nonhydrostatic Meso-Scale Modelling) for large areas covering parts of or the entire continents, for which a complete forecast is calculated. This kind of models reduce the investment cost compared with weather stations per coffee plot.
- The properties of coffee plots are relevant information. We propose the development or the use of mobile applications for information recording of the coffee farms management as: application of fungicides, fertilizations, properties of crop sowing, etc. The main task is to encourage the farmers in the use of this kind of applications.
- Mobile technologies nowadays have a vast potential in multiple domains. To increase the measurements of IRR, we propose use mobile applications to compute the Incidence Rate of Rust based on computer vision approaches. The mobile application would analyze the pictures taken from the coffee plots, search the yellow spores on coffee trees leaves.

Acknowledgments. We thank Centro Nacional de Investigaciones de Café (Cenicafé) and Mr. Alvaro Gaitan Bustamante, PhD, for his knowledge. We also thank the Telematics Engineering Group (GIT) of the University of Cauca, and the Control Learning and Systems Optimization Group (CAOS) of the Carlos III University of Madrid, for the technical support. Finally, this work has been partially supported by Agro-Cloud project of the RICCLISA Program, and the Spanish Government (under projects TRA2011-29454-C03-03 and TRA2015-63708-R), and Colciencias for PhD scholarship granted to MsC. David Camilo Corrales.

References

1. Arcila, J., Farfan, F., Moreno, A., Salazar, L., Hincapie, E.: Sistemas de produccion de cafe en colombia. Cientific bot036, Cenicafe (2007)
2. Avelino, J., Cristancho, M., Georgiou, S., Imbach, P., Aguilar, L., Bornemann, G., Läderach, P., Anzueto, F., Hruska, A.J., Morales, C.: The coffee rust crises in colombia and central america (2008–2013): impacts, plausible causes and proposed solutions. Food Secur. **7**(2), 303–321 (2015)
3. Breiman, L., Friedman, J., Olshen, R., Stone, C.: Classification and Regression Trees. Wadsworth and Brooks, Monterey (1984)
4. Carrizosa, E., Olivares-Nadal, A.V., Ramírez-Cobo, P.: Time series interpolation via global optimization of moments fitting. Eur. J. Oper. Res. **230**(1), 97–112 (2013)
5. Cintra, M.E., Meira, C.A.A., Monard, M.C., Camargo, H.A., Rodrigues, L.H.A.: The use of fuzzy decision trees for coffee rust warning in Brazilian crops. In: 2011 11th International Conference on Intelligent Systems Design and Applications (ISDA), pp. 1347–1352, November 2011

---

[4] https://www.meteoblue.com.

6. Corrales, D.C., Corrales, J.C., Figueroa-Casas, A.: Towards detecting crop diseases and pest by supervised learning. Ingeniería y Universidad **19**, 207–228 (2015)
7. Corrales, D.C., Figueroa, A., Ledezma, A., Corrales, J.C.: An empirical multi-classifier for coffee rust detection in Colombian crops. In: Gervasi, O., Murgante, B., Misra, S., Gavrilova, M.L., Rocha, A.M.A.C., Torre, C., Taniar, D., Apduhan, B.O. (eds.) ICCSA 2015. LNCS, vol. 9155, pp. 60–74. Springer, Cham (2015). doi:10.1007/978-3-319-21404-7_5
8. Corrales, D.C., Casas, A.F., Ledezma, A., Corrales, J.C.: Two-level classifier ensembles for coffee rust estimation in Colombian crops. Int. J. Agric. Environ. Inf. Syst. (IJAEIS) **7**, 41–59 (2016)
9. Corrales, D.C., Ledezma, A., Peña, A.J., Hoyos, J., Figueroa, A., Corrales, J.C.: A new dataset for coffee rust detection in Colombian crops base on classifiers. Sistemas y Telemática **12**(29), 9–23 (2014)
10. Corrales, D.C., Peña, A.J., León, C., Figueroa, A., Corrales, J.C.: Early warning system for coffee rust disease based on error correcting output codes: a proposal. Revista IngenierÂas Universidad de MedellÂn **13**, 57–64 (2014)
11. Friedman, J.H.: Multivariate adaptive regression splines. Ann. Stat. **19**(1), 1–67 (1991)
12. Gunn, S.R.: Support vector machines for classification and regression. Technical report (1998)
13. Hall, M., Frank, E., Holmes, G., Pfahringer, B., Reutemann, P., Witten, I.H.: The weka data mining software: an update. SIGKDD Explor. Newsl. **11**(1), 10–18 (2009)
14. Haykin, S.: Neural Networks: A Comprehensive Foundation, 2nd edn. Prentice Hall, Upper Saddle River (1998)
15. Lasso, E., Thamada, T.T., Meira, C.A.A., Corrales, J.C.: Graph patterns as representation of rules extracted from decision trees for coffee rust detection. In: Garoufallou, E., Hartley, R.J., Gaitanou, P. (eds.) MTSR 2015. CCIS, vol. 544, pp. 405–414. Springer, Cham (2015). doi:10.1007/978-3-319-24129-6_35
16. Luaces, O., Rodrigues, L.H.A., Meira, C.A.A., Bahamonde, A.: Using nondeterministic learners to alert on coffee rust disease. Expert Syst. Appl. **38**(11), 14276–14283 (2011)
17. Luaces, O., Rodrigues, L.H.A., Alves Meira, C.A., Quevedo, J.R., Bahamonde, A.: Viability of an alarm predictor for coffee rust disease using interval regression. In: García-Pedrajas, N., Herrera, F., Fyfe, C., Benítez, J.M., Ali, M. (eds.) IEA/AIE 2010. LNCS, vol. 6097, pp. 337–346. Springer, Heidelberg (2010). doi:10.1007/978-3-642-13025-0_36
18. Mohanty, P.K., Reza, M., Kumar, P., Kumar, P.: Implementation of cubic spline interpolation on parallel skeleton using pipeline model on CPU-GPU cluster. In: 2016 IEEE 6th International Conference on Advanced Computing (IACC), pp. 747–751, February 2016
19. Nutman, F.J., Roberts, F.M., Clarke, R.T.: Studies on the biology of *Hemileia vastatrix* Berk and Br. Trans. Br. Mycol. Soc. **46**(1), 27–44 (1963)
20. Perez-Ariza, C., Nicholson, A., Flores, M.: Prediction of coffee rust disease using Bayesian networks, pp. 259–266. DECSAI University of Granada (2012)
21. Quinlan, R.J.: Learning with continuous classes. In: 5th Australian Joint Conference on Artificial Intelligence, pp. 343–348. World Scientific, Singapore (1992)
22. R Core Team: R: A Language and Environment for Statistical Computing. R Foundation for Statistical Computing, Vienna, Austria
23. Rivillas, C., Serna, C., Cristancho, M., Gaitan, A.: La Roya del Cafeto en Colombia. Impacto, manejo y costos de control. Cientific bot036, Cenicafe (2011)

24. Smola, A.J., Schölkopf, B.: A tutorial on support vector regression. Stat. Comput. **14**(3), 199–222 (2004)
25. Vapnik, V., Golowich, S.E., Smola, A.J.: Support vector method for function approximation, regression estimation and signal processing. In: Jordan, M.I., Petsche, T. (eds.) Advances in Neural Information Processing Systems, vol. 9, pp. 281–287. MIT Press, Cambridge (1997)
26. Vapnik, V.N.: Statistical Learning Theory. Wiley-Interscience, New York (1998)
27. Zivot, E., Wang, J.: Modeling Financial Time Series with S-PLUS®. IFIP. Springer, New York (2007). doi:10.1007/978-0-387-32348-0

Urban Sprawl, Labor Incomes and Real Estate Values

Massimiliano Bencardino[1(✉)] and Antonio Nesticò[2]

[1] Department of Political, Social and Communication Sciences,
University of Salerno, Via Giovanni Paolo II, 132, 84084 Fisciano, SA, Italy
mbencardino@unisa.it
[2] Department of Civil Engineering, University of Salerno, Fisciano, SA, Italy
anestico@unisa.it

Abstract. The analysis of the processes of urban growth and sprawl should be conducted taking into account the temporal evolution of a plurality of parameters: economic, demographic and socio-cultural. These factors are so related that the complex territorial system sometimes seems indecipherable. Thus, as the quantitative modeling suggests, the real phenomenon is simplified, identifying a limited number of exogenous variables and researching the effect that these variables generate on the simplersystem, object of the study. The goal is to find the functional relationships that govern the events.

In the present study, the temporal evolution of the built in a metropolitan area is investigated, in order to determine the effect that the age of construction and geographical location have on the market value of the property for residential use. Then, correlations are recorded with both the labor or capital income that the population perceives on the territory, and with the spatial distribution of the unemployment rate.

The construction of the reference dataset is conducted on the basis of official information provided in Italy by the National Institute of Statistics (ISTAT) and the Property Market Observatory (OMI) of Territorial Agency.

Finally, the implementation of Geographic Information Systems (GIS), at the level of Census and OMI units, allows to draw thematic maps, from which comparisonthe results of the study are derived (Both the authors conceived and designed the research and the paper. Massimiliano Bencardino collected data on the evolution of the built-up area and Antonio Nesticò collected data on real estate values, both wrote parts in every section.).

Keywords: Urban sprawl · Territorial planning · Appraisal · Real estate values · Spatial economic analysis · Geographic Information Systems

1 Introduction and Aim of the Paper

Since the postwar period, many Italian and European cities were characterized by urban sprawl and chaotic and fragmented expansion, so as to give rise to large urban agglomerations [1, 2]. In Italy, these phenomena have affected all the largest cities, such as Milan, Rome and Naples. The expansion started from the old historic center

© Springer International Publishing AG 2017
O. Gervasi et al. (Eds.): ICCSA 2017, Part II, LNCS 10405, pp. 17–30, 2017.
DOI: 10.1007/978-3-319-62395-5_2

and has progressively invested suburbs, semi-rural and rural areas, leading to a mixture of residential, commercial and productive functions [3–6].

The evolution of the territory anyway tends to respect economic laws, albeit sometimes conditioned by unclear speculative logics. They respond to the pursuit of differential rents in land use [7–9]. Thus, by invoking a principle of David Ricardo, we tend to cultivate at first the most fertile lands and gradually the less fertile, favoring a differential rent of the first than the others. More, as proposed by agricultural economist Johann Heinrich von Thünen, we proceed in order to create positional differential rents, because the shortest distance from the market of production areas produces lower costs for moving goods, and thus a higher income against the soils further away. Similarly, in Urban Economics, the shortest distance from the business center generates extra income and, consequently, higher merchant appreciation compared to what happens for properties located in peripheral areas [10].

The present study intends to conduct diachronic analysis of the built in a metropolitan area, which is that of Naples and its surrounding municipalities. In the same area, properly circumscribed, the values of urban property for residential purposes are recognized. These values are examined both with reference to their specific spatial distribution and with regard to the age of construction which characterizes each micro-zone in which the entire area is divided. This aims to establish the existence of functional relationships between the identified parameters.

Given that the urban growth is strongly related to income distribution in the territory, or – diametrically – to its unemployment rate, the urban sprawl and the spatial sequence of real estate values are read in the light of income and unemployment maps. These maps are built by developing spatially ISTAT dataset with GIS.

2   Methodology of Analysis and Data Processing

With regard to the methodology of study and numerical elaborations, the diachronic analysis on the built up areas are developed on the basis of the National Information Institute of Statistics (ISTAT) at the Census units scale. However, this scale is redefined both because of errors that may occur in the geo-coding of ISTAT data [11] and because these should be compared with the corresponding real estate values taken by the Observatory of the Real Estate Market (OMI) of Territorial Agency, which adopts a different reference scale.

So, the OMI micro-zone is assumed as spatial units of analysis, expressing uniform levels of the local real estate market and thus of urban, socio-economic and environmentalcharacteristics, and of service facilities. Spatial processing of large datasets allow the conversion from ISTAT Census scale to that of OMI microzones.

The following Table 1 summarizes the elaborations developed on the datasets.

The investigation area is divided according to the age of edification, distinguishing between the buildings before the 1919, those realized in the periods 1919–1945, 1946–1960, 1961–1970, 1971–1980, 1981–1990, 1991–2000, 2001–2005, and finally those of a later period to 2005.

Table 1. Extracted from dataset of the study area.

OMI zones	Area (km²)	Unemploym. rate		Before 1919	Age of construction of the buildings					After 2005	Values
		2001	2011		1919–1945	1946–1960	1961–1970	...	2001–2005		2013
ACERRA - D1	42.57	11.9%	11.4%	30	39	34	62	...	44	5	710
ACERRA - B1	1.12	17.8%	12.7%	186	321	270	316	...	36	18	980
ACERRA - C1	11.21	14.5%	11.3%	21	81	171	438	...	395	196	860
AFRAGOLA - D1	1.61	17.9%	11.9%	0	3	26	65	...	52	19	700
AFRAGOLA - B2	0.66	27.7%	13.4%	953	151	60	65	...	63	50	800
AFRAGOLA - C2	3.57	28.0%	12.5%	30	46	221	380	...	78	113	620
AFRAGOLA - B1	0.14	24.5%	9.2%	112	30	15	11	...	10	8	980
AFRAGOLA - C1	2.70	17.0%	10.8%	145	218	860	760	...	87	74	830
AFRAGOLA - D3	2.55	16.4%	9.9%	0	2	5	10	...	3	2	595
AFRAGOLA - D2	6.75	19.2%	10.4%	0	0	0	6	...	19	1	595
ARZANO - D1	0.80	36.4%	11.9%	1	0	0	3	...	0	1	1,300
ARZANO - B2	0.53	17.4%	13.5%	552	98	124	200	...	0	0	1,425
ARZANO - B1	0.85	15.9%	11.9%	63	25	115	165	...	3	1	1,520
ARZANO - C1	2.14	15.8%	13.0%	1	15	32	129	...	0	1	1,110
ARZANO - C1	0.35	18.0%	10.4%	0	1	0	12	...	0	0	1,110
BACOLI - D1	2.20	12.1%	10.1%	11	8	19	88	...	1	0	875
BACOLI - B2	2.99	13.2%	9.3%	36	201	351	336	...	14	0	1,010
BACOLI - C2	0.55	14.9%	10.4%	8	17	50	46	...	2	0	960
BACOLI - C2	5.17	15.8%	10.6%	108	119	187	160	...	36	16	960
BACOLI - B3	0.03	17.0%	11.1%	4	2	3	7	...	0	0	1,110
BACOLI - B3	0.31	11.7%	12.4%	13	9	14	68	...	3	0	1,110
BACOLI - B3	0.24	12.7%	11.5%	8	12	19	28	...	2	4	1,110
BACOLI - C3	0.11	10.7%	11.3%	2	15	12	4	...	0	0	875
BACOLI - C4	1.40	15.5%	11.3%	33	18	55	70	...	1	0	725
BACOLI - B1	0.11	11.1%	13.3%	4	9	8	6	...	2	0	1,295
BACOLI - C1	0.36	9.6%	8.2%	3	4	11	34	...	0	0	1,160
CAIVANO - D1	0.26	25.5%	14.4%	1	1	0	4	...	0	0	535
CAIVANO - D1	1.48	22.3%	15.7%	2	3	1	26	...	10	1	535
CAIVANO - B2	0.18	33.4%	14.9%	30	12	19	38	...	2	0	770
CAIVANO - B2	0.14	13.4%	18.8%	16	21	222	71	...	0	0	770
CAIVANO - C2	0.28	35.6%	10.5%	1	2	6	14	...	1	0	680
CAIVANO - C3	1.79	26.7%	15.4%	51	113	7	128	...	4	0	680
CAIVANO - B1	1.43	19.4%	16.1%	232	148	654	580	...	11	3	1,010
CAIVANO - C1	2.54	17.1%	13.8%	18	27	36	158	...	91	20	710

(*continued*)

Table 1. (*continued*)

OMI zones	Area (km²)	Unemploym. rate		Before 1919	Age of construction of the buildings					After 2005	Values
		2001	2011		1919–1945	1946–1960	1961–1970	...	2001–2005		2013
CAIVANO - D3	15.79	23.2%	13.3%	3	6	3	11	...	18	1	420
CAIVANO - D2	3.20	18.6%	17.3%	31	9	10	10	...	4	0	480
CALVIZZANO - D1	2.11	20.0%	14.6%	5	1	3	1	...	6	2	670
CALVIZZANO - C2	0.97	17.2%	12.4%	33	4	16	60	...	4	0	620
...	...	...	...	...	...	...	...	...	...	...	...

For each age and for the i-th micro-zone, the third quantile Q(3/4) of the distribution is represented (1):

$$Q\left(\frac{3}{4}\right) = \left|\frac{Built_{\Delta t}}{Built_{Tot}}\right|_i \tag{1}$$

It is the distribution of the ratio between the number of buildings constructed in the micro area i-th in a given period Δt and the total number of buildings realized there.

3   Urban Growth on a Metropolitan Area

The study area is the «Napoli *de facto*», which includes a less large area of the whole metropolitan cities but more homogeneous regarding the territorial functions [12, 13]. It goes from Monte di Procida to Giugliano and from Acerra up to Herculaneum (cfr. Fig. 2 below), forming an urban area which differs from the excluded areas of Nola, Vesuvius, Castellammare di Stabia and Sorrento peninsula. These areas gravitate around other centroid, i.e. Nola, Torre del Greco or Castellammare.

The age of construction of the built for each of the 313 micro-zones is reflected in the nine synthesis thematic maps (Fig. 1).

The urban growth process that, from the historic center of Naples progressively engages the outside areas, is immediately evident. In particular, before 1919, the city is concentrated in the strip that goes from Herculaneum to Chiaia Posillipo, to a portion of Fuorigrotta and up to Poggioreale. Settlements are found in Frattamaggiore, Afragola, Pomigliano d'Arco, Arzano, Marano, Pozzuoli and Monte di Procida.

In the following period, until 1945, the buildings extend around Herculaneum, Pomigliano d'Arco, Caivano and Monte di Procida. A strong development was also recorded in Bagnoli and Fuorigrotta. Here, during the Fascism, important urban interventions redefine the urban setting, with the realization of the Avenue of Augustus and theMostra d'Oltremare (headquarters of Neapolitan exhibition company).

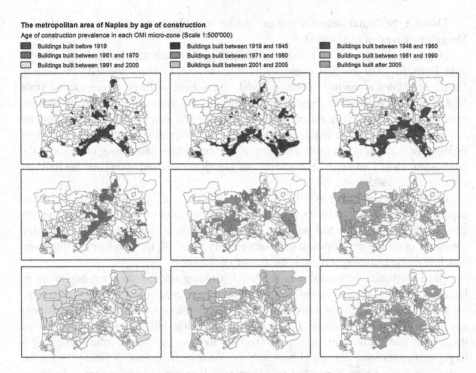

Fig. 1. The metropolitan area of Naples by age of construction.

The Fig. 2 shows the evolution of the built in the metropolitan city of Naples.

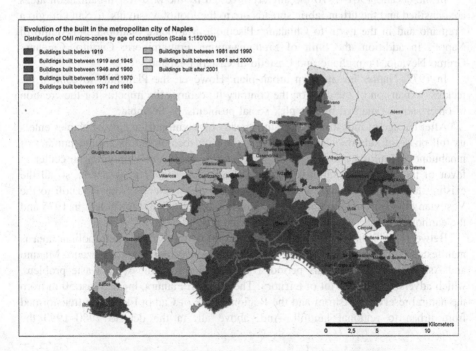

Fig. 2. Evolution of the built in the metropolitan city of Naples.

Thus, a profound transformation of the neighborhood placed «outside the Mergellina cave» is manifested.

In the neighborhoods of Fuorigrotta and Bagnoli urban transformation continues after 1946. In 1959 the San Paolo stadium is opened. In 1963, the RAI Production Centre of Naples is opened. Then, in 1964, the School of Engineering moved from Mezzocannone Avenue in Fuorigrotta, alongside the National Institute of motors, existing since 1940. This redevelopment plan for the neighborhood, promoted by Engineer Luigi Cosenza, definitely marks the transformation of the suburb area of Fuorigrotta in the new centrality and area of services. At the same time, in Bagnoli, the establishments of Cementir, Montecatini and Eternit are realized, alongside the Italsider existing since 1905.

In the 15 years between 1946 to 1960, Naples is in a phase of strong urban expansion and connection to the national infrastructure system. In its boundary the city concentrates services for the province and for the region and proposing itself asthe connector of all the development of Southern Italy. In 1964, the «highway of the Sun» was inaugurated, joining Milan at the center of Naples.

Since the fifties, economic and territorial planning documents were beginning to highlight the strategies of rebalancing the regional territory [14–17]. The city was expanding disorderly in the hills, generating neighborhoods with very different characteristics from the urban core. It transships by the same hills and completely fills all those areas that are the current suburbs, without form and without a urban design, in a large agglomeration. The neighborhoods of Pianura, Soccavo, Vomero, Arenella, Stella, San Carlo Arena, Secondigliano, Ponticelli, Barra and San Giovanni in Teduccio were born. Furthermore, the urban centers of Casalnuovo, Afragola Caivano are expanded [18–20].

In the decade 1961–1970 the already invested by the post-war urbanization areas are densified and the urban fabric spreads out to the foot of Vesuvius, to San Giorgio a Cremano and in the north to Chiaiano, Piscinola, Marinella, Scampia and Melito di Napoli. In addition, the built of Frattamaggiore, Frattaminore, Cardito, Caivano, Grumo Nevano, Pomigliano and Castello di Cisterna expand.

In 1972, Naples has its own urban plan. However, the Plan fails to contain the reckless expansion of the city, on the contrary it becomes the impetus for the creation of other suburbs marked by complex social problems, as Scampia.

After the age of industrialization, the already highlyurbanized City of Naples enters its full phase of suburbanization. In 1971, the City records the maximum number of inhabitants (1 million and 226 thousand), and the process of emptying the center in favor of the many new suburbs began. The agglomeration process starts, so all the existing towns form a single metropolitan area, which extends from Pozzuoli to the Vesuvian towns, from Giugliano up to Caivano. The city lives the cholera in 1975 and the earthquake of 1980, which will have prolonged effects.

Between 1980 and 1990 further densification of the formed metropolitan area is manifested, with significant effects in Volla, Casoria, Villaricca, Calvizzano, Marano and Astroni area. During this period, Naples begins to address the waste problem, which adversely affects a lot of territory. The dump in Pianura, located just 50 m from the natural reserve of Astroni and the Regional Park of Campi Flegrei, is transformed from urban to regional landfill. And, above all, in the decade 1980–1990 the metropolitan area expands to the Domitian Coast, where the construction of holiday homes was replaced by a widespread and illegal housing, which has devoured natural areas [21].

After 1990 further densification occurs in Giugliano, Villaricca and Quarto and to the east in Acerra. More recently, the first suburbs are again involved by densification associated sometimes to urban regeneration. These processes involve Naples, Pozzuoli, Quarto, Villaricca, Sant'Antimo, Afragola e Acerra.

So, the temporal evolution of the built can be read in relation to the spatial distribution of the market value of urban residential properties.

4 Spatial Distribution of Real Estate Values

The collection and processing of data provided by the Observatory of the Real Estate Market (OMI) for the entire province of Naples, allow to draw the map of the values of the residential units in the area of study. Figure 3 shows the average unit value of the housing market for each OMI micro-zone.

Most profiles of interest result. At first we observe that the highest market values are recorded in the core of ancient construction age. In Fig. 3 the areas with the most pronounced red color, with corresponding values between 6,151 and 10,300 €/m^2, or between 4,151 and 6,150 €/m^2, refer to the ancient center of Naples (Chiaia, Posillipo, San Ferdinando, Avvocata, Montecalvario, Fuorigrotta) or even to the first post-war

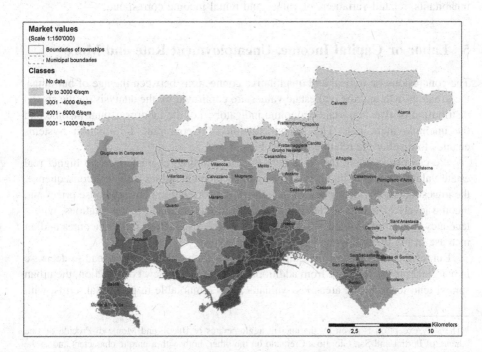

Fig. 3. Market value of urban residential properties for each OMI zone. (Color figure online)

build, until 1960 (Vomero, Arenella, Stella, San Carlo all'Arena). It means that the ancient heart of Naples remains the center of business and professional activities, so with higher incomes, benefiting from positional advantages, with corresponding higher market prices.

This condition occurs not only in Naples, but also in many cities and metropolitan areas, both Italian and European, which have an ancient foundation and which preserve the old town, often surrounded by peripheral and semi-peripheral areas. It appears that urban real estate values highlight the centers with the oldest buildings.

Figure 3 shows that the distribution of property values tend to gradually cluster going from the old center of the city of Naples to the outside[1], with a temporal sequence of building phases that responds to the economic criterion of maximizing the urban rent in building activities. So if the ancient urban centers are characterized by optimal location and with top ratings by market players, gradually less profitable soils are gradually built up, according to the economic principles of Ricardo and Von Thunen – among others – that arise in agricultural economics but obviously are valid even in urban economics. So that, the urban real estate values are able to explain the logic of spatial distribution of urban income.

As argued, the displayed spatial distribution of property values retraces the phenomena of urban growth. By an analysis on the population, taking the entire Naples *de facto* as a single urban system, the study area is affected by contemporary disurbanization and suburbanization [22–25]. So, the survey area is divided into three areas (core, first and second ring), as already been demonstrated [12]. To these areas, respectively characterized by stability, reduction and increase in the number of inhabitants, related variations of values and rental income correspond.

5 Labor or Capital Income, Unemployment Rate and More

The conclusions on logical and quantitative connections between the age of buildings, the urban growth and the real estate values are confirmed by the analysis on the spatial distribution of two important economic indicators, i.e. the laboror capital income and the unemployment rate. The implementation of Geographic Information Systems enables to display the relative maps in Figs. 4 and 5.

We can observe that the areas with the old building, characterized by higher real estate values, show the highest labor or capital income and, as a logical consequence, the lowest unemployment rates. In fact, the direct correlation between house prices and income levels of residents clearly emerges from the graphical representations, with a tendency to decline in values in those areas where a reduction in the incomes and an increase in the unemployments are recorded.

But, as the evolution of the buildings follows economic logic that lead to decrease in marginal profits resulting from additional investments in new construction, the urban sprawl tend to go to the areas less valuable, less remarkable in positional terms, with

---

[1] The only exception is given by the ancient agglomerates of Bacoli and Monte di Procida on one hand, of Portici and San Giorgio a Cremano on the other, both with a unique characters and in the coastal area.

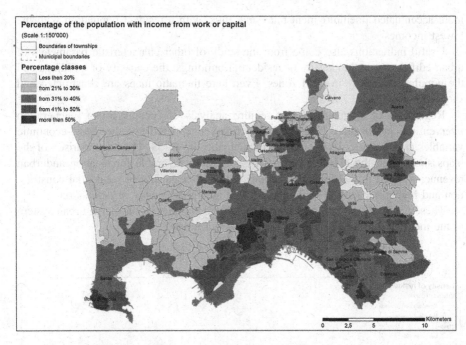

Fig. 4. Distribution of labor or capital income in Naples *de facto*.

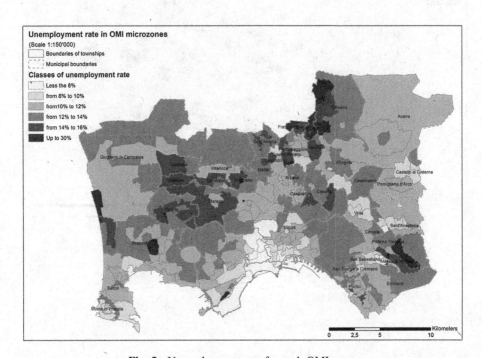

Fig. 5. Unemployment rate for each OMI zone.

more accentuated unemployment rates and, thus, with a resident population with the lowest incomes.

Useful indications also come from the study of other characteristic parameters of urban edification as the density of residential buildings, the capacity of buildings and the population density in OMI zones. Even here thematic maps are constructed and shown in the Figs. 6, 7 and 8.

Regardless of singularities that require analysis of specific urban regeneration interventions put in place [26–32] or of actions engendered by other socio-economic variables at the district level or at the municipality level [33–41], a comparison of the maps shows that the older construction areas, more favorable for location andurban revenue, resulted in the timestratifications with higher density of residential construction and higher capacity of buildings, as well as higher density of population.

These results are not obvious, especially if the complexity of a large urban system as the metropolitan area of Naples is taken into account.

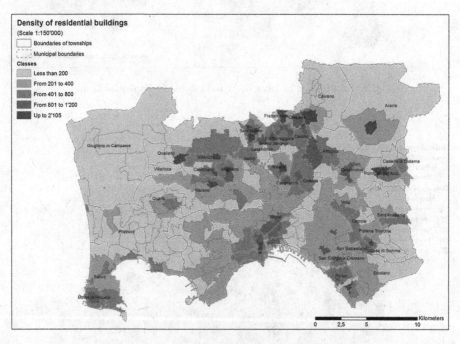

Fig. 6. Density of residential buildings.

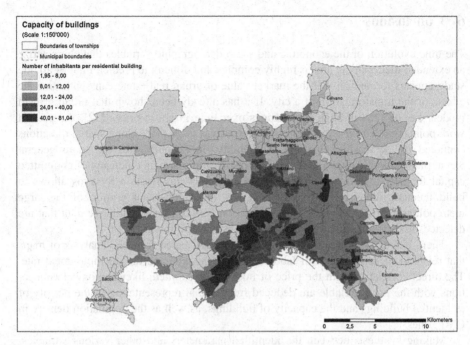

Fig. 7. Capacity of buildings for each OMI zone.

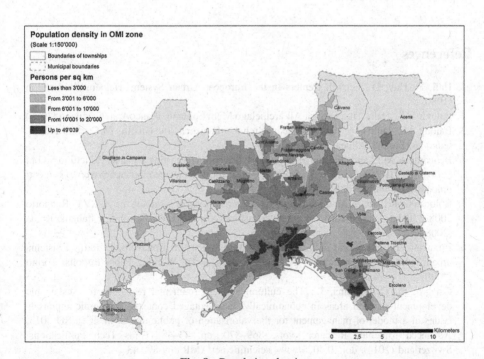

Fig. 8. Population density.

6   Conclusions

The time evolution of the economic and socio-demographic variables that characterize an extended metropolitan area is highly complex and difficult to predict. First of all, the study appears to confirm that the market value of urban real estate can synthesize the effects of the transformation of a city. If it has already been shown that urban growth models can be read through the changes over time in house prices [12], the present work points out that the expanding of built up areas takes place along spatial directions connoted by decreasing marginal profitability, in keeping with the more general Keynesian economic model that returns a decreasing marginal efficiency of committed capital for new investments. The use of Geographic Information Systems allows to build thematic maps for the study area, i.e. for a consistent portion of the large metropolitan fabric of Naples. The comparison between the quantitative data that are detected by the representations returns what is here exhibited.

Then, interesting functional correlations are descended from the analysis of maps that relate to the distribution of labor or capital income, and the unemployment rate. The direct correlation with the price of housing is recorded, likewise logical connections with the price variable are deduced from spatial representation of the density of residential buildings and the capacity of buildings, as well as the population density in OMI zones.

Making synthesis between the identified parameters and other various variables, with the intent to deduce a single function of the urban property value, constitutes the purpose of future research.

References

1. Hall, P., Hay, D.: Growth Centers in the European Urban System. Heinemann, London (1980)
2. Indovina, F.: Dalla Città Diffusa All'arcipelago Metropolitano. FrancoAngeli, Milano (2009)
3. Baioni, M.: Diffusione, dispersione, anarchia urbanistica. In: Gibelli, M.C., Salzano, E. (eds.) No Sprawl. Alinea, Firenze (2006)
4. Bencardino, M.: Land take and urban sprawl: drivers e contrasting policies. Bollettino S.G.I., Serie XIII **VII**(2), 217–237 (2015). SGI, Roma. http://societageografica.net/wp/wp-content/uploads/2016/08/bencardino_eng.pdf
5. Bolocan Goldstein, M.A.: Città senza confini, territori senza gerarchie. In: AA.VV. Rapporto 2008. L'Italia delle città, tra malessere e trasfigurazione. Società Geografica Italiana, Roma (2008)
6. Bonavero, P.: Traiettorie della ricerca urbana europea. In: Bonavero, P. (ed.) Il sistema urbano europeo fra gerarchia e policentrismo, pp. 8–14. Working Paper Eupolis, Torino (2000)
7. Calabrò, F., Della Spina, L.: The cultural and environmental resources for sustainable development of rural areas in economically disadvantaged contexts. Economic-appraisals issues of a model of management for the valorisation of public assets. In: ICEESD 2013, Advanced Materials Research, vols. 869–870, pp. 43–48. Trans Tech Publications, Switzerland (2014). doi:10.4028/www.scientific.net/AMR.869-870.43

8. Derycke, P.H.: Economia Urbana. Il Mulino, Bologna (1972)
9. Orefice, M.: Estimo. UTET, Torino (1984)
10. Barber, W.: Storia del Pensiero Economico. Feltrinelli, Milano (1982)
11. Bencardino, M.: Demographic changes and urban sprawl in two middle-sized cities of campania region (Italy). In: Gervasi, O., et al. (eds.) ICCSA 2015. LNCS, vol. 9158, pp. 3–18. Springer, Cham (2015). doi:10.1007/978-3-319-21410-8_1
12. Bencardino, M., Granata, M.F., Nesticò, A., Salvati, L.: Urban growth and real estate income. a comparison of analytical models. In: Gervasi, O., et al. (eds.) ICCSA 2016. LNCS, vol. 9788, pp. 151–166. Springer, Cham (2016). doi:10.1007/978-3-319-42111-7_13
13. Calafati, A.: Nuova perimetrazione e nuove funzioni per le Città metropolitane. Il caso di Napoli. Centro Studi Unione Industriali di Napoli (2014). http://www.lavoce.info/archives/17288/citta-metropolitane-delrio-province/. Accessed
14. Bencardino, F., Greco, I.: Politiche di sviluppo regionale e assetto territoriale in Campania. Alcune riflessioni. In: Rivista Documenti Geografici, Dipartimento di Scienze Storiche, Filosofico-sociali, dei Beni culturali e del Territorio, Università di Roma, Tor Vergata, no. 2 (2016). doi:10.19246/docugeo2281-7549/201602_02
15. Bencardino, F., Cresta, A., Greco, I.: Le «città medie» nello sviluppo territoriale della Campania: alcune riflessioni. In: Viganoni, L. (a cura di) A pasquale Coppola. Raccolta di scritti, Collana «Memorie della Società Geografica Italiana», LXXXIX, pp. 385–396. SGI, Roma (2010)
16. Bencardino, M., Greco, I., Ladeira, P.R.: The comparative analysis of urban development in two geographic regions: the state of Rio de Janeiro and the Campania region. In: Murgante, B., Gervasi, O., Misra, S., Nedjah, N., Rocha, Ana Maria A.C., Taniar, D., Apduhan, B.O. (eds.) ICCSA 2012. LNCS, vol. 7334, pp. 548–564. Springer, Heidelberg (2012). doi:10.1007/978-3-642-31075-1_41
17. Greco, I.: Dalla città resiliente alla campagna resiliente: gli spazi aperti e rurali come luogo di riequilibrio città-campagna al tempo della crisi. In: Capineri et al. (a cura di) Memorie Geografiche, vol. XII. Oltre la globalizzazione Resilienza/Resilience, Società di Studi Geografici, Firenze (2014)
18. Acierno, A.: Periferie napoletane: recinti di insicurezza. J. Urbanism, 1–8 (2007). Planum
19. Amato, F.: La periferia italiana al plurale: il caso del Napoletano. In: Sommella, R. (a cura di) Le città del Mezzogiorno. Politiche, dinamiche, attori, pp. 219–242. FrancoAngeli, Milano (2008)
20. Talia, I.: Appunti su Napoli, 1951–2010, "Tante Italie una Italia. Dinamiche territoriali e identitarie". In: Muscarà, C. et al. (eds.) Mezzogiorno, la modernizzazione smarrita, vol. II. FrancoAngeli, Milano (2011)
21. Porcaro, S.: Castelvolturno, intervista a Mario Luise. Napoli Monitor 59 (2014). http://napolimonitor.it/old/2014/07/16/26233/castelvolturno-intervista-mario-luise.html
22. Berry, B.J.L.: Urbanization and counterurbanization in the United States. Ann. Am. Acad. Polit. Soc. Sci. 451, 13–20 (1980). Changing Cities
23. Champion, T.: Urbanization, suburbanization, counterurbanization and reurbanization. In: Padison, R. (ed.) Handbook of Urban Studies. SAGE Publication, London (2001)
24. Norton, R.D.: City Life-Cycles and American Urban Policy. Academic Press, New York (1979)
25. Van den Berg, L., et al.: Urban Europe, A Study of Growth and Decline. Elsevier Ltd., London (1982)
26. Bencardino, M., Greco, I.: Smart Communities. Social Innovation at the service of the smart cities. I: TeMA. J. Land Use Mob. Env. SI/2014, 39–51 (2014). University of Naples "Federico II" Print, Napoli

27. De Mare, G., Granata, M.F., Nesticò, A.: Weak and strong compensation for the prioritization of public investments: multidimensional analysis for pools. Sustainability **7** (12), 16022–16038 (2015). doi:10.3390/su71215798. MDPI AG, Basel, Switzerland

28. Murgante, B., Misra, S., Carlini, M., Torre, Carmelo M., Nguyen, H.-Q., Taniar, D., Apduhan, B.O., Gervasi, O. (eds.): ICCSA 2013. LNCS, vol. 7974. Springer, Heidelberg (2013). doi:10.1007/978-3-642-39649-6

29. Guarini, M.R., Battisti, F.: Benchmarking multi-criteria evaluation methodology's application for the definition of benchmarks in a negotitation-type public-private partnership. A case of study: the integrated action programmes of the Lazio Region. Int. J. Bus. Intell. Data Mining **9**(4), 271–317 (2014). doi:10.1504/IJBIDM.2014.068456

30. Nesticò, A., Mare, G.: Government tools for urban regeneration: the cities plan in Italy. A critical analysis of the results and the proposed alternative. In: Murgante, B., et al. (eds.) ICCSA 2014. LNCS, vol. 8580, pp. 547–562. Springer, Cham (2014). doi:10.1007/978-3-319-09129-7_40

31. Nesticò, A., Pipolo, O.: A protocol for sustainable building interventions: financial analysis and environmental effects. Int. J. Bus. Intell. Data Mining **10**(3), 199–212 (2015). doi:10.1504/IJBIDM.2015.071325. Inderscience Enterprises Ltd., Genève, Switzerland

32. Tajani, F., Morano, P.: An evaluation model of the financial feasibility of social housing in urban redevelopment. Prop. Manag. **2**, 133–151 (2015). doi:http://dx.doi.org/10.1108/PM-02-2014-0007

33. Battisti, F., Guarini, M.R.: Public Interest Evaluation in Negotiated Public-Private Partnership. Int. J. Multicrit. Dec. Making **7**(1) (2017). Inderscience Enterprises Ltd.

34. Bencardino, M., Nesticò, A.: Demographic changes and real estate values. A quantitative model for analyzing the urban-rural linkages. Sustainability **9**(4), 536 (2017). doi:10.3390/su9040536

35. Greco, I., Bencardino, M.: The paradigm of the modern city: *SMART and SENSEable cities* for smart, inclusive and sustainable growth. In: Murgante, B., et al. (eds.) ICCSA 2014. LNCS, vol. 8580, pp. 579–597. Springer, Cham (2014). doi:10.1007/978-3-319-09129-7_42

36. Guarini, M.R., Battisti F.: Benchmarking multi-criteria evaluation methodology's application for the definition of benchmarks in a negotitation-type public-private partnership. A case of study: the integrated action programmes of the Lazio Region. Int. J. Bus. Intell. Data Mining **9**(4), 271–317 (2014)

37. Guarini, M.R., Buccarini, C., Battisti, F.: Technical and economic evaluation of a building recovery by public-private partnership in Rome (Italy). In: Stanghellini, S., Morano, P., Bottero, M., Oppio, A. (eds.) Appraisal: From Theory to Practice. GET, pp. 101–115. Springer, Cham (2017). doi:10.1007/978-3-319-49676-4_8

38. Della Spina, L., Ventura, C., Viglianisi, A.: A multicriteria assessment model for selecting strategic projects in urban areas. In: Gervasi, O., et al. (eds.) ICCSA 2016. LNCS, vol. 9788, pp. 414–427. Springer, Cham (2016). doi:10.1007/978-3-319-42111-7_32

39. Morano, P., Tajani, F.: The break-even analysis applied to urban renewal investments: A model to evaluate the share of social housing financially sustainable for private investors. Habitat Int. **59**, 10–20 (2017). doi:10.1016/j.habitatint.2016.11.004

40. Nesticò, A., Galante, M.: An estimate model for the equalisation of real estate tax: a case study. Int. J. Bus. Intell. Data Mining **10**(1), 19–32 (2015). doi:10.1504/IJBIDM.2015.069038. Inderscience Enterprises Ltd., Genève, Switzerland

41. Nesticò, A., Macchiaroli, M., Pipolo, O.: Costs and benefits in the recovery of historic buildings: the application of an economic model. Sustainability **7**(11), 14661–14676 (2015). doi:10.3390/su71114661. MDPI AG, Basel, Switzerland

OLAP Analysis of Integrated Pest Management's Defense Rules: Application to Olive Crop in Apulia Region

Claudio Zaza[1(✉)], Sandro Bimonte[2], Crescenzio Gallo[3],
Nicola Faccilongo[1], Piermichele La Sala[1], and Francesco Contò[1]

[1] Department of Economics, University of Foggia,
1, Largo Papa Giovanni Paolo II, Foggia, Italy
{claudio.zaza,nicola.faccilongo,piermichele.lasala,
francesco.conto}@unifg.it
[2] TSCF, Irstea, 9 Av. Blaise Pascal, Aubiere, France
sandro.bimonte@irstea.fr
[3] Department of Clinical and Experimental Medicine,
University of Foggia, 1, Viale Luigi Pinto, Foggia, Italy
crescenzio.gallo@unifg.it

Abstract. The Agri-Food sector is facing global challenges. The first concerns feeding a world population that in 2050, according to UN projections, will reach 9.3 billion people. The second challenge is the request by consumers for high quality products obtained by more sustainable, safely and clear agri-food chains. The Integrated Pest Management (IPM) could be an important instrument to help farmers to face these challenges. The IPM requires the simultaneous use of different crop protection techniques for the control of pests through an ecological and economic approach. This work explores the possibility to develop a framework that combines the Information and Communication Technologies (ICTs) with the IPM principles, in order to support the farmers in the decisional process, improving environmental and production performances. The proposed ICT tool is On-Line Analytical Processing (OLAP), which allows performing analysis in the domain of time and space verifying for a single farm the respect of the IPM technical specifications.

Keywords: Data warehouse · OLAP · Integrated Pest Management · Olive crop

1 Introduction

The Agri-Food sector is facing global challenges. The first issue concerns feeding a world population that in 2050, according to UN projections, will reach 9.3 billion people [28]. The second challenge is the request by consumers for high quality products obtained by more sustainable, safely and clear agri-food chains [15, 27]. To answer these requests, farmers need to increase quality and quantity of the production, reducing the environmental impact through new management strategies and tools. In this context, Farm Management Information Systems (FMISs) play an important role.

O. Gervasi et al. (Eds.): ICCSA 2017, Part II, LNCS 10405, pp. 31–44, 2017.
DOI: 10.1007/978-3-319-62395-5_3

FMIS has been defined as a "planned system for collecting, processing, storing, and disseminating data in the form needed to carry out a farm's operations and functions." According to Fountas [13], FMISs provide functionalities for field operations management, best practice tools[1], finance, inventory, traceability, reporting, site-specific tools, sales, machinery management, human resource management, and quality assurance. In particular, incorporating the Integrated Pest Management (IPM) framework into FMISs appears as a mandatory instrument to help farmers facing to the challenges of sustainable agriculture. The IPM requires the simultaneous use of different crop protection techniques for the control of insects, pathogens, weeds and vertebrates, through an ecological and economic approach [24]. The aim is to combine different techniques to control pest populations below an economic damage threshold [6]. The relevance of the IPM is underlined by the EU that has recognized to IPM a central role to reduce the reliance on the use of conventional pesticides, in the context of the Framework Directive 2009/128/EC [12] on Sustainable Use of Pesticides.

Unlucky, the IPM is manually fulfilled by farmers, and IPMs of different campaigns and farms are not shared and stored. Therefore, it appears difficult to investigate for the best sustainable practices according to past IPMs. This is an important limitation in the support of the farmers in the decisional process for improving environmental and production performances. To overcome this limitation, in this work we explore the usage of Data Warehouse (DW) and On-Line Analytical Processing (OLAP) systems for IPMs analysis. DW and OLAP system are Business Intelligence technologies allowing the analysis of huge volume of data at different spatial, temporal and thematic levels (i.e. scales) [19]. Decision-makers (e.g. farmers, researchers, etc.) explore warehoused data by means of user-friendly OLAP tools, such as pivot tables and graphic displays. In the last years, OLAP systems have been successfully applied in several domains such as marketing, health, etc., and also in agriculture.

Therefore, in this work we propose an OLAP (i.e. multidimensional) model, and its implementation, aimed at improving the environmental performance of farms. These will be assessed verifying the application of the IPM technical specifications. With the OLAP analysis, it is possible to estimate the level of sustainability at various farm layers and different time intervals.

The paper is organized in the following way: Sect. 2 presents a state of art of the agricultural sustainable indicators and OLAP applications in the agricultural domain; Sect. 3 presents our OLAP model for the analysis of IPMs; Sect. 4 details its implementation, and Sect. 5 concludes the paper.

2 Related Work

In this section we review the main sustainable agriculture production indicators, and the OLAP application in the agriculture domain.

---

[1] Best practices tools: based on (complex) agronomy models suggest to farmers some particular technical operations (such as respecting the organic standards) [14].

2.1 Sustainable Agriculture Indicators

The worldwide-accepted definition of sustainable development was introduced by the World Commission on Environment and Development in the Brundtland Report [30] as "development that meets the needs of the present without compromising the ability of future generations to meet their own needs". In the last years, many indicators to assess the sustainability of agricultural systems have been developed (for a complete review see [10]).

For this work, we consider only indicators that support the following requirements:

1. *Farm-oriented.* The indicator must be tailor made to evaluate the farm sustainability, in particular working on the management aspects related to phytosanitary treatments, due to the documented hazards to human health, environment, food commodities, etc. associated to the extensive usage of these substances [1].
2. *Easy to implement.* The acquisition of agri-environmental data is time and money consuming. Therefore, the indicator must exclusively use data that are easy to collect by farmers without any additional (money and time) costs.
3. *Understandable.* To encourage the farmer to use the indicator in his/her agricultural practices, s/he has to trust it. Therefore, we believe that the farmer has to understand how the indicator is calculated to adopt it.

In literature there are many works focused on evaluating the sustainability at farm level. In many of the studied works, data required for the evaluation are too extensive for our purpose, for example [9, 21, 26]. In other cases, the assessments are specific on restricted topics, such as nitrogen management [20]. In other works, the models developed are too complicated to satisfy the third requirement, for example [11, 25].

2.2 OLAP for Agriculture

Recently, some publications have investigated the usage of OLAP for the agricultural data. Most of works address the analysis of agriculture production under different points of view (see [3] for a systematic review). Chaudhary [8] proposed a system for the analysis of economic factors affecting the yield of cereal crops. Gupta [16] study the crop production according to irrigation, weather, seeding and the soil type. Chaturvedi [7] studies the cotton production, defining 13 multidimensional models including different analysis axes such as soil resources, etc. In [4] authors have used a Spatial OLAP architecture to analyze the data collected using sensors networks. In particular we studied the energy consumptions of farms for wheat.

Environmental aspects of agriculture productions are studies in [23, 29]. Pinet [23] analyze the spread quantity per commune and fertilizer. Vernier [29] study the quantity of pesticides in lakes and rivers of the south of France.

To conclude no work proposes the historical and multidimensional analysis of sustainable practices using information systems.

3   OLAP Model for IPM

In this section, we describe the Integrated Pest Management for Olive tree (Sect. 3.1) that we use in this work, and the OLAP model that we propose to analyze it (Sect. 3.2).

3.1   Integrated Pest Management for Olive Tree

The Framework Directive 2009/128/EC [12] on Sustainable Use of Pesticides obliges all professional users of pesticides to apply the IPM approach in the European member countries from 1 January 2014. The task of the Member States is to combine the eight IPM principles [2], in order to achieve crop technical specifications. For example, in Italy each Region develops its own guidelines, taking into account the agri-environmental conditions, the present crops and the related pests.

Therefore, in this work we use the IPM like a sustainability indicator. The main idea is to verify that farmers apply the IPM. The IPM is not an indicator, but the application of its specifications ensures to the farms a better performance, by sustainable and economic points of view. Furthermore, in some regions, like Apulia Region, with the Measure 10.1.1 of the Rural Development Programme (RDP) 2014–2020, the diffusion of certified integrated production was stimulated by an economic incentive. To get incentives, farmers must meet all the requirements included in the crop specific Disciplinary provided by the Region. Moreover, producers are also checked by an independent control authority, who verifies that all the farm's operations are compliant with the IPM.

Each crop's specific IPM is composed of two parts:

1. *General:* a set of recommendations regarding all the aspects of the management of crop, such as fertilization restrictions, soil management, irrigation frequency, etc.
2. *Defense rules:* all the allowed operations (agronomic and chemical) to protect the crop from the pests listed (Table 1).

The IPM satisfies the requirements described in Sect. 3. In particular, it is designed for arable farms and it is mandatory for the European community (*farm-oriented requirement*); the data required to verify the respect of IPM are all recorded in the "Farm's Notebook" (*easy to implement requirement*); the farmers understand easily the requirements and the results, because they are based on daily agricultural management practices (*understandable requirement*).

In particular, in this work we focus our attention to Defense rules included in the Apulia Region technical specification for the cultivation of the Olive Tree, which is the most important crop in the selected area. Furthermore, according to the Italian Institute of Statistics [18], Apulia is the first olive-producing region in Italy. In particular, in 2016 the agricultural surface invested in Apulia for olive production was equal to 0.380 Mha, thus representing 33% of the corresponding national one (1.17 Mha). The obtained production was 0,79 Mt, corresponding to 38% of the national amount (2,09 Mt) [18]. Therefore, in Apulia region there is an important number (227.245) [17] of farms working on Olive tree, which generate huge amount IPM data. Actually, this data is quite unexploited.

Table 1. Excerpt of the IPM for the protection of Olive crop in Apulia region

Pest	Intervention criteria	Active substance and auxiliary	Maximum numb. of intervention	Usage limitations and notes
Olive fruit fly (*Bactrocera oleae*)	Intervention threshold: • Tables olive: presence of the first punctures; • Oil olives: 10-15% presence of active infestation (summary of eggs and larvae)	*Beauveria bassiana*	/	
		Dimethoate	2	
		Imidacloprid	1*	* allowed only after blossoming

Table 1 shows an excerpt of the IPM for the protection of Olive crop in Apulia region. In the following, we describe each column:

- *Pest*: It is the pest for Olive crop. In the example, the pest is the *Bactroceraoleae*, which is most dangerous pest of Olive tree worldwide [22].
- *Intervention criteria*: The intervention threshold represents the limit before that the pest does not cause economic damage to the cultivated crop. In the example, the value is different between table and oil olive cultivars. In the first case, the damage is higher because a single puncture strongly reduces the value of the product, so it is necessary to control immediately the pest. Instead, in olives for oil, fly presence lower than 10–15%, in function of different cultivars, is economically acceptable, so the control starts when the pest pressure exceeds this value.
- *Active substance and auxiliary:* It reports the list of authorized Active Substances (ASs) and auxiliaries that could be used to control the pest. In the example, phytosanitary products containing *Beauveriabassiana*, entomopathogenic fungus, Dimethoate or Imidaclopridare allowed.
- *Maximum number of treatments*: It is the maximum number of treatments admitted for a specific molecule. For example, products based on *B. bassiana* are not limited, because it is a natural organism without environmental impact. In contrast, Dimethoate and Imidacloprid are limited to 2 and 1 treatment per year respectively, due to the toxicity of the molecules.
- *Usage limitations and notes*: It represents the list of eventual other restriction for specific molecule, such as temporal restriction. For example, Imidacloprid products are allowed only after the phenological phase of blossoming.

3.2   OLAP Model

Warehoused data is stored according to the OLAP (i.e. multidimensional) model [14, 19]. It organizes data according to dimensions and facts. *Dimensions* represent the analysis axes and they are organized in hierarchies. A *hierarchy* is composed of levels that define the analysis granularities. *Facts* are the analysis subjects, and are described by *measures*. Usually, measures are numerical values that are aggregated using numerical aggregation functions over the dimensions levels. Warehoused data is then explored using the OLAP operators, which allow navigating into hierarchies (i.e. Roll-Up and Drill-Down), and selecting a subset of warehoused data (i.e. Slice and Dice).

The OLAP model we propose for the analysis of Olive fruit IPM, previously described, is shown in Fig. 1 using the UML Profile ICSOLAP [5]. ICSOLAP is a UML profile for the conceptual design of spatial DW (SDW). It makes possible to conceptually represent classical and advanced aspects of spatio-multidimensional modeling, such as multiple and complex hierarchies, etc. The profile defines a stereotype or a tagged value for each multidimensional element, for example: <<Fact>> for the fact and <<TemporalAggLevel>> for temporal dimension levels.

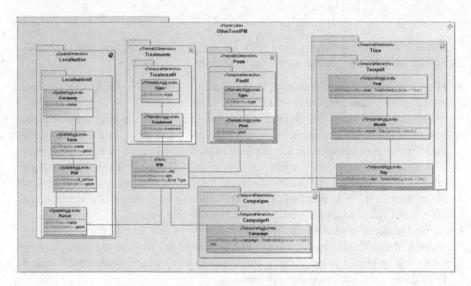

Fig. 1. Olive fruit IPM multidimensional model

The multidimensional model presents 5 dimensions:

- *Time*: it is the temporal dimension with the hierarchy: day<month<year
- *Campaign*: it represents the crop campaign (for example 2009–2010).
- *Company*: it describes the organization of the farms. It groups parcels in plots, and farms in companies. This hierarchy allows analyzing all the treatments performed at different spatial scales. For example, the same farm could treat two plots of the same crop in different ways during a single campaign, in order to evaluate the best management strategy for the future years.
- *Pests*: It represents the pest of the IPM (first column of Table 1). Pests are organized in type, for example weed, fungi, insects, bacteria and so on. Therefore, there is the scientific name of pest, e.g. *Bactrocera oleae* (Table 1), *Spilocea oleagina*, etc.

- *Treatments*: It represents the different types of interventions allowed in the IPM technical specification. They are grouped in types, such as Agronomic, e.g. Pruning, and Chemical, e.g. Dimethoate, Imidacloprid, etc.

  Measures and their aggregations are:

- *Number of treatments*: It represents the number of performed treatments (for example the application of Dimethoate during the campaign 2009–2010). This measure is aggregated along all dimensions using the sum function.
- *Respect of IPM*: This is a Boolean measure describing when the performed action respects the IPM rules. For example, the usage of the Glyphosate 30.8% is not allowed. Therefore, the measure value for this treatment is "NO". *Respect of IPM* is aggregated using the AND operator.
- *Error type*: Describes the treatment according to the IPM rules. It is an enumeration value:
  - *Allowed*. The treatment is allowed (for example the usage of Copper to control the pest *S. oleagina*).
  - *A.S. not Allowed*. The treatment is not allowed (for example the usage of glyphosate 30.8%).
  - *Exceeded nb inter*. The number of treatments exceeds the maximum amount of treatment allowed (fourth column, Table 1), for example the third intervention based on Dimethoate to control *B. oleae*is not allowed.
  - *A.S. not allowed in time*. The treatment performed is not allowed in that time period. For example, Imidacloprid products are forbidden before the phenological phase of blossoming.
  - *Errors*. It represents whenmore errors occur, so it is necessary a drill-down operation to obtain the specific error type for single treatment.

For the aggregation of *Error type*we use a user-defined aggregation that returns *Errors* when more than one of: *A.S. not Allowed* OR *Exceeded nb inter* OR *A.S. not allowed in time are aggregated*, otherwise it outputs *Allowed*.

Thanks to this model, the decision-makers can trigger different kinds of questions:

(i) *Technical specification*: Compliance with the Olive Crop IPM technical specification provided by Apulia Region. In this case, the model allows investigating the respect of the technical specification. These queriescan be grouped in two classes:

  - *"inter-farms"* queries allow to compare treatments among different farms over historical data. Examples of this kind of queries are:
    - **Q1:** *"Visualize all chemical treatments of farms"* (Fig. 4),
    - *"Visualize the treatments for Bactroceraolea of farms that respect the technical specifications"*,
    - *"Visualize all the not allowed chemical treatments of farms that not respect the technical specification"*,
    - ...

For privacy reason deeper information are available only for the user's farm. In this way, the model allows to perform light inter-farm analysis and deeper intra-farm analysis as shown below.

- *"intra-farm"* queries allow matching treatments between different plots or parcels of the same farm. For example:
  - **Q2:** *"Visualize all the chemical treatments performed for plots of the Bimonte farm"* (Fig. 5),
  - *"Visualize the treatments that not respect the technical specification for plots of the farm"*;
  - *"Visualize the allowed chemical treatments for all the parcels"*,
  - ...

In this way, it is possible to analyse the compliance with technical specification at lower scale, allowing identifying exactly where a wrong treatment was performed.

(ii) *Management*: Detailed analysis of crop protection strategies. In the second group of query, detailed analyses are provided at intra-farm level. In particular, the different crop protection strategies are investigated through different point of view, such as the number of treatment for plot or parcel, the type of interventions for each pests, etc. Farmers and/or decision-makers can answer to queries like these:

  - **Q3:** *"Show the number and type of treatments performed at farm level for each pest"* (Fig. 6),
  - *"Show the number of agronomical and chemical treatments performed to control fungi pests"*,
  - *"Visualize all the treatment for all the pests at plot scale"*,
  - ...

This type of queries enabling the user to carry out different analyses, for example comparing different strategies of crop protection realized in different plot, or evaluating which pest caused more damage in order to control better in the next campaign.

These queries can be triggered at different spatial, temporal and thematic levels using the dimensions hierarchies and the OLAP operators.

For example, it is possible to:

- **Q4:** *"Visualize the number and type of treatment performed to control pests grouped per type at plot scale"*, applying a Drill-Down operator on the spatial dimension,
- *"Visualize the allowed chemical treatments per parcel and day"*,
- ...

4 Implementation

In this section we describe the implementation of the OLAP model presented in Sect. 3.

Data used in this work concerns two real Apulian farms, involved by the Department of Economics of University of Foggia during the development of the AFC-WDM FMIS. Then, in this paper the farms were renamed respectively *"Bimonte"* and *"Zaza"* for privacy reasons.

The OLAP system is integrated into a FMIS, called AFC-WDM (Agri-Food Chain Web Data Management). AFC-WDM is an integrated cloud platform working on a

Web 2.0 environment on multiple devices (PCs, tablets, smartphones) in the model SaaS - Software as a Service. The data collected by the platform concern all the managing aspects of the agricultural farms and can be updated in every moment by the users themselves. Data is stored in a database structured in two layers:

- *Company and Farm*: All the basic information related to the Company and the Farm are collected (VAT number, address, director, etc.);
- *Resources*: All the possible resources (Buildings, Warehouses, Machines, Plots) that could be assigned to a single farm starting from a specific date are stored. For example, on a single plot (group of Parcels), all the technical operations (such as irrigations, phytosanitary treatments, tillage operations, etc.) could be managed.

These technical operations are used to feed the DW tier of the OLAP system.

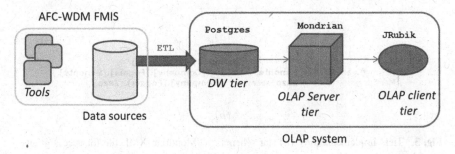

Fig. 2. BI architecture

The OLAP system is a three-tier Relational OLAP system composed of (Fig. 2):

- The Data Warehouse tier. It is responsible for the storage of data and is implemented using the Postgres DBMS. Data is modeled using the star schema logical model [19]. The star schema denormalizes dimension tables to avoid expensive join operations.
  The star schema implemented in our case study is shown in Fig. 3a. Let us note that we have introduced an additional attribute "farmid". It represents farms using simple numerical identifiers. It allows for an anonymous navigation in the company dimension (see Fig. 4 for an example).
- The OLAP Server tier. The OLAP server implements the OLAP operators (Drill-Down, Roll-Up, etc.), and access grants. In particular, as shown in Fig. 3b we define a set of access grants to allow only the farm owner to visualize data about his/her plots.
- The OLAP client tier. It is in charge of the visualization and exploration of warehoused data and is implemented using the OLAP client JRubik.

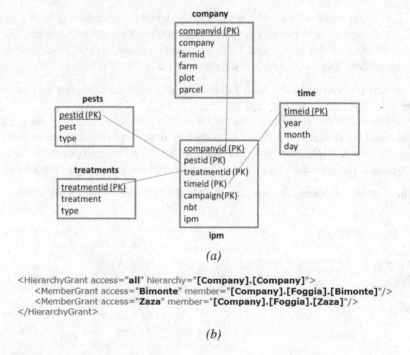

(a)

```
<HierarchyGrant access="all" hierarchy="[Company].[Company]">
    <MemberGrant access="Bimonte" member="[Company].[Foggia].[Bimonte]"/>
    <MemberGrant access="Zaza" member="[Company].[Foggia].[Zaza]"/>
</HierarchyGrant>
```

(b)

Fig. 3. Tiers implementations: *(a)* star schema, *(b)* Mondrian XML file for access grants

In the rest of the section, a set of examples of OLAP queries results are shown using JRubik.

In Fig. 4, an example of *Inter-farm Technical specification* query is represented.

Company	Treatments	Time	Error type	Respect of IPM	Number of Treatments
+Bimonte	Copper	+2016	Allowed	OK	2
	Deltamethrin	+2016	Errors	NO	2
	Dimethoate	+2016	Exceeded nb inter.	NO	5
	glyphosate 30,8%	+2016	A. S. not allowed	NO	1
	Pyraclostrobin	+2016	A. S. not allowed in time	NO	1
+2	Copper	+2016	Allowed	OK	5
	Dimethoate	+2016	Allowed	OK	3

Fig. 4. Example of light inter-farm Technical specification query

In particular, **Q1:** *"Visualize all chemical treatments of farms"* is shown, allowing the comparison between the user's farm (Bimonte) and another farms (FarmID = 2). The user can compare his own farm with others for the measures "Error type", "Respect of IPM" and "Number of Treatments" aggregated at farm level, but for privacy reason he could not know deeper information. As showed in Fig. 4, many errors were detected on Bimonte farm.

Company	Treatments	Time	Error type	Respect of IPM	Number of Treatments
+Plot1	Copper	−2016	Allowed	OK	1
		+09/2016	Allowed	OK	1
	Deltamethrin	−2016	A. S. not allowed	NO	1
		+10/2016	A. S. not allowed	NO	1
	Dimethoate	−2016	Allowed	OK	2
		+06/2016	Allowed	OK	1
		+09/2016	Allowed	OK	1
	glyphosate 30,8%	−2016	A. S. not allowed	NO	1
	Pyraclostrobin	−2016	A. S. not allowed in time	NO	1
+Plot2	Copper	−2016	Allowed	OK	1
		+09/2016	Allowed	OK	1
	Deltamethrin	−2016	A. S. not allowed	NO	1
		+10/2016	A. S. not allowed	NO	1
	Dimethoate	−2016	Exceeded nb inter.	NO	3
		+06/2016	Allowed	OK	1
		+09/2016	Allowed	OK	1
		+10/2016	Exceeded nb inter.	NO	1

Fig. 5. Example of deeper intra-farm Technical specification query

To obtain more information, an *Intra-farm Technical specification* query is launched (Fig. 5). The example provided is the result of the query **Q2**: *"Show all the chemical treatments performed for plots of the Bimonte farm"*. In this way the farmer can obtain detailed information for his/her own farm on where (Plot level and deeper) and when (Month level and deeper) s/he has not respected the IPM. Furthermore, the column "Error type" allows you to easily understand the explanation of eventual errors, for a single treatment. For example in the Fig. 5 in the Plot1 there are two A.S. not allowed, Deltamethrin and Glyphosate 30,8%, and one A.S. not allowed in time, Pyraclostrobin. For the Plot2 the wrong interventions are only two, again Deltamethrin and the third Dimehoate treatment, which exceed the threshold of maximum number of treatments allowed per year (Table 1).

The *Management* query enables performing detailed analyses at intra-farm level. For example, in Fig. 6 is the result of following query: *"Show the number and type of treatments performed at farm level for each pest"*. As showed, at farm level (in this case the Zaza's farm) for each pest were reported the type of intervention, Agronomic or Chemical, and detailed information, such as Active Substance used and the number of treatments. Like in the previous case, this type of query allows you to visualize an overview of the treatment performed at farm level.

Figure 7 shows the result of the query **Q3**: *"Visualize the number and type of treatment performed to control pests grouped per type at plot scale"*. This type of query allows comparing the different crop protection strategies carried out by the farmer in different plots of same farm. For example, in Plot11 the farmer executes only one agronomic treatment to control bacteria pest, while in the other plots/he does not. The total number of chemical treatments is five in both plots, but in the first one the farmer has used more fungicides than insecticides, while in the second plot the situation

Company	Treatments	Pests	Number of Treatments
+Zaza	—agronomic	P.syringae pv.savastanoi	1
	Pruning	P.syringae pv.savastanoi	1
	—chemical	Spilocea oleagina	5
		Bactrocera olea	5
	Copper	Spilocea oleagina	5
	Dimethoate	Bactrocera olea	3
	Imidacloprid	Bactrocera olea	2

Fig. 6. Example intra-farm management query

		Number of Treatments			
		Pests			
Company	Treatments	+ bacteria	+ fungi	+ insects	+ weed
+Plot11	—agronomic	1			
	Pruning	1			
	+chemical		3	2	
+Plot12	+agronomic				
	+chemical		2	3	

Fig. 7. Example of management query at plot scale

is reversed. Based on this information, the farmer could decide the best future crop strategy. For example in the next campaign he will pay particular attention to fungi protection in the Plot11, due to the high number of treatments performed.

5   Conclusions

The Agri-Food sector is facing global challenges. The first concerns feeding a world population that in 2050, according to UN projections, will reach 9.3 billion people. The second challenge is the request by consumers for high quality products obtained by more sustainable, safely and clear agri-food chains. The Integrated Pest Management (IPM) could be an important instrument to help farmers to face these challenges. Motivated by the lack of a decision-support tool for the historical analysis of IPM, in this work we present an OLAP model, and its implementation, that allows verifying the respect by farmers of Olive Crop technical specification defense rules provided by Apulia Region, which encourage the diffusion of the certified integrated production with an economic incentive. As shown, the user could obtain by the system detailed spatial and temporal information related to his/her farm, thanks to the different hierarchies implemented. Moreover, two groups of queries have been developed and tested. In the first group there are queries concerning the Technical Specification compliance, investigating the respect of the defense rules at inter and intra-farm level. The second group includes the Management queries, allowing the user to analyse the crop protection strategies of his/her farm through different points of view, such as the number of treatment for plot or parcel, the type of interventions for each pests, etc.

Our on-going work is the integration of more real datasets, the introduction of spatial data to provide a Spatial OLAP system, and finally as future work we will investigate indices to be able to compare similar (from a climatic and agronomy points of view) farms.

References

1. Aktar, W., Sengupta, D., Chowdhury, A.: Impact of pesticides use in agriculture: their benefits and hazards. Interdisc. Toxicol. **2**, 1–12 (2009)
2. Barzman, M., Bàrberi, P., Birch, A.N.E., Boonekamp, P., Dachbrodt-Saaydeh, S., Graf, B., Hommel, B., Jensen, J.E., Kiss, J., Kudsk, P., Lamichhane, J.R., Messéan, A., Moonen, A.-C., Ratnadass, A., Ricci, P., Sarah, J.-L., Sattin, M.: Eight principles of integrated pest management. Agron. Sustain. Dev. **35**, 1199–1215 (2015). doi:10.1007/s13593-015-0327-9
3. Bimonte, S.: Current approaches, challenges, and perspectives on spatial OLAP for agri-environmental analysis. IJAEIS **7**, 32–49 (2016)
4. Bimonte, S., Pradel, M., Boffety, D., Tailleur, A., André, G., Bzikha, R., Chanet, J.P.: A New sensor-based spatial OLAP architecture centered on an agricultural farm energy-use diagnosis tool. Int. J. Dec. Supp. Syst. Technol. **5**, 1–20 (2013)
5. Boulil, K., Bimonte, S., Pinet, F.: Conceptual model for spatial data cubes: A UML profile and its automatic implementation. Comput. Stand. Interfaces **38**, 113–132 (2015)
6. Chandler, D., Bailey, A.S., Tatchell, G.M., Davidson, G., Greaves, J., Grant, W.P.: The development, regulation and use of biopesticides for integrated pest management. Philos. Trans. Roy. Soc. Lond. B: Biol. Sci. **366**, 1987–1998 (2011)
7. Chaturvedi, K., Rai, A., Dubey, V., Malhotra, P.: On-line analytical processing in agriculture using multidimensional cubes. Int. Indian Soc. Agric. Stat. J. **62**, 56–64 (2008)
8. Chaudhary, S., Sorathia, V., Laliwala, Z.: Architecture of sensor based agricultural information system for effective planning of farm activities. In: IEEE International Conference on Services Computing 2004, pp. 93–100 (2004)
9. de Olde, E.M., Oudshoorn, F.W., Bokkers, E.A., Stubsgaard, A., Sørensen, C.A., de Boer, I.J.: Assessing the sustainability performance of organic farms in Denmark. Sustainability **8**, 957 (2016). doi:10.3390/su8090957
10. Deytieux, V., Munier-Jolain, N., Caneill, J.: Assessing the sustainability of cropping systems in single-and multi-site studies. A review of methods. Eur. J. Agron. **72**, 107–126 (2016)
11. Dong, F., Mitchell, P.D., Colquhoun, J.: Measuring farm sustainability using data envelope analysis with principal components: the case of Wisconsin cranberry. J. Environ. Manage. **147**, 175–183 (2015)
12. European Parliament. Directive 2009/128/EC of the European Parliament and of the Council of 21 October 2009 establishing a framework for Community action to achieve the sustainable use of pesticides. Off. J. Eur. Union **309**, 71–86 (2009)
13. Fountas, S., Carli, G., Sørensen, C.G., Tsiropoulos, Z., Cavalaris, C., Vatsanidou, A., Liakos, B., Canavari, M., Wiebensohn, J., Tisserye, B.: Farm management information systems: Current situation and future perspectives. Comput. Electron. Agric. **115**, 40–50 (2015)

14. Gallo, C., De Bonis, M., Perilli, M.: Data Warehouse Design and Management: Theory and Practice. Department of Economics, University of Foggia, Italy, Quaderno n. 07/2010, pp. 1–18 (2010). http://econpapers.repec.org/paper/ufgqdsems/07-2010.htm
15. Grunert, K.G.: Food quality and safety: consumer perception and demand. Eur. Rev. Agric. Econ. **32**, 369–391 (2005)
16. Gupta, A., Mazumdar, B.: Multidimensional schema for agricultural data warehouse. Int. J. Res. Eng. Technol. **2**, 245–253 (2013)
17. ISTAT. Italian Institute of Statics (ISTAT) Agricultural Census (2010). http://dati-censimentoagricoltura.istat.it/. Accessed 20 Mar 2017
18. ISTAT. Italian Institute of Statics (ISTAT) "Superfici e Produzioni" (2016). www.istat.it/it/. Accessed 20 Mar 2017
19. Kimball, R., Ross, M.: The Data Warehouse Toolkit: The Definitive Guide to Dimensional Modeling. Wiley, New York (2013)
20. Langeveld, J.W.A., Verhagen, A., Neeteson, J.J., Van Keulen, H., Conijn, J.G., Schils, R.L.M., Oenema, J.: Evaluating farm performance using agri-environmental indicators: recent experiences for nitrogen management in The Netherlands. J. Environ. Manage. **82**, 363–376 (2007). doi:10.1016/j.jenvman.2005.11.021
21. Meul, M., Van Passel, S., Nevens, F., Dessein, J., Rogge, E., Mulier, A., Van Hauwermeiren, A.: MOTIFS: a monitoring tool for integrated farm sustainability. Agron. Sustain. Dev. **28**, 321–332 (2008)
22. Navrozidis, E.I., Vasara, E., Karamanlidou, G., Salpiggidis, G.K., Koliais, S.I.: Biological control of Bactroceraoleae (Diptera: Tephritidae) using a Greek Bacillus thuringiensis isolate. J. Econ. Entomol. **93**, 1657–1661 (2000)
23. Pinet, F., Schneider, M.: Precise design of environmental data warehouses. Oper. Res. Int. J. **10**, 349–369 (2010)
24. Prokopy, R.J.: Two decades of bottom-up, ecologically based pest management in a small commercial apple orchard in Massachusetts. Agr. Ecosyst. Environ. **94**, 299–309 (2003)
25. Reig-Martínez, E., Gómez-Limón, J.A., Picazo-Tadeo, A.J.: Ranking farms with a composite indicator of sustainability. Agric. Econ. **42**, 561–575 (2011). doi:10.1111/j.1574-0862.2011.00536.x
26. Rodrigues, G.S., Rodrigues, I.A., de Almeida Buschinelli, C.C., de Barros, I.: Integrated farm sustainability assessment for the environmental management of rural activities. Environ. Impact Assess. Rev. **30**, 229–239 (2010). doi:10.1016/j.eiar.2009.10.002
27. Seuring, S., Müller, M.: From a literature review to a conceptual framework for sustainable supply chain management. J. Clean. Prod. **16**, 1699–1710 (2008)
28. United Nation. World Population Prospects: The 2010 Revision. Department of Economic and Social Affairs. Population Division, New York (2010)
29. Vernier, F., Miralles, A., Pinet, F., Carluer, N., Gouy, V., Molla, G., Petit, K.: Pesticides: An environmental information system to characterize agricultural activities and calculate agro-environmental indicators at embedded watershed scales. Agric. Syst. **122**, 11–21 (2013)
30. WCED U: Our Common Future. World Commission on Environment and Development Oxford University Press, Oxford (1987)

Adaptive Prediction of Water Quality Using Computational Intelligence Techniques

Iván Darío López[1]([⊠]) [iD], Apolinar Figueroa[2] [iD],
and Juan Carlos Corrales[1] [iD]

[1] Grupo de Ingeniería Telemática (GIT), Universidad del Cauca,
Popayán, Colombia
{navis,jcorral}@unicauca.edu.co
[2] Grupo de Estudios Ambientales (GEA), Universidad del Cauca,
Popayán, Colombia
apolinar@unicauca.edu.co

Abstract. Water is not only vital for ecosystems, wildlife and human consumption, but also for activities such as agriculture, agro-industry, and fishing, among others. These are some of the activities that require water to be developed. However, in the same way as their water use has increased, it has also been detected an accelerated deterioration of its quality. In this sense, to have predictive knowledge about Water Quality (WQ) conditions, can provide a significant relevance to many socio-economic sectors. The premise of this paper is to predict the water quality for different uses of water (aquaculture, irrigation and human consumption) represented by several datasets. This approach is based on Support Vector Regression (SVR) technique configured with the Pearson VII Universal Kernel (PUK), and an evolutionary algorithm of Particle Swarm Optimization (PSO). Experimental results show that the proposed predictive mechanism provides an acceptable prediction accuracy on different water-use datasets. These results indicate that bio-inspired techniques improve the adaptive capacity of a prediction algorithm.

Keywords: Computational Intelligence · Forecasting · Particle Swarm Optimization · Support Vector Regression · Water Quality

1 Introduction

Ecological Quality of Water (EQW) refers to the set of physical, chemical, biological and radiological characteristics of surface and underground waterbodies [1]. Moreover, experts in Integrated Water Resources Management (IWRM) show that there is no single definition of EQW because it strictly depends on its use [2]. Thus, for example, water that cannot be used for human consumption may be used for other activities such as irrigation or aquaculture, among others, whereas it has specific characteristics that make it suitable for such use [3]. Some of this water is used by infrastructure installed by human, and most of the extracted water is then returned to the environment after it has been used. However, during the use, changes on its features may occur [4].

© Springer International Publishing AG 2017
O. Gervasi et al. (Eds.): ICCSA 2017, Part II, LNCS 10405, pp. 45–59, 2017.
DOI: 10.1007/978-3-319-62395-5_4

Furthermore, is important to clarify that this study focuses mainly on the physico-chemical component of EWQ.

In IWRM, there is a dynamic relationship between basin stakeholders and governmental institutions, which must work together to ensure the viability of their decisions in order to achieve the objectives of sustainable development. Traditionally, respective watershed management authorities and the responsible staff, collect samples from different water body points and different water bodies to determine the current status of water quality. This is possible by implementing plans for improvement and water management. However, this task involves time, both sampling and construction of quality data series, and in most cases, the management response is given below. This, occasionally could represent an economic and hugely environmental lost which affect territorial sustainability on a regional or global scale [5].

In this sense, to control suitable WQ conditions is not enough to establish monitoring activities which provide corrective actions to certain pollution types. Thus, predicting water quality plays an important role for many socio-economic sectors which depend on this liquid use. This scenario highlights the need of models or mechanisms to anticipate the materialization of pollution risk with plenty range of time, preventing negative effects that impact water resources quality.

Nevertheless, the reviewed literature focused the prediction process on a specific water use, leaving aside the generalization capacity of models, which in this study has been called "*adaptive capacity*". In view of this, the present study plans to guide the problem of low adaptive capacity of models for water quality prediction through the implementation of an adaptive WQ prediction system. Thus, if a prediction model gets accurate results when it is applied on a water-use, e.g. treatment for human consumption, same results should be acceptable if the model is applied to other uses such as water for fish farming, crops irrigation, river-based recreation, among others. Considering that WQ variables in these uses are not the same for drinking water treatment, the mechanism should have the capacity to generalize new situations, providing reliable predictions in each one of them.

Predicting water quality conditions in a watershed allows formulating management alternatives associated to the land use for agricultural or agro-industrial activities developed there; additionally, the planning processes and IWRM can be facilitated. Thus, measures or decisions to reduce the pollution risk of surface and ground water can be supported. This proposal is important because it would allow government authorities to take decisions about different productive sectors, which use water like an input for all their processes. In this way, these organisms would have a greater ability to forecast, in an integral and systemic capacity associated with the reality of ecosystems, land use, and human activities which makes it more flexible and useful.

This paper is organized as follows: Sect. 2 introduces the related studies to the main topics addressed around water quality adaptive prediction, Sect. 3 describes the study area and data sources for testing, Sect. 4 explain the proposed adaptive prediction mechanism, Sect. 5 shows the experimental evaluation results. And finally, Sect. 6 presents the conclusions of this research.

2 Related Studies

Recent developments in Computational Intelligence (CI) [6] are focused in methods inspired by biological processes. These techniques have the ability to learn and adapt to new situations, generalize, abstract, discover, and associate; therefore, these have become a promising alternative in many applications.CI defines different techniques such as Artificial Neural Networks (ANN) [7], Support Vector Machines (SVM), Evolutionary Computation (EC) [8], among others. Authors of [9] establish a list of squares regression method supported by SVM for prediction of WQ parameters. In the same direction, Guo [10] propose a hybrid methodology improved through SVM and ARIMA models in ordér to increase the predictability of WQ problems from time series of crops irrigation. On the other hand, Džeroski [11] proposes a regression tree for each variable in order to predict physicochemical parameters from biological features; while Poor [12] uses a regression tree to improve predictions of low-flow nitrate and chloride in Willamette river basin watersheds.

Moreover, predicting WQ parameters has been widely discussed within techniques like ANN [13, 14]. This paradigm is one of the most widely explored fields when addressing the above problem. Pop [15] presents a biologically inspired method to predict the time series values and describes how to apply it to the specific case of water quality monitoring for irrigation. Hatzikos [16] uses the ANN with activation neurons as a modeling tool for seawater quality prediction. Furthermore, in [17] a neural network is proposed to predict biological parameters from physicochemical features. Finally, several studies were developed on water quality data in Colombian, Spanish, and Mexican rivers respectively [18–20].

On the other hand, Huang and Liu [21] aim to couple a Hybrid Genetic Algorithm (HGA) and ANN for multi-objective calibration of surface water quality model. At the same time, a prediction model of water quality by SVM and an optimization of SVM parameters using GA is proposed in [22]. While Baltar and Fontane [23] propose a Multi-Objective Particle Swarm Optimization algorithm (MOPSO), which is applied to find solutions to minimize deviations from target variables such as total dissolved solids, dissolved oxygen, pH, and temperature.

According to the previous literature review, it is important to highlight the advances and contributions in the development of hybrid predictive models, primarily to cover the disadvantages of some techniques and improve them with the addition of other algorithms. It is in this regard that CI has introduced algorithms and predictive techniques which have the ability to estimate future water conditions based on the analysis of collected historical data. However, none of them considers the application of prediction model for different uses of water. The studies found are limited to make the prediction process with reference to data from a certain water-use. Sometimes skewing results in that particular scenario, regardless of the applicability that the same prediction model may have to other uses, which may bring benefits to different sectors and their productive processes. Moreover, there are techniques that despite of not having been thoroughly explored within an adaptive component, they have important features such as the ability to adapt to changing environments, as in the case of EC techniques: PSO and GA.

3 Data and Study Area

In order to develop an accurate predictive mechanism for different water-use datasets; it is important to have several data sources which represent different uses of water. To ensure an acceptable behavior of the mechanism, three datasets from two data sources were employed in this study. These were selected according to the following criteria: availability (free access data), quality (data reviewed by experts in water resources management), quantity (sufficient data for experimental tests), and finally the use of water in the sampling site (information about the productive processes of each site).

3.1 United States Geological Survey (USGS)

USGS is a research organization of the United States federal government [24]. This agency provides reliable water quality data in the United States for public access. The accessed data are labeled into two general categories: data approved for publication (A), which have been processed and thoroughly reviewed by the USGS staff; and provisional data subject to revision (P), which have no approval of the review staff. In this study, data with label (A) was selected.

These datasets belong to two sampling sites in the state of California. The first one comprising the territory of Alviso; Guadalupe River and Coyote Creek end up in Alviso wetland through an estuary that flows into the San Francisco bay, and one of the main activities is fishing. Variables such as temperature, conductivity, turbidity, suspended sediment concentration, and dissolved oxygen were considered in this dataset, with a periodicity of 15 min (96 samples daily for 5 days). The second one is Don Pedro Lake, located in Mariposa County, which covers an area of 32.56 km^2 where one of the main uses is the irrigation. This dataset was composed of the following variables: temperature, conductivity, pH, dissolved oxygen, turbidity, colored dissolved organic matter, chlorophyll a, and nitrate. The sampling interval was 1 h for 5 days.

3.2 Cauca River Modeling Project Phase II (PMC II)

PMC II is a project for monitoring the water quality of Cauca River [25]. The study area was the stretch *Hormiguero - Mediacanoa*, specifically the monitoring station of *Puente Juanchito*, near Santiago de Cali city, Colombia, where the main water use is human consumption. Cauca River Basin is the second largest waterway of Colombia and crosses around 183 municipalities, representing approximately 41% of the Colombian population. WQ variables in this dataset are as follows: temperature, conductivity, dissolved oxygen, biochemical oxygen demand 5, chemical oxygen demand, pH, and total suspended solids. The total sampling interval was 5 days, with 24 samples per day.

4   Water Quality Adaptive Prediction

To implement the adaptive prediction mechanism, CRISP-DM methodology (Cross Industry Standard Process for Data Mining) [26] was used, which is the reference guide most widely used in the development of data mining projects. Additionally, in order to guide the mechanism construction and modeling, a methodology oriented to prediction process was chosen. This methodology is called SVM-UP (Support Vector Machine - Unified Process) [27] and it was developed to implement regression models for time series forecasting. Adaptive prediction mechanism is comprised of components seen in Fig. 1. Below, parameters selection, predictive and adaptive components are described, including CI algorithms that compose them.

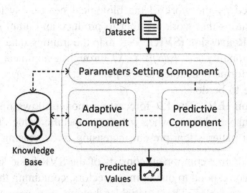

Fig. 1. General scheme of the adaptive prediction mechanism.

4.1   Parameters Setting Component

The purpose of this component is to obtain a relevant subset of data from the original dataset, which is associated with a specific use of water. It is based on two selection strategies; the first one uses methods defined by other authors. Guyon and Elisseeff [28] classify them into three categories: filter, wrapper, and embedded methods; in this study, wrapper selection method (Sequential Forward Selection - SFS) was chosen considering the high performance according to the survey made by Hamon [29]. The second strategy was to use a Knowledge Base (KB) of different uses of water.

To better understand the operation of this component, an example of parameters selection is shown below. Considering an aquaculture-use data set D, where n is the number of variables, in this case $n = 5$. V is the set of variables in D, where $V = \{temperature, conductivity, turbidity, suspended sediment concentration, dissolved oxygen\}$. By SFS selection method, V' is generated from V, i.e., $V' = \{temperature, conductivity, turbidity, dissolved oxygen\}$. On the other hand, KB contains, among others, the set R, which stores information from experts about the relevant variables in aquaculture, i.e., $R = \{temperature, conductivity\}$. Finally, P is the set of selected variables, which is the output of parameters setting component (intersection between the sets V' and R), i.e., $P = \{temperature, conductivity\}$.

4.2   Predictive Component

As its name implies, predictive component is responsible for performing prediction operations. This component uses the subset of data returned by parameters setting component. To determine the CI algorithm that offers better results in predicting water quality, it is necessary to have a theoretical basis to establish the reference algorithms for comparison. Studies refer to the use of ANN (Multilayer Perceptron) and Support Vector Regression (SVR) as strategies for obtaining good results in predicting water quality from time series values [9, 30, 31]. Additionally, Linear Regression method (LR) was evaluated as being traditionally used to solve such problems (evaluation results are shown in Results and Discussion section).

- **Artificial Neural Networks (ANN).** ANN is a paradigm of learning and automatic processing inspired by the work of the biological nervous system. It is an interconnection of neurons that work together to produce an output stimulus.
- **Support Vector Regression (SVR).** SVR map the training data $x \in X$, into a space of higher dimensionality $F = \{\alpha(x) | x \in X\}$ through a nonlinear mapping where a linear regression, formally denoted as $x = \{x_1, x_2, ..., x_n\} \rightarrow \alpha(x) = \{\alpha(x)_1, \alpha(x)_2, ..., \alpha(x)_n\}$, is possible to be done.
- **Linear Regression (LR).** The LR forecasting model allows finding the expected value of a random variable, a, when b takes a specific value. This forecast is an optimal model for patterns that present increasing or decreasing trends.

Formally, the predictive component consists of the SVR basic architecture, except for its inputs, which correspond to one or more vectors, containing the data of each one of the water quality variables. These variables belong to a particular use and have previously been selected by the parameters setting component. These vectors are denoted by $x = \{x_1, x_2, ..., x_n\}$ and they are connected directly to one or more *Pearson VII Universal Kernel (PUK)* functions, responsible for mapping the input space X to a new space with higher dimensional features. In addition, and according to the literature review, PUK kernel has the possibility to easily change through parameter adaptation, starting from a Gaussian to a Lorentzian function. It means that this property uses PUK as a generic Kernel able to replace other methods [32, 33].

The output of this component is a vector for each water quality variable which contains predicted values using SVR. These values are then used as input by the adaptive component. As a summary, this component is the main part of the proposed mechanism and was selected because it gets a better performance in accuracy with a significant reduction in the predicted values error percentage (results are presented in Results and Discussion section).

4.3   Adaptive Component

The previous component has a close relationship with the adaptive component, which aims to implement a technique to adjust the predicted to the actual values for a specific water-use, i.e., the optimization aim for this study is focused on minimizing the difference between the two curves (predicted values and actual values). Consequently, the adaptive component also requires information from the knowledge base to adjust the input vectors to the predicted values. For each of these vectors, a set of particles is created, which by their interaction and cooperation try to find new values more accurate than predicted values obtained in the predictive component. Each set of particles is based on an adaptive function (objective function) that is responsible for obtaining the prediction value adapted to each of the predicted values.

In order to optimize the corresponding predicted values to the water quality parameters of different uses of water, two of the most used and best performing evolutionary techniques were selected, considering the studies presented in [34–36]: Genetic Algorithms (GA) and Particle Swarm Optimization (PSO) (experimental evaluation is shown in Sect. 5).

- **Genetic Algorithms (GA).** GA is inspired by biological evolution and genetic-molecular base. These algorithms make a population of individuals evolve by subjecting it to such random actions that act in biological evolution (mutations and genetic recombination). GA select individuals according to some criterion for deciding which individuals are more adapted to survive and which of them are the less fit, which are discarded.
- **Particle Swarm Optimization (PSO).** PSO is a CI technique based on population seeking to emulate the social behavior of certain groups of individuals such as flocks of birds in flight or some fish families. In this sense, the main focus of this technique is to provide a simple and effective optimization algorithm. Individuals are known as particles moving through a hyper-dimensional search space and each one has a velocity and a position. Therefore, changes in a particle within the group are influenced by experience or knowledge of their neighbors.

5 Experimental Results

As a starting point, the selected variables used as input for each dataset are shown in Table 1 (variables obtained from parameters setting component). Furthermore, to select the most accurate regressor, different tests were performed to adaptive prediction mechanism. The evaluation process was developed by 10-fold cross-validation method following the experimental baseline shown in [9, 14, 31]. In this way, it was divided into two phases; the first one was the algorithm selection (prediction of time series with a water quality dataset to obtain the most accurate results). In the second phase, it was defined and set an algorithm to adapt the predicted values on different water-use datasets. Tests conducted on each of these phases are detailed below.

Table 1. Input variables for each dataset (USGS and PMC II).

Dataset	Sampling site	Instances	Parameter
USGS	Alviso estuary	480	Water temperature
			Water conductance
	Don Pedro Lake	120	Water temperature
			Water conductance
PMC II	Puente Juanchito station	192	Water temperature
			Water conductance
			Dissolved oxygen
			pH

5.1 Regressor Selection for Predictive Component

Initially, a dataset of water temperature values corresponding to 96 instances was obtained (daily recordings with measurements every 15 min). This dataset was the entry for each of the three prediction algorithms, and a predicted value for the same variable was the expected output for each one. The software used to implement the algorithms was Weka 3.7.12, and the metrics to determine the algorithms accuracy are given in Table 2.

Table 2. Accuracy metrics for the three prediction algorithms (one predicted value).

Precision metric	LR	ANN	SVR
Mean Absolute Error (MAE)	0.051	0.058	0.044
Root Mean Square Error (RMSE)	0.069	0.075	0.081
Mean Absolute Percentage Error (MAPE)	0.239	0.274	0.206

The results presented in Table 1 show that the SVR technique is more accurate compared to the ANN and LR since the Mean Absolute Error was less than the others (0.44). Furthermore, the Mean Absolute Percentage Error was 20.6%. However, this error rate was still considerably high for a good prediction model. On this basis, in this paper, a new assessment of the three algorithms was performed, but in this case increasing the size of input data (480 instances) and the number of values to predict (five values). Accuracy metrics for this new test are reported in Table 3.

Table 3. Accuracy metrics for the three prediction algorithms (five predicted values).

Precision metric	LR	ANN	SVR
Mean Absolute Error	0.053	0.058	0.050
Root Mean Square Error	0.077	0.078	0.084
Mean Absolute Percentage Error	0.281	0.309	0.263

On average, SVR performed better than ANN and LR predicting 5 values. As a result, SVR with PUK kernel was selected as the predictive component regressor to be used in this study.

5.2   Algorithm Selection for Adaptive Component

From previous section, predicted values for each variable were obtained initially by SVR configured with the PUK kernel. In a second evaluation phase, different tests were conducted applying SVR-PUK model with three different datasets: Alviso estuary, which represents the aquaculture use; Don Pedro Lake, for irrigation; and Puente Juanchito station for human consumption. These datasets were preprocessed inside the parameters setting component by a features selection process. In addition, conductance values were normalized considering the different levels or/and scales caused by the salinity of certain water sources; as a consequence, noise could be added to the model. In this case, Alviso estuary is the confluence between the river and the sea, so it has a higher salinity. Similarly, Don Pedro Lake shows high levels of salinity, which contrasts with the presence of low salt levels in Cauca River at Puente Juanchito station.

Figure 2 shows predicted values of water temperature in (a) Alviso estuary, (b) Don Pedro Lake, and (c) Puente Juanchito station (predicted values of conductance, dissolved oxygen and pH are referenced in [37]). In the same way, results obtained from the preliminary analysis of predictive performance for each variable, are summarized in Table 4.

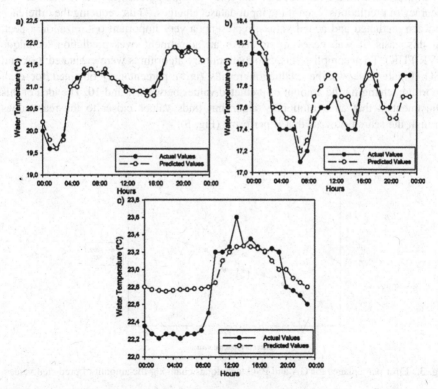

Fig. 2. Temperature predicted values using SVR with PUK kernel for three datasets. (a) aquaculture, (b) irrigation, (c) human consumption.

Table 4. Accuracy of SVR-PUK algorithm for each predicted variable defined in two datasets (USGS and PMC II).

Dataset	Sampling Site	Parameter	MAE	RMSE	MAPE
USGS	Alviso estuary	Water temperature	0.0216	0.0398	0.1138
		Water conductance	0.3280	0.3912	0.1689
	Don Pedro Lake	Water temperature	0.1971	0.2423	0.1938
		Water conductance	0.2174	0.2591	0.3552
PMC II	Puente Juanchito station	Water temperature	0.3612	0.4368	0.5868
		Water conductance	0.0741	0.1432	0.6523
		Dissolved oxygen	0.3759	0.4589	0.6931
		pH	0.0992	0.1362	0.5554

Comparing the above curves, predicted values of temperature for the first curve in Alviso estuary (Fig. 2(a)) were approximated to the actual values for this variable. This is evidenced by the low error percentage obtained (11.38% and 16.89% for water temperature and conductance respectively). However, when tests were applied using the other two datasets, the predicted values showed a different behavior (Fig. 2(b) and (c)). In the same way, the error rate increases considerably for each variable.

In this scenario, the objective of this study is highlighted, which is to maintain the accuracy of predictions, even if the input dataset changes. Thus, reducing the difference between predicted and actual values, represents a very important optimization aspect. On this basis, it was necessary to make an adjustment over prediction technique (SVR-PUK). To accomplish this, two evolutionary algorithms were evaluated: GA and PSO. Figure 4 shows the evaluation results (error percentages) obtained for each algorithm, changing the amount of predicted values between 1 and 10. The idea of this adjustment is that an evolutionary algorithm finds values closer to the real values through the search process that it performs (Fig. 3).

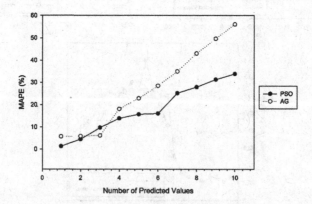

Fig. 3. Error percentages for GA and PSO techniques changing the amount of predicted values.

As can be seen from the previous figure, PSO is more accurate than GA, particularly when the number of predicted values increases. For this reason, PSO algorithm was selected as the main element of the adaptive component, and it was setting with the following parameters for all datasets (based in [38]): number of particles = 10, Vmax = 20, learning factors = 2, and maximum number of iterations = 2000. Each individual in the population tries to find a local solution and at the end of the number of iterations, find a global one. The search space is composed of values stored in the knowledge base, which was previously used in the parameter setting component. Finally, the global solution represents the new predicted value for a given physico-chemical variable.

Based on these considerations, Fig. 4 shows water temperature predicted values (conductivity, dissolved oxygen, and pH predicted values are presented in [37]) using the proposed hybrid mechanism (SVR-PUK-PSO) with three water-use datasets (a) aquaculture, (b) irrigation, and (c) human consumption). Table 5 summarizes the SVR-PUK-PSO algorithm accuracy for each variable of the three sampling sites.

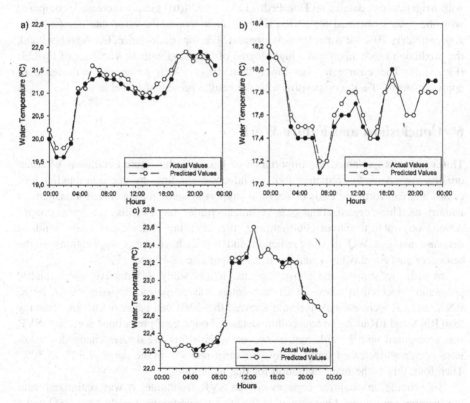

Fig. 4. Temperature predicted values using SVR-PUK and PSO techniques for three datasets. (a) aquaculture, (b) irrigation, (c) human consumption.

Table 5. Accuracy of SVR-PUK-PSO techniques for each of the predicted variables defined in two datasets (USGS and PMC II).

Dataset	Sampling site	Parameter	MAE	RMSE	MAPE
USGS	Alviso estuary	Water temperature	0.1672	0.2556	0.2438
		Water conductance	0.1952	0.2375	0.2239
	Don Pedro Lake	Water temperature	0.0586	0.0789	0.3095
		Water conductance	0.1521	0.2043	0.2862
PMC II	Puente Juanchito station	Water temperature	0.0108	0.0499	0.0473
		Water conductance	0.0205	0.0480	0.0889
		Dissolved oxygen	0.004	0.0079	0.0901
		pH	0.0038	0.0179	0.0514

Similar results of accuracy were obtained for the predicted values using SVR-PSO; however, the accuracy decreases with the aquaculture dataset in Alviso estuary (Fig. 4(a)). This decrease is not significant considering that predicted values accuracy with irrigation-use dataset on Don Pedro Lake (Fig. 4(b)) greatly increased, compared with the values obtained for SVR-PUK. In other words, the error rate decreased by approximately 70% for water temperature and 50% for conductance. On the other hand, the prediction mechanism had a high degree of precision with all variables of PMC II (Fig. 4(c)). The error rate was lower than 10%, which represents a decrease of approximately 50% in comparison with the results presented in Table 4.

6 Conclusions and Future Work

This paper has explained the importance of knowing the future conditions of water quality. An adequate predictive tool would allow guiding the decision-making processes and can form the basis of decision-support systems for watershed management authorities. These organizations or government entities may establish integrated control actions on water resources. Furthermore, the sampling and lab-test time could be decrease using a WQ in situ-prediction and it will have direct implications in the resources cost for this tasks making it less expensive and faster [5].

In order to support the above aspects, in this study an adaptive water quality prediction mechanism based on CI techniques was proposed. Experimentally, SVR, RNA, and LR were evaluated; results showed that SVR had better prediction accuracy than RNA and LR using an aquaculture dataset. To increase predictions accuracy, SVR was configured with PUK kernel. However, when water-use dataset changed, predictions accuracy decreased significantly. The error rate was in the range of 30% to 70%. Therefore, this is the main limitation for this technique.

To provide an adaptive capacity to the SVR technique, it was optimized with bio-inspired algorithms. On average, in the adaptive component evaluation, PSO had a better performance than GA technique. In a new iteration of prediction process, the error rate was lower than 31% for each of water-use datasets. These results suggest that SVR-PUK and PSO combination can predict acceptable water quality values regardless of usage. Returning to the hypothesis posed at the beginning, it is now possible to state that bio-inspired techniques improve the adaptive capacity of a prediction algorithm. Additionally, by incorporating water quality information through a knowledge base allowed to reinforce the algorithms adaptation process. In this sense, the predictive mechanism can be applied to any water quality dataset. However, the prediction accuracy could decrease if the actual values for some variables have different scales depending upon their water-use dataset. To address this problem, it was necessary to normalize these values.

Finally, future work aims to complement the prediction mechanism with two specific improvements: in the first place, research is also needed to determine automatic water-use detection from a specific dataset. In order to improve mechanism adaptability, it must be tested with more datasets of uses of water, distinct to the three datasets analyzed in this proposal. Additionally, it would be interesting to identify possible uses of water from a specific dataset and their characteristics. In the second place, provide recommendations to the staff responsible for water resources management, it is suggested automatic causes detection of water pollution from a context analysis.

Acknowledgements. The authors are grateful to the Telematics Engineering Group (GIT) and Environmental Studies Group (GEA) of the University of Cauca, Institute CINARA of the University of Valle, RICCLISA Program and AgroCloud project for supporting this research, and the Administrative Department of Science, Technology and Innovation (Colciencias) for PhD scholarship granted to MSc. Iván Darío López.

References

1. La gestión del Agua en Castilla-La Mancha. http://pagina.jccm.es/agenciadelagua/. Accessed 11 Mar 2015
2. Comunidad Autónoma de Extremadura: Agentes Forestales de Extremadura (ForestryAgents of Extremadura). MAD-Eduforma, Sevilla, Spain (2003)
3. Carbó, C.: Genética, patología, higiene y residuos animales (Genetics, pathology, hygiene and animal waste), vol. 4. Mundi-Prensa Libros, Madrid (1995)
4. AQUASTAT - FAO's Information System on Water and Agriculture. http://www.fao.org/nr/water/aquastat/water_use/indexesp.stm. Accessed 16 Apr 2015
5. López, I.D., Valencia, C.H., Corrales, J.C.: Predicting water quality based on multiple classifier systems. In: 21st Century Watershed Technology Conference and Workshop Improving Water Quality and the Environment Conference, pp. 1–10. ASABE, Quito, Ecuador (2016)
6. Engelbrecht, A.P.: Computational Intelligence: An Introduction, 2nd edn. Wiley, England (2007)
7. Gurney, K.: An Introduction to Neural Networks, 1st edn. Taylor & Francis, Bristol (2003)
8. Goldberg, D.E.: Genetic Algorithms in Search, Optimization, and Machine Learning, 1st edn. Addison-Wesley, Boston (1989)
9. Tan, G., Yan, J., Gao, C., Yang, S.: Prediction of water quality time series data based on least squares support vector machine. Procedia Eng. **31**, 1194–1199 (2012)

10. Guo, Y., Wang, G., Zhang, X., Deng, W.: An improved hybrid ARIMA and support vector machine model for water quality prediction. In: Miao, D., Pedrycz, W., Ślęzak, D., Peters, G., Hu, Q., Wang, R. (eds.) RSKT 2014. LNCS, vol. 8818, pp. 411–422. Springer, Cham (2014). doi:10.1007/978-3-319-11740-9_38

11. Džeroski, S., Demšar, D., Grbović, J.: Predicting chemical parameters of river water quality from bioindicator data. Appl. Intell. 13(1), 7–17 (2000)

12. Poor, C.J., Ullman, J.L.: Using regression tree analysis to improve predictions of low-flow nitrate and chloride in Willamette river basin watersheds. Environ. Manag. 46(5), 771–780 (2010)

13. Romero, C.E., Shan, J.: Development of an artificial neural network-based software for prediction of power plant canal water discharge temperature. Expert Syst. Appl. 29(4), 831–838 (2005)

14. Palani, S., Liong, S.-Y., Tkalich, P.: An ANN application for water quality forecasting. Mar. Pollut. Bull. 56(9), 1586–1597 (2008)

15. Pop, F., Ciolofan, S., Negru, C., Mocanu, M., Cristea, V.: A bio-inspired prediction method for water quality in a cyber-infrastructure architecture. In: 2014 Eighth International Conference on Complex, Intelligent and Software Intensive Systems (CISIS), pp. 367–372. IEEE, Birmingham, UK (2014)

16. Hatzikos, E., Hätönen, J., Bassiliades, N., Vlahavas, I., Fournou, E.: Applying adaptive prediction to sea-water quality measurements. Expert Syst. Appl. 36(3), 6773–6779 (2009)

17. He, L.-M.L., He, Z.-L.: Water quality prediction of marine recreational beaches receiving watershed baseflow and stormwater runoff in Southern California, USA. Water Res. 42(10–11), 2563–2573 (2008)

18. Gutiérrez, J., Riss, W., Ospina, R.: Bioindicación de la calidad del agua con macroinvertebrados acuáticos en la sabana de Bogotá, utilizando redes neuronales artificiales. Caldasia 26(1), 151–160 (2004)

19. Saint-Gerons, A.I., Adrados, J.M.: Desarrollo de una Red Neuronal para estimar el Oxígeno Disuelto en el agua a partir de instrumentación de E.D.A.R. In: XXV Jornadas de Automática, Navarra, España (2004)

20. García, I., Rodríguez, J.G., López, F., Tenorio, Y.M.: Transporte de contaminantes en aguas subterráneas mediante redes neuronales artificiales. Información Tecnológica 21(5), 79–86 (2010)

21. Huang, Y., Liu, L.: Multiobjective water quality model calibration using a hybrid genetic algorithm and neural network-based approach. J. Environ. Eng. 136(10), 1020–1031 (2010)

22. He, T., Chen, P.: Prediction of water-quality based on wavelet transform using vector machine. In: 2010 Ninth International Symposium on Distributed Computing and Applications to Business Engineering and Science (DCABES), pp. 76–81. IEEE (2010)

23. Baltar, A.M., Fontane, D.G.: A generalized multiobjective particle swarm optimization solver for spreadsheet models: application to water quality. In: Proceedings of the Twenty Sixth Annual American Geophysical Union Hydrology Days, pp. 20–22 (2006)

24. USGS - U. S. Geological Survey. http://www.usgs.gov/. Accessed 14 Apr 2015

25. CVC: Segunda campaña de muestreo con propositos de calibracion del modelo de calidad del agua del rio cauca. Corporacion Autonoma Regional del Valle del Cauca Caracterizacion Modelamiento Matemático del Río Cauca -PMC- Fase II Convenio Interadministrativo 0168, 27 November 2002, vol. 15 (2005)

26. Chapman, J.C.: CRISP-DM 1.0: Step-by-Step Data Mining Guide. SPSS, USA (1999)

27. Guajardo, J., Weber, R., Miranda, J.: A forecasting methodology using support vector regression and dynamic feature selection. J. Inf. Knowl. Manag. 5(4), 329–335 (2006)

28. Guyon, I., Elisseeff, A.: An introduction to variable and feature selection. J. Mach. Learn. Res. 3, 1157–1182 (2003)

29. Hamon, J.: Combinatorial optimization for variable selection in high dimensional regression: Application in animal genetic. Université des Sciences et Technologie de Lille - Lille I (2013)

30. Faruk, D.Ö.: A hybrid neural network and ARIMA model for water quality time series prediction. Eng. Appl. Artif. Intell. 23(4), 586–594 (2010)

31. Thoe, W., Wong, S.H.C., Choi, K.W., Lee, J.H.W.: Daily prediction of marine beach water quality in Hong Kong. J. Hydro-Environ. Res. 6(3), 164–180 (2012)

32. Abakar, K.A., Yu, C.: Performance of SVM based on PUK kernel in comparison to SVM based on RBF kernel in prediction of yarn tenacity. Indian J. Fibre Text. Res. 39(1), 55–59 (2014)

33. Üstün, B., Melssen, W.J., Buydens, L.M.C.: Facilitating the application of support vector regression by using a universal pearson VII function based kernel. Chemom. Intell. Lab. Syst. 81(1), 29–40 (2006)

34. Hassan, R., Cohanim, B., De Weck, O., Venter, G.: A comparison of particle swarm optimization and the genetic algorithm. In: 46th AIAA/ASME/ASCE/AHS/ASC Structures, Structural Dynamics and Materials Conference. American Institute of Aeronautics and Astronautics, Austin, Texas, USA (2005)

35. Chiu, C.-C., Cheng, Y.-T., Chang, C.-W.: Comparison of particle swarm optimization and genetic algorithm for the path loss reduction in an urban area. J. Appl. Sci. Eng. 15(4), 371–380 (2012)

36. D'heygere, T., Goethals, P.L.M., De Pauw, N.: Genetic algorithms for optimization of predictive ecosystems models based on decision trees and neural networks. Ecol. Model. 195 (1–2), 20–29 (2006)

37. López, I.D., Corrales, J.C.: Predicción Adaptativa de la Calidad del Agua mediante Técnicas de Inteligencia Computacional (Adaptive Prediction of Water Quality using Computational Intelligence Techniques). MSc. Thesis, Telematics Engineering Department, University of Cauca, Popayán, Colombia (2016)

38. Kennedy, J., Eberhart, R.C.: Particle swarm optimization. In: Proceedings of IEEE International Conference on Neural Networks, pp. 1942–1948, Piscataway, NJ, USA (1995)

A Tool for Classification of Cacao Production in Colombia Based on Multiple Classifier Systems

Julián Eduardo Plazas ⓘ, Iván Darío López$^{(\boxtimes)}$ ⓘ,
and Juan Carlos Corrales ⓘ

Grupo de Ingeniería Telemática (GIT), Universidad del Cauca,
Popayán, Colombia
{jeplazas,navis,jcorral}@unicauca.edu.co

Abstract. Cacao is one of the central crops that support the agrarian production in Colombia. In some areas, it is the main source of income for about 25.000 families. Frequently, its production is affected by several factors such as climate, soil, water, wind, among others. This paper presents a Machine Learning approach for classifying cacao production in the region of Santander, Colombia. The proposed system aims to link climate and cacao production data to develop the classification task. In this sense, several techniques were experimentally evaluated in order to determine the algorithm that generates the best model to classify new climate instances on the cacao production dataset. Experimental results showed a better precision for Random Forest in comparison with other evaluated techniques.

Keywords: Cacao production · Classifiers · Multiple Classifier Systems · Classification · Random Forest · Supervised learning

1 Introduction

Cacao culture is an agricultural activity that involves the cultivation of *Theobroma cacao* (cocoa tree) to obtain cocoa beans, which are used in the elaboration of chocolate. It is important to mention the difference between the terms cacao and cocoa. Cacao refers to the tree, its pods and the beans inside; while cocoa refers to two by-products of the cacao bean: cocoa powder and cocoa butter; both are extracted from the bean when it is processed in the factory [1]. In this study, we employ the term "cacao", because the data used belong to the production of this crop in its natural state.

Cacao is an evergreen tree that is always in flowering and growing up to 10 m high. To ensure a proper growth and development, several factors are required like a humid climate with a temperature between 20 °C and 30 °C, shade, wind protection, and a porous soil rich in nitrogen, magnesium, and potassium. Additionally, its ideal altitude for cultivation is of 400 m above sea level [1]. Colombia is an agrarian country and its economy largely depends on crop productivity. In this sense, cacao is one of the most important crops; in some areas, it is the main source of income for about 25000 families. Departments of Santander, Huila, Norte de Santander, Arauca, and Tolima are

© Springer International Publishing AG 2017
O. Gervasi et al. (Eds.): ICCSA 2017, Part II, LNCS 10405, pp. 60–69, 2017.
DOI: 10.1007/978-3-319-62395-5_5

the main producers with a contribution of approximately 75% of total national cacao production, which in 2004 amounted to 41703 tons [2].

Production is affected by several factors such as climate, which influences crop growth. In order to be economically sustainable, cacao cultivation must have yields above 1500 kg/ha of dry grain annually from the fifth year of establishment. To guarantee this, it is not enough that the crop is established in areas with appropriate conditions of climate and soil, or to plant certified seed of good quality; it is also necessary the timely and adequate application of practices for the profitability of both the crop and the harvest. In view of this, it is necessary to have models or mechanisms to anticipate the production for a crop in a short-term (cacao in this case), preventing negative effects that impact agricultural activities for sustainability of the crop production stakeholders.

Some studies propose strategies around modelling crops for management under certain climatic conditions [3, 4]. In other cases, works are based on methods used for the analysis of statistical data over a period of time (time series and statistical analysis). Recently, Machine Learning (ML) has introduced algorithms and predictive techniques, which have the ability to classify different values of sowing and production, based on the analysis of collected historical data. Techniques like Artificial Neural Networks (ANN), Decision Trees (DT), K-Means clustering, K Nearest Neighbor (KNN), Support Vector Machine (SVM), among others, have been applied in the domain of agriculture with acceptable results [5]. In the same way, classifications made by Multiple Classifier Systems (MCS) are often more accurate than classifications made by the best single classifier, taking into account the combination of a set of classifiers [6, 7]. Finally, in [8], *AdaSVM* and *AdaNaive* are proposed as an ensemble model to project the rice paddy, cotton, sugarcane, groundnut, and black gram production over a period of time.

This paper presents a web interface to determine a possible cacao production value from different climatic conditions on a specific area. In this development, various architectures for combining classifiers were tested with the aim of establishing the most accurate one for improving the classification precision. The remainder of this paper is organized as followed: Sect. 2 defines the study area and the data sources for testing. Subsequently, in Sect. 3, the general method for preprocessing the climatic and cacao production data sets is described. Section 4 presents the model selection process, and the experimental evaluation results. Section 5 shows the model deployment by a web application. Finally, Sect. 6 highlights the conclusions of this study.

2 Data and Study Area

To acquire cacao yield data, this study focused its search on data belonging to official entities. One of these is the National Federation of Cacao Growers of Colombia (FEDECACAO) [9], which is an organization for the continuous improvement of cacao cultures. It provides a free database of the cocoa economy in Colombia with three main sections: Annual production of cocoa beans by department (departmental records from 2002 to 2014); Area, production and annual yield of cocoa (national records from 2005 to 2011); and Monthly production of cocoa beans (national records from 2006 to 2010).

In order to define the time period and the geographical area where the data were obtained; information about annual production of cocoa beans by departments was used. It was detected that the most important department for the production of cacao in Colombia is Santander, which contributes about 50% of the annual national production. This region is located in the central northern part of the country, in the Andean region; it has an area of 30537 km^2 and a population of 2061 million people. In the same way, there was a greater amount of cacao production data between 2006 and 2010 (monthly information).

After the previous process, it was proceeded to look for climatological information of Santander between 2006 and 2010. The first source of climatological data was tutiempo.net [10], which contains data collected at the climatic stations of the Colombian main airports. This source includes data from two regions of Santander: Barrancabermeja (1964–2016) and Bucaramanga (1975–2016); the second region was the most complete for the period established (2006–2010). However, this climate information belongs to a single point, while production is at the departmental level; thus, it was necessary to look for another source of climatic data. NOAA's NCEP/NCAR Reanalysis [11] collects satellite data on climate values since 1948 and the information of the entire department was obtained at the dates defined. Therefore, NOAA data source was used instead of the climatic data of tutiempo.net.

3 Data Preprocessing

In order to obtain an appropriate dataset for cacao culture support systems, raw data was cleaned, constructed, integrated and formatted by an iterative process before having an acceptable version of the dataset [12]. The first step was data selection. For cacao production dataset, the monthly data was complete; however, its spatial scale was in a national level. On the other hand, climate data source contained about 48 climatic variables from around the world, however, only some of these attributes like *daily average temperature (°C)*, *maximum daily temperature (°C)*, *daily minimum temperature (°C)*, *mean daily relative humidity (%)*, and *total daily precipitation rate (mm/h)* were available for Santander region.

In the second step, datasets were cleaned. Here, only the climate data in Santander were modified. *Daily minimum temperature* and *daily maximum temperature* were discarded, since these attributes had constant values throughout the dataset (these were considered as atypical values). In addition, the negative values found in the *daily total precipitation rate* were replaced by zero, assuming that any value below zero represents a day without rain. Furthermore, no negative values were found in the other variables.

Data construction corresponds to the third step. In this phase, aiming to determine the monthly production of cacao in the region of Santander, the annual contribution percentage of this region from 2006 to 2010 (Table 1) was used to scale the monthly values of national production to monthly values of regional production; these percentages are based on [9].

Table 1. Annual contribution of Santander to the national cacao production.

Year	Annual contribution
2006	50%
2007	50%
2008	48%
2009	48%
2010	46%

Once the approximate values of cocoa production in Santander were obtained for the 2006–2010 period, the instances were multiplied in a new subset. The total production in a given month was assigned to each day of that month; this multiplication was carried out precisely, including even the leap years. Moreover, variables like *average daily temperature* and *average daily relative humidity* were selected; however, *total daily precipitation* rate was converted into *average daily cumulative precipitation* by the recommendation of a machine-learning expert.

The fourth step was data integration; considering both data subsets (climate and production) contain the same number of instances (one instance per day in the period 2006–2010), these were integrated into a preliminary dataset with about 1800 instances, and four attributes: *month, average daily temperature, average daily relative humidity* and *average daily cumulative precipitation* in Santander. In addition, a numerical class (*cacao production* in Santander) was defined.

Subsequently, three datasets with different time scales (daily, weekly, fortnightly) were selected, considering the climate data have a daily scale, the production data have a monthly scale, but the cacao harvests occur every two weeks:

- **Daily scale dataset.** It is the same as the preliminary dataset, so all instances are preserved.
- **Weekly scale dataset.** A new dataset was constructed based on the preliminary dataset, but taking the average climate of the last seven days as instances (or accumulated in case of rainfall). In the same way, *average weekly minimum temperature* and *average maximum weekly temperature* were selected (preserving the production of the corresponding month, and maintaining a number of instances close to 1800).
- **Fortnightly scale dataset.** Same as the weekly scale dataset, but extending the sampling range to fourteen days.

In order to define which of these datasets is most appropriate to continue with the modeling phase, experimental tests were conducted by three supervised learning algorithms implemented in WEKA environment [13]: M5P, M5Rules, and kNN [14]. Experimental results performed with cross-validation [15] are shown in Table 2 (the comparison criterion was the correlation coefficient).

As shown in Table 2, when the time scale is changed from daily to weekly, there was a large increase in the efficiency of all tested algorithms (by comparing the correlation coefficient). This increase is much more significant on a fortnightly scale. From this preliminary comparison, it is possible to identify that the last dataset presents a

Table 2. Comparison of algorithms and data sets by the correlation coefficient criterion.

Data set	Algorithm		
	M5P	M5Rules	kNN
Daily	0.8648	0.8605	0.867
Weekly	0.9255	0.8992	0.9636
Fortnightly	0.9619	0.9631	0.9908

better behavior when performing the modeling with supervised learning algorithms, and therefore can be considered as the best of the three.

In the fifth step (data formatting), the numerical target class "*cacao production*" was discretized. Five ranges of values were defined as "*very low production*" (0–876 tons), "*low production*" (876–1274 tons), "*medium production*" (1274–1673 tons), "*high production*" (1673–2071 tons), and "*very high production*" (>2071 tons).

In the final data set, the monthly production level (very low, low, medium, high, very high) can be identified according to the average climate of the last week. This dataset has 1813 instances (all of these without missing values), 6 attributes (five continuous and one discrete), and the class to be predicted, which was discretized with 5 different values (very low, low, medium, high, and very high), with a maximum imbalance of 50%. At the end of the initial data preparation, an acceptable version of the dataset is available using the fortnightly scale. It is possible to continue with the modeling phase to define the algorithm that best allows to classify new instances in the dataset.

4 Model Selection

In order to train classifiers, it is necessary to describe the main characteristics of the final data set, the method used to test the resulting models, metrics to evaluate them, and the algorithms applied in the modeling process. Thus, to generate an accurate data model from the described dataset, the following algorithms were used (implemented in WEKA software).

- **IBk**. Implements the k-nearest neighbor algorithm. It was tested with 1 to 100 neighbors, finally k = 1 was selected [14].
- **J48**. Implements the C4.5 Decision Tree.
- **Naïve Bayes**.
- **Random Tree**.
- **Random Forest**.
- **Bagging**. Bootstrap Aggregating.
- **AdaBoost**. Adaptive Boosting.
- **Stacking**.
- **Grading**.
- **DECORATE**. Diverse Ensemble Creation by Oppositional Relabeling of Artificial Training Examples.

The generated models were tested applying the *10-fold cross validation* method, one of the most reliable mechanisms to estimate the accuracy of a model [15]. Furthermore, the following metrics were used to compare the models: *Relative Absolute Error (RAE), Precision, Recall, Receiver Operating Characteristics Area (ROC Area)* [16], and the model training time. The obtained results are presented in Table 3.

Table 3. Experimental results of each algorithm of the modeling process.

Algorithm	RAE	Precision	Recall	ROC area	Training time
IB1	2.47%	0.983	0.983	0.990	0 s
J48	7.08%	0.952	0.951	0.979	0.09 s
Naïve Bayes	58.54%	0.598	0.602	0.866	0.01 s
Random tree	6.03%	0.953	0.953	0.970	0.003 s
Random Forest	9.26%	0.982	0.982	0.999	0.33 s
Bagging	9.34%	0.978	0.978	0.997	0.06 s
AdaBoost	6.77%	0.947	0.946	0.966	0.01 s
Stacking	2.15%	0.983	0.983	0.997	2.19 s
Grading	2.52%	0.980	0.980	0.987	0.15 s
DECORATE	11.26%	0.987	0.987	0.999	0.43 s

From the above table, it can be concluded that the preprocessed dataset presents an excellent performance with each algorithm, with the exception of *Naïve Bayes*; for this reason, it is certainly difficult to determine which is the best technique. However, considering the models with a RAE less than 10%, Precision and Recall of over 0.980, and a training time less than 0.5 s; *Random Forest* differs from other for having a higher ROC area (Table 3). In consequence, this technique is the algorithm that generates the best model in order to classify new instances in the cacao production dataset. The main features of the selected model are described below.

- A correctly classified instances rate of 98.23% (1781 of 1813 instances).
- A RAE less than 10% (it is greater than an average classification by more than 90%).
- Precision and Recall greater than 0.980 (the probability of poor ranking is less than 2%).
- A ROC area of 0.999 (the model has an almost ideal specificity-sensitivity ratio).
- A training time less than 0.5 s (if the user needs to train the model again, it will not feel a noticeable latency).

5 Model Deployment

From the previous modeling process, an application was developed, and java platform was used in two ways; web service and web client deployment. The general architecture is shown in Fig. 1.

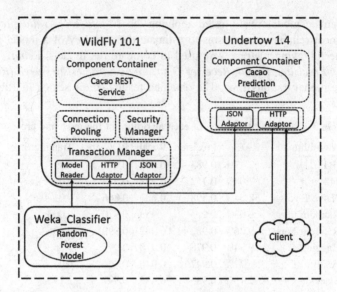

Fig. 1. General architecture for the classification module.

WildFly. It is the open-source application server offered by Red Hat, which supports REST services [17].

- Component Container: contains the different parts of the cacao REST service, which allows to classify new fortnightly climate instances in order to predict the monthly cacao production in Santander, Colombia, organizing the information and recommending mitigation actions.
- Connection Pooling: allows to pool the connections between server and the cacao REST service.
- Security Manager: handles the security of all the server transactions.
- Transaction Manager: allows the server to be accessed and the applications to connect with external components, applications, or servers through different API and protocols, leveraging the resource adaptors.
  - Model Reader: communicates with the *Weka_Classifier* in order to classify a new instance.
  - HTTP Adaptor: allows the server and the cacao REST service to communicate via HTTP protocol.
  - JSON Adaptor: allows the server and the cacao REST service to send JSON data.

Undertow. It is the open-source web server offered by Red Hat, which supports HTTP and JSON [18].

- Component Container: contains all the different parts of the cacao classification client, which offers the farmers and crop managers a basic interface to consume the cacao rest service; organizing the service response in a simple way.
- Resource Adaptors: allow the server to process JSON and HTML requests.

Weka Classifier. It is the conceptual model implementation represented by the climate and cacao production data in Santander, Colombia. In this case, the classifier was a java library generated with Random Forest algorithm in Weka tool software, which is an open-source machine learning tool, offered by the University of Waikato [13].

The system response is given through discrete values corresponding to the level of expected cocoa production in Santander, the estimated maximum and minimum production, and the precision percentage for classification. In addition, a number of useful recommendations for farmers are presented to the user considering the optimization of their logistics according to the expected production level. These recommendations were taken mainly from [2, 19], and the knowledge of experts in crop management, who placed a set of recommendations in a given category (Table 4).

Table 4. List of cacao production classes and recommendations for farmers.

Production class	Recommendation
Very low	- Save the minimum space to store the production - Hire the minimum staff to function - If you have spare machinery, rent it to other businesses - Foresee the cancellation of sales agreements with your partners
Low	- Save less space to store the production - Hire less staff to function - If you have spare machinery, rent it to other businesses
Medium	- Keep the usual space to store the production - Hire only the essential staff - Use only the essential machinery
High	- Save extra space to store the production - Hire extra staff - Rent out extra machinery
Very high	- Save extra space to store the production - Hire extra staff - Rent out extra machinery - Establish sales agreements with more partners

6 Conclusions and Future Work

This paper has presented a machine learning process for predicting the monthly cacao production in Santander, Colombia, from the fortnightly climate conditions on the ripening. This process produced a quality dataset, a precise model, and a web service and client. The quality of the dataset was tested by modeling it through various algorithms, obtaining satisfactory results. The precision of the model was over 98% when tested through a ten-fold cross validation. The web service and client were deployed on a server and are accessible for different governmental and non-governmental organizations of Santander department via Internet.

These outcomes will allow farmers, technicians and farmers' federations to optimize their courses of action when handling unexpected levels of cacao yield. According to the yield predicted amount, farmers or technicians cloud contract only the essential storage space, staff, and machinery; farmers' federations could prepare only the essential sales agreements. This way all the yield will be harvested, stored and sold without wasting resources or loosing production.

Similarly, the quality and precision of our outcomes would allow different organizations to develop their own products, underpinned on the dataset, the data model or the web service, in order to offer specialized agricultural services to farmers, technicians and farmers' federations. The organizations leveraging these outcomes could be private companies, research and development institutes, research groups, universities or even governmental organizations.

However, the process of gathering climate and yield data from cacao crops was strenuous and exiguous since there is almost no open data about the crops, and in Colombia only some secretive cacao farms are sufficiently monitored. Notwithstanding a complete department had to be selected (including uncultivated land) in order to link climate and yield in the cacao production, the process of data preparation was thorough enough to build a quality dataset without missed or anomalous data; obtaining very good results for almost all the algorithms in the data modeling (Table 3) and meeting the selection criteria stated previously. Moreover, the best algorithm was selected in order to obtain the most precise data model.

As future work, we propose to gather location, climate, and yield data from different cacao experimental farms in order to develop a complete classification system capable of precisely forecasting a cacao farm production from its location and climate conditions during the ripening; not only in Santander, Colombia but anywhere in the world.

Acknowledgements. The authors would like to thank Universidad del Cauca, AgroCloud project of the RICCLISA program for supporting this research, and Colciencias (Colombia) for PhD scholarship granted to MSc. Iván Darío López and MSc(c). Julián Eduardo Plazas.

References

1. Leon, J.: Botánica de los cultivos tropicales (Botany of tropical crops), 2nd edn. IICA, San José (2000)
2. FEDECACAO - Cacaocultura en el departamento de Cundinamarca (Cacao culture in Cundinamarca). https://www.fedecacao.com.co/site/images/recourses/pub_doctecnicos/fedecacao-pub-doc_08B.pdf. Accessed 22 Nov 2016
3. Jones, J.W., Hoogenboom, G., Porter, C.H.: The DSSAT cropping system model. Eur. J. Agron. **18**(3), 235–265 (2003)
4. Dias, M., Navaratne, C., Weerasinghe, K., Hettiarachchi, H.: Application of DSSAT crop simulation model to identify the changes of rice growth and yield in Nilwala River Basin for mid-centuries under changing climatic conditions. Procedia Food Sci. **6**, 159–163 (2016)
5. Mishra, S., Mishra, D., Santra, G.H.: Applications of machine learning techniques in agricultural crop production: a review paper. Indian J. Sci. Technol. **9**(38), 1–14 (2016)
6. Hansen, L.K., Salamon, P.: Neural network ensembles. IEEE Trans. Pattern Anal. Mach. Intell. **12**(10), 993–1001 (1990)

7. Zhou, Z.-H.: Ensemble Methods: Foundations and Algorithms. CRC Press, New York (2012)
8. Balakrishnan, N., Muthukumarasamy, G.: Crop production-ensemble machine learning model for prediction. Int. J. Comput. Sci. Softw. Eng. **5**(7), 148–153 (2016)
9. FEDECACAO - Economía nacional del cacao (Cacao National Economy). http://www.fedecacao.com.co/portal/index.php/es/2015-02-12-17-20-59/nacionales. Accessed 28 Sept 2016
10. Tutiempo Network S.L - Clima en Bucaramanga/ Palonegro (Bucaramanga/ Palonegro Climate). http://www.tutiempo.net/clima/Bucaramanga_Palonegro/800940.htm. Accessed 28 Sept 2016
11. NOAA - NCEP/NCAR Global Reanalysis Products. http://rda.ucar.edu/datasets/ds090.0/. Accessed 30 Nov 2016
12. Chapman, J.C.: CRISP-DM 1.0: Step-by-Step Data Mining Guide. SPSS, New York (1999)
13. Machine Learning Project at the University of Waikato in New Zealand. http://www.cs.waikato.ac.nz/ml/weka/. Accessed 01 Dec 2015
14. Hassanat, A.B., Abbadi, M.A., Altarawneh, G.A., Alhasanat, A.A.: Solving the problem of the K parameter in the KNN classifier using an ensemble learning approach. Int. J. Comput. Sci. Inf. Secur. **12**(8), 1–7 (2014)
15. Kohavi, R.: A study of cross-validation and bootstrap for accuracy estimation and model selection. In: Proceedings of the 14th International Joint Conference on Artificial Intelligence, vol. 2, pp. 1137–1143. Morgan Kaufmann Publishers Inc. (1995)
16. Ćosović, M., Obradović, S., Trajković, L.: Performance evaluation of BGP anomaly classifiers. In: 2015 Third International Conference on Digital Information, Networking, and Wireless Communications (DINWC), pp. 115–120. IEEE (2015)
17. Red Hat - About WildFly. http://wildfly.org/about/. Accessed 13 Jan 2017
18. Red Hat - Undertow JBoss Community. http://undertow.io/. Accessed 13 Jan 2017
19. FEDECACAO - Manual de buenas prácticas agrícolas - BPA en el cultivo del Cacao. http://www.fedecacao.com.co/portal/images/recourses/pub_doctecnicos/fedecacao-pub-doc_11B.pdf. Accessed 16 Jan 2017

Decision Support System for Coffee Rust Control Based on Expert Knowledge and Value-Added Services

Emmanuel Lasso$^{(\boxtimes)}$, Óscar Valencia, and Juan Carlos Corrales

Grupo de Ingeniería Telemática, Universidad del Cauca,
Campus Tulcán, Popayán, Cauca, Colombia
{eglasso,ovalencia,jcorral}@unicauca.edu.co

Abstract. In coffee production, the quality and quantity of harvests depends on the diseases treatment. One of the diseases with the most negative impact is the Coffee Rust. It causes losses around 30% in Colombian coffee crops. Sever-al studies propose the use of computer sciences techniques for the automat-ic detection of conditions that trigger epidemics. On the other hand, the knowledge of experts in the control of the disease needs to be disseminated in a more agile way so that the farmers can make correct decisions to avoid great losses in the production. This paper presents a Decision Support System (DSS) for Coffee Rust Control based on expert knowledge. It recommends the best alternative for the application of fungicides and proposes some value-added services based on the integration of its functionalities with those offered in the services of an Early Warning System (EWS) for Coffee Rust. Our proposal represents a highly scalable and flexible solution for the disease management at farmer level.

Keywords: Decision support · Service bus · SOA · Expert knowledge · Coffee rust · Agriculture · Disease management

1 Introduction

The detection and control of diseases in crops are some of the most important tasks in agricultural sector. In coffee production, the Coffee Rust is a disease that causes low productivity and quality in harvests (production reduced by about 30% [1]), generated by the *Hemileia Vastatrix* fungus. The development of the fungus depends on two main factors: weather conditions and crop properties [2]. In this sense, several studies from the computer sciences [3–9] have been carried out in order to obtain models to detect favorable conditions for coffee rust, making use of machine learning techniques. The application of these techniques is based on the analysis of monitored weather conditions in crops. The data monitored is used as the training dataset for supervised learning algorithms, where the records of past disease epidemics constitute the target variable (variable to predict). In this way, the results obtained in the approaches mentioned can be used as main element for the generation of coffee rust alerts. However, following the generation of an alert, it is necessary to take control measures so that the disease does not cause a great impact on the crops. The combination of methods for coffee rust

© Springer International Publishing AG 2017
O. Gervasi et al. (Eds.): ICCSA 2017, Part II, LNCS 10405, pp. 70–83, 2017.
DOI: 10.1007/978-3-319-62395-5_6

detection and recommendation of the appropriate control measures represents a scenario of better production volume and quality. To achieve an effective disease control, it is necessary for decision makers (in this case farmers) to know the fungicides to use according to the characteristics of the area where a coffee crop is planted, the plant phenology, as well as the state of the rust infection in the trees.

A Decision Support System (DSS) can be presented as a solution to the diseases management in crops, allowing to develop less intensive and integrated agricultural systems with a fewer application of fertilizers and pesticides, and restricted use of natural resources (water, soil, energy, etc.) [10]. These systems are a combination of individual intellectual resources and the capabilities of a computer to improve the decisions quality [11]. The DSS compiles, organizes and integrates all kinds of information necessary about a problem to be resolved, analyze and interpret this information, and finally use the analysis to recommend the most appropriate action to take. In an scenario like the agriculture, the DSS needs to update the execution flows for the decision-alternatives generation in order to address different events or problems and adapt its response to a language/presentation of easy understanding for any user [12]. The development of a DSS must consider interrelated domain factors, such as institutional resources, processes and policies [13]. As presented in [10, 14], decision-making can be based on environmental monitoring data, such as weather and crop properties, as it allows the generation of recommendations for disease control automatically and in line with climate variability. A well-known DSS in the agriculture is the DSSAT [15], which incorporates models of 16 different crops, available through a software for the evaluation and application of these models in agricultural production. Due to the advances of researchers in the different domains covered by DSSAT, the models have been designed to allow the incorporation of new knowledge and the knowledge base update.

On the other hand, in computer sciences, the integration of Expert Systems(ES) with DSS is proposed, in order to create more useful and accurate computer-based systems. Expert Systems are a computational tool that combine a knowledge base for a specific area with an inference capacity. The different advantages in the use of expert knowledge can be taken as a valuable element in decision making [16]. The integration od the expert knowledge in the DSS is possible in a total way or by components, either in data analysis, user interaction or in the inference model building [17]. For agriculture, expert knowledge can be integrated into a DSS through a rule repository [18, 19]. An inference engine can interpret these rules, identify risk scenarios and generate suggestions for problem solving in crop production (diseases, pests, low production, etc.).

In this scenario, a large amount of information is handled, requiring a set of services to consult or update it from different user applications (web, mobile, desktop, among others). For this, a Service Oriented Architecture (SOA) represents a solution for the management, exhibition and invocation of these services [20]. Some studies have made use of SOA for the design of systems applied in the agricultural environment [21–23] in order to take advantage of the maximum reuse of services from their low coupling. In this sense, each service can be modified in a simple way without affecting to a great extent the business processes. In addition, the Enterprise Services Bus (ESB) allows communication between applications services in an agile and flexible way, enabling the generation of value-added services that reuse features offered by heterogeneous services [24].

This paper presents a Decision Support System(DSS) for Coffee Rust Control based on expert knowledge for the recommendation of the best alternative for the application of fungicides and value-added services for the disease control management from the integration of the DSS with an Early Warning System (EWS) for Coffee Rust. This proposal represents a highly scalable and flexible solution for the end user (in this case a farmer). The remainder of this paper is organized as follows: Sect. 2 describes the DSS for Coffee Rust Control; Sect. 3 refers to the integration of the DSS with the EWS; Sect. 4 relates the evaluation of the DSS and Sect. 5 address the conclusions.

2 Decision Support System for Coffee Rust Control: DSS-CRC

The objective of the Decision Support System for Coffee Rust Control (DSS-CRC) is to support to the farmer for the management of the disease, from the recommendation of different control process aspects, such as the type of fungicide required according to the state of the infection, the appropriate dates of sprinkling and the technology (equipment and dosage) that should be used to achieve good protection against the rust. The correct articulation of these aspects leads to a reduction of economic losses due to the impact of the disease and an improper use of the fungicides. Thus, our proposal starts from the inclusion of expert knowledge in the DSS for the construction of the rules within the inference model that generates a recommendation. In particular, the expert knowledge was extracted from the Colombian Coffee Research Center (Cenicafé) documentation for the management and control of Coffee Rust in Colombia [25]. Additionally, our system was implemented within the AgroCloud project, which belongs to the Interinstitutional Network of Climate Change and Food Security of Colombia[1] (RICCLISA). AgroCloud is focused on providing support services for threat situations in coffee crops located in the upper Cauca River basin.

In this sense, the DSS is able to provide recommendations for three decision-making scenarios related to Coffee Rust Control: fungicide type, application time and technology. Each section can be consulted separately or in order as a major decision-making process (called Control Management). The sections are detailed below.

2.1 Fungicide Type

The fungicide to be applied for the control of Coffee Rust depends on the disease stage and rainfall conditions in the crop area. The input values required to the user are:

- Disease stage: germination, colonization or sporulation.
- Location of the crop: municipality to which it belongs. It is used to get the rain forecast from forecasting services provided by AgroCloud.

From these data, the rules generated from expert knowledge determine the type of fungicide to be used, which can be cupric, systemic or with strobirulins. With this

---

[1] www.ricclisa.org .

information the user can continue with the Control Management process to know the moment of fungicide application and the technology recommended for the applications.

2.2 Moment of Fungicide Application

The time to start chemical control of Coffee Rust can be determined based on three criteria.

- Fixed calendar: is based on the phenological development of the crop, the evaluation of the disease in different coffee growing areas and the distribution of the crop. In this way, the user must enter the department(state) where the crop is located and the type of fungicide that will be used. In response, the number of fungicide sprays per year and the exact dates for the applications are obtained.
- Flowering date: starts from the farmer's knowledge of the date of flowering in his crop. From this date, it is possible to determine some characteristics of the coffee production taking into account the quantity and concentration of the flowering. The productive potential gives an indication of how the rust epidemic will be. As input elements, the user must specify the flowering present in his crop (abundant or sparse) and, if it is abundant, specify the date in which it occurred. The system response will be the number of fungicide sprays per year and the exact dates for the applications.
- Infection levels: the number of sprays vary depending on the evolution of the disease. This method is based on the study of the impact on the production of certain levels of leaves affected in the plant and allows to use rationally the systemic fungicides, without creating conditions that favor the presence of new races of the fungus and to diminish the number of sprays. This reduces the costs of disease control in the crop.

2.3 Fungicide Application Technology

The technology in this case refers to the process for determining and relating factors for the correct execution of a fungicide application. For this, the user must enter the following parameters:

- Age (in months) of the crop.
- Fungicide to be used (cupric, systemic, with strobirulins).
- Fungicide sub-type (in the case of systemic fungicides).
- Type of terrain (flat or slope).
- Seed characteristics (in grooves or irregular).
- Density: can be less than or equal to 5000 plants per hectare, or greater than that amount.

The response of the system will be the spray volume of the selected fungicide, the dose to be used, the spray equipment to be handled and finally the spray system.

To automate the decision process, we conceived each decision-making scenarios as service units that represent independent functionalities. Precisely, through SOA concepts we developed the DSS-CRC from this perspective. The services functionalities in each scenario were taken to compose some value-added service for Coffee Rust Control Management.

2.4 Knowledge Base Definition

From expert knowledge found in [25], a set of rules for each decision-making scenarios were defined. The rules constitute the knowledge base of the proposed DSS and are modeled as decision trees. In this way, each branch of the tree can be seen as an IF-THEN rule. The conceptualization of our knowledge base as decision trees allows the consistency verification of the rules and their improvement by an expert in the treatment of the disease, since its structure represents a level of abstraction of easy understanding and modification.

As an example, in the Fig. 1, the Decision Tree (left side) and rules (right side) associated with each branch of the tree are shown. The example corresponds to the scenario of fungicide application technology, specifically the recommendation of fungicide spray volume and dose (Sect. 2.3). The nodes in the three represents an evaluation of the parameters submitted by a user and the conditions determining the tree branch to follow are expressed in the IF-THEN type rules.

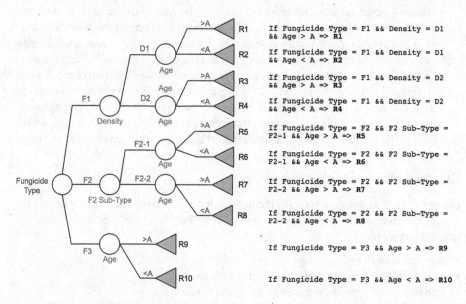

Fig. 1. Decision Tree for recommendation of fungicide application technology

3 DSS-EWS Integration

The objective The DSS described above represents a useful tool for farmers in decision-making scenarios, such as Coffee Rust control. In the AgroCloud project, an EWS to detect favorable conditions for this disease, using the patterns obtained by Lasso et al. [3], was developed. This system obtains the weather conditions around coffee crops and, from this information, a data graph that characterizes the state of a crop at a given time is constructed. The patterns for the disease are searched within the data graph, obtaining the crops where disease conditions are presented. In this way, an alert is generated and it is necessary that farmers take measures to avoid large losses in their crops. Thus, a DSS-EWS integration represents a complete support for Coffee Rust problem in crops, from the detection to its chemical control.

The AgroCloud EWS functions and process are available as web services. Therefore, we take them and integrate their responses to the proposed DSS-CRC as an input parameter in the *Coffee Rust Control Management* value-added service. It is composed by three main services that correspond to the DSS sections related in Sect. 2 (Get fungicide, Get Moment of Application and Get Application Technology). Each service can be consulted separately or in order as a major decision-making process. The BPMN diagram [26] in the Fig. 2 shows the integration of the EWS and DSS services. The BPMN allows to represent the invocation between services, the value-added service containing different services of the DSS, the messages shared, the actions performed, and the services responses. The execution flow is described below.

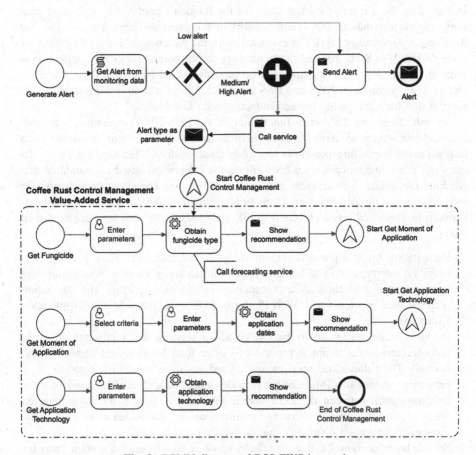

Fig. 2. BPMN diagram of DSS-EWS integration

The *Generate Alert* service in the EWS execute a script for the generation of alerts from weather data. The alert level is evaluated by the gateway and two scenarios can be identified:

– Continuity to preventive treatment: for low alert level.
– Corrective treatment for immediate application: for medium and high alert levels.

If the alert is low, this service only shows its details. In case of being medium or high, the crop conditions are favorable for the accelerated disease development, which influences in the fungicide type that must be used in the control. The details of the alert are shown and a call to the DSS value-added service is made. This call starts the process for obtaining the fungicide type categorized as a corrective treatment scenario. In addition, the service makes a call to the forecasting services, since the rain affects the fungicide application. In this case, the user gets a recommendation characterized from the disease alert level. Next, the service *Get Moment of Application* is started. The user selects the criteria that adjust to his needs and enters the parameters required (described in Sect. 2.2). As a result, the ideal dates for the fungicide application is showed (this fungicide corresponds to that recommended in the previous step). Finally, the *Get Application Technology* service is executed, with the parameters specified by the user (as is described in Sect. 2.3) and the technology for the correct fungicide application is showed. In this way, in a scenario of crop hazards caused by Coffee Rust, the articulation and integration of EWS and DSS services represent a complete information and support tool for high quality coffee production with low losses.

In order to ensure the correct functioning of the DSS-EWS integration, a flexible and accurate communication between them is necessary. The main objective is to respond to the events flow present in favorable situations for Coffee Rust epidemics. To carry out this communication, an ESB allows the integration based on standards that combine messaging, web services and intelligent routing. The system considered for the integration is divided into four levels or layers, and the ESB is located as mediator between business and service layers (Fig. 3). The elements of each layer are described below:

– Presentation layer: represents the interfaces to the end user. They provide easy usability, interpretation and human-machine interactivity, receive input parameters in user queries and show system responses for decision-making. The interaction with the user is offered as a Web Platform (PHP based) or a Mobile Application (Android).
– Business layer: composed by two value-added services that represents the main tasks to carry out a complete process of Coffee Rust Management (detection and control). The value-added services are defined based on WS-BPEL notation.
– Enterprise Service Bus (Mule ESB): provides a flexible and reliable channel for the communication between the DSS and EWS services. It is a mediator between business and service layer, allowing the formation of value-added services from the services orchestration defined in the business process model.
– Service layer: contains the fine granularity services, which perform specific tasks for decision alternatives and alert generation. The services are hosted in a Glassfish application server and are SOAP based.

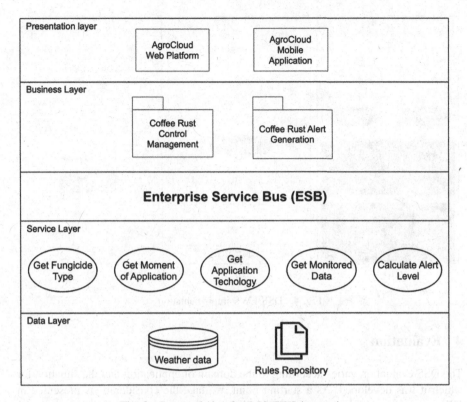

Fig. 3. System design for DSS-EWS integration

- Data layer: includes a graph database (Neo4j) to store the weather data monitored in crops zones, used to generate the disease alerts; and a rules repository created from expert knowledge about Coffee Rust control and management. The rules are extracted from the conceptual decision tree mentioned in Sect. 2.4 and are codified based on RuleBook, a rule abstraction for java.

Each layer described above contains elements with a specific role within the integration. The ESB allows the reuse of functions in the services, facilitate scalability and integration with different tools of the AgroCoud project. In this way, the DSS and EWS can be seen as individual and complementary tools to provide full support to the farmer to face a disease such as Coffee Rust.

The system implementation corresponds to a mobile application developed for Android and an example is shown in the Fig. 4. The alerts section (Fig. 4a) contains a form to consult the generated alerts by locations. For high (or medium) alerts, the link presented in the alert description redirect to the fungicide type recommendation (Fig. 4b). This section shows the possible fungicides related to the type recommended. The SOA approach allows to integrate different resources and services, while the front-end (web, desktop, or mobile application) communicates with the value-added services in a simple way. Additionally, in the Fig. 4c an example of application moment recommendation is presented.

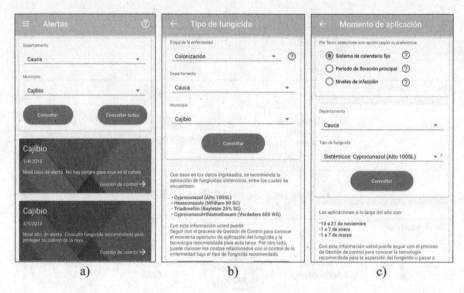

a)   b)   c)

Fig. 4. DSS-EWS implementation

4   Evaluation

The DSS evaluation varies according to the domain of application and the situation for which it was developed. As a starting point, we take the considerations presented in [27], such as decision (outcome) quality, decision process efficiency and the decision-maker's satisfaction, to define our measurement variables. Coffee Rust epidemics generate a great impact on the production quality and quantity. In this case, the decision-makers are the farmers responsible for carrying out disease control measures. The results are presented below.

4.1   Decision Quality

The possible recommendations obtained from the DSS were evaluated by an expert in Coffee Rust in Colombia, and an agronomist. Given that the inference rules of the DSS are constructed from expert knowledge, the responses are appropriate for the disease control. The system represents a great support for the farmer, since it allows him to access the knowledge about the control elements in a fast and interactive way, without being dependent on a communication with the advisory service of the Colombian Federation of Coffee Growers. In addition, the DSS-EWS joint work is a great success, linking the alert generation with the control measures that must be carried out in a short time. As suggested changes, the expert recommends adding the possibility of specifying the percentage of disease incidence in the fungicide type consultation, because the farmer may not know the disease stage.

4.2   Decision Process Efficiency

The decision process for DSS-EWS integration was considered since the generation of the disease alert. From a dataset of weather and cropping properties obtained by Corrales et al. [4], 124 alerts were generated, and 56 of these corresponded to situations of corrective treatment for immediate application (described in Sect. 3). The integration was successful, since it correctly identified the type of disease situation that was present and established the link between DSS and EWS in the 56 mentioned situations. It's necessary to stress that the precision in the alerts generation is measured in the EWS and this evaluation does not concern the present proposal. On the other hand, the different DSS processes without integration with the EWS presented a correct functioning, as separate processes (recommendation of type of fungicide, dates of applications and technology of application) and as a process in order (Control Management).

Given the time of development of the disease in coffee crops may take several weeks and the observation of treatment results takes another additional time, in the present work we have left aside the on-site evaluation of the effect of the proposed system. For this reason, only DSS-EWS integration tests based on historical information collected on a farm were presented.

4.3   Decision Maker's Satisfaction

End-user satisfaction is one of the most important DSS evaluations in agriculture. The end users in this case are farmers that interact with the product on a daily and constant basis. For this, we used the User Acceptance Testing (UAT) [28] as the last phase of the software solution validation process. Farmers made use of the DSS-EWS integration in order to be able to support and respond to real situations in specific scenarios for which it was developed, according to the specifications given. The concept of "black box" was addressed, which is commonly categorized as a functional test but can also be used for UAT. The application functionalities analysis was performed and the user only knew and interacts with the system inputs and outputs, without being able to see the code and internal flow of operation. In addition, the end user knew the business requirements. The principal UAT elements and results obtained are presented in Table 1.

We carried out the UAT process including, in addition to farmers, an agronomist and a technical assistance group of the National Federation of Coffee Growers, in order to obtain different points of view. Our proposal generated a great interest on the end users in using information systems as support to face a problem like coffee rust. Further, the UAT allows us to know specific needs at the farmer level, taking their comments as a starting point to improve the proposed system.

The Fig. 5 shows the acceptance of the system proposed. The total of users (53 users) were able to complete the decision-making processes without problems. In some cases, farmers needed assistance in handling the device through which the system was accessed (39 users). However, the forms were understood to a greater extent (45 users). Although the disease control procedures have been disseminated by the authorities of the coffee sector for many years, the users of the system consider that the tool is useful as a way to obtain a personalized recommendation based on the characteristics of their crops and the environmental conditions present in them.

Table 1. UAT results.

UAT for DSS-EWS integration for coffee rust control management	
Scenarios	• Fungicide type recommendation • Application moment recommendation • Application technology recommendation • Coffee rust control management (complete process)
Validation queries	• Does the process provide you with relevant information regarding coffee rust control? • Is the process clear and easy to follow? • Is the information shown clear? • What improvements could the tool have?
Users	• 5 medium-sized producers from Cajibío (Cauca) • An association of organic coffee producers (Popayán) (12 producers) • An association of coffee growers from Los Andes - Corinto (ASPROCCAN) (20 producers) • An agronomist • Technical assistance group of the National Federation of Coffee Growers (Colombia) (15 users)
General perception	The system is understandable by the user and there is a great motivation to start using it. There were no problems in the DSS execution in the different scenarios. On the other hand, technical assistance group and agronomist agree that while it is a useful tool, however it should not completely replace the technical assistance of the Federation of Coffee Growers
Main recommendations	• Show the fungicide type that should be applied in the recommendation based on main flowering • The recommendation of application moment should be more didactic (with graphics) • Be able to visualize images of the different elements needed in the control (fungicides, spray equipment, etc.) • Recommend other options of spray systems in case the producer does not have the capacity to acquire the recommended one • Provide fungicides specifications (composition, picture, laboratory, environmental impact) • Ability to print or mail recommendations
New features or suggested functions	• Take into account other coffee diseases, in addition to pests • Take into account disease control processes related to organic coffee production • Recommend chemical-free procedures

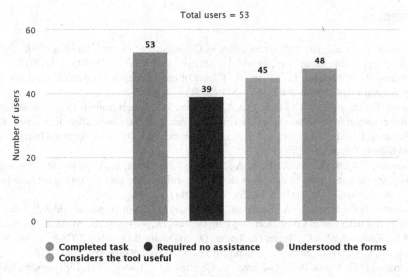

Fig. 5. Representation of acceptance

5 Conclusions

This paper addresses one of the most common problems in coffee production: Coffee Rust. In the management of this disease, both detection and treatment are the main factors. We took the expert's knowledge in order to formalize the recommendations for the disease control obtained from their studies, and a set of inference rules were defined. From these rules as main element, a Decision Support System for Coffee Rust Control was built under a Service Oriented Architecture approach. This allowed us to get a set of independent and interoperable services to cover each stage of the disease control management. In this sense, some value-added services were defined to integrate the DSS with an Early Warning System (EWS) for Coffee Rust developed in the AgroCloud project. Thus, after an alert generation, a support for the disease control characterized according to the alert level is provided. In the evaluation of our proposal, it represents a high scalable and flexible solution for the end user (in this case a farmer). Testing at the end user level allowed us to know the opinions and additional requirements from the decision makers.

As future work, we will seek to follow up on crops where the recommendations of our system have been followed, in order to observe the impact of the decision process. Also, to apply our proposal for other diseases or pests, as well as other crops. The ESB can be used to integrate other services with the DSS services, such as weather services, harvest and flowering prediction, cost management and climate change models.

Acknowledgements. The authors are grateful to the University of Cauca and its Telematics Engineering Group (GIT), the Colombian Administrative Department of Science, Technology and Innovation (Colciencias), AgroCloud project of The Interinstitutional Network of Climate Change and Food Security of Colombia (RICCLISA) for supporting this research and the *InnovAccion Cauca* project of the Colombian Science, Technology and Innovation Fund (SGR-CTeI) for PhD scholarship granted to MSc. Emmanuel Lasso.

References

1. Avelino, J., et al.: The coffee rust crises in Colombia and Central America (2008–2013): impacts, plausible causes and proposed solutions. Food Secur. **7**(2), 303–321 (2015)
2. Avelino, J., Willocquet, L., Savary, S.: Effects of crop management patterns on coffee rust epidemics. Plant. Pathol. **53**(5), 541–547 (2004)
3. Lasso, E., Thamada, T.T., Meira, C.A.A., Corrales, J.C.: Graph patterns as representation of rules extracted from decision trees for coffee rust detection. In: Garoufallou, E., Hartley, R.J., Gaitanou, P. (eds.) Metadata and Semantics Research, pp. 405–414. Springer International Publishing, Cham (2015)
4. Corrales, D.C., Ledezma, A., Peña, A.J., Hoyos, J., Figueroa, A., Corrales, J.C.: Un nuevo conjunto de datos para la detección de roya en cultivos de café Colombianos basado en clasificadores. Sist. Telemática **12**(29), 9–23 (2014)
5. Corrales, D.C., Figueroa, A., Ledezma, A., Corrales, J.C.: An empirical multi-classifier for coffee rust detection in Colombian crops. In: Gervasi, O., Murgante, B., Misra, S., Gavrilova, M.L., Rocha, A.M.A.C., Torre, C., Taniar, D., Apduhan, B.O. (eds.) ICCSA 2015. LNCS, vol. 9155, pp. 60–74. Springer, Cham (2015). doi:10.1007/978-3-319-21404-7_5
6. Corrales, D.C., Casas, A.F., Ledezma, A., Corrales, J.C.: Two-level classifier ensembles for coffee rust estimation in Colombian crops. Int. J. Agric. Environ. Inf. Syst. IJAEIS **7**(3), 41–59 (2016)
7. Meira, C.A.A., Rodrigues, L.H.A., Moraes, S.A.: Analysis of coffee leaf rust epidemics with decision tree. Trop. Plant Pathol. **33**(2), 114–124 (2008)
8. Meira, C.A.A., Rodrigues, L.H.A., de Moraes, S.A.: Modelos de alerta para o controle da ferrugem-do-cafeeiro em lavouras com alta carga pendente. Pesqui. Agropecuária Bras. **44**, 233–242 (2009)
9. Cintra, M.E., Meira, C.A.A., Monard, M.C., Camargo, H.A., Rodrigues, L.H.A.: The use of fuzzy decision trees for coffee rust warning in Brazilian crops. In: 2011 11th International Conference on Intelligent Systems Design and Applications (ISDA), pp. 1347–1352 (2011)
10. Magarey, R.D., Travis, J.W., Russo, J.M., Seem, R.C., Magarey, P.A.: Decision support systems: quenching the thirst. Plant Dis. **86**(1), 4–14 (2002)
11. Keen, P.G., Morton, M.S.S.: Decision Support Systems: An Organizational Perspective, vol. 35. Addison-Wesley, Reading (1978)
12. Greer, J.E., Falk, S., Greer, K.J., Bentham, M.J.: Explaining and justifying recommendations in an agriculture decision support system. Comput. Electron. Agric. **11**(2), 195–214 (1994)
13. Rath, B.: Decision support system for drafting of sustainable agriculture production policy for Odisha. Int. J. Supply Chain Manag. **2**(2), 92–98 (2013)
14. Rossi, V., Meriggi, P., Caffi, T., Giosué, S., Bettati, T.: A Web-based Decision Support System for Managing Durum Wheat Crops (2010)
15. Jones, J.W., et al.: Decision support system for agrotechnology transfer: DSSAT v3. In: Tsuji, G.Y., Hoogenboom, G., Thornton, P.K. (eds.) Understanding options for agricultural production. Systems Approaches for Sustainable Agricultural Development, vol. 7, pp. 157–177. Springer, Dordrecht (1998)
16. Turban, E., Frenzel, L.E.: Expert Systems and Applied Artificial Intelligence. Prentice Hall Professional Technical Reference, Upper Saddle River (1992)
17. Turban, E., Watkins, P.R.: Integrating expert systems and decision support systems. MIS Q. **10**, 121–136 (1986)
18. Mansingh, G., Reichgelt, H., Bryson, K.-M.O.: CPEST: an expert system for the management of pests and diseases in the Jamaican coffee industry. Expert Syst. Appl. **32**(1), 184–192 (2007)

19. Chevalier, R.F., Hoogenboom, G., McClendon, R.W., Paz, J.O.: A web-based fuzzy expert system for frost warnings in horticultural crops. Environ. Model Softw. **35**, 84–91 (2012)
20. Erl, T.: Service-Oriented Architecture (SOA): Concepts, Technology, and Design. Prentice Hall, Upper Saddle River (2005)
21. Xu, L., Chen, L., Chen, T., Gao, Y.: SOA-based precision irrigation decision support system. Math. Comput. Model. **54**(3), 944–949 (2011)
22. Han, W., Yang, Z., Di, L., Mueller, R.: CropScape: a web service based application for exploring and disseminating US conterminous geospatial cropland data products for decision support. Comput. Electron. Agric. **84**, 111–123 (2012)
23. Murakami, E., Saraiva, A.M., Ribeiro, L.C., Cugnasca, C.E., Hirakawa, A.R., Correa, P.L.: An infrastructure for the development of distributed service-oriented information systems for precision agriculture. Comput. Electron. Agric. **58**(1), 37–48 (2007)
24. Papazoglou, M.P., Traverso, P., Dustdar, S., Leymann, F.: Service-oriented computing: state of the art and research challenges. Computer **40**(11), 38–45 (2007)
25. Rivillas, C., Serna, C., Cristancho, M.A., Gaitán, A.L.: Roya del Cafeto en Colombia: Impacto, Manejo y Costos del Control, Chinchiná Bol. Téc. (36) (2011)
26. White, S.A.: Introduction to BPMN, vol. 2. IBM Coop., Foster City (2004)
27. Rhee, C., Rao, H.R.: Evaluation of decision support systems, Handbook on Decision Support Systems, 2nd edn., pp. 313–327. Springer, Heidelberg (2008)
28. Cimperman, R.: UAT defined: a guide to practical user acceptance testing. Addison-Wesley Professional, Reading (2006)

Impact of Temporal Features of Cattle Exchanges on the Size and Speed of Epidemic Outbreaks

Aurore Payen[1,2]([✉]), Lionel Tabourier[2], and Matthieu Latapy[2]

[1] AgroParisTech, Paris, France
[2] Sorbonne Universités, UPMC Univ Paris 06, CNRS, LIP6 UMR 7606,
Paris, France
{aurore.payen,lionel.tabourier,matthieu.latapy}@lip6.fr

Abstract. Databases recording cattle exchanges offer unique opportunities for a better understanding and fighting of disease spreading. Most studies model contacts with (sequences of) networks, but this approach neglects important dynamical features of exchanges, that are known to play a key role in spreading. We use here a fully dynamic modeling of contacts and empirically compare the spreading outbreaks obtained with it to the ones obtained with network approaches. We show that neglecting time information leads to significant over-estimates of actual sizes of spreading cascades, and that these sizes are much more heterogeneous than generally assumed. Our approach also makes it possible to study the speed of spreading, and we show that the observed speeds vary greatly, even for a same cascade size.

1 Introduction

Production of dairy and meat products is a major economic field in France. Early detection of disease outbreaks is thus a key issue for the protection of economic interests, as well as animal welfare. Among the various routes to infect holdings, such as contamination by wildlife or contacts between different herds in pastures, cattle trade movements spread pathogens at national and international level, and are thus a major way of infection.

At least three approaches can be used to study epidemic spreading occurring because of these exchanges: agent-based models, generating functions, and network approaches. The first two approaches require assumptions either on the agent behavior or about the exchanges, such as statistical equivalence of nodes of same degree for Degree-Based Mean-field theory [15] or a local treelike structure for generating functions [13,15]. In this work, we would like to study the cattle trade movements without making any assumption about the nodes' behavior and the structure of the network. That is why using network modeling seems more appropriate, as it does not require any assumption about the dynamics of exchanges.

Thanks to the creation of databases recording animal exchanges, modeling cattle trade movements for animal health purposes has started developing. Since

O. Gervasi et al. (Eds.): ICCSA 2017, Part II, LNCS 10405, pp. 84–97, 2017.
DOI: 10.1007/978-3-319-62395-5_7

the Bovine spongiform encephalopathy crisis of 1996, each state of the European Union has an obligation to identify every farm animal on its territory and to register cattle trade movements. The *Base de Données Nationale d'Identification* (BDNI) database is the French enforcement of this decision. The year 2005 has been described in [17], and [8] described in details years 2005 to 2009.

When studying cattle trade movements, data are often aggregated in a static network, where an edge links two holdings if at least one exchange occurred, no matter how many times and when they interact. Yet, advances in the field of temporal networks show the importance of considering the temporality of interactions among nodes to describe propagation phenomena, especially when the edges are not stable through time. Indeed, the order and the frequency of interactions are of great importance to estimate the final number of reached nodes [19].

Consequently, many recent studies also use sequences of static graphs (sequence of snapshots) to observe the evolution of networks over time [5, 8, 12, 14]. For instance, [8] mainly focused on describing aggregated data for different time windows: the authors study monthly, quarterly and yearly networks. They measure the stability of several static features over successive snapshots, and also used dynamical measures, such as the reachability ratio, for which we express interest in the following. A significant asset of sequences of snapshots is that they allow the use of graph theory. However, despite the fact that each snapshot represents the aggregation of interactions that take place in a given time window, the precise dates of interactions are lost, and with them, their order and frequency. One might wonder if using a temporal network model, where the date of occurrence of a link is specified [9], could improve the evaluation of the potential size of an infection.

To answer this question, we study the French cattle trade network during the year 2005, by transforming it into a static network, sequences of monthly or quarterly networks, and a model of temporal network called link stream. After describing the network basic properties and how to measure the potential infection size of a disease in a static and a dynamical context, we compare the estimations of potential sizes of disease outbreaks, obtained using the different data representations. Finally, we propose a refinement not only to measure an outbreak size but also to take into account its propagation speed. Our contribution is then mainly methodological, comparing outbreak sizes estimations with static and temporal graph models, and proposing an intrinsically temporal measure to enhance our understanding of the measure.

2   Dataset

Our work relies on a data export from the BDNI which contains all cattle trade movements from 2005 to 2015 inside France. This makes approximately 148 millions cattle transfers in the following format: date of the transfer, origin holding, destination holding, animal identification number. We do not use animal identification numbers here, and focus on the fact that an exchange occurred between two given holdings at a given date.

In this paper, we focus on year 2005 because it already received much attention in previous works [8,17]. This choice makes it easier to compare our results to existing ones, and we checked that they still hold for recent years. We therefore consider 2005 as representative of all years, to this regard. In this subsection, we only present here a few basic properties of the dataset that help us analyze the measurements implemented in the rest of this work.

During 2005, 245,821 holdings exchanged at least one animal with another holding, leading to a collection of approximately 10 million cattle transfers. We display in Fig. 1 the distributions of the number of outgoing and incoming movements for each holding. We also show in Fig. 2(left) the total number of transfers occurring per week during the whole year, and in Fig. 2(right) the total number of transfers occurring per day, for several weeks.

The movement distributions clearly confirm the heterogeneity between holdings, already observed in previous works [8,17]: although some holdings are the origin or destination of up to approximately one thousand in- or out-going movements, the vast majority of holdings are involved in only a few movements, with all intermediary behaviors present in the data. Interestingly, there is a cutoff in the out-going movement distribution indicating that holdings involved in more than 100 out-going movements are exceptions (less than 2% of the nodes).

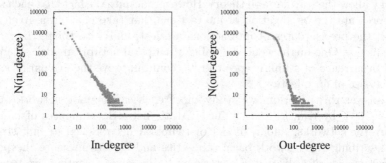

Fig. 1. Left: distribution of incoming movements per holding in 2005, in log-log scales. For each value x on the horizontal axis, we plot the number of holdings that were the destination of x cattle movement during 2005. Right: distribution of outgoing movements per holding in 2005.

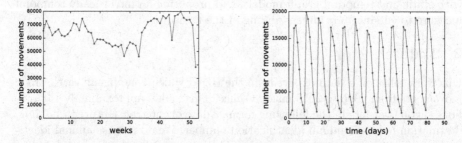

Fig. 2. Left: weekly number of transfers in the database in 2005. Right: daily number of transfers in the database for the first 12 weeks of 2005.

The number of movements per day (Fig. 2, right) is also interesting in the context of investigating the dynamical properties of the dataset. As already discussed in previous works on different animal trade networks [3,6,11], it shows that the short-term activity strongly changes with time, with peaks of activity on Mondays that progressively vanish through the week. Such variations may have an important impact on spreading phenomena.

It has also been observed that the French cattle trade network is asymmetric [8], in the sense that the existence of a link from i to j does not systematically involve the existence of a link from j to i. This property may be measured using the *reciprocity ratio*, which is the fraction of reciprocated links during a given period. We measure that this ratio is indeed 0.52 for the yearly network, and of the same order of magnitude for monthly networks (from 0.51 to 0.53). It stems from the fact that most of the holdings sell or transfer animals to other holdings, while a much lower amount receive or buy animals. This property also has direct consequences on the potential outbreak sizes, that will be discussed later.

3   Data Modeling

The data presented above may be represented in several ways. In spreading phenomena studies, the main approach consists in modeling relations between holdings using networks (i.e., sets of nodes and links between them), which capture the structure of exchanges. The direction of exchanges plays a key role in spreading phenomena, as a cattle transfer from holding A to holding B is different from a transfer from B to A, to this regard. Likewise, the time-ordering of cattle transfers is important: a transfer from A to B followed by a transfer from B to C is different from a transfer from B to C followed by a transfer from A to B, regarding spreading.

After presenting how the problem of estimating the potential size of an outbreak is usually addressed in the static and snapshots case, we propose an adaptation to link streams, a dynamical model of networks. In all cases, we consider directed links. Previous studies already insisted on these temporal aspects [7,14,18], and introduced notions such as the in/outgoing infection chains in order to better capture temporal features of cattle transfers. Our approach is related to these ones, focusing on the comparison that can be made with the snapshot case.

The most basic approach ignores temporal information: nodes represent holdings and there is a directed link from i to j if there was at least one cattle transfer from holding i to holding j in 2005. We call this model the **static network**. On 2005 dataset, it has 245,821 nodes and 1,646,510 directed links. Its degree distributions are the quantities plotted in Fig. 1.

In order to take into account the temporal features of exchanges while still using network formalisms, one may divide the data into time slices, each slice corresponding to a static network called **snapshot**. This leads to sequences of snapshots, each capturing the structure of exchanges during a time slice. Choosing appropriate time slices is a difficult question in general, but monthly and quarterly time slices are generally used for cattle trade exchanges, leading to 12 and 4 snapshots respectively, see for instance [8]. Choosing small time slices leads to a better conservation of the dynamics but also to snapshots with less information, while choosing larger time slices gives more complex snapshots but the dynamics within each time slice is lost.

In our case, the number of active nodes, that is to say holdings involved in at least one exchange, varies from 105295 (43% of the total) to 138907 (57%) per month (see Fig. 3), or from 180930 (74%) to 196166 (80%) quarterly.

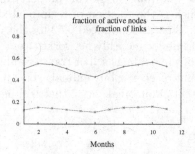

Fig. 3. Fraction of active nodes and links for each monthly network

Finally, in order to capture all the dynamics in the data, one may use the **link stream** or temporal network formalisms [10,20]. A link stream is a sequence of triplets (t, i, j) indicating that an interaction (here at least one cattle transfer) occurred at time t from node i to node j (here holdings). In our case, the obtained link stream involves 245,821 nodes and 3,355,680 temporal links.

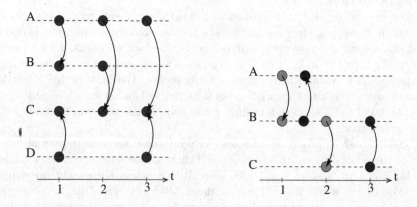

Fig. 4. Left: Link stream representation of interactions, A, B, C and D are holdings and a directed link at t represents a movement from the departure holding to the arrival holding at time t. Right: SI cascade on a link stream, starting from A at time 1, and reaching nodes B and then C following directed links.

4   Infection Modeling

Accurate modeling of infections is a challenge in itself [16]. However, most models rely on the assumption that infections spread over a population through the links of a network representing contacts between individuals. This leads in particular to the Susceptible-Infected (SI) model, which represents a worst case scenario: the disease spreads from a node A to all its neighbors as soon as A is infected, and nodes never recover. The size and speed of obtained spreading cascades are therefore upper bounds of what one may expect in reality. The SI model does not claim to be realistic, yet, as we are interested here in quantifying the size and speed of *potential* outbreaks, without any assumption on the disease under concern, we consider this model as a baseline and study its behavior in static and dynamical settings.

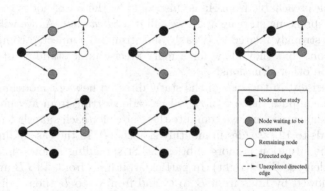

Fig. 5. SI cascade in the static snapshot case starting from the leftmost node: at each step, the neighbors (gray) of an infected node (black) are reached, until all the nodes are processed.

Many studies – e.g. [1] – consider the *undirected* version of the static graph defined above: they consider that a link from A to B also induces a link from B to A. Then, the key property for spreading phenomena is the size of the connected components. Indeed, a connected component is a maximal set of nodes such that there is a path from any node in the set to any other node in the set. As a consequence, any spreading starting in a connected component will eventually reach all the nodes of this connected component, and so connected component sizes give the sizes of epidemic outbreaks according to the SI model. In particular, most real-world networks have a connected component that contains most nodes of the network, called the Giant Connected Component, or GCC. Using such a representation, most epidemic outbreaks therefore reach all the nodes in the GCC, *i.e.* most nodes of the network.

Considering *directed* static networks is more realistic and leads to more detailed insights. The connected components of the undirected network are then called the Weakly Connected Components (WCC) of the directed network, and a Strongly

Connected Component (SCC) is a set of nodes such that there is a *directed* path from any node in the set to any other node in the set. Strongly connected components are therefore included in weakly connected components, and an infection starting in a weakly connected component will not in general reach all nodes of this component. Instead, an infection starting in a strongly connected component will spread in the whole strongly connected component. Therefore, the size of strongly connected components, and in particular the size of the largest one, are key features for epidemic studies [17]. However, an infection starting in a strongly connected component will also spread to all nodes reachable from it, which are not necessarily in the component. Likewise, some infections starting outside a strongly connected component may reach it and all its nodes. Therefore, considering connected components misses much information.

As a consequence, we decided to conduct our study by truly simulating SI spreading phenomena (Fig. 5): we start from a given node A and compute the set of all nodes reachable from A; as this would be the set of nodes reached by a SI epidemic spreading starting at A, we call it a *SI cascade*. As we will see, such cascades are strongly related to (weakly and strongly) connected components in some situations, but they allow for a more precise understanding of spreading phenomena in other situations.

More precisely, in the case of the static directed network modeling of cattle mobility, we compute the size of the SI cascade starting from any node. In the case of snapshot modeling, we compute these sizes for each snapshot (each node therefore leads to 12 cascades in monthly snapshots). In the case of link streams modeling, the situation is more subtle: the SI spreading follows links in their temporal order, see Fig. 4(right). In particular, if links from A to B and from B to A are followed by links from B to C and from C to B then a disease may spread from A to B and then to C but not the converse (although the directed links from C to B and from B to A do exist, they are not in the appropriate temporal order). In this setting, SI cascades are therefore constrained not only by link directions but also by time ordering. Equivalent objects were previously defined and used in the literature, they are called outgoing infection chains in [21] or in [7].

In order to compare SI cascades in link streams to the ones in snapshots, we add the following constraint: for a given time duration (one or three months in general), we consider only the nodes reached within this duration after the cascade starts. Indeed, cascades in snapshots represent epidemics that have a limited time to spread; adding this time limit to spreadings in link streams makes the comparison fair. In addition, we have to choose a starting time in the link stream context. In our experiments, we choose for each node a random time and then observe SI cascades starting from each node at this time.

In terms of computational complexity, we can compare the cascading process on link streams to the one on static graphs. Setting a propagation duration have an impact on the time complexity of calculations. Let $\mathscr{L}$ be the set of temporal links (t, i, j), with t between the departure time and the propagation duration. The time complexity is $\mathcal{O}(n|\mathscr{L}|)$, with n the number of nodes in the network.

It only depends on the propagation duration and not on the total duration of the dataset. In the static case, it depends on the length of snapshots and the number of active nodes and links in this period, so the time complexity is smaller than $\mathcal{O}(nm)$, with m the total number of links of the network. Experimentally we observed that the computation of SI cascades is faster in the sequence of snapshots case than in the link stream case on a standard workstation: results on link streams are obtained in a couple of days, while some minutes to a few hours are enough on sequences of snapshots. In both cases, it is possible to make a sample of source nodes to speed up the calculations, but it raises the question of representativity of the results. Concerning memory complexity, computing SI cascades is in $\mathcal{O}(n)$, as it only requires to store their state (susceptible or infectious) over time. The link stream is read link by link in increasing value of time, to update the infectious state of the nodes. Thus, it is not needed to load it in memory, contrary to sequences of snapshots. The required memory space is then smaller in the case of link stream.

5   Size of Cascades

This section is devoted to the study of cascade sizes (*i.e.* the total number of nodes reached by SI cascades) in the frameworks described above. Our goal is to gain insight on the effect of temporal features of cattle mobility on cascade sizes.

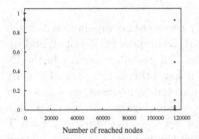 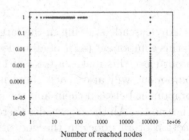

Fig. 6. Inverse cumulative distribution of cascades sizes for the 2005 static directed network. Left: lin-lin scales. Right: log-log scales.

Let us first consider the static directed network. The distribution of cascade sizes is given in Fig. 6. It clearly displays two different situations: either cascade sizes are over 100,000 nodes (93% of them reach more than 41% of active nodes), or they are much smaller, between 1 and 300 nodes. This is due to the fact that the network has the structure depicted in Fig. 7. According to this representation, commonly used as a large-scale map of the world wide web [4], the nodes are spread over different groups: a central core, the GSCC, where any node can reach any other node following a directed path; the *IN component*, where some nodes can reach nodes of the GSCC following a directed path, but cannot be reached by the nodes of the GSCC. Reciprocally, some nodes can be reached by the nodes of the GSCC following a directed path, but cannot reach it, they are located downstream to it, and constitute the *OUT component*. The remaining nodes are either isolated in small connected components, or part of structures called *tendrils* and *tubes*. Tendrils are subgraphs going either out from the in-component without reaching the GSCC, or subgraphs going into the out-component without coming from the GSCC. Tubes connect directly the in- to the out-component, without going through the GSCC.

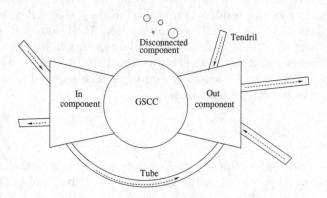

Fig. 7. Bow-tie structure of a directed graph with a giant strongly connected component.

Any cascade starting in the largest strongly connected component will reach all this component (44% of all nodes) and the OUT component (3% of all nodes). In our case, this makes a total of 115,250 nodes. All cascades starting in the IN component will also reach these nodes, and a few others: the part of the IN component between them and the strongly connected component, as well as few other nodes. All these cascades therefore have very similar sizes, and reach a large fraction of all nodes in the network. Finally, a few cascades start in the OUT component, or at other nodes that do not reach the largest strongly connected component (7% of them), they reach only a small part of the network and thus lead to much smaller cascade sizes. These results are consistent with previous results on the size of giant connected components, and confirm that taking into account the direction of links is crucial: SI cascades would otherwise reach 99% of nodes [7,17]. Considering the distribution of cascade sizes however highlights the fact that two very focused regimes co-exist, and that other cascade sizes never occur.

Now, we display in Fig. 8 the cascade size distribution for the monthly snapshots and the corresponding experiments for the link stream modeling. Quarterly snapshots give very similar results so we do not plot them here. Similarly to the static case, the cascade size distribution obtained for monthly snapshots display two distinct regimes, and for the same reason: each snapshot has a bow-tie large-scale shape. However, the largest cascades are much smaller than in the static network: they only reach 8% (July) to 13% (April) of nodes. This is also true for quarterly snapshots, in which cascades reach at most 23% of nodes. This may be due to the fact that spreading time is bounded, but it also shows that the cattle mobility is far from being a monthly or quarterly repetition of movements similar to the ones performed at the yearly level. This confirms the fact that time information is crucial to have an estimate of the potential epidemic outbreak sizes, as pointed out previously by several works [1,7,19]. It is the main motivation for turning to link stream modeling, that captures this information much more precisely.

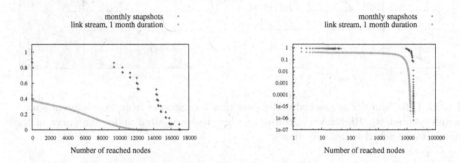

Fig. 8. Inverse cumulative distribution of cascades sizes for the 2005 monthly snapshot networks (in green, x points) and for the link stream modeling (in red, + points). Left: lin-lin scales. Right: log-log scales. The vertical axis is normalized to account for the fact that different numbers of cascades are considered in the two cases. (Color figure online)

The results using link stream modeling first show that maximal SI cascades lasting one month actually reach 5% (i.e. 13225) of all nodes only (13% with a 3 months propagation duration). This confirms that neglecting temporal information leads to significant over-estimates of the possible size of epidemic outbreaks. The distribution even shows that the number of negligible SI cascades is much larger than indicated by snapshots: respectively 60% of SI cascades are of size smaller than 5 nodes in the link stream, compared to 13% in monthly snapshots.

Figure 8 also displays another striking fact: whereas static network, quarterly snapshot and monthly snapshot modelings consistently lead to two different kinds of SI cascades (a few very small ones and a huge majority reaching a large part of the network, with nothing in between), link streams lead to a continuum in the observed cascade sizes. Although many cascades are large and many are very small, there are also all kinds of cascade sizes in between, and the heterogeneity of cascade sizes is much higher.

6   Speed of Cascades

One key advantage of the link stream modeling is that it accounts for the speed
of spreading, and not only for the number of nodes eventually reached, which is
a key feature for fighting epidemic outbreaks. We therefore dedicate this section
to the study of the speed of SI cascades in link streams.

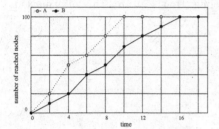 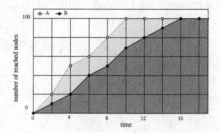

Fig. 9. Left: Number of reached nodes over time by cascades from two different source
nodes (A and B). Right: Visualization of the virulence (area under the curve of the
number of reached nodes over time).

In Fig. 9(left), the SI cascades from A and B reach the same number of nodes
at the end of the propagation period. Yet, the number of nodes reached at a
given time is higher for the propagation from source A than from source B.
As a consequence, if the cascade represents the spreading of a disease, the first
situation certainly implies a higher level of risk than the second one. Measuring
the number of reached nodes does not allow to make this distinction. That is
why we also calculated what we called the *virulence* of a cascade, that is simply
the area under the curve of the number of reached nodes over time (Fig. 9, right),
using the following definition. Let t_i be the beginning of the spreading, d the
propagation duration (parameter of the model), and $n(t)$ the number of reached
nodes at time t:

$$virulence = \sum_{t=t_i}^{t_i+d} n(t) \tag{1}$$

Using this measure, the SI cascade starting from B gets a lower virulence score
than the spreading starting from A. It allows us to differentiate the case described
in Fig. 9(left), where two cascades reach the same number of nodes with different
speeds.

We plot in Fig. 10 the virulence as a function of the final number of reached
nodes. We observe that for a given number of infected nodes, the corresponding
scores of virulence cover a broad range of values, especially for large spreading
cascades. For instance, for a score of 6000 reached nodes, the virulence spreads
from about 25000 to about 75000, that is to say three times as much. This means
that cascades of the same size may correspond to very different virulences, in
other words very different propagation speeds. Similar observations can be made

with a propagation duration of 3 months. Refining the comparison on the number of reached nodes and the virulence and their impact on the propagation sizes is a perspective of this study.

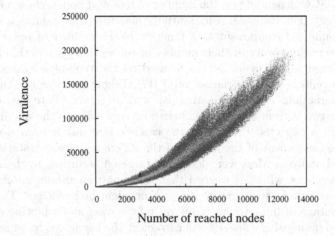

Fig. 10. Virulence as a function of the final number of reached nodes. In green, average virulence values for a fixed number of reached nodes. (Color figure online)

7 Conclusion

In this study, we evaluate the potential epidemic outbreak sizes, by taking into account temporal information in a network-based representation of cattle trade movement. For this purpose, we used data extracted from the year 2005 of the French database of cattle trade movements, and compared the sizes of SI cascades as produced using snapshot-based representations, to the closest possible protocol on a link stream representation of the data. We observed that cascades on static networks are systematically larger than the ones on link streams and thus that not taking into account the temporality of contacts leads to overestimating potential infection sizes. Moreover, while the cascades on static snapshots display a clear 2-mode behavior, where cascades are either very small or close to the maximum size, cascades on link streams can have any kind of size between these two extremes. Besides that, using SI cascades on link stream offers the possibility to take into account the spreading speeds, and not only the total number of reached nodes. Our results show that a same number of infected nodes can correspond to a wide range of propagation speed, which suggests that our measure should be a better tool to estimate the impact of an outbreak.

This study leads to several relevant considerations for future work. In the *BDNI* dataset, information about the type of holding (farms, markets, assembling centers...) is available. A special focus on the proportion of holdings being reached by SI cascades depending on their type would be interesting, in order to gain insight on whether the category they belong to leads to different risk exposures to disease outbreak.

In the context of controlling pathogen spread, targeted control is a key issue. One of the common procedures to test an intervention strategy is to measure the size of the GSCC depending on the number of removed nodes, chosen according to the strategy. For instance, some authors measured the reduction in size of the giant connected component as a function of the number of removed node, chosen in decreasing order of their number of connections, strength, or centrality measurements [2,8], while others focused on the type of holdings removed (assembling centers, markets, farms, etc.) [17]. Taking into account the temporality of interactions allows to distinguish various intermediate cascades sizes between the two extremes (very small sizes or very close to the maximal size). Therefore, it is expected that using the cascade size distribution would be a more precise assessment of the impact of the different control strategies than a GSCC-based analysis. Moreover, nodes are targeted according to their features in static snapshots, while [1] warned that using past snapshots information to devise efficient control strategies in the present might be inefficient. That is why taking advantage of link stream temporal information and adapting strategies to focus on dynamical features in order to select the key nodes to be removed is also a perspective of great interest.

Finally, we only considered the cattle trade movements as propagation routes. Thanks to the location of holdings, infection by contact in pastures may be modeled by adding links between holdings in the same neighborhood. Providing that we circumvent privacy issues related to the geographical location, such an approach would allow a more accurate modeling of contacts, but would also complexify the studied network.

Acknowledgements. This work is funded in part by the European Commission H2020 FETPROACT 2016–2017 program under grant 732942 (ODYCCEUS), by the ANR (French National Agency of Research) under grants ANR-15-CE38-0001 (Algo-Div) and ANR-13-CORD-0017-01 (CODDDE), by the French program "PIA - Usages, services et contenus innovants" under grant O18062-44430 (REQUEST), and by the Ile-de-France program FUI21 under grant 16010629 (iTRAC).

References

1. Bajardi, P., Barrat, A., Natale, F., Savini, L., Colizza, V.: Dynamical patterns of cattle trade movements. PLoS One **6**(5), e19869 (2011)
2. Bajardi, P., Barrat, A., Savini, L., Colizza, V.: Optimizing surveillance for livestock disease spreading through animal movements. J. R. Soc. Interface **9**(76), 2814–2825 (2012)
3. Belik, V., Fiebig, F., Lentz, H.H.K., Hövel, P.: Controlling contagious processes on temporal networks via adaptive rewiring (2015)
4. Broder, A., Kumar, R., Maghoul, F., Raghavan, P., Rajagopalan, S., Stata, R., Tomkins, A., Wiener, J.: Graph structure in the web. Comput. Netw. **33**(1), 309–320 (2000)

5. Büttner, K., Krieter, J., Traulsen, A., Traulsen, I.: Efficient interruption of infection chains by targeted removal of central holdings in an animal trade network. PLoS One **8**(9), e74292 (2013)
6. Büttner, K., Salau, J., Krieter, J.: Quality assessment of static aggregation compared to the temporal approach based on a pig trade network in Northern Germany. Prev. Vet. Med. **129**, 1–8 (2016)
7. Dube, C., Ribble, C., Kelton, D., Mcnab, B.: Comparing network analysis measures to determine potential epidemic size of highly contagious exotic diseases in fragmented monthly networks of dairy cattle movements in Ontario, Canada. Transboundary Emerg. Dis. **55**, 382–392 (2008)
8. Dutta, B.L., Ezanno, P., Vergu, E.: Characteristics of the spatio-temporal network of cattle movements in France over a 5-year period. Prev. Vet. Med. **117**(1), 79–94 (2014)
9. Holme, P., Liljeros, F.: Birth and death of links control disease spreading in empirical contact networks. Sci. Rep. **4**, 4999 (2014)
10. Holme, P., Saramäki, J.: Temporal networks. Phys. Rep. **519**(3), 97–125 (2012)
11. Lentz, H.H.K., Selhorst, T., Sokolov, I.M.: Unfolding accessibility provides a macroscopic approach to temporal networks. Phys. Rev. Lett. **110**(11), 1–5 (2013)
12. Natale, F., Giovannini, A., Savini, L., Palma, D., Possenti, L., Fiore, G., Calistri, P.: Network analysis of Italian cattle trade patterns and evaluation of risks for potential disease spread. Prev. Vet. Med. **92**(4), 341–350 (2009)
13. Newman, M.E.J.: Spread of epidemic disease on networks. Phys. Rev. E Stat. Nonlinear Soft Matter Phys. **66**(1), 1–11 (2002)
14. Nöremark, M., Hakansson, N., Lewerin, S.S., Lindberg, A., Jonsson, A.: Network analysis of cattle and pig movements in Sweden: measures relevant for disease control and risk based surveillance. Prev. Vet. Med. **99**(2–4), 78–90 (2011)
15. Pastor-Satorras, R., Castellano, C., Van Mieghem, P., Vespignani, A.: Epidemic processes in complex networks. Rev. Mod. Phys. **87**(3), 925–979 (2015)
16. Pellis, L., Ball, F., Bansal, S., Eames, K., House, T., Isham, V., Trapman, P.: Eight challenges for network epidemic models. Epidemics **10**, 58–62 (2015)
17. Rautureau, S., Dufour, B., Durand, B.: Vulnerability of animal trade networks to the spread of infectious diseases: a methodological approach applied to evaluation and emergency control strategies in cattle, France, 2005. Transbound. Emerg. Dis. **58**(2), 110–120 (2011)
18. Schärrer, S., Widgren, S., Schwermer, H., Lindberg, A., Vidondo, B., Zinsstag, J., Reist, M.: Evaluation of farm-level parameters derived from animal movements for use in risk-based surveillance programmes of cattle in switzerland. BMC Vet. Res. **11**(1), 149 (2015)
19. Vernon, M.C., Keeling, M.J.: Representing the UK's cattle herd as static and dynamic networks. Proc. R. Soc. Biol. Sci. **276**(1656), 469–476 (2009)
20. Viard, J., Latapy, M.: Identifying roles in an ip network with temporal and structural density. In: 2014 IEEE Conference on Computer Communications Workshops (INFOCOM WKSHPS), pp. 801–806. IEEE (2014)
21. Webb, C.R.: Investigating the potential spread of infectious diseases of sheep via agricultural shows in Great Britain. Epidemiol. Infect. **134**(1), 31–40 (2006)

Creating Territorial Intelligence Through a Digital Knowledge Ecosystem: A Way to Actualize Farmer Empowerment

Monica Sebillo[1(✉)], Giuliana Vitiello[1], and Athula Ginige[2]

[1] University of Salerno, Fisciano, Italy
{msebillo,gvitiello}@unisa.it
[2] Western Sydney University, Sidney, Australia
a.ginige@westernsydney.edu.au

Abstract. When properly combined territorial intelligence and ICT may contribute to strengthen the skills of a territory, to understand its phenomena, to interpret local dynamics involving targeted citizens, institutions and organizations. Among the others, the domain of sustainable agriculture production could benefit from mobile technology both in terms of single user data collector and support in daily activities, and in terms of user community data processing. In this way, the territory itself could contribute to the construction of a collective meaning, take advantage of collective knowledge, produce innovation in the management activities, and create cross-fertilization in terms of added value for different domains. This paper presents a technological solution conceived to increase the automation process for knowledge creation and sharing. On the basis of a digital knowledge ecosystem, users are allowed to query thematic datasets and visualize (geolocated) information. Each new information produced by the ecosystem through aggregating and mining processes can be in turn integrated as a resource belonging to the heritage of the spatially enabled territory.

Keywords: Digital knowledge ecosystem · Territorial intelligence · User's empowerment

1 Introduction

The information technology (IT) available today enables citizens to contribute in a new form of participatory democracy where private actors, associations and individuals, play a joint role in making decisions and bringing together local resources. This form of citizen engagement is currently known as collective intelligence and represents an indispensable step towards the actualization of the three pillars of the Open Government (OG) paradigm, namely transparency, participation and collaboration [19]. Examples of how people are involved in these community-based activities are now becoming common enough and they mainly refer to civic consultations in defense of cultural and environmental heritages, such as preserving a historical good, as well as to the urban context, such as approving plans to rebuild urban areas. However, the scope where the collective intelligence may result fruitful is quite wider, ranging from natural

© Springer International Publishing AG 2017
O. Gervasi et al. (Eds.): ICCSA 2017, Part II, LNCS 10405, pp. 98–111, 2017.
DOI: 10.1007/978-3-319-62395-5_8

environment to healthcare, from safety to mobility, due to the pervasive impact of the unifying element of those domains, namely the territory underlying a community and the information deriving from it. Actually, the geographic information itself plays a relevant role within the collective intelligence building process since it synthesizes and aggregates homogeneous collections of data thus preparing a base to enable people cooperate to a problem solution. When such shared collections of data heritage generated by a territory are then regularly organized through a system of models, methods, processes, people and tools, the collective intelligence assumes a specific territorial connotation and it becomes Territorial Intelligence (TI) [3, 6, 21]. As reported in [8, 10], the term Territorial Intelligence (TI) relates the development of a territory and communities acting on it. This implies that pursuing the territorial valorization and quality means acting so that a territory can meet the needs of its community (consumers) through continuous improvement of its economic, social, environmental and cultural (sustainability) dimensions. This claim explains the important role that TI is gaining for organizations and companies belonging to the same geographic area: it works as a collector of information deriving from multiple sources and, by processing it, it products knowledge relevant to the underlying territory and support the exchange of know-how among local actors of different cultures [3]. Indeed, territory underlies every activity where location plays a role in reaching such a goal and it performs a twofold action in terms of information producer and consumer (prosumer). Moreover, when this information prosumerism is supported by the mobile technology, the extent of TI is limited only by the real capability of the underlying infrastructure.

Handling the multidisciplinary complexity of data coming from a spatially enabled territory and supporting a fruitful exchange of it represent the goals of the research we are conducting in this field. In particular, we are investigating a digital knowledge ecosystem, as an open and shared environment with properties of scalability and sustainability, capable to realize services through the integration of four basic territory-oriented elements, namely content, communities, practices and policy, and technology.

In [4, 24] we presented some initial results of the research we are conducting on these topics applied to the specific domains of mobile health (m-health) and sustainable agriculture. We showed that both domains represent fields where TI and ICT can be combined to strengthen the skills of a territory, to understand its phenomena, to interpret local dynamics involving patients as well farmers, institutions and organizations. In particular, we also described how the management of healthcare as well as of sustainable agriculture production could benefit from mobile technology both in terms of single user data collector and support in daily activities, and in terms of patient/farmer community data processing. Moreover, we sketched how relationships among interested actors could be strongly beneficial when they are interpreted on the basis of the underlying territory. In this way, the territory itself could contribute to the construction of a collective meaning, take advantage of collective knowledge, produce innovation in the management activities, and create cross-fertilization in terms of added value for different domains.

Our current efforts are now focused on the latter feature that is the role that territorial intelligence may play in enabling knowledge creation for the development of different-sized communities. To support the achievement of this goal, in this paper we present a technological solution conceived to increase the automation process for knowledge creation and sharing. On the basis of a digital knowledge ecosystem, we discuss how users are allowed to query thematic datasets and visualize (geolocated) information. Each new information produced by the ecosystem through aggregating and mining processes can be in turn integrated as a resource belonging to the heritage of the spatially enabled territory.

The paper is organized as follows. Section 2 recalls the mobile-based information system implemented to support farmers' decision making for the crop cultivation activities. In Sect. 3, the rationale that stimulated the rethinking of the information system is discussed and the approach used to outline the information flow pattern is presented. Section 4 presents the architectural principles adopted to design the digital knowledge ecosystem, with special focus on both the Metadata Collection Framework and the module addressed to the engagement of the mobile users. Conclusions are drawn in Sect. 5.

2 A Mobile-Based Information System for Farmers Empowerment

The digital ecosystem we present in this paper evolved from a research aiming at developing mobile-based information systems to address special needs of targeted users, namely diabetes patients [24] and farmers from emerging countries [2, 4]. These represent two completely different scopes that helped us to recognize both specific necessities derived for communities with individual requests, and general-purpose requirements common to different typologies of users.

In the following, we briefly recall results deriving from studies performed within the sustainable agriculture domain and summarize both the proposed solution evolved.

A major agricultural problem in Sri Lanka is the overproduction of vegetables. The reason is twofold. Many farmers grow the same crop in the same area without awareness of what others famers are cultivating. This causes over production, waste and price fluctuations. Moreover, the open loop control system adopted so far to match supply and demand is based on many unpredictable variables, such as soil type and weather that represent contextual parameters on which the agricultural production strongly depends. This rarely led to an optimal balancing between total production and market demand.

In August 2011, an international collaborative research team consisting of researchers from seven universities in four continents started to explore a possible solution for farmers in Sri Lanka to face the problem associated with the crop cultivation [22]. An initial analysis showed that the major reason causing such an issue was users not getting the necessary information at the right time to make extensively informed decisions. Moreover, analyzing the surrounding environment, it came out that some useful information that could be of interest to farmers was available with various other stakeholders. Preliminary results derived by providing farmers with a mobile-based solution inspired by both the wide spread use of mobile phones and the user empowerment provided by Social Networking applications such as Facebook and Twitter. In particular, the awareness that these social networks represent an effective means to make information available to users in real time in order to detect evolving situations, motivated the adoption of the concept of Social Life Networks (SLN), as discussed in [26].

The initial physical solution evolved on the basis of field tests carried out with farmers in Sri Lanka, and a Design Science Research (DSR) methodology was selected to create a complete artefact as a solution to this research problem on which the Mobile-based Information System (MBIS) could be based. The paper [4] exhaustively describes steps followed to achieve this goal while Fig. 1 illustrates the DSR cycle evolved to develop the MBIS.

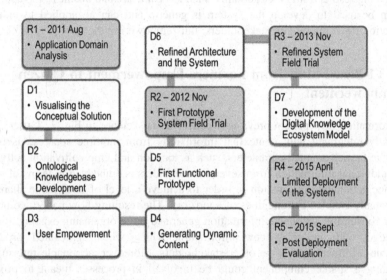

Fig. 1. The order of DSR cycles that evolved to develop the MBIS [4]

The study described in [4] has emphasized some relevant aspects. First, the information needs of farmers can be divided into two broad categories: Quasi Static and Dynamic or Situational. The former includes information that evolves slowly with time, such as crop varieties, fertilizer, pests and diseases. The latter refers to information that changes rapidly, such as local weather, market prices and current demand.

The second relevant aspect concerns the introduction of the "actionable information" concept that is the information that enables a stakeholder to act with least amount of further processing [23]. Providing farmers with such a context specific actionable information then represented the right support to help farmers make optimal decisions. In particular, using a user-centered design let the actionable information be useful for many other activities, thus connecting all major stakeholders in a complex information flow model. This allowed researchers to recognize relationships among consumers and producers, namely how to aggregate and disaggregate actionable information to generate dynamic content. As an example, the possibility to create an online market stimulated two banks to approach the researcher team to investigate how provide farmers with micro finance in case of cash flow difficulty. Analogously, the

Government could issue import permits for different food types on the basis of the information resulting from a cultivation planning functionality shared among farmers and agriculture department officers.

Generalizing this approach the architecture underlying the MBISs evolved towards a more complex system that, starting from a portion of collected personal information, shares it with other stakeholders (e.g., through dedicated social networks) and allows to discover new patterns and additional facts built also by performing spatio-temporal analyses. The derived collective knowledge could be in turn convenient for the user himself/herself when it is offered through, e.g., location based services. Moreover, investigating results from other domains, such as healthcare and business, the approach that can be used to develop the system is generic and can be applied to enhance information flow among key stakeholders, thus empowering the users.

3   A TI-Based Shift from Farmers Empowerment to Citizen Empowerment

The information flow pattern previously conceived is sketched in Fig. 2. A user query is initially integrated with contextual information from multiple sources, useful to instantiate farmer-oriented parameters, such as location and crop cultivation activities. Then, additional information from several stakeholders is embedded in terms of slices of ad hoc actionable information in order to achieve a level of knowledge shareable with spatially enabled communities of a territory. The resulting flow pattern combines initially consuming actionable information generated by re-organising existing domain knowledge, and producing new information which when aggregated produces actionable information to some other stakeholders. Moreover, actionable information embedding a spatial component could be involved in processes meant to produce territorial intelligence that can be provided to communities on different platforms. The resulting empowerment is then addressed both to the initial community of interest and to the wider one of citizens. Such a generalization in terms of both architecture and behavior motivated us to re-think the whole design and re-conceptualize both mobile-based systems as a digital knowledge ecosystem [5]. As a matter of fact, the information flow pattern mimics energy flow patterns in a biological ecosystem made up of living (biotic) and non-living (abiotic) components, which interact and give life to a dynamically stable, self-perpetuating system [2]. As a biological ecosystem, where all living components evolve especially depending on the environment and spatial interactions represents a key factor in the maintenance of diversity, in our vision, transforming the traditional client-server architecture so as to build a digital knowledge ecosystem, entails creating stronger links between the territory actors and the community of targeted users through information sharing and cooperative exploitation. Moreover, information flowing across individuals and the surrounding territory may be aggregated and transformed into knowledge, which can support decision-making procedures and can be ultimately delivered back to individuals in the form of services.

Such a conceptualization also fits Flichy's model of contemporary society named "connected individualism," which emphasizes the role of communities and networks for individual's growth and empowerment. Within this transformation, the goal of

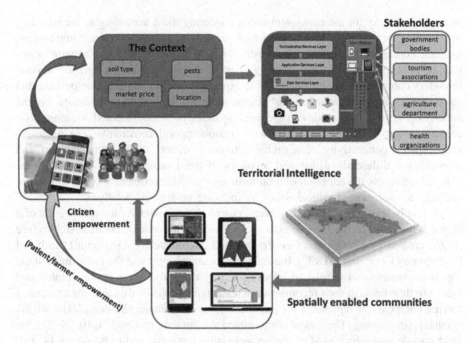

Fig. 2. The information flow pattern resulting from the enhancement of the mobile-based information system

information sharing and cooperative exploitation can be achieved by a profitable combination of territorial knowledge with personal data and events organized through a software infrastructure conceived for the development of special-purpose applications meant to improve users' experience while creating public value for services. In order to design and develop such a platform, factors of primary importance that should be taken into account are the need of a shared communication protocol among all the involved entities, extensibility (i.e., the ability to add new features without affecting the existing components), and the opportunity to hide the format differences of data coming from heterogeneous data sources. In particular, to this aim it is necessary to identify people who produce information, to aggregate/disaggregate it, and to include slices of them as part of the system, thus satisfying the need of accessing and providing both prior knowledge (static) and information generated in real time (dynamic).

4   The Digital Knowledge Ecosystem for Spatially-Enabled Farmer Communities

In a digital context, an ecosystem is a distributed, adaptive, open socio-technical system with properties of self-organization, scalability and sustainability. Basically, it embeds a set of satellite applications, Web portals, externals systems, legacy systems, and corporate DBs. These elements represent the Technology and Infrastructure component that integrates the Service Providers and the IT Infrastructure and Hardware meta-elements. As for the ecosystem users, clustering them according to the role they play allows for a simpler management of services allocated to them. Single users, thematic communities, organizations/institutions, regulatory bodies are generic examples of clusters of users who access and produce information through the Service Providers component that coordinates the aspects related to fruition, competition and collaboration for resources among diverse entities. This facility is based on the underlying spatial interactions that constitute a set of interconnected locations in a continuous space at multiple scales with spatio-temporal constraints. The heterogeneous spatial connectivity of a digital ecosystem affects the search strategies and determines a differently distributed topology of the landscape. This characteristics, although obstructs the adoption of a uniform approach for problem solving, allows for speeding the solution of critical issues by focusing on specific problems.

Generally speaking, there exist several proposals for designing the architecture of a digital ecosystem. In [6] Briscoe et al. propose an Ecosystem-Orientated Architecture (EOA) created by extending a Service-Oriented Architecture (SOA) with Distributed Evolutionary Computing (DEC). Basically, applications represent the individuals of the digital ecosystem and consist of groups of services. Those services are created and evolve in time through their recombination and aggregation. To this aim, the automated service composition represents an important issue to manage the complexity of the manual programming. There exist efforts aimed at achieving this goal [4, 10, 20–23], the most prevalent of which is SOA and its associated standards and technologies [7, 19].

In the following Subsection, basic architectural principles are discussed.

4.1   Architectural Principles of the Digital Knowledge Ecosystem

The involvement of advanced technological solutions conceived to produce and consume information represents the essential component to actualize services for territorial prosumerism in a digital ecosystem. On the basis of the requirements analysis, ICT is supposed to pursue three main goals, namely

a. to support decision makers in their daily activities and strategies adoption, by providing them with the territorial knowledge distributed among heterogeneous (virtual) sites,
b. to define easy-to-use solutions which help users to handle information of interest through integrated and sharable applications, and
c. to capitalize distributed territorial information creating processes of participatory democracy which involve intelligent communities.

The above goals could be translated into a set of well-designed technological objectives that include the definition of standards and shared guidelines about data and processes, and the design of an architecture addressed to implement a digital ecosystem where actors involved in any net of interest could share data, extract information, perform interoperable processes, and provide users with advanced services. Once data from possible heterogeneous sources are collected, they have to be transformed into information, they have to be preserved, made available, and presented into a simple, flexible and effective form to constitute a support for strategic, tactical and operating decisions. To this aim, TI represents a suitable and appropriate means to support data processing, analyses and aggregations. Datasets from users have to be modeled according to international standards and associated with appropriate metadata thus guaranteeing their interoperability and contributing to the construction of a collective meaning. Domain data and metadata standards should be investigated in order to identify a sharable common model capable to capture properties of data also associated with geographic components, and express their semantics in a complete and extensible manner. Once data is properly modeled and managed, the goal is to provide target users with a collaborative open environment that facilitates interaction and data sharing to integrate heterogeneous services and improve their delivery. The focus of such an ecosystem is on the usage of any information content within atomic applications arranged by diverse actors. Generally speaking, the digital ecosystem is addressed to combine visualization/provision of information as it usually happens through a virtual site, and acquisition of knowledge inferred from multiple sources (also human activities, users' actions and interactions) by applying appropriate processing functionality. Figure 3 shows a high-level overview of the entire backend architecture.

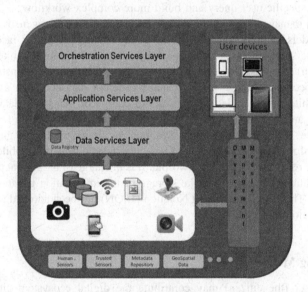

Fig. 3. The information flow pattern resulting from the enhancement of the mobile-based information system

Although the detailed specification of its components is still an ongoing process, the design is mainly influenced by two requirements concerning the seamless integration of heterogeneous data sources into the overall computational process, and the ability to dynamically extend the list of available functionality.

The digital ecosystem is organized in terms of users, data and services. An initial set of architectural principles has been presented in [8] where the above requirements have been guaranteed through a backend service oriented architecture, as specified in [20]:

- the data services layer,
- the application services layer,
- the orchestration services layer.

The main role of the data services layer is to encapsulate the different data formats and encodings of various information sources into a unique representation, and provide the higher levels with a standardized interface for their retrieval and exploitation. In addition, the data services layer offers support for the extension of the available data sources.

The intermediate level, the application services layer, collects the diverse software applications that make up the building blocks of the entire system functionality, ranging from basic reporting to geo-processing functions.

The orchestration services layer represents the architecture level that supports the definition of user queries and complex business processes, realized by properly combining both the functionality offered by lower layers services and the functionality provided by previously defined services composition. The composition is implemented by using the Business Processes Execution Language (BPEL) [21] that can be used to both answer a specific user query and build more complex workflow.

As for data managed through the digital ecosystem, it is heterogeneous in terms of structures, models, sources and semantics. Such a heterogeneity is faced by using different ad hoc solutions properly integrated in a general-purpose architecture where one of the abstraction levels is organized in terms of semantic relationships. Besides components specifically addressed to manage different datasets, such as spatial and media, the architecture includes components in charge of managing data harmonization, provision and visualization.

The Mobile Devices Management Module, as shown in Fig. 3 guarantees the bidirectional information exchange between the backend and the mobile devices of final users. The role of this module is twofold. It is in charge of the format translation of the input and output messages and the storage of the information generated from the mobile device. The JavaScript Object Notation (JSON) has been adopted as underlying data representation language [22].

4.2   Engaging Mobile Users

The mobile user (the citizen) may contribute the digital ecosystem either through activities performed on his/her smartphone (e.g., calls, photos, messages, web browsing, etc.) or through metadata collected by device sensors. For a mobile user to benefit from the digital ecosystem, his/her device must include some functionality that allows him/her to interact and exchange information with the ecosystem. Besides the traditional components of a mobile architecture, the device is expected to embed a Metadata Collection Famework, which is responsible to aggregate and manage metadata either directly generated by user-performed activities or coming from the various sensors usually available on an Android device.

Besides the collection and management of user-generated metadata, the framework offers a high level API that simplifies the development of mobile applications designed to extract information from the retrieved metadata. The main components of the framework are the Metadata Collection and Aggregation module, the Background module and the High Level Libraries, as shown in Fig. 4.

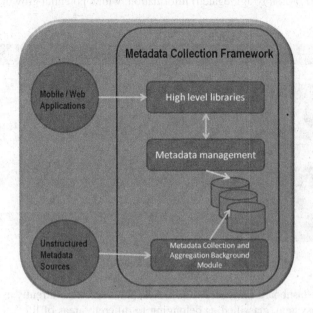

Fig. 4. The metadata collection framework

The Background Module performs the metadata retrieval activity, exploiting an Android service that scans (at regular or user-defined intervals) all metadata sources available on the mobile device. Figure 5 illustrates the Background module interface (entitled 'Monitor') which can be used to set the interval rates at which the scanning happens, to run/stop the service and to enable its background activity.

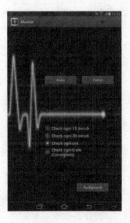

Fig. 5. Setting the background module through the monitor interface.

As a proof-of-concept, we have developed a prototypal digital ecosystem, built on a web platform, which supports individual citizens and allow service providers to create, monitor and share different services. The whole community may leverage such services and the derived (possibly aggregated) information, with a potential growth in territorial knowledge.

Fig. 6. The dashboard digital knowledge ecosystem.

The initial dashboard illustrated in Fig. 6 displays services currently available from the digital ecosystem, classified as belonging to different areas of life:

- culture;
- education;
- food;

Fig. 7. The initial overview of the digital knowledge ecosystem.

- health;
- sports;
- transport.

Figure 7 shows the initial interface of the Web site of the digital knowledge ecosystem.

5 Conclusion

The focus of the research we are carrying out is mainly addressed to investigate how territorial intelligence could support and boost sustainable development actions by different-sized communities. To achieve this goal, both methodological and technological solutions have been designed and specifically conceived to increase the automation process for knowledge creation and sharing among several territorial actors.

In this paper, a digital knowledge ecosystem is described as a collaborative open environment that facilitates interaction and data sharing to integrate heterogeneous services and improve their delivery. Each new information produced by it through aggregating and mining processes can be integrated as a resource belonging to the heritage of the spatially enabled territory.

The initial prototype of the digital knowledge ecosystem allows information systems from public and private organizations communicate and interact. Information provided by different providers are geographically and semantically aggregated, thus enhancing awareness by territorial communities.

The ability to obtain aggregated information related to the territory in nearly real-time opened up the possibility for different companies, organizations and government agencies to enroll in the digital knowledge ecosystem and adjust their processes as well as their services dynamically, thus bringing back to the farmers knowledge they could use to improve their quality of life.

References

1. Alves, A., Arkin, A., Askary, S., Barreto, C., Bloch, B., Curbera, F., Ford, M., Goland, Y., Guízar, A., Kartha, N., Liu, C.K., Khalaf, R., König, D., Marin, M., Mehta, V., Thatte, S., van der Rijn, D., Yendluri P., Yiu, A.: Web Services Business Process Execution Language Version 2.0. OASIS (2007). http://docs.oasis-open.org/wsbpel/2.0/wsbpel-v2.0.pdf
2. Begon, M., Harper, H., Townsend, C.: Ecology: Individuals, Populations and Communities. Blackwell Publishing, Boston (1996)
3. Bertacchini, Y., Rodriguez-Salvador, M., Souari, W.: From territorial intelligence to compositive & sustainable system. Case studies in Mexico & in Gafsa University. In: Proceedings of International Conference of Territorial Intelligence, Huelva, Spain, pp. 106–124, October 2007
4. Bertolotto, M., Giovanni, P., Sebillo, M., Vitiello, G.: Standard-based integration of W3C and geospatial services: quality challenges. In: Casteleyn, S., Rossi, G., Winckler, M. (eds.) ICWE 2014. LNCS, vol. 8541, pp. 460–469. Springer, Cham (2014). doi:10.1007/978-3-319-08245-5_31

5. Briscoe, G.: Complex adaptive digital ecosystems. In: Proceedings of the International Conference on Management of Emergent Digital EcoSystems, pp. 39–46 (2010)

6. Briscoe, G., Marinos, A.: Digital ecosystems in the clouds: towards community cloud computing. In: 3rd IEEE International Conference on Digital Ecosystems and Technologies, pp. 103–108 (2009)

7. Curbera, F., Duftler, M., Khalaf, R., Nagy, W., Mukhi, N., Weerawarana, S.: Unraveling the web services web: an introduction to SOAP, WSDL, and UDDI. IEEE Internet Comput. **2**, 86–93 (2002)

8. Damy, S., Gerardin, P., Girardot, J.J., Herrmann, B., Masselot, C.: Catalyse: an approach and a tool of governance help for territorial intelligence. In: INTI Huelva 2013 XII International Conference in Territorial Intelligence Social Innovation and New Ways of Governance for the Socio-Ecological Transition, 22 November 2013, Huelva (2013)

9. Silva, L., et al.: Interplay of requirements engineering and human computer interaction approaches in the evolution of a mobile agriculture information system. In: Ebert, A., Humayoun, S.R., Seyff, N., Perini, A., Barbosa, Simone D.J. (eds.) UsARE 2012/2014. LNCS, vol. 9312, pp. 135–159. Springer, Cham (2016). doi:10.1007/978-3-319-45916-5_9

10. Di Giovanni, P., Bertolotto, M., Vitiello, G., Sebillo, M.: Web services composition and geographic information. In: Pourabbas, E. (ed.) Geographical Information Systems - Trends and Technologies, pp. 104–141. Taylor and Francis Group, New York (2014)

11. Dumas, P., Gardere, J.-P., Bertacchini, Y.: Contribution of socio-technical systems theory concepts to a framework of territorial intelligence. In: Proceedings of International Conference of Territorial Intelligence, Huelva, Spain, pp. 92–105, October 2007

12. Erl, T.: Service-Oriented Architecture: Concepts, Technology, and Design. Prentice Hall PTR, Upper Saddle River (2005)

13. European Network of Territorial Intelligence (ENTI). http://www.territorial-intelligence.eu. Accessed Dec 2014

14. Ginige, A.: Social life networks for the middle of the pyramid (2011). http://www.sln4mop. org//index.php/sln/articles/index/1/3. Accessed 25 Mar 2012

15. Ginige, A.: Digital knowledge ecosystems: empowering users through context specific actionable information. Presented at the 9th International Conference on ICT, Society and Human Beings (ICT 2016), Madeira, Portugal (2016)

16. Ginige, A., Anusha, W., Ginige, T., de Silva, L., Di Giovanni,.P., Mathai, M., Goonatillake, M.D.J.S., Wikramanayake, G., Vitiello, G., Sebillo, M., Tortora, G., Richards, D., Jain, R.: Digital knowledge ecosystem for achieving sustainable agriculture production: a case study from Sri Lanka. Special Session on Data Science for Agricultural Decision Support Systems (DS4ADSS) at the 3rd IEEE International Conference on Data Science and Advanced Analytics, Montreal, Canada, 17–19 October 2016

17. Girardot, J.J., Masselot, C.: Specifications for the territorial intelligence community systems (TICS). In: International Conference of Territorial Intelligence Tools and Methods of Territorial Intelligence, MSHE, Besançon, France, October 2008, pp. 273–282 (2009)

18. JavaScript Object Notation. http://www.json.org. Accessed Jan 2017

19. Lathrop, D., Ruma, L., Open Government: Collaboration Transparency, and Participation in Practice. O'Reilly Media, Sebastopol (2010)

20. McIlraith, S., Son, C., Zeng, H.: Semantic web services. IEEE Intell. Syst. **16**, 46–53 (2001)

21. Narayanan, S., McIlraith, S.: Simulation, verification and automated composition of web services. In: International Conference on World Wide Web, pp. 77–88. ACM Press, New York (2002)

22. Milanovic, N., Malek, M.: Current solutions for web service composition. IEEE Internet Comput. **8**, 51–59 (2004)

23. Rao, J., Su, X.: A survey of automated web service composition methods. In: Cardoso, J., Sheth, A. (eds.) SWSWPC 2004. LNCS, vol. 3387, pp. 43–54. Springer, Heidelberg (2005). doi:10.1007/978-3-540-30581-1_5

24. Sebillo, M., Tucci, M., Tortora, G., Vitiello, G., Ginige, A., Di Giovanni, P.: Combining personal diaries with territorial intelligence to empower diabetic patients. J. Vis. Lang. Comput. **29**, 1–14 (2015). doi:10.1016/j.jvlc.2015.03.002. Elsevier

25. Violino, B.: How to navigate a sea of SOA standards. http://www.cio.com/article/104007/How_to_Navigate_a_Sea_of_SOA_Standards2007

26. Di Giovanni, P., et al.: Building social life network through mobile interfaces - the case study of Sri Lanka farmers. In: Spagnoletti, P. (ed.) Organizational Change and Information Systems. Lecture Notes in Information Systems and Organisation, vol. 2, pp. 399–408. Springer, Heidelberg (2013). ISBN 978-3-642-37227-8

Workshop on Advanced Methods in Data Mining for Applications (AMDMA 2017)

The Classification of Turkish Economic Growth by Artificial Neural Network Algorithms

Yeliz Karaca[1(✉)], Şengül Bayrak[2], and Emrullah Fatih Yetkin[3]

[1] Visiting Engineering School (DEIM), Tuscia University, Viterbo, Italy
yeliz.karaca@ieee.org
[2] Department of Computer Engineering, Halic University, Istanbul, Turkey
[3] Department of Computer Engineering, Kemerburgaz University, Istanbul, Turkey

Abstract. The development of globalization means that economies of the world are placing more importance on international trade. The increase in the variety of goods or services and the implementation of deregulation policies by many countries have made increases in international trade unavoidable. This study investigates the relationship between Turkey's international trade balance and economic growth between the years of 1960 and 2015 by using data obtained by the Turkish Statistical Institute. Two data sets were used in this study to identify the factors that affect the Turkish international trade balance and economic growth. Data Set 1 was formed by combining the parameters that make up international trade balance and those that determine economic growth. Data Set 2 consists only of parameters that define international trade balance. The aim of this study is to be able to identify the relationship between the parameters that define international trade balance and economic growth in a computerized system. In order to achieve this, Data Set 1 and Data Set 2 were subjected to various Artificial Neural Network methods such as Feed Forward Back Propagation and Cascade Forward Back Propagation algorithms and were classified in accordance with the international trade volume parameters. At the conclusion of the experimental work, the accuracy of the Feed Forward Back Propagation and Cascade Forward Back Propagation algorithms obtained from the test operation in the classification process was calculated. As a result, the study has classified the factors that influence the growth of the economy and international trade in Turkey.

Keywords: Artificial neural networks · Feed Forward Back Propagation algorithm · Cascade Forward Back Propagation algorithm · Economic growth

1 Introduction

International trade is defined by the volume of imports and exports of a country within a specific time period. The increase of deregulation initiatives between countries increases imports and exports, which results in an increasing volume

O. Gervasi et al. (Eds.): ICCSA 2017, Part II, LNCS 10405, pp. 115–126, 2017.
DOI: 10.1007/978-3-319-62395-5_9

of international trade. References cited [1–5] present valuable information on the details and contents concerning the growth of economy and international trade.

There are many analyses based on various methods on components of international trade. Jacop [6] used the Attraction Model in an attempt to measure the developmental and diversionary effect from the perspective of the member countries involved on trade of the free trade zones established between African nations. As a conclusion, it was demonstrated that the African Nations Economic Communities (ECOWAS) had more of a creative effect on development of trade. It was also determined that the trade diversionary effect was of a weak nature in the free trade zones agreement. Aguilar [7] used the Attraction Model to evaluate trade determinants with respect to specific products defined in the 9+ class, Standard International Trade Classification (SITC). As a result, it was found that an increase of the population of the importing country resulted in a corresponding increase in trade, thus population was found to have a positive correlation. Dimitri et al. [8] used the Attraction Model to measure Direct Foreign Investment (DFI) in Southeastern European countries. The importance of policies on labour unit costs, corporate tax burdens, foreign currency rates and trade regimes of the host country were emphasised as important factors in facilitating DFI. Tenreyro [9] used Panel Data Analysis (pseudo maximum-likelihood and instrumental variable methods) to estimate an Attraction Model to determine whether currency exchange rate volatility had a significant influence on bilateral exports. As a result of the work carried out, it emerged that for the period between 1970 and 1977, currency exchange rate volatility did not have a significant impact on bilateral exports. Allen [10] made use of the Attraction Model to investigate the relationship between DFI and exports. At the conclusion of the experimental work carried out, it was determined that America possessed a complementary DFI-Export relationship in both high and low capital/labour ratio industries. It was also concluded that DFI had a small but a positively significant impact on Mexican exports. Arize et al. [11] investigated the effect of currency exchange rate volatility on export volumes for 8 Latin American countries. Autoregressive conditional heteroscedasticity (ARCH) Model was used in calculation of the currency exchange rate volatility. For all the investigated countries, it was determined that currency exchange rate volatility had a negative influence on exports in both long and short range periods. Kayuma [12] researched the effect of currency exchange rate volatility for the 1996–2009 period on Exports in Uzbekistan. Three different volatility measurement techniques were used and these were the variance of the real currency exchange rate in respect of the trend, its relative change in respect of the previous period value, and Arch based modelling. According to the model that uses the variance of the real currency exchange rate in respect of the trend, currency exchange rate volatility has a negative impact not only on total exports but also on chemical substances, machines, energy, foodstuffs, and metal exports. However, volatility does not have any significant impact on exports of cotton products.

The work carried out analyzed the significant relationship between various parameters and statistical analysis methods. However, a study that classifies the relationship between international trade balance and economic growth using machine learning methods has never been carried out in Turkey or the rest of the world before. Based on the experience in the past and upcoming years, it can be stated that there is a need for a decision making mechanism with a high level of accuracy in analyzing the relationship between Turkish international trade volume and economic growth parameters. In this study, a machine learning approach with a high level of accuracy, which has never been used in previous literature and which classifies the economic growth related to Turkish international trade has been used in this study. Thus, a system is being presented to economists that can aid them while they are making inferences on the growth of international trade volume.

This study is based on the data obtained from the Turkish Statistical Institute, covering the period between 1960 and 2015. The study is based on parameters that affect international trade balance and economic growth. There exist a total of twelve parameters, which are: time period, international trade volume, exports, imports, tourism, exchange rate, foreign capital investments, gross domestic product (GDP), public free capital investment, private fixed capital investment, labour and population density. The aim of the study is to classify the relationship between Turkish international trade balance and economic growth for the period 1960 to 2015 period, using Artificial Neural Network (ANN) methods. International Trade Volume has been selected as the parameter which delimits this classification. In identifying how the parameters that ensure economic growth influence the increase in international trade volume, Feed Forward Back Propagation and Cascade Forward Back Propagation algorithms out of ANN methods were used on two different data sets (as can be seen on Table 1).

2 Materials and Methods

2.1 Materials

Out of ANN methods, Feed Forward Back Propagation and Cascade Forward Back Propagation algorithms were used in order for the impact of economic growth parameters on international trade balance for the period 1960 to 2015 to be classified with regard to international trade volume.

Data Details. In Table 1, Data Set 1 consists of the parameters (time period, international trade volume, exports, imports, tourism, exchange rate, foreign capital investments, gross domestic product (GDP), public free capital investment, private fixed capital investment, labour and population density) which determine international trade balance and economic growth.

Again in Table 1, Data Set 2 consists of the parameters (international trade volume, exports, imports, tourism, exchange rate, foreign capital investments, time period) which determine international trade balance.

In Fig. 1, the increase in Turkish International Trade Volume is shown according to data obtained from the Turkish Statistical Institute for the period 1960 to 2015.

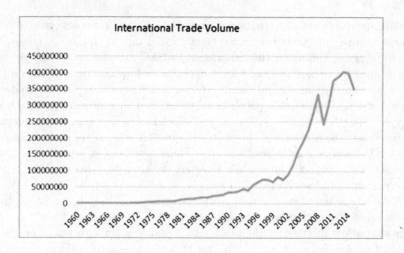

Fig. 1. Increase in Turkish International Trade Volume between 1960 and 2015.

Table 1. Parameters for Data Set 1 and Data Set 2.

Data Sets	Data set matrix size	Data set parameters description
Data Set 1	56 × 12	International trade volume, exports, imports, tourism, exchange rate, foreign capital investments, gross domestic product (GDP), public free capital investments, private fixed capital investment, labour, population density, time period
Data Set 2	56 × 7	International trade volume, exports, imports, tourism, exchange rate, foreign capital investments, time period

2.2   Methods

Data Set 1 and Data Set 2 have been classified by applying the Feed Forward Back Propagation algorithm and the Cascade Forward Back Propagation algorithm out of ANN algorithms.

Artificial Neural Networks. Human brains are made up of neurons. Neurons interconnect with each other to create a network. ANN has been developed using the operation of biological neuron cells in human brain as a model. In recent years, ANN has been modelled on computer based digital systems. In an ANN model, learning is achieved during learning process when the data arriving at the entry level has its weighting minimized by means of cyclical updates. Thus, at the end of a learning cycle, it can duplicate a human kind of prediction using the same working principles [13–15].

Feed Forward Back Propagation Algorithm. This is a feed forward, ANN system where one or more hidden layers are used in the entry and exit layers of a Multilayer Perceptron (MLP) network. In multilayer networks, the data is accepted by the input layer (x). The result value created at the output layer (d), at the end of operations within the network is compared to the target value. The error value between the calculated and target values is updated so as to minimize the weightings. Thus, in systems that possess a multiple hidden layered architecture, each system variances are deducted from the previous layers corrected values; and the operation is iterated. As a result, weighting correction operation starts from the weightings connected to the exit layer and the operation continues in reverse direction until it reaches the input layer [16].

In this study, Data Set 1 and Data Set 2 was trained using the Feed Forward Back Propagation (FFBP) algorithm. The multilayer network architecture which is essential for both Data Sets is provided in Fig. 2a for Data Set 1, and in Fig. 2b for Data Set 2.

Algorithmic flow for the training process is as follows:

Step 1: Algorithms network architecture is defined and the weightings are assigned using small random numbers and the first input is presented to the network [17].

Step 2: when m-dimension (m; for Data Set 1 (56×12), for Data Set 2 (56×7) input values are entered, $x_i = [x_1, x_2, ..., x_m]^T$ is yielded. Similarly, the required n-dimension output values are given by $d_k = [d_1, d_2, ..., d_n]^T$ If the x_i is the output values of neurons at layer i (for Data Set 1 and Data Set 2 n; 4×1), then the total input arriving to a neuron at the j level is assigned according to Eq. 1 [17].

$$net_j = \sum_{i=1}^{m} w_{ji}.x_i (from \ \ i.node \ \ to \ \ j.node) \tag{1}$$

Step 3: The output of the j neuron at the hidden layer (transfer function output) is

$$y_j = f_j(net_j), \quad j = 1, 2, ..., j \tag{2}$$

and here f_j is the transfer function.

Step 4: The total output to arrive to k neuron at output layer is calculated according to Eq. 3.

$$net_k = \sum_{j=1}^{j} w_{kj}.y_j \tag{3}$$

Step 5: The nonlinear output of k neuron at the output layer is calculated.

$$o_k = f_k(net_k), \quad becomes \ \ k = 1, 2, ..., n \tag{4}$$

Step 6: The output obtained from the network is compared to the actual data and e_k error is calculated.

$$e_k = (d_k - o_k) \tag{5}$$

Step 7: d_k and o_k are the outputs from a random k neuron at the output layer and represent the target (required) value and the actual one, respectively, obtained from the network.

Step 8: Weighting assignment has to be taken into account for the connections at the output layer. For each pattern, the total of square of the errors is calculated.

$$E = \frac{1}{2} \sum_k [(d_k) - o_k]^2 \tag{6}$$

Step 9: When there is a hidden layer, error level is classified as of local and global minimums. Local minimums may be more than one. Global minimum is where the error is the smallest minimum. The aim is to achieve global minimum during training process. Weightings are updated as can be seen in [18].

$$\Delta w_{kj} = -\varepsilon \frac{\delta E}{\delta w_{kj}} \tag{7}$$

According to Fig. 2, for Data Set 1 and Data Set 2, training data set was carried out for a 56-year-period using parameters.

Cascade Forward Back Propagation Algorithm. Cascade Forward Back Propagation (CFBP) algorithm is similar to the FFBP algorithm. The only difference is in the connection between the neurons at the input layer, hidden layer and the output layer [19–21]. Neurons that follow each other are connected; and training is carried out in this manner. Training process can be applied at two or more levels.

In this study, Data Set 1 and Data Set 2 were trained by applying the CFBP algorithm. The model structure used for Data Set 1 is given in Fig. 3.

The algorithmic flow of the training process is as follows:

Step 1: The initial values of the weightings are assigned.

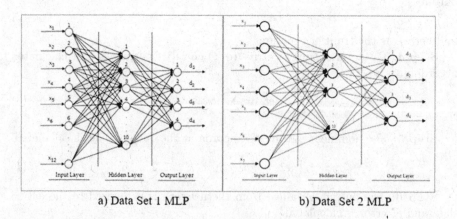

a) Data Set 1 MLP   b) Data Set 2 MLP

Fig. 2. Multilayer Perceptron (MLP) architecture for Data Set 1 and Data Set 2.

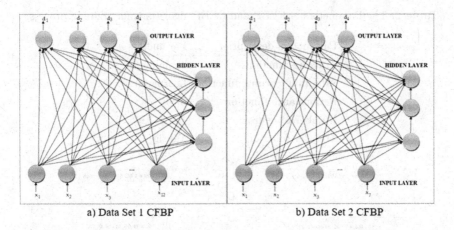

a) Data Set 1 CFBP   b) Data Set 2 CFBP

Fig. 3. Cascade Forward Back Propagation (CFBP) model architectures.

Step 2: for each (x_k, d_k) data pattern x_k inputs are propagated forward via the neural network layers.

$$p_k = f_j(w_j a_{j-1} - b_j), \quad j = 1, ..., J \tag{8}$$

Sensitivities and results are calculated.

$$\delta_j = f_j(d_j - w_{j+1})^T \delta_{j+1}, \quad j = J - 1, ..., 1 \tag{9}$$

If the error value is higher than estimated, weightings (weights and biases) are recalculated.

$$\Delta w_j = -\eta \delta_j (d_{j-1})^{(T)}, \quad j = 1, ..., J \tag{10}$$

$$\Delta b_j = \eta \delta_j, \quad j = 1, ..., J \tag{11}$$

Step 3: When the estimated error value is reached, the training process is terminated. If this value is not reached minimum, then step 2 is repeated.

3   Results and Discussion

The results of the FFBP and CFBP algorithm analyses used in this study have been obtained through Matlab 2016a software. Two data sets have been applied to FFBP and CFBP algorithms input layer. The first data set (Data Set 1) consists of parameters that determine international trade balance and economic growth. The second data set (Data Set 2) consists of parameters that determine international trade balance only. At the output layer of the FFBP and CFBP algorithms neurons have been represented according to the international trade balance volume (1: Small, 2: Medium, 3: Good, 4: Excellent).

The parameters used during the application of Data Set 1 and Data Set 2 to the FFBP and CFBP algorithms are provided in Table 2.

Table 2. The Parameters of FFBP and CFBP algorithms for training process.

Parameters description	Parameters
Training function	Trainlm
Adaption learning function	Learngdm
Performance function	MSE
Number of neurons	10
Transfer function	Tansig

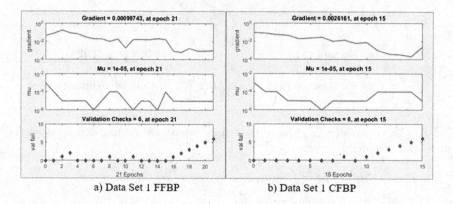

a) Data Set 1 FFBP   b) Data Set 1 CFBP

Fig. 4. According to Data Set 1 FFBP and CFBP training state.

For the FFBP algorithm, Data Set 1 was trained for 21 iterations with an error rate of 0.00102. As seen in Fig. 4a, when the best classification performance was achieved at Epoch 21 in the training process, a gradient ratio of 0.00099, a momentum value of 1e−05, and a validation check of 6 were obtained.

For CFBP algorithm, Data Set 1 was trained with 15 iterations of Data Set 1 with an error rate of 0.00236. As seen in Fig. 4b, when the best classification performance was achieved at Epoch 15 in the training process, a gradient ratio of 0.00261, a momentum value of 1e−05, and a validation check of 6 were obtained.

As can be seen from Fig. 5a, in the international trade volume-based classification of Data Set 1 in the FFBP algorithm, the best validation ratio obtained was 0.0014944 at Epoch 15.

As can be seen from Fig. 5b, in the same classification of Data Set 1 in the CFBP algorithm, the best validation ratio obtained was 0. 0012049 at Epoch 9.

For FFBP algorithm, Data Set 1 was trained for 16 epochs with an error rate of 0.0219. According to Fig. 6a, when the best classification performance was reached at Epoch 16, the gradient ratio was 0.00503, momentum value was 1e−05, and a validation check value of 6 was obtained. For CFBP algorithm, Data Set 1 was trained for 45 epochs with an error rate of 3.37e−09. According to Fig. 6b), when Data Set 2 CFBP reached best classification performance

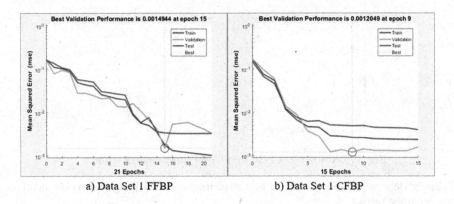

a) Data Set 1 FFBP   b) Data Set 1 CFBP

Fig. 5. FFBP and cascade FFBP performance graphs for Data Set 1.

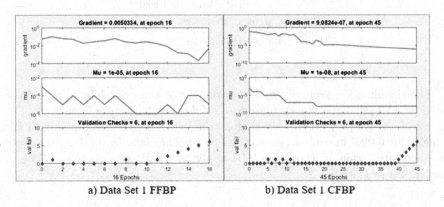

a) Data Set 1 FFBP   b) Data Set 1 CFBP

Fig. 6. According to Data Set 2 FFBP and CFBP algorithms training state.

at epoch45, a gradient ratio of 9.0824e−07, momentum value of 1e−08 and a validation check 6 was obtained.

As can be seen from Fig. 7a, in the classification process of Data Set 2 in accordance with international trade volume using the FFBP algorithm, the best validation ratio was obtained at epoch 10, with a value of 0.0023377.

As can be seen from Fig. 7b, in the classification process of Data Set 2 in accordance with international trade volume using the CFBP algorithm, the best validation ratio was obtained at epoch 39, with a value of 1.2715e−05.

As can be seen from Table 3, the following can be stated:

During the training of Data Set 1, which consists of parameters relating to economic growth and international trade volume, the best classification accuracy rate was obtained from the CFBP algorithm with a value of 99.87%.

During the training of Data Set 2, which consists of parameters international trade volume, the best classification accuracy rate was obtained from the FFBP algorithm with a value of 99.76%.

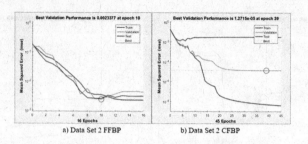

a) Data Set 2 FFBP   b) Data Set 2 CFBP

Fig. 7. Data Set 2, FFBP and CFBP performance graphs.

Table 3. Data Set 1 and Data Set 2 FFBP algorithm and CFBP algorithm classification test accuracy rates.

	FFBP	Cascade FFBP
Data Set 1	99.85%	99.87%
Data Set 2	99.76%	99.14%

The classification accuracy rate performance of Data Set 1 by applying the FFBP and CFBP algorithms is more successful in comparison to Data Set 2. The classification accuracy rate obtained after training Data Set 1 by applying FFBP is higher by 0.09% in comparison to Data Set 2. The classification accuracy rate obtained after training Data Set 1 by applying CFBP is higher by 0.73% in comparison to Data Set 2.

4   Conclusion

In this study, in order to classify the effect of economic growth and international trade balance on each other, two data sets were used. Data Set 1 consists of parameters relating to economic growth and international trade volume, while Data Set 2 consists only of parameters that define international trade balance. The aim of this study is to be able to classify the effect of parameters that influence economic growth in international trade by using the FFBP and CFBP methods out of ANN algorithms. At the end of the experimental study, classification test accuracy rate obtained from Data Set 1 by applying the FFBP algorithm is 0.09% higher than such rate obtained from Data Set 2 by applying the same algorithm. Classification test accuracy rate obtained from Data Set 1 by applying the CFBP algorithm is 0.73% higher than such rate obtained from Data Set 2 by applying the same algorithm. According to experimental results economic growth parameters have an influence on the development of international trade. The statistical data obtained from Turkish Statistical Institute are of a nature that supports this experimental result. It can be deduced that international trade volume increases in relation to the growth of the components that impact economic growth, and this is supported by the test accuracy rates. Results of the experimental study is compatible with the inferences of the experts in real life. In future work, the number of parameters that measure international trade volume will be increased to create new models that are able to perform predictions for the future. Results from this study are in relation to the parameters affecting the foreign trade balance amidst the development of the Turkish economy in the years from 1960 to 2015. Such analysis results obtained in our study could be applied to such fractal time series as spatial and temporal fractals [22], differential equation of fractional order, response of a fractional system, fractional filter driven with a white noise [23], Hurst exponent [24] and other methods for the estimation of the change or activities in the economies of countries in the upcoming years. Statistical models to be so obtained could provide valuable insight, and thus guide the literature.

References

1. Jorgenson, D., Gollop, F.M., Fraumeni, B.: Productivity and US Economic Growth, p. 169. Elsevier, Amsterdam (2016)
2. Solow, R.M.: Resources and economic growth. Am. Econ. **61**(1), 52–60 (2016)
3. Lu, X., Guo, K., Dong, Z., Wang, X.: Financial development and relationship evolvement among money supply, economic growth and inflation: a comparative study from the US and China. Appl. Econ. **49**(10), 1032–1045 (2017)
4. Greiner, A., Semmler, W., Gong, G.: The Forces of Economic Growth: A Time Series Perspective. Princeton University Press, Princeton (2016)
5. Srinivasan, A., Jayalakshmi, G.: Probabilistic analysis on time to recruitment for a single grade man power system when the breakdown threshold has two components using a different policy of recruitment. Indian J. Appl. Res. **5**(7), 1–3 (2016)
6. Jacop, M.W.: The intensity of trade creation and trade diversion in COMESA, ECCAS and ECOWAS: a comparative analysis. J. Afr. Econ. **14**(1), 117–141 (2005)
7. Aguilar, C.A.: Trade analysis of specific agri-food commodities using a gravity model, michigan state university department of agricultural economics. Master of Science Thesis, Michigan (2006)
8. Dimitri, D.G., Balazs, H., Elina, R.: Foreign direct investment in European transition economies - the role of policies. IMF Working Papers **20431**, 26–42 (2007)
9. Tenreyro, S.: On the trade impact of nominal exchange rate volatility. J. Dev. Econ. **82**(2), 4 (2007)
10. Allen, J.T.: The foreign direct investment-exports relationship: a US-Mexico analysis using the gravity model. Northern Illinois University, Doctor of Philosophy Dissertation, Dekalb, Illinois (2007)
11. Arize, A.C., Osang, T., Slottje, D.J.: Exchange-rate volatility in Latin America and its impact on foreign trade. Int. Rev. Econ. Finan. **17**, 33–44 (2008)
12. Kayumova, N.O.: How Exchange Rate Volatility Affects on the Main Export Goods of Uzbekistan? (2013). doi:10.12955/ejbe.v5i0.166
13. Wang, S.C.: Artificial neural network. In: Wang, S.C. (ed.) Interdisciplinary Computing in Java Programming, vol. 743, pp. 81–100. Springer, US (2003)
14. Akerkar, R., Sajja, P.: Knowledge-Based Systems. Jones & Bartlett Publishers (2010)
15. Milačiĉ, L., Joviĉ, S., Vujoviĉ, T., Miljkoviĉ, J.: Application of artificial neural network with extreme learning machine for economic growth estimation. Phys. A Stat. Mech. Appl. **465**, 285–288 (2017)

16. Skiba, M., Mrówczyńska, M., Bazan-Krzywoszańska, M.: Modeling the economic dependence between town development policy and increasing energy effectiveness with neural networks. Case study: the town of Zielona Gra. Appl. Energy **188**, 356–366 (2017)
17. Han, J., Pei, J., Kamber, M.: Data Mining: Concepts and Techniques, pp. 398–406. Elsevier, Amsterdam (2011)
18. Wang, L., Zeng, Y., Chen, T.: Back propagation neural network with adaptive differential evolution algorithm for time series forecasting. Expert Syst. Appl. **42**(2), 855–863 (2015)
19. Karaca, Y., Hayta, Ş.: Application and comparison of ANN and SVM for diagnostic classification for cognitive functioning. Appl. Math. Sci. **10**(64), 3187–3199 (2016)
20. Qiao, J., Li, F., Han, H., Li, W.: Constructive algorithm for fully connected cascade feedforward neural networks. Neurocomputing **182**, 154–164 (2016)
21. Qiu, M., Song, Y., Akagi, F.: Application of artificial neural network for the prediction of stock market returns: the case of the Japanese stock market. Chaos Solitons Fractals **85**, 1–7 (2016)
22. Peters, E.E.: Fractal Market Analysis Applying Chaos Theory to Investment and Economics. Wiley, Hoboken (1994)
23. Li, M.: Fractal time series—a tutorial review. Math. Probl. Eng. **2010** (2009). doi:10.1155/2010/157264
24. Loffredo, M.I.: Testing chaos and fractal properties in economic time series. In: International Mathematica Symposium (1999)

Standardized Precipitation Index Analyses with Wavelet Techniques at Watershed Basin

Funda Dökmen[1(⊠)], Zafer Aslan[2,3], and Ahmet Tokgözlü[4]

[1] Food and Agricultural Vocational School, Kocaeli University,
Campus of Arslanbey, 41285 Kartepe, Turkey
f_dokmen@hotmail.com, funda.dokmen@kocaeli.edu.tr
[2] Faculty of Engineering, Istanbul Aydın University, 34295 Istanbul, Turkey
zaferaslan@aydin.edu.tr
[3] International Centre for Theoretical Physics (ICTP), Trieste, Italy
[4] Science and Literature Faculty, Süleyman Demirel University, Isparta, Turkey
tokgozlu68@gmail.com

Abstract. Climatic changes play an important role on agricultural production. Their impacts on agricultural sciences are significantly different form one region to another. There are many factors on the impact of climate change in farmlands. SPI (Standardized Precipitation Index) is one of the drought indices. SPI values show similar variation in relatively drought periods and different amounts of precipitation. The main aim of this study is to understand different factors and their role at different scales with wavelet analysis of long-term precipitation and the SPI data. All data is being evaluated under the catchment basin area. Data records cover Big Menderes Catchment area at Aegean Region in Turkey between 1960 and 2015. The results of this study underline in general small scale influences have more important role in all selected stations. Spatial variation of entropy at watershed basin has been associated with micro, meso or large scale influences. In general, gradually increasing trends of drought and wet conditions are resulted in small scale factors.

Keywords: Drought index · Rainfall · Sustainable agriculture · Signal analysis

1 Introduction

A fractal time series has been taken one of the solutions to a different equation of fractional order or a response of a fractional system or a fractional filter driven with a white noise in the domain of stochastic (Li 2009). It gives a tutorial review about conventional time series, solution to a differential equation of integer order with the excitation of white noise in mathematics.

Adequate knowledge of spatial connections in rainfall is important for reliable modeling of catchment processes and water management. This study applies the ideas of network theory to examine and interpret the spatial connections in rainfall in Turkey. Jha and his group studied that the clustering coefficient values are interpreted in terms of topographic factors (latitude, longitude, and elevation) and rainfall properties (mean, standard deviation, and coefficient of variation) in Australia. After this paper, reliable

© Springer International Publishing AG 2017
O. Gervasi et al. (Eds.): ICCSA 2017, Part II, LNCS 10405, pp. 127–141, 2017.
DOI: 10.1007/978-3-319-62395-5_10

information about rainfall is the driving force in most hydrologic studies. Rain-gages are common devices for rainfall observations, providing point measurements. In addition to estimation of stream flow in space, knowledge of spatial connections in rainfall in a network of monitoring stations is crucial for various purposes, including for; (1) applying interpolation methods to obtain rainfall at locations where rain-gage measurements are not available; (2) filling gaps in historical rainfall record using available rainfall observations at neighboring stations and; (3) determining optimal density and locations for installing new rain-gages (Jha et al. 2015).

The increasing global water demand is aggravating the competition between agriculture and the environment for fresh and good quality water. Drainage water, containing agrochemicals and salts, may degrade water quality in receiving water bodies (Moyano et al. 2015).

A new approach to characterize geographical areas with a Drought Risk Index (DRI) is suggested, by applying an Artificial Neural Network (ANN) classifier to bioclimatic time series for which Operational temporal Units (OtUs) are defined. DRI = 1 is represented by the reciprocal OtUs to former one. The classification of the cells based on DRI time profiles showed that, at the scale used in this work, DRI has no dependence on land cover class, but is related to the location of the cells. The methodology was integrated with GIS (Geographic Information System) software, and used to show the geographic pattern of DRI in a given area at different periods (Incerti et al. 2007). Variation of climatic conditions, occurrence of extreme drought in recent years and increasing of human activity convert measured discharges of hydrometric stations in the future. Available data cannot predict next variations (Adib 2011). Historically, precipitation in the region is highly variable with dry years alternating with wet ones as, for example, during the 1970–2010 period. The effects of the dry period were exacerbated by land use changes in the catchment area over the last 50 years. Mediterranean environments are particularly sensitive to climate variations due to the hot dry summer season, year-round open-water conditions, and limited rainfalls (Giadrossich et al. 2015).

Standardized drought indices computed from precipitation, evapotranspiration, stream-flow or soil moisture have been used to monitor drought and to quantify drought severity. The Standardized Precipitation Index (SPI) is the most widely used. SPI values quantify deviations from 'normal precipitation' (McKee et al. 1993). SPI values are classified in drought (wetness) categories, with the more negative values indicating a more severe drought category (McKee et al. 1995). Shorter time scales are appropriate to monitor the effects of precipitation shortages in soil water storage and agriculture, while longer time scales are used to monitor drought effects on surface and ground water resources (Paulo et al. 2016). Generally, agricultural studies prefer to use Palmer Drought Severity Index (PDSI, Palmer 1965). PDSI combines precipitation and evapotranspiration to define a deviation from normal through a soil water balance, and a modification of the PDSI to better perform the soil water balance using an olive crop as standard perennial crop (MedPDSI, Pereira et al. 2007; Paulo et al. 2012). Due to its standardized nature, the same negative SPI value computed in different locations or time periods corresponds to the same relative drought severity but, to different amounts of precipitation. The relative measure provided by the SPI would be more transparent if accompanied by an absolute value of the monthly precipitation thresholds relative to the SPI drought categories as shown by Paulo and Pereira (2008).

Massel (2001), Dökmen and Aslan (2013) studied on water quality and management subjects by using wavelet analyses. Moosavi (2017) examined that modelling of runoff by analyses of wavelet packet on water resources management.

The present paper discusses spatio – temporal variation of SPI in Southwestern part of Turkey.

2 Materials and Methods

The following part of this paper covers some statistical analyses, spatio-temporal variations and wavelet applications on SPI.

2.1 Study Area and Data

Study area and geographic situation are given in Fig. 1. Monthly rainfall rate at watershed region has been analyzed.

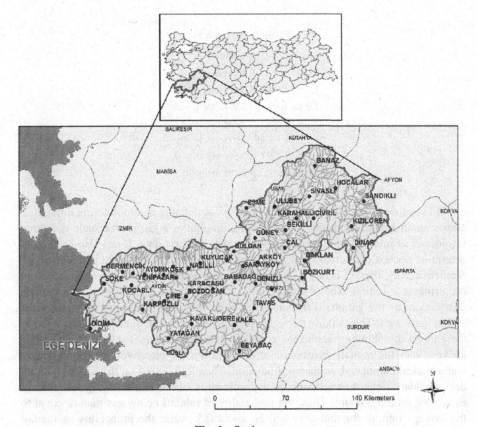

Fig. 1. Study area

2.2 Standardized Precipitation Index

Calculation of meteorological indices, SPI is one of the indices that were presented for drought monitoring. It is calculated in short-term (3, 6 and 9 months) and long-term (12, 24 and 48 months) periods. In any time scale, the SPI mean may reach zero in a location and its variance become equal to 1. Using the SPI, quantitative definition of drought can be established for each time scale. A drought event for time scale "i" is defined here as a period in which the SPI is continuously negative and the SPI reaches a value of −1.0 or less. The drought begins when the SPI first falls below zero and ends with the positive value (Mckee et al. 1993). The SPI is calculated by using the following equation:

$$SPI = \frac{P_i - \bar{P}}{s} \tag{1}$$

where P_i and $-\bar{P}$ are precipitation and average precipitation at selected period, s is standard deviation of precipitation (http://www.ijmer.com/papers/Vol3_Issue4/DP3424072411.pdf, 6.03.2017). Table 1 shows characteristics of SPI values (Han and Kamber 2006).

Table 1. SPI values

SPI	Class
2.0+	Extremely wet
1.5 to 1.99	Very wet
1.0 to 1.49	Moderately wet
−.99 to .99	Near normal
−1.0 to −1.49	Moderately dry
−1.5 to 1.99	Severely dry
−2 and less	Extremely dry

SPI is a tool which was developed primarily for defining and monitoring drought. It allows an analyst to determine the rarity of a drought at a given time scale (temporal resolution) of interest for any rainfall station with historic data. It can also be used to determine periods of anomalously wet events. The SPI is not a drought prediction tool. Mathematically, the SPI is based on the cumulative probability of a given rainfall event occurring at a station. The historic rainfall data of the station is fitted to a gamma distribution, as the gamma distribution has been found to fit the precipitation distribution quite well. This is done through a process of maximum likelihood estimation of the gamma distribution parameters, α and β. In simple terms, the process described above allows the rainfall distribution at the station to be effectively represented by a mathematical cumulative probability function. Therefore, based on the historic rainfall data, an analyst can then tell what is the probability of the rainfall being less than or equal to a certain amount. Thus, the probability of rainfall being less than or equal to the average rainfall for that area will be about 0.5, while the probability of rainfall being less than or equal to an amount much smaller than the average will be also be lower (0.2, 0.1, 0.01 etc., depending on the amount). Therefore if a particular rainfall event gives a low probability on the cumulative probability function, then this is indicative of a likely drought event. Alternatively, a rainfall event which gives a high probability on the cumulative probability function is an anomalously wet event (http://drought.unl.edu/portals/0/docs/spi-program-alternative-method.pdf, 6.03.2017).

2.3 Wavelet Analyses

Dimensionality Reduction

In *dimensionality reduction*, data encoding or transformations are applied so as to obtain a reduced or "compressed" representation of the original data. If the original data can be *reconstructed* from the compressed data without any loss of information, the data reduction is called lossless. If, instead, we can reconstruct only an approximation of the original data, then the data reduction is called lossy. There are several well-tuned algorithms for string compression. Although they are typically lossless, they allow only limited manipulation of the data. In this section, we instead focus on two popular and effective methods of lossy dimensionality reduction: *wavelet transforms* and *principal components analysis*.

Wavelet Transforms

The discrete wavelet transform (DWT) is a linear signal processing technique that, when applied to a data vector X, transforms it to a numerically different vector, $X0$, of wavelet coefficients. The two vectors are of the same length. When applying this technique to data reduction, we consider each tuple as an n-dimensional data vector, that is, $X = (x1; x2; : : : ; xn)$, depicting n measurements made on the tuple from n database attributes. *"How can this technique be useful for data reduction if the wavelet transformed data are of the same length as the original data?"* The usefulness lies in the fact that the wavelet transformed data can be truncated. A compressed approximation of the data can be retained by storing only a small fraction of the strongest of the wavelet coefficients.

For example, all wavelet coefficients larger than some user-specified threshold can be retained. All other coefficients are set to 0. The resulting data representation is therefore very sparse, so that operations that can take advantage of data scarcity are computationally very fast if performed in wavelet space. The technique also works to remove noise without smoothing out the main features of the data, making it effective for data cleaning as well. Given a set of coefficients, an approximation of the original data can be constructed by applying the *inverse* of the DWT used. In our notation, any variable representing a vector is shown in bold italic font; measurements depicting the vector are shown in italic font.

The DWT is closely related to the *discrete Fourier transform (DFT)*, a signal processing technique involving sines and cosines. In general, however, the DWT achieves better lossy compression. That is, if the same number of coefficients is retained for a DWT and a DFT of a given data vector, the DWT version will provide a more accurate approximation of the original data. Hence, for an equivalent

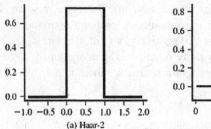
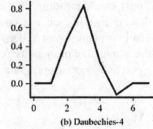

(a) Haar-2   (b) Daubechies-4

Fig. 2. Examples of wavelet families. The number next to a wavelet name is the number of *vanishing moments* of the wavelet. This is a set of mathematical relationships that the coefficients must satisfy and is related to the number of coefficients, (Han and Kamber, 2006).

approximation, the DWT requires less space than the DFT. Unlike the DFT, wavelets are quite localized in space, contributing to the conservation of local detail.

There is only one DFT, yet there are several families of DWTs. Figure 2 shows some wavelet families. Popular wavelet transforms include the Haar-2, Daubechies-4, and Daubechies-6 transforms. The general procedure for applying a discrete wavelet transform uses a hierarchical *pyramid algorithm* and iteration, resulting in fast computational speed. The method is as follows (Han and Kamber 2006):

1. The length, L, of the input data vector must be an integer power of 2. This condition can be met by padding the data vector with zeros as necessary (L_n).
2. Each transform involves applying two functions. The first applies some data smoothing, such as a sum or weighted average. The second performs a weighted difference, which acts to bring out the detailed features of the data.
3. The two functions are applied to pairs of data points in X, that is, to all pairs of measurements ($x2i$; $x2i + 1$). This results in two sets of data of length $L = 2$. In general, these represent a smoothed or low-frequency version of the input data and the high frequency content of it, respectively.
4. The two functions are recursively applied to the sets of data obtained in the previous loop, until the resulting data sets obtained are of length 2.
5. Selected values from the data sets obtained in the above iterations are designated the wavelet coefficients of the transformed data.

Equivalently, a matrix multiplication can be applied to the input data in order to obtain the wavelet coefficients, where the matrix used depends on the given DWT. The matrix must be orthonormal, meaning that the columns are unit vectors and are mutually orthogonal, so that the matrix inverse is just its transpose. Although we do not have room to discuss it here, this property allows the reconstruction of the data from the smooth and smooth-difference data sets. By factoring the matrix used into a product of a few sparse matrices, the resulting *"fast DWT"* algorithm has a complexity of $O(n)$ for an input vector of length n. Fourier transforms are spherical transforms. On contrary, wavelet transforms are local transforms. They present analyses at both frequency and time domains. In this paper, $f(t)$ is SPI values (Aslan et al. 2011):

$$\int_{-\infty}^{\infty} |f(t)|^2 \, dt < \infty \tag{2}$$

$\psi(t)$ is a continuous wavelet function and given at Eqs. 3 and 4.

$$\int_{-\infty}^{\infty} |\psi(t)|^2 \, dt = 1 \tag{3}$$

$$\int_{-\infty}^{\infty} |\psi(t)| \, dt = 0 \tag{4}$$

Wavelet transforms can be applied to multidimensional data, such as a data cube. This is done by first applying the transform to the first dimension, then to the second, and so on. The computational complexity involved is linear with respect to the number of cells in the cube. Wavelet transforms give good results on sparse or skewed data and on data with ordered attributes. Lossy compression by wavelets is reportedly better than JPEG compression, the current commercial standard. Wavelet transforms have many real-world applications, including the compression of fingerprint images, computer vision, analysis of time-series data, and data cleaning.

3 Analyses

3.1 Time Series Analyses of SPI

Figures 3(a)–(g) show temporal variation of SPI in watershed area (Denizli, Afyonkarahisar, Muğla, Uşak, Manisa, Kütahya and Aydın). SPI values show a decreasing trend in Muğla, Manisa and Kütahya. Drought index is gradually increasing in Denizli, Afyonkarahisar, Uşak and Aydın. There is a sufficient evidence of linear trends with = 0,05 and they are associated with rapid urbanization in that regions.

Gradually a decreasing trend are changing from normal to moderate dry conditions between 1960 and 2014. Very wet and extremely wet conditions have been observed in Uşak in recent years

3.2 Wavelet Analyses

Figures 4a, 4b, 4c, 4d, 4e and 4f show wavelet analyses at two selected stations (Aydın, Muğla) in respectively. Amplitudes of small scale fluctuations (d1) increase in the second part of study period. In this period, combined effects of meso and small scale fluctuations have been recorded. Role of large scale influences are decaying in Aydın in the recent period. Increasing trend of d1 coefficients are resulted in rapid urbanizations at study areas.

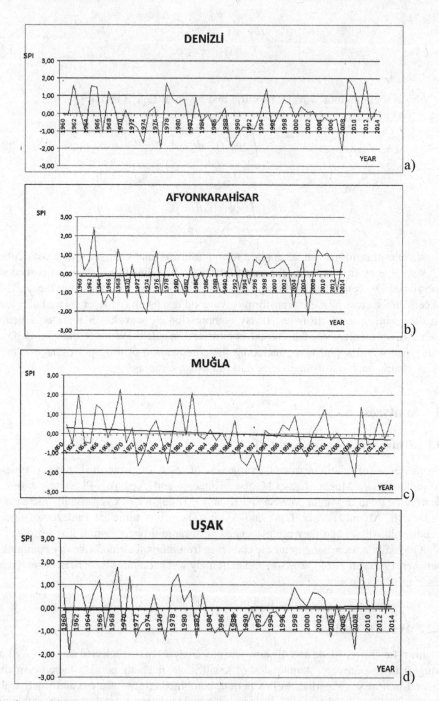

Fig. 3. (a)–(g) Temporal variations of SPI in selected stations at the watershed area (1960–2014)

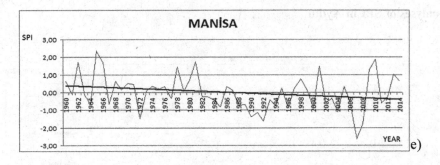

e)

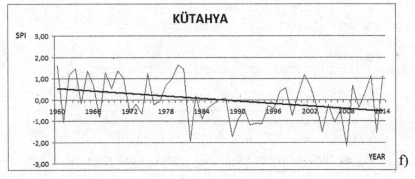

f)

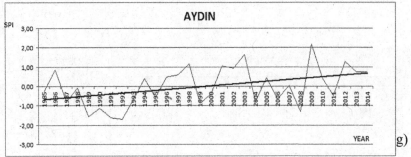

g)

Fig. 3. (continued)

Analyses of SPI in Aydın

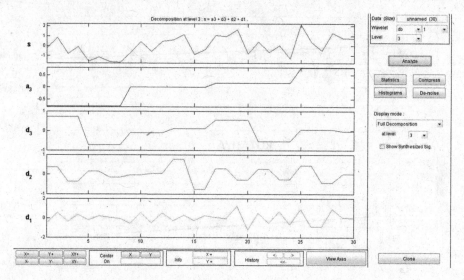

Fig. 4a. Wavelet 1D, DMeyer, Level 3, SPI in Aydın (1985–2014)

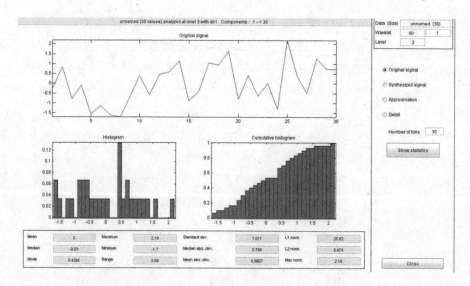

Fig. 4b. Wavelet 1D, DMeyer, Level 3, frequency distribution, SPI in Aydın (1985–2014)

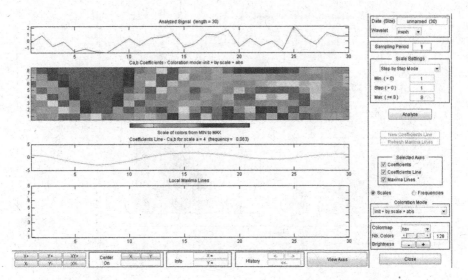

Fig. 4c. Continuous wavelet, 1D, Maxican Hat, SPI in Aydın (1985–2014)

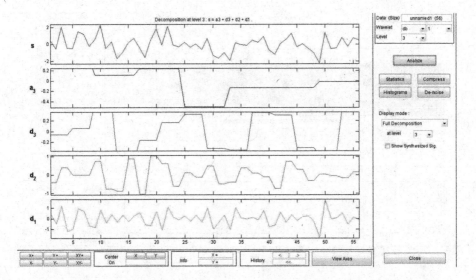

Fig. 4d. Wavelet 1D, DMeyer, Level 3, SPI in Muğla (1960–2014)

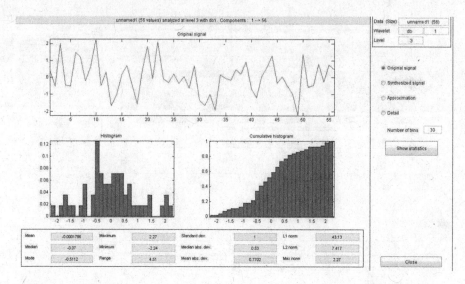

Fig. 4e. Wavelet 1D, DMeyer, Level 3, frequency distribution, SPI in Muğla (1960–2014)

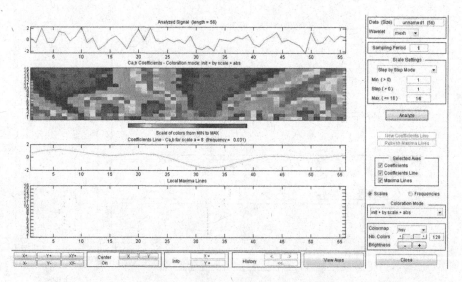

Fig. 4f. Continuous wavelet, 1D, Maxican Hat, SPI in Muğla (1960–2014)

Analyses of SPI in Muğla

Role of small scale factors on SPI variation in Aydın and Muğla show an increasing trend in last decade. Periodicity of large scale events have been changed from three months up to 16 months.

Table 2 shows spatial variation of entropy at watershed basin. In this study instead of d coefficients square of d values have been compared. Small scale factors and their role at all stations are important. Most of the energy is decaying through the higher levels.

Table 2. Spatial variation of square of entropy values at watershed basin

d^2	Muğla	Aydın	Kütahya	Manisa	Uşak	Afyon	Denizli
d1	**27,84135**	**0,413692**	**0,553781**	**0,45508**	**0,47817**	**0,50804**	**0,40785**
d2	11,40818	0,342813	0,354805	0,29023	0,29098	0,44789	0,18130
d3	4,346863	0,018134	0,182704	0,20269	0,15635	0,07463	0,12655
d4	1,7278	**0,056364**	0,116754	0,10386	0,09811	0,04664	0,03684
d5	0,28455	0,022877	0,032549	0,032549	0,04657	0,04162	0,02003

It explains gradually decreasing energy components at meso and larges scales at all stations except Aydın. Energy is transferred from lower levels to larger levels. It explains the combined role of urbanization and synoptic factors in Aydın. Role of small scale factors which are mainly associated with urbanization has less importance than other stations in Denizli. Not only the role of large scale factors but role of small and meso scale fluctuations are more important in Muğla if it was compared with the other stations. Gradually trend of Drought is mainly resulted in small, scale factors in Kütahya, Manisa and Muğla. Uşak, Aydın ad Denizli tends from moderate to very and extremely wet classes.

4   Results and Conclusion

Specific results of the study are below:

- "Büyük Menderes Watershed" and Aydın is under the near normal to moderate wet conditions in forthcoming years.
- Minimum drought risk is observed in Aydın.
- 2008 is a dry year in general.
- Short term dry conditions have been recorded in 1990's. Extremes were recorded before 1990.
- Action Plan in Afyon was carried out by the Governmental State Organization. Energy levels and distribution explain this local factor.
- Role of High Population.
- Farmlands, agricultural activities are mainly affected by drought.

Akcknowledgements. The authors gratefully appreciate and acknowledge the International Centre for Theoretical Physics (ICTP)-Associateship Program.

References

Adib, A.: A new approach for determination of peak flood hydrograph. J. Food Agric. Environ. **9**(3&4), 1129–1130 (2011)

Aslan, Z., Siddiqi, A.H., Manchanda, P.: Temporal and spatial variation of temperature and precipitation indices. J. Food Agric. Environ. **9**(3–4), 912–922 (2011)

Dökmen, F., Aslan, Z.: Evaluation of the parameters of water quality with wavelet techniques. Water Resour. Manag. **27**, 4977–4988 (2013). doi:10.1007/s11269-013-0454-5. Springer

Giadrossich, F., Niedda, M., Cohen, D., Pirastru, M.: Evaporation in a Mediterranean environment by energy budget and penman methods, Lake Barats, Sardania, Italy. Hydrol. Earth Syst. Sci. **19**, 2451–2468 (2015). doi:10.5194/hess-19-2451

Han, J., Kamber, M.: Data Mining: Concepts and Techniques, 2nd edn. Morgan Kaufmann Publishers, San Francisco (2006)

Incerti, G., Feoli, E., Salvati, L., Brunetti, A.: Analysis of bioclimatic time series and their neural network-based classification to characterise drought risk patterns in south italy. Int. J. Biometeorol. **51**, 253–263 (2007). doi:10.1007/s00484-006-0071-6

Jha, S.K., Zhao, H., Woldemeskel, F.M., Sivakumar, B.: Network theory and spatial rainfall connections: an interpretation. J. Hydrol. **527**(2015), 13–19 (2015)

Li, M.: Fractal time series. Math. Probl. Eng. **2010**, 1–26 (2009). doi:10.1155/2010/157264. Article ID 157264. Hindawe Publishing Corporation

Massel, S.R.: Wavelet analysis for processing of ocean surface wave records. Ocean Eng. **28**, 957–987 (2001)

McKee, T.B., Doesken, N.J., Kleist, J.: The relationship of drought frequency and duration of time scales. In: 8th Conference on Applied Climatology, pp. 179–186. American Meteorological Society, Boston (1993)

McKee, T.B., Doesken, N.J., Kleist, J.: Drought monitoring with multiple time scales. In: 9th 1999 Conference on Applied Climatology, pp. 233–236. American Meteorological Society, Boston (1995)

Moosavi, V., Talebi, A., Hadian, M.R.: Development of a hybrid wavelet packet-group method of data handling (WPGMDH) model for runoff forecasting. Water Resour. Manag. **31**, 43–59 (2017). doi:10.1007/s11269-016-1507-3

Moyano, M.C., Tornos, L., Juana, L.: Water balance and flow rate discharge on a receiving water body: application to the B-XII irrigation district in Spain. J. Hydrol. **527**(2015), 38–49 (2015)

Palmer, W.C.: Meteorological drought. Research Paper No. 45, US Department of Commerce, Weather Bureau, Washington, DC (1965)

Paulo, A.A., Pereira, L.S.: Stochastic prediction of drought class transitions. Water Resour. Manag. **22**, 1277–1296 (2008)

Paulo, A.A., Rosa, R.D., Pereira, L.S.: Climate trends and behaviour of drought indices based on precipitation and evapotranspiration in Portugal. Nat. Hazards Earth Syst. Sci. **12**, 1481–1491 (2012)

Paulo, A., Martins, D., Pereira, L.S.: Influence of precipitation changes on the SPI and related drought severity. An analysis using long-term data series. Water Resour. Manag. **30**(15), 5737–5757 (2016)

Pereira, L.S., Rosa, R.D., Paulo, A.A.: Testing a modification of the palmer drought severity index for Mediterranean environments. In: Rossi, G., Vega, T., Bonaccorso, B. (eds.) Methods and Tools for Drought Analysis and Management, pp. 149–167. Springer, Dordrecht (2007)

http://drought.unl.edu/portals/0/docs/spi-program-alternative-method.pdf. Accessed 6 Mar 2017
http://www.ijmer.com/papers/Vol3_Issue4/DP3424072411.pdf. Accessed 6 Mar 2017

Comparison of Deep Learning and Support Vector Machine Learning for Subgroups of Multiple Sclerosis

Yeliz Karaca[1]([⊠]), Carlo Cattani[2], and Majaz Moonis[3]

[1] Visiting Engineering School (DEIM), Tuscia University, Viterbo, Italy
yeliz.karaca@ieee.org
[2] Engineering School (DEIM), Tuscia University, Viterbo, Italy
[3] University of Massachusetts Medical School, Worcester, USA

Abstract. Machine learning methods are frequently used for data sets in many fields including medicine for purposes of feature extraction and pattern recognition. This study includes lesion data obtained from Magnetic Resonance images taken in three different years and belonging to 120 individuals (with 76 RRMS, 6 PPMS, 38 SPMS). Many alternative methods are used nowadays to be able to find out the strong and distinctive features of Multiple Sclerosis based on MR images. Deep learning has the working capacity pertaining to a much wider scaled space (120×228), less dimension (50×228) (also referred to as distinctive) feature space and SVM (120×228). Deep learning has formed a more skillful system in the classification of MS subgroups by working with fewer sets of features compared to SVM algorithm. Deep learning algorithm has a better accuracy rate in comparing the MS subgroups compared to multiclass SVM algorithm kernel types which are among the conventional machine learning systems.

Keywords: Deep learning · Support vector machines kernel types · Multiple Sclerosis subgroups · MRI

1 Introduction

Multiple Sclerosis (MS) is a disorder of central nervous system afflicting the brain, cerebellum and spinal cord [1]. One of the most distinctive features of the disorder is the damage seen in myelin which is a matter with fat surrounding the neural fibers. Demyelization, damage of the myelin, occurs in this process and many damaged parts called plaque form which are not so functional as the normal tissue. Plaque is not like the normal tissue and it is hard [2,3]. This study examines three of the MS subgroups among Clinical Progress Types which are: Remitting Relapsing Multiple Sclerosis (RRMS), Secondary Progressive Multiple Sclerosis (SPMS) and Primary Progressive Multiple Sclerosis (PPMS). In Primary Progressive Multiple Sclerosis (PPMS) improvement is recorded from the beginning of the disease, characterized by slow or rapid progression. The emergence of Secondary Progressive Multiple Sclerosis (SPMS) is quite similar to that

© Springer International Publishing AG 2017
O. Gervasi et al. (Eds.): ICCSA 2017, Part II, LNCS 10405, pp. 142–153, 2017.
DOI: 10.1007/978-3-319-62395-5_11

of the Relapsing Remitting MS type. Half of the patients belong to this category within the first ten years of the disorder followed by attacks and improvements. The early phase lasts 5 to 6 years; following the early phase, the disease enters a secondary progressive period. Although the number of attacks decreases and healing is relatively slow, impairment becomes worse. Relapsing Remitting Multiple Sclerosis (RRMS) sees attacks and the attacks either have a full or partial recovery. These attacks can last for days, weeks or even months. It is not possible to estimate how frequent such attacks will occurs. At times, some attacks might be seen in some patients even after 15 or 20 years [4,5].

Over the past decade, medical doctors have been using Deep learning algorithm for facilitating the diagnosis and follow-up the course of various disorders. Liu et al. [6] applied Deep Learning algorithm on 3 dimensional MR images for the early diagnosis of Alzheimer disease. A comparison was made through SVM algorithm to emphasize the classification accuracy of deep learning algorithm compared to machine learning algorithms. Using Auto encoder and Softmax Regression models in Deep Learning algorithm, the existence of NC, ncMCI, cMCI, AD was examined for the disorder diagnosis based on MR images. It was concluded that compared to SVM algorithm, classification performance with deep learning algorithm proved to be higher. Xie et al. [7] used gene data and performed gene expression inference with deep learning model. The method proposed taking the genetic variations in genotype into training procedure through Stacked Denoising Auto Encoder method. Classification was made based on useful features in gene data. Deep learning model was applied on 7085 gene data and reduced to 6611 genes. When the genes eliminated were examined, it was seen that they were unimportant genes. Wang et al. [8] proposed the use of deep learning model to solve the trigger identification and argument detection problem in biomedical text data. Window size around candidate triggers was fed in deep neural network. The classification was made afterwards in Softmax Regression model. It was emphasized by the authors that deep learning proved to be more effective compared to SVM in the extraction of biomedical features. Sarraf et al. [9] worked on 15 healthy subjects and 28 individuals with Alzheimer. They used fMRI data and their estimation was based on Convolutional Neural Network. This method enabled them to extract the important features from the fMRI dataset, which proved to be useful for the diagnosis of the disorder. The test accuracy was 96.85% pertaining to patient or healthy individuals. Petersson et al. [10] did classification through deep learning model for the Hyperspectral image data (the class of which was not certain. By applying the label mapping method on these features, they managed to get a classification with high accuracy rate.

In order to make the interaction between mathematical model and multiple sclerosis more understandable and smooth, researchers have attempted to design different methodologies. For example, Brosch et al. [11] performed an accurate classification with deep learning method having modeled the morphological features and lesion distribution of MR images taken from patients of SPMS. Brosch et al. [12] did a classification through Deep 3-D Convolutional Encoder Network through the segmentation of lesion entities in MR images for the diagnosis of MS. Karaca et al. [13] developed a clinical decision support system to classify the patients diagnostics based on features obtained from Magnetic Resonance Imaging (MRI) and Expanded Disability Status Scale (EDSS). For the kernel trick, efficient performance in non-linear classification, the Convex Combination of Infinite Kernels model was developed to measure the health status of patients. They showed that the proposed model classified MS diagnosis level with a better accuracy compared to that of single kernel, artificial neural network and other machine learning methods.

This study includes 120 MS patients' MR image data obtained over three different years. Lesion diameters seen from the MR images of the MS patients included three different parts of brain which are brain stem, corpus callossum periventricular and upper cervical parts. The dataset includes a large scale of 120×228 in the application. This means the deep learning method is rich in terms of the features and the dataset is comprised of multiple classes. The classification of MS subgroups was performed accordingly. The purpose in this study is to provide an alternative for the accuracy of MS subgroups classification through Deep learning method and contribute to literature in this way. This study will make a contribution to literature by being the first study performed, having classified the MS subgroups by applying deep learning method on the MR images of the patients belonging to three different years.

Deep learning algorithm works with fewer feature sets compared to those in SVM algorithm. Deep learning therefore can form a system which is much more capable in classifying the subgroups of MS. In addition, deep learning has yielded much more accurate results compared to kernel types of multiclass SVM algorithm, which is one of the classical machine learning methods.

2   Material and Method

2.1   Materials

The dataset pertaining to MR images of 76 RRMS, 6 PPMS, 38 SPMS patients, taken over a period of three different years was used. Based on this dataset, the MS subgroups were classified by means of Deep Learning and Support Vector Machine kernels. Their performance was compared in terms of accuracy.

Patient Details. MR image data was obtained from a total of 120 subjects with MS subgroups. The subjects were aged between 18–65. Hacettepe University Medical Faculty Neurology and Radiology Department in Primer Magnetic Resonance Imaging Center, Ankara provided the data and the diagnosis for the Multiple Sclerosis disorder was made in accordance with McDonald Criteria [14].

Magnetic Resonance Imaging Features. Magnetic Resonance (MR) enables imaging through nuclear magnetic resonance forming the images based on the density of the hydrogen atoms in the tissue and movements thereof. It is particularly used for the imaging of soft tissues. It proves to be an important method used in the diagnosis of the central nervous system disorders (including the brain and the spinal cord) [15]. The MR images of the patients were taken from the device called 1.5 tesla device (Magnetom, Siemens Medical Systems, Erlangen, Germany). Magnetic Resonance Imaging is the most important method for MS diagnosis [14]. MR images taken over a period of three different years enabled the study. The lesion diameters formed the basis of application including three different parts which are brain stem, corpus callossum periventricular and upper cervical parts.

3 Method

If the data is not linearly separable like our proposed features extracted from MRI, a statistical methods cannot prove to be adequate to classify the nonlinear data. For this reason, non-linear systems are preferred. For example, Kernel methods are appropriate methods for these kinds of problems. They are widely used to deal with a variety of learning tasks such as classification, regression, ranking, clustering, and dimensionality reduction. The appropriate selection of a kernel is often a decision to be made by the user. Yet, poor choices may cause sub-optimal performances. Searching for an appropriate kernel manually tend to be an art which might be time-consuming and imperfect. In this study, the researchers proposed a combination of nearly infinite kernel learning for the classification of MS subgroups.

Using the lesion diameter data obtained from MR images of 120 MS patients, dataset of 120×228 (initial phase) was used and Deep learning as well as SVM algorithms were applied. The comparison of the MS subgroups was performed by Matlab 2016a software tool.

Deep Learning Regression Model. Encoding structure consists of a neural network which has multiple hidden layers as it is shown in Fig. 1. Neurons of the input layer signify the original input vector. Each hidden layer is to be seen as a higher level representation of the previous layer. However, it is not important to define the exact meaning of each layer. The output layer is a sparse representation of the input layer which has the same dimensionality as the input [16].

Figure 1 provides the depiction of the deep learning structure with multilayered neural network. The input layer feed-forwards the pre-computed features from MRI data to the hidden layers. Each hidden layer gains the nonlinear transformations from the previous layer. It is optimized to restructure the original instance. The softmax layer takes the activations of the last hidden layer as inputs, presenting the probability of each MS Subgroups.

The activation signals are propagated forward iteratively through the network based on Eqs. (1) and (2) until the output layer is reached. The neuron activation of each layer can be calculated by

$$\begin{cases} a_l^i = x^{(i)}, & l = 1 \\ a_{(l)}^{(i)} = \sigma(W^l a + b), & l > 1 \end{cases} \tag{1}$$

$$h(W, b, x) = a^{(N)} \tag{2}$$

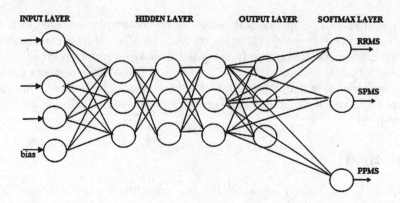

Fig. 1. MS data set deep learning structure.

where x is the unlabeled data $\{x^{(i)}\}_{i=1}^{m}$; W is the weight matrix controlling the activation effect between the neurons on neighbor layers. b is the bias term and σ is the activation function which can be set to sigmoid function or hyperbolic tangent function for the purpose of introducing the non-linearity for network to be able to model complex relationships. $h(W, b, x)$ is the illustration of the input data and the activations at the output layer [17].

It is possible to perform the feature dimensionality reduction or over - completion when one changes the number of the neurons at each hidden layer. One can also combine features of different views or modalities through concatenating features into one input vector. The sparse auto-encoder is said to work well in data fusion by catching the synergy between different modalities [18].

The representation loss is used as the objective function for optimization as shown in Eq. (3). To train this unsupervised model, the following is used:

$$L\left(W, b, x, z\right) = \min_{W, b} E\left(W, b, x, z\right) + \gamma||W||_2^2 + \beta K\left(W, b, x\right) \tag{3}$$

where $E\left(W, b, x, z\right) = ||h\left(W, b, x\right) - z||_2^2$ is the representation loss with squared error. The second term is the weight decay that causes small weights. The third item is the sparsity penalty that is regularized by β with target activation of ρ close to 0, enforcing a sparse representation penalizing the objective function through the use of Kullback-Leibler divergence across n training samples in Eq. (4) [19].

$$K\left(W, b, x\right) = \sum_{j}^{n} ID_{KL}(\rho||\frac{1}{m}\sum_{i=1}^{m} h_j(x^{(i)}; W, b)||) \tag{4}$$

In this study the hidden layers of sparse auto-encoder was trained one at a time. They were stacked to form a complete neural network by removing the temporary output layer. It was proven beneficial that the first and last few hidden layers yielded more than propagating the entire network.

Softmax Regression. Softmax output layer is added on the top of the trained auto - encoder stack that contains only previous hidden layers regarding the classification of MS subgroup [13,18]. A different activation function is used by the softmax layer. This may have nonlinearity that is different from the one which has been applied in previous layers. The softmax activation function is Eq. (5).

$$h_i^l = \frac{e^{w_i^l h^{l-1} + b_i^l}}{\sum j e^{w_i^l h^{l-1} + b_j^l}} \tag{5}$$

where w_i^l is the i-th row of w^l and b_i^l is the i-th bias term of last layer. h_i^l can be used as an estimator of $P(Y = a|x)$, where Y is the associated label of input data vector x. In such a case, three neurons at the softmax layer can be classified as RRMS, SPMS, PPMS.

Akin to the procedure of training the Deep Belief Net (DBN) [20], further fine-tune for all the parameters in the network as regards the overall classification loss is seen through the unfolding of all the auto-encoders and also the application of back propagation algorithm on the entire network [15,20].

Support Vector Machines (SVM). SVM has the principle of being separable linearly. It matches the input space with a kernel space. Afterwards, it forms a linear model over this kernel space [21]. SVM algorithm can be used for the classification and regression problems.

The main idea in the regression method is to be able to reflect the characteristics of the training data available close to the actual value as much as possible. It is also important to find the linear distinctive function that is in line with statistical learning theory. Kernel functions are used for the processing of nonlinear cases in regression as well similar to classification [22,23].

There are two instances encountered in SVM algorithm. One is that the data are on a plane where they can be separated linearly or on plane where they cannot be separated linearly. In cases when the data cannot be separated linearly, nonlinear classifiers can be used instead of linear counterparts. A non-linear feature space transforms $x \in R^n$ observation vector to z vector from a higher degree. Linear classifiers are calculated based on Eq. (6) in this new space. The feature space where z vector is included is shown through F.

θ denotation makes matching with $R^n \to R^F$ and it is represented as $z = \theta(x)$ [21].

$$x \in R^n \to z(x) = [a_1, \theta_1(x), ..., a_n, \theta_n(x)^T] \in R^F \tag{6}$$

when it is the case of being separable non-linearly, the training sets cannot be separated linearly in the original input space. Through the aid of non-linear mapping function, SVM performs transformation to feature space from the original input space where it can perform linear classification. Support Vector machines do such transformation denoted mathematically as follows: $K(x, x_j) = \phi(x)\phi(x_j)$ The kernel function is depicted by Eqs. (7), (8) and (9). Thus, this process enables the separation of data in a linear way [24,25].

$$Cubic\ Kernel : K(x_i, x_j) = (x_i.x_j + 1)^d \tag{7}$$

$$Gaussian\ Radial\ Basis\ Function\ \ Kernel: K(x_i, x_j) = e^{-\frac{||x_i - x_j||^2}{2\sigma^2}} \qquad (8)$$

$$Quadratic\ Kernel: K\left(x_i, x_j\right) = \tanh\left(x_i, x_j - \delta\right) \qquad (9)$$

In this study, the classification of MS subgroups, namely RRMS, SPMS, PPMS was performed by 10-fold cross validation method based on classification accuracy method with three different kernels.

4   Results and Discussion

This study uses lesion diameter data obtained from MR images of 120 individuals 76 RRMS, 6 PPMS, 38 SMPS), taken from a period of three different years. Many alternative methods are used these days to be able to obtain the most distinctive and strong features of MS disorder from MR images. Deep learning has the working capacity relating to a much wider scaled space (120×228), less dimension (50×228) (also referred to as distinctive) feature space and SVM (120×228). Working with fewer feature sets, Deep learning algorithm has proven to be a more capable system compared to SVM algorithm in making the subgroup classification of MS. Deep learning algorithm has had a better accuracy rate in classifying the MS subgroups compared to kernel types of multiclass SVM algorithm. A data matrix of 120×228 was used in this study. The dataset is comprised of MR images belonging to subgroups of MS which are RRMS, SPMS, PPMS. Matlab 2016a software was used to obtain the results of this experimental study. It has been seen through this study that the accuracy rate of deep learning model for classifying RRMS, SPMS, PPMS subgroups is superior to that obtained by three different kernels of SVM Algorithm (Gaussian Radial Basis Function Kernel, Linear Kernel, Cubic Kernel).

Application of Deep Learning. The dataset with a dimension of 120×228 including lesion diameter data of MR images belonging to MS patients has been applied on the Deep Learning model through the steps provided below: This study reveals the functionality of Neural Network Toolbox autoencoders for training a deep neural network to classify the subgroups of RRMS, SPMS, PPMS MR based on image data.

Neural networks with multiple hidden layers can be helpful for the solution of classification problems through complex data like data of MR images. Each layer can learn features at a different level of abstraction. Yet, training neural networks with multiple hidden layers may prove to be a difficult task in practice.

One way to train a neural network with multiple layers effectively is to do the training one layer at a time. This study has been able to achieve this goal by training a special type of network known as an autoencoder for the each desired hidden layer.

This study reveals the fact about how one can train a neural network with two hidden layers to classify the subgroups of MS based on the MR images. As a first phase, MRI data set has been trained on the hidden layers individually in an unsupervised fashion by using autoencoders. Next, MRI data set has been trained through final softmax layer, and the layers has been joined together to form a deep network, which requires training one final time in a supervised fashion.

This study has used synthetic data throughout the training and testing stages. The MR images have been generated by applying random affine transformations to MS Subgroups with MR images created using different fonts.

Step 1: Training the first autoencoder. This step is initiated by training a sparse autoencoder on the training data without using the labels.

An autoencoder is a neural network that attempts to replicate its input at its output. Hence, the size of its input will be the same as its output size. When the number of neurons in the hidden layer is less than the size of the input, the autoencoder learns a compressed representation of the input.

Neural networks have weights randomly initialized before training. For this reason, the results from training are different each time. To eliminate such a behavior, it is important that one set the random number generator seed in an explicit manner.

Step 2: The size of the hidden layer is set for the autoencoder. The autoencoder has been trained to make this smaller than the input size.

Step 3: The type of autoencoder has been trained as a sparse autoencoder. This autoencoder utilizes regularizers to enable the learning of a sparse representation in the first layer, which can control the influence of these regularizers through the setting of various parameters:

- L2 Weight Regularization [18] controls the impact of an L2 regularizer for the weights of the network (and not the biases). This should normally be quite small.
- Sparsity Regularization [18] controls the impact of a sparsity regularizer that attempts to exert a constraint on the sparsity of the output from the hidden layer. This is a different procedure compared to applying a sparsity regularizer to the weights.
- Sparsity Proportion [17,18] is a parameter of the sparsity regularizer. It controls the sparsity of the output from the hidden layer. A low value for Sparsity Proportion usually causes each neuron in the hidden layer "specializing" by only giving a high output for a small number of training examples.
- The autoencoder is trained and the values for the regularizers described above are specified.

The autoencoder contains an encoder that is followed by a decoder. Figure 2 depicts that the encoder maps an input to a hidden representation and the decoder attempts to reverse this mapping to restructure the original input.

After training the first autoencoder, the second autoencoder is also trained in a similar fashion. The main difference in terms of use of the features is that they were generated from the first autoencoder as the training data in the second autoencoder. The original vectors in the training data had a dimension of 120×228. After passing them through the first encoder, this was reduced to

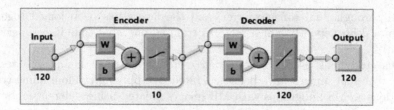

Fig. 2. Application of Autoencoder on MS Data.

10 dimensions. After using the second encoder, this was reduced again to 10 dimensions. As can be seen Fig. 3, a final layer is trained to classify these 50-dimensional vectors into different subgroups of MS.

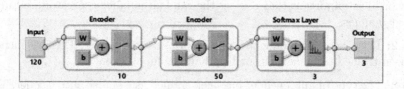

Fig. 3. Application of Softmax Regression on MS Data.

Step 4: Training the final softmax layer. A softmax layer is trained in order to classify the 50-dimensional feature vectors. Unlike the autoencoders, as can be seen in Fig. 3. the softmax layer is trained in a supervised manner, using labels for the training data. As has been explained previously, the encoders from the autoencoders have been used to extract the features. As can be seen from Fig. 3, the encoders from the autoencoders are stacked together with the softmax layer so as to form a deep network.

Mean square error is obtained as 0.0021114 at Epoch 1000 for Deep learning method as can be seen from Fig. 4.

No matter how wide scaled data matrix is applied on the Deep Learning Model, the dimension of the data matrix is reduced in order to be able to do accurate classification in the deep learning model. MS dataset with a dimension of 120×228 was applied on the deep learning model in this study. As a result of the application, the MS dataset feature dimension was reduced to 50×228 and the classification of the MS subgroups (RRMS, SPMS, PPMS) was made accordingly.

Application of SVM. The matrix with a dimension of 120×228 including the lesion diameter data of MR images of MS patients was applied on the SVM algorithm. The classification accuracy obtained from the Gaussian, Linear and Cubic Kernel training procedures of SVM algorithm is obtained through ROC Curve [24] as can be seen from Fig. 5. Table 1 presents the classification accuracy rates of SVM kernel types.

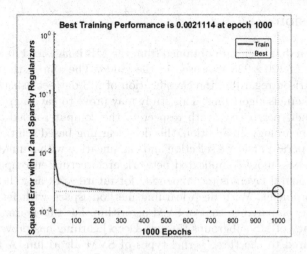

Fig. 4. MS data set deep neural network performance graph.

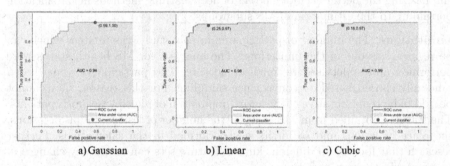

a) Gaussian   b) Linear   c) Cubic

Fig. 5. MS Data Set SVM Kernels ROC Curve.

As can be seen from Table 1, a higher classification accuracy has been obtained in Deep learning method compared to three different kernel methods of SVM by working on space with fewer features.

Deep learning algorithm has ensured the most accurate classification by working on the least number of features compared to SVM algorithm.

Table 1. Deep learning and SVM kernels classification accuracy rate.

Algorithms	Accuracy rate	Data size
Gaussian kernel SVM	78.30%	120×228
Linear kernel SVM	88.30%	
Cubic kernel SVM	90.80%	
Deep learning model	99.78%	50×228

5 Conclusion

Using the lesion diameter data obtained from the MR images of 120 MS patients a data matrix of 120×228 was used in the study. The aim of the study is to make a comparison regarding the classification of MS dataset using deep learning and SVM kernels algorithms. This study may prove to have a great potential in forming a new perspective with respect to the computer-aided diagnosis in other fields of neurology. In addition, this deep learning-based solution can allow researchers to perform feature selection and classification with a unique architecture. Nevertheless, more complicated network architecture encompassing more convolutional neural layers is recommended for future work as well as for more complicated problems. With deep learning method, space with fewer features was worked on compared to that of SVM algorithm. For this reason, the classification accuracy of MS subgroups through Deep Learning has proved to be the highest compared to the three kernel types of SVM algorithm. A larger scale of dataset shall be obtained in the future from the MR images. It is aimed at emphasizing the better accuracy power of deep learning methods in classification pertaining to the estimation of MS subgroups.

Limitations. The number of participants in this study was limited, which can be regarded as one of the limitations. The analyses for MS health status were performed and supervised by medical doctors. The participants were chosen randomly chosen based on convenience sampling. For this reason, they cannot be said to be representative of the entire population of multiple sclerosis patients. The features obtained from MR images (the number of lesions, radius of lesions, the place of lesions and so forth) was performed in the calculations. Further research that focuses on different kind of feature sets can reveal any changes in terms of effectiveness over time.

References

1. Duncan, I.D., Franklin, R.J.: Myelin Repair and Neuroprotection in Multiple Sclerosis, pp. 23–47. Springer Science and Business Media, Heidelberg (2012)
2. Murray, T.J., Saunders, C., Holland, N.J.: Multiple Sclerosis: A Guide for the Newly Diagnosed, pp. 1–39. Demos Medical Publishing, New York (2012)
3. Scalfari, A., Lederer, C., Daumer, M., Nicholas, R., Ebers, G.C., Muraro, P.A.: The relationship of age with the clinical phenotype in multiple sclero sis. Multiple Scler. J. **22**(13), 1750–1758 (2016)
4. McAlpine, D., Compston, A.: McAlpine's Multiple Sclerosis, pp. 2–10. Elsevier Health Sciences, Amsterdam (2005)
5. Hirst, C., Ingram, G., Swingler, R., Compston, D.A.S., Pickersgill, T., Robertson, N.P.: Change in disability in patients with multiple sclerosis: a 20-year prospective population-based analysis. J. Neurol. Neurosurg. Psychiatry **79**(10), 1137–1143 (2008)
6. Liu, S., Liu, S., Cai, W., Pujol, S., Kikinis, R., Feng, D.: Early diagnosis of Alzheimer's disease with deep learning. In: IEEE 11th International Symposium Biomedical Imaging (ISBI), pp. 1015–1018 (2014)

7. Xie, R., Quitadamo, A., Cheng, J., Shi, X.: A predictive model of gene expression using a deep learning framework. In: IEEE International Conference Bioinformatics and Biomedicine (BIBM), pp. 676–681 (2016)
8. Wang, A., Wang, J., Lin, H., Zhang, J., Yang, Z., Xu, K.: Biomedical event extraction based on distributed representation and deep learning. In: 2016 IEEE International Conference on Bioinformatics and Biomedicine (BIBM), p. 775 (2016)
9. Sarraf, S., Tofighi, G.: Deep learning-based pipeline to recognize Alzheimer's disease using fMRI data. In: Future Technologies Conference, pp. 816–820 (2016)
10. Petersson, H., Gustafsson, D., Bergstrom, D.: Hyperspectral image analysis using deep learning: a review. In: 2016 6th International Conference Image Processing Theory Tools and Applications (IPTA), pp. 1–6 (2016)
11. Brosch, T., Yoo, Y., Li, D.K.B., Traboulsee, A., Tam, R.: Modeling the variability in brain morphology and lesion distribution in multiple sclerosis by deep learning. In: Golland, P., Hata, N., Barillot, C., Hornegger, J., Howe, R. (eds.) MICCAI 2014. LNCS, vol. 8674, pp. 462–469. Springer, Cham (2014). doi:10.1007/978-3-319-10470-6_58
12. Brosch, T., Tang, L.Y., Yoo, Y., Li, D.K., Traboulsee, A., Tam, R.: Deep 3D convolutional encoder networks with shortcuts for multiscale feature integration applied to multiple sclerosis lesion segmentation. IEEE Trans. Med. Imaging **35**(5), 1229–1239 (2016)
13. Karaca, Y., Zhang, Y., Cattani, C., Ayan, U.: The differential diagnosis of Multiple Sclerosis using convex combination of infinite kernels. CNS Neurol. Disord. Drug Targets **16**(1), 36–43 (2017)
14. Karaca, Y., Osman, O., Karabudak, R.: Linear modeling of multiple sclerosis and its subgroups. Turk. J. Neurol. **21**, 7–13 (2015)
15. Rajan, S.S.: MRI A Conceptual Overview. Library of Congress Catalog in Publication Data. Springer, New York (1998)
16. Mamoshina, P., Vieira, A., Putin, E., Zhavoronkov, A.: Applications of deep learning in biomedicine. Mol. Pharm. **13**(5), 1445–1454 (2016)
17. Deng, L., Yu, D.: Deep learning methods and applications. Found. Trends Sig. Process. **7**(3–4), 230–239 (2014)
18. Graupe, D.: Deep Learning Neural Networks: Design and Case Studies, pp. 23–53. World Scientific Publishing, Singapore (2016)
19. Galas, D.J., Dewey, T.G., Kunert-Graf, J., Sakhanenko, N.A.: Expansion of the Kull back-Leibler divergence, and a new class of information metrics. Entropy Inf. Theor. MDPI **6**(2), 8 (2017)
20. Goodfellow, I., Bengio, Y., Courville, A.: Deep Learning, pp. 155–194. MIT Press, Cambridge (2016)
21. Suthaharan, S.: Machine Learning Models and Algorithms for Big Data Classification: Thinking with Examples for Effective Learning, vol. 36. Springer (2015)
22. Han, J., Kamber, M., Pei, J.: Data Mining Concepts and Techniques, 3rd edn., pp. 408–413. Elsevier, Amsterdam (2012)
23. Amari, S.I., Wu, S.: Improving support vector machine classifiers by modifying kernel functions. Neural Netw. **12**(6), 783–789 (1999)
24. Fung, G.M., Mangasarian, O.L.: Multicategory proximal support vector machine classifiers. Mach. Learn. 1–21 (2004)
25. Karaca, Y., Hayta, Ş., Karabudak, R.: Case study application for C-support vector classification: the estimation of MS subgroup classification with selected kernels and parameters. Eur. J. Pure Appl. Math. **9**(2), 196–215 (2016)

Workshop on Advanced Smart Mobility and Transportation (ASMAT 2017)

Preliminary Investigation on a Numerical Approach for the Evaluation of Road Macrotexture

Mauro D'Apuzzo[1], Azzurra Evangelisti[1(✉)], and Vittorio Nicolosi[2]

[1] University of Cassino and Southern Lazio,
Via G. Di Biasio 43, 03043 Cassino, Italy
dapuzzo@unicas.it, aevangelisti.ing@gmail.it
[2] University of Rome Tor Vergata, Via del Politecnico 1, 00133 Rome, Italy
nicolosi@uniroma2.it

Abstract. Safety aspects, as dry and wet friction and splash and spray phenomena, and environmental aspects, as rolling noise, in-vehicle noise and rolling resistance are highly affected by the pavement surface macrotexture. For these reasons, predicting macrotexture, is a crucial aspect for pavement engineers. In this connection, several statistical empirical models have been proposed in the scientific literature. However, none of them seems to be effective in predicting macrotexture. For these reason, in order to better understand relationship between grading and volumetric properties of bituminous mixes and the corresponding macrotexture level a more theoretical approach has to be pursued.

In this paper a preliminary analysis toward the theoretical prediction of road macrotexture is presented. A numerical model by means of a Discrete Element Method (DEM) approach has been developed in order to simulate compaction of a bituminous mix in gyratory compactor. Several DEM simulation speciements have been examined and different simulation strategies have been investigated in order to highlight strengths and weaknesses of each tool.

Simulation have been performed on mono-granular mixes and a comparison with experimental specimen prepared in laboratory has been performed. Preliminary results seem rather encouraging showing that this approach may provide useful information for the development of a theoretical prediction model of pavement macrotexture.

Keywords: Macrotexture · Road pavement · Discrete Element Method (DEM) · Predicting model · Numerical model · Numerical simulation

1 Introduction

The highway agencies and the vehicles/tyres industrial groups are called to analyze and control the highway conditions and the vehicle-related aspects, respectively. For this reason, significant efforts have been spent at both National and International levels for developing and administering effective design, construction, maintenance, and management practices and policies.

© Springer International Publishing AG 2017
O. Gervasi et al. (Eds.): ICCSA 2017, Part II, LNCS 10405, pp. 157–172, 2017.
DOI: 10.1007/978-3-319-62395-5_12

Road crash investigations have consistently shown a proved relationship between crashes and pavement surface features, as friction and texture [6, 8, 13, 20, 21, 24]. According to ISO 13473, the pavement texture is defined as "the deviation of the pavement surface from a true planar surface" [23] and the texture depth was found to stronger affect accident frequency than low-speed skid resistance, in fact the risk of both wet and dry weather crashes were found to increase greatly for low texture level surfaces [22, 25]. In particular, according to ISO 13473, the road texture can be divided into four subdomains, characterized by different wavelength scales. The macrotexture domain is correlated to particle size distribution of the aggregates and on the compaction energy exerted to the mixture; it contributes to the friction by means of tire's hysteresis, and its presence ensures the effective drainage of rainwater. A clear relationship between crash occurrence and pavement macrotexture is demonstrated [9].

The most used technique to measure road macrotexture is the volumetric technique ASTM E965 [5]. The Sand Patch method consists of spreading a known volume of sand on a dry surface, which previously has been cleaned. The sand then is spread in an even circular motions into a circular area until the sand is level with the tops of the cover aggregate. The measure of the diameter of the circle is performed and repeated four times in 45° angles from one another and the mean average of diameters (Di) is then calculated. The Mean Texture Depth (MTD), that according to ASTM E965 [5], is defined as average pavement macrotexture depth, is the synthetic index used to describe the macrotexture value, obtained with the Sand Patch method by means of the Eq. 1.

$$MTD = 4V/(\pi Di^2) \tag{1}$$

Where:

MTD is the macrotexture value [mm];
V is the fixed volume of the sand [mm^3];
Di is the measured diameter [mm].

With the aim to develop a theoretical model for predicting road macrotexture, several efforts have been performed in the past [15] and in this study, the numerical simulation has been selected. In particular, the Distinct Element Method (DEM) has been chosen and used for the numerical simulations. In fact the DEM Method, due to the discrete nature of the elements that are used in the simulations, is the most appropriate method to investigate the particles packing phenomenon. In particular, the DEM model of a gyratory compactor specimen has been used to simulate the field compaction phase, which reduces the volume of a mixture of hot asphalt, aggregates and filler materials to form the required final density.

2 Objective

This paper has the main aim to investigate the possibility to use Discrete Element Method (DEM) to simulate the behavior of pavement's upper layer in terms of aggregates' arrangement, compaction and macrotexture.

Due to the computational burden, which is required to simulate the field compaction phase, a smaller sample has been selected. In particular, the gyratory compaction, which is considered to be one of the best methods of laboratory compaction, has been chosen.

To investigate the effects of the laying and the compaction methods, on the macrotexture, the simulations of nine monogranular mixtures, with diameters from 4 mm to 20 mm, have been performed and two monogranular specimens with diameter of 8 mm have been made.

3 Laboratory Test: Gyratory Compactor Technique and Experimental Campaign

The construction of asphalt roadway requires that pavements be laid down with a compaction process, which consists of the release of direct and indirect compacting forces exerted by rollers passing over the loose mix to produce dense layers materials.

In the laboratory tests the gyratory compactor is used to simulate and reproduce the real compaction conditions under actual road paving operations, hence determining the compaction properties of the asphalt. In fact, during the test, the bituminous specimen is subjected to shearing forces similar to those encountered under the action of a roller during field compaction.

According to the Standards [4, 17, 18], the gyratory compaction method is used to prepare cylindrical specimens of hot mix asphalt (HMA) for determining the volumetric and physical properties of compacted HMA mix.

This test method is useful for monitoring the density of test specimens at any point in the compaction process, and it is suited for the laboratory design and field control of volumetric and physical HMA properties.

The gyratory compactor allows the application of an axial static pressure and of a dynamic shear, at the same time on an asphalt mix specimen and the energy applied by the gyratory compactor should represent the energy applied in the field.

In fact, according to European standard [18], the compaction is achieved by the application of a vertical stress (compaction pressure of 600 kPa) to a cylindrical sample of asphaltic mixture inserted in a mould with an internal diameter of 150 mm. The longitudinal axis of the mould is rotated at a fixed angle of 1.25° respect to the vertical axis. The mould performs 30 rev/min around the sloping axis, applying a dynamic shear to all the samples faces.

Two different specimens with steel spheres of diameter 8 mm, by means of the gyratory compactor, have been realized (see Fig. 11a, b, c and d). Both have been characterized by a spheres weight of 8700 g, a bitumen weight of 70 g and a diameter (D) of 150 mm. Due to the weight of the spheres, the compaction was found to be difficult and several attempts have been performed, in fact after the construction of the specimens and during the extraction phase, usually the lateral collapses have been occurred.

According to the ASTM E965 [5], on each face of the specimens, the macrotexture measurements have been performed. In particular the Eq. 1 has been used. Due to the strong irregularity of the faces' surface of the specimen 1, the macrotexture values for both top and down faces, have been high.

The volumetric features and the measured macrotexture values of the mix have been summarized in the Table 1.

Table 1. Steel spheres specimens' features

Specimen	Final weight [g]	H [mm] after 50 rev.	% Voids	MTD [mm] Top	MTD [mm] Down
1	8770	103,7	36,245	5,3	6,5
2	8737	103,5	35,994	4,97	2,75

4 Numerical Modeling

The distinct element method (DEM) is a numerical solution used to describe the mechanical behavior of discontinuous bodies. It was introduced by Peter Cundall in the 1971 with the scope to analyzing rock mechanic problems using discrete spherical particles and then applied to soils [14]. Since that moment, several Distinct Element Code have been developed.

The discrete element method (DEM) is an intuitive method in which discrete particles collide with each other and with other surfaces using the force-displacement law.

In order to simplify the calculation process, the following assumption are usually used for DEM simulations:

- is suitable to model particular material behavior or other systems in which large numbers of discrete particles contact each other;
- each particle is modeled as a single-node element that is a rigid body and it can be represented as a rigid spherical shape;
- can be used with finite elements for modeling discrete particles interacting with deformable continua or other rigid bodies.

In particular, the single-node elements are rigid spheres with fixed radii and represent individual grains of a particular media. Their features can be summarized as:

- discrete particle elements interacting with each other and with other bodies;
- each particle has uniform density and radius;
- each element can have only one node;
- the coordinates of the node define the center location of the physical grain;
- each node has three displacement and three rotational degrees of freedom.

In DEM applications, the materials' constitutive behavior is simulated by associating the constitutive model with each contact. For this reason the contact behavior is a critical factor and it can strongly influence the effectiveness of the DEM simulation.

In the past, due to the simple application, for simulating asphalt mixture, the elastic contact model has been used [7, 11, 12, 16].

However, asphalt mixture is a thermal rheological material which depends upon the temperature, loading frequency, and level of strain. For this reason, viscoelastic contact model have been used in the DEM simulation of asphalt mixture.

In rheological model theory, the basic mechanical behaviors of materials are elastic, viscous and plastic and the basic mechanical elements which describe mechanical behaviors by means of mathematics models are Hooke's spring and dashpot models for elasticity and viscosity, respectively.

More complicated models through parallel and series connection with each other have been proposed and thus they can reflect the real mechanical behaviors of materials. In particular, for studying the rheological behavior of asphalt mixture, the viscoelastic model, called Hertz-Mindlin Model, is widely used [1–3, 12]. The rheological representation is reported in the Fig. 1.

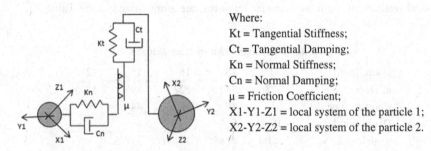

Where:
K_t = Tangential Stiffness;
C_t = Tangential Damping;
K_n = Normal Stiffness;
C_n = Normal Damping;
μ = Friction Coefficient;
X1-Y1-Z1 = local system of the particle 1;
X2-Y2-Z2 = local system of the particle 2.

Fig. 1. Hertz-Mindlin mechanical constitutive model

The Hertz-Mindlin Model is a system composed by two Kelvin's constitutive model and in this study, it was used to simulate the time and temperature dependent property of asphalt mixture.

The spring stiffness K_n acts in the contact normal direction and it may represent a simple linear or a nonlinear contact stiffness. The dashpot C_n represents contact damping in the normal direction. The tangential spring stiffness K_t along with the friction coefficient μ represent friction between the particles. The dashpot C_t represents contact damping in the tangential direction. Basing on this system, the forces acting on particle surfaces cause moments at particle centers that are moved across the interface.

4.1   The DEM Simulation of the Gyratory Compactor

The gyratory compactor is used for mixing and compacting granular materials' specimens. Several factors, including shape, size, density, and contact stiffness of particles, friction, speed of rotation and grade of the mould axis, influence the level of mixing that will be achieved in a given amount of time. These factors also affect the amount of energy required to mix and to compact the specimens.

Geometry and Materials. The mould geometry is cylindrical: the height H is 150 mm, the circular diameter D is 150 mm, and the compaction angle α is 1.25°. The wall of the mould is assumed to be without depth, and the drum inner radius is 75 mm. The geometry conforms the real laboratory size of the compacted specimens. The drum is made of steel but for this simulation, the drum is assumed to be a rigid body (meshed with shell elements). The plate has the same internal diameter of the mould (D = 150 mm), it has six degree of freedom (d.o.f.) and the compaction of the spheres occurs moving down with a pressure of 600 kPa.

The granular media consist of spherical particles, which are characterized by a density of $2.5 \cdot 10^{-9}$ ton/mm^3, a changing radius between 20 and 4 mm and are represented by rigid body. The friction coefficient for contact between particles is 0.5. For contact between particles and the mould wall and plate, the friction coefficient is 0.2. Mass proportional damping is used in the analysis to reduce the analysis noise.

In order to investigate the DEM applicability for the macrotexture prediction and evaluation, nine monogranular mixtures have been generated and analyzed. All the physical features of the monogranular mixtures, are summarized in the Table 2.

Table 2. Monogranular mixture features.

Diam. [mm]	20	18	16	14	12
n. elem	273	341	527	795	1275
Comput. time [h]	8	10	15,5	44	70
Diam. [mm]	10	8	6	4	
n. elem	2109	4285	10107	34125	
Comput. time [h]	117	238	561,5	1080	

Simulation Steps. The total virtual duration for the analysis is 31.2 s and it is composed by four steps that are summarized in the Table 3 and in to the Fig. 3.

The result of the DEM simulation is the final position of the particles after the compaction. A table with the number code of the particles, the x coordinate, the y coordinate and the z coordinate of the center of the particles is the final output. In order to evaluate the macrotexture, by means of the MTD index, the calculation of the surface voids is required. For this reason a code to convert the coordinates of the particles node into a grid-surface has been developed.

The surface is composed of a grid with a sample spacing of 0.5 mm. The Fig. 3 represents the difference between single spherical particle specimen (Fig. 3a), and the final surface (Fig. 3b). On the final surface (Fig. 3b), the macrotexture value, by means of the, Eq. 1, have been evaluated.

5   Analysis of Results

The repeatability analysis to investigate if the macrotexture value of the specimens can be affected by the fixed initial position of the spheres; the wall effect analysis, with the aim to characterized the effect of the mould wall on the sphere arrangement; and the time history analysis to evaluate the evolution of the compaction during the simulation, have been performed.

Table 3. Simulation step

Step 0		
Boundary conditions:	Initial position for mould, plate and aggregates	Fig. 2a
Loads:	NaN	
Interactions:	NaN	
Time:	0 s	
Step 1		
Boundary conditions:	Pinned to mould and plate; 6 d.o.f. to aggregates	Fig. 2b
Loads:	Gravity load to the model: acceleration of −9810 mm/s^2 in the vertical direction	
Interactions:	dem-dem/dem-mould	
Time:	1,1 s	
Step 2		
Boundary conditions:	Inned to the mould; 1 d.o.f., in the horizontal direction to the plate; 6 d.o.f. to aggregates	Fig. 2c
Loads:	Gravity load to the model; velocity type load, in the horizontal direction, to the plate	
Interactions:	dem-dem/dem-mould	
Time:	0,1 s	
Step 3		
Boundary conditions:	Rotation around z-axes to the mould; vertical direction to the plate; 6 d.o.f. to aggregates	Fig. 2d
Loads:	Gravity load to the model; velocity type load, rotation around z-axes at 30 rev/min, to the mould; vertical load, of 600 kPa, to the plate	
Interactions:	dem-dem/dem-mould/dem-plate	
Time:	30 s	

5.1 Repeatability Analysis

A fixed position of the particles in the model, by means of layers of non-overlapping bodies, can be required for the DEM modeling technique. In order to guarantee a random mix of the aggregates, as has been done in previous studies [10], the initial position of the particles (Step 0) has been forced in a cylinder (see Fig. 2a).

This cylinder is not a part of the mould, but it allowed, in the Step 1, to direct the particles into the mould in a stable random configuration, under gravity force.

The particles are initially positioned next to each other and at a certain initial height from the interior wall of the cylinder. Each layer of the cylinder grid is rotated respect to both above and below next layers, of a random angle (different from each layer). In this way the initial order of the grid nodes is not a priori fixed (see Fig. 2a). In the Step 1, the particles are dropped inside the cylinder and the mould, which are fixed in the initial position, during this step (see Fig. 2b).

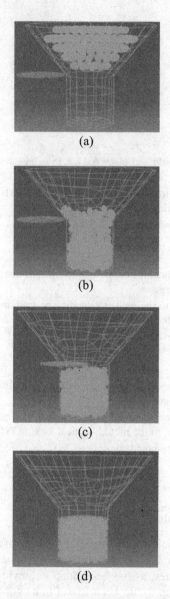

Fig. 2. (a) Initial position. (b) Casual arrangement of particles. (c) Plate positioning. (d) Compaction and mould rotation.

In order to investigate, if the initial position of each aggregate, affects the final macrotexture value, a repeatability analysis on the monogranular mix with 20 mm of diameter, has been performed. 25 simulations with different initial cylinder positions have been analyzed. The results of these simulations at time of 10 s, are summarized in the Fig. 4.

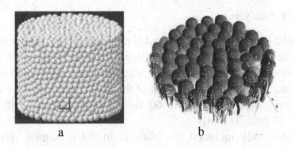

a   b

Fig. 3. a. Specimen of single spherical particles and b. Final surface.

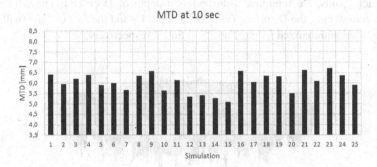

Fig. 4. Repeatability analysis results

A statistical analysis on this population has been performed and the best probability distribution for this population has been resulted "Wakeby". The Features of the analyzed population of simulations has been summarized in the Table 4.

Table 4. Repeatability Analysis results at 10 s of simulation

Mean	6,03	Standard deviation	0,463
Median	6,088	Coeff. of variation	0,077
Min	5,081	Standard error	0,092
Max	6,714	Skewness	−0,501
Range	1,634	Excess Kurtosis	−0,743

As it is possible to see, the Coefficient of Variation is 7, 7%. It means that the value of the macrotexture is not affected by the initial fixed position of the particles in the model, but, according to previous evidences [19], the value of macrotexture is highly variable itself.

5.2   Wall Effect Analysis

Due to the limited dimension of the mould, the wall effect can strongly affect the arrangement of the particles along the boundary line. In order to better understand the wall effect related to the aggregate dimensions (compared to the mould dimension), for each mono-granular mixture, the evaluation of the MPD is performed reducing the grid dimension (75 mm, 70 mm, 65 mm, 60 mm, 55 mm, 50 mm, 45 mm, 40 mm, 35 mm, 30 mm, 25 mm 20 mm).

In addition the study included the analysis in the specimen vertical direction, per-forming a cut at two different depth with respect to the upper boundary surface: D and D/2. In fact, according to the spherical packing theory (27), to evaluate the inter-particle voids, the reference volume has a depth of D or D/2. The whole final surface (without cut), the D surface (with cut a depth D) and the D/2 surface (with cut a depth D/2) are summarized in the Figs. 5, 6 and 7 respectively.

Fig. 5. Whole surface of the D6 mix with a grid radius of 60 mm.

Fig. 6. D surface of the D6 mix with a grid radius of 60 mm.

Fig. 7. D/2 surface of the D6 mix with a grid radius of 60 mm.

In general, the spherical packing theory considers two main theoretical packing arrangement of uniform spheres: the Cubical packing model and the Tetrahedral packing model which represent the loose and the dense configurations respectively.

The first configuration can be obtained with a very low compaction energy applied to the mixtures and the second produces the smallest voids content which can be achieved by the spherical particles configuration (infinite compaction energy applied to

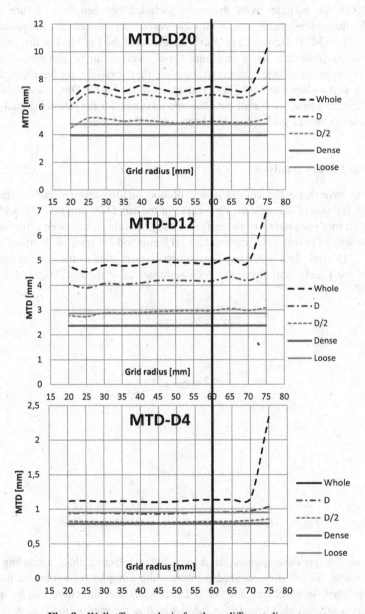

Fig. 8. Wall effect analysis for three different diameters.

the mixtures). Generally the theoretical MTD evaluations have been performed considering D/2 as reference depth.

By way of example, the wall effects analysis results evaluated on the three surfaces ("Whole", "D" and "D/2") for the monogranulare mixtures, are summarized in the Fig. 8. In addition, the MTD values obtained from the loose and dense theoretical configurations have been added to the diagram.

Due to their significant dimensions respect to the mould size (ratio between 7.5 ÷ 18.75), the particles with diameters included between 20 ÷ 8, are strongly affected by the wall effect, in fact, in all these cases both theoretical configurations are not reached (see MTD-D20 and MTD-D12 diagrams of the Fig. 8). For this reason it is possible to conclude that the 6 mm and 4 mm models can be considered effective models for the macrotexture evaluation, because they seem do not be affected by the wall effect, as it is shown in the MTD-D4 diagram of the Fig. 8. Due to the stabilization of the macrotexture values, for all the observed mixtures, for the following analysis, the threshold at 60 mm of grid radius has been selected and used.

5.3   Time History Analysis

In order to investigate the compaction evolution, a time history analysis has been performed. By way of example the time curve of the mix D 20 mm for 30 s, with a grid radius of 60 mm (belonging to the range of MTD stability), has been reported in the Fig. 9. By way of example, for the mixture D12 mm and D4 mm, on the three surfaces ("Whole", "D" and "D/2"), a zoom on the first five seconds of the compaction phase (Step 3 of the DEM simulation), has been summarized in the Fig. 10.

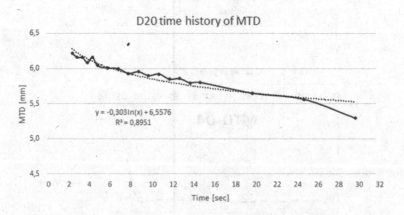

Fig. 9. Time history curve for diameter 20 mm mixture

Looking the previous figures, it is possible to observe that, according to the physical sense, for all the analyzed mixtures, the compaction evolution follows a decreasing trend. In particular, for the D 20 mm mixture, the curve can be approximated with a logarithmic curve ($R^2 = 0{,}89$) which is reported in the Fig. 9.

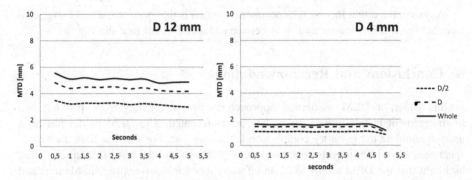

Fig. 10. Zoom on the first five seconds on time history curve for diameter 12 mm and 4 mm.

The compaction phenomenon is more evident for mixtures with larger diameters (from 20 mm to 10 mm). On the other hand, the mixtures characterized by smaller diameters (from 8 mm to 4 mm), seem to achieve high level of compaction immediately. It can be due to the size of the particles diameters respect to the grid radius: the arrangement of the particles with larger diameters can be more difficult and time consuming, respect to the smaller ones, as shown in the wall effect analysis.

5.4 Real Specimens and DEM Comparison

Finally a comparison between the real specimens and the DEM results of the D 8 mm mixtures has been performed. As it is possible to see in the Fig. 11e, in this case, due to the high weight of the steel spheres and the bitumen presence, the real specimens could not achieve the compaction level that the DEM simulations had. In fact the decompaction phenomenon is occured during the extraction phase of the specimen from the cylindrical mould. Probably the aluminum particles, which have similar weight to the aggregates, instead of steel ones, should been used.

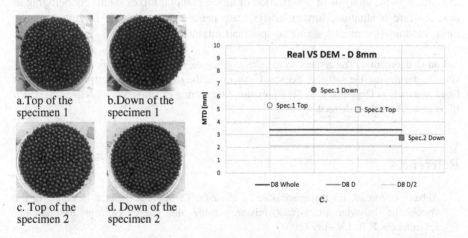

a.Top of the specimen 1

b.Down of the specimen 1

c. Top of the specimen 2

d. Down of the specimen 2

e.

Fig. 11. Comparison between real specimens and DEM results of the D 8 mm.

Anyway, the down face of the specimen 2, which is the more stable and compacted face, achieved DEM values and, it is between D and D/2 cut (see the Fig. 11e).

6 Conclusions and Recommendations

In this paper, a DEM numerical approach to investigate aggregate packing and re-arrangement in bituminous mixes has been presented. The DEM model has been used to simulate the gyratory compaction laboratory test. For the first time the virtual specimens have been used to macrotexture evaluations and the results obtained so far, highlight that the DEM seems to be an effective tool for macrotexture simulations and predictions. In fact the results obtained on the large amount of monogranular simulations, seem to be perfectly in agreement with the physical sense.

The repeatability analysis demonstrated that the final macrotexture value is not affected by the fixed initial position of the DEM nodes and that the simulations are repeatable. The wall effect analysis and time history analysis demonstrated that the sphere particles dimensions affect the speed and capability of the virtual mixture to compact itself.

So far, the more satisfactory results have been obtained with small-size particles mixtures (6 mm and 4 mm), but in the same time, they results the highest time consuming in the computational process. Nevertheless, some aspects have to be highlighted:

- the bitumen film is considered in the constitutive behavior of contact model but the whole bituminous matrix and the volume occupied by the film have not been taken into account for the macrotexture evaluation;
- the spherical form of the particles helps the achievement of the maximum compaction (in the cases where it has been achieved), but it is far away from the real aggregates arrangement because the aggregates indentation phenomenon is neglected.

Although the analysis of compaction of mono-granular mixes seems encouraging in macrotexture evaluation, further analysis are needed to investigate bi-granular mixtures, random-size mixtures and no-spherical mixtures.

Acknowledgements. The authors would like to thank the Silesian University of Technology di Gliwice, for the use the software licence. Furthermore the support of the staff of Laboratory of Road Materials at Department of Transportation Engineering of University of Naples "Federico II" is gratefully acknowledged.

References

1. Abbas, A., Masad, E., Papagiannakis, T., Harman, T.: Micromechanical modeling of the viscoelastic behavior of asphalt mixtures using the discrete-element method. Int. J. Geomech. **7**(2), 131–139 (2007)

2. Adhikari, S., You, Z.: 3D discrete element models of the hollow cylindrical asphalt concrete specimens subject to the internal pressure. Int. J. Pavement Eng. **11**(5), 429–439 (2010)

3. Di Renzo, A., Di Maio, F.P.: Comparison of contact-force models for the simulation of collisions in DEM-based granular flow codes. Chem. Eng. Sci. **59**, 525–541 (2004)

4. ASTM D6925-09: ASTM D6925-09, standard test method for preparation and determination of the relative density of hot mix asphalt (HMA) specimens by means of the superpave gyratory compactor. ASTM International, West Conshohocken (2009). www.astm.org

5. ASTM E965 Standard: Standard test method for measuring pavement macrotexture depth using a volumetric technique. American Society for Testing and Materials International, West Conshohocken (2006). www.astm.org

6. Bray, J.: The role of crash surveillance and program evaluation: NYSDOT's skid accident reduction program (SKARP). In: Presented at 28th International Forum on Traffic Records and Highway Information Systems, Orlando, Florida (2002)

7. Buttlar, W., You, Z.: Discrete element modeling of asphalt concrete: a micro-fabric approach. Geomaterials, No. 1757, pp. 111–118. National Research Council, Washington, D.C. (2001)

8. Cairney, P.: Skid resistance and crashes – a review of the literature. Research report No. 311. ARRB Transport Research Ltd., Vermont South Victoria, Australia (1997)

9. Cairney, P., Styles, E.: A pilot study of the relationship between macrotexture and crash occurrence. Australian Transport Safety Bureau, Victoria, Australia (2005)

10. Chen, J., Huang, B., Chen, F., Shu, X.: Application of discrete element method to superpave gyratory compaction. Road Mater. Pavement Des. **13**, 480–500 (2012). ISSN 1468-0629print/ISSN 2164-7402online

11. Collopa, A., McDowellb, G., Lee, Y.: Use of the distinct element method to model the deformation behavior of an idealized asphalt mixture. Int. J. Pavement Eng. **5**(1), 1–7 (2004)

12. Collopa, A., McDowellb, G., Lee, Y.: Modelling dilation in an idealised asphalt mixture using discrete element modelling. Granular Matter **8**(3–4), 175–184 (2006)

13. Craus, J., Livneh, M., Ishai, I.: Effect of pavement and shoulder condition on highway accidents. Transportation Research Record 1318. Transportation Research Board (TRB), Washington, D.C. (1991)

14. Cundall, P., Strack, O.: A discrete numerical model for granular assemblies. Dans Geotech. **29**(1), 47–65 (1979)

15. D'Apuzzo, M., Evangelisti, A., Nicolosi, V.: Preliminary findings for a prediction model of road surface macrotexture. Procedia Soc. Behav. Sci. **53**, 1110–1119 (2012). doi:10.1016/j. sbspro.2012.09.960. ISSN: 1877-0428

16. Dai, Q., You, Z.: Prediction of creep stiffness of asphalt mixture with micro-mechanical finite-element and discrete-element models. J. Eng. Mech. **133**(2), 163–173 (2007)

17. EN 12697-10: Bituminous mixtures. Test methods for hot mix asphalt. Compactibility, 28 January 2002. ISBN 0 580 38964 2

18. EN 12697-31: Bituminous mixtures. Test methods for hot mix asphalt. Specimen preparation by gyratory compactor, 31 May 2007. ISBN 978 0 580 50689 5

19. Evangelisti, A., D'Apuzzo, M., Nicolosi, V., Flintsch, G., De Leon Izeppi, E., Katicha, S., Mogrovejo, D.: Evaluation of variability of macrotexture measurement with different laser-based devices. In: Airfield & Highway Pavement Conference 2015 (2014)

20. Gandhi, P., Colucci, B., Gandhi, S.: Polishing of aggregates and wet-weather accident rates for flexible pavements. Transportation Research Record 1300. Transportation Research Board (TRB), Washington, D.C. (1991)

21. Giles, C., Sabey, B., Cardew, K.: Development and Performance of the Portable Skid Resistance Tester. ASTM Special Technical Publication No. 326. American Society for Testing and Materials (ASTM), Philadelphia (1962)

22. Gothie, M.: Relationship between surface characteristics and accidents. In: 3rd International Symposium on Pavement Surface Characteristics, Christchurch, New Zealand (1996)
23. ISO 13473: Characterization of pavement texture by use of surface profiles – determination of mean profile depth. International Organization for Standardization (1997)
24. Kamel, N., Gartshore, T.: Ontario's Wet Pavement Accident Reduction Program. ASTM Special Technical Publication 763. American Society of Testing and Materials (ASTM), Philadelphia (1982)
25. Larson, R.: Consideration of Tire/Pavement Friction/Texture Effects on Pavement Structural Design and Materials Mix Design. Office of Pavement Technology. Federal Highway Administration (FHWA), Washington, D.C. (1999)
26. Lees, D.Y.: Rational design of continuous and intermittent aggregate grading for concrete. HRR No. 441-1973 (1973)

Novelty Detection for Location Prediction Problems Using Boosting Trees

Khaled Yasser[✉] and Elsayed Hemayed

Computer Engineering Department, Faculty of Engineering,
Cairo University, Cairo, Egypt
khaled.yasser@engl.cu.com.eg, hemayed@ieee.org

Abstract. Due to the enormous use of mobile applications and the wide spread of location-based services, as Foursquare, google maps, Facebook check-ins, it became a must to focus on studying these data and its impact on our social norms. In this paper, we are tackling the location novelty problem, which evaluates the user's curiosity to explore new places. In order to maintain a better service and offer new services, such as recommending new places, optimizing marketing campaigns, we conducted these experiments to classify the next check-ins to be either Novel or regular. We can predict the novelty of the next Point of Interest (POI) up to 82%, by extracting different types of features, in space and time, and using boosting trees.

Keywords: Location based analytics · Boosting trees · Location mining

1 Introduction

As the mobile technologies are widely spread, new lines of research are emerging. One track is the location analytics, which exploit the smartphones as sensors to collect the users' locations. The era of the social networks are thriving, which enhanced the techniques of collecting location data by offering check-ins features, like Facebook and Twitter, or even building networks based on the location check-ins, like Foursquare and previously Gowalla. We used this latter network Gowalla to conduct our research. These Social networks are in need to enhance their services by applying analyses to their enriched data. They offer interfaces to their platforms for third party applications, so researchers use these interfaces to collect data and publish it to the open source community, for research.

Location analytics are now solving lots of problems such as avoiding traffic congestions, optimizing retail campaigns and analyzing human mobility patterns. Novelty prediction is one type of location based analytics. It can play an important role on products' and services' campaigns, if coupled with location semantics and next location predictions. For example, if a user is expected to visit a novel place to eat at specific area of the city, the system can recommend a more optimized list of new restaurants with tempting offers. Moreover, the results can be enhanced by adding the social connections between the users to the system, as they are well studied, and proved that the social relationships influence the user mobility behavior.

© Springer International Publishing AG 2017
O. Gervasi et al. (Eds.): ICCSA 2017, Part II, LNCS 10405, pp. 173–182, 2017.
DOI: 10.1007/978-3-319-62593-5_13

In this paper we address the novelty prediction of the next location (exploration prediction) as defined by [1]. In our experiments, we focus on two types of datasets, a social network dataset Gowalla [2], a check-ins dataset made by the users, and another location-based social-networking service Geolife [3–5], a project conducted by a team from Microsoft Asia. In this project, we collect the users' geolocations periodically every 1 to 5 s, so the data considered as dense dataset. We run through the process of reshaping the data, extracting temporal, spatial, and historical features of the data to be able to determine the novelty of the next Point of interest (POI). We apply a set of classifiers to these features to measure the error rate of classifying the novelty of the places to be visited by the users. Gradient boosting trees has the minimum error rate. In our experiments, we used two well-known boosting algorithms. The first algorithm is the extreme gradient boosting (XGBoost) [6]. It uses boosting trees with additive training to minimize the error of a bunch of problems at both classification and regression. We also used lightGBM boosting trees algorithm which is developed by Microsoft [7]. This paper is organized as follows. Related work is discussed in Sect. 2. The proposed system is presented in Sect. 3. Section 4 presents and discusses the experimental results. Finally the paper is concluded in Sect. 5.

2 Related Work

Location based predictions is becoming at the focus of many research projects. It has great impact on advertising, traffic congestion control and many other fields. There are three main types for tackling location prediction problems:

First, Markovian models, as the data can be treated as sequences of states, the users are moving between them, [8–10]. Applied a Markov model to predict the next location, based on the frequency of the locations found in the user's history [11]. Used Markov chains to figure out the next POI, also [12] proposed a hidden Markov model for location prediction, using time intervals as observed states, and locations as the hidden states. These techniques model single user behavior or small group of users with common interests or relations. They failed to generalize for all users' behaviors. Also these models can't recommend new places. Furthermore, they can mistakenly predict private location of a user to others. If this location has high visiting frequency.

Second, recommender systems. Zheng et al. [5], used collaborative recommendation filters, to model the travel pattern of the users. These models can recommend novel POIs, using the social connections, and also the spatial and temporal features. They model the problem for all users, they failed to personalize the analysis, so they can't predict a private place to a specific user (home, work).

Third, is a hybrid technique, which separates the predictions of the novel places from the predictions of the regular ones [1]. They used features to check if the next place is novel or regular. If novel, then the inputs passed to a collaborative recommendation engine, otherwise, if the next location is regular, the input data passed to a hidden Markov model. They exploit the power of HMMs to discover the regularities, and the collaborative filters to offer novel POIs. They combined the result based on the novelty of the POIs [13], also separated the problem into two smaller ones, one for regular places and the other for novel ones. They combined the results linearly.

We followed these latter hybrid techniques to solve the problem, and we found that discovering the novelty of the next location, is one cornerstone for a more comprehensive model for human mobility behavior.

3 Proposed System

We applied a sequential pipeline as depicted in Fig. 1. This sequence helped us reproduce the baseline problem solution and the flexibility to change the classifier and to add more features. The pipeline has the following stages:

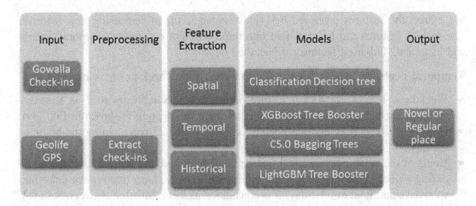

Fig. 1. System architecture

- We load the datasets and prepare the naming and formats.
- Then we apply check-in extraction process to Geolife dataset.
- Then we extract the features.
- Then we apply five different classification and regression trees, regular decision trees, then bagging and boosting trees.
- Finally to classify the new observation as a regular place or a novel one.
- The details of each stage are discussed in the following sections.

3.1 Preprocessing

For Geolife dataset, we need to aggregate the GPS points concentrated at one place to be a stay point. A stay point is the place where the user spent more time than a threshold. We applied the algorithm, proposed by [14], to detect the stay points. Briefly, the algorithm checks if the GPS points are within a spatial threshold and more than temporal threshold. In our implementation, we set these thresholds as follows: spatial threshold $\theta_s = 200$ m and temporal threshold $\theta_t = 20$ min.

Once having these stay points, we cluster them into POIs by their geolocation. Then we calculate the distance between stay points using Haversine formulae [15]. Then we use Density-based spatial clustering of applications with noise (DBSCAN) algorithm [16]. Unlike k-means algorithm, which cluster the data based on predefined number of clusters, DBSCAN can generate dynamic number of clusters based on the density of the data points. Thus we configure it to cluster the stay points based on the distance of 50 m to be considered as the same cluster, with minimum number of points in each cluster.

For Gowalla dataset, it doesn't need these preprocessing tasks, as it already has check-ins (POI) with labels. We added the novelty labels for both datasets.

3.2   Features Extraction

We extract the features proposed by [1], we enhance the system using new features. The overall features can be classified into three different categories; temporal, spatial and historical. The used features are detailed below:

Temporal Features. These features are extracted from check-in timestamp:

- Weekdays: It is a categorical feature between 1 at Sunday up to 7 as Saturday representing the days of the week. The effect of this feature is shown in Fig. 2(a).
- Hour of the day: A categorical feature between 0 and 23 representing the hours of the day.
- Hours of the week: a categorical feature between 0 and 167 representing the hours during a week. This feature is better differing the novel places and regular ones than weekdays and hours of the day. It is considered as a combination between both [1]. The effect of this feature is shown in Fig. 2(b).
- Time interval between consecutive check-ins: it's the difference in time between the current check-in and the previous check-in in minutes.

Spatial Features. These features depend on the location of the check-in and the surrounding locations (neighborhood). We used Haversine formulae [15], to calculate the distance between two check-ins, using their GPS locations. We used DBSCAN to build the neighborhoods:

- The distance between consecutive locations per user. We calculated the distance between the current and previous locations visited by the user.
- Location entropy: using Shannon's formula:

$$H = -\sum p(x) \log p(x) \tag{1}$$

Where $p(x)$ is the ratio between the visiting frequency of the location's neighborhood to the visiting frequencies to other territories.

- Visiting ratio: it's the ratio between the number of locations visited by the user to the number of all check-ins in this area within the location neighborhood (a geographical boundary)
- Visited locations to neighborhood locations ratio: we found that the ratio between the number of locations visited to the locations of the neighborhood correlated to

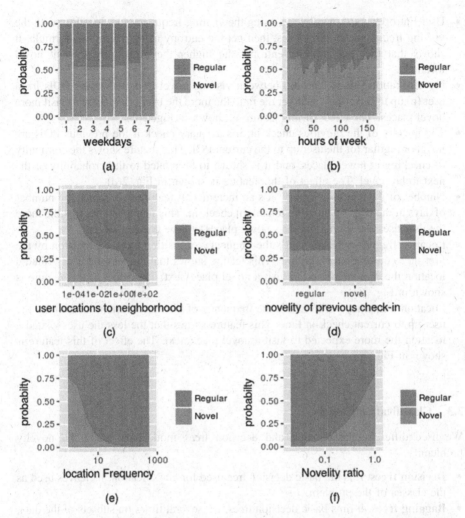

Fig. 2. The different features and its effect on the probability of novelty to the regularity of places. The x-axis of each graph is the feature values, and the y-axis is the conditional probability of novelty given the corresponding feature.

the explorations made by the user, the less the user uses the service by adding check-ins the more the novel places to be visited. The effect of this feature is shown in Fig. 2(c).

Historical Features. These features are extracted from the user's check-in history and they are as follows:

- Number of POIs the user had visited: the different locations the user visited up to the current check-in. This feature indicates the user behavior of using the check-in service in general and novelty tendency.

- User Entropy: As shown by [1], using the visiting frequency to a location, with the visiting frequencies to all places, then get the entropy using Shannon's formula. It shows that the less the user entropy, the higher the probability to visit novel locations.
- Novelty ratio: This is the ratio between visiting novel POIs to regular POIs for a user (u) up to time (t). The larger the ratio the more the user is expected to visit more novel places. The effect of this feature is shown in Fig. 2(f).
- The novelty of the previous check-in: it's a binary check if the previous POIs are novel or regular for the user up to the current POI. This feature shows the continuity of checking at novel places, and it is shown to be related to the probability of the next to be novel. The effect of this feature is shown in Fig. 2(d).
- Number of days in user history: it's an incremental feature that counts the number of days in the user's history up to current check-in. This feature shows that the more the user uses the service the less novel places will be checked-in.
- Location frequency per user: it's the frequency of visiting a certain location by the user up to current check-in time. This feature shows that the less the user visited a location the more expected to visit a novel place next. The effect of this feature is shown in Fig. 2(e).
- Location frequency per user: it's the frequency of visiting a certain location by the user up to current check-in time. This feature shows that the less the user visited a location the more expected to visit a novel place next. The effect of this feature is shown in Fig. 2(e).

3.3   Classification Models

We used different classification and decision trees models to tackle the novelty problem:

- **Decision trees.** It is the basic decision tree used for classification, its leaves used as the classes of the problem.
- **Bagging trees.** It runs basic decision trees, for several times on subsets of the data, using bootstrap technique. Then takes the average results, as the output.
- **Boosted trees (Xgboost)** [6]. Using decision trees, with additive training trees. Every iteration tries to correct the error of the previous one. Using an objective function (loss and regularization).
- **Boosted trees (Lightgbm)** [7]. It is another boosted tree implementation. It uses histograms to convert continuous features to discrete ones. This technique speeds up the training process, also, it uses leaf-wise tree growing, which reduces the model memory footprint.

We used a random search cross validation [17], it used for model selection, it applies the whole process, training, predicting and scoring. To optimize the algorithm parameters (hyper-parameters). For example, the number of trees, learning rate and maximum depth. This search technique runs the algorithm with random combinations of hyper-parameters. These combinations can be set with intervals, we set, as an example, learning rate from 0.5 to 0.05. This search can lead to the best tuning of the hyper-parameters, so we can select the best hyper-parameters, which produce the minimal error rate.

4 Experimental Results

4.1 Datasets

We conducted the experiments using two types of datasets:

- Geolife which is a GPS dataset [3–5]: it consists of users' trajectories. These trajectories are the GPS locations related to the users' trips per day. This dataset contains 17,621 trajectories for 182 users and 24,876,978 GPS points collected from April 2007 to Aug. 2012 in the city of Beijing. We applied clustering preprocessing for these raw GPS points to extract the POI at which the users spent most of their time.
- Gowalla data [2]: Gowalla is a location based social network, in which users check-in at the places they visited. The dataset extracted from the website with 6,442,890 check-ins over the period of Feb. 2009–Oct. 2010 for 107,092 users. We consider the check-ins of the users as the POIs, so this dataset does not need preprocessing as Geolife dataset, though we process these data to extract the features, and to set the novelty label.

4.2 Experiments

We used both datasets Gowalla and Geolife. To evaluate the proposed architecture. We divided both datasets into training, validation and test sets. We used the default features used by [1]. We extracted them from Gowalla as (feature set 1) and from Geolife as (feature set 3). Then, we added more features proposed in the previous section. For Gowalla as (feature set 2), and for Geolife as (feature set 4). We applied four experiments:

1. Gowalla (feature set 1) to the algorithms separately. We applied the random search for hyper-parameters. To fine-tune every algorithm parameters. The optimized hyper-parameters change from algorithm to another, with different set of features.
2. Gowalla (feature set 2) to the algorithms. To test the accuracy enhancements when adding new features.
3. Geolife (feature set 3) to the algorithms with different tuning from Gowalla, as the data structure differs from Gowalla.
4. Geolife (feature set 4) to the algorithms, to get the best results with boosted trees algorithm Xgboost. The prediction accuracies are shown in Table 1.

Table 1. The ratio of the correct predictions to all the test set examples, columns are the algorithms' results and rows are the used features

Features\Technique	Decision trees	Xgboost	LightGBM	Bagging trees
Feature set 1[a]	0.7276	0.8017	0.795	0.7844
Feature set 2[b]	0.7585	0.8202	0.82	0.8166
Feature set 3[c]	0.7783	0.8192	0.7748	0.8041
Feature set 4[d]	0.8451	0.8558	0.8537	0.8485

[a] Feature set 1: Gowalla dataset with the 12 features.
[b] Feature set 2: Gowalla dataset with all proposed features.
[c] Feature set 3: Geolife dataset with the 12 features.
[d] Feature set 4: Geolife dataset with all proposed features.

4.3   Results

By applying the proposed architecture to Gowalla dataset, we showed that using (feature set 1) with basic decision trees, we got low accuracy level of 72%. We consider it as the baseline classification, as the other models are enhancements (bagging, boosting) over these basic decision trees. When using feature set 1 with boosted trees, we got a result of 80%, which was reported by [1]. By using Gowalla (feature set 2) with the boosted trees, we were able to enhance the accuracy up to 82%. By applying the boosted trees to Geolife dataset, we got accuracy up to 85%. Geolife exceeded Gowalla by 3% accuracy. This conclusion maybe because of the way the data collected. For Geolife, the data fetched at the time of the visit. When the user visits a

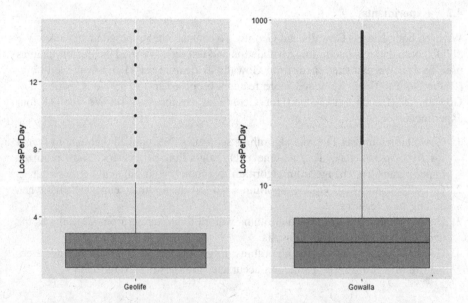

Fig. 3. To the left, Geolife number of locations per day. Its mean is 2.5 and variance is 3.4. To the right, Gowalla scaled using log10 for depiction. It spread up to a maximum of 731 locations per day. Its mean is 6.5 and variance is 332.7.

location and spent enough time. But for Gowalla, the user can check-in at places he/she didn't visit. Or he/she check-in on different time from the actual visiting time. As shown in Fig. 3, the mean of Geolife 2.5, variance of 3.5 and maximum of 15. For Gowalla, it has a maximum of 731 locations per single day, with 332.7 variance and 6.2 mean. These reasons make Gowalla dataset noisy.

5   Conclusion

In this paper, we highlighted the importance of the next place novelty, an exploration prediction problem. We proposed new features to enhance the prediction accuracy. We were able to reduce the prediction error of about 8% over decision trees, and about 2% to 4% over bagged trees using boosting trees. We compared different classification models and we found out that, Xgboost exceeds other tree-based algorithms. Lightgbm model is a very fast to solve such problems, with high accuracy levels. In general, the proposed architecture for solving the next location novelty, can be coupled with location semantics and arrival time predictions, to build a comprehensive model, for tracking user mobility behavior.

References

1. Lian, D., Xie, X., Zheng, V.W., Yuan, N.J., Zhang, F., Chen, E.: CEPR: a collaborative exploration and periodically returning model for location prediction. ACM Trans. Intell. Syst. Technol. 6(1), 1–27 (2015)
2. Cho, E., Myers, S.A., Leskovec, J.: Friendship and mobility: user movement in location-based social networks. In: ACM SIGKDD International Conference on Knowledge Discovery and Data Mining (2011)
3. Zheng, Y.: Mining interesting locations and travel sequences from GPS trajectories. In: WWW, Madrid (2009)
4. Zheng, Y.: GeoLife: a collaborative social networking service among user, location and trajectory. In: UbiComp, Seoul, Korea (2008)
5. Zheng, Y.: GeoLife: a collaborative social networking service among user, location and trajectory. IEEE Data Eng. Bull. 33, 32–40 (2010)
6. Chen, T., Guestrin, C.: XGBoost: a scalable tree boosting system. In: Proceedings of the 22nd ACM SIGKDD International Conference on Knowledge Discovery and Data Mining - KDD 2016 (2016)
7. Distributed machine learning toolkit. http://www.dmtk.io/. Accessed 10 Jan 2017
8. Pan, J.-S., Polycarpou, M.M., Woźniak, M., Carvalho, A.C.P.L.F., Quintián, H., Corchado, E. (eds.) Hybrid Artificial Intelligent Systems. Springer, Heidelberg (2013)
9. Cho, S.-B.: Exploiting machine learning techniques for location recognition and prediction with smartphone logs. Neurocomputing 176, 98–106 (2016)
10. Ashbrook, D., Starner, T.: Using GPS to learn significant locations and predict movement across multiple users. Pers. Ubiquit. Comput. 7(5), 275–286 (2003)
11. Gambs, S.: Next place prediction using mobility Markov chains. In: Eurosys' MPM (2012)

12. Mathew, W., Raposo, R., Martins, B.: Predicting future locations with hidden Markov models. In: Proceedings of the 2012 ACM Conference on Ubiquitous Computing - UbiComp 2012 (2012)
13. Wang, Y., Yuan, N.J., Lian, D., Xu, L., Xie, X., Chen, E., Rui, Y.: Regularity and conformity: location prediction using heterogeneous mobility data. In: Proceedings of the 21th ACM SIGKDD International Conference on Knowledge Discovery and Data Mining - KDD 2015 (2015)
14. Zheng, Y., Zhang, L., Ma, Z., Xie, X., Ma, W.-Y.: Recommending friends and locations based on individual location history. ACM Trans. Web 5(1), 1–44 (2011)
15. Haversine formula. Wikipedia https://en.wikipedia.org/wiki/Haversine_formula. Accessed 17 Jan 2017
16. Ram, A., Jalal, S., Jalal, A.S., Kumar, M.: A density based algorithm for discovering density varied clusters in large spatial databases. Int. J. Comput. Appl. 3(6), 1–4 (2010)
17. James Bergstra, Y.B.: Random search for hyper-parameter optimization. J. Mach. Learn. Res. 13, 281–305 (2012)

Estimation of an Urban OD Matrix
Using Different Information Sources

Asma Sbaï[1], Henk J. van Zuylen[2,3,4], Jie Li[4(✉)], Fangfang Zheng[3],
and Fattehallah Ghadi[1]

[1] Faculty of Science, Ibn Zohr University, Agadir, Morocco
{a.sbai,f.ghadi}@uiz.ac.ma
[2] Delft University of Technology, Delft, The Netherlands
h.j.vanZuylen@tudelft.nl
[3] South West Jiaotong University, Chengdu, China
[4] Hunan University, Changsha, China
Ljlj369@msn.com

Abstract. An Origin Destination matrix for urban trips is more difficult to develop than for interurban long and medium distance trips. The socio-economic characteristics are valuable parameters to estimate trip attractions and destinations, but often the distance does not have a significant effect on the distribution of urban trips. Since the 1980s methods are developed to estimate the trip matrix from traffic volumes. The problem is underdetermined: the information in the OD matrix is more than the information contained in the traffic volumes. Nowadays there are more information sources like probe vehicles, Automated Number Plate Recognition cameras, mobile phone data etc. This article discusses the possibilities and limitations of these additional information sources. Use is made of traffic data collected in Changsha, a town in middle-south China.

Keywords: Origin Destination matrix · Probe vehicles · Information minimization · Urban traffic

1 Introduction

The Origin Destination matrix (OD) is a concept that is needed for planning, monitoring and management of traffic. It contains the number of trips made during a certain time period between origin zones and destination zones. Often the OD matrix is determined for trips with all modes. The fraction of trips per mode is derived from the total OD matrix using a modal choice model. In such a way the demand for public transport and the traffic volumes for cars and bicycles can be determined [1].

For the determination of the full OD matrix often socio-economic data are used: the trip production and attraction of zones are estimated based on the characteristics of the zone: number of jobs, shops, inhabitants, schools etc. The trip production and attraction are then estimated based on such figures and model parameters. The distribution of the trips over different destinations is done based on the distance concept: the trips over a short distance are more likely than long distance trips. This calculation step creates an OD matrix, where each element gives the number of trips from origin zone to

© Springer International Publishing AG 2017
O. Gervasi et al. (Eds.): ICCSA 2017, Part II, LNCS 10405, pp. 183–198, 2017.
DOI: 10.1007/978-3-319-62395-5_14

destination. The sum of all trips from one origin is constrained by the fact that it should be equal to the total trip production as estimated in the socio-economic model. The same applies to the sum of all trips with the destination in a zone. The split of trips over different modes is calculated by a modal choice model where the perceived costs and values of travel modes are assumed to determine the choice of travelers. This modeling procedure is the well-known and standard four-step model [1].

There are several criticisms on the four step model, for instance the assumption that trip production and attraction are not influenced by the accessibility or zones is rather weak. Furthermore, in urban areas the distance between origin and destination has not always a strong influence on the elements of the OD matrix.

An alternative method to determine an OD matrix is the use of questionnaires, asking people living in the different zones about their travel behavior. Such processes to obtain an OD matrix are complex and expensive and in general not suitable to be repeated on a regular basis.

Alternatives have been sought to estimate the OD matrix using readily available data. For instance, travel patters with public transport can be obtained from the smart card data, if the travelers use electronic payment with traceable debit cards. Something similar can be done with data from toll stations where the entry and exit of a vehicle are registered and give information about the part of the journey that was made over the tolled roads.

Traffic volumes that are observed are the consequences of trips and the route choice. When the OD matrix for car trips and the route choice are known, the traffic volumes are uniquely determined. From the 1980s methods were developed to 'reverse engineer' an OD matrix from observed traffic volumes. Even when route choice is known, the reverse calculation of the OD matrix from the traffic volumes is not unique. The number of constraints for the a-posteriori OD matrix (i.e. the number of road links with known volumes) is nearly always less than the number of elements of the matrix [2, 3]. A possible way to obtain still an estimate for the OD matrix is to choose the matrix that satisfies the constraints imposed by the traffic volumes and contains as little information as possible [2]. Other approaches have been developed like: minimizing the difference between an *a-priori* OD matrix and a matrix that satisfies the volume constraints. Ease of calculation was often more leading in the choice of the estimation methodology than logic. Several researchers assume a normal distribution of the elements of the OD matrix which makes a least square estimation method possible. In reality the OD matrix has many elements with a small value, for which a normal distribution is not realistic.

At present, many more sources of information become available, next to traffic volumes. Probe vehicles with GPS are driving in many cities, reporting their positions on regular basis. Cameras with Automated Number Plate Recognition (ANPR) are installed at intersections registering the number plates of passing cars, which makes it possible to follow cars, measure their travel times and determine their path through a part of the network. Bluetooth scanners can register Bluetooth devices and determine their path and travel times. Furthermore, mobile phones can be traced and tracked when they move through the network.

For the monitoring of car traffic the option to use ANPR as information source is most useful. The ANPR cameras can count passing vehicles and identify them. Mobile phone data are less representative for the traffic: one car can have no, one or more mobile phones, mobile phones can also be carried by pedestrians and public transport passengers. While ANPR gives information about the movement pattern of all cars, mobile phone give information about movement of people, but restricted to people carrying a phone. Bluetooth scanners have the same limitation.

ANPR has the limitation that the cameras are often installed on a limited number of places. Several roads are not monitored with cameras and a route of a journey contains roads that are not all monitored by cameras. For the estimation of an OD matrix the data from ANPR should be extended with traffic flows on the roads without cameras. Many cities have traffic control systems that use detectors measuring the traffic flow, like SCATS and SCOOT. However, these data are not complete, partially because the failure rate of detectors is rather high [4]. Therefore a third source of information is useful: the probe vehicles. Often the probe vehicles with GPS are taxis. That means that traffic patterns of probe vehicles are not necessarily representative for the Origin Destination traffic flows [5]. In Sect. 6, this subject will be illustrated with empirical data.

2   The Empirical Data

The methodology for the estimation of an OD matrix from multiple data sources is illustrated with a real life situation. For Changsha, the capital of Hunan province in China, traffic information for the central city area is collected by loop detectors, taxis with GPS and ANPR cameras. Data for three days in April 2015 were collected and available for analysis.

The loop detectors are a part of the SCOOT traffic control system. On each controlled lane of more than 200 intersections (situation 2015) a loop detector is installed on a short distance from the stop line. The traffic counts from the detectors are aggregated in 5 min periods. In total there are 2529 loop detectors. However, the quality of the detectors is low [4], only 24% of the detectors give traffic counts that are reliable for 95% (Fig. 2), 47% of the detectors give data that are more than 90% reliable. Error checking and data completion is necessary and, in general, the information from loop detectors are only useful when additional information is added.

All taxis in Changsha have GPS and they transfer their position and speed every 30 s by GSM to a central computer system. There are 7200 taxis driving through the network, most of them are continuously driving around, transporting passengers and finding new passengers along the roads. A specific property of taxi routes is that taxis often stop to alight passengers, to get new passengers and to follow instruction of their passengers, e.g. to stop at shopping centers or schools. Tracing the path of taxis through the network shows that more than about 20% is not following the road straight from intersection to intersection, but have intermediate activities: they stop in between or disappear from a road and emerge again after some time. Furthermore, as visible in Fig. 3, the taxi drivers are faster than other drivers during day time but at night they use more time traveling over a road (probably while searching for passengers).

The ANPR cameras register number plates of cars passing the stop line. The quality of these data is good, for most intersections less than 5% of the cars are not well recognized, only at night a larger fraction of cars is not recognized.

Important is that not only taxi drivers but also about 15% of the drivers of other cars make stops half way between intersections. Often such stops have a short duration. In the framework of OD matrix estimations it is a question whether such intermediate stops represent an intermediate destination. In fact, trips are often not a movement between the place of origin to the destination, but a chain of displacements combined with activities [6, 7]. So each stop between two intersections should be considered as a node in a chain of displacements. The whole trip should be cut into pieces that are driven without special interruptions. This gives a constraint to the use of ANPR data: exceptional stops between monitored intersections should be identified and trajectories on which such interruptions occur should be split.

3 State of the Art of OD Estimations

Estimation of OD matrices from traffic count has received much interest from researchers for the last decades. Especially, to find methods for situations where data is limited and can lead to non-unique OD matrices [3]. A number of methods were developed based on Bayesian statistics [8–10], Generalized least squares [11, 12], and Maximum likelihood [13]. All these methods are based on prior information and thus are very sensitive to this input. When the prior information has a low validity, the estimated OD matrix has a low quality.

3.1 Categories of Sensors

In order to provide information about the traffic state, different sources of data are used. Those technologies differ on the type of data collected, their usage cost and their reliability. In the literature, authors used different types of data depending on the ability of the model to process it. Antoniou [14] distinguishes traffic sensor technologies according to their functionality in three categories:

- Point sensors: Inductive loop detectors are the most widely used. They are cheap but not fully reliable. Additionally, installation and maintenance may cause disturbance to traffic. This equipment provides traffic flow information, time mean speed and vehicle length. Weight In Motion (WIM) systems allow continuous collection of vehicle weight information in a measurement zone.
- Point to point sensors: Can detect vehicles at single locations and thus for some case provide travel time, route choice fractions, paths and OD flows. AVI Automated Vehicle Identification System is a technology based on the identification of vehicles on various points of the network. Another tool in this category is the CCTV cameras that records license plates and send this information to a center system to process it and match subsequent detections of particular vehicles. But the reliability depends on the quality and sensitivity of the camera to weather conditions.
- Area-wide sensors: A number of new generation sensors are having the attention of researchers for their ability to improve dynamic estimation of traffic. Smartphone, GPS are typical examples of such technologies, including Airborne sensors that continuously fly over a study zone surveying the traffic and cameras at high

positions along roads that observe traffic over longer stretches of the road. Cars equipped with GPS devices collect information about its location and transmit this data to a central for processing. The inconvenience of this technology is the need for a wireless telecommunication connection to send data from the vehicle to the taxi management center. Another limitation is the inability to recognize the type of vehicles unless the registration of data is linked to a vehicle specification.

3.2 Multiple Data Source

While using single data source provide interesting outputs, the fusion of all available resources often leads to significant better results. However, to use these diverse data, transformations such as scaling, aggregation and reinterpretation are required to provide readable information for traffic managers.

Byon et al. [16] join different data sources to estimate traffic conditions as soon as any single sensor becomes active using a single constraint at a time.

Bugeda et al. [21] have used several ad-hoc linear Kalman filter approaches to estimate the OD matrix based on Bluetooth and Wifi data and proved that this approach works in the case of both congested and uncongested network but tuning on the initialization points and matrices is critical in some situation.

In 2012, Ma et al. [17] developed an approach based on the extension of entropy maximization and an extension of Bayesian networks to estimate freight OD matrix using data from loop detectors data, WIM data and Bluetooth. Ma integrated the different data sources and showed that especially data from number plate cameras improves the quality of the OD matrix estimation [18]. In the same year, Morimura and Kato [19] use Tokyo multiple data sets from: probe-car, road traffic census and landmark data, to estimate a microscopic OD matrix using an L1-regularized Poisson regression with an adaptive baseline where the data from probe vehicles is propagated to the entire area along with landmark information. Iqbal et al. [20] use mobile phone Call Detail Record (CDR) and limited traffic counts from video recording. They combine the two data sources by generating an OD pattern from CDR data and scale this matrix up to match the traffic counts. Bugeda et al. [21] have used several ad-hoc linear Kalman filter approaches to estimate the OD matrix based on Bluetooth and Wifi data and proved that this approach works in the case of both congested and uncongested networks. They found that tuning on the initialization points and matrices is critical in some situation.

Kostic and Gentile [21] treats the calibration as an optimization problem and use derivative free optimization algorithm (SPSA, Nelder-Mead and CMA-ES). It consists on a constrained minimization problem of two parts: distance from initial demand and distance from observed traffic data. The framework uses an a-priori matrix from previous studies to initiate the algorithms and to measure the deviation from estimated results. The objective function is on one hand a comparison between historic and current demand and simulated and measured data. On the other hand, measures are used for calibration and each type has its weight factor. Their approach is similar to the method proposed longer ago by Willumsen [23]. To test simulations in optimization,

Kostic and Gentile use traffic real time equilibrium. They implemented four termination criterions: Minimum relative gap, number of iterations, maximum number of function evaluations and maximum optimization time. They started the simulation with only flow data then, add other types: speed, density and number of vehicles. The solutions obtained are within 20% errors and shows that combining different type of data reproduces better the known true solution.

The use of sensors raises the problem of its location, which involves the right choice of number and links where to install sensors. Some authors aimed through their researches to minimize the number of sensors and maximize the precision of their methods [25, 27].

4 The Methodological Framework

The distribution of traffic over the roads is determined by travel times. Drivers choose the route with the shortest perceived travel time. Travel times depend, among others, on the traffic volumes: on busy roads the delays at intersections are longer than on roads with less traffic. We will assume that we know the volumes on the most important links in the network and that we can determine the travel times. Then we know which routes are chosen by the drivers. The methodology used in this section is based on the assumption that the whole journey is started and completed in the same time period. The analysis given in this section applies to a certain time period. The time index is omitted in the formulas (Fig. 4).

A journey from zone i to zone j passes over several links. We denote the number of trips between i and j with T_{ij}. The traffic flow V_a over a link a is

$$V_a = \sum p_{ij}^a \cdot T_{ij} + E_a \tag{1}$$

Where p_{ij}^a is the fraction of trips from i to j that pass link a and E_a represents the errors which come from observation errors and inaccuracy of the route choice parameters p_{ij}^a. Different approaches can determine the proportion of traffic assigned to each link: All or Nothing assignment, Deterministic Equilibrium assignment, Stochastic assignment or Stochastic User Equilibrium. Basically all assignments depend on the route travel time [1]. When the volumes V_a and the route fractions p_{ij}^a are known, the information contained in Eq. (1) is still not sufficient to determine the OD matrix uniquely. Several methods have been developed to estimate an OD matrix that satisfies the volume constraints, as described in Sect. 3. Van Zuylen and Willumsen [3] defined the most logical estimate as the OD matrix that contains the minimum information. This is equivalent of the maximization of the entropy of the estimated OD matrix, subject to the constraints imposed by the traffic volumes. That gives the following formula

$$\begin{cases} T_{ij} = t_{ij} \prod_a X_a^{p_{ij}^a/g_{ij}} \\ \text{Where: } g_{ij} = \sum_a p_{ij}^a \end{cases} \tag{2}$$

And t_{ij} is an a-priori estimate of the OD matrix, e.g. a historical OD matrix that has to be updated. Other authors used other approaches, such as the minimization of the difference between T_{ij} and t_{ij}. The factors X_a are determined by making the T_{ij} matrix consistent with the volumes as given in Eq. (1). If the counted traffic flows V_a are mutually consistent and the appropriate algorithm is chosen to determine X_a, a perfect match can be found between the measured traffic flows and the result of the assigned OD matrix according to Eq. (1). The information contained in the traffic volumes might be inconsistent. The most obvious is the case where incoming flows on an intersection are not equal to the outgoing flows. Such inconsistencies are often due to counting errors or to the fact that counts were obtained at different times. The elimination of these inconsistencies can be done in different ways, e.g. by maximizing the likelihood of the volumes or by fusing the data [28].

Often the traffic volumes are not available on all roads. For the Changsha case a lot of detectors were not reliable. We merged the traffic volumes with the taxi data. Taxis are driving on all roads, but they are only a small proportion of the traffic. The number of taxis passing a road can be determined from the GPS records. By determining the proportion of taxis in the total flow, the volumes on roads without valid loop detectors can be estimated by up scaling the taxi counts.

Furthermore, taxi OD patterns can be determined from the GPS tracks where we can assume that a taxi that stops for more than a certain time (we used 2 min based on observations) is changing a passenger. In the case that there is no a-priori OD matrix available, the taxi OD offer a possibility to choose t_{ij} in Eq. (2).

If we have ANPR data, the information about the OD matrix increases considerably. A matrix can be made of the flow between intersections with an ANPR camera. This matrix contains more specific information than the traffic flow. If two cameras numbered k and l count D_{kl} vehicles recognized by both cameras, this imposes the following constraint to the OD matrix:

$$D_{kl} = \sum_{ij} T_{ij} q_{ij}^{kl} \tag{3}$$

Where q_{ij}^{kl} is the fraction of the trips between i and j that pass the cameras k and l. The generalization of the solution based on information optimization (Eq. 2) is now

$$T_{ij} = t_{ij} \prod_a X_a^{p_{ij}^a / g'_{ij}} \prod_{kl} Y_{kl}^{q_{ij}^{kl} / g'_{ij}} \tag{4}$$

Where

$$g'_{ij} = \sum_a p_{ij}^a + \sum_{lm} q_{ij}^{kl} \tag{5}$$

The situation in Changsha is that each camera is observing a single lane, so that left turning and straight on movements can be distinguished. However, right turning traffic is not controlled by signals on most intersections and no cameras are installed at right turning lanes. There are also no loop detectors on right turning lanes, so that these flows have to be estimated from taxi data.

This method to estimate traffic flows from taxi flows gives reliable results only when the fraction of taxis is sufficiently well known. For flows passing ANPR cameras the ratio of taxis can be directly determined. The question is whether such fractions can be extrapolated to flows that are not monitored by cameras. In the next section we describe how we verified the validity of this extrapolation.

In Fig. 5 it is clear that the OD matrix element for the trips between zone 7 and zone 14 can directly be determined from the number of number plates recognized by cameras at intersection 51 and 59. Still formulas (4) and (5) can be applied as generic expressions to determine OD flows.

5   Some Preliminary Results

The data that were available in Changsha were traffic counts on the lanes of signalized intersection, excluding the right turning lanes. These counts are aggregated in 5 min periods and the missing data (due to defect detectors) were identified. Furthermore, a check was made whether the distribution of the flows over parallel lanes was logical [4]. Counts where the traffic was too much on one lane are an indicator that there might be something wrong with the detectors. Figure 6 gives an example for two parallel lanes. The distribution of traffic between both lanes should be between the 5 and 95 percentile limits.

After the check of the detector data we kept the valid ones. In the right hand part Fig. 1, the roads where reliable counts could be obtained are shown as white lines.

The GPS taxi data are first map-matched. The roads in the study area were digitized and the taxis are assigned to the location on the link that is the closest to its GPS position. The trajectories of the taxis over the links of the digitized network are used to determine taxi flows and travel times. Taxi trajectories often show that they deviate from the main roads, probably to go to a minor side road or parking place that is not part of the digitized network. For the determination of the travel time such trajectories

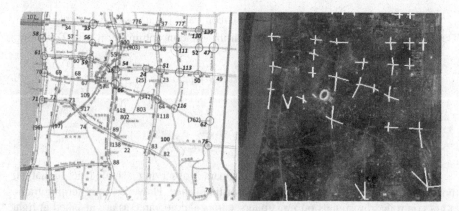

Fig. 1. The network of the central area of Changsha. The intersections with circles have ANPR cameras. Right: the intersections in the SCOOT traffic control system. The white links have reliable loop detectors.

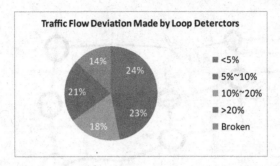

Fig. 2. The unreliability statistics of the loop detectors [4]

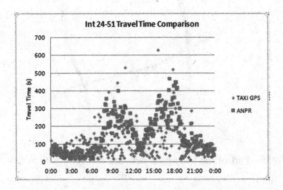

Fig. 3. Comparison of travel times measured for taxis and other cars

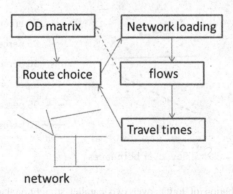

Fig. 4. Structure of the estimation of the OD estimation

have to be ignored. However, for the taxi OD matrix such deviations from the main street shows that they have a destination in a minor zone at the side of the main road, as e.g. zone 51 in the upper left corner of the network in Fig. 5.

For the ANPR several procedures have to be followed. Some double counting occurs; such records can easily be identified and eliminated. Unrecognized number

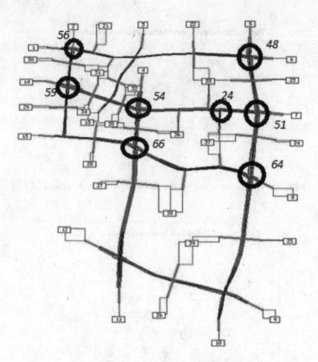

Fig. 5. Part of the Changsha network of Fig. 1 with zones.

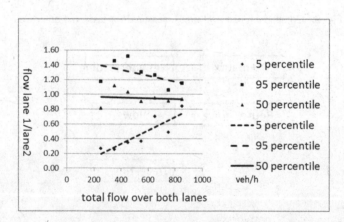

Fig. 6. Observed distribution of traffic over two parallel straight-on lanes for different total volumes (lane 1 is the right lane, lane 2 the left one, the 5, 95 and 50 percentiles for the left lane (lane 1). Only counts distributions between the 5 and 95 percentiles are considered as reliable [4].

plates count for the determination of the flows, but are useless for the OD matrix between intersections. The situation that drivers stop between two consecutive intersections has to be identified (Fig. 7).

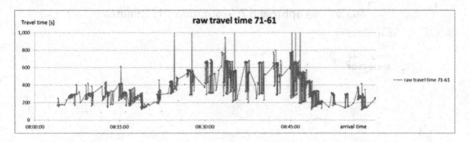

Fig. 7. Travel times of a series of cars, showing several outliers

If a car numbered i arriving at time t_i, has a travel time tt_i and the car before has a travel time tt_{i-1} and the next car tt_{i+1}, we can assume that when $tt_i > (tt_{i-1} + tt_{i+1})/2 + 120$ s, the car has made an intermediate stop, so it has an intermediate destination on that link or at a side road of that link.

An important question is whether taxi flows can be used to estimate the total traffic flows. Taxis are driving everywhere in a city while loop detectors and ANPR cameras are only present and operational on a limited number of spots.

A first analysis was done to see whether taxis flows can be considered as representative for the total traffic flow. First of all the proportion of taxis were analyzed to see whether these proportions are sufficiently stable over the network and over time to be used for the conversion of taxi flows to total flows.

Figure 8 shows that the percentages on consecutive links are similar for consecutive links with a rather high R^2 values for time periods of one hour. This similarity is less when we look at 15 min values Fig. 9.

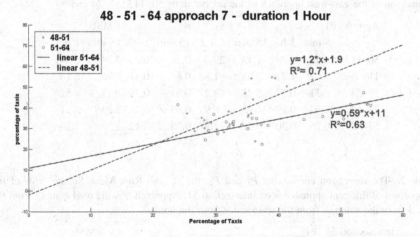

Fig. 8. The relation between the percentages of taxis on the south approach of intersections 48 and 51 (blue line) and 51 and 64 (red line). Approach 7 is the link coming from the south. The intersections numbers are as given in Fig. 1. (Color figure online)

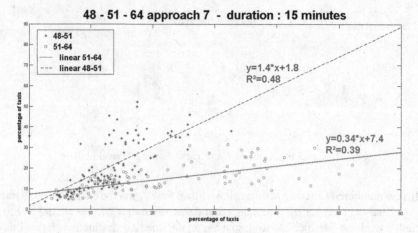

Fig. 9. The relation between percentages of taxis for 15 min averages.

The overview of some more links are shown in Tables 1 and 2. It is clear that links that go in the same direction are stronger correlated than links in different directions, so that a single percentage of taxis for a whole network is not justifiable. Therefore, we can use taxi flows to estimate missing traffic volumes, when we use the taxi percentages as are found from the ANPR data for the directions that are similar to the flows that are to be estimated. We have to realize that such estimated flows are still relatively uncertain.

Table 1. The regression coefficients P_1 and P_2, the R^2 and Root Mean Square error of the comparison of the east-west approach of the intersections 50, 113, 51, 24 and 59

Appr 6	P1		P2		R^2		RMSE	
	15 min	1 h	15 min	1 h	15 min	1 h	15 min	1 h
50–113	0.45	0.23	1.13	29.3	0.36	0.23	7.14	6.347
113–51	1.42	1.21	0.85	−1.83	0.67	0.71	9.70	8.16
51–24	0.31	0.57	5.28	8.27	0.61	0.79	4.17	4.52
24–54	1.03	0.79	1.73	8.9	0.63	0.70	5.23	5.11
54–59	−0.26	0.30	10.47	17.29	0.15′	0.12	5.56	7.33

Table 2. The regression coefficients P_1 and P_2, the R^2 and Root Mean Square error of the comparison of different approaches on intersection 51. Approach 6 is the road coming from the east, 7 from the south, 8 from the west and 9 from the north.

Intersection 51	P1		P2		R^2		RMSE	
	15 min	1 h	15 min	1 h	15 min	1 h	15 min	1 h
51.6–51.9	0.6	0.53	0.06	9.64	0.21	0.34	7.91	7.39
51.6–51.8	0.25	0.49	3.76	13.24	0.18	0.47	3.66	5.17
51.6–51.7	1.35	1.09	2.71	2.83	0.55	0.75	8.29	6.20

Looking at the OD matrix between intersections we investigated whether the taxi OD matrix has a similar structure as the full OD matrix, i.e. the matrix for all traffic. We did that for the OD matrix for 2 h periods (which covers time periods like a morning peak, midday and afternoon peak) and for 15 min periods. OD matrices for periods of 15 min are especially interesting for Dynamic Traffic Assignment programs [26]. For the 2 h OD matrix the correlation between the matrix elements of the full OD matrix and the taxi OD matrix is very high: $0.67 < R^2 < 0,94$. For the 15 min OD matrices this is slightly lower: $0.23 < R^2 < 0.91$. In only 7% of the 15 min periods the R^2 was lower than 0.5.

For the comparison of the taxi and full OD matrices we used also the information content as introduced originally by Shannon [30]:

$$I = \sum_{ij} \frac{T_{ij}}{T} \ln \left(\frac{T_{ij}}{T} \middle/ \frac{t_{ij}}{t} \right) \tag{6}$$

Where T is the sum of all elements of the full OD matrix T_{ij} and t is the sum of all elements of the a-priori (taxi) OD matrix.

If we take the taxi OD matrix as a-priori for the full OD matrix, we find that the information in the full OD matrix for all traffic is much less than in the case that we don't assume that the taxi OD matrix contains the a-priori information (e.g. the a-priori matrix is uniform $t_{ij} = 1$ for all i and j). This difference in information is larger for time periods where the correlation between t_{ij} and T_{ij} is large.

As a result of the analysis of the empirical data, we found the following procedure to estimate an OD matrix:

1. Select the loop detector data that are valid,
2. Map match the taxi GPS data and determine the taxi link flows,
3. Determine the taxi percentage and estimate missing traffic flows by using the taxi flows and the taxi percentage,
4. Determine the taxi trajectories and split trajectories if a taxi uses more than 2 min additional time to drive over a link compared to other traffic (i.e. consider such links as destinations and origins),
5. Determine the taxi OD matrix for the area of interest,
6. Determine the flows from ANPR data, additionally fill in gaps in traffic flows from the detector data and up scaled taxi flows,
7. Calculate the route choice using the actual link travel times (often quickest routes are adequate).
8. Use the traffic flows, ANPR counts, route matrix to estimate the OD matrix, for instance according the information minimization method (Eqs. 4 and 5).

Since not all input data have the same reliability, the estimation method like the information minimization should be adapted to take the reliability of traffic flows and route matrix into account.

6 Discussion and Conclusions

The availability of more data of urban traffic makes it possible to monitor the status of the traffic in urban road networks in ever increasing detail. However, errors in the data collection systems and incompatibility of the meaning and temporal and spatial variation in the data makes it necessary to carefully analyze the quality, semantics and semiotics of the data (semantics stands for the interpretation, semiotics stands for the conclusion that can be drawn from the data). In this article we analyzed loop detector data, probe vehicle data and ANPR registration. We found that the detector failures leave many gaps in the available traffic data. The probe vehicles (taxis) give very useful and rather accurate information about taxi flows, the taxi OD matrix and link travel times. The extrapolation from taxi flows to total traffic flows is still tedious, because the proportion of taxis in the total traffic flow is not constant. For time periods of one hour or longer, we can assume that taxi flows on corridors in one direction are rather constant, but on other directions significant differences occur.

The data from Automated Number Plate Recognition (ANPR) cameras are also very useful and accurate. Due to the fact that these cameras are not present on all intersections makes it necessary to use the other information sources (loop detectors and probe vehicles) to fill in the gaps.

The use of ANPR, probe vehicles and loop detectors together gives the possibility to estimate the Origin Destination matrix. Still much care should be given to data cleaning. The fact that many drivers have some activity alongside the road and in side roads makes it necessary to select these vehicles and give these a special place in the estimation of the OD matrix. The comparison of the OD matrix between intersections with ANPR for all traffic and taxis, shows that the taxi OD matrix is well suited to be used as a-priori matrix for the estimation of the full OD matrix.

References

1. Hensher, D.A., Button, K.J.: Handbook of Transport Modelling. Pergamon, Oxford (2000)
2. Van Zuylen, H.J.: Some improvements in the estimation of an OD matrix from traffic counts. In: The 8th Internaional Symposium on Transportation and Traffic Theory, Toronto (1981)
3. Van Zuylen, H.J., Willumsen, L.G.: The most likely trip matrix estimated from traffic counts. Transp. Res. **14**(3), 281–293 (1980)
4. Li, J., van Zuylen, H.J., Wei, G.: Loop detector data error diagnosing and interpolating with probe vehicle data. Transp. Res. Rec. J. Transp. Res. Board (2014)
5. Li, J., van Zuylen, H.J., Liu, C., Lu, S.: Monitoring travel times in an urban network using video, GPS and Bluetooth. In: 14th Meeting of the Euro Working Group on Transportation, Poznan (2011). http://www.sciencedirect.com/science/article/pii/S1877042811014509
6. Jones, P. M., Koppelman, F.S., Orfeuil, J.P.: Activity analysis: state of the art and future directions. In: Developments in Dynamic and Activity-Based Approaches to Travel Analysis, pp. 34–55 (1990)
7. Ma, Y., Kuik, R., van Zuylen, H.J.: Day-to-day origin destination tuple estimation and prediction with hierarchical Bayesian networks using multiple data sources. Transp. Res. Rec. J. Transp. Res. Board **2342**(1), 51–61 (2013)

8. Maher, M.J.: Inferences on trip matrices from observations on link volumes: a Bayesian statistical approach. Transp. Res. Part B Methodol. **17**(6), 435–447 (1983)
9. Tebaldi, C., West, M.: Bayesian inference on network traffic using link count data. J. Am. Stat. Assoc. **93**(442), 557–573 (1998)
10. Li, B.: Bayesian inference for origin-destination matrices of transport networks 1210 using the EM algorithm. Technometrics **47**(4), 399–408 (2005)
11. Cascetta, E.: Estimation of trip matrices from traffic counts and survey data: a generalized least squares estimator. Transp. Res. **18B**, 289–299 (1984)
12. Toledo, T., Kolechkina, T.: Estimation of dynamic origin–destination matrices using linear assignment matrix approximations. IEEE Trans. Intell. Transp. Syst. **14**(2), 618–626 (2013)
13. Spiess, H.A.: Maximum likelihood model for estimating origin-destination matrices. Transp. Res. **21B**(5), 395–412 (1987)
14. Constantinos, A., Balakrishna, R., Koutsopoulos, H.N.: A synthesis of emerging data collection technologies and their impact on traffic management applications. Eur. Transp. Res. Rev. **3**(3), 139–148 (2011)
15. Bachmann, C., Abdulhai, B., Roorda, M.J., Moshiri, B.: A comparative assessment of multi-sensor data fusion techniques for freeway traffic speed estimation using microsimulation modeling. Transp. Res. Part C Emerg. Technol. **26**, 33–48 (2013)
16. Byon, Y., Shalaby, A., Abdulhai, B., Elshafiey, S.: Traffic data fusion using SCAAT Kalman filters. In: Transportation Research Board 89th Annual Meeting Compendium of Papers DVD, Washington, DC (2010)
17. Ma, Y., van Zuylen, H., Kuik, R.: Freight origin-destination estimation based on multiple data source. In: 2012 15th International IEEE Conference on Intelligent Transportation Systems (ITSC), pp. 1239–1244. IEEE (2012)
18. Ma, Y.: The use of advanced transportation monitoring data for official statistics. Thesis ERIM Series Research in Management, Rotterdam (2016)
19. Morimura, T., Kato, S.: Statistical origin-destination generation with multiple sources. In: 2012 21st International Conference on Pattern Recognition (ICPR), pp. 3443–3446. IEEE, November 2012
20. Iqbal, M.S., Choudhury, C.F., Wang, P., González, M.C.: Development of origin–destination matrices using mobile phone call data. Transp. Res. Part C Emerg. Technol. **40**, 63–74 (2014)
21. Bugeda, B., Montero Mercadé, J., Marqués, L., Carmona, C.: A Kalman-filter approach for dynamic OD estimation in corridors based on Bluetooth and Wi-Fi data collection. In: 12th World Conference on Transportation Research, WCTR (2010)
22. Kostic, B., Gentile, G.: Using traffic data of various types in the estimation of dynamic OD matrices. In: 2015 International Conference on Models and Technologies for Intelligent Transportation Systems (MT-ITS), pp. 66–73. IEEE, June 2015
23. Willumsen, L.G.: Estimating time-dependent trip matrices from traffic counts. In: Volmuller, J., Hamerslag, R. (eds.) Proceedings of the Ninth International Symposium on Transportation and Traffic Theory. VNU Science Press, Utrecht (1984)
24. Dunlap, M., et al.: Estimation of origin and destination information from Bluetooth and Wi-Fi sensing for transit. Transp. Res. Rec. J. Transp. Res. Board **2595**, 11–17 (2016)
25. Antoniou, C., Barceló, J., Breen, M., Bullejos, M., Casas, J., Cipriani, E., Montero, L.: Towards a generic benchmarking platform for origin–destination flows estimation/updating algorithms: design, demonstration and validation. Transp. Res. Part C Emerg. Technol. **66**, 79–98 (2016)
26. Astaria, V.: A continuous time link model for dynamic network loading based on travel time function. In: Lesort, J.-B. (ed.) Transportation and Traffic Flow Theory, pp. 79–102. Pergamon, Oxford (1996)

27. Hadavi, M., Shafahi, Y.: Vehicle identification sensor models for origin–destination estimation. Transp. Res. Part B Methodol. **89**, 82–106 (2016)
28. Kikuchi, S., Miljkovic, D., van Zuylen, H.J.: Examination of methods that adjust observed traffic volumes on a network. Transportation Research Board Meeting, TRB Paper Number 00-1378 (2000)
29. Ye, P., Wen, D.: Optimal traffic sensor location for origin-destination estimation using a compressed sensing framework. IEEE Trans. Intell. Transp. Syst. (2016)
30. Shannon, C.E.: A mathematical theory of communication. Bell Syst. Tech. J. **27**, 379–423, 623–656 (1948)

Workshop on Advances in Information Systems and Technologies for Emergency Preparedness, Risk Assessment and Mitigation (ASTER 2017)

Flood Hazard Assessment of the Fortore River Downstream the Occhito Dam, in Southern Italy

Ciro Apollonio[(⊠)] ⓘ, Gabriella Balacco ⓘ, Andrea Gioia ⓘ,
Vito Iacobellis ⓘ, and Alberto Ferruccio Piccinni ⓘ

Dipartimento di Ingegneria Civile, Ambientale, del Territorio,
Edile e di Chimica, Politecnico di Bari, 70125 Bari, Italy
{ciro.apollonio,gabriella.balacco,andrea.gioia,
vito.iacobellis,albertoferruccio.piccinni}@poliba.it

Abstract. In recent decades the effects generated by human activity that have led to the reduction of the natural water retention capacity of the soil and those generated by climatic factors rapidly changing, have exacerbated the hydrogeological instability and amplified the national territory vulnerability. Despite the flood risk in Italy is a topic of great interest, there are several aspects, especially in the management of flood events, that require further investigation; examples are floods caused by dam discharge. In this paper the distribution of hazard areas in downstream of Occhito Dam, has been carried out. Several hazard scenarios were considered related to only dam discharge or relating to the simultaneous presence of the dam contribution and contributions from river basins downstream of the Occhito dam. The results, validated using flooding photo, are presented and discussed, providing flow-rate in order to predict potential floods.

Keywords: Floodplain mapping · Dam · Hydraulic model · Hazard assessment · Flood risk

1 Introduction

European programs of scientific research as Horizon 2020 focused their attention about that and duties for governments and institutions like the Basin Authorities and Civil Protection agencies, consistently with the requirements dictated by the recent European legislation (Flood Directive 2007/60/EC) on flood risk protection and management of the territory.

In face, floods are one of the major threats to human life and assets. Despite the flood risk is still high in many areas of the world, several aspects regarding flood prediction and management, require further investigation.

Recently, noteworthy efforts have been spent for the prevention and mitigation of floods, focusing the attention on risk assessment and management. Numerous researchers spent time and energy on this topic with the aim to identify tools for flood risk evaluation and/or guidelines for risk management [1–6] or to propose alternative procedure with the aim to provide better results for the peak discharge [5] and, consequently, for floods mapping.

© Springer International Publishing AG 2017
O. Gervasi et al. (Eds.): ICCSA 2017, Part II, LNCS 10405, pp. 201–216, 2017.
DOI: 10.1007/978-3-319-62395-5_15

Recently many scientific works have investigated the hydrological basin behaviors using the remote sensing imagery extraction of land use parameters and the geographical information systems (GIS), particularly in ungauged basins [7–12].

Flooding risk is also expected to increase significantly in the 21st century, according to accredited climate change scenarios [13]. Changes in flood frequencies and intensities as a result of climate change have been identified as real issue that need to be factored into national and local planning [14].

In recent times, Chen et al. [15] verified that the increasing city area did not automatically generate more runoff. In particular Chen studied the consequence of urbanization on inundation behaviors at basin scale, in a national (China) case study characterized by an severe town development. These results demonstrate that runoff increase depends also by other features, for example climate conditions, soil type, basin slope, and groundwater [16–18].

Winsemius et al. [19] proposed recently, a framework for global flood risk assessment for river floods, which can be applied in current conditions, as well as in future conditions due to climate and socio-economic changes. The framework's goal was to establish flood hazard and impact estimates at a high enough resolution to allow for their combination into a risk estimate, which can be used for strategic global flood risk assessments.

At the same time, flood prediction and management are also important tasks for governments and basin authorities. Considering the serious effects of several flood events in Italy and consequent damages, a rigorous policy of enforcement of prevention and mitigation of flood risk is required. As already stated by the European Community Flood Directive (2007/60/CE) [20] and by Italian laws [21], the use of best practices and best available technologies is necessary in order to correctly identify areas exposed to flooding.

For these areas, technical flood defense is being complemented or replaced by measures for reducing effects of flooding, such as warning systems, emergency measures, spatial planning regulation, flood-proofing of buildings or insurance solution. In recent years, several hydrological models derived from studies on artificial intelligence have found increasing use at different scale [22].

Historically, the preferred flood management options are engineered structural solutions, such as dams and embankments [23]. A change in flood risk reduction policy can be observed, in the European Flood Directive (2007/60/CE) on the assessment and management of flood risks (European Commission 2007). It requires developing management plans for areas with significant flood risk, focusing on the reduction of the probability of flooding and of the potential consequences to human health, the environment and economic activity.

Flood risk management plans should be integrated in the long term with the river basin management plans of the Water Framework Directive, contributing to integrated water management on the scale of river catchments. It follows that the reduction of potential damage should have top priority: risk reduction through spatial planning must be strengthened. Although technical flood defense is essential for reducing extreme flooding, it is only effective up to certain return periods, so the residual risk must be accounted for and disclosed to increase flood awareness and precaution against the false sense of security.

Whilst the policies of flood risk management are still being assimilated into practice, the need of a review and an improvement of flood risk management is increasing in the last few years. In this context, the availability of accurate and efficient flood modeling tools is crucial to assist engineers and managers dealing with flood risk assessment, prevention and mitigation. Flood hazard mapping is a fundamental instrument for the protection of human lives, goods and activities.

Researchers have done many efforts to identify innovative and simplified procedures to provide a preliminary delineation of the flood-prone areas useful for the planning of further detailed studies like a numerical analysis. Manfreda et al. [24] investigated the role of different morphological descriptors, derived by a DEM derived by satellite images, in the identification of flood-prone area of an Italian river basin, adopting a linear binary classifier and ROC analysis at basin and local scale.

The analysis results have been compared with a flood map derived by a one-dimensional simulation for the basin scale and one derived by a two-dimensional model for the local and flat scale. Good results have been observed in particular for two index: the distance from the nearest stream (D) and the elevation of the nearest stream (H), both showed to be less sensitive to the change of resolution in the adopted DEM. While DEM-based methods have been recently adopted for preliminary floodplain characterization at basin scale [24–26], detailed inundation maps are generally based on advanced hydrologic and hydraulic modeling [27, 28]. In particular, the hydraulic analysis is often performed by means of a 1D model or a 2D model. Nevertheless, whenever the river stage overtops levees and the flood covers areas external to the river network a 2D simulation should be preferred in order to account for a detailed representation of the terrain surface.

In practice, 2D models represent the most sophisticated feasible approach to predict inundation, using topographical and land cover information and computational constraints and considering finite element and finite volume approaches for predicting shallow water hydraulics. Horrit et al. [29] compared inundation extents obtained by remotely sensed observations with flooding maps predicted using both a finite volume model (SFV) and a finite element model (TELEMAC-2D). Predicted inundation extents are compared with four airborne synthetic aperture radar images and these are used to calibrate the values of Manning's roughness for the channel and floodplain. The results showed good performances of both models, which decreases with the increase of the flow discharge; moreover, they highlighted a very interesting topic that requires future investigation, that is the different roughness parameter sensitivity; in particular optimum roughness also depends on the model, its resolution and flow rate.

On the other hand, two-dimensional hydraulic models require highly accurate digital terrain models [30]. In fact, flat areas are a critical issue for the characterization of drainage patterns using digital elevation models (DEM). Although notable attempts of representing fluvial channels, using triangulated irregular networks or implementing morphological indices, have been applied in hydrogeomorphic simulations, the majority of stream network identification approaches are based on the estimation of flow directions (FDs) using digital elevation models (DEMs). Several research projects have been developed in the last few years, aimed to investigate raster-based FD model performances, showing the importance, for hydrogeomorphic modelers, to consider the use of more detailed approaches that include the implementation of flow direction algorithms and introduce a physically-based DEM correction model, for investigating their influence on the topological properties of the channel network [31].

Nowadays, the availability of high-resolution images and satellite products allows a considerable improvement in model parameterization [32–34]. In particular, the acquisition of high-resolution topographic data, provides a considerable advance in the use of numerical models for flood mapping and inundation extent evaluation [29, 35, 36]. More in detail, airborne LiDAR (Light Detection and Ranging) systems [37] can provide elevation and/or bathymetry measurements at a horizontal sampling interval of the order 1 to 5 m, with vertical accuracies in the region of ±0,05 to ±0,15 m depending on the system configuration and flight profile [38, 39]. In Italy LiDAR campaigns, undertaken by the Environmental Agency, cover actually significant coastal strips, as well as a number of river floodplains.

Flood maps can be considered the most incisive form of presentation of that natural phenomenon, regardless of the categories in which they can be classified: flood hazard maps, flood vulnerability maps and flood risk maps [40]. Maps of the inundation area, generated using rainfall events with different return periods, are the most used to illustrate flood hazard [41], even if, recently, this approach has been criticized [36] by comparing the inundation maps obtained using either deterministic or probabilistic approaches.

Limits highlighted about the deterministic approaches are real and incontrovertible, especially considering that they renounce to take any account of the uncertainties in the modelling process [42]; however, in absence of efficient alternative, this kind of approach represents an useful tool for institutions or professionals in the context of goods safety. Probably the time is not yet mature to accept a new probabilistic point of view or maybe the end-users (like environmental agencies, Basin Authorities or engineers) are not yet able to understand it. This is partly due to the fact that the transfer of relative know-how from scientist to end-users is still difficult notwithstanding the relevant research activity developed in last decades [43].

Currently a strong policy of prevention and mitigation of flood risk is required and, to achieve that, it is necessary to adopt methodologies that, despite their limitations, can be easily applied and understood by end-users. This need becomes even more evident when particularly intense rainfall events, occurring on river basins characterized by the presence of hydraulic structures such as dams, can induce significant dangers to human lives and goods. With this aim, this paper analyzed several hazard scenarios for the Fortore river basin (Southern Italy) in order to identify the potential flooding area for different timescales events and to define best strategies for Occhito dam management. An hydraulic two-dimensional numerical model, FLO-2D [44], has been applied for the definition of Fortore flood propagation and, consequently, the analysis of the hazard scenarios downstream of the Occhito Dam was achieved exploiting the dam contribution together with that deriving from the catchment located downstream of the dam. Field observations and high-resolution aerial photographs of flooding have been used for the validation and calibration of the numerical model results, in absence of updated tail water rating curve and flow rate measures for the Fortore river.

2 The Case Study Description

The Occhito dam, among the largest in Europe in the category of earth-filled dams, is located on the stream network of the Fortore river. The interregional Fortore river basin crosses over the Southern Italy regions of Campania, Puglia and Molise, as shown in Fig. 1. The Occhito dam, located near Carlantino town, in Puglia, serves as a natural border between Puglia and Molise regions, for about 10 km. The dam is managed by the "Consorzio di Bonifica di Capitanata", and provides an irrigated land of about *143,000* ha located along the Fortore river valley and on a wide plain called "Tavoliere delle Puglie". The Puglia Region is mainly dominated by agriculture, that is a vital economic resource for the region, with more than 70% of the total area occupied by cropped land [45].

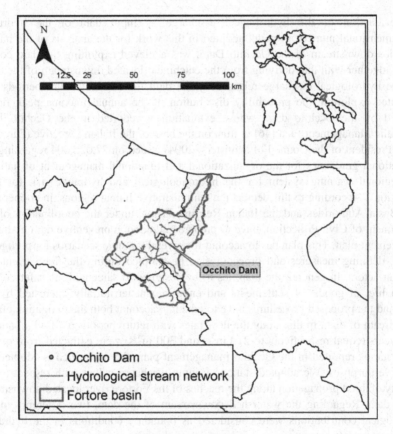

Fig. 1. Study area

A critical flood event occurred on the 4 and 5 March 2005 on the Fortore river basin and on its nested sub-catchment upstream of the Occhito dam, caused enormous damages and disasters in the locality called Ripalta located about 50 km downstream of the dam. In the night between the 4th and 5th of March a huge quantity of rain has fallen on the basin areas placed both upstream and downstream the dam. The reservoir water level, which had never reached the overflow level, provided the activation of the overflow spillways for the first time after their construction occurred in 1966. The total peak discharge at Ripalta reached about 630 m³/s considering also the runoff contribution arising from the basin downstream the dam. The total discharge caused the flooding of Fortore river that, also due to the reduced hydraulic conveyance, invaded areas located on the right and left sides of the river banks. It was declared the alert state by the Authorities, for the area around the dam and for communities downstream of the Occhito dam; the most important local transport structures were closed.

3 Evaluation of the Hydrological Discharges

The evaluation of the hydrological discharge as input data of the hydraulic two-dimensional numerical model adopted in this work for the analysis of the hazard scenarios downstream of the Occhito Dam, was achieved exploiting the dam contribution together with that deriving from the catchment located downstream of the dam.

The hydrological discharge released by the dam for different return periods was computed exploiting the probability distribution of the annual maxima peak floods released by the Occhito dam, whose evaluation is reported in the Occhito Dam Hydraulic Management Plan [46] written on the base of the Italian Directive (Directive of the President of the Council of Ministers, 2004) issued on 27.02.2004 regarding the "operational guidelines for the organizational and functional management of national and regional warning system for the hydrogeological and hydraulic risk for civil protection". According to this decree, the administrative Italian regions, in cooperation with Basin Authorities and the Italian Registry Dams, under the coordination of the Department of Civil Protection, have to prepare and adopt a preventive dam hydraulic management plan. This plan has to account for several possible scenarios for prefigured events, defining measures and procedures to be adopted during the flood events by different actors. In general, the plan has to be designed to safeguard the safety of the human life, of goods, of settlements and environment territorially interested by the event and the proposed procedures have to take into account both the mitigation effects downstream of dam. In this study the flow rates with return periods of 10 (T_{10}) and 30 (T_{30}) years (equal respectively to 210 m³/s and 300 m³/s) were extracted from one of the scenarios reported in the Occhito management plan, the one which is adopted for operative purposes. We adopted stationary input discharge values with respect to the velocity of flood propagation along the reaches of the Fortore river located downstream of the dam. Regarding the watershed downstream of the dam, two different type of hydrological contributions were considered as boundary conditions in the hydraulic numerical model:

- the first one from the tributary sub-basins which are hydrologically independent from the main river network;
- the second one from the surface areas belonging to the catchments (called residuals) located between two sequential river basins.

In the first case a triangular hydrograph was introduced in each cross sections (identified with numbers 9, 11, 13, 15 and 17, in Fig. 2) located on the tributary rivers upstream of the junctions; in particular the triangular flood hydrograph is characterized by a duration equal to three times the corrivation time of the river basin and peak flood, for a fixed return periods, extracted from the plan for hydrogeological risk of the interregional basin of the Fortore river (here after called PAI Fortore) written by the Basin Authority of Trigno, Biferno, Minori, Saccione and Fortore [47].

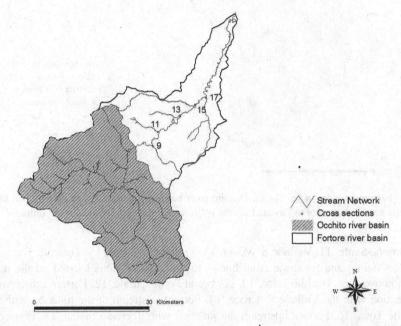

Fig. 2. Fortore river basin (in black), stream network (in blue), Occhito river basin (in green) and cross sections (red points) located on the tributaries. (Color figure online)

The flow contribution of each residual river basin was introduced as a triangular hydrograph in a section inside the pertinent reach. In this case the peak flood was computed exploiting the discharge per unit (contributing) area extracted from the PAI Fortore [47]. In particular seven residuals river basins were identified downstream of the dam, relatively to seven cross section located in the main river and reported in Fig. 3 as red points (n.8, n. 10, n. 12, n.14, n. 16, n.18, n.19, n.20); for each of these cross-sections the surface area of the residual river basin was evaluated as the difference between the surface area of the river basin contributing to the considered river section and the river basins located upstream.

In this work for each residual river basin the highest discharge per unit (contributing) area was considered between the different values (corresponding to different return periods) reported on the PAI Fortore [47].

The values of peak flood for different return periods and lag-times (t_r) are calculated both for the river basins contributing to the cross sections located on the tributary rivers

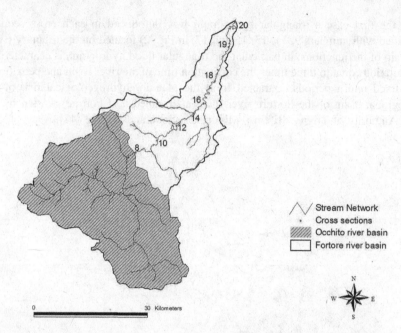

Fig. 3. Fortore river basin (in black), Occhito river basin (in green), stream network (in blue) and cross sections (red points) located on the main stream network. (Color figure online)

(9. Torrente Sente, 11. Vallone S. Maria, 13. Vallone S. Croce, 15. Torrente Tona, 17. Torrente Staina) and for those contributing to the cross sections located on the main stream network (8. Occhito Dam, 10. Fortore at Ponte Casale, 12. Fortore upstream of the junction with the Vallone S. Croce, 14. Fortore upstream of the junction with the Torrente Tona, 16. Fortore upstream the junction with Torrente Staina, 18. Fortore at Ponte Civitate, 19. Fortore at Ripalta and 20. Fortore at the outlet).

In particular, for the estimation of the peak floods (Q_T = 30 years) with return period T = 30 years, the values reported on the PAI Fortore were adopted in this study for each cross section; instead the peak floods with return period T = 10 years (Q_T = 10 years) were obtained by linear interpolation of Gumbel reduced variable obtained starting from the probabilities values of peak floods reported on the PAI Fortore. The computed peak discharges were adopted as boundary conditions in the hydraulic two dimensional model; in particular for a fixed return period (equal to 10 and 30 years) a constant value released from the dam was introduced in the hydraulic model, while the contribution of the river basin downstream of the dam was introduced as a triangular flood hydrographs characterized by a duration equal to three times the corrivation time (evaluated from the lag-time exploiting the relationship proposed by Ward and Elliot [48]) of the river basin and peak flood extracted from the PAI Fortore. For residual river basins, the evaluation of the lag-time was achieved exploiting the relationship between the lag-time and surface area for the tributary rivers which values, extracted from the PAI Fortore.

In this work five scenarios were investigated for the delineation of the flood hazard maps (as shown in Table 1). In the first three scenarios only a constant value released by the Dam was considered as boundary condition (equal respectively to 50, 100 and 300 m³/s); for the remaining two scenarios the boundary conditions were characterized by the overlapping of the discharge released from the dam with return periods of 10 and 30 years and the contribution of the river basins downstream of the dam for the same return periods. The five scenarios just defined were placed in three different computational domains, constructed in series, so that the output of the previous domain becomes the inputs of the next.

Table 1. Hazard scenarios considered

$\psi_d(\%)$	Discharge released from the dam [cm/s]	Contribution of the basin downstream of the dam
Scenario 1	50	–
Scenario 2	100	–
Scenario 3	300	–
Scenario 4	210	T10
Scenario 5	300	T30

The decision to divide the stream of the Fortore river in three computational domains is due to the stability criteria of the numerical model [43]. The three domains are shown in Fig. 4:

- Domain 1: From the Occhito Dam to the Torremaggiore Viaduct.
- Domain 2: From the Torremaggiore Viaduct to the Ripalta Bridge.
- Domain 3: From the Ripalta Bridge to the Fortore River mouth.

The flood modeling have been performed by a 2D commercial hydraulic model FLO-2D. FLO-2D is a volume conservation model that numerically routes a flood hydrograph while predicting the area of inundation and flood wave attenuation over a system of square grid elements. In this case, the DEM adopted to simulate flow on a floodplain topography was carried out by LiDAR data. The model considers river channel flow as one-dimensional and reproduces river overbank flow and floodplain flow as 2-dimensional. The flooding simulation is developed in two dimensions, through the integration of following the conservative equation and the motion equation [43].

The FLO-2D features used in this work also included, roughness shape, flow obstructions and hydraulic structures (e.g. bridges). The friction slope parameter is calculated by Manning's equation.

The numerical model was run with a uniform Manning's coefficient of 0.04 m$^{-1/3}$s, representative of whole floodplain. The computational domains were defined by a square grid with resolution of 15 m, which assures reduced long simulations times; while the time step was set to 0.5 h, to ensure the model stability.

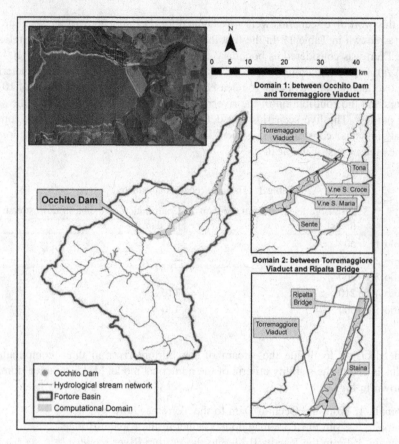

Fig. 4. Computational domains

4   Results

In this section, the results of the application of a two-dimensional model, in particular the maximum water level and flow velocity, were analyzed for each spatial domain investigated in relation to the different flow scenarios hypothesized. Maps of floodplain areas for the selected five scenarios were obtained by filtering the initial results, considering the inundated area only the surface characterized by a water level major than 0.05 m.

The spatial representation of the simulated water level for Domain 1 shows that downstream of the Occhito Dam the flood area is embanked in a section wide of about 300 m. In this area the most vulnerable zones are those with the presence of extensive floodplains used for agriculture and some minor roads (located near the Fortore Bridge).

In the reach downstream of River Tona (Fig. 4) confluence, flood areas greatly remain within the alluvial terraces also for flood characterized by return period of 30 years. In particular, the area at risk, for all scenarios considered, is that located upstream of Torremaggiore Viaduct (Fig. 4) where is located a large structure for aggregates production. All of hydraulic structures located in Domain 1 are never overtopped for all hypothesized scenarios.

Examining the results for Domain 2, slight criticality has been identified for some agricultural areas (scenarios 4 and 5), located upstream of Civitate Bridge, that is the only hydraulic structure that occurs in this domain.

Downstream of Civitate Bridge the valley is larger; on the right bank the river flows in adherence to the embankment, while on the left bank it is possible to recognize a cultivated plain with scattered houses. It can be observed a marked reduction in slope and the presence of dense vegetation in the channel. Civitate Bridge, however, never resulted overtopped for all hypothesized scenarios.

By analyzing the results of Domains 3 and 4, it is evident that these areas are the part characterized by the major flood hazard.

In particular, the transport infrastructure of considerable importance as railway Bologna-Lecce and S.S. 16 are subjected to floods and overtopped for scenarios 4 and 5, the same infrastructure was frequently exposed to flood events of the Fortore river in recent years.

The first critical area is on the way to Ripalta at the hydraulic left of the bridge, this road is already at hazard for scenario 3.

Downstream of Ripalta Bridge, the discharge is no longer contained in the alluvial terraces, involving the plain until arriving to the embankment of S.S 16, of A14 highway, the road to Colle Arena Bridge and Bologna-Lecce railway line.

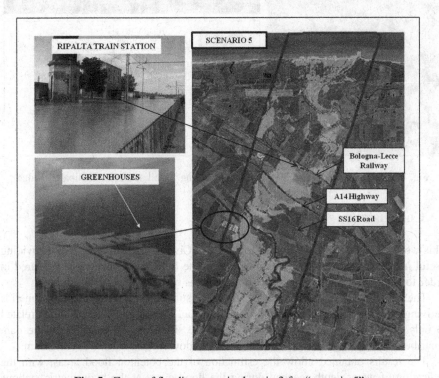

Fig. 5. Zoom of flooding areas in domain 3 for "scenario 5"

There is the involvement of some greenhouses in hydraulic left of the river upstream of the S.S. 16 as shown in Fig. 5.

Downstream of the railway, flood areas continue to expand both in the right and in left bank concerning secondary roads and scattered houses. These floodplains are already present for the scenario 1 and increase for the scenarios 4 and 5 where the S.S. 16 and the railway, are overlapped. In this domain several crossing can be found; they are not overlapped for all hypothesized scenarios. Hydrometric stations of Puglia Civil Protection are placed downstream of the Ripalta and Colle Arena bridges (Fig. 6). In the Fig. 6 the maximum values of water levels (H_{max}) obtained from the model for each investigated scenario are plotted for the section corresponding to the position of the hydrometric station of the Ripalta Bridge. The water levels were calculated as sum between the topographic elevations of the cell and flow depths calculated by the numerical model. The hydraulic model calibration has been done on the basis of flooding photographs (Fig. 5). In this way, the outflow stairs are obtained, without the Flow Measurements on the Fortore river. This system has finally provided a precious tool for Civil Protection in terms of the flooding forecast.

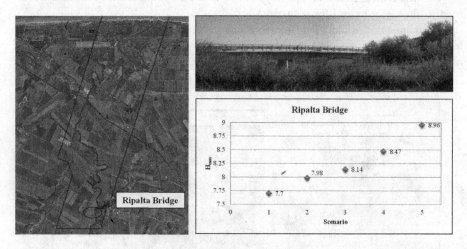

Fig. 6. Water level obtained by numerical modeling for all scenarios at Ripalta Bridge

5   Conclusions

This work represents a management tool for the Civil Protection as well as providing useful insights in view of the activities that in the future will have to be prepared in order to define safety interventions for these areas recently affected by flooding.

The distribution of hazard areas has been studied with reference to scenarios deriving from the application of the Mitigation Plan. Scenarios were considered related to only release from the Dam or relating to the simultaneous presence of the dam contribution and contributions from river basins downstream of the Occhito dam.

The results obtained allow the estimation of the potential flood areas, with the indication of water depth reached during the flood event at the hydrometric stations recently installed by the Civil Protection agency of Puglia Region. In the context of hazard and risk management this tool allows the real time monitoring of flooding, exploiting the data recorded by the gauged river station.

On the other hand, the numerical analysis conducted downstream of the Occhito dam, have highlighted some critical points in relation to the flow capacity in safety conditions.

It appears clear that preventing certain situations of particular risk, already highlighted in scenarios characterized by minor flow entities, would greatly enhance benefits of the real-time management of flood risk downstream the dam.

Acknowledgments. The authors are grateful to the Regione Puglia – Settore Protezione Civile for funding this research.

References

1. Mo, C.X., Feng, J., Sun, G.K., Zhang, C., Ma, R.Y., Wu, X.R.: Study on flood control risk management for reservoir earth dam. In: 2009 1st International Conference on Information Science and Engineering, ICISE, Nanjing, China, 26–28 December 2009. IEEE (2009)
2. Ahmad, F., Yahaya, A.S., Ali, M.M., Hussain, W.N.A.M.: Environmental risk assessment on hill site development in Penang, Malaysia: recommendations on management system. EJSR **40**(3), 318–340 (2010)
3. Koutroulis, A.G., Tsanis, I.K.: A method for estimating flash flood peak discharge in a poorly gauged basin: Case study for the 13–14 January 1994 flood, Giofiros basin, Crete, Greece. J. Hydrol. **385**, 150–164 (2010)
4. Baker, M.E.: Dodging a bullet: Lily Lake Dam and the 2013 Colorado flood. In: Association of State Dam Safety Officials Annual Conference 2014, Dam Safety 2014, San Diego, California – USA, pp. 512–522. Curran Associates, New York (2014)
5. Manfreda, S., Nardi, F., Samela, C., Grimaldi, S., Taramasso, A.C., Roth, G., Sole, A.: Investigation on the use of geomorphic approaches for the delineation of flood prone areas. J. Hydrol. **517**(863–876), 1694 (2014)
6. Sharif, H.O., Al-Juaidi, F.H., Al-Othman, A., Al-Dousary, I., Fadda, E., Jamal-Uddeen, S., Elhassan, A.: Flood hazards in an urbanizing watershed in Riyadh, Saudi Arabia. Geomat. Nat. Hazard Risk **7**(2), 702–720 (2016)
7. Novelli, A., Tarantino, E., Caradonna, G., Apollonio, C., Balacco, G., Piccinni, A.F.: Improving the ANN classification accuracy of landsat data through spectral indices and linear transformations (PCA and TCT) aimed at LU/LC monitoring of a River Basin. In: Gervasi, O., Murgante, B., Misra, S., Rocha, A.M.A.C., Torre, C.M., Taniar, D., Apduhan, B.O., Stankova, E., Wang, S. (eds.) 16th International Conference on Computational Science and Its Applications, Beijing, China, 4–7 July 2016 (2016)
8. Tarantino, E., Novelli, A., Laterza, M., Gioia, A.: Testing high spatial resolution WorldView-2 imagery for retrieving the leaf area index. In: Proceedings SPIE 9535, Proceedings of Third International Conference on Remote Sensing and Geoinformation of the Environment (RSCy2015), p. 95351N (2015)
9. Skakun, S., Kussul, N., Shelestov, A., Kussul, O.: Flood hazard and flood risk assessment using a time series of satellite images: a case study in Namibia. Risk Anal. **34**(8), 1521–1537 (2014)

10. Nardi, F., Annis, A., Biscarini, C.: On the impact of urbanization on flood hydrology of small ungauged basins: the case study of the Tiber river tributary network within the city of Rome. J. Flood Risk Manage. (2015). doi:10.1111/jfr3.12186

11. Apollonio, C., Balacco, G., Novelli, A., Tarantino, E., Piccinni, A.F.: Land use change impact on flooding areas: the case study of Cervaro Basin (Italy). Sustainability **8**(10), 996 (2016)

12. Tarantino, E., Novelli, A., Aquilino, M., Figorito, B., Fratino, U.: Comparing the MLC and JavaNNS approaches in classifying multitemporal LANDSAT satellite imagery over an ephemeral river area. Int. J. Agric. Environ. Inf. Syst. **6**, 83–102 (2015)

13. IPCC: Managing the risks of extreme events and disasters to advance climate change adaptation. A special report of working groups I and II of the Intergovernmental Panel on Climate Change, pp. 1–594. Cambridge University Press, Cambridge, UK (2012). https://www.ipcc.ch/pdf/special-reports/srex/SREX_Full_Report.pdf

14. Lawrence, J., Quade, D., Becker, J.: Integrating the effects of flood experience on risk perception with responses to changing climate risk. Nat. Hazards **74**, 1773–1794 (2014)

15. Chen, X., Tian, C., Meng, X., Xu, Q., Cui, G., Zhang, Q., Xiang, L.: Analyzing the effect of urbanization on flood characteristics at catchment levels. Proc. IAHS **370**, 33–38 (2015)

16. Giordano, R., D'Agostino, D., Apollonio, C., Scardigno, A., Pagano, A., Portoghese, I., Lamaddalena, N., Piccinni, A.F., Vurro, M.: Evaluating acceptability of groundwater protection measures under different agricultural policies. Agric. Water Manage. **147**, 54–66 (2015)

17. Portoghese, I., D'Agostino, D., Giordano, R., Scardigno, A., Apollonio, C., Vurro, M.: An integrated modelling tool to evaluate the acceptability of irrigation constraint measures for groundwater protection. Environ. Model Softw. **46**, 90–103 (2013)

18. Giordano, R., D'Agostino, D., Apollonio, C., Lamaddalena, N., Vurro, M.: Bayesian belief network to support conflict analysis for groundwater protection: the case of the Apulia region. J. Environ. Manage. **115**, 136–146 (2013)

19. Winsemius, H.C., Van Beek, L.P.H., Jongman, B., Ward, P.J., Bouwman, A.: A framework for global river flood risk assessments. Hydrol. Earth Syst. Sci. **17**, 1871–1892 (2013)

20. Flood Directive 2007/60/EC of the European Parliament and of the Council of 23 October 2007 on the assessment and management of flood risks. http://eur-lex.europa.eu/legal-content/EN/TXT/?uri=CELEX:32007L0060

21. Directive of the President of the Council of Ministers: Indirizzi operativi per la gestione organizzativa e funzionale del sistema di allertamento nazionale, statale e regionale per il rischio idrogeologico e idraulico ai fini di protezione civile. Official Journal of 11 Marzo 2004 n. 59, 1–17, (in Italian). http://www.protezionecivile.gov.it/jcms/it/view_prov.wp?contentId=LEG21144

22. Granata, F., Gargano, R., de Marinis, G.: Support vector regression for rainfall-runoff modeling in urban drainage: a comparison with the EPA's storm water management model. Water **8**(3), 69 (2016)

23. Zhang, Y., Zhao, Y., Wang, Q., Wang, J., Li, H., Zhai, J., Zhu, Y., Li, J.: Impact of land use on frequency of floods in Yongding River Basin, China. Water **8**, 401 (2016)

24. Manfreda, S., Samela, C., Gioia, A., Consoli, G.G., Iacobellis, V., Giuzio, L., Cantisani, A., Sole, A.: Flood-prone areas assessment using linear binary classifiers based on flood maps obtained from 1D and 2D hydraulic models. Nat. Hazards **79**, 735–754 (2015)

25. Nardi, F., Vivoni, E.R., Grimaldi, S.: Investigating a floodplain scaling relation using a hydrogeomorphic delineation method. Water Resour. Res. **42**, W09409 (2006)

26. Kim, J., Kuwahara, Y., Kumar, M.: A DEM-based evaluation of potential flood risk to enhance decision support system for safe evacuation. Nat. Hazards **59**(3), 1561–1572 (2011)

27. Grimaldi, S., Petroselli, A., Arcangeletti, E., Nardi, F.: Flood mapping in ungauged basins using fully continuous hydrologic–hydraulic modeling. J. Hydrol. **487**, 39–47 (2013)
28. Iacobellis, V., Castorani, A., Di Santo, A.R., Gioia, A.: Rationale for flood prediction in karst endorheic areas. J. Arid Environ. **112**(A), 98–108 (2015)
29. Horritt, M.S., Di Baldassarre, G., Bates, P.D., Brath, A.: Comparing the performance of a 2-D finite element and a 2-D finite volume model of floodplain inundation using airborne SAR imagery. Hydrol. Process. **21**, 2745–2759 (2007)
30. French, J.R.: Airborne LiDAR in support of geomorphological and hydraulic modelling. Earth Surf. Proc. Land. **28**, 321–335 (2003)
31. Nardi, F., Grimaldi, S., Santini, M., Petroselli, A., Ubertini, L.: Hydrogeomorphic properties of simulated drainage patterns using digital elevation models: the flat area issue. Hydrol. Sci. J. **53**, 1–18 (2008)
32. Frappart, F., Seyler, F., Martinez, J.M., León, J.G., Cazenave, A.: Floodplain water storage in the Negro River basin estimated from microwave remote sensing of inundation area and water levels. Remote Sens. Environ. **99**(4), 387–399 (2005)
33. Domeneghetti, A., Tarpanelli, A., Brocca, L., Barbetta, S., Moramarco, T., Castellarin, A., Brath, A.: The use of remote sensing-derived water surface data for hydraulic model calibration. Remote Sens. Environ. **149**, 130–141 (2014)
34. Iacobellis, V., Gioia, A., Milella, P., Satalino, G., Balenzano, A., Mattia, F.: Inter-comparison of hydrological model simulations with time series of SAR-derived soil moisture maps. Eur. J. Remote Sens. **46**, 739–757 (2013)
35. Cobby, D.M., Mason, D.C., Davenport, I.J.: Image processing of airborne scanning laser altimetry data for improved river flood modeling. ISPRS J. Photogrammetry Remote Sens. **56**(2), 121–138 (2001)
36. Di Baldassarre, G., Schumann, G., Bates, P.D., Freer, J.E., Beven, K.J.: Flood-plain mapping: a critical discussion of deterministic and probabilistic approaches. Hydrol. Sci. J. **55**(3), 364–376 (2010)
37. Measures, R.M.: Laser Remote Sensing: Fundamentals and Applications. Krieger Publishing, Melbourne (1991)
38. Lillycrop, W.J., Irisch, J.L., Parson, L.E.: SHOALS system: three years of operation with airborne lidar bathymetry - experiences, capability and technology advancements. Sea Technol. **38**, 17–25 (1997)
39. Gomes Pereira, L.M., Wicherson, R.J.: Suitability of laser data for deriving geographical information. A case study in the context of management of fluvial zones. ISPRS **54**, 105–114 (1999)
40. Merz, B., Hall, J., Disse, M., Schumann, A.: Fluvial flood risk management in a changing world. Nat. Hazards Earth Syst. Sci. **10**, 509–527 (2010)
41. Bates, P.D.: Remote sensing and flood inundation modelling. Hydrol. Process. **18**, 2593–2597 (2004)
42. Bates, P.D., Horritt, M.S., Aronica, G., Beven, K.: Bayesian updating of flood inundation likelihoods conditioned on flood extent data. Hydrol. Process. **18**, 3347–3370 (2004)
43. Montanari, A.: What do we mean by uncertainty? The need for a consistent wording about uncertainty assessment in hydrology. Hydrol. Process. **21**, 841–845 (2006)
44. O'Brien, J.: FLO-2D Users Manual. FLO-2D Inc., Nutrioso (2001)
45. Giordano, R., Milella, P., Portoghese, I., Vurro, M., Apollonio, C., D'Agostino, D., Lamaddalena, N., Scardigno, A., Piccinni, A.: An innovative monitoring system for sustainable management of groundwater resources: objectives, stakeholder acceptability and implementation strategy. In: Proceedings of Environmental Energy and Structural Monitoring Systems (EESMS), IEEE Workshop, Taranto, Italy, pp. 32–37 (2010)

46. Piccinni, A.F., Iacobellis, V., Laricchia, V., Apollonio, C., Gigante, V., Gioia, A.: Piano di laminazione della diga di Occhito sul fiume Fortore – Final report (2009)

47. Autorità di Bacino dei Fiumi Trigno, Biferno e Minori, Saccione e Fortore, Piano Stralcio per l'Assetto Idrogeologico del Bacino Interregionale del fiume Fortore - Relazione generale (2005). http://adbpcn.regione.molise.it/autorita_old/pdf/rg/fortore.pdf

48. Ward, A.D., Elliot, W.J.: Environmental Hydrology. CRC Press and LLC, Boca Raton and Florida (1995)

Hierarchical Spatial Distribution of Seismic Risk of Italian RC Buildings Stock

Marco Vona, Monica Mastroberti^(✉), and Federico Amato

University of Basilicata, Potenza, Italy
{marco.vona, monica.mastroberti,
federico.amato}@unibas.it

Abstract. Due to both the huge amount of strategic and residential buildings that require interventions and limited economic availability, the definition of mitigation strategies based on retrofit priority intervention are a fundamental step. Interventions priority ranking based on concept of seismic risk should be defined. They should be able to address the economic resources by the areas more seismic risky to less ones. In this work, an effective hierarchical spatial distribution of seismic risk of Italian RC buildings stock has been defined based on a spatial georeferencing of Italian seismic hazard, vulnerability, building types exposure, and the employment of Analytical Hierarchical Method. Moreover, a novel and reliable definition of seismic risk index has been employed in building type characterization.

Keywords: Seismic risk analysis · Existing RC buildings · Hierarchical spatial distribution · GIS

1 Introduction

In seismic risk mitigation strategies the prioritization of intervention plays a key role [1]. This issue has become a fundamental problem especially with regard to the limited economic resources and their allocation on national, regional or sub-regional territories. This is a typical political and administrative problem and it is a fundamental topic for decision makers. Moreover, as reported in other work [2], the seismic losses is mainly due to physical damage on buildings. This latter influence also the recovery ability, e.g. the resilience of communities, negatively conditioned from high vulnerability of the residential buildings.

For public or strategic buildings several mitigation strategies have been defined and applied in the last years, due to their important social role, in order to improve their expected performance during seismic events and to reduce the emergency time. For example, in Italy several laws (OPCM 3274/2003) have been defined a wide and ambitious plan for the strategic buildings. The seismic vulnerability has been evaluated for these buildings and based on results seismic risk mitigation strategies have been defined in order to allocate the economic resources through a simply procedures.

However, for existing residential buildings similar procedures have been defined only in the last years, but the high number of existing buildings requires a significant economic recourse, the citizen's cooperation, and mainly accurate mitigation strategies.

© Springer International Publishing AG 2017
O. Gervasi et al. (Eds.): ICCSA 2017, Part II, LNCS 10405, pp. 217–229, 2017.
DOI: 10.1007/978-3-319-62395-5_16

In recent Italian law (D.L. 28/2/2017, [3]), seismic risk mitigation of residential buildings, the economic funding has been allocated based on an Seismic Risk Index. It is based on the simple capacity - demand ratio (α), both expressed in terms of seismic intensity. A new proposal for seismic risk level quantification has been defined. In addition, the EAL (Expected Annual Losses) has been proposed as global evaluation parameters of seismic quality or resilience of existing building in seismic region.

In order to improvement the resources allocation, due to both the huge amount of residential buildings that require interventions and the limited economic resources, it would be necessary as first step the definition of priority of interventions (for residential buildings) on a wide territorial scale. In particular, based on seismic risk concepts, the priority interventions strategies should be defined. Seismic risk could be quantified as the probability of losses due to an earthquake, where these losses may be human, social, or economic. Thus, effective seismic risk mitigation strategies should be based on propriety lists, which are directed by more risky areas to less ones.

Priority lists based only on the concept of seismic risk would address the resources only to the more hazardous zones, without any consideration about the building vulnerability and exposure. In this way, minor importance is given to less hazardous zones but characterized by more vulnerable buildings and an important exposure.

On the other hand, priority lists based just on concept of vulnerability would neglect the effects of seismic hazard and exposition. In this way, for same vulnerability the difference in damage levels resulting to intense and continue aftershocks typical of more hazardous zones are neglected. In the same manner the exposure effects.

Rigorously, in definition of effective seismic risk mitigation strategies all the factors influencing seismic risk, such as seismic hazard, geotechnical soil characteristics, vulnerability of structures, repair costs of damaged buildings, and exposed population must be considered and combined [2]. In order to obtain the optimal result, these variables should be gathered and analyzed with GIS (Geographical Information System). The GIS constitutes a highly useful working tool that facilitates data interoperability and joint analysis at territorial scale. Therefore, it is a power-full tool to develop seismic risk study that requires the integration of a large set of heterogeneous information (data of different nature, size and format), and their spatial representation to an efficient monitoring and interpretation of all stages of risk analysis. Moreover, in definition of seismic risk priority list the Multi Criteria Decision Analysis methods (MCDA) have a key role [1, 4].

In Italy, after recent earthquakes, the existing buildings have been showed high more vulnerable. In particular, significant part of Reinforced Concrete (RC) buildings has been designed only to gravity load, without any anti-seismic details and precautions, and with old seismic codes [5]. Moreover, actually, due to the low maintenance state they are close to the end of their effective life.

The amount of exposure (human, economic system etc.) may be directly connected with the amount of anthropic elements (buildings). These latter could be damaged or destroyed during an earthquake, producing serious losses (human, economic, financial, social), as result of hazard level of vulnerability. The Italian buildings exposition, especially refereed to RC buildings, is strongly linked to social-economic-historical reasons. Moreover, buildings exposition is influenced by the past seismic history.

In this work, a map of seismic risk of Italian residential RC-existing buildings has been defined. This map could be a valid support tool for decision makers, politicians operating at national scale in redistribution of economic resources. The definition of seismic risk mitigation strategies reducing the future economic losses is the purpose of map. It has been defined on the basis on a rational hierarchical process, addressed to maximize the effect of vulnerability and hazard, checking the influence of buildings amount. A preliminary study of Italian hazard, vulnerability and existing RC-buildings exposure has been performed; finally, information have been joined by the GIS platform Q-GIS. Local information for Italian hazard provided by [6] have been exploited. Then, the PGA is the reference seismic intensity parameter, while the Poissonian model is the reference hazard model. The RC existing buildings exposition has been characterized based on information provided by the last population and buildings census of ISTAT 2011 [7]. Coherently with the aim of work, the RC building exposition has been evaluated at typological scale. The amount of two RC existing building types has been defined (RC-Pre Code, RC-Old Code). Their vulnerability level has been quantified in terms of seismic risk index for life safety performance level, as quantitatively defined by NTC08 Italian Code [8]. A new definition of Seismic Risk Index has been applied, as proposed by the same authors in previous work [9].

The GIS implementation has allowed the synthetic characterization of each Italian municipality in terms of seismic hazard, amount of two building types considered and their level of seismic risk.

Finally, an Analytical Hierarchical Process based on a pair wise comparisons among of different seismic risk criteria (seismic hazard, vulnerability, building type exposure) has been performed. Then, a hierarchical spatial distribution of seismic risk of existing Italian RC buildings has been defined.

2 Elements for Italian Existing RC Buildings Seismic Risk Map

In this work, Q-GIS has been used to define a geographical information system, while the municipality is the geographical unit. The data used in geo-referenced process of seismic risk at municipal level have been provided by different sources, and in different format. In following, further information about the original source of hazard data, characteristic and amount of exposed building types have been reported. Moreover, information about the transformation process applied in reporting the information at municipal level have been provided. Finally, a novel approach to characterize the vulnerability level of identified types is briefly reported and applied.

2.1 Italian Seismic Hazard

The hazard information for each municipality have been defined starting from the hazard grid (0.05 grade) of median PGA on an horizontal and free bedrock characterized by a probability of 10% in 50 (reference hazard level for Life Safety Performance Level) years published in [6]. The hazard grid consisting in 104567 points characterized in terms of latitude and longitude.

Based on a geographic correspondence between the administrative borders of municipality and the hazard grid, a mean and variance value of reference PGA for life safety PL (characterized by a probability of occurrence of 10% in 50 years) has been associated at each municipality. Moreover, according with Italian hazard map [6] that splitted the Italian territory in four seismic regions characterized by a range of PGA; each municipality has been characterized as belonging by a specific seismic region.

In Fig. 1, the Italian Hazard Map based on the mean value of PGA (probability of 10% in 50 years) assigned at each municipality is reported.

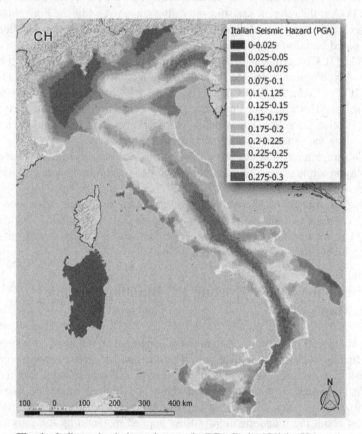

Fig. 1. Italian seismic hazard maps (in PGA Prob. 10% in 50 years).

2.2   Building Types

The amount of existing RC buildings have been provided by Italian population and buildings census ISTAT 2011 [7]. The census variable provided at census section level (different census sections for each locality of municipality) are 134.76 related to population, 7 to homes, 27 to buildings, 9 to family and 15 to foreigners. In the study only the information related to buildings have been considered.

For each census section the amount of masonry, reinforced concrete and others types of buildings have been provided. In a quick way, a simple and rational typological desegregation process (reinforced concrete, masonry) based on construction age has been performed for this first application. Moreover, the information provided for each census section of municipality have been aggregated; consequently, municipal information have been obtained.

The amount of RC existing building designed according with Pre seismic codes and Old codes [5] have been defined at municipal level. In follow, the identified building types will be respectively called as RC-Pre Code and RC-Old Code. The number of storey is not considered. This is coherently with previous studies [5, 10] that have been highlighted the key role of the construction age in structural performance; this choices is coherent with the main goal of this work which is a first application of a new definition for Seismic Risk Index (SRI) for large scale assessment [9]. Based on the amount of RC Pre-Code and RC Old-Code buildings for each Italian municipality the exposure map reported in Fig. 2 have been defined.

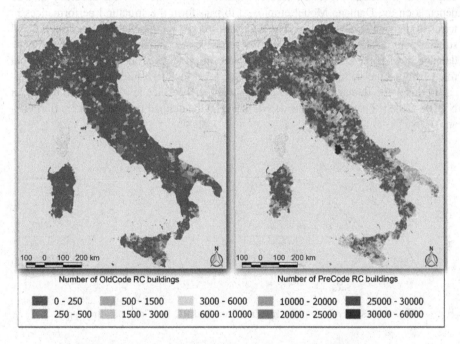

Fig. 2. Italian exposure for RC Old-Code and RC Pre-Code maps.

The exposure maps highlight as the number of RC Old-Code buildings is less than that of RC Pre-Code. Moreover, the most important Italian cities, which have had a major development in recent years, such as the cities of Po Valley, of Veneto, of Central Apennine, of Puglia, Sicily are characterized by the high number of buildings. Finally, the Rome city is characterized by the higher amount of RC buildings.

In previous studies of authors [5, 9, 10], the structural behavior of the most widespread in Italy and others European areas RC-MRF building types have been studied. More information about building types characterisation, analysis methods, structural modelling, and seismic actions have been provided in [5, 11]. In particular, the building types analysed have been mainly characterized in term of designed code (Old Code and Pre Code building types), number of storey (low-mid-high rise), effectiveness and distribution of infill panel (BareFrame, Infillframe, PilotisFrame).

The numerical results (of NLDAs) of each investigated building types have been employed as reference in characterization of seismic behaviour of two identified building types (RC-Pre Code, RC-Old Code).

In this way, in order to define the Fragility Curves (FCs) of types, the numerical results have been aggregated in only two FCs set. The FCs allow a probabilistic characterization of vulnerability in terms of probability of exceedance of specific Performance Level varying the seismic intensity. It is to be highlighted that the FCs effectiveness strongly depends by the accuracy of building model and analysis method employed in structural performance evaluation [10]. Moreover, the accuracy of FCs depends on the Damage Model employed to transform the structural performances in terms of damage level achieved. The relationship between qualitative description, local structural limit state, and global engineered response parameter threshold for each damage level is the core of a Damage Model. In previous work [10] the fundamental role of limit state definition for each damage level has been highlighted.

In this work, damage levels coherently with performance levels proposed by Italian NTC08 Code [8] have been considered. In Fig. 3 the defined Fragility Curves (in PGA) are shown.

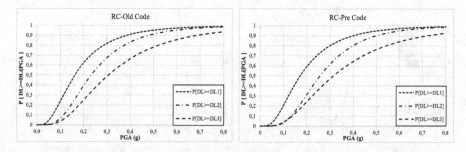

Fig. 3. Fragility Curves for RC Old Code and RC Pre Code building types.

2.3 Seismic Risk Index

Bases on defined FCs and Italian Hazard characterization, carrying on a probabilistic way, a novel probabilistic definition of Seismic Risk Index (SRIs) has been applied [9]:

$$SRI_{PLi} = 1 - \frac{\sum_{I_{PLi,i}}^{I_{PLi,f}} P[PL \geq PLi|I] \Delta I}{\sum_{I_{PLi,i}}^{I_{PLi,f}} \Delta I}$$

The new SRI definition quantifies the level of seismic risk in terms of probability that the building capacity no fulfill the required demand by each limit state. It is based on Fragility Curves and an accurate intensity characterization of Hazard Level required by each PL. $\sum_{I_{PLi,i}}^{I_{PLi,f}} P[PL \geq PLi|I]\Delta I$ is the area identified between the fragility curve for a specific performance level and the reference seismic range $\sum_{I_{PLi,i}}^{I_{PLi,f}} \Delta I$. This area is normalized to the total area identified by the intensity range; and measures the seismic risk level regard to a specific PL (as probability of exceedance of PL).

Based on SRI equation each considered building type has been characterized by different seismic risk indexes for life safety PL, according with PGA range that characterize each seismic region identified by Italian hazard map [6].

In Table 1, the SRI indexes obtained for two building types and seismic regions are reported. The reported values highlight as two considered buildings types are characterized by similar seismic risk level. Really, these results are conditioned from simply aggregation considered for FCs definition, but is coherent with the goal of the work. Based on the SRI values, and the mean value of reference PGA of each Italian municipality, the considered types for each municipality have been further characterized in term of seismic risk level for life safety PL.

Table 1. RC Old Code and RC Pre Code building types: seismic risk index for life safety PL.

	SRI RC - Old Code	SRI RC - Pre Code
S.R. 4	1,00	1,00
S.R. 3 (0,05–0,1 g)	0,99	0,99
S.R. 3 (0,1–0,15 g)	0,92	0,92
S.R. 2 (0,15–0,2 g)	0,81	0,81
S.R. 2 (0,2–0,25 g)	0,68	0,69
S.R. 1 (0,25–0,275 g)	0,59	0,60
S.R. 1 (0,275–0,3 g)	0,53	0,55

In Fig. 4 an Italian map of seismic risk index for two considered building types (RC Old-Code and RC Pre-Code) have been reported.

The maps highlight a comparable safety level of two building types for the reason previously explained. Moreover, for the same building type the seismic risk level strongly depends by hazard level (the seismic risk index contour plots are similar to seismic hazard one).

3   A Novel Italian RC Buildings Seismic Risk GIS Map

Once defined the basic items of seismic risk for each Italian municipality (hazard, vulnerability, and exposure information), a first Italian RC building seismic risk spatial distribution has been defined. In first analysis, the seismic risk at municipality and building type level has been quantified as a simple convolution between the three basic items (mean PGA, SRI|, and amount of types). The obtained spatial distribution of seismic risk of Italian RC building is reported in Fig. 5.

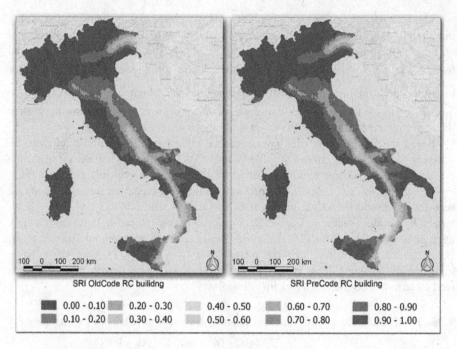

Fig. 4. Italian seismic risk index (RC Old-Code and RC Pre-Code) maps.

Figure 5, the map seems provide seismic risk quantification as good balance between seismic hazard and exposure. For example the internal zones of central Italy (Molise, Abruzzo), even if are characterized by a high hazard are less risky than the coast zones as result of low exposure.

A second map of spatial distribution of seismic risk of Italian RC building has been defined based on the Analytical Hierarchical Process (AHP). The reason for this choice is the different format (mean PGA, seismic risk index, number of buildings) of the seismic risk elements.

In fact, they assume fairly different values that could ill conditioned the importance relation between different criteria and the final result. AHP is one of the Multi Criteria Decision Analysis methods (MCDA) and allows a prioritization of a range of decision-making alternatives. Moreover, it permits the comparison of qualitative and quantitative evaluation criteria, which are normally not directly comparable, by combining multidimensional scales of measurements into a single priority scale [12, 13]. AHP is based on a set of pairwise comparisons between the criteria by assigning them a relative importance score and ending with the assignment of a percentage weight. In this study, the scores to be used at each comparison are chosen in a range of 1 to 9, where the lowest value indicates that the criteria compared are equally important, while the highest value indicates that the first criterion is absolutely more important than the second one.

Let A_i be the single alternative and a_{ij} the value resulting from the comparison between criterion i and j, and consider a number of criteria equal to n. The result of all

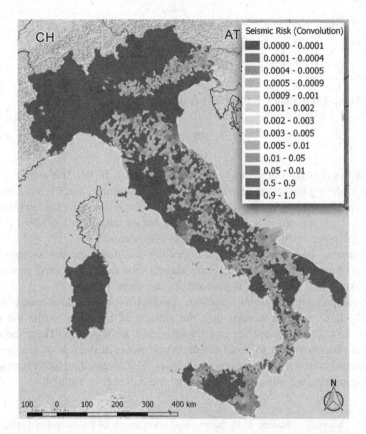

Fig. 5. Italian RC building seismic risk spatial distribution (first approach).

the $n(n-1)/2$ comparisons will generate the $A_{n \times n}$ matrix, which will be used to evaluate the percentage weight vector of each single criterion. Once the matrix A has been obtained, the percentage weight vector to be assigned to each stimulus is calculated by determining the maximum eigenvalue λ and its relative v_λ eigenvector. By normalizing the eigenvector v_λ so that the sum of its elements is equal to 1, the vector of percentage weights (or priority) for the A_i stimuli is obtained.

Once the priority vector has been determined, it is important to understand if the comparison matrix is consistent or not. Therefore, a Consistency Index (CI) is defined as the value obtained from:

$$CI = \frac{\lambda - n}{n - 1}$$

where λ is the maximum eigenvalue of the matrix A and n represents the size of the matrix itself. Moreover, the Random Consistency Index (RI) is defined according to Table 2.

Table 2. Random Consistency Index for different sizes of the matrix A.

RI	0	0	0.58	0.9	1.12	1.24	1.32	1.41	1.45	1.49
n	1	2	3	4	5	6	7	8	9	10

Finally, the Consistency Ratio (CR) is defined as:

$$CR = \frac{CI}{RI}$$

The matrix A is considered as consistent if $CR < 0.1$. In this study AHP has been applied considering the seismic hazard, the seismic risk index, and the amount of each building types as criterion to be weighted. Due to their different format and size (PGA values (0–0.3 g), seismic risk index (0–1), number of building (0–60000)) a rational relative importance relation between them has been defined.

The pairwise comparison was developed by weighing the assessments of the importance of each criterion and the influence that the different criteria format could have in seismic risk evaluation as proposed by the three authors.

In the comparison both seismic risk index and seismic region have been considered respectively fairly more important than the amount of buildings, while the seismic region and seismic risk index are equally important among them. Thus, the defined hierarchical process is able to avoid the underestimation of the role of seismic region and seismic risk criterion than the exposure values. The obtained A matrix is reported in Table 3. Moreover, the others AHP indicators are reported in Table 4.

Table 3. Matrix A obtained from the pairwise comparison of the considered criterion.

	Number of buildings	Seismic risk index	Seismic region
Number of buildings	1.00	0.25	0.128
Seismic risk index	4.00	1.00	1.00
Seismic region	5.50	1.00	1

Table 4. The AHP indicators obtained from the pairwise matrix.

Number of buildings	0.096
Seismic risk index	0.428
Seismic region	0.475
λ	3.01
CI	0.005
CR	0.009

Based on the obtained results, the hierarchical spatial distribution of seismic risk for considered types has been defined (Fig. 6). The seismic risk map reported in Fig. 6 seems provide a seismic risk quantification strongly influenced by the buildings exposure. For example, the most part of highly hazardous regions are characterized by

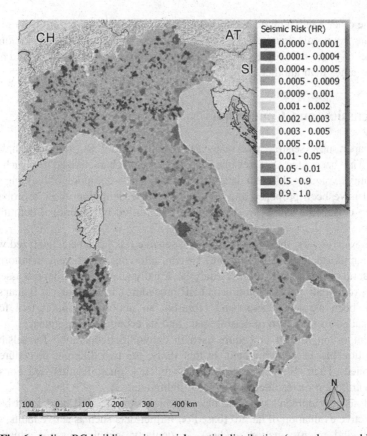

Fig. 6. Italian RC building seismic risk spatial distribution (second approach)

less seismic risk due to the lower values of exposure. On the other hand, a greater seismic risk characterize the most densely RC building populated regions, even if little hazardous.

The City of Rome is a clearly example; it is characterize by a medium-low seismic hazard but the highest number of existing RC Old-Code and RC Pre-Code buildings. In general, according to seismic risk map of Fig. 6 the Italian cities that had have a high expansion in decades '70–'90, are characterized by the most risk, in relation to which priority should intervene.

The comparison between the obtained spatial seismic risk distributions (Figs. 5 and 6) highlights as the procedure for seismic risk quantification strongly influence the final result, providing discrepancies in subsequent decisional phases.

The choice of a procedure rather than another strongly depends by decision maker's profile, strategies objective, and final purpose.

A map of seismic risk influenced by the spatial buildings exposure would be more useful in reducing of future economic losses, giving priority to areas with a high number of low-medium vulnerable buildings that globally are responsible of a huge amount of seismic losses.

On the other hand, a map linked to vulnerability and seismic hazard is more able to reduce the economic losses linked to strong but limited damage conditions. The evaluation of economic consequence of seismic risk have a key role in the choice of approach rather than another.

4　Discussion and Conclusion

In this paper, new seismic risk maps for Italian existing RC buildings have been defined. The basic items of seismic risk have been characterized for each Italian municipality exploding different data. In particular, a new definition of building seismic risk level have been proposed and applied. The new definition allow a probabilistic evaluation of seismic safety level, completely neglected by traditional definition and procedure.

In a powerful way, the variable of Italian seismic risk have been analyzed within a GIS (geographical information system) system, allowing an efficient monitoring and interpretation of all stages of risk analysis. Finally, based on different assembling procedure (conventional convolution and HR procedure) hazard maps of Italian seismic risk have been defined. These latter represent an important support tool for each subjects called in reduction of seismic risk, and its economic consequence.

The important role of procedure used in convolution operation for seismic risk quantification has been highlighted. In this study, the two different procedures have been applied. Their different impact in seismic risk quantification and consequent spatial distribution of resources have been highlighted.

In future developments of work, new seismic risk maps will be defined based on more effective evaluation of hazard (based on further data such as active faults, seismic catalogue, and micro-zonation studies). Particularly, peculiar hazard situation not emphasized by Poisson approach, but important in a correct collocation of resources will be considered. Moreover, more accurate building information will be used, avoiding too conservative types aggregation. Finally, a spatial distribution directly connected with the economic effects of building seismic risk will be defined.

References

1. Vona, M., Anelli, A., Mastroberti, M., Murgante, B., Santa-Cruz, S.: Prioritization strategies to reduce the seismic risk of the public and strategic buildings. Disaster Adv. 10(4), 21–34 (2017)
2. Vona, M., Harabaglia, P., Murgante, B.: Thinking about resilient cities studying Italian earthquakes. Urban Des. Plan. 169(4) (2016). http://dx.doi.org/10.1680/udap.14.00007
3. D.M. 28 febbraio 2017, n. 58 recante: Sisma Bonus - Linee guida per la classificazione del rischio sismico delle costruzioni nonché le modalità per l'attestazione, da parte di professionisti abilitati, dell'efficacia degli interventi effettuati. Ministero delle infrastrutture e dei trasporti (2017). (in Italian)

4. Rivas-Medina, A., Gaspar Escibano, J.M., Benito, B., Bernabè, M.A.: The role in GIS in urban seismic risk studies: application to the city of Almerìa (southern Span). Nat. Hazard Earth Syst. Sci. **13**, 2717–2725 (2013)
5. Vona, M.: Fragility curves of existing RC buildings based on specific structural performance levels. Open J. Civil Eng. **4**, 120–134 (2014)
6. Meletti, C., Montaldo, V.: Stime di pericolosità sismica per diverse probabilità di superamento in 50 anni: valori di ag. Progetto DPC-INGV S1, Deliverable D2 (2007). http://esse1.mi.ingv.it/d2.html
7. Descrizione dei dati geografici e delle variabili censuarie delle Basi territoriali per i censimenti: anni 1991, 2001, 2011 Versione definitiva 25/02/2016 (2016)
8. NTC 2008: Norme Tecniche per le Costruzioni. Decreto del Ministero delle infrastrutture, Supplemento Ordinario n.30 alla Gazzetta Ufficiale della Repubblica italiana, 29, 4/02/2008, Italy (2008). (in Italian)
9. Mastroberti, M., Vona, M.: New seismic risk index for existing buildings. In: 6th ECCOMAS Thematic Conference on Computational Methods in Structural Dynamics and Earthquake Engineering, COMDYN 2017, Rhodes Island, Greece, 15–17 June 2017
10. Vona, M., Mastroberti M.: A critical review of fragility curves for existing RC buildings. In: VII European Congress on Computational Methods in Applied Sciences and Engineering, ECCOMAS Congress, Crete Island, Greece, 5–10 June 2016
11. Masi, A., Vona, M.: Vulnerability assessment of gravity-load designed RC buildings, evaluation of seismic capacity through non linear dynamic analyses. Eng. Struct. **45**, 257–269 (2012)
12. Saaty, T.L.: The Analytic Hierarchy Process. McGraw Hill, New York (1980)
13. Figueira, J.R., Greco, S., Ehrgott, M.: Multiple Criteria Decision Analysis: State of the Art Surveys. International Series in Operations Research and Management Science, vol. 233. Springer, New York (2016)

Resilience Modification and Dynamic Risk Assessment in Hybrid Systems: Study Cases in Underground Settlements of Murgia Edge (Apulia, Southern Italy)

Roberta Pellicani[1(✉)], Ilenia Argentiero[1], Alessandro Parisi[2],
Maria Dolores Fidelibus[2], and Giuseppe Spilotro[1]

[1] Department of European and Mediterranean Cultures, DiCEM,
University of Basilicata, Via Lazazzera, 75100 Matera, Italy
{roberta.pellicani,ilenia.argentiero,
giuseppe.spilotro}@unibas.it
[2] Department of Civil, Environmental, Land, Building Engineering and
Chemistry, Polytechnic University of Bari, Via Orabona, 70100 Bari, Italy
{alessandro.parisi,mariadolores.fidelibus}@poliba.it

Abstract. The resilience of natural system, not affected by anthropic modifications, can be altered by many natural drivers (e.g. geological conditions, climate, etc.) and their spatial modifications. Over time, human activities have modified many natural systems generating "hybrid systems" (both human and natural), in which natural and anthropic drivers changed their vulnerability, in order to decrease or increase their resilience. Potential emerging signals of the resilience variation are difficult to assess because of wrong risk perception and lack of communication. In this context of soft crisis, it would be appropriate a dynamic risk assessment of hybrid systems in order to avoid disaster when hazardous phenomena occur, but it is a quite complex issue. The aim is to define the relationship between the hybrid system resilience, referring to study cases located in Apulia region, and some emerging signals and their records over time. Furthermore, the aim is to understand how human and natural drivers were involved in the shift.

Keywords: Hybrid system · Soft crisis · Resilience · Risk · Urban settlements · Instability · Emerging signals

1 Introduction

The ultimate goal of risk studies is generally an accurate assessment of the level of threat from hazarduous phenomena, i.e. an objective, reproducible, justifiable, and meaningful measure of risk [1]. The process of establishing such a measure of risk is referred to as risk estimation. The estimated level of risk can then be evaluated (risk evaluation) in the light of the consequences accrued from being exposed to that risk and as a result, decisions can be made on whether that level of risk is intolerable, tolerable, or acceptable [2]. An objective measure of risk can also be employed in terms of

© Springer International Publishing AG 2017
O. Gervasi et al. (Eds.): ICCSA 2017, Part II, LNCS 10405, pp. 230–245, 2017.
DOI: 10.1007/978-3-319-62395-5_17

cost/benefit (or cost/risk reduction) analysis of proposed risk treatment measures. The full range of procedures and tasks that ultimately lead to the implementation of rational policies and appropriate measures for risk reduction are collectively referred to as risk management [3]. Figure 1 summarizes the components that constitute risk management and their hierarchical relationships. The hazard and risk identification stage essentially identifies those factors that should be further investigated and taken into consideration in risk estimation. The process of risk estimation integrates the behaviour of the hazard (hazard analysis) with the elements at risk and their vulnerability (consequence analysis) in order to allow risk calculation [4]. The risk/benefit ratio and the absolute level of risk strongly influence not only the acceptability of risk but also the nature of the response. High levels of risk may warrant a legislative or sophisticated engineering response while low levels of risk may be accepted or treated by common sense and education. The characteristics of hazard factors and elements at risk can also dictate the type of treatment measures to be adopted. The range of options available for reducing the risk can be subdivided as follows:

- Hazard modification: usually engineering solutions aimed at modifying the characteristics of hazard factors and reducing the phenomenon occurrence.
- Behavior modification: reducing the consequences and the impact on the elements at risk, by options such as warning systems, reduction of vulnerability, development planning, education, regulations and economic incentives.
- Loss sharing: including systems for insurance, disaster relief, development aid and compensations.

Resilience, as risk component, indicates a dynamical characteristic of a generic system, which estimates how the system varies under the modifications of natural or anthropic drivers (Fig. 2). According to the dynamical systems theory, stable states and repelling points can be identified in a complex system. Repelling points, that are also known as tipping points or catastrophic bifurcations, are located between stable states, which are also called attractors or basin of attractions: the system can have alternative stable states associated to several configurations [5, 6].

If natural or anthropic drivers alter the system configuration over time, the system resilience can lessen, and the considered system can tip from a basin of attraction to another through a repelling point, causing a critical transition (Fig. 3) [6].

This work aims at describing three case studies located in Apulia region (Southern Italy), where natural systems have been modified by natural and anthropic factors over time. These modifications highlight a risk variation in function of natural and human drivers and their dynamic interactions with the impacted system. Moreover, emerging signals of risk variation and methodology for susceptibility assessment are described in order to a preliminary definition toward a dynamic risk assessment.

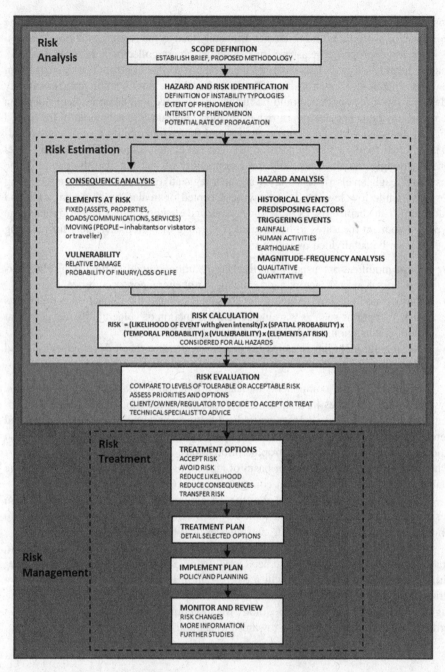

Fig. 1. Flow chart showing all the stages involved in landslide risk management, based on [3], modified.

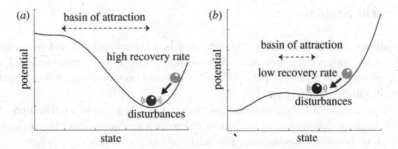

Fig. 2. Far from the tipping point (a), the basin of attraction is steep and the rate of recovery from perturbations is relatively high. When the system is closer to the tipping point (b), the basin of attraction shallows and the rate of recovery from small perturbations is lower. Modified by [10].

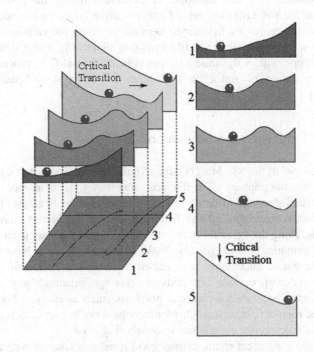

Fig. 3. Both human and natural drivers can vary the resilience of a system. On the right, the five sections represent qualitative configurations of the system related to several resilience values. Modifications can reduce the resilience of the system, as shown from the 1st to the 5th section, in which it is shown a critical transition thought a bifurcation point. Modified by [11].

2 Hybrid Systems

The resilience of a generic natural system, that is a system not affected by anthropic modifications, varies over time according to natural drivers (e.g. environmental, ecological, geological, climatic, etc.). If human-induced modifications affect a natural system, it will be defined hybrid system.

If the resilience of hybrid systems lessen, because of anthropic modifications, they can shift from an equilibrium to another in an easier way than without human-induced variations of boundary conditions. The term soft crisis, which is a *"situation with several elements malfunctioning (i.e., potentially leading to a crisis) and "waiting for" another element that leads, that tips the situation in an active crisis"* [7, p. 89], defines the resilience reduction of hybrid systems before the tipping point. Human-induced modifications of natural systems have interested areas in Altamura and Ginosa, as it will be later described, where anthropic modifications reduce the resilience of the hybrid systems: the soft crisis has turned into an active crisis, causing critical transitions and instability processes. In order to estimate the resilience variation of a generic system over time, it is possible to identify emerging signals [8]. These signals can be a better way to cope with a dynamic risk assessment of instability processes than to estimate the variation of resilience boundaries of the system because of their uncertainly [9].

3 Apulian Underground Settlements

Along the hillslopes of the SW Murgia edge (Apulia region, South Italy), the presence of water and the outcropping of soft calcareous stones, such as calcarenites, has favoured the establishment and development of rupestrian civilizations. These settlements, which often represent cultural, historical and artistic heritage, are located in natural systems, along the "gravine" (canyons due to karst deep karstic circulation and subsequent undermining) or old coastline flanks and date back to the Neolithic in some cases [12]. The native inhabitants have started a slowly excavation of the soft calcareous stones outcropped since thousands of years ago, creating many underground spaces. These have been used for several purposes, such as caverns, house-caverns, rainwater tanks, rock cave churches, silos for the wheat storage, aqueducts and caves to extract block of calcarenites as building materials (Fig. 4).

Despite to their excellent characteristic, good resistance coupled with an excellent workability [13–15], the structural weakening of the calcarenite leads to the possibility of brittle fracture of relevant parts of naturally or artificially carved mass, often in contexts where the actual geometry and the perception of risk cannot be properly evaluated. The soft crisis of these fragile contexts, where natural systems excavated overtime interact with build-up areas constructed by using the same materials, can lead to disastrous evolutions of the instability processes.

In Apulia, the main rupestrian settlements are located in Matera, Gravina, Massafra, Ginosa, Laterza e Grottaglie (Fig. 5), while the main underground quarries for mining calcarenites are sited in Altamura, Gravina, Cutrofiano, Gallipoli, Montescaglioso, Canosa e Andria (Fig. 5).

Fig. 4. Example of hybrid system: rupestrian settlements and overlying buildings along the "gravina" natural system of Gravina in Puglia.

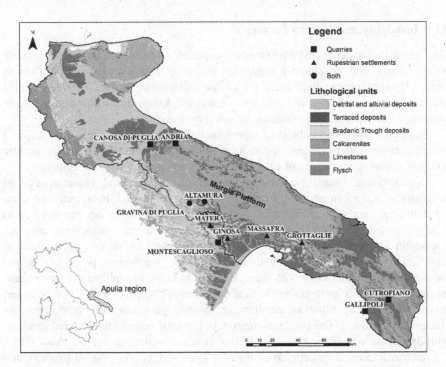

Fig. 5. Geolithological map of Apulia with highlighting of calcarenite outcrops and main rupestrian civilitazion sites.

Nowadays these underground or surface cavities are often abandoned and damaged, as characterized by collapse of the vaults and degradation of the calcarenite material. In fact this material is affected by typical pathologies, regarding both the excavated and the constructed. These pathologies may create instability and risk conditions to people frequenting rupestrian buildings or to structures and infrastructures above the underground cavities.

The damage processes affecting calcarenites, associated to a progressive degradation of their mechanical properties, are mainly due to weathering processes involving the material upon the saturation. In particular, the rupestrian habitats/spaces, excavated within this kind of rock, do not offer fruibility conditions, in terms of human health and hygienic standard, due to the high humidity conditions deriving from the air saturation, fed by the capillary saturation of the cavity calcarenite walls.

The will is to define the relationship between the hybrid system resilience, referring to some study cases located in Apulia region, and some emerging signals and their records over time. Furthermore, the aim is to understand how human and natural drivers were involved in the shift.

This analysis is supported by failure events, occurred in some urban settlements, such as Altamura, Ginosa and Gravina, and which allowed codifying susceptibility provisional models.

3.1   Instability and Hazard Factors

The presence of underground environments, especially if located at little depth beneath built-up areas, may constitute a significant threat for structure integrity and human safety. Upward propagating cavities and the consequent ground subsidence may severely damage buildings and may even cause life losses by swallowing people or causing the collapse of the structures occupied by them.

The conservation and stability of rupestrian underground settlements is endangered by different environmental and human hazards such as weathering agents, syngenetic and post-depositional history of calcarenites and uncontrolled urban activities.

Environmental hazards essentially depend on the material characteristics of underground environments. Calcarenites, as already mentioned, have excellent characteristics but their good resistance is influenced by a series of factors that can lead to weakening and instability phenomena. The exposure to the atmosphere, and the interaction with infiltrating rainfall can be at the origin of various geomechanical instabilities [16]. The most important factor of weathering processes is the chemical-physical interaction with water, due to infiltration or capillarity or atmosphere transport, that induce a chemo-mechanical degradation of the rock mass. In fact, water, in its movement by filtration, capillary or hygroscopic absorption, may cause the transport or dilution of fine particles (matrix), or the weakening of the material structure with desquamation effects, aided by climatic conditions (thermal cycles, wind) [15].

Another hazard factor depends on the rock mass and its geographical location, that contains the elements of its syngenetic and post-depositional history, such as longitudinal variations of granulometric composition, sedimentary discontinuities, levels or organogenic beds, post-sedimentary discontinuities, internal (karst) and external erosion. For example, in an environment with tectonic and slope decompression joints the frequency of fractures is influenced by its geographical location, being in fact controlled by climatic exposure: northern exposure leads to a greater intensity of jointing

[13]. The presence of discontinuities or joints influences the mechanical behaviour of the calcarenites, indeed they represent preferential pathways of water flow, that triggers weathering phenomena as discussed above.

Another frequent type of evolutive process is deformation under constant load or tension release (surface degradation or alteration), when the stress sedimentary field has been altered: this is a true creep. Unfortunately, creep phenomena associated with the concentration of stresses prelude progressive failure, coming from fracture propagation along cementation broken bonds.

The spatial recurrence of these phenomena can be rhythmic, arithmetic or random, thus introducing into the calcarenite mass discontinuity elements not always evident to the fresh cut of rock, which are very visible after exposure to the atmosphere for many years and most likely completely absent in the size of the laboratory specimens [17]. In addition to environmental hazard factors, human activities have contributed to situations of weakening in underground environments. The same construction technique of these environments, often developed on multiple levels, has resulted in an alternation of full and empty gaps in rock masses, that also causes significant concentration of strains. This architecture is the characterizing element of rupestrian settlements, but coexistence of rupestrian with the built is a widespread situation. In fact, urban agglomerations have grafted above, near and within most of the main rupestrian settlements (those of Gravina in Puglia, Matera Sassi, Ginosa, Palagianello and Massafra). The penetration and the perfect osmosis among the different types of artifacts have made the rupestrian-built complex grow in symbiosis [18].

Therefore, there has been an expansion of both existing and constructed without any consideration of the existing cavities, due also to the lack of specific technical regulations and adequate control of competent authorities. This has in some cases led to a decrease in the thickness of the calcarenite walls and consequently to the loss of lateral confining pressure (unloading) and to significant concentrations of strains, due also to the strains transmitted by overlying elevation building.

Another negative factor of this rupestrian-built complex is the lack of adequate drainage systems of rainwater with the consequent problems of absorption and infiltration with the above-mentioned problems. All these situations represent hazard factors that modify the resilience of the system and can led to instability phenomena, such as collapses and sinkholes.

3.2 Risk Factors

The *risk* associated to instability phenomena depends on hazard factors, and their spatial and temporal probability of occurrence, on vulnerability of the threatened system, assumed as the level of damage or loss caused by the instability event, and, finally, on the exposure, that is the value (economic, social, cultural, artistic, etc.) of the assets potentially at risk.

As already mentioned, *hazard* factors for underground rupestrian settlements are associated both to natural phenomena – natural degradation of the rock due to weathering, presence of weakness zone inside rocky mass due to syngenetic sedimentary gaps, etc. – and human activities – intense urbanization without considering the presence of underground cavities, lack of superficial water regulation, etc. The spatial and temporal combination of natural and antropogenic hazard factors generates often the most severe risk conditions.

Vulnerability is a fundamental component in the evaluation of risk associated to instability phenomena. It may be defined as the degree of potential damage or loss (expressed on a scale from 0 - no loss to 1 - total loss) produced by an instability phenomenon of a given intensity to an element exposed to risk [19]. The underground rupestrian and hybrid enviroments are affected by specific vulnerabilities, in terms of physical, functional, and technological obsolescence, as well as in terms of historic and architectural values, whose conservation and enhancement might conflict with the performance improvement [20]. The physical vulnerability represents the degree of loss of an element due to external stresses, deriving from hazard scenario of a given intensity. The functional vulnerability represents the tendency of an element to lose the original working capability due to external pressure. Hybrid systems, such as underground settlements below built-up areas, can be characterized by systemic vulnerability. It indicates the loss of functionality of the system due to a hazard scenario. It comprises physical and functional vulnerability of each single element at risk, but also of their relative interconnections. The vulnerability of a complex system cannot be assessed as sum of singular vulnerabilities of each element at risk (underground settlements, buildings, roads, etc.). The damage to a system must be evaluated in terms of loss of system functionality and organization.

Finally, *exposure* is representative of the value or amount of elements at risk. Therefore, it refers to: (i) socio-economic, cultural and artistic value of rupestrian heritage, (ii) people visiting the rupestian churches or living in hypogea used for housing or cellars, etc., (iii) buildings, above the underground environments, and their inhabitants. Each of these elements has its own characteristics, which may be spatial (the location in relation to the hazard) or temporal (such as persons, which will differ in time at a certain location).

3.3   Resilience and Emerging Signals

The term *resilience* can take various meanings depending on fields (e.g. psychological, ecological, social and technical) in which it is employed [21]. In disaster risk reduction field, the resilience is defined as *"the ability of a system, community or society exposed to hazards to resist, absorb, accommodate to and recover from the effects of a hazard in a timely and efficient manner, including through the preservation and restoration of its essential basic structures and functions"* [7, p. 24]. In ecology, it identifies a property of a system, that is *"[...] a measure of the persistence of systems and of their ability to absorb change and disturbance [...]"* [5, p. 14]. In the risk formulation, even with the relative caution, it is plausible considering resilience as the inverse of vulnerability [22].

As represented in the theoretical resilience state model (Fig. 3) and taking its cue from ecological systems theory [3, 5], the resilience represents an indicator in order to identify the evolution of hybrid systems over time. Therefore, it would be necessary identifying both human-induced modifications of a natural system and emerging signals of resilience variation in the hybrid system consequent to anthropic modifications of the natural one. As first step towards the definition of thresholds, they can permit to understand if the hybrid system is near to the tipping point because they represent the information about the variation of the attractors over time.

Referring to human settlements along the "gravine", the emerging signals of resilience variation can be waterfall in the undergrounds, the presence of cracking in calcarenites and their evolution over time. In order to cope with the resilience variation, the main goal would be improving local inhabitants and policy-makers perception and communication of potential emerging signals, in order to collect data related to human-modifications over time and to define a qualitative approach of dynamic risk system assessment.

4 Methodologies and Study Cases

The assessment of the short-term and long-term stability of underground quarries, in terms of spatial probability of future instability phenomena (susceptibility) can be carried out by applying statistic previsional models. Deterministic procedures generally require the expensive and time-consuming acquisition of a great deal of information on the rock properties and the geometry of the cavities, and generally simplify the complexity of a non-static phenomenon.

The first study case regards the urban area of Altamura, which together with Canosa, Cutrofiano, Gravina and Ginosa represent the towns in Apulia region with the greater concentrations of man-made underground cavities. Here, extensive underground mining was carried out in the northeastern sector of the city, aimed at extracting calcarenite blocks, thus producing large and irregular cavity systems. The underground mining zone covers an area of approximately 6,490 m^2 (Fig. 6) and the depth of the cavity roofs ranges from 5 m to 25 m. These mine galleries have been progressively abandoned and knowledge on their spatial distribution has vanished through time. Currently, due to recent urban expansion of Altamura, the largely cavities underlie a densely developed built-up area. Over the last few years, several sinkholes occurred, determining the need of assessing the hazard associated with mine galleries. The instability of these voids is adversely affected by both the natural degradation of the rock mass - development of unloading cracks, natural creep leading to strength loss, weathering processes - and various human activities - construction of deep foundations, absence of ventilation for closure of cavities, water infiltration.

The susceptibility to instability processes within the mine galleries, expressing the relative spatial probability of having instability processes within the cavities, which may eventually result in the development of sinkholes at the surface, was assessed and mapped by means of two statistical multivariate methods: discriminant analysis and logistic regression [23]. These methods are based on the assumption that the factors which controlled previous instabilities in the quarries are the same as those that will govern instability processes in the future. These approaches calculate the percentage or probability of having instability through a statistical relationship between the known instability phenomena (cavity sections with evidence of instability) and a number of environmental and anthropogenic predisposing factors weighted by correlation coefficients. The predisposing factors

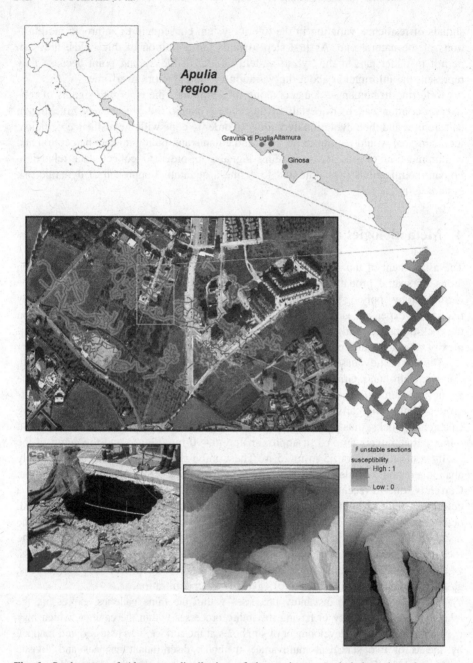

Fig. 6. Study case of Altamura: distribution of the cavity network below the urban area, instability phenomena inside cavities and on the ground surface, and susceptibility map.

considered in these analyses, because having a relevant influence on the instability process, are the following: (a) geometric parameters, i.e. maximum and minimum height, width (in vertical sections) and extension (in plan-view) of the cavity, (b) lithostratigraphic parameters, i.e. thickness of lithological units above the cavity roof (calcarenite, clay and vegetated soil), (c) structural parameters, (d) loading parameter and (e) presence of buildings above the cavity. The susceptibility maps (Fig. 6), resulted from of the two statistical models, can contribute to identify the areas on the ground surface with the greatest potential for mining subsidence.

The practical interest of the presented approach is that it offers the possibility of producing spatially distributed susceptibility models for cavity instability in a more rapid and inexpensive way than deterministic models. The methodology should be considered as a general first level analysis for near surface underground cavities aimed at identifying critical sectors that can be the focus of more detailed analyses, especially those associated with critical sites related to the presence of buildings, roads and frequent presence of human activity.

The second study case is Ginosa, whose historic center is a good example of rupestrian-built complex (Fig. 7a), where rupestrian cavities have been completely integrated into the most recent buildings. Unfortunately, this symbiosis has developed in years where specific technical regulations were absent. In fact, there are houses with structural elements obtained directly from the calcarenite mass, sometimes of reduced thickness, some are the extension of cavities or the cavity functions as cellar, others are founded on boulders of old collapses or on multiple levels of cavities with reduced thickness of the vaults. All this is added to the lack of adequate drainage systems of rainwater often causing flooding of cavities.

Fig. 7. View of rupestrian/build complex of Ginosa (a) and photos of the collapse of January 2014 (b, c).

All these anthropic modifications represent hazard factors that has reduced the resilience of this hybrid systems. In fact, in January 2014 the soft crisis that afflicted this fragile context led to a disastrous evolution of the instability processes, causing the collapse of a section road and several housing units with their cavities, fortunately without victims (Fig. 7b, c). Only during reconstruction of the section road have been discovered multiple levels of cavities (Fig. 8). Many were the emerging signals; in fact measures were taken to avoid the loss of human lives, but a better assessment of hazard factors useful for evaluating the structural vulnerability would have allowed to safe such an important cultural heritage.

Fig. 8. Cavities close to the collapse: each colour corresponds to a different level (a) and section of reconstruction project (b).

The third study case regards the whole historic center of Gravina in Puglia. Here, the most of the last century constructions have been built on calcarenites, which is the main outcropped Plio-Pleistocene carbonate Formation on the Apulia Foreland in this area. Since Neolithic period, local inhabitants have excavated outcropped calcarenite and limestone in order to create caverns along the "gravine" [12, 24]. From Roman period to the middle of 1950s, the modification of the natural system continued [24, 25].

Nowadays, the local speleological federation has identified around sixty underground cavities [26], whereas a local association has identified more than one-hundred underground spaces (such as rock churches, cellars, cistern and silos), most of them located in the historic center under houses, churches, streets and squares.

Often, there is not consciousness about the presence of underground spaces in the historic center and in its neighborhood. Moreover, some of these undergrounds spaces do not respect the perimeter of upper spaces: they can have a spatial underground extension, which underlie the space occupied by constructions, squares and streets.

These human-induced modifications of the natural system has generated a hybrid system with a modified resilience. In some cases, the interaction between drainage system, aqueducts and underground caves [27], related to the modification of the upper construction with a low awareness of spatial extension of underground, has led to a resilience reduction. Some emerging signals are the track of waterfalls during rainy days and the presence of cracking in calcarenites. The perception of these emerging signals requires the awareness of the existence of underground voids and the communication of these signals to managers in order to cope with the resilience reduction in a dynamical way, avoiding critical transition towards alternative stable state of hybrid system.

5 Conclusions

The "gravine" along the Murgia edge represented natural systems before rupestrian settlements and the urbanistic evolution of settlement. The main elements that constituted these natural systems were outcrops of soft calcareous stones, water, native vegetation and fauna. The resilience related to the variation of the system configuration, which originates instability processes, was associated only to the spatial variation of natural drivers over time. Rupestrian settlements along the "gravine" modified these natural systems creating hybrid systems, in which the resilience is related to both natural factors and the human-induced modifications over time.

The assessment of the risk, related to the exposure of both assets and human lives, is aimed at safeguarding, safe using and retraining the cultural heritage of underground rupestrian civilizations. Moreover, the dynamic risk assessment can be helpful in understanding emerging signals, indicating a variation of system resilience, and in prioritizing current and future actions for preserving the underground and rupestrian assets.

A further development of this study could be the definition of a dynamic risk assessment procedure, able to evaluate and compare several risks and their changes on rupestrian and underground assets due to different combinations of hazard factors and exposed elements. The goal could be to build a framework able to incorporate also multiple social, institutional and cultural dimensions for enhancing resilience capacities of rupestrian heritage in risk management.

References

1. Glade, T., Anderson, M., Crozier, M.J.: Landslide Hazard and Risk. Wiley, Chichester (2005)
2. Fell, R.: Landslide risk assessment and acceptable risk. Can. Geotech. J. **31**, 261–272 (1994)
3. AGS (Australian Geomechanics Society): Landslide risk management concepts and guidelines. Aust. Geomech. **35**, 49–92 (2000)
4. Dai, F.C., Lee, C.F., Ngai, Y.Y.: Landslide risk assessment and management: an overview. Eng. Geol. **64**, 65–87 (2002)
5. Holling, C.S.: Resilience and stability of ecological systems. Ann. Rev. Ecol. Syst. **4**, 1–23 (1973)

6. Scheffer, M.: Critical Transitions in Nature and Society. Princeton University Press, Princeton (2009)

7. UNISDR (United Nations International Strategy for Disaster Reduction): Terminology on Disaster Risk Reduction, Geneva, Switzerland (2009)

8. Siegel, W., Schraagen, J.M.: Developing resilience signals for the Dutch railway system. In: 5th REA Symposium on Managing Trade Offs, Soesterberg, Netherlands, pp. 191–196, 24–27 June 2013

9. Cook, R., Rasmussen, J.: "Going solid": a model of system dynamics and consequences for patient safety. Qual. Saf. Health Care **14**(2), 130–134 (2005). doi:10.1136/qshc.2003. 009530

10. Scheffer, M., Bascompte, J., Brock, W.A., Brovkin, V., Carpenter, S.R., Dakos, V., Held, H., van Nes, E.H., Rietkerk, M., Sugihara, G.: Early-warning signals for critical transitions. Nature **461**, 53–59 (2009). doi:10.1038/nature08227

11. Scheffer, M., Carpenter, S., Foley, J.A., Folke, C., Walker, B.: Catastrophic shifts in ecosystems. Nature **413**, 591–596 (2001)

12. Vinson, S.P.: Neolithic pottery of Inland Apulia: field work and speculation. Am. J. Archaeol. **82**(4), 449–459 (1978)

13. Spilotro, G., Fidelibus, M.D., Fidelibus, C., Zinco, M.R.: Lithological and geotechnical features of the calcarenites in the west of the Murgian platform. In: Anagnostopoulos, A., et al. (eds.) Proceedings of the International Symposium on Hard Soils - Soft Rocks, Athens, Balkema, Rotterdam, pp. 293–300, September 1993

14. Spilotro, G., Qeraxhiu, L., Pellicani, R., Argentiero, I.: Caratteristiche tecniche delle rocce calcarenitiche e loro variabilità in relazione all'ambiente di esposizione, pp. 81–84. Laboratorio di pratiche della conoscenza nei Sassi di Matera, Ediz. Archivia (2015)

15. Spilotro, G., Fidelibus, M.D., Pellicani, R., Qeraxhiu, L., Argentiero, I., Pergola, G.: Il patrimonio architettonico di Matera e i matetiali naturali da costruzione: nel tufo e col tufo: Caratterizzazione tecnica delle calcareniti e variazioni per condizioni ambientali. Geologia Territorio Ambiente **25**, 10–24 (2016)

16. Ciantia, M.O., Castellanza, R., Crosta, G.B., Hueckel, T.: Effects of mineral suspension and dissolution on strength and compressibility of soft carbonate rocks. Eng. Geol. **184**, 1–18 (2015)

17. Cotecchia, V., Calò, G., Spilotro, G.: Caratterizzazione geolitologica e tecnica delle calcareniti pugliesi. 3° Conv. Naz. Attività estrattiva dei minerali di 2a categoria, Bari, 17–19 January1983

18. Grassi, D., Grimaldi, S., Simeone, V.: Localizzazione e problemi di stabilità dei siti rupestri dell'area pugliese. Giornale di Geologia Applicata **4**, 65–72 (2006)

19. Wong, H.N., Ho, K.K.S., Chan, Y.C.: Assessment of consequence of landslides. In: Cruden, R., Fell, R. (eds.) Landslide Risk Assessment, pp. 111–149. Balkema, Rotterdam (1997)

20. Fatiguso, F., De Fino, M., Cantatore, E., Caponio, V.: Resilience of historic built environments: inherent qualities and potential strategies. In: International High-Performance Built Environments Conference – A Sustainable Built Environment Conference. Procedia Engineering (2017)

21. Alexander, D.E.: Resilience and disaster risk reduction: an etymological journey. Nat. Hazards Earth Syst. Sci. **13**, 2707–2716 (2013). doi:10.5194/nhess-13-2707-2013

22. Pescaroli, G., Alexander, D.: Critical infrastructure, panarchies and the vulnerability paths of cascading disasters. Natural Hazards, 1–18 (2016). doi:10.1007/s11069-016-2186-3

23. Pellicani, R., Spilotro, G., Gutiérrez, F.: Susceptibility mapping of instability related to shallow mining cavities in a built-up environment. Eng. Geol. **217**, 81–88 (2017). doi:10.1016/j.enggeo.2016.12.011
24. Nardone, D.: Notizie storiche sulla città di Gravina (455–1860). Ed. Gravina, L. Attolini (1922)
25. Capuzzi, L.: Gravina: un paese del Sud. Quaderno di storia urbanistica n.1 (1981)
26. Barnaba, F., Caggiano, T., Castorani, A., Delle Rose, M., Di Santo, A.R., Dragone, V., Fiore, A., Limoni, P.P., Parise, M., Santaloia, F.: Sprofondamenti connessi a cavità antropiche nella regione Puglia. 2° International Workshop "I Sinkholes: gli sprofondamenti catastrofici nell'ambiente naturale ed in quello antropizzato", session 5, pp. 635–672 (2009)
27. Fidelibus, M.D., Pellicani, R., Argentiero, I., Spilotro, G.: The geoheritage of the water intake of Triglio ancient aqueduct (Apulia Region, Southern Italy): a lesson of advanced technology insensitive to climate changes from an ancient geosite. Geoheritage. doi:10.1007/s12371-017-0238-z

Preventive Approach to Reduce Risk Caused by Failure of a Rainwater Drainage System: The Case Study of Corato (Southern Italy)

Ciro Apollonio[(⊠)] [iD], Roberto Ferrante,
and Alberto Ferruccio Piccinni [iD]

Dipartimento di Ingegneria Civile, Ambientale, del Territorio,
Edile e di Chimica, Politecnico di Bari, Bari 70125, Italy
{ciro.apollonio,albertoferruccio.piccinni}@poliba.it,
r.ferrante2@studenti.poliba.it

Abstract. The presence of ancient underground urban drainage system in cities entails serious risks to public safety. Often, the presence of these hydraulic structures is forgotten and their projects missing in public offices. In this work the critical issues of the old urban drainage system of Corato, a city in the south of Italy, are described. In particular, through a targeted identification, acquisition and spatialization of key variables, and subsequent processing in a GIS software, a method for risk assessment has been provided. The choice of key variables was carried out through visual inspections and acquisition of "historical criticalities" of the urban drainage system. Finally, this paper presents a methodology, calibrated on the old city of Corato, to evaluate the risk caused by the presence of this hydraulic infrastructure, in order to help technicians to define the correct order of operations to be carried out.

Keywords: Risk map · Structural engineering · Risk assessment · Hydraulic infrastructure protection · Urban drainage system

1 Introduction

The drainage system represents an important infrastructure for urban areas. For this reason these hydraulic infrastructures are increasing in Italy. However, each urban settlement cannot be said to be immune to risks related to the malfunction or failure of these hydraulic infrastructures. Due to a complex series of factors, the ancient drainage systems are very vulnerable to failure, with serious consequences for public safety caused by structural damage and local floods leading to inflow of water into basements, threatening traffic safety, or causing streets and surface erosion and pollution of water supply system [1].

Stresses to which these structures have been subjected by time events (earthquakes or also urban planning modifications), leaks of the water mains that alter the Geotechnical conditions of underlying soils [2] have, as a natural consequence, an uncontrolled change (and short-term uncontrollable) of stress state in masonries. All this contributes to the triggering of the inexorable deterioration process that, thus

© Springer International Publishing AG 2017
O. Gervasi et al. (Eds.): ICCSA 2017, Part II, LNCS 10405, pp. 246–260, 2017.
DOI: 10.1007/978-3-319-62395-5_18

affecting structure safety and functionality (i.e. respectively, under the profile of the structural safety and of the adequate regime of the hydraulic operation), it can result in sagging of the overlying structures, the collapse of the work itself and hygiene & health issues for the urban areas [3].

Some historical Italian cases of sudden collapses were documented, such as the "*Emissario di Cuma*" (Pozzuoli, Naples, 1976), causing a chasm of about ten meters deep. The collapse has triggered a series of events chain: the complete obstruction of the sewer system due to the debris of collapse [4]. As an obvious consequence, the stretch of sea facing the hydraulic infrastructure and the small lake of Averno were polluted (for six months), and for as long as required to remedy the sites and to recover the collapsed network. Economic damage of various types, (i.e. fishing and sea bathing) were also recorded. Even the bimillennial Cloaca Maxima in Rome, still operating, is inexorably subject to failures and collapses involving a protected archaeological heritage; the last event occurred in April 2013 near the "*Arco di Giano*" [4].

All collapsing modes of urban water drainage system (mechanisms, causes and warning signs) have been extensively analysed in detail by *Water Research council* WRc [5], by collecting, examining and comparing data of 250 collapses reported in literature, with particular attention to the loss of bearing capacity of the soil and the formation of voids. The mode of failure and collapse of the practicable sewers, especially those in tunnels, are generally associated with an inadequate tensile strength of the structure. This may be caused by mortar joints weakening due to corrosion or abrasion, or by infiltration, and excessive hydraulic load on the overhead structure ones.

Other factors to take into account about the underground drainage system are road traffic level and loads, water level, soil type, terrain, weather conditions and the groundwater system.

In many cases, the rank of the road service undergoes changes over time, so that the underground drainage system differs to the original load pattern considered during the project phase. This is very uncommon in an ancient city sewer, thanks to the constructive technique in archway. In particular, *egg-shaped* pipes show higher resistance against traffic loads than the conventional circular pipes and also present a better hydraulic performance in normal flow conditions [6].

Collapse repair costs are greater than new construction, but costs associated with refurbishing may be one-sixth of the cost of pipe replacement/renewal. Major breaks and stoppages occur more frequently in older systems than in new ones, then is worthwhile to carry out a preventative maintenance of the older parts of the water drainage system [7].

It is therefore desirable sharing knowledge and best practices on water management issues within urban areas, with local approaches for modelling and mapping urban storm water [8].

Although several computational models to simulate urban hydrology and hydraulics are readily available and accepted [9], the risk assessment of urban flooding elements has been little researched. Some studies developed a model to quantify flood risk by applying a digital elevation model, using the relationship between flood stage and the corresponding damage cost [10].

In recent years, statistical approaches in the study for rehabilitation of underground systems were developed; in particular a way to determine the progression in time of the rainwater pipeline deterioration and to plan the restoration work, considering their performance decay factor over time, was used by Abraham et al. [11]. This forecast is based on the Markov probabilistic approach. It is based on the assumption that since the behaviour of the sewer lines (i.e., the rate of deterioration) is probabilistic, the choice of an appropriate intervention strategy is an uncertain process [12]. Deterioration models featured by a semi-Markov chain using transition probabilities based on user-defined survival functions show that the probability of failure of buried sewer pipes is subjected to combined effect of corrosion and stresses.

A Stochastic Finite Element Method offers many advantages over traditional probabilistic techniques since it does not use any empirical equations in order to determine the failure of pipes [12]. Flood hazard management is a problem incorporating aspects of the natural sciences (hydrology, ecology, etc.), the social sciences (economics, politics, psychology, culture, etc.) and engineering [13].

The inclusion of risk assessment and cost benefit analyses provides indeed a solid basis for decision-making in the sewer sector, by viewing comparatively various considerations of a drainage system [14]. Another research topic for a rational framework sewer network rehabilitation decision-making is represented by the Computer Aided Rehabilitation of Sewer and Storm water networks (CARE-S). It is aimed to improve the structural and functional reliability of the wastewater networks [1] and can be a useful tool to support decision-making at the local level, and facilitate the assessment and monitoring of the process within its comprehensive context [13].

However, does exist a solution most greatly reducing the risk of critical asset failure? The application of a multi-objective genetic algorithm optimization model, combined with an enhanced critical risk methodology, has successfully answered these questions when applied to a data set case study [15].

Existing hydraulic reliability based approaches have focused on quantifying the functional failure caused by extreme rainfall or increase in dry weather flows that lead to an hydraulic overloading of the system. This requires a new approach: the Global Resilience Analysis in which a wide range of structural failure scenarios results from random cumulative failures. Failure representing loss of system functionality are characterized by total flood volume (failure magnitude) and average flood duration (failure duration). A new resilience index combines the failure magnitude and duration into a single metric to quantify system residual functionality at each considered failure level [16]. In literature, "hazard" is defined as the occurrence of a flood event with a definite exceedance probability. "Risk" is defined as the potential damages associated with such an event, expressed in monetary losses. It becomes clear that hazard and risk quantification depend on spatial specifications (e.g., area of interest, spatial resolution of data) [17].

Recently, many scientific works have mapped the risk using the geographical information systems (GIS) [18]. In fact, the primary phase concern before a potential disaster is to mitigate the impact of a hazard. Here GIS is gaining favour in risk assessment and the development of long-term mitigation strategies [19].

In fact, to illustrate the risk distribution in an urban zone the use of a geographic information system (GIS) can result very useful, providing a georeferenced criticality database [20–24]. GIS, furthermore, can be an important tool for involving the public in the different stages of the planning process of risk alleviation [13]. The literature suggests to assign weight and rank values to the GIS layers and to the classes of each layer, respectively. The assignment of the weight/rank values and their analysis is usually realized by the application of the Analytic Hierarchy Process (AHP) method [25].

In this work a criticality study of the water drainage system of Corato city (in southern Italy) is proposed. In particular, using a GIS, a criticality forecast is provided, taking into account several parameters and data such as: sewer diameter, slope, material, age and depth. This approach provides an essential tool to validate a decision support system (DSS) to decide the priority measures for the citizens' security.

2 The Case Study

The city of Corato (48273 inhabitants in 2016) is located in Apulia Region, in southern Italy, on the north-eastern foothills of the Murge Baresi, at an altitude of 232 m above sea level.

It is bordered by two streams, the smallest in the North-West and the most relevant to the East, respectively classified with hydraulic risk [26] in *hydro-geological assessment plane* written by Apulian Basin Authority.

Geologically, the town is located on a limestone karst valley filled with clay, sand and tuff, forming the plugging of a suspended aquifer. The drainage works focused in this study are located in the "old town", built between 1876 and 1926. The large part of its main path, including channels that form the lateral adductions, is almost unknown.

Some data, largely related to its general planimetry, are collected in the Municipal Technical Office (as well as at the State Archives), however there are no detailed information that would illustrate the morphology and function.

In this context, the study and research cannot be separated from its general mapping, which takes into account several features, in particular the masonry conditions.

The greatest danger the city of Corato goes to face, due to the presence of that infrastructure, is the sudden collapse in high-traffic vehicular areas (see Fig. 1).

On April 21th, 2016 in *Corso Cavour*, during routine road maintenance, a wide chasm suddenly opened, without giving any warning signs of collapse [27].

The collapse, however, has allowed us to verify the regular hydraulic functioning of the drainage network.

In June 1985 in *via Duomo,* during the excavation for installation of gas pipelines, a collapse occurred. A similar event, in the same urban area already described, occurred on July 20th, 2006 in *via Asproni*, caused by the passage of a truck engaged in a nearby construction site.

This tunnel of drainage system, built in the 20s, appeared in bad conditions, in a poor state of preservation, ruined at several points, partially filled and interested by the presence of flowrate coming by urban waste water network [28].

As a result of speleological inspection after the collapse of 2016, on the walls it has been possible to identify the flood average levels and the levels of maximum storage reached in time. The survey allowed to verify the good hydraulic behaviour of this part of the urban drainage network.

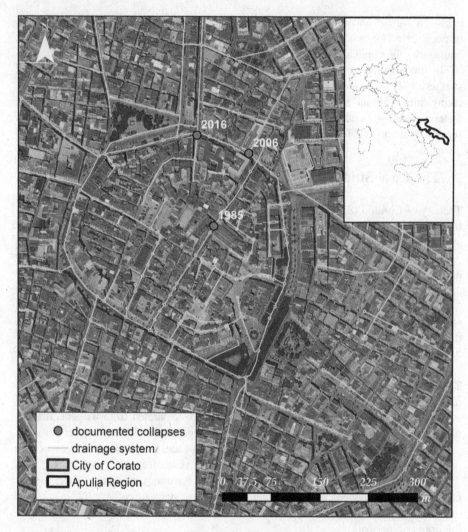

Fig. 1. Documented collapses of drainage system in the "old town" of Corato.

A thick layer of rough plaster has provided the size of the aggregates used (structural state information); in several points have been registered (in a database) signs of erosion due to water flow.

In particular, this inspection signaled some sections to be verified, since the state of deformation internally inflected of some wall portions suggests a propensity to ruin for squashing.

In areas with no plasterwork, it can be appreciated the thickness of the same (about 2 cm) and a substantial lack of mortar between the junctions, probably eroded from the direct contact with the water flow (defect named "mortar missing" in WRc MSCC) [29].

The city is not new to sinking episodes of the road surfaces due to the collapse of underground structures, a circumstance that makes it more and more topical the need for understanding, monitoring and in-depth maintenance of the existing drainage network. The city has suffered from severe episodes of hydrogeological instability in 1922, with collapses and victims, for the infiltration and rising of groundwater in the basements and in local houses of the old town buildings. This represents a further aggravating factor in the risk status assessment.

Such actions, planned or constructed, have resulted in benefits often only temporary [30], so much so that today it is reported a frequent raising of the groundwater level in case of intense meteoric precipitations, even more so if the drainage structures are not constantly object of maintenance [26].

The irregularity of the impermeable geological surface, made up of concavity depth and variable extension, also creates stagnation in the areas of major depression, and changes in the runoff under varying groundwater excursions [22].

In light of the above, it seems essential to avoid the risk that the meteoric hydraulic load, usually drained by rain sewer, adversely affect the precarious aquifer, after the collapse (or loss of efficiency) of the sewerage network. Consequences, comparing the dramatic historical circumstances, could be catastrophic and without any clear and tangible warning signs.

In Corato, it appears evident the necessity to know the state of groundwater hydraulic infrastructures, in particular of drainage system that presents serious criticisms due especially to the age (see Fig. 2).

Knowing how the network unravels below the city centre will also help the municipal technicians to draw up a protocol for excavation procedures regarding to the buried channels.

The structural diagnosis will facilitate the identification of critical points of incipient collapse (to establish before any damage may occur to the community in the event of collapse) and the areas requiring an ordinary maintenance. Furthermore, it is desirable to adopt solutions implementing the hydraulic characteristics.

Due to the special hydrogeological nature of the city, it is always wise to have an efficient underground drainage network for the timely evacuation of the most critical flood events. It is worthy to note that the old town has no such problems of flooding (instead existing in the "new areas" of the city). It can possibly be attributed to the activity of this silent and old drainage system.

3 Data and Method

It is well known that geographic information systems (GIS) are designed to analyse the interrelation between different layers of spatially distributed data. GIS is the tool of choice when integrated information from various sources to tackle complex problems [31, 32] are needed. The novelty of this work is represented by the use of geospatial information to provide a link between urban water drainage system and human risk. In the case of Corato drainage system, GIS has been applied to delineate the area with major risk, using several variables. On the whole, this procedure seems as a low-cost and rapid method of data collection based on the combination of remote sensing (RS), global positioning systems (GPS), digital video (DV) and geographic information systems (GIS) [33].

Fig. 2. (a) collapse occurred in April 2016; (b) a junction of the Corato drainage system; (c) a tunnel of drainage system without mortar; (d) mortar missing in a part of drainage system.

In order to quantify the risk status, this paper proposes three steps closely linked together. The first one is intended to collect and to organize the available information into a database to identify the "key variables", that are the most important parameters to classify risk in the case study. The second one is targeted to collect and to organize the available data into a GIS database. In the third step, using collected data, is proposed a map of risk depending on the drainage system, to provide a tool for risk management in the city of Corato.

3.1 Key Variables and Data Collected

The choice of key variables represents the most complex step in the GIS database construction. For the analysed case study, the following key variables are considered: the exposition, the caused inconvenience and the pipe status.

The exposition takes into account the data concerning all types of traffic: pedestrians, vehicles, (trucks included), and the presence of buildings near the underground drainage network (where exists a danger for building substructures).

The caused inconveniences involve all possible consequences caused by a failure of urban drainage system. In particular flooding, structural collapse and contamination (i.e. of groundwater system) are considered.

The last key variable considered is the pipe status, in which the structural status, the hydraulic pipe behaviour and the presence of external infiltration (i.e. by urban sewer) are collected. On the basis of the proposed method, it can be noted that hydrological and hydraulic considerations are used to evaluate flood risk, socio-economic data is assessed for damage estimation and regulatory constraints on urban development or environmental protection are taken into consideration [13].

The urban planning data (see Fig. 3) are collected using the available public database and by consulting the several projects of Corato drainage system.

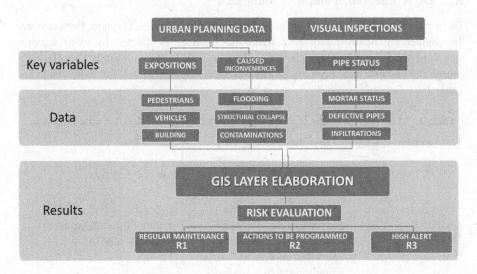

Fig. 3. Structure of the GIS database, including key variables.

In particular in "caused inconveniences" the critical events illustrated in literature are considered [26, 30] (i.e. historical documents and scientific paper).

Instead, the data on the "pipe status" are obtained by survey on field and visual inspection.

In fact a specific inspection was conducted by speleologists and engineers in order to understand the drainage system status. The survey on field was completed collecting photographic documentation of the hydraulic behaviour of the system.

It was also possible, using GPS, to locate position of all accesses and special points (for example manholes, outfall, interferences) and geo-referencing them (coordinates in UTM 33 system datum WGS-84).

This operation, in addition to provide a first idea of the trend of the network, has allowed to report each underground measurement minimizing possible errors.

Another issue of visual inspection was to carry out an analysis of used materials, not only about their state of preservation, but also, as far as possible, on their composition.

Regarding the hydraulic data, the survey field analysis was mainly aimed to measure the flow in the pipelines, identify incorrect connections and to measure the depth groundwater.

Finally, in order to identify and classify the structural criticality, this working group has chosen to consider, in the parameter called "pipe status", the scheme proposed by WRc in the "Manual of Sewer Condition Codification" (MSCC) [29] in which any defect is encoded with an alphanumeric code (i.e. deformation; roots, infiltration, sedimentary deposits, attached deposits, other obstacles; defects in manholes and dissolution of the mortar) in relation to its gravity.

3.2   Layer Integration and Risk Mapping

There are many kinds of potential failure of an old urban drainage system. Furthermore, the engineering behaviour of each hydraulic structure is affected by many factors, including flood control capacity, seepage situation, and structure stability [34]. To each considered variable have been assigned a weighted value equal to those indicated in the Table 1.

Table 1. Main features of the spatial database, including range values for used layer data.

Key variables	Layer data	Range value	Obtained from
EXPOSITIONS (E)	Pedestrians (P)	0–1	Urban planning data
	Vehicles (V)	0–1	
	Buildings (B)	0–1	
CAUSED INCONVENIENCE (CI)	Flooding (F)	1–3	Literature (historical documents and scientific paper)
	Structural Collapse (SC)	1–3	
	Contaminations (C)	0–1	
PIPE STATUS (PS)	Mortar Status (MS)	1–3	Visual Inspections
	Defective Pipes (DP)	1–3	
	Infiltrations (I)	0–1	

In particular, the parameters which are supposed to be closely related to failure of drainage system were used to establish a comprehensive evaluation system of the risk status. This entails both direct and indirect damage, where the indirect one is induced by the drainage system failure side effects, but occurs – in space or time – outside the collapse event. Examples are disruption of traffic, trade and public services [17].

The spatial database consists overall of nine layer, a number chosen referencing to literature [25] according to which nine objects are the maximum that an individual can simultaneously compare and consistently rank for an immediate analysis. Judgments are based on the best information available, and the engineers/speleologists knowledge and experience.

When the value range is 0–1, that means the key variables can only be considered absent or present; while for value range included in the interval 1–3 is considered a growing probability. The flooding variable values depend on the return time (T) of critical events; while for "structural collapse" the value depends on the number of occurred events. Finally, for mortar status and defective pipes, the scale value depends on the results of the visual inspections. In particular, in following table are reported the flooding variable values depending on the return time (T) according to the technical requirements provided by the Hydrographic District of Southern Apennine – Apulia basin authority [35] (Table 2).

Table 2. Value scale for the "flooding variable"

Flooding value	Return time (years)
1	>200
2	30–200
3	<30

The risk status R can be calculated by the following formula (1):

$$R_i = \sum E_i \cdot \sum (CI_i + PS_i) \tag{1}$$

Three classes of risk can be derived from this formula, corresponding to areas with high risk, areas with moderate risk and areas with low risk, as indicated in Table 3.

Table 3. Risk degree classification

Risk class	Risk degree	Range value
R1	Low	0–7
R2	Medium	8–21
R3	High	22–42

4 Results

The Fig. 4 presents some results of the risk analysis conducted. In general terms, the most dangerous areas are located in the "old town". The map in Fig. 4 shows how the drainage system in the old town is affected by risk, classified as *R1*, *R2* and *R3*. Within this context, the "pipe status" data play a defining role in the final score. These data are confirmed in the "old town" by the visual inspections. The high risk is the result of the presence of several factors reported in Fig. 4. Actually, many factors associated with the study area provide an increasing risk. The old town area was affected by the hydrogeological instability in 1922; this information represents an important aspect to take into account in the evaluation of risk degree. In Table 4, is illustrated a risk evaluation for the urban drainage system indicated in Fig. 4.

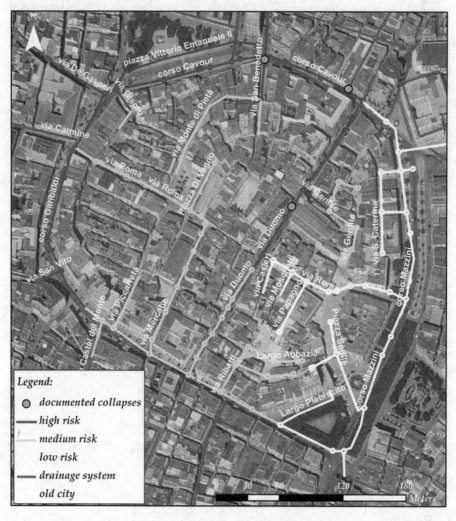

Fig. 4. Map of risk in "old town" confirmed by field survey.

Table 4. Used values for the key variables to estimate the risk status of the drainage network

Street name	Expositions (E)			Caused inconvenience (CI)			Pipe status (PS)			Risk value	Risk class
	(P)	(V)	(B)	(F)	(SC)	(C)	(MS)	(DP)	(I)		
corso Mazzini	0	1	0	1	1	0	1	1	0	4	R1
corso Cavour	1	1	0	1	3	1	3	2	1	22	R3
corso Garibaldi	1	1	1	2	2	1	2	2	0	27	R3
corso Garibaldi	1	1	1	1	2	0	2	1	0	18	R2
piazza Vittorio E.	1	1	0	2	3	0	3	2	1	22	R3
via Duomo	1	1	1	2	3	1	2	1	0	27	R3
via Duomo	1	1	1	1	2	1	2	1	0	21	R2
via Roma	1	1	1	3	1	1	1	1	0	21	R2
via Roma	1	0	0	1	1	0	1	1	0	4	R1
via Mercato	0	1	1	2	2	1	1	1	1	16	R2
via Piccarreta	1	0	1	3	3	1	1	1	1	20	R2
via Ribatti	0	0	1	1	1	0	3	3	0	8	R2
largo Abbazia	1	0	0	2	2	0	2	2	1	9	R2
piazza Sedile	1	0	0	1	1	0	1	1	0	4	R1
largo Plebiscito	0	1	0	1	1	0	1	1	0	4	R1
via Casieri	0	0	1	1	1	0	2	1	0	5	R1
via Moschetti	0	0	1	1	1	0	3	1	0	6	R1
via Papagno	0	0	1	1	1	0	1	2	0	5	R1
via Gentile	1	0	1	3	2	0	3	2	1	22	R3
via Gentile	0	0	1	2	1	0	2	2	1	8	R2
v. Santa Caterina	0	0	1	1	1	1	1	1	0	5	R1
v. Monte di pietà	0	0	1	3	2	1	2	1	1	10	R2
piazza Di Vagno	1	0	1	3	2	0	1	1	1	16	R2
v. Castel d Monte	1	1	1	1	1	1	2	1	1	21	R2
via San Vito	1	1	1	1	1	1	2	2	1	24	R3
via Carmine	1	1	1	2	1	1	2	2	0	24	R3
via Gisotti	1	0	1	3	2	0	3	2	1	22	R3
via Aldo Moro	1	1	1	1	1	0	3	3	0	24	R3
via S. Benedetto	1	0	1	2	3	1	3	2	0	22	R3

In the boundary conditions is included the internal status of the drainage system, detected by visual inspections. For example, under the road called "Corso Cavour" sections with missing bricks and mortar, with deformation on the walls were identified.

5 Conclusions

This paper has presented an approach to draw up a rapid risk assessment of ancient urban drainage networks, located in the old town of Corato.

Indeed, the drainage system criticalities (collapses, floods, environmental disasters) expose the cities to significant risks. It is possible to act in several technical ways. In this work is carried out a risk map using a GIS software.

The novelty of this analysis is represented by the fastest method to map the risk caused by criticalities of urban drainage system. Some key parameters to identify actions are collected, in order to avoid the collapse. The parameters required for the analysis, obtained by overlaying urban/topographical data with visual inspections, are therefore strictly necessary to define the priorities for action in order to guarantee the safety of people. In the case of Corato, on which the proposed method has been successfully tested, it was possible to identify the most critical areas where to begin the rational planning of interventions.

The need of a risk mapping is highlighted by the circumstance that random local repairs, with emergency character, in a critical urban drainage system, are not convenient and normally do not solve the real problem, sometimes contributing to worsen the actual scenario. It is also not negligible the economic factor.

Furthermore, according to the literature, as hazard, risk and vulnerability are not constant in time, corresponding analyses and maps must be updated after significant changes with a minimum of additional expenses.

This work, in accordance with the most recent literature guidelines, highlights the need for further developments based on site-specific considerations in order to improve new methods to monitor the drainage system conditions.

References

1. Saegrov, S., Schilling, W.: Computer Aided Rehabilitation of Sewer Water Networks ASCE Global Solutions for Urban Drainage (2002)
2. Antognoli, L., Pelagalli, L., Bianchi, E., Maxima, C.: Un recente intervento di restauro nel tratto sottostante il Giano del Velabro nel 2013. Bullettino della Commissione Archeologica Comunale di Roma numero unico annuale, 322–325 (2014)
3. Filetici, M.G., Scaroina, L.: Sicurezza e conoscenza: un binomio indispensabile per la conservazione. La distrazione della conservazione - Il rischio idrogeologico dell'area archeologica centrale di Roma. In: Bianchi, E. (ed.) La Cloaca Maxima e i sistemi fognari di Roma dall'antichità ad oggi, pp. 231–239. Palombi Editore, Roma (2014)
4. D'Elia, E.: Sulle origini storiche e sull'evoluzione della fognatura di Napoli. L'Acqua, Associazione Idrotecnica Italiana, pp. 19–41, Novembre/Dicembre 2015

5. Serpente, R.F.: Understanding the Modes of Failure for Sewers in William A. Macaits. Urban Drainage Rehabilitation Programs and Techniques. ASCE, USA (1994)
6. Regueiro-Picallo, M., Naves, J., Anita, J., Puertas, J., Suarez, J.: Experimental and numerical analysis of egg-shaped sewer pipes flow performance. Water **8**(12), 587 (2016)
7. Randrup, T., McPherson, B., Gregory, E., Costello, E., Laurence, R.: Tree root intrusion in sewer systems: review of extent and costs. J. Infrastruct. Syst. **7**(1), 26 (2001)
8. Blanksby, J., Boogard F.C.: Modelling and mapping of urban storm water flooding - using simple approaches in a process of triage. Skintwater (2010)
9. Granata, F., Gargano, R., de Marinis, G.: Support vector regression for rainfall-runoff modeling in urban drainage: a comparison with the EPA's storm water management model. Water **8**(3) (2016)
10. Ryu, J., Butler, D., Makropoulos, C.: Assessing sewer flood risk. In: Proceedings of the 2nd IMA International Conference on Flood Risk Assessement, Plymouth, UK (2007)
11. Abraham, D.M., Wirahadikusumah, R.: Development of prediction models for sewer deterioration. In: Lacasse, M.A., Vanier, D.J. (eds.) Durability of Building Materials and Components, pp. 1257–1267. Institute for Research in Construction - National Research Council of Canada, Ottawa (1999)
12. Alani, A.M., Faramazi, A.: Predicting the probability of failure of cementious sewer pipes using stochastic finite element method. Int. J. Environ. Res. Public Health **12**, 6641–6656 (2015)
13. Nunes Correia, F., Francisco, et al.: Flood hazard assessment and management: interface with the public. Water Resour. Manage. **12**(3), 209–227 (1998)
14. Fenner, R.A.: Approaches to sewer maintenance: a review. Urban Water **2**(4), 343–356 (2000)
15. Ugarelli, R., Di Federico, V., Sveinung, S.: Risk Based asset management for wastewater systems. In: NOVATECH, p. 917 (2007)
16. Johanses, N.B., et al.: Risk assessment of sewer systems. In: NOVATECH, p. 925 (2007)
17. Büchele, B., et al.: Flood-risk mapping: contributions towards an enhanced assessment of extreme events and associated risks. Nat. Hazards Earth Syst. Sci. **6**(4), 485–503 (2006)
18. Novelli, A., Tarantino, E., Caradonna, G., Apollonio, C., Balacco, G., Piccinni, F.: Improving the ANN classification accuracy of landsat data through spectral indices and linear transformations (PCA and TCT) aimed at LU/LC monitoring of a river basin. In: Gervasi, O., et al. (eds.) ICCSA 2016. LNCS, vol. 9787, pp. 420–432. Springer, Cham (2016). doi:10.1007/978-3-319-42108-7_32
19. Cova, T.J.: GIS in emergency management. Geograph. Inf. Syst. **2**, 845–858 (1999)
20. Ward, B., Savic, D.A.: A multi-objective optimisation model for sewer rehabilitation considering critical risk of failure. Water Sci. Technol. **66**, 2410–2417 (2013)
21. Mugume, S.N., et al.: A global analysis approach for investigating structural resilience in urban drainage systems. Water Res. **81**, 15–26 (2015)
22. Mair, M., et al.: GIS-based applications of sensitivity analysis for sewer models. Water Sci. Technol. **65**(7), 1215 (2012)
23. Halfawy, M., Dridi, L., Baker, S.: Integrated decision support system for optimal renewal planning of sewer networks. J. Comput. Civil Eng. **22**(6), 360–372 (2008)
24. Catasto Cavità Artificiali Pugliesi. Il progetto. http://www.catasto.fspuglia.it/df/il-progetto.php. Last accessed 26 Nov 2016
25. Fernández, D.S., Lutz, M.A.: Urban flood hazard zoning in Tucumán Province, Argentina, using GIS and multicriteria decision analysis. Eng. Geol. **111**(1), 90–98 (2010)
26. Nicotera, G., Abruzzini, E.: Il sovralzamento della falda freatica di Corato (Bari). Geo-tecnica, 69–77 (1962)

27. Ferrante, R.: Cunicoli o fogne bianche? L'affascinante sottosuolo coratino a un bivio. Lo Stradone - Il Giornale di Corato, 22–24, Aprile 2016
28. Vernice, S.: Una tragedia sfiorata - CORATO: il dissesto e la voragine in via Asproni forse provocati dai lavori del cinema Kursaal. La Gazzetta del Mezzogiorno 20 Luglio 2006, 16 (2006)
29. Thornhill, R., Wildbore, P.: Sewer Defect Codes - Origin And Destination. Underground Construction. U-Tech: Underground Technology Cutting Edge Technical Information for Utility Construction & Rehabilitation, pp. 32–36 (2005)
30. Cherubini, C., Romanazzi, E.: The problem of groundwater rise in Corato. Giornale di Geologia Applicata, 383–386 (2005)
31. Portoghese, I., D'Agostino, D., Giordano, R., Scardigno, A., Apollonio, C., Vurro, M.: An integrated modelling tool to evaluate the acceptability of irrigation constraint measures for groundwater protection. Environ. Modell. Softw. **46**, 90–103 (2013)
32. Giordano, R., Milella, P., Portoghese, I., Vurro, M., Apollonio, C., D'Agostino, D., Lamaddalena, N., Scardigno, A., Piccinni, A.F.: An innovative monitoring system for sustainable management of groundwater resources: objectives, stakeholder acceptability and implementation strategy. In: Proceedings of Environmental Energy and Structural Monitoring Systems (EESMS), IEEE Workshop, Taranto, Italy, pp. 32–37 (2010)
33. Montoya, L.: Geo-data acquisition through mobile GIS and digital video: an urban disaster management perspective. Environ. Modell. Softw. **18**(10), 869–876 (2003)
34. Jiang, C., Sheng, J., Zhang, G., Xu, G.: Calculation of failure probability of hydraulic structures for rural hydropower. In: Proceedings of 2012 International Conference on Modern Hydraulic Engineering. Procedia Eng. **28**, 161–164 (2012)
35. Apulian Regional Authority Watershed. Piano Stralcio di Assetto Idrogeologico (PAI). (Italian only). http://www.adb.puglia.it. Last accessed 02 Feb 2017

Earthquake's Rubble Heaps Volume Evaluation: Expeditious Approach Through Earth Observation and Geomatics Techniques

Sergio Cappucci[(⊠)] [iD], Luigi De Cecco, Fabio Gemerei,
Ludovica Giordano [iD], Lorenzo Moretti, Alessandro Peloso,
Maurizio Pollino [iD], and Riccardo Scipinotti [iD]

ENEA - Italian National Agency for New Technologies,
Energy and Sustainable Economic Development, Roma, Italy
{sergio.cappucci,luigi.dececco,fabio.geremei,
ludovica.giordano,lorenzo.moretti,alessandro.peloso,
maurizio.pollino,riccardo.scipinotti}@enea.it

Abstract. When an earthquake of a certain magnitude hit a populated zone, a huge amount of data have to be collected in order to address the typical hazard and emergency actions for rescue, assistance, viability, etc. In this paper we concentrate our attention on a particular dataset used to estimate the amount (and location) of rubbles generated by partial or total collapse of buildings/structures, which strongly influences the consequent environmental hazard and reconstruction phase. Despite this information is particularly valuable for optimizing the emergence response, for example by improving the management of their prompt removal, there are not many methods to estimate the amount of the rubbles in terms of volume/weight. Here, a procedure to estimate the volume of rubble heaps through earth observation data and Geomatics techniques is presented and preliminary results discussed.

Keywords: Earthquake · Rubbles · Geomatics · LIDAR · Post-emergency management · Emergency Management System COPERNICUS

1 Introduction

The central Apennines is one of the most seismically active areas in Italy and Europe. Central Italy is characterized by a rural and mountain landscape with several small ancient villages, teeming with tradition and cultural assets, characterized by poorly engineered buildings and structures.

Generally, the bigger and closer the earthquake, the stronger the shaking. There have been significant earthquakes with extraordinarily modest damage either because they caused little shaking or because the buildings were construct to withstand that level of acceleration on the land surface. In other cases, modest earthquakes have caused considerable damage either because the trembling was locally amplified, or more likely because the structures were poorly engineered. The latter case is what happen in central Italy on the borders of Umbria, Lazio, Abruzzo and Marche regions

© Springer International Publishing AG 2017
O. Gervasi et al. (Eds.): ICCSA 2017, Part II, LNCS 10405, pp. 261–277, 2017.
DOI: 10.1007/978-3-319-62395-5_19

on 24 August 2016 when a 6.0-magnitude scale earthquake occurred, followed by thousands of localized aftershocks (more than 60,000 on April 2017), whose some powerful ones (5.3–6.5 magnitude; see Sect. 1.1).

After the rescue activities and beside the typical actions taken to restore the transport and communication networks and to ensure safety and assistance to the local population, a parallel emergency immediately appears to be tackled: the management of the huge amount of rubbles heaps. Their weight is always unknown and their volume has to be estimated in order to plan the best possible management options.

Many studies related to waste management generated by catastrophic events have been conducted by using weight dataset either measured before or after transportation of waste to treatment or recycling plants [1, 2]. But, typically, this evaluation is possible only after a long time from the event.

The volume and variety of disaster waste are often difficult to assess. For the institution in charge of their removal is difficult and risky to rely only on what is possible to observe from the ground perspective, in order to deal with the organisation and management of the necessary actions (e.g., building demolition, rubbles removal [3]). After the 2010 Haiti earthquake, Talbot and Talbot [4] demonstrated the post-event performance of a fast rubble detection algorithm applied at the server, which enhances mono-temporal optical imagery in less than one second/plat, indicating that an end-to-end prototype designed to meet the needs of first responders after disaster is needed.

In our study, requested by the Italian National Civil Protection Department (CPD), we outline a specific and straightforward approach to estimate the rubble heaps volume. The study is mainly based on the LIDAR and Earth Observation (EO) dataset provided by the CPD acquired in the immediate post-event period within the framework of Emergency Management System (EMS) COPERNICUS.

In the Italian disaster management framework, such approach represents an innovative way to deal with post-emergency tasks. Nevertheless, it still needs a massive effort to make it reliable and adoptable by National and Local Authorities as an operative protocol. At the same time, we specify that we do not present or discuss any data related to building stability, earthquake resonance on building or reasons why buildings of different characteristics (e.g., height or construction materials) responded differently during the earthquake. Finally, it is worth highlighting that in some cases apparently undamaged buildings (at time of LIDAR and orthophotos surveys and from the perspective of visual photointerpretation) may have been partially or totally destroyed later on during following aftershocks.

1.1 Earthquake Events

The first relevant shock occurred at 3:36 A.M. of August 24th, 2016 and had a magnitude of 6.0, with a hypocentre at a depth of 8 km situated along the Valle del Tronto, between the towns of Accumoli (RI) and Arquata del Tronto (AP). Other two powerful earthquakes took place on October 26th and 30th, 2016 with epicentres in the Umbria-Marche border (the strongest shock, with moment magnitude of 6.5 having its epicentre between the towns of Norcia and Preci). On January 18th, 2017 a new sequence of four strong events (more than magnitude 5 shocks) took place between the towns of L'Aquila Montereale and Capitignano. The earthquakes and aftershocks were felt in most of Central and Northern Italy. The seismic event area is located in a very active seismological part of Central Italy that also includes L'Aquila, where the earthquake of April 6th, 2009 caused more than 300 deaths and about 65,000 displaced people. Further back in time, another relevant event happened in Marche and Umbria regions (September 26th, 1997). A detailed list of the main earthquake events recently occurred in Central Italy (since August 24th, 2016) is retrievable from the INGV (Italian National Institute of Geophysics and Volcanology) website [5].

2 Study Area

Immediately after the event, the CPD entrusted Helica Ltd. company (http://www.helica.it/) with LIDAR and orthophotos data acquisition over seven - geographically separated - areas hit by the earthquakes (Fig. 1).

Fig. 1. Overview LIDAR surveyed areas (Source: CPD and Helica Ltd.).

2.1 Geology

Apennines are affected by a multiphased contractioned and extensional tectonics [6]. A fault and thrust system belt, formed by African European continental margins collision during Neogene which result in development of Umbria–Marche northern Apennines and Lazio-Abruzzo Central Apennines [7, 8]. Then, Apennines belt was interested by a post orogenic extension. Such extension is expressed by superficial faulting [9]. The fault system (N) NW - (S) SE trending and usually 10–30 km long is the uppermost part of a complex normal-oblique system (Fig. 2). Many studies described and quantified the Holocene displacement of faults [10]. In Fig. 3, a simplified geological section show the complex geometric pattern along a WSW-ENE transect [11].

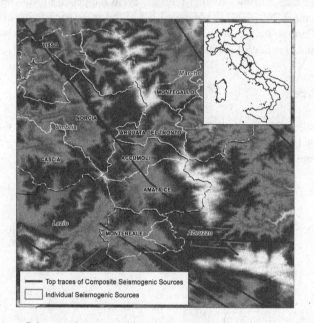

Fig. 2. Study area. Seismogenic sources (blue) and some of their top traces (red) are indicated in the areas interested by the first earthquake. (Color figure online)

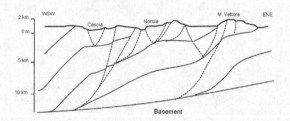

Fig. 3. Schematic geological section of the complex geometric pattern WSW-ENE oriented across Norcia-M. Vettore. Modified from [11].

Amatrice and many other villages interested by spread collapses and damaged buildings are predominantly located on ridges and incoherent deposit [12]. This circumstance may have influenced local increase of acceleration, but no data analysis is available at present to prove such hypothesis. The two most essential variables influencing earthquake damages are (1) the intensity of ground shaking caused by the event tied with (2) the quality of the engineering of structures in the area. The level of trembling itself is strongly influenced by the proximity of the earthquake source to the affected region and the types of rocks (mostly those at or close to the land surface) that seismic waves pass through during their propagation.

3   Data and Methods

3.1   Dataset

The core dataset is made up of:

1. LIDAR Point Clouds (2016, *post-event* flight- provided by CPD);
2. 1 m-pixel Digital Terrain Model (DTM; derived from LIDAR data acquired during 2016 flight, hereinafter indicated as DTM_{post});
3. 1 m-pixel Digital Surface Model (DSM; derived from LIDAR data acquired during 2016 flight, hereinafter indicated as DSM_{post});
4. RGB orthophotos (15 cm ground spatial resolution, 2016 flight);
5. LIDAR Point Clouds (2008-2012 acquisition flights, *pre-event*, provided[1] by the Ministry of the Environment and Protection of Land and Sea of Italy (MATTM) and made available through the Italian National Geoportal);
6. 1 m-pixel DTM (derived from above mentioned 2008–2012 LIDAR flights, hereinafter indicated as DTM_{pre});
7. 1 m-pixel DSM (derived from above mentioned 2008-2012 LIDAR flights, hereinafter indicated as DSM_{pre});
8. Damaged/collapsed buildings: delimitation and classification (provided by Copernicus Emergency Management Service [13]).

As listed above, two LIDAR datasets were available: one acquired few days after the August 24th earthquake and the other referred to some years before the event (and available only for a portion of the investigated areas). All the DSM and DTM data, in raster format, pre and post-earthquake, are at the same spatial resolution: 1 pixel per square meter. In the following Section, different approaches are presented, using either only post-earthquake data or exploiting the pre-existing DSM and DTM data, too.

Specifically, data listed from 1 to 4 were provided by the CPD: in particular, these data were acquired by the Helica Ltd. by means of a ~ 1000 m altitude flights, during which the RGB orthophotos were acquired too. Helica pre-processed the 10 dots/m^2 LIDAR dataset (about 9 million points clouds) and generated the DSM and DTM layers, by mosaic them in 892 tiles (700 × 700 m each). In Fig. 4, an example of different data layers available for one of the hit zones is reported.

Data listed from 5 to 7 were provided by the MATTM and refer to a pre-earthquake status as data were acquired during flights occurred between 2008 and 2012. Elevations data are in this case orthometric heights (measured above the geoid). However, it has to be highlighted that these data didn't cover the whole hit zones, since they were acquired for different purpose.

---

[1] http://www.pcn.minambiente.it/geoportal/catalog/search/resource/details.page?uuid=%
7BB8A39D4E-D7DF-4621-9318-4EEE3B1511CF%7D (Source: Italian National Geoportal).

Fig. 4. Orthophoto, post-event LIDAR Point Cloud (3-D rendering) and DSM extracted from LIDAR data.

The geodetic system used is ETRS89, implementing the realization ETRF2000 for 2008-12 LIDAR dataset. The coordinate system is UTM-WGS84-ETRF2000 Zone 33, according to the reference system of map cartography used in this part of Italy. It is worth underlining that this dataset is characterized by ellipsoidal heights, since it is not possible to transform them into orthometric heights until the new after-shock reference point's height will be upgraded by the *Istituto Geografico Militare Italiano* (IGMI). In some case, Google street view data (e.g. buildings and streets characteristics such as building's elevation, street's path tendency) as well as data provided by Social Network/other web sources were considered - where available - as ancillary information (Fig. 5). The damaged buildings map (Fig. 10b) derives from Copernicus EMS (Activation EMRS177[2]).

Fig. 5. A building before the earthquake and the same collapsed building pointed out in the orthophotos acquired after the event (blue line indicates the field of view). (Color figure online)

3.2 Methodology

In order to store and share all data among the multiple users, an HW/SW architecture based on the specific Spatial Data Infrastructure (SDI), already available and running in the ENEA [14, 15], was exploited (Fig. 6). The SDI was developed in a FOSS4G (Free

---

[2] http://emergency.copernicus.eu/mapping/list-of-components/EMSR177.

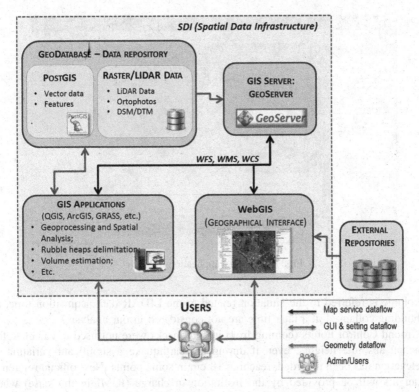

Fig. 6. HW/SW architecture of the exploited SDI

and Open Source Software for Geospatial) environment [16, 17]: a PostGIS-based GeoDatabase (GeoDB) [18] for data storage and management was realized. The GeoDB can be directly accessed by means of different desktop GIS software suites (e.g., QGIS). Then, by means of the server stratum, composed by GeoServer [19], data can be accessed via web by means of OGC standards (WMS, WFS and WCS services). Moreover, a restricted WebGIS application (accessible using appropriate credentials) and developed in the GeoPlatform environment (a free/open source suite developed by geoSDI [20]), has been implemented [21]. A specific layout, thus, was created so as to further support our mapping activities (Fig. 7).

In order to take into account different issues, typical of such a kind of emergency situations (e.g., the availability/lack of data, the specific needs of stakeholders, etc.), and to tackle appropriately them, two different methodologies, indicated as M1 and M2, respectively, were conceived and applied. The first one (M1) has been defined in order to be adopted in case of no pre-event data (e.g. LIDAR) are available and/or exploitable; the second one (M2) is based on the availability and exploitability of previously acquired (pre-event) LIDAR data (i.e., DSM_{pre}, DTM_{pre}).

Whatever the methods used, the following assumptions were made on rubble heaps volume evaluation:

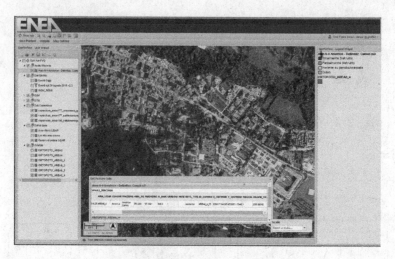

Fig. 7. WebGIS application interface

- volume is referred to the situation related to the LIDAR data acquisition time, as building collapsed at a later time are not considered in our analysis;
- Ground Control Points (coming from in-situ check) were not used to calculate the actual absolute heights, even if during the earthquake a significant variation in elevation has been recorded, reaching 18 cm in some points. New official reference heights will be provided by the Institution in charge (IGMI in this case) when emergency will be over;
- volume assessment was carried-out without considering the void ratio, which can be extremely variable as many ruins are composed by a multiple continuous and discontinuous wreckage of different materials (bricks, tiles, reinforced concrete columns and curbs mixed to fragments of metals, wood, home appliances and other bulky wastes).

Additionally, only in M1, the following assumption was done, too:

- variation of terrain elevation underneath the rubble heap perimeter was not considered. It means that a flat horizontal surface (points laying at a constant elevation) was always considered as the base of the rubble heap's volume.

Method 1. The implemented methodological approach, outlined in Fig. 8, is structured into the following steps:

1. The collapsed buildings (rubble heaps perimeters) were delimited by means of visual photo-interpretation, taking advantage of orthophotos' high resolution (15 cm) as well as using the already available Copernicus EMS data [13]. The polygons (P) were traced considering the displacement caused by the oblique photo acquisition (e.g., by integrating the DTM_{post} and DSM_{post} in the assessment). Wherever possible, i.e. when a single building was effectively recognizable as a well-defined single structure, a unique rubble heap was digitized for each single

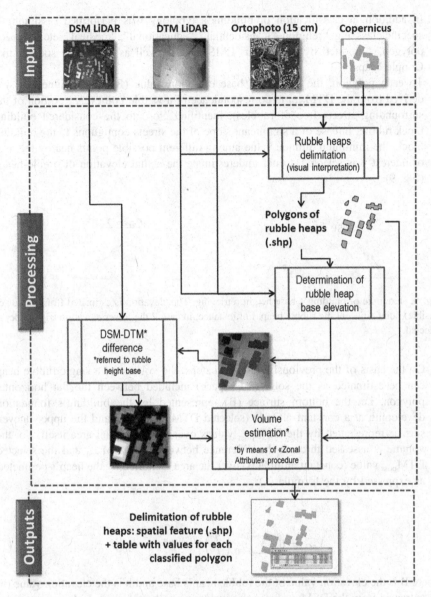

Fig. 8. Flow chart of the methodological approach pursued.

building; a pooled polygon referring to a group of (not distinguishable) buildings was sketched otherwise.

2. A suitable classification of "level of damage" has been reported, by assigning specific codes to each polygon: (1) "totally collapsed"; (2) "partially collapsed", (3) "rubbles piled over a slope"; (4) "uncertainty in defining the level of damage".

3. In addition, a classification of building's typology was performed (assigning a specific class, e.g. "religious", "residential", "educational", "medical", etc.), to each polygon, supported by Copernicus EMS data, as well as other web sources like Google Maps.
4. To each polygon, the building's base elevation value (B) was assigned: it was derived from the DTM_{post} in correspondence of a visible point (e.g., portion of the surrounding streets free of rubbles), identified close to the considered building block/rubble. In case of a significant slope of the streets contiguous to the building block, the minimum altitude value among different possible points nearby the area of interest was selected, to not underestimate the actual elevation of rubble heaps (Fig. 9).

Fig. 9. Example of building's base height retrieving. The elevation is estimated from the lower (visible) point adjacent the rubble heap. For instance, in case 2 the lower elevation (right side) is selected.

5. On the basis of the previously described steps, the volume of a single rubble heap can be estimated as the solid (3-D space) included between the flat horizontal polygon, i.e. the bottom surface (B) represented by the building's foundation developing at a constant altitude (selected DTM_{post} value), and the upper uneven surface represented by the DSM_{post} values over the polygonal area itself. So, the volume is assessed through the difference between the DSM_{post} and the selected DTM_{post} value (constant altitude B) over the area identified by the heap's perimeter, as expressed by the formula (1):

$$V_{M1} = B \cdot \sum_{k=0}^{n} \binom{n}{k} h^{DSMpost} \quad [m^3] \tag{1}$$

where B is the elevation value of the rubble heap's base and h is the elevation value (in m) extracted from the DSM_{post} for each pixel, n = number of pixel. Volume values are expressed in m^3.

This calculation was carried out in GIS environment by means of the ERDAS Imagine "zonal attribute" function[3], which allows to assign the calculated volume value

---

[3] Zonal Statistics To Polygon Attributes: this ERDAS Imagine (by Hexagon Geospatial) operator extracts the zonal statistics of the background image of a vector feature layer and save them as vector attributes.

as an attribute of the corresponding GIS polygon representing the rubble heap (in shapefile format).

Method 2. Method 2 follows the same steps 1 and 2 of M1 but different step 3, since the assignment of the elevation values of rubble's base is not manually performed. Rather, the base values are automatically extracted by the pre-event DTM_{pre}, which can be assumed as the bottom surface.

Therefore, in M2 processing chain, the volume calculation is directly obtained by the difference in elevation between post-event DSM_{post} and pre-event (DTM_{pre} over the rubbles' perimeter (step 3), as described in Eq. (2):

$$V_{M2} = \sum_{k=0}^{n} \binom{n}{k} h^{DSMpost} - \sum_{k=0}^{n} \binom{n}{k} h^{DTMpre} \qquad [m^3] \qquad (2)$$

However, particular attention is needed in order to make the two datasets (post- and pre-event) consistent with each other in case of differences in the absolute elevation reference system among them, and the variation in height, which can occur over some of the areas hit by earthquakes.

Ultimately, we investigated an additional approach limited to the exploitation of LIDAR data after and before the earthquake, thus without using any orthophotos and the manual extraction of height values from DTM. In fact, the raster resulting from the difference $DSM_{post} - DSM_{pre}$ may provide an overview and a rapid assessment of collapsed areas, ideally returning a null value wherever nothing happened, negative values where buildings collapsed and positive values where rubble heaps piled up over previous clear area (e.g., streets, courtyards, etc.). We do not have obtained consolidated results yet and we can argue that the idea deserves to be further developed. However interesting preliminary results were obtained at least in terms of automatic location (and geometrically definition) of rubbles (fair agreement with results manually obtained in step 1 (in M1/M2).

4   Results

The present study shows the possible results, in terms of rubble's volume estimation, which can be provided in a rather short time starting from LIDAR and orthophotos datasets. In particular, a total of about 15,000 m^2 of rubble heaps were identified and outlined in the area of Amatrice, corresponding to a total volume of about 75,000 m^3. In Table 1 a summary of the total amount of rubble volumes assessed by using the two

Table 1. Estimated rubble heaps volume in Amatrice area

Municipality	Settlement type	Volume M1 (m^3)	Volume M2 (m^3)
Amatrice	residential	54,827.1	56,021.8
	religious	18,112.3	19,655.,4
Total vol. (residential + religious)		73,826.0	75,764.0

different methodologies (as described in Sect. 3.2) is reported. It is worth highlighting that those heaps are the majority but not the total amount of rubbles in this area, as this is an ongoing research (aiming at testing the approach) and thus only the preliminary results are here presented.

As an example, selected perimeters of rubble heaps identified in Amatrice's area are reported in Fig. 10a. The same polygons are classified according to the level of damage (Fig. 10b), to the settlement typologies (Fig. 10c) and to the estimated volume of each single rubble heap (Fig. 10d): this helps to understand the potential applicability of the results obtained in the frame of our research.

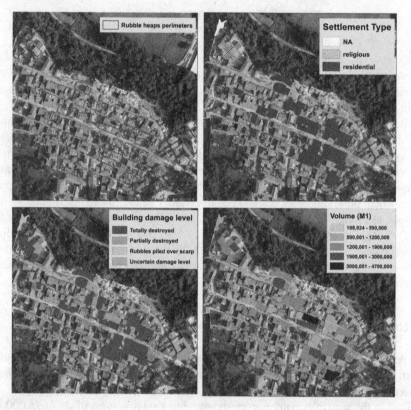

Fig. 10. Polygons: (a) representing rubble heaps delimitation in Amatrice; (b) classified according to the level of damage; (c) referring to the settlement typologies (residential, religious, etc.); (d) volume values (m^3).

In Fig. 11 the raster images whose values represent the difference between the DSM$_{post}$ and DSM$_{pre}$ are reported. This additional approach offers a viable solution only in case a pre-existing DSM$_{pre}$ is available in the affected area: those datasets may also be utilized when orthophotos are not (or not yet) available.

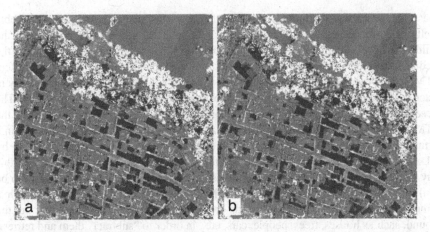

Fig. 11. Raster images resulting by the difference ($DSM_{post} - DSM_{pre}$): (a) darker areas indicate the collapsed buildings; (b) blue polygons are the automatically selected areas for rubbles volume estimation. (Color figure online)

5 Discussion and Conclusions

For each rubble heaps it was possible to delineate the perimeter (that is the surface extension) with a high degree of accuracy, thanks to the high resolution of orthophotos obtained immediately after the earthquake event. In order to support specific management needs, such as the restoration of a particular site, viability, etc. for each polygon (rubble heap) a corresponding address was associated, too. Tracing of the polygons (rubble heap) is a time consuming task especially if the aim is to associate each rubble heaps to a single source (building generating the rubble). To this aim we tried to subdivide into multiple polygons what was actually a single heap. However, especially in areas interested by many contiguous collapsed building, it was not easy, sometimes even not feasible, to retrieve the exact rubble-building correspondence.

In other cases, some uncertainty remains about the actual grade of a building's damage: for example, the combination of rubbles and an apparently intact building leaves doubts about the real stability of the structure and the consequent prediction about the management of such specific situation (i.e. building to be demolished or restored).

In the present study, we would like to stress that different methods may be applied depending on the available resources (i.e. time/budget/data).

Therefore, two methodologies (M1 and M2, see Sect. 3.2) have been implemented in order to provide effective support to the CPD operators and to cooperate with the existing Copernicus EMS service. M1 uses a single, consistent set of data (collected immediately after the event) whereas the other, M2, needs pre-existing LIDAR data (specifically DTM) to be available and the consequent adjustment to be performed in order to obtain consistent datasets.

As described in the Results Section, in M1 a flat surface is considered as the bottom side of rubble's volume, whose z-position (height) needs to be manually retrieved by operators (which have to look for a rubble-free point as near as possible to the rubble itself). In some cases, it was not so easy to find a heap-free point to retrieve the building's base height due to the ruins covering all the surrounding streets. In the second method (M2), the elevations are automatically derived from the pre-existent DTM_{pre}, hence overcoming such a drawback, though adjustment are likely necessary in order to make the two datasets (pre- and post-event) coherent one with the other. The presence of the rubble heaps themselves can also interfere with the production of the DTM from LIDAR data, leading to less reliable DTM (DTM_{post}) values (resulting in a virtual intermediate value between the actual DTM and DSM). Data acquired by LIDAR surveys, indeed, are laser pulses returning to the sensor and measuring in this way the distance to a target. Therefore, by using the flight parameters these data can be easily converted into DSM data. The DTM is instead derived by DSM through a complex algorithm that strongly depends on the slope of the object protruding out the ground, such as houses, trees, people, cars, etc., in order to "subtract" them and retrieve the underlying terrain. For that reason, when the acquisition is performed in a "normal" situation and not after a catastrophic event, the elevation of the terrain (i.e. the resulting DTM) is very accurate and reliable since there are no errors due to rubbles and/or scattered debris. On the contrary, such debris (commonly present after an earthquake) can be processed as ground level elevation in some spot (see Fig. 12) and some discrepancies can be found comparing the pre-earthquake DTM_{pre} with the post-earthquake DTM_{post}.

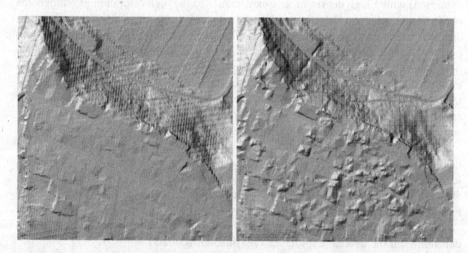

Fig. 12. Amatrice centre. On the left, DTM before earthquake (LIDAR flight 2008–2012); on the right DTM after Earthquake 1 (LIDAR flight 2016)

In general, we can argue that whether to use the first or the second proposed methodology (M1 or M2) depends more on the specific case (data, time, skilled manpower available) as well as on the requested result's accuracy. Despite the total volume is basically comparable, as the overall percentage difference is within 2,6%, specific differences, up to 50% in volume, can be found when single rubble heaps are compared. In fact, significant different results were obtained utilizing the two approaches on small hotspots of collapsed buildings, however weight/volume comparison is still an ongoing challenging activity. For an assessment of the weight must be taken into account some parameters not yet known in the current state and linked to each construction, such as the type of material used, the number of floors and also the housing presence [22, 23].

Clearly, the bigger the surface extension of a rubble, the higher the influence of using a single elevation as the base of the rubble, especially if the actual base is a significantly sloping or uneven surface (e.g., typically along streets). In this case it is better to limit the polygon size at minimum extension with a relevant increase in time consuming processing chain. A specific discussion is needed for when pre and post LIDAR datasets are available. The M2 procedure, using only post event LIDAR dataset, certainly offers a faster procedure to locate the area characterized by buildings' collapses and rubble heaps, along with their volume's estimates. The M1 procedure shows the advantage to consider also differences in landscape that may be occurred between the different data acquisitions.

In fact, the time-span between the two LIDAR datasets can reach more than a few years (in this case 2016 versus 2008–2012), possibly resulting in new housing, roads, changing in vegetation, etc. Such change in detection are not a direct consequence of the earthquake, but only the common dynamics of territories [24] or the emerging issue of urban sprawl [25], which need to be accounted for in some way. It is a challenge we are going to tackle in the future developments of our approach, indeed.

It is worth noting that changes referring to the removal and displacement of debris, collapses or demolition occurred after the date of the survey (post-event dataset) and hence not accounted for in the dataset provided by the National and Local Authorities were not taken into account in the present study.

Ultimately, preliminary results are showing an interesting correspondence between the geology of the area, the level of damages and the volume of the identified rubble heaps which will hopefully produce a research development.

We finally would like to stress the fact that the possibility to use the Copernicus EMS has been vital for the success of our research. Following the European Flood Awareness System (EFAS [26]) and European Forest Fire Information System (EFFIS [27]) we hope that a GIS-based Web-application that provides near real-time and historical information on buildings will be considered by the European Commission (EC) and that the acquisition of LIDAR data will be constantly carried out in order to cover the entire territory at European level.

Acknowledgments. The present research has been funded by ENEA and requested by DPC. EO data were provided under COPERNICUS by European Union and ESA. Thanks to Eng. F. Campopiano, P. Marsan and P. Pagliara of National Civil Protection for the fruitful discussion and suggestions and to all staff involved during the emergency action. ENEA is not responsible for any use, even partial, of the contents of this document by third parties and any damage caused to third parties resulting from its use. The data contained in this document is the property of ENEA.

References

1. United States Environmental Protection Agency: Household Hazardous Waste Management: A Manual for One-Day Community Collection Programs. EPA530-R-92-026, Washington, D.C., August 1993
2. Lund, H.F.: Household Hazardous Wastes. The McGraw-Hill Recycling Handbook (1993)
3. Memon, M.A.: Disaster waste recovery and utilization in developing countries-Learning from earthquakes in Nepal. In: 15th Asian Regional Conference on Soil Mechanics and Geotechnical Engineering, ARC 2015: New Innovations and Sustainability, pp. 143–147 (2015)
4. Talbot, B.G., Talbot, L.M.: Fast-responder: Mobile access to remote sensing for disaster response. Photogramm. Eng. Remote Sens. **79**(10), 945–954 (2013)
5. Amatrice, Norcia, Visso Seismic Sequence, INGV. http://terremoti.ingv.it/it/ultimi-eventi/1023-sequenza-sismica-in-italia-centrale-aggiornamenti.html. Last access: 28 Mar 2017. (In Italian)
6. Pizzi, A., Galadini, F.: Pre-existing cross-structures and active fault segmentation in the northern-central Apennines (Italy). Tectonophysics **476**, 304–319 (2009). doi:10.1016/j.tecto.2009.03.018
7. Coltorti, M., Farabollini, P.: Quaternary evolution of the Castelluccio di Norcia Basin (Umbria-Marchean Apennines, Italy). Il Quaternario **8**, 149–166 (1995)
8. King, G.C.P.: Speculations on the geometry of the initiation and termination processes of earthquake rupture and its relation to morphology and geological structure. Pure Appl. Geophys. **124**(3), 567–585 (1986)
9. Hernandez, B., Cocco, M., Cotton, F., Stramondo, S., Scotti, O., Courboulex, F., Campillo, M.: Rupture history of the 1997 Umbria-Marche (Central Italy) main shocks from the inversion of GPS, DInSAR and near field strong motion data. Ann. Geophys. **47**, 1355–1376 (2004)
10. Valensise, G., Pantosti, D.: The investigation of potential earthquake sources in peninsular Italy: a review. J. Seismol., 5287–5306 (2001)
11. Calamita, F., Pizzi, A.: Tettonica quaternaria nella dorsal appenniniica umbro-marchigiana e bacini intrappenninici associate. Studi Geologici Camerti, SI, 17–25 (1992)
12. Cacciuni, A., Centamore, E., Di Stefano, R., Dramis, F.: Evoluzione morfotettonica della conca di Amatrice. Studi Geologici Camerti **2**, 95–100 (1995)
13. Copernicus Emergency Management Service (EMS). http://emergency.copernicus.eu/
14. Pollino, M., Modica, G.: Free web mapping tools to characterize landscape dynamics and to favor e-participation computational science and its applications. In: Murgante, B., et al. (eds.) ICCSA 2013, LNCS 7973, Part III, 566–581. Springer, Heidelberg (2013)
15. Modica, G., Pollino, M., Lanucara, S., Porta, L., Pellicone, G., Fazio, S., Fichera, C.R.: Land suitability evaluation for agro-forestry: definition of a web-based multi-criteria spatial decision support system (MC-SDSS): preliminary results. In: Gervasi, O., et al. (eds.) ICCSA 2016. LNCS, vol. 9788, pp. 399–413. Springer, Cham (2016). doi:10.1007/978-3-319-42111-7_31
16. Steiniger, S., Hunter, A.J.S.: Free and open source GIS software for building a spatial data infrastructure. In: Bocher, E., Neteler, M. (eds.) Geospatial Free and Open Source Software in the 21st Century, pp. 247–261. Springer, Heidelberg (2012)
17. Modica, G., Laudari, L., Barreca, F., Fichera, C.R.: A GIS-MCDA based model for the suitability evaluation of traditional grape varieties: the case-study of 'mantonico' grape (Calabria, Italy). IJAEIS **5**(3), 1–16 (2014)
18. PostgreSQL with PostGIS extension. http://postgis.net

19. Geoserver. http://geoserver.org/
20. GeoPlatform by geoSDI. https://github.com/geosdi/geo-platform
21. Pollino, M., Caiaffa, E., Carillo, A., Porta, L., Sannino, G.: Wave energy potential in the mediterranean sea: design and development of DSS-WebGIS "waves energy". In: Gervasi, O., Murgante, B., Misra, S., Gavrilova, Marina L., Rocha, A.M.A.C., Torre, C., Taniar, D., Apduhan, B.O. (eds.) ICCSA 2015. LNCS, vol. 9157, pp. 495–510. Springer, Cham (2015). doi:10.1007/978-3-319-21470-2_36
22. CNR-ITC, C.N.VV.FF.: Sisma Abruzzo 6 Aprile 2009 - Stima quantificazione macerie, Activity Report (2010). (In Italian)
23. Marghella, G., Marzo, A., Moretti, L., Indirli, M.: Uso del GIS per il Piano di Ricostruzione di Arsita (TE). In: ANIDIS 2013 – XV Convegno (2013). (In Italian)
24. Modica, G., Zoccali, P., Fazio, S.: The e-participation in tranquillity areas identification as a key factor for sustainable landscape planning. In: Murgante, B., Misra, S., Carlini, M., Torre, C.M., Nguyen, H.-Q., Taniar, D., Apduhan, B.O., Gervasi, O. (eds.) ICCSA 2013. LNCS, vol. 7973, pp. 550–565. Springer, Heidelberg (2013). doi:10.1007/978-3-642-39646-5_40
25. Fichera, C.R., Modica, G., Pollino, M.: Land Cover classification and change-detection analysis using multi-temporal remote sensed imagery and landscape metrics. Eur. J. Remote Sens. **45**, 1–18 (2012)
26. European Flood Awareness System (EFAS). https://www.efas.eu/
27. European Forest Fire Information System (EFFIS). http://effis.jrc.ec.europa.eu/

A Geospatial Decision Support Tool for Seismic Risk Management: Florence (Italy) Case Study

Luca Matassoni[1], Sonia Giovinazzi[2(✉)], Maurizio Pollino[3],
Andrea Fiaschi[1], Luigi La Porta[3], and Vittorio Rosato[3]

[1] Fondazione Parsec (Former Prato Ricerche), Prato, PO, Italy
{l.matassoni,a.fiaschi}@pratoricerche.it
[2] University of Canterbury, Christchurch, New Zealand
sonia.giovinazzi@canterbury.ac.nz
[3] ENEA Italian National Agency for New Technologies,
Energy and Sustainable Economic Development, DTE-SEN-APIC Lab,
Casaccia Research Centre, Rome, Italy
{maurizio.pollino,luigi.laporta,
vittorio.rosato}@enea.it

Abstract. Seismic risk assessment, which attempts to predict earthquake-induced physical impacts on structures and infrastructures, casualties and losses can be a powerful tool to support emergency response planning as well as the development of effective mitigation strategies. The Civil Protection (CP) Department of Florence Municipality commissioned this study as historical earthquakes showed an appreciable seismic risk for the city that needed careful civil protection planning. A Decision Support System DSS (*CIPCast-ES*) developed by ENEA, APIC Lab, in the framework of the EU-funded project CIPRNet, was used to simulate the seismic and impact scenarios for two major historical earthquakes felt in Florence, to assess the earthquake-induced damage at single building level, and the relative expected consequences on population. The possibility to account for the seismic microzonation (i.e. the possible amplification of the seismic hazard and therefore of the expected impacts due to soil conditions) was also included within DSS. The results of the scenario analysis, presented in the paper in tabular format, were provided to the CP of Florence Municipality as queryable, interactive and end-user friendly web-version maps.

Keywords: Seismic scenario · Civil Protection · Macroseismic vulnerability · Decision Support System

1 Introduction

The estimation of possible impacts induced by seismic events, including damages to the built environment, consequences for the population, direct and indirect losses, is a critical aspect for planning and preparing mitigation actions pre-event and for managing the emergency post-disaster.

© Springer International Publishing AG 2017
O. Gervasi et al. (Eds.): ICCSA 2017, Part II, LNCS 10405, pp. 278–293, 2017.
DOI: 10.1007/978-3-319-62395-5_20

In Italy the Civil Protection (CP) Department is responsible for promoting and enhancing the community resilience to earthquake disasters, at different levels, including national, regional and municipal levels. As far as the municipal level is concerned the Mayor has the primary responsibility for the protection of the territory under his jurisdiction. Aiming to conduct an informed planning for post-earthquake emergency management, the CP Office of Florence Municipality commissioned this study to get an estimation of the possible earthquake-induced impacts in the City. In recent years significant progresses have been made in developing software platforms for seismic risk assessment to provide decision-makers with tools to assess, at regional or local level, and almost in real time, earthquake-induced impact and damage scenarios. The most well-known and internationally used ones include, among others: [1–8].

This study is based on a web-based Decision Support System DSS [9, 10], developed in the framework of a EU-funded project *CIPRNet, Critical Infrastructures Preparedness and Resilience Research Network* (https://www.ciprnet.eu/), and named *CIPCast* (https://www.ciprnet.eu/ads.html) [11], by the *APIC Lab*, of *ENEA*. The initial version of the *CIPCast* DSS (named DSS 1.0) was conceived as a combination of free/open source software environments [12, 13] including GIS features. These peculiar features of DSS 1.0 made it is particularly suitable for the final aim of this study, that was to provide the CP of Florence Municipality with queryable, interactive and end-user friendly web-version maps of selected earthquake scenarios, that they could use to plan for and test alternative emergency management strategies and resource allocations. In the context of this implementation DSS 1.0, has been upgraded to become a specific *Earthquake Simulator* module of the *CIPCast* DSS (hereafter named *CIPCast-ES*) [11]. It was customised to assess the earthquake-induced damage at single building level, and the relative expected consequences on the residents in term of casualties and displaced population. The possibility to account for the seismic micro-zonation (i.e. the possible amplification of the seismic hazard and therefore of the expected impacts due to soil conditions) was also included within *CIPCast-ES*, and used in this implementation thanks to the availably of a seismic microzonation study providing the relative amplification factors for Florence Municipality. Furthermore, a WebGIS application has been developed, as the natural geographical interface of the *CIPCast-ES*: basic information, maps and scenarios can be visualized and queried via web, by means of standard Internet browsers and, consequently, the main results can be easily accessible to the users and exploitable for further analyses [11].

The implementation of *CIPCast-ES* to Florence case-study allowed the estimation of the expected physical damage to the built environment (limited to the analysis of residential buildings) for two major historical earthquakes felt in Florence in 1895 and 1919, along with the estimation of the consequences to the population in terms of homeless and expected casualties (injured and dead people).

After giving a brief overview on general and peculiar aspects of platforms for seismic risk analysis (a more comprehensive overview is out of the scope of this paper and can be found elsewhere: [5, 14, 15]) the paper presents in Sect. 3 the specific method implemented within ENEA *CIPCast-ES* for the analysed case study. Section 4 of the paper presents and discusses the obtained results. Potentialities and limitations of the study, along with the necessary steps to advance and support the dissemination of the use of DSS for CP purposes are discussed in the conclusions of the paper (Sect. 5).

2   A Short Overview on the Steps

Generally, essential components of platforms for seismic risk assessment include:

(a) *Hazard Component*: for the assessment and representation of the seismic hazard;
(b) *Exposure Component*: for the inventory and characterisation of the exposed assets (e.g. building, infrastructures) and communities;
(c) *Fragility/Vulnerability and Damage assessment component*: for the assessment of the physical assets vulnerability and the propensity to sustain damage when subjected to earthquakes;
(d) *Impacts/Loss assessment component*: for the estimation of the expected direct and indirect impacts and losses due to the earthquake-induced physical damage to structures and infrastructures.

Hazard component should provide, using either deterministic or probabilistic approaches, an estimation of the expected ground-motion considering the specific characteristics of the local/regional seismicity. Deterministic approaches estimate, for the city/area under analysis the expected ground motion, and related consequences, for one selected seismic event deterministic earthquake (e.g. the maximum historical event from a pertinent seismogenic source, or the maximum earthquake compatible with the known tectonic framework); probabilistic approaches estimates the probability of occurrence of a certain ground motion, and relative consequences, in a certain time frame, due to all the possible seismogenic sources that might generate an earthquake in the area under analysis. The selection of one of the two approaches should consider the goal of the study [16–18]. For both deterministic and probabilistic approaches *attenuation relationship* are used to describe how the ground motion attenuates as the distance away from the fault rupture increase. The earthquake ground motion, can be described in term of either qualitative parameters. Examples of quantitative/engineering parameters include, among others, the peak ground acceleration (PGA), elastic spectral quantities (S_a) that have been observed to be more efficient predictors of the structural performance than PGA. Macroseismic Intensity I is a qualitative description of the effects of the earthquake at a particular location, as evidenced by observed damage on the natural and built environment, and by the human and animal reactions at that location. Thanks to its qualitative nature it is the oldest measure of the earthquake Macroseismic Intensity scales have been extensively used when adopting empirical-based model for assessing the building stock seismic vulnerability and for estimating the possible earthquake-induce damage to it (see below).

To account for geological and geophysical characteristics of the sites that might modify the seismic ground motion, or might generate earthquake-induced geotechnical secondary hazards (e.g. potential liquefaction susceptibility, landslide and rock fall hazard, as well as earthquake-related flooding), seismic microzonation should be considered. Empirical observation or analytical simulation (for example, site response analysis) [19] can be used for estimating how the predicted ground motion at from either deterministic or probabilistic approaches should be modified to account for the local seismic microzonation.

Exposure Component: Should provide an accurate classification of the exposed assets, including structures and infrastructures. As far as buildings are concerned, the final aim would be to classify the building into groupings that are expected to behave similarly, sustaining a comparable level of damage, following an earthquake. To this aim, relevant data and information on the building stock should be collected. Preparation of a building stock inventory is a critical part of a seismic risk analysis study and can be both time and resourcing consuming. A compromise needs to be sought between the time and cost required for higher quality data on the building stock and the uncertainties arising from low quality data. For example, in the event of insufficient for a direct classification of the building in terms of the construction types inferences might be established between larger groups of buildings (referred later as categories), recognized on the basis of more general information and building types. Inferences may take the form of "for masonry buildings built before 1919, 40% belong to rubble stone typologies and 60% are old brick masonry buildings".

Fragility/Vulnerability and Damage assessment component: The damage that might affect a building stock, when subjected to a seismic event, can be predicted via different approaches, including: (a) actual damage observation (b) expert judgment, (c) simplified-mechanical and analytical models; or a combination of the three aforementioned approaches. Methods based on actual damage observations are derived statistically processing the damage data from different earthquakes and summarising the outcomes in term damage of: (a) damage probability matrices, DPM (e.g. [20, 21]); (b) vulnerability curves (e.g. Eq. 5); or fragility curves (e.g. [22]).

Impacts/Loss assessment component: Earthquake-induced structural and non-structural damage to building is the root cause for several losses and impacts [23]. Consequences to buildings and their contents, direct economic losses (e.g. cost of repair/reconstruction of building structural or non-structural elements, or replacement of machinery and content), consequences to inhabitants and building occupants can be estimated once the physical damage sustained by the buildings has been assessed. Different empirical correlations, translating structural and non-structural damage into percentage of losses are proposed by different authors and in the framework of different seismic risk analysis methods. These correlations apparently different can be, actually, represented according to a common formulation (Eq. 6).

3 Case Study Description

Florence, the capital of the Tuscany Region, in Italy, is a city, which bustles with industry and craft, commerce and culture, art and science. The population of Florence municipality is about 400,000 inhabitants spread over a territory of just over 100 km^2. The territory pertains mainly to the plain of fluvial-lacustrine basin of Florence-Prato-Pistoia hosting a conurbation of about 1,000,000 inhabitants (Fig. 1). Population growth was impressive especially in the last two centuries, reaching the maximum in the seventies of the last century.

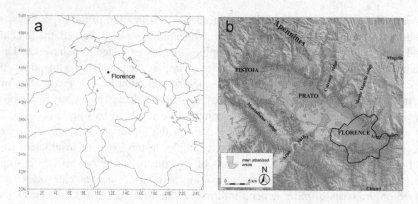

Fig. 1. Florence (Northern Tuscany, central Italy): localization (a) and morphological setting of the area and boundary of the Municipality (b)

3.1 Historical Seismicity and Selection of the Earthquake Event for the Scenarios

Despite the low-moderate historical seismicity presents in the area, the seismic risk in Florence should be regarded as a serious issue nowadays, considering the increased urbanisation and related population and the fact that mostly of the buildings were not designed according to seismic criteria. The national rules recognized Florence as affected by seismic hazard only starting from 1981, currently identifying a low seismic hazard (class 3 to 4 with 1 as higher hazard) with acceleration peak values spanning the 0,125 g–0,150 g range with a return period of 475 years (10% probability of exceedance in 50 years [24]).

Historically, 229 earthquakes were felt in the territory of Florence [25]. Out of these, 20 had intensity larger or equal than the damage threshold i.e. $I_{MCS} = 5-6$. The largest felt historical earthquake, among those known, occurred on May 18, 1895 with an estimated magnitude M = 5.5, intensity at epicentre $I_{0MCS} = 8$ located near Florence, in an area without known seismogenic sources [26]. The felt intensity in Florence was $I_{MCS} = 7$ causing minor to moderate damages to several buildings [27].

The closest seismogenic source is *"Mugello east"* (about 25 km far from Florence) that originated the June 29, 1919 earthquake with an estimated M = 6.3, and $I_{0MCS} = 10$ [26]. The felt intensity in Florence was $I_{MCS} = 6$ causing minor damages to the building stock. Further seismogenic sources include *"Mugello west"*, which is linked to the 1542 event M = 5.9 and *"Poppi"* in Casentino, for which there is only one record of a strong historical earthquake (i.e. 1504, M = 5.1). All the aforementioned individual areas (i.e. *Mugello east, Mugello weast* and *Poppi*) belong to a same composite seismogenic source, referred to as *"Mugello-Città di Castello– Leonessa"* running for 200 km along the Apennines from the latitude of Pistoia about (north-west) to the upper valley of the Nera River in Umbria (south-east). The Prato-Fiesole fault system is a further seismogenic source that is closer to the city, but is a debated one since strong earthquake have not been reported/measured for that (see Fig. 2 [27]).

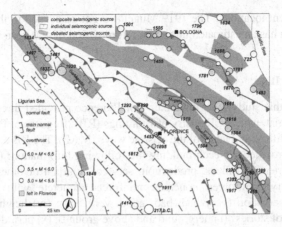

Fig. 2. Tectonics features (adapted from [28]), main seismicity (from [21]), and seismogenic sources (from [26])

The historical seismic events with significant felt macroseismic intensity in Florence, i.e. $I_{MCS} \geq 6$, were collected and collated [25], to identify the maximum credible earthquake: it is possible to notice the occurrence of four events with felt intensity $I_{MCS} = 7$ in Florence (years 1414, 1453, 1895), with epicentre location either in Florence or its surrounding areas (i.e. 1414 epicentre Wester Tuscany). These events were characterized by epicentral intensity $6 \geq I_{0MCS} \geq 8$, estimated magnitude $5.1 \geq M_w \geq 5.7$; their seismogenic sources is unfortunately unknown. The events where the epicentre was attributable to a seismogenic source (i.e. 1542, 1729, 1919) were characterized by a lower felt intensity in Florence, i.e. $I_{MCS} = 6$ despite the higher magnitude at the epicentre, (i.e. $M_w = 5.95$ and $M_w = 6.29$ for the earthquake generated by the Mugello seismogenic source in 1542 and 1919 respectively). Due to the aforementioned gaps of knowledge in the historical seismicity, it seemed unfeasible to establish a maximum credible earthquake for Florence area. Therefore, the decision was made, for the sake of this study, to analyse the damage scenarios generated by two historical events with epicentre location one in the Florentine area and the other in the Mugello area, respectively: (i) 18 May 1895, estimated magnitude $M_w = 5.50$ (epicentre located only few km outside the municipal boundaries); 29 June 1919, estimated magnitude $M_w = 6.38$.

The epicentral locations of the selected events were assumed (in WGS84 from [21]) as: 43.7 N − 11.267 E for the 1895 Florentine event; 43.95 N − 11.483 E for the 1919 Mugello event. The appropriateness of the selected events is supported by [29] showing that the higher percentage contribution (25–30%) to the probabilistic seismic hazard in Florence, by disaggregation, is related to a $M_w = 4.5$–5 earthquake occurring within 10 km from the city boundary; while a Mw = 6–6.5 earthquake occurring between 20 and 30 km, is contributing by less than 5%. Moreover the selected scenario events were preferred to others causing comparable levels of impacts on the build environment, being more recent and therefore supported by a greater number of information on the observed effects for different locations. The 1919 Mugello event associated to the *Mugello east* seismogenic area has an estimated minimum return period of 450 years [26].

3.2   Available Data

For the sake of this study the Municipality of Florence provided three detailed databases in GIS format, namely: Registry of Buildings (RB); Registry of Population (RP); Map of the Amplification Factor of the seismic wave (MAF). The RB database (version of April 2015) provided for each building the following information: (i) function (i.e. residential, commercial, offices, production, services, tourism or other); (ii) building material (masonry, reinforced concrete, reinforced concrete with pilotis, other); (iii) state of preservation (very bad, bad, good, very bad); (iv) height (meter above ground)/number of stories[1] (underground and above ground); (v) period of construction (<1918; 1918–1945; 1946–1960; >1960). The RP database provided the distribution of residents within the city (377,139 inhabitants at 31 December 2014), in term of number of people localized at point level at their residence address.

For the sake of this study, RB and RP databases had to be spatially joined to assign the population to the residential buildings. People of each street number were assigned to the closest residential building (characterized by a minimum area of 25 m^2 to exclude many smaller accessory building). The automatic GIS operation for integrating the two databases introduced some uncertainties. For 4.100 buildings (14% of the total) it was not possible to associate some resident: some were actually vacant; but for others that were far from the related street number point (e.g. building without direct access to the main street) the inhabitants were assigned to other closer building, This error might affect the results related to the assessment of the impacts on the population, in particular when conducted at single building level.

Finally, the MAF database provided values of the local seismic amplification factor AF. The University of Florence, Department of Earth Sciences [30] elaborated MAF database on the basis of the available geological and geophysical data (1,220 borehole and down-hole tests overall) using the 1-D software Shake-91 [31] for the numerical analysis and the Kriging method for the spatial interpolations of the results. The maximum value obtained for the local seismic amplification factor resulted AF = 2.4; values AF < 1.2 were regarded as negligible.

4   Method

4.1   Deterministic Seismic Hazard and Site Effects Analysis

In this study, the deterministic approach was preferred to the probabilistic one since the main focus was on planning the emergency management and the response for a single event [18]. For the assessment of deterministic hazard scenarios, DSS 1.0 [10]

---

[1] When the information on the number of stories above the ground was not available it was estimated as the height divided by 3, being 3 m the standard inter-story height for residential building in Italy.

implemented by default the Sabetta and Pugliese attenuation law [32] that calculates the attenuation of the ground shaking, a function of the distance, in terms of PGA for a given earthquake defined in term of epicentral location and estimated magnitude M at the epicentre. In the present study, the DSS 1.0 was modified, aiming to evaluate the deterministic hazard in terms of macroseismic intensity, I instead of PGA. Macroseismic intensity provides a qualitative description of the seismicity in relation to the damage observed on the built environment. In this respect it can be easily communicated and understood by the end-users and easily handled for managing post-disaster emergencies. Moreover, since DSS 1.0 implemented the macroseismic method for the vulnerability assessment, CIPCast-ES [11] is able to provide a hazard scenario directly represented in terms of I avoided the introduction of large uncertainties when converting instrumental measures e.g. PGA, into I [33].

Three attenuation laws were selected for implementation within CIPCast-ES, from a review available for the Italian territory [34, 35], namely: Eq. (1), by Faccioli and Cauzzi [36]; Eq. (2), by Pasolini et al. [37]; Eq. (3), by Allen et al. [38].

$$I_{MMI} = 1.0157 + 1.2566M_w - 0,6547\ln\sqrt{R^2+4} \qquad (1)$$

$$I_{MCS} = 5.862 + 2.460M_w - 0.0086\left(\sqrt{R^2+3.91^2} - 3.91\right)$$
$$- 1.037\left(\ln\sqrt{R^2+3.91^2} - \ln 3.91\right) \qquad (2)$$

$$I_{MMI} = 2.085 + 1.428M_w - 1.402\ln\sqrt{R^2+h^2+(2.042e^{(M_w-5)} - 0.209)^2} + S \qquad (3)$$

where M_w is the moment magnitude of the earthquake, R (km) is the epicentral distance, h (km) is the hypocentral depth (set as 3.91 km in Eq. (2)), S in Eq. (3) is a site topographic factor (not considered here because absent in the other equations). Equation (1) is validated for $M_W < 5.5$ (otherwise R in the equation is not the epicentral distance but the shortest distance from site to surface projection of the ruptured fault), Eq. (3) is validated for hypocentral distance ($R^2 + h^2$) less or equal to 50 km (otherwise see [38]). Macroseimic intensity I was given referring to Mercalli-Cancani-Sieberg, MCS, scale (I_{MCS}) or Modified Mercalli (I_{MMI}) scale.

According to Margottini et al. [39], Eq. (1) assumed MCS intensity equal to Medvedev-Sponheuer-Karnik (MSK) intensity for strong earthquakes [36] while Eq. (2) considered only the MCS intensity [37] and Eq. (3) the MMI [38]. MMI and MSK macroseismic scale have a direct correspondence [40] with the European Macroseismic Scale, EMS-98 [41], universally recognized in Europe as the standard macroseismic intensity scale and implemented in the Macroseismic vulnerability method. Grades of MCS macroseismic scale tend to be higher than those assigned using other scales, the relationship between the two scales it is often solved setting EMS intensity a degree level lower than the MCS intensity, but in realty more complex than this; as such there is not a clear correspondence between MCS and EMS-98 [42, 43]. CIPCast-ES was customised to allow the end-users to select one of the three aforementioned attenuation law and to set the parameters defining the selected scenario earthquake (e.g. for Eq. 3: epicentre coordinates; magnitude; and hypocentral depth).

CIPCast-ES was able to calculate, after that, at each building level, R [km], as the distance between the epicentre of the earthquake scenario and the centroid of each building footprint, and therefore the macroseismic intensity, for each one of the three equations, assumed to be equivalent to I_{EMS-98}.

The DSS 1.0 could not account for the effects of amplification due to soil conditions. *CIPCast-ES* was modified so that, the I_{EMS-98} values assessed at the bedrock according to Eqs. (1) or (2) or (3) could be locally increased taking into account the local amplification factor AF, available from microzonation studies. The increase in the macroseismic intensity to account for soil condition (ΔI_{soil}) was calculated, according to the macroseismic method [44], as:

$$\Delta I_{soil} = \frac{\ln(AF)}{\ln 1.6} \qquad (4)$$

For individual building, the related amplification factor was evaluated as the average between the maximum and minimum AF range identified within the boundaries of each building.

4.2 Building Vulnerability and Damage Assessment

For the assessment of the seismic vulnerability of the building stock at census tract-level DSS 1.0 implemented a method referred to as Macroseismic vulnerability method [40, 45] that uses a vulnerability index V and a ductility index Q to characterize the vulnerability of a single building or group of buildings, accounting for the building type, constructive and geometrical features. DSS 1.0 assessed the building stock vulnerability, as a function of the building typology and height, by sourcing the data on the building stock from the 2001 national census of population and houses [10].

In this implementation *CIPCast-ES* was customised to allow for the vulnerability assessment at single building level, as a function of the following information: construction material (e.g. masonry or reinforce concrete, RC); period of construction; number of stories; and state of maintenance. The estimation of V, at the individual building, was carried out implementing the procedure described in [45]. Table 1 shows the basic vulnerability index V*, and the vulnerability modifiers ΔV (i.e. increment or decrement to V*), assigned to each building or group of buildings as a function, respectively, of: (i) building material and period of construction; (ii) state of maintenance, number of stories, possible aggregation with adjacent buildings, presence of weak plan (referred to as pilotis) for RC buildings.

Reference was made to the information included in the RB database to calculate the ΔV values by assuming:

– for the maintenance - good and very good as *good*; bad and very bad as *bad*;
– for the number of stories - 1 to 2 floors as *low*; 3 to 5 as *medium*, >5 as high;
– aggregation status and presence of pilotis.

For buildings where information about the period and/or material of building and/or state of preservation were vacant, the worst case scenario (i.e. higher value of V* and/or

Table 1. Assumed V* values by period and building material, and ΔV values by state of preservation, number of stories, aggregation status and presence of pilotis

Category	Building period Masonry	V^*	State of maintenance		Number of storeys			Aggregate building		Pilotis
			good	bad	low	med.	high	no	yes	
I	< 1919	0.79	0	0.08	−0.08	0	0.08	−0.04	0.04	-
II	1919 - 1945	0.73	0	0.06	−0.08	0	0.08	−0.04	0.04	-
III	1945-1971	0.69	0	0.04	−0.08	0	0.08	−0.04	0.04	-
IV	> 1971	0.65	0	0.04	−0.08	0	0.08	−0.04	0.04	-
	RC		good	bad	low	med.	high	no	yes	
V	< 1971	0.59	0	0.04	−0.03	0	0.03	0	0.04	0.12
VI	1971–1981	0.55	0	0.04	−0.03	0	0.03	0	0.04	0.12
VII	> 1981	0.42	0	0.04	−0.03	0	0.03	0	0	0.06

ΔV was assumed). According to the Macroseismic method the correlation between the seismic input and the expected physical damage is expressed in terms of a curve of vulnerability, described by a closed analytic function:

$$\mu_D = 2.5\left[1 + \tanh\left(\frac{I + 6.25V - 13.1}{Q}\right)\right] \tag{5}$$

where:

- μ_D is the mean damage expected for the individual building or group of buildings;
- I is the seismic hazard described as a continuous parameter, according to the EMS-98 [41], calculated as $I_{EMS-98} = I_{badrock} + \Delta I_{soil}$;
- V is the value of the vulnerability index calculated as $V = V^* + \Delta V$;
- Q is the index of ductility assumed equal to $Q = 2.3$ [46];

From the resulting mean damage μ_D CIPCast-ES allocated the level of damage to each building, according to the EMS-98 5 level damage scale [41], plus the absence of damage D_0 as clarified in Table 2.

Table 2. Degrees of damage by the average damage μD value according to EMS-98 scale [41].

Damage degree	None	Slight	Moderate	Heavy	Very heavy	Destruction
	D_0	D_1	D_2	D_3	D_4	D_5
μ_D range values	<0.5	≥0.5; <1	≥1; <2	≥2; <3	≥3; <4	≥4; ≤5

4.3   Expected Consequences on Population and Buildings

The DSS 1.0 could not estimate the expected consequence on the population due to earthquake-induced physical damage to the buildings. This feature was included in CIPCast-ES, as for the approach proposed in [40] The occurrence probability of a certain consequence p_c was derived as a function of the likelihood of the building suffering the 5 different EMS-98 damage damages, p_{Dk} k from 0 to 5 as follow:

$$p_c = \sum_{k=0}^{5} w_{ck} p_{Dk} \qquad (6)$$

The assessment of the weighting factors $w_{c,k}$ was obtained from empirical relationships derived from the statistical analysis of the damage and the consequences observed after past earthquakes [47, 48]. In particular, in this work reference was made to the correlations proposed by [49] calibrated on Italian earthquake events, recognizing however few limitations of the same, as far as the assessment of the consequences on the population are concerned, including the facts that it does not account for and distinguish night and day, seasonal and weather conditions, different behaviours for different class of residents (e.g. due to age, gender, particular physical conditions, etc.) [50]. Table 3 reports the weighting factors $w_{c,k}$ for the consequences on buildings.

Table 3. Weighting factors wc, k for assessing the expected consequences on buildings [49].

Damage degree		None	Slight	Moderate	Heavy	Very heavy	Destruction
		D_0	D_1	D_2	D_3	D_4	D_5
Collapsed buildings (CB)	w_{CB}	0	0	0	0	0	1
Unusable buildings (UB)	w_{UB}	0	0	0	0.4	1	1

The consequences on the population are assessed as follow: Casualties (CA), including serious injuries and deaths 30% of the residents living in collapsed buildings; Displaced People (DP) all the residents living in Unusable Buildings (UB) minus the number of casualties, estimated as above.

5 Results and Discussion: Scenarios for Selected Earthquakes

The simulations of the two selected earthquake scenarios run with *CIPCast-ES* provided, for different attenuation laws (Eq. from 1 to 3) and in the hypothesis of considering or neglecting site amplification (AF = none, AF = average respectively) the results summarised in Table 4, in terms of consequences on buildings, and people, including: collapsed buildings (CB); buildings with damage level D4 and D5; displaced population (DP); and casualties (CA).

As a general comments, it can be noted that the implementation of the different attenuation laws brought to fairly similar results (Eq. (2) was not used for the 1919 Mugello case study as the assumption of $I_{MCS} = I_{EMS-98}$ did not seem to be acceptable) and that the role of soil amplification can be very relevant and should not be therefore neglected when estimating earthquake-induced impacts. The analysis showed that strong earthquakes originated in the Mugello area (such as the 1919 Mugello earthquake) could cause in Florence an overall damage to the built environment similar to the one caused by weaker earthquakes with epicentre close to the City (such as the 1895 Florentine earthquake).

Table 4. Impacts on buildings and population resulting from *CIPCast-ES* for the two selected earthquake scenarios

	AF	Attenuation Eq.	Consequences on buildings		Consequences on people	
			CB	D4-D5	DP	CA
1895 Florentine	None	(1)	–	97	852	–
		(2)	–	120	1081	–
		(3)*	–	123	1091	–
	Mean	(1)	–	1681	17894	–
		(2)	–	2379	26308	–
		(3)*	–	2458	27139	–
1919 Mugello	None	(1)	–	140	1364	–
		(3)*	–	100	916	–
	Mean	(1)	–	3701	41788	–
		(3)*	–	1782	20007	–

*the hypocentral depth for Eq. (3) is assumed h = 3.91 km, as for Eq. (2).

Reading the result from a CP end-user perspective, Table 4 shows that: casualties should not be a major issue in the event of an earthquake in Florence; similarly complete collapse of buildings should not be probable. However the CP Department of Florence Municipality might have to deal with a number of displaced people that might range between 800 and 28,000 approximately. A number of buildings in the range of 100 to 3,700 approximately might suffer severe/moderate damage, potentially requiring the need for activating rescue operations and for cleaning rubbles from the urban road network.

In term of the possibility to validate the results on observations from historical events, for the first selected earthquake, i.e. 1895 Florentine, the results obtained seem

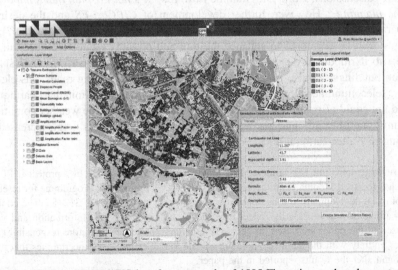

Fig. 3. *CIPCast-ES.* WebGIS interface: example of 1895 Florentine earthquake scenario

coherent with the few available historical information, although a direct comparison is not feasible. The 1895 Florentine earthquake, caused in Florence a minor to moderate widespread damage to the residential buildings (Fig. 3) and to all the monuments, including churches and historical buildings [27, 51, 52]. Four people died, while the number of injured is unknown [21, 52]. No casualties were estimated by the *CIPCast-ES* for the 1895 Florentine earthquake, however it is fair to say that the scenarios evaluated the consequences only on residential buildings, excluding monuments and strategic buildings, such as schools and hospitals.

6 Conclusion

The study presented in this paper provided a good opportunity for upgrading the DSS 1.0 tool to become the *CIPCast - Earthquake Simulator* and showed the effectiveness of the tool to bridge the gaps between scientists and end-users, providing a platform well understood and manageable by both the groups. As seen from this and similar implementations several uncertainties might affect the final results of a seismic risk analysis including: the completeness and reliability of the data available for charac-terising the exposed asset and population; the assumption made to estimate the hazard, the building vulnerability and expected damage along with the reliability of the adopted models; the several limitations when assessing the possible impacts on people due to the difficulties to account for population dynamics and several further variables that might be influential.

Despite that, and recognising all the aforementioned limitation, this implementation showed that it is very useful to provide to CP Authorities at least with a rough estimation, possibly supported by historical evidences, that could inform their planning and it is very useful to provide them with queryable, interactive and end-user friendly maps, as the ones generated with *CIPCast-ES*, of possible scenarios that could prompt "what-if" discussions; citing Dr. Gramme Edwards *"It's not the plan that's important, it's the planning"*. However, further development of *CIPCast-ES* and the desirable comparison and integration of potentialities with other platforms such as [5, 7], for example, will be promoted to enhance the tool and possibly reduce the uncertainties. Thanks to its modularity and capability, *CIPCast-ES* allows to perform analysis not only on buildings but also on distributed infrastructures, including roads, gas electric power, telecommunication, etc. Moreover, *CIPCast-ES* is not limited to earthquake hazard but could also simulate impacts from flooding and extreme weather conditions, allowing multi-hazard analysis and cascading hazard analysis (e.g. increased of flooding risk after earthquakes).

Acknowledgement and Disclaimer. This article was derived from the FP7 project CIPRNet, which has received funding from the European Union's 7th Framework Programme for research, technological development and demonstration under grant agreement no. 312450. The contents of this article do not necessarily reflect the official opinion of the EU. Information and views expressed in the paper are based on personal research experiences, therefore responsibility for that lies entirely with the authors. There are several variables and data uncertainties that might affects and alter the results reported in the paper.

References

1. Daniell, J.E.: Open source procedure for assessment of loss using global earthquake modelling software (OPAL). Nat. Hazards Earth Syst. Sci. **11**, 1885–1900 (2011)
2. Erdik, M., Şeşetyan, K., Demircioğlu, M.B., Hancılar, U., Zülfikar, C.: Rapid earthquake loss assessment after damaging earthquakes. In: Garevski, M., Ansal, A. (eds.) Earthquake Engineering in Europe. GGEE, vol. 17, pp. 523–547. Springer, Cham (2010). doi:10.1007/978-90-481-9544-2_21
3. Spence, R.J.S., So, E.K.M.: Human casualties in earthquakes: modelling and mitigation. In: Proceedings of the Ninth Pacific Conference on Earthquake Engineering Building an Earthquake-Resilient Society 2011, p. 224 (2011)
4. Strasser, F.O., Bommer, J.J., Şeşetyan, K., Erdik, M., Çağnan, Z., Irizarry, J., Goula, X., Lucantoni, A., Sabetta, F., Bal, I.E., Crowley, H., Lindholm, C.: A comparative study of European earthquake loss estimation tools for a scenario in Istanbul. J. Earthq. Eng. **12**, 246–256 (2008)
5. GEM: The OpenQuake-engine User Manual. Global Earthquake Model (GEM) Technical report 2017-02. (2017)
6. NIBS: HAZUS MR4 Technical Manual. National institute of Building Sciences, Washington, DC (2004)
7. MAE, M.-A.E.C.: Mid-America Earthquake Centre Seismic Loss Assessment System - MAEviz v3.1.1 (2007). http://mae.cee.illinois.edu/software/software.html
8. Molina, S., Lang, D.H., Lindholm, C.D.: SELENA – an open-source tool for seismic risk and loss assessment using a logic tree computation procedure. Comput. Geosci. **36**, 257–269 (2010)
9. Pollino, M., Fattoruso, G., Rocca, A.B., Porta, L., Curzio, S.L., Arolchi, A., James, V., Pascale, C.: An open source GIS system for earthquake early warning and post-event emergency management. In: Murgante, B., Gervasi, O., Iglesias, A., Taniar, D., Apduhan, Bernady O. (eds.) ICCSA 2011. LNCS, vol. 6783, pp. 376–391. Springer, Heidelberg (2011). doi:10.1007/978-3-642-21887-3_30
10. Pollino, M., Fattoruso, G., La Porta, L., Della Rocca, A.B., James, V.: Collaborative open source geospatial tools and maps supporting the response planning to disastrous earthquake events. Futur. Internet. **4**, 451–468 (2012)
11. Pietro, A., Lavalle, L., La Porta, L., Pollino, M., Tofani, A., Rosato, V.: Design of DSS for supporting preparedness to and management of anomalous situations in complex scenarios. In: Setola, R., Rosato, V., Kyriakides, E., Rome, E. (eds.) Managing the Complexity of Critical Infrastructures. SSDC, vol. 90, pp. 195–232. Springer, Cham (2016). doi:10.1007/978-3-319-51043-9_9
12. Pollino, M., Modica, G.: Free web mapping tools to characterise landscape dynamics and to favour e-participation. In: Murgante, B., Misra, S., Carlini, M., Torre, Carmelo M., Nguyen, H.-Q., Taniar, D., Apduhan, B.O., Gervasi, O. (eds.) ICCSA 2013. LNCS, vol. 7973, pp. 566–581. Springer, Heidelberg (2013). doi:10.1007/978-3-642-39646-5_41
13. Rosato, V., Pietro, A., Porta, L., Pollino, M., Tofani, A., Marti, José R., Romani, C.: A decision support system for emergency management of critical infrastructures subjected to natural hazards. In: Panayiotou, C.G.G., Ellinas, G., Kyriakides, E., Polycarpou, M.M.M. (eds.) CRITIS 2014. LNCS, vol. 8985, pp. 362–367. Springer, Cham (2016). doi:10.1007/978-3-319-31664-2_37
14. GFDRR: Understanding Risk - Review of Open Source and Open Access Software Packages Available to Quantify Risk from Natural Hazards (2015)

15. Kongar, I., Giovinazzi, S.: Damage to Infrastructure: Modeling. In: Beer, M., Kougioumtzoglou, I.A., Patelli, E., Siu-Kui Au, I. (eds.) Encyclopedia of Earthquake Engineering, pp. 1–14. Springer, Heidelberg (2014). doi:10.1007/978-3-642-36197-5_356-1

16. Bommer, J.J.: Deterministic vs. probabilistic seismic hazard assessment: an exaggerated and obstructive dichotomy. J. Earthq. Eng. **6**, 43–73 (2002)

17. Krinitzsky, E.L.: Deterministic versus probabilistic seismic hazard analysis for critical structures. Eng. Geol. **40**, 1–7 (1995)

18. McGuire, R.K.: Deterministic vs. probabilistic earthquake hazards and risks. Soil Dyn. Earthq. Eng. **21**, 377–384 (2001)

19. Kramer, S.L.: Geotechnical Earthquake Engineering. Prentice Hall, Upper Saddle River (1996)

20. Whitman, R.V., Reed, J.W., Hong, S.T.: Earthquake damage probability matrices. In: Proceedings of 5th European Conference on Earthquake Engineering, Roma, pp. 25–31 (1973)

21. Guidoboni, E., Ferrari, G., Mariotti, D., Comastri, A., Tarabusi, G., Valensise, G.: CFTI4Med - Catalogue of Strong Earthquakes in Italy (461 B.C.-1997) and Mediterranean Area (760 B.C.-1500). http://storing.ingv.it/cfti4med/

22. Rossetto, T., Elnashai, A.: Derivation of vulnerability functions for European-type RC structures based on observational data. Eng. Struct. **25**, 1241–1263 (2003)

23. Tiedemann, H.: Casualties as a function of building quality and earthquake intensity. In: Proceedings of the International Workshop on Earthquake Injury Epidemiology for Mitigation and Response, pp. 420–434. Johns Hopkins University, Baltimore (MD)

24. Stucchi, M., Meletti, C., Montaldo, V., Crowley, H., Calvi, G.M., Boschi, E.: Seismic hazard assessment (2003–2009) for the Italian building code. Bull. Seismol. Soc. Am. **101**, 1885–1911 (2011)

25. Locati, M., Camassi, R., Rovida, A., Ercolani, E., Bernardini, F., Castelli, V., Caracciolo, C.H., Tertulliani, A., Rossi, A., Azzaro, R., D'Amico, S.: DBMI15, the 2015 version of the Italian Macroseismic Database. http://emidius.mi.ingv.it/

26. DISS Working Group: Database of Individual Seismogenic Sources (DISS), Version 3.2.0: A compilation of potential sources for earthquakes larger than M 5.5 in Italy and surrounding areas. http://diss.rm.ingv.it/diss/

27. Boccaletti, M., Corti, G., Gasperini, P., Piccardi, L., Vannucci, G., Clemente, S.: Active tectonics and seismic zonation of the urban area of Florence, Italy. Pure Appl. Geophys. **158**, 2313–2332 (2001)

28. Martini, I.P., Sagri, M.: Tectono-sedimentary characteristics of Late Miocene-Quaternary extensional basins of the Northern Apennines, Italy. Earth-Sci. Rev. **34**, 197–233 (1993)

29. Barani, S., Spallarossa, D., Bazzurro, P.: Disaggregation of probabilistic ground-motion hazard in Italy. Bull. Seismol. Soc. Am. **99**, 2638–2661 (2009)

30. Dipartimento di Scienze della Terra - DST: Studi geologici e sismici dell'area fiorentina: approfondimenti per una zonazione sismica [Geological and seismic studies in the Florence area: insights for a seismic zonation], Florence, Italy (2013)

31. Idriss, I.M., Sun, J.I.: User's Manual for Shake91 – A Computer Program for Conducting Equivalent Linear Seismic Response Analyses of Horizontally Layered Soil Deposits. University of California - Center for Geotechnical Modeling, Department of Civil & Environmental Engineering (1992)

32. Sabetta, F., Pugliese, A.: Estimation of response spectra and simulation of nonstationary earthquake ground motions. Bull. Seismol. Soc. Am. **86**, 337–352 (1996)

33. Douglas, J., Climent, D.M., Negulescu, C., Roullé, A., Sedan, O.: Limits on the potential accuracy of earthquake risk evaluations using the L'Aquila (Italy) earthquake as an example. Ann. Geophys. **58**, 1–17 (2015)

34. Mak, S., Clements, R.A., Schorlemmer, D.: Validating intensity prediction equations for Italy by observations. Bull. Seismol. Soc. Am. **105**, 2942–2954 (2015)
35. Mak, S., Schorlemmer, D.: Erratum to validating intensity prediction equations for Italy by observations. Bull. Seismol. Soc. Am. **106**, 2409–2413 (2016)
36. Faccioli, E., Cauzzi, C.: Macroseismic intensities for seismic scenarios estimated from instrumentally based correlations. In: Proceedings of the First European Conference on Earthquake Engineering and Seismology - Geneva, Switzerland, 3–8 September 2006, ECEES (2006)
37. Pasolini, C., Gasperini, P., Albarello, D., Lolli, B., D'Amico, V.: The attenuation of seismic intensity in Italy, Part I: theoretical and empirical backgrounds. Bull. Seismol. Soc. Am. **98**, 682–691 (2008)
38. Allen, T.I., Wald, D.J., Worden, C.B.: Intensity attenuation for active crustal regions. J. Seismol. **16**, 409–433 (2012)
39. Margottini, C., Molin, D., Serva, L.: Intensity versus ground motion: a new approach using Italian data. Eng. Geol. **33**, 45–58 (1992)
40. Giovinazzi, S.: The Vulnerability Assessment and the Damage Scenario in Seismic Risk Analysis (2005). http://www.digibib.tu-bs.de/?docid=00001757
41. Grünthal, G., Musson, R., Schwarz, J., Stucchi, M.: European Macroseismic Scale 1998 (EMS-98). Centre Europèen de Géodynamique et de Séismologie, Luxembourg (1998)
42. Spence, R.: Intensity, damage and loss in Earthquakes in seismic damage to masonry buildings. In: Bernardini, A. (ed.) Seismic Damage to Masonry Buildings, pp. 27–40. Balkema, Rotterdam (1999)
43. Musson, R.M.W., Grünthal, G., Stucchi, M.: The comparison of macroseismic intensity scales. J. Seismol. **14**, 413–428 (2010)
44. Giovinazzi, S.: Geotechnical hazard representation for seismic risk analysis. Bull. New Zeal. Soc. Earthq. Eng.. **42**, 221–234 (2009)
45. Lagomarsino, S., Giovinazzi, S.: Macroseismic and mechanical models for the vulnerability and damage assessment of current buildings. Bull. Earthq. Eng. **4**, 415–443 (2006)
46. Lagomarsino, S.: Vulnerability assessment of historical buildings. In: Oliveira, C.S., Roca, A., Goula, X. (eds.) Assessing and Managing Earthquake Risk: Geo-scientific and Engineering Knowledge for Earthquake Risk Mitigation: Developments, Tools, Techniques. Geotechnical, Geological and Earthquake Engineering, vol. 2, pp. 135–158. Springer, Netherlands (2006). doi:10.1007/978-1-4020-3608-8_7
47. Coburn, A., Spence, R.: Earthquake risk modelling. In: Earthquake Protection, pp. 311–352. Wiley, Chichester (2006)
48. Franchin, P.: A computational framework for systemic seismic risk analysis of civil infrastructural systems. In: Pitilakis, K., Franchin, P., Khazai, B., Wenzel, H. (eds.) SYNER-G: Systemic Seismic Vulnerability and Risk Assessment of Complex Urban, Utility, Lifeline Systems and Critical Facilities. GGEE, vol. 31, pp. 23–56. Springer, Dordrecht (2014). doi:10.1007/978-94-017-8835-9_2
49. Bramerini, F., Di Pasquale, G., Orsini, G., Pugliese, A.: Rischio sismico del territorio italiano. Proposta di una metodologia e resultati preliminari [Seismic risk of the Italian territory. Proposal for a methodology and preliminary results] - Technical report n. SSN/RT/95/01, Rome, Italy (1995)
50. Anhorn, J., Khazai, B.: Open space suitability analysis for emergency shelter after an earthquake. Nat. Hazards Earth Syst. Sci. **15**, 789–803 (2015)
51. Cioppi, E.: 18 maggio 1895: storia di un terremoto fiorentino [18 May 1895: history of a Florentine earthquake]. Osservatorio Ximeniano, Firenze (1995)
52. Guidoboni, E., Guidoboni, E., Ferrari, G.: Historical cities and earthquakes: Florence during the last nine centuries and evaluations of seismic hazard. Ann. Geophys. **38**, 617–647 (1995)

Enhancing Creativity in Risk Assessment of Complex Sociotechnical Systems

Alex Coletti[1][ID], Antonio De Nicola[2]([✉])[ID], and Maria Luisa Villani[2][ID]

[1] SMRC, Ashburn, VA, USA
acoletti@smrcusa.com
[2] ENEA-CR Casaccia, Via Anguillarese 301, 00123 Rome, Italy
{antonio.denicola,marialuisa.villani}@enea.it

Abstract. We propose the CREAM (CREAtivity Machine) software system to enhance the creativity of experts during vulnerability and risk assessment of complex sociotechnical systems. Our assumption is that a new idea related to a risk can be represented as a fragment of a conceptual model, here named *risk mini-model*, that can be generated by means of an ontology-based approach for computational creativity. In our solution risk mini-models activate a creative process for stakeholders to identify and understand risks. The whole set of risk mini-models for a specific risk constitutes a risk conceptual model. Such models are included in a knowledge base together with a domain ontology and a set of rules.

Keywords: Risk assessment · Computational creativity · Water system · Ontology

1 Introduction

Infrastructures, such as those for transport, telecommunication, water and energy services, are prime examples of sociotechnical systems. These systems are complex because they represent major communal investments, they are increasingly interconnected, and failures can severely impact communities livelihood and prosperity (from [30]). The analysis of likely system mishaps is essential in the definition of risk preventive plans. However, since failure rates can change with time, societal resilience, systems designs and re-allocation of resources, risk assessments often need critical reviews. Risks assessment protocols specific to each system types exist but the present trends of critical infrastructures to be increasingly interconnected makes them harder to validate because their endorsements require a wider range of specialized types of expertise.

Generally, the situations that hazards can create are not obvious to be conceived in advance. Risks elicitation is difficult and different views and types of expertise are needed when, for example, duration and intensity are unknown for a likely natural hazard. In this context, system actors, such as managers and operators, need a risk assessment management tool that enables them to access other work-group experiences while exploring new plausible scenarios.

© Springer International Publishing AG 2017
O. Gervasi et al. (Eds.): ICCSA 2017, Part II, LNCS 10405, pp. 294–309, 2017.
DOI: 10.1007/978-3-319-62395-5_21

Our objective is to enhance the creativity of experts by means of a software system that can be used during vulnerability and risks assessments workshops. We automatically generate input for risk models (i.e., conceptual descriptions) to be analyzed by a community of users. A risk model describes how stakeholders and experts perceive a certain risk for a system (or a subsystem), and the threats capable to exploit system vulnerabilities.

Our main assumption is that a (new) idea concerning a possible risk of a sociotechnical system can be represented as a fragment of a conceptual model, called *risk mini-model*. As such, it is possible to generate it by means of ontology-based techniques. The whole set of risk mini models, concerning some risk for a system, forms a risk model, that is, a formal conceptual representation of that risk. We propose to support the risks elicitation process through a suite of semantics-based tools for computational creativity. Indeed, risk mini-models would drive a creative process for stakeholders to identify and comprehend risks by incrementally constructing a shared knowledge base.

Our solution is based on a domain ontology together with a set of rules, on risk models and on the CREAM (CREAtivity Machine) software system [9,10]. This work is a follow-up to the results of an on-line experiment of a decision support system for Community Water Systems.

The paper is organized as follows. Section 2 presents related work in the area. Section 3 describes the water systems case study to which the creative approach is applied. Section 4 presents the risk models and their generation through CREAM, to the aim of enhancing creativity in risk assessment of complex sociotechnical systems. Finally Sect. 5 draws conclusions.

2   Related Work

Following [20], creativity is the ability to generate original ideas, concepts and objects. Our approach for computational creativity addresses the problem to enhancing human creativity by means of technology. In this respect, we propose a process and tools towards satisfaction of creativity needs in the activity of risks analysis for a complex sociotechnical system. As far as we know, this kind of application for computational creativity techniques is novel. Other works use ontologies for inventive problem solving [32] and for ontology-based identification of innovation challenges [33]. With these works we share the adoption of domain ontologies but we have a different objective as we aim at retrieving new "ideas" by means of patterns-based ontological queries.

Concerning the state of the art of technological frameworks for risk analysis, they generally implement quantitative methods, but there is an increasing interest in providing on-line decision systems realized by linking available quantitative resources (e.g., models, historical data and standards) to semantics to allow for integration of knowledge with qualitative local experiences [14,18]. Our work is also positioned in this line as our creative process is a means to building a new knowledge base for risk analysis of water systems, as described in Sect. 3. Another relevant work in this area is the CORAS platform for risk assessment of security critical systems [21], result of a EU-funded project, which introduces the field of model-driven risk assessment. Similarly to the works in model-driven engineering [24], CORAS proposes modelling methods and high-level graphical notations to support risk analysts in describing risks and related aspects, such as threat scenarios, to the purpose of analysis.

Several works addressed the creativity problem by proposing a paradigm shift from closed innovation to open innovation, where the boundaries of institutions are open to exploit also external ideas [5]. Along this line current research is focusing in defining collaboration frameworks for supporting innovation [1,13]. With these works we share the goal of involving a group of people in brainstorming activities but with respect to them we provide also automatically built ideas to stimulate the debate.

3   The Water Systems Case Study

3.1   Vulnerability Assessment Support System Experiment

In the United States, the Environmental Protection Agency [15] estimated that over one third of the total population gets some or all of their drinking water from highly vulnerable public systems. In Europe, over two thirds of all river and lake habitats and inland water species are in unfavorable conservation status, and about 25% of all groundwater bodies across Europe are in poor chemical status [31].

Therefore, small water system managers need easy access to local information that enables them to prepare for risks in ways that are most suited to the needs and priorities of their communities [25]. In this Section we use a Vulnerability Assessment Support System (VASS) experiment aimed at identifying the risks of water systems by means of focus groups of experts stakeholders in coastal communities. During the various phases of the experiment, the focus groups had the opportunity to compare their findings against reliable information sources and the data resulted to be crucial for the formulation of realistic action plans suitable for improving the supply of safe water to small communities.

The VASS experiments demonstrated that by coordinating focus group activities according to rather strict protocols, it is possible to obtain clear and transferable lists of risks descriptions in natural language.

The major finding of the experiment was that, to be useful, software tools need to provide common collaborative platforms for discussing vulnerability according facts and data, and facilitate the comparison of assessments according to standardized metrics. Therefore, as long as suitable software tools are available, qualitative risk-ranking approaches can provide managers, policymakers, and stakeholders key understanding of the risks of a system, and deliver the knowledge necessary for the formulation of realistic resilience plans.

These findings lead to a conceptual representation of the general definition of system risks into a Vulnerability Upper Model (VUM) [6]. This enabled the creation of an ontology (that is a formal, explicit specification of a shared conceptualization as per [3,17]) on the risks related to the water systems thus extending the ontological representation of the VUM.

3.2 VUM Ontology

As discussed in [4], ontological upper models can be a means to facilitate engineering of multi-disciplinary ontologies [23], such as those representing environmental and technological aspects of the knowledge concerning complex domains. Therefore, the generality of the VUM can be extended to represent risks for all types of technological systems (e.g., telco and energy networks), and could be suitable for establishing a unifying link of system specific risk ontologies. In this research work, the VUM was used to develop an ontology of risks and vulnerabilities specific to water systems as they relate to the effect of climate change on the frequency distribution of natural hazards.

A detailed analysis of the ontologies associated with the vulnerability of the assessment experiment was performed on the VASS as a preliminary step toward the validation of the viability of the semantic web solution for use in other cases. This analysis led to development of a new ontology from the mentioned Vulnerability Upper Model (VUM). The ontology consists of concepts linked to each other by a somewhat lower number of properties [7]. Interestingly, the definition of the VUM required to organize all the elements addressed in the vulnerability assessment into hierarchical lists. This demonstrated the ability of the VUM to support experts needs in building individual vulnerability ontologies. Given the variety of different concepts needed in the ontology, it is appropriate for sake of clarity, to focus first on the core high-level aspects of a system services (Core VUM), and provide the blueprints on how a detailed VUM ontology may be built.

Since our VUM ontology is mainly focused on natural hazards, the (natural) Environment Aspect at the top of Core VUM organizes all Hazards entities according to suitable general categories (e.g. Climatological, Ecosystem/Habitats, Geological, Hydrological and Weather). In turn, all Hazards are characterized by their set of Severity metrics defined to specify properties such as time, duration, location, and intensity. Individual Hazards of defined Severity property can be associated to collections of likely threats meaningful for selected System aspects of the assessment.

In the VUM, Threat and System Risk differentiate between the threats a hazard can pose to a system when likely Vulnerability conditions can make the threats significant to risk estimates. This way the System Risk can be defined according to vulnerability and the severity parameters within the System aspect. In turn, both System Risk and Vulnerability are also a property of the system. The System defines all the users interfaces with the participating Stakeholders. However, Stakeholders not only control the System through its interfaces, but, when the system fails to function within operational conditions, they are themselves affected by its Risk.

Figure 1 shows a pictorial representation of the VUM. As a proof of concept of the Vulnerability Upper Model, we constructed a VUM ontology[1] by extending it. The basic mechanism to extend the VUM is generalization/specialization [2], which is expressed by the relationship IS-A that allows the introduction of new concepts whenever they can be associated with more abstract ones. IS-A, is generally shortened to ISA, and the idea is that an ontology is built by first modeling the most general concepts characterizing the domain of interest (core concepts), and then sub-cases through more specialized concepts [2].

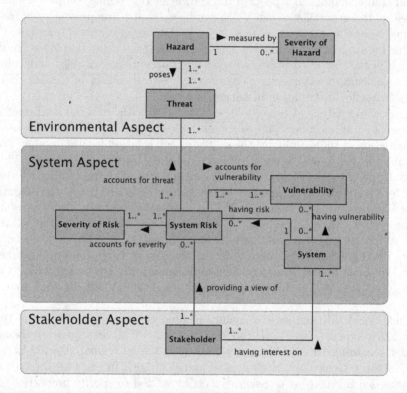

Fig. 1. The diagram of the Core VUM illustrates the conceptual relationships occurring between System Aspect and Environmental Aspect at one side and Stakeholder Aspect at the other side. A double conceptual pathway of relationships occurs between System and System Risk. System properties such as design or location affect its vulnerability and System Risk. However, System Risk affects also the System performance. Likewise, Stakeholders act as actors in their relationship with the System but are also impacted by the System Risk. For the sake of clarity the VUM is presented by means of the Unified Modeling Language (UML) graphical notation (www.uml.org).

By means of the ISA relationship, it is possible to organize a second level of logical units where the new concepts further specify the well-defined

---

[1] The VUM ontology is available at https://tinyurl.com/VUM-ontologyICCSA2017.

entities of the core VUM by providing more information content. This extended VUM model facilitates logical subgrouping of well-organized ontologies. The other upper level concepts (e.g., Stakeholder) belonging to the VUM can be specialized by means of the same technique (e.g., Manager ISA Stakeholder, User ISA Stakeholder, Regulator ISA Stakeholder, Operator ISA Stakeholder).

The ontology formally represents the risks descriptions identified in the VASS experiment.

Interestingly, while the semantic framework based on the VUM succeeded in organizing all the risks descriptions identified by the stakeholders of the VASS experiment into an ontology, it has also provided the methodology for knowledge enrichment. Machine generated risks are valuable to both verifying correctness and completeness of the VUM ontology, and in identifying new real risk situations.

For this reason we built a software system devoted to generating such (potentially) creative insights.

4   CREAM-Based Risk Models Generation

A risk model is a conceptual tool to support vulnerability assessment of complex sociotechnical systems. As mentioned in the introduction, it describes how stakeholders and experts perceive a certain risk for a system (or a subsystem). In our approach a risk model consists of a set of risk mini-models fully describing the specific risk. A risk mini-model structure is derived from the VUM. One example of risk mini-model is that including the following concepts: system risk, system, hazard, threat, severity, vulnerability and stakeholder, with the aim of identifying threats and system vulnerabilities related to a given system risk, from the perspective of different stakeholders. Figure 2 shows a simple template to provide a visual presentation of the risk mini-model just described, facilitate its exploration and understanding and support systems stakeholders in the risk assessment process of complex sociotechnical systems [22].

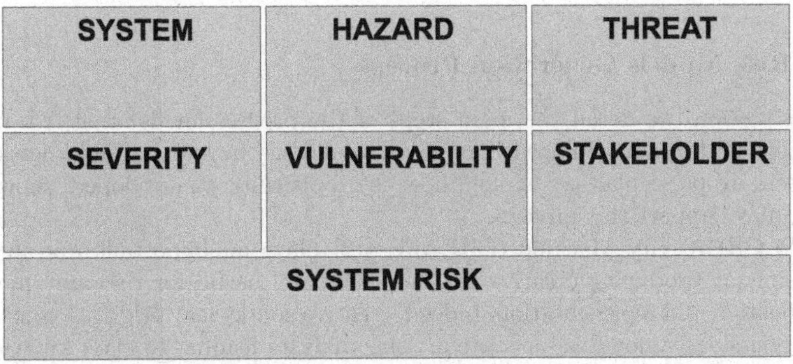

SYSTEM	HAZARD	THREAT
SEVERITY	VULNERABILITY	STAKEHOLDER
SYSTEM RISK		

Fig. 2. Risk mini-model template

The process to define a risk model is cyclic. Starting from a given knowledge on risks collected in a knowledge base, including a VUM ontology and a set of contextual rules, a set of *creative sparks* are automatically generated. We call *creative spark* a risk mini-model that has not been evaluated yet by humans. After a validation process, a subset of creative sparks may give rise to risk mini-models. These risk mini-models are stored in the knowledge base and used to enrich the risk model for a particular sociotechnical system. The aim is both to support the sharing of ideas among focus groups by leveraging on a multidisciplinary knowledge base concerning different aspects (e.g., technical and socio-economical) and different geographical areas and to overcome personality barriers that could arise in brainstorming activities by automatically proposing discussion hints (i.e. creative spark). This process is sketchily represented in Fig. 3 and will be deeply analyzed in the next subsection.

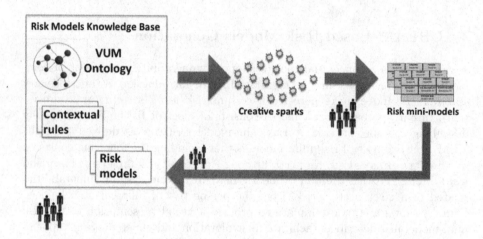

Fig. 3. Risk model lifecycle

4.1 Risk Models Generation Process

In this section, we detail the main steps of the process for defining risk models, based on formalized knowledge on risks collected by experts and knowledge engineers. In particular, we explain how the tools suite we developed, shown in Fig. 4, may support this process.

The CREAtivity Machine (CREAM) [12] plays an important role in this as it aims at producing creative sparks potentially useful for risk mini-models identification and representation. Indeed, creative sparks may originate constructive discussions among water system risks analysts leading to risks knowledge enhancement.

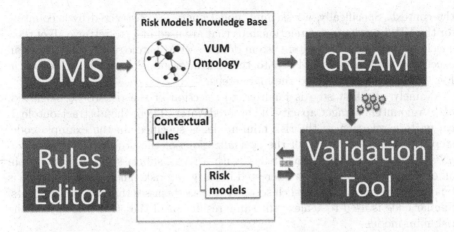

Fig. 4. A CREAM-based system for risk assessment consisting of the Ontology Management System (OMS), the rules editor, the CREAtivity Machine (CREAM), the validation tool, and the risks models knowledge base

Knowledge Base Building and Enrichment. This step consists of the following three activities: VUM Ontology engineering, contextual rules specification, and risk models definition. Generally, ontology engineering and evolution [11] aims at providing structural knowledge concerning the domain of complex sociotechnical systems by means of an ontology management system (OMS), as Protégé [29]. Rules writing aims at formally representing contextual knowledge to be applied on top of the ontology to represent specific aspects (e.g., nuclear power plants must be closed after 40 years of operation in State X). Risk models definition aims at identifying relevant risks by means of interviews and focus groups activities.

Creative Sparks Generation. This step is an automatic activity of the process, performed through the CREAtivity Machine. This is a software component aiming at realizing semantic bindings of abstract ontological models with domain-specific concepts, by exploiting entities and object properties specialization relations of an ontology. For this application, we consider bindings of risk mini-models structures, corresponding to patterns of the VUM, with the most specific concepts of the water system VUM ontology, preserving the semantics of the pattern. Thus, the creative sparks are the result of such binding function.

Indeed, in our computational creativity approach, we consider the risk models detection as a search process within a space consisting of the VUM ontology constrained by the contextual rules, which is updated and refined at each iteration of the process described in Subsect. 4.1, as a result of the validation activity by the risk analysts.

Semantic binding is realized through a SPARQL [27] select query by using three types of knowledge: from the risk mini-model, from the domain and from the context. Specifically, a risk mini-model query is constructed by accounting for the VUM concepts and relationships that are used in it, to retrieve all of their specializations from the water system domain VUM ontology. Figure 5 shows an excerpt of such a query related to the risk mini-model template presented in Fig. 2, where we highlighted the main steps.

Namely, the first step is finalized to selection of the risk to be discussed, e.g., AmountOfStorageCapacityOfThreatedWater. Then, the abstract ontological pattern referring to the risk mini-model is specified. In the example code, step 2 includes retrieving all the specializations of the object property *havingRisk*, connecting a `System` subclass with a `Risk` subclass, in order to obtain all of the `System` subclasses interested in the given risk. The subsequent code is finalized to retrieving, from such set, just those subclasses that are leaf concepts. Similar code is used to achieve the same result for all the other concepts of the risk mini-model.

The effect of a VUM pattern-base query is a set of risk mini-models resulting by linking each semantic relationship of the pattern to one of the object properties specializing it, and each VUM construct involved in that relationship to a leaf subclass of the corresponding VUM concept. This pattern-base query can be refined with contextual rules, each providing a filter SPARQL statement. The final code block of Fig. 5 shows an example of implementation of contextual rules, encoding the situation where the specific geographical location of the water system under analysis excludes tropical-types hazards.

CREAM was implemented in Java, based on the Apache Jena framework including the ARQ library [19], which implements a SPARQL 1.1 engine. The application is configurable with respect to the ontology, contextual rules and query patterns (risk mini-models), by means of a XML file.

Risk Mini-Models Definition. This step takes as input the creative sparks generated by CREAM and aims at selecting a (sub)set of them as components of the envisaged risk model. It includes an analysis of the candidate creative sparks. Our main assumption is that a creative spark can be promoted from its candidate status to a "full-fledged" risk mini-model if it is creative and, hence, novel and valuable [26]. Along these lines, we envisage that the brainstorming activity devoted to accept or modify the suggested creative spark also foster discussions and ideas exchange in a community of users.

However whereas assessing relevance of a creative spark can be done (at least in principle) by assigning, for instance, a relevance weight to its elements and assessing its novelty can be done by measuring how much it is different from risk mini-models already considered in the risk model [26], the main difficulties concern its plausibility. The issue here is to determine how much the risk mini-model is reasonable and likely. This would require availability of a formal specification of common-sense knowledge that is currently an open issue in the artificial intelligent research field [28]. We envisage also to use creative sparks that were discarded in the previously mentioned approaches as hints for identifying new risk mini-models. Different methods can be adopted to this purpose.

```
PREFIX : <http://www.../VUMOnto#>
................
SELECT DISTINCT ?risk ?system
?vulner ?hazard ?threat ?stakehldr
WHERE {

# selection of risk
?risk rdfs:subClassOf
AmountOfStorageCapacityOfTreatedWater .

# pattern definition
?ipros a owl:ObjectProperty;
   rdfs:domain ?system1 ;
   rdfs:range ?risk ;
   rdfs:subPropertyOf havingRisk .
   FILTER (! (?system1 = System))
..................
# selection of leaf concepts
?system rdfs:subClassOf System  .
   FILTER NOT EXISTS {
   ?sub_sy1 rdfs:subClassOf ?system  .
   ?sub_sy2 rdfs:subClassOf ?system .
   FILTER (! (?sub_sy1 = ?sub_sy2))   }
   FILTER EXISTS {
   ?system rdfs:subClassOf ?system1  .}
.......................

# contextual rules
FILTER (
!(?hazard in (TropicalStorm,
   TropicalDepression))  )
.......................
}
```

Fig. 5. Pattern-query-excerpt

A promising one is the appreciative inquiry (AI) methodology [8]. This seems to be particularly suitable in our case as the first step of the AI methodology is to recognize the best in the item to be discussed (and, hence, even the best in the unlikely creative sparks) as baseline to imagine something completely new (as new risk mini-models) to be determined in subsequent steps of the brainstorming activity. Automatizing the process of generating creative sparks could support the group in "thinking outside the box" as no one would be ashamed in proposing apparently absurd and unlikely situations to be discussed.

Figure 6 shows an example of how participatory vulnerability assessment can be done in practice by a focus group supported by a collaboration tool that we are defining. A creative spark produced by CREAM and selected for evaluation is represented as a fragment of conceptual model in the center of a panel. The participants are required to contribute to a discussion on that basis, possibly leading to a new definition of risk mini-model. To this aim, a set of pre-made sticky notes referring to concepts retrieved from the ontology, and representing alternative or additional concepts for the risk description, can be moved by the focus group coordinator over each box in the diagram during the brainstorming. The ontology allows this activity be supported and guided by the tool through reasoning techniques, for example exploiting semantic dependence relationships between the concepts, application of contextual rules and performing redundancy checks.

The result of this activity will be a number of risk mini-models, each contributing to assess the overall level of risk of the corresponding system. A detailed summary of best practices for risk ranking methods is available in [16]. This level could depend, for instance, on information pre-loaded in the system as the level of threat posed by weather and climate on system or component, the vulnerability of a system or component, the severity of the failure or the likelihood of the event. Figure 6 shows the case where the level of risk derives from likelihood and severity according to the confusion matrix depicted on the bottom right of the image.

It should be noted also that new knowledge elicited from the users can be used to enrich the knowledge base. In fact new risk models arising from brainstorming activities could be an input for ontology enrichment whereas discarded creative sparks could originate new contextual rules or ontology updates. Hence such knowledge will be available also to other focus groups with different competences and from different geographical regions.

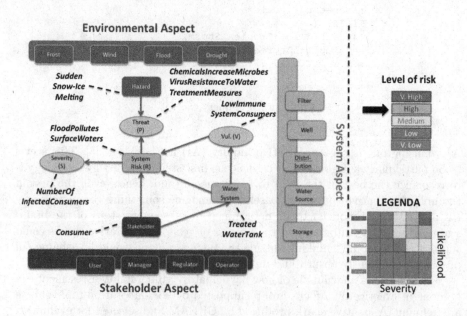

Fig. 6. Partecipatory vulnerability assessment of a creative spark.

Finally the risk mini-models are added to the risk model in order to enrich the knowledge base.

4.2   Risk Mini-Models Examples

In this section we present some examples of risk mini-models generated by means of the CREAM software and validated by experts in the field. We provide also a short natural language description of them to show how they can be interpreted by experts.

Risk Mini-Model 1
SystemRisk : FloodPollutesSurfaceWaters.
Hazard : SuddenSnow-IceMelting.
Threat : ChemicalsIncreaseMicrobes/VirusResistanceToWater TreatmentMeasures.
System : TreatedWaterTank.
Vulnerability : LowImmuneSystemInConsumers.
Stakeholder : Consumer.
Severity : NumberOfInfectedConsumers.

Description of Risk Mini-Model 1. A flood caused by unpredictably fast snow melting rate, transports surface pollutants (possibly including antibiotics from farming) into surface waters. Germs with resistance to water treatment measures in the drinking water system, colonize the treated water tank. Due to the presence of unsanitary water in the system, low immune system consumers are affected.

Risk Mini-Model 2
SystemRisk : DieOutOfWildlifePollutesSurfaceWaters.
Hazard : UndeterminedViralInfection.
Threat : WaterTreatmentOxidationProcessCreatesSecondary Carcinogens.
System : TreatedWaterTank.
Vulnerability : UnmonitoredTraceChemicalsExceedSafeLimits.
Stakeholder : Consumers.
Severity : NumberOfInfectedConsumers

Description of Risk Mini-Model 2. Significant water pollution due to wildlife die-out, the carcinogens, normally produced in untraceable amounts during the oxidation occurring in bio-chemical water treatment, sudden surges in concentration. Safety margins are exceeded and consumers might get exposed.

Risk Mini-Model 3
SystemRisk : RuptureOfAgeingWaterDistributionPipes.
Hazard : UndergroundWaterFlow.

Threat : SinkHoleWeakensHighwayStructures.
System : PilastersOnSandstone.
Vulnerability : VehicleSafety.
Stakeholder : HighwayUsers.
Severity : NumberOfAccidents

Description of Risk Mini-Model 3. Rupture of ageing water distribution pipes running near highway pilasters built on sandstone could trigger formation of sink holes thus weakening the highway structure. This affects vehicle safety which is a relevant issue for highway users.

Risk Mini-Model 4
SystemRisk : DisgruntledEmployee.
Hazard : PollutedWaterSource.
Threat : WrongAlternateWaterSourceInjectsPollutantsIntoDistributionSystem.
System : AlternateWaterSources.
Vulnerability : InsufficientPeriodicWaterQualityMonitoringProtocols.
Stakeholder : Consumers.
Severity : NumberOfInfectedConsumers

Description of Risk Mini-Model 4. Disgruntled employees could purposely connect a polluted alternate water source to the system. Inadequate periodic water quality monitoring could expose consumers to polluted water.

When using a domain specific ontology (like the VUM ontology) and a set of rules, CREAM is enabled to bound in a risk mini-model domain entities and object properties residing in related but different informational spaces. In the hypothetical examples, CREAM applies documented and factual threats of system risks to the water system and creates risk mini-models that may be plausible even though they may be neither relevant nor likely. Therefore, different risks observed in circumstances that may not have impacted water systems in the past, can be included for consideration in a risk assessment as long as they comply with the conceptual framework of the vulnerabilities of the system. In the examples, CREAM generated risk mini-models are envisioned as prompting complete assessments of system risks in order to reduce the need of direct participation of vulnerability experts to focus groups.

5   Conclusions

Risk assessment of complex sociotechnical system requires both a deep knowledge of the infrastructure and a creative attitude in envisioning system mishaps. In this context we provided an ontology-based method and the CREAM software system to support such process. Our approach is conceived to be supported also by human activities aimed at validating the results, stimulating discussions and enriching the knowledge base to improve the system performance.

CREAM improves system performance by leveraging two of its primary functionalities. Firstly, the semantic binding of risk mini-models facilitates the sharing of experiences among focus groups in spite of geographical and socio-economical separation of the systems. Secondly, the machine generated creative spark, while driving the discussion toward added creativity and original thinking, also cuts across the personality barriers and the biases that often affect focus group deliberative objectivity.

At the best of our knowledge the presented approach is novel for two reasons. From a scientific perspective, our ontology-based approach to stimulate and generate new ideas for knowledge enhancement is novel. From an application perspective it represents one of the first works to apply computational creativity methods to the engineering problem of risk assessment in complex sociotechnical systems.

As future work we intend to further detail the risk mini-models definition step of our method by focusing on the creative brainstorming approach.

Acknowledgements. The work of Alex Coletti was conducted at SMRC, with partial funding from a NOAA-CPO research grant. The work of Antonio De Nicola and Maria Luisa was conducted at ENEA and was partially supported by the Italian Project ROMA (Resilience enhancement Of Metropolitan Area) (SCN_00064). We kindly acknowledge Bálint Bálazs and Michele Melchiori for stimulating discussions.

References

1. Bálazs, B.: Social innovation and science cafe in the digital age. Keynote speech at the second international workshop on mobile and social computing for collaborative interactions (MSC 2015), October 2015
2. Borgida, A., Mylopoulos, J., Wong, H.K.T.: Generalization/specialization as a basis for software specification. In: Brodie, M.L., Mylopoulos, J., Schmidt, J.W. (eds.) On Conceptual Modelling, pp. 87–117. Springer, New York (1984)
3. Borst, W.N.: Construction of engineering ontologies for knowledge sharing and reuse. Universiteit Twente (1997)
4. Camporeale, C., De Nicola, A., Villani, M.L.: Semantics-based services for a low carbon society: an application on emissions trading system data and scenarios management. Environ. Model. Softw. **64**, 124–142 (2015)
5. Chesbrough, H.: From open science to open innovation. Institute for Innovation and Knowledge Management, ESADE (2015)
6. Coletti, A., De Nicola, A., Villani, M.L.: Building climate change into risk assessments. Nat. Hazards **84**(2), 1307–1325 (2016)
7. Coletti, A., Howe, P., Yarnal, B.: Local resilience of community water systems and severe weather patterns. In: 94th AMS Conference, Atlanta, GA (2014)
8. Cooperrider, D., Whitney, D.D: Appreciative Inquiry: A Positive Revolution in Change. Berrett-Koehler Publishers (2005)

9. De Nicola, A., Melchiori, M., Villani, M.L.: A semantics-based approach to generation of emergency management scenario models. In: Mertins, K., Bénaben, F., Poler, R., Bourrières, J.-P. (eds.) Enterprise Interoperability VI. PIC, vol. 7, pp. 163–173. Springer, Cham (2014). doi:10.1007/978-3-319-04948-9_14

10. De Nicola, A., Melchiori, M., Villani, M.L.: A lateral thinking framework for semantic modelling of emergencies in smart cities. In: Decker, H., Lhotská, L., Link, S., Spies, M., Wagner, R.R. (eds.) DEXA 2014. LNCS, vol. 8645, pp. 334–348. Springer, Cham (2014). doi:10.1007/978-3-319-10085-2_31

11. De Nicola, A., Missikoff, M.: A lightweight methodology for rapid ontology engineering. Commun. ACM **59**(3), 79–86 (2016)

12. De Nicola, A., Villani, M.L.: Creative sparks for collaborative innovation. In: Rossignoli, C., Virili, F., Za, S. (eds.) Digital Technology and Organizational Change: Reshaping Technology, People, and Organizations Towards a Global Society. LNISO, vol. 23. Springer, Cham (2018). doi:10.1007/978-3-319-62051-0

13. Diamantini, C., Missikoff, M., Potena, D.: Open innovation in virtual enterprises: an ontology-based approach. In: Poler, R., Doumeingts, G., Katzy, B., Chalmeta, R. (eds.) Enterprise Interoperability V: Shaping Enterprise Interoperability in the Future Internet, pp. 165–175. Springer, London (2012)

14. Dow, A.K., Dow, E.M., Fitzsimmons, T.D., Materise, M.M.: Harnessing the environmental data flood: a comparative analysis of hydrologic, oceanographic, and meteorological informatics platforms. Bull. Am. Meteorol. Soc. **96**(5), 725–736 (2015)

15. EPA2013: Geographic information systems analysis of the surface drinking water provided by intermittent, ephemeral and headwater streams in the U.S. (2014)

16. Florig, H.K., Granger Morgan, M., Morgan, K.M., Jenni, K.E., Fischhoff, B., Fischbeck, P.S., DeKay, M.L.: A deliberative method for ranking risks (i): overview and test bed development. Risk Anal. **21**(5), 913–913 (2001)

17. Gruber, T.R.: A translation approach to portable ontology specifications. Knowl. Acquis. **5**(2), 199–220 (1993)

18. Hunt, J.R., Baldocchi, D.R., van Ingen, C.: Redefining ecological science using data. In: Hey, T., Tansley, S., Tolle, K. (eds.) The Fourth Paradigm: Data-Intensive Scientific Discovery, pp. 21–26. Microsoft Research, Redmond, Washington (2009)

19. Apache Jena, version 2.11.1 (2013). http://jena.apache.org

20. Kusiak, A.: Put innovation science at the heart of discovery. Nature **530**(7590), 255 (2016)

21. Lund, M.S., Solhaug, B., Stølen, K.: Model-Driven Risk Analysis: The CORAS Approach. Springer Science & Business Media, Heidelberg (2010)

22. Lurie, N.H., Mason, C.H.: Visual representation: implications for decision making. J. Mark. **71**(1), 160–177 (2007)

23. Noy, N.F.: Semantic integration: a survey of ontology-based approaches. ACM Sigmod Rec. **33**(4), 65–70 (2004)

24. OMG-MDA: MDA guide, version 1.0.1 (2003). http://www.omg.org/mda/presentations.htm. Accessed 4 Mar 2016

25. Pagano, A., Giordano, R., Portoghese, I., Fratino, U., Vurro, M.: A bayesian vulnerability assessment tool for drinking water mains under extreme events. Nat. Hazards **74**(3), 2193–2227 (2014)

26. Pease, A., Winterstein, D., Colton, S.: Evaluating machine creativity. In: Workshop on Creative Systems, 4th International Conference on Case Based Reasoning, pp. 129–137 (2001)

27. Pérez, J., Arenas, M., Gutierrez, C.: Semantics and complexity of SPARQL. In: Cruz, I., Decker, S., Allemang, D., Preist, C., Schwabe, D., Mika, P., Uschold, M., Aroyo, L.M. (eds.) ISWC 2006. LNCS, vol. 4273, pp. 30–43. Springer, Heidelberg (2006). doi:10.1007/11926078_3

28. Speer, R., Havasi, C., Lieberman, H.: AnalogySpace: reducing the dimensionality of common sense knowledge. In: Proceedings of the 23rd National Conference on Artificial Intelligence, AAAI 2008, vol. 1, pp. 548–553. AAAI Press (2008)

29. Stanford: Protégé ontology management system (2016). http://protege.stanford.eduabout.php. Accessed 24 Feb 2016

30. Weijnen, M.P.C., Herder, P.M., Bouwmans, I.: Designing complex systems: a contradiction in terms. In: Eekhout, M., Visser, R., Tomiyama, T. (eds.) Delft Science in Design^2. A Congress on Interdisciplinary Design, pp. 235–254 (2008)

31. Werner, B.: European waters - current status and future challenges - synthesis. EEA report No. 9/2012 (2012)

32. Yan, W., Zanni-Merk, C., Cavallucci, D., Collet, P.: An ontology-based approach for inventive problem solving. Eng. Appl. Artif. Intell. **27**, 175–190 (2014)

33. Zanni-Merk, C., Cavallucci, D., Rousselot, F.: An ontological basis for computer aided innovation. Comput. Ind. **60**(8), 563–574 (2009)

Spatial Analysis and Ranking for Retrofitting of the School Network in Lima, Peru

Angelo Anelli[1], Sandra Santa-Cruz[1], Marco Vona[2(✉)],
and Michelangelo Laterza[2]

[1] Pontifical Catholic University of Peru, Lima, Peru
{angelo.anelli,ssantacruz}@pucp.edu.pe
[2] University of Basilicata, Potenza, Italy
{marco.vona,michelangelo.laterza}@unibas.it

Abstract. Retrofitting and management strategies of existing buildings are actually a crucial topic. In this work, an approach based on GIS and MCDM method has been used in order to define a retrofitting ranking for seismic risk management. The main goal is to define a framework based on a multidisciplinary approach. The proposed procedure can be used in different ways and applications. It can be the basis of seismic risk mitigation strategies which are a typical problem of public administrations. Due to both the amount of essential buildings that require seismic retrofitting and the restricted economic availability, it is necessary to prioritize interventions on large territorial scale in order to optimize the allocation of available economic resources and ensure an efficient seismic risk mitigation. The paper provides a simple and rational seismic risk mitigation policy in order to consider the possible variables or disciplines that are not generally integrated in studies of risk (not only seismic) on a large territorial scale.

Keywords: Seismic risk mitigation strategies · Seismic losses evaluation · GIS · Ranking list

1 Introduction

In last years, several work have been carried out about seismic risk assessment and mitigation of urban areas (for example but not only [1–4]). The most recent seismic risk studies have been extensively referred to the resilience concept [5–7]. The operative implementation of resilience concept is a hard work, particularly in complex and strongly heterogeneous system. Several applications have been carried out on a different territorial scale and in some cases seem that the territorial scale could affect the methods and expected results.

In fact, there are several cases in which many of these approaches could fail in order to obtain realistic scenarios and set up effective mitigation or management strategies. These topics are relevant when megalopolises (or megaregions) and their sub systems are investigated. In particular, in modern earthquake engineering, the megalopolises should be considered with their fundamental and significant characteristics. In fact, the best cases where modern networks (as lifelines, rail, and highway, etc.) have

O. Gervasi et al. (Eds.): ICCSA 2017, Part II, LNCS 10405, pp. 310–325, 2017.
DOI: 10.1007/978-3-319-62395-5_22

been accurately designed in order to guarantee optimal development, and other cases in which the strong and very fast growth is often based on chaotic and uncontrolled processes. In these last cases, most of validated models and methods could be not applicable due to the poor level and reliability of available information.

On the other hand, the strong and continuous improvement of tools and instruments, the availability of new and more powerful technologies have made possible significant progress. In particular, using open source tools and database, and low cost information is possible to define accurate and always updated investigation. In this way, the developing trend and its strongly variability can be monitored and investigated.

In this study, a concrete application to school network of Lima, Peru, is reported. The work is based on intensive use of database on GIS integrated with seismic risk and resilience studies. Consequently, the complex system of Lima has been treated considering it on a smaller territorial scale. The analyses have been carried out on single school, census section, and finally district. In fact, on the basis of the carried out investigations, the interaction between different parts of the megalopolis are actually limited. The study is strongly in progress and the strategies to link the different parts of the megalopolis will be considered a result.

2 The Case Study

2.1 Description of the Lima Metropolitan Area

Lima is one of the oldest cities in Latin America. In Lima several issues are overlapped. The Lima Metropolitan Area comprises the 43 districts of Lima and the 6 districts of the close province of Callao. The metropolitan area has an area of about 2800 km^2 and a population of 10,000,000 according to INEI statistics [8–10]. Monumental and cultural heritage are quite diffuse in the town and could condition the planning strategies.

In Lima, the several neighborhoods (called districts) are strongly different in terms of average income per citizen, services, network, lifelines, and security [11]. From economic point of view, the differences between districts are also strong and these differences play a fundamental role on all other aspects which influence the school management, security and vulnerability, i.e. their resilience.

In fact, several problems in management of the school are closely linked with the socio - economic aspects as well as the economic available resources. In the poorest districts, the basic services (water, electricity, sewerage and sanitation) are available for only a few hours a day or unavailable and the lifelines are inefficient, insufficient and highly vulnerable [12].

Moreover, it is to be highlighted that a fundamental question is the security of the school, its surrounding area, and the district where each school is located. Several problems are related to the perception of the security and real security. A significant part of population has been victim of theft, in particular the younger population. In any case, several districts with low average income have strong problem of security, as Rimac, El Agustino, Villa El Salvador, San Juan de Lurigancho, La Victoria, Callao and Los Olivos. Security in each district is in charge of National Police service, municipal police and private security companies. The number of policemen is dependent on the available economic resources of each district (ranging approximately from about 5000 to 100 citizens per police officer official).

The security problem is visible in the enclosures, gates and defensive walls of the different districts. In school complexes, the same condition is visible and it is important to highlight that a significant part of the management and employment of the economic resources is used for security problems.

The current status of the city is strongly dependent form the substantial absence of a global or local urban planning. In the last decades, the city has grown in an uncontrolled and often in an illegal way (see Fig. 1). As consequence, significant risks anthropogenic are present in the urban area.

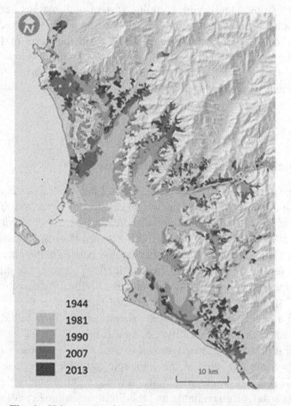

Fig. 1. Urban expansion of Lima metropolitan area [11].

The above descriptions are useful for the future resilience study and their influence can also be seen from the analysis of the available database.

The research project SIRAD [13] on the seismic risk in the metropolitan area of Lima and Callao has divided the territory of Lima in five subsoil category:

- Zone I (S1) corresponds to the hard ground or rock. This area has not local amplification;
- Zone II (S2) consists of fine granular soils and colluvial and alluvial clay soils on gravel. This area has moderate amplification effects;
- Zone III (S3) corresponds to sandy soil without the presence of water. The soils in this area are very durable but have important effects of amplification;
- Zone IV (S4) is formed by sandy soil with water;
- Zone V comprises filler soil.

In zone IV and V, there are very important amplifications that may lead to complications such as instability of the structure and the phenomenon of liquefaction. In the Zone V there are not schools.

2.2 Seismic Risk of Public Schools in the City

The Lima schools were often built with poor construction practices due to the lack of regulated procedures and quality supervision [14]; these problems are very common in the city especially on illegal constructions. Most of the school buildings in Lima were built in a staged construction process and additional capacity has been added when there has been money from either families or the government [15].

In many schools the safety standards are not respected (i.e. narrow corridors, evacuation routes that include staircases with low parapets, unsupervised alcoves, dead ends, incorrect definition and maintenance of safety areas and emergency exits, etc.). Moreover, government entities, academic staff, teachers, and healthcare workers lack the proper training required to provide information or humanitarian aid to students affected by seismic events [16].

Based on the available data of previous studies [15, 17, 18], 1825 schools have been analyzed, for a total of 7428 pavilions. The original data have been opportunely elaborated and integrated coherently with the goal of this work.

The pavilions built area is approximately 2.055 million square meters and they house about 665 thousand students.

As reported in [15], based on the collected geo-referenced information, for the city of Lima 6 types of school buildings have been defined: modular 780-PRE (PRE), great school unit (GUE), modular 780-POST (POST), unconfined masonry (MS-PRE-B), adobe walls (A-PRE-B) and prefabricated lightweight material (PREF). In this study 1512 public schools were visited (a total of 1870 public schools are located in the field of study) in order to collect information, characterize their building types and calculate their probable losses.

The PRE building type (4401 pavilions) is a modular system of reinforced concrete frames and confined masonry, designed and built in the years preceding the entry into force of the anti-seismic regulations of 1997. It consists of adjacent classrooms with longitudinal corridor along the building's facade. The staircase is located next to the pavilions. The modular 780-PRE was used since 1960, with small architectural and structural variations according to the construction year. These modifications do not change the main structural system based on reinforced concrete structures. They have produced an addition of columns in the corridors, changes in the dimensions of the beams and columns and changes in the classrooms and in the windows. For the purposes of this evaluation, all these variants have been considered within the modular 780-PRE. The modular buildings 780-PRE suffered significant damage in the earthquakes of the past, mainly because of limited lateral rigidity that cause shear failure of short column, problems of connection, etc.

The GUE building type (57 pavilions) consists of reinforced concrete frames and masonry walls, which vary in thickness depending on the location and year of construction. These school buildings are very old; on average, these were built fifty years ago. They were designed to accommodate a large number of students. Until 1960, the buildings of the type GUE were principally in masonry with thickness of 25 cm. They had openings of small size and a good amount of high walls, which have conferred an adequate stiffness and resistance to the structures. After 1960 the GUE system was changed. Currently, in Lima and in other cities of Peru, some buildings GUE have undergo an expansion and modernization process.

The building type currently used in Peru for the school buildings is the modular 780-POST (997 pavilions in Lima Metropolitan Area). It is a modular system of reinforced concrete frames and confined masonry. The distribution of classrooms, stairwells and corridor locations is similar to the modular system 780-PRE, but only this building type is designed and built through the Peruvian seismic design code of 1997 or later editions. These regulations increased the requirements of rigidity of the structures. Seismic behavior of POST building type was tested in two recent earthquakes in southern Peru. Pavilions of this type were not damaged.

The MS-PRE building type (1388 pavilions) consists of unconfined masonry walls, similar to the modules 780 but with lightweight roofs, usually covered with calamine in the last level. This configuration is more vulnerable to seismic action of the type GUE, Modular 780-PRE and Modular 780-POST, because they do not have the rigid diaphragm which binds and limits the walls. In an earthquake, the structures could collapse for the overturning of the walls. This system was built according to previous codes to 1997, and usually its construction was carried out without any technical supervision.

The A-PRE-B building type (147 pavilions) are made of adobe walls. They are generally older buildings that have performance undesirable in case of earthquakes, low strength and brittle fracture. The Peruvian seismic design code of 1997 prohibited the use of this construction type for the strategic buildings.

Finally, 438 pavilions of the considered school buildings are PREF type. They were built with light prefabricated materials. Due to their low mass they are not particularly vulnerable to earthquakes but in most cases these structures do not offer adequate lighting conditions, and not even adequate protection against rain or other environmental phenomena.

In Figs. 2, 3 and 4 are respectively reported the number of buildings, building area and students of each building type related to the subsoil category e.g. to the seismic amplification factor.

According to the aforementioned study, only the pavilion built after 1997 have acceptable risk levels, because for the remaining structures, the main source of risk is the lack of modern seismic criteria in the design and construction of buildings. It shows

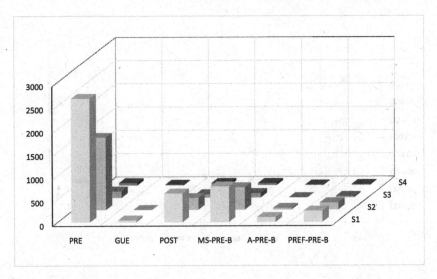

Fig. 2. Synthetic view of the buildings number, in terms of buildings types and type of soil (hazard).

that there is a possibility to generate critical situations due to lack of operability in the educational constructions even under medium intensity earthquakes. This would mean that for the school buildings analyzed in this paper, only 14.58% of their built area is safe and 86.33% of the student population is exposed to high risk.

Anyway, according to the Ministry of Education in Peru [9], more than 50% of the school buildings in Lima require total replacement to bring them into conformance with the Peruvian building code.

These alarming conditions are due to the large percentage of buildings built before 1997, with structural systems less efficient to seismic loads than those built in later years.

As reported in [15], the results of that work confirm the presence of a high risk in the educational buildings of all districts of Lima and Callao, and reveal the urgent need to strengthen their structures. These results refer to each pavilion and represent their repair cost for an occasional earthquake (magnitude $M_w = 8.1$), a frequent earthquake (magnitude $M_w = 6.8$) and the probable annual loss determined with a probabilistic analysis of seismic risk, considering several occasional earthquakes. These repair costs are normalized to the reposition cost of the pavilions.

In this paper, the aforementioned cost ratios have been processed and associated to each school building, weighting their built area. In particular, for each school, three different weighted normalized repair cost indices I_R are determined according to the following type of formula:

$$I_R = \sum_{i=1}^{n} \frac{A_i \cdot R_i}{A_T} \qquad (1)$$

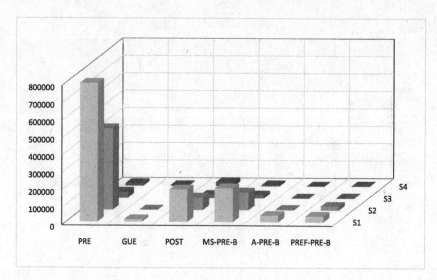

Fig. 3. Synthetic view of the built area, in terms of buildings types and type of soil (hazard).

where n is the number of school pavilions, A_i is the built area of each pavilion, A_T is the total built area of the school pavilions, and R_i is the aforementioned normalized repair cost for an occasional earthquake ($M_w = 8.1$), for a frequent earthquake ($M_w = 6.8$), and the normalized probable annual loss for several occasional earthquakes, respectively.

In this way, for each school, different levels of risk have been determined based on the available data and the importance of the exposed structures. The Fig. 5 shows the

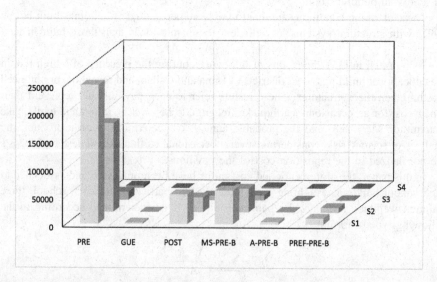

Fig. 4. Synthetic view of the student's number, in terms of buildings types and type of soil (hazard).

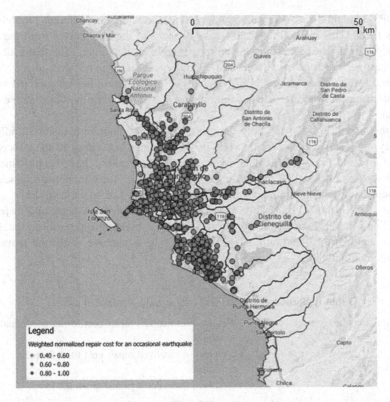

Fig. 5. Public schools in Lima Metropolitan Area with a high weighted normalized repair cost index (>0.40).

public schools with a high weighted normalized repair cost index (>0.40). This index is referred to the occasional seismic event that generates an acceleration of 0.25 g on the coast of Peru (earthquake of magnitude $M_w = 8.1$) with a return period T_R between 75 and 100 years. As shown, the school buildings have a high risk in all districts of Lima Metropolitan Area.

Due to the remarkable amount of school buildings that are extremely vulnerable and the limited economic availability, a prioritization strategy of interventions is required.

3   Retrofitting Prioritization Ranking

3.1   Methodology

In the present work, to develop rational and optimal prioritization rankings of interventions, multiple conflicting criteria have been elaborated through the TOPSIS method [19] and all the collected information have been processed in a geographic information system (GIS) for its analysis and computational management.

TOPSIS method has been already applied successfully for different problems of various fields. This method is based on the distance of the alternatives from the ideal solutions, namely the ideal solution and the negative ideal solution. The first is defined by the best performance of all attributes. On the other side, the negative-ideal solution is defined by the worst performance values. The best solution is characterized by the shortest distance from the ideal solution and the farthest from the negative-ideal solution.

In this paper, in order to apply the proposed methodology and ensure a multidisciplinary approach with a holistic view of the problem, according to the concept of resilient city and the available data, fourteen evaluation criteria have been defined.

The authors believe that the city resilience is based on three main aspects: seismic risk parameters, emergency management and integration and social cohesion.

The Table 1 shows the relationships between the resilience aspects and their corresponding criteria. Four of the considered criteria – building type, subsoil category, lifelines, and historical and cultural significance – are qualitative, while the remaining ones are quantitative.

Table 1. Resilience aspects and corresponding judgment criteria.

Seismic risk parameters	C_1	Weighted normalized repair cost for an occasional earthquake
	C_2	Weighted normalized repair cost for a frequent earthquake
	C_3	Weighted normalized probable annual loss
	C_4	Building type
	C_5	Subsoil category
	C_6	Density of school population
Emergency management	C_7	Lifelines
	C_8	Free usable area
	C_9	Quantity of roads in a radius of 500 m from the school
	C_{10}	Distance from the nearest main road
	C_{11}	Density of population in the surroundings
Integration and Social cohesion	C_{12}	Historical and cultural significance
	C_{13}	Social economic condition
	C_{14}	Community organization

In the first resilience aspect, six sub-criteria have been considered to analyze the factors that determine the seismic risk of the buildings (i.e. hazard, vulnerability and exposure). Three of these six criteria represent the weighted normalized repair cost indices discussed above, while the remaining three criteria analyze the collapse modes of the structures and their age, the problems of liquefaction and instability in the constructions, and the number of pupils exposed to the risk, respectively.

In the second group, five pertinent criteria have been analyzed in order to highlight how the lifelines, green areas, streets and density of the population surrounding the schools are important to manage and recover the emergency.

Finally, in the last aspect, the importance of integration and social cohesion on the city resilience has been underlined through three relevant criteria that consider the historical, cultural, socio-economic and organizational conditions of the community. These criteria safeguard the spirit of the community, try to mitigate the socio-economic inequality in order to ensure a more equal distribution of primary goods (security, justice, health, basic services, etc.) and promote the cooperation of the community with the schools, respectively.

As regards the case study, the importance of a multidisciplinary approach to define the priorities of interventions on large territorial scale, has also been highlighted in another studies [17, 18, 20]. In these works the need to use the analyzed criteria is validated. Based on each criterion, the schools have been evaluated in order to construct a decision matrix (1825 × 14).

Using Geographic Information Systems (GIS), according to each criterion, the schools have been analyzed separately to consider their real geographical distribution and identify the spatial and territorial relationships of the school buildings with their surroundings. Moreover, for each criterion has been assigned a weight.

In order to evaluate quantitatively the qualitative criteria and determine the weights of the criteria is necessary a conversion of the qualitative variables of judgment, in quantitative terms. The method used for the transformation has been the Analytic Hierarchy Process (AHP) of Saaty [21, 22].

Using the procedure proposed by Saaty, for each qualitative criterion and criteria weight vector, a matrix of preferences has been constructed. These matrices have been built through simple binary comparisons. For each comparison has been attributed a corresponding judgment of relative importance between two alternatives (for the qualitative criteria) or between two criteria (for the weight vectors) using the Saaty's scale.

The sought numerical values have been evaluated through the principal eigen-vectors of the matrix of preferences so that their sum is equal to 1. While through the use of principal eigenvalues is possible to do a consistency check in order to exclude unacceptable conflicts. In this way, the final solution is a logical solution and not a random order.

In particular, for each qualitative criterion, a square matrix 1825 × 1825 (equal to the number of the schools) has been defined in order to calculate its principal eigen-value and the corresponding eigenvector. Then the data of the quantitative criteria and the calculated eigenvectors have been inserted directly into the decision matrix.

Through the assignment of the criteria weights, thirty-eight political scenarios have been considered. They want to simulate some seismic risk mitigation strategies of policymakers and consider the possible uncertainties involved in planning. In this way, these scenarios could predict in advance the trends and changes in the urban system in respect of possible future disasters. In the proposed procedure, they represent the criteria weight vectors.

The number of scenarios considered has been conditioned by the thirty-four consistent scenarios (based on the possible comparisons of the three resilience aspects), the three extreme scenarios (where for every resilience aspect is assigned a weight equal to 100%) and a particular case (where all the sub-criteria have the same weight). In this way, for each resilience factor a weight between 0 and 1 has been assigned.

In particular, with the exception of the aforementioned particular case (where its principal eigenvector is easily calculable), for each scenario a preference matrix based on the resilience aspects has been defined. The resulting weight has been subsequently distributed between the relevant criteria through the definition of a new preference matrix in order to determine the concerning weight vector of the criteria. For each aspect, to simplify the proposed procedure, the same relative importance for the relevant criteria has been assigned. In this simple case, all terms of the preference matrices are equal to one and are perfectly consistent.

All scenarios are being developed, but in order to illustrate the proposed methodology, below just four political scenario has been analyzed. They want to show some different political philosophies and how the results and the possible choices of the decision-maker can vary.

In the first scenario (W_1), the decision maker considers only the seismic risk factors of the school buildings. In the second scenario (W_2), the decision maker considers only the criteria of management and recovery of the emergency. In the third scenario (W_3), the decision maker considers only the criteria of integration and social cohesion. Finally, in the fourth scenario (W_4), the decision maker gives equal importance to the various aspects.

Based on each political scenario indicated, the relevant criteria weight vector $W^T = \{w_1, ..., w_{14}\}^T$ is shown below in the Table 2. After the construction of the decision matrix and criteria weight vectors, the TOPSIS methods has been applied.

Table 2. Criteria weight vector for each considered scenario.

$W_1^T = \{0.17, 0.17, 0.17, 0.17, 0.17, 0.17, 0.00, 0.00, 0.00, 0.00, 0.00, 0.00, 0.00, 0.00\}^T$
$W_2^T = \{0.00, 0.00, 0.00, 0.00, 0.00, 0.00, 0.20, 0.20, 0.20, 0.20, 0.20, 0.00, 0.00, 0.00\}^T$
$W_3^T = \{0.00, 0.00, 0.00, 0.00, 0.00, 0.00, 0.00, 0.00, 0.00, 0.00, 0.00, 0.33, 0.33, 0.33\}^T$
$W_4^T = \{0.06, 0.06, 0.06, 0.06, 0.06, 0.06, 0.07, 0.07, 0.07, 0.07, 0.07, 0.11, 0.11, 0.11\}^T$

3.2 Results

Applying the TOPSIS method, through the C_i^* values a ranking of the alternatives has been defined in accordance with each scenario (Figs. 6, 7, 8 and 9).

Using a geographic information system, the distribution of the priorities of the interventions in the various districts according to the analyzed criteria can be effectively and quickly visualized. With this territorial approach, it is easier to identify and assess the "weaknesses" of the various districts.

The results show that the schools are ranked in a different manner depending on the considered scenario. These decision scenarios have been defined through the assignment of a weight for each analyzed criterion and represent the limit and intermediate conditions in which a decision maker could define its prioritization strategies. The criteria weights play a fundamental role in the decision-making processes. Because of

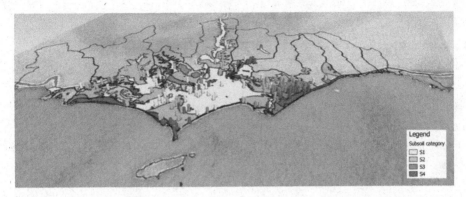

Fig. 6. The first two hundred C_i* values (see the heights) of the political scenario that considers only the seismic risk factors of buildings.

Fig. 7. The first two hundred C_i* values (see the heights) of the political scenario that considers only the management and recovery of the emergency.

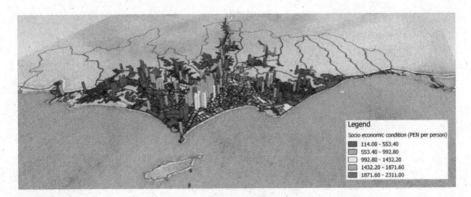

Fig. 8. The first two hundred C_i* values (see the heights) of the political scenario that considers only the integration and social cohesion.

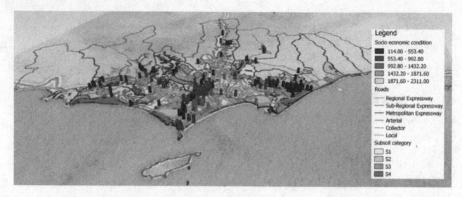

Fig. 9. The first two hundred C_i^* values (see the heights) of the political scenario that considers the previous three factors with an equal importance.

the large number of schools and the considerable differences of the considered scenarios, some estimates of the alternatives are amplified, others are overturned for small changes of the weights value.

It is a hard work to choose the political scenario more appropriate due to the possible uncertainties involved in the planning, the available economic resources and the urban system's ability to absorb, adapt and respond to future adverse changes.

In Fig. 10, assuming to strengthen the first two hundred schools of each considered political scenario, their repair cost (without reinforcements) for an occasional earthquake is shown. On the basis of these repair costs, the political scenario that seems to have the lowest costs is the number 3. In this political scenario, the decision maker considers only the criteria of integration and social cohesion.

Policy-makers could always be tempted to choose the most economical political scenario or the scenario that allows to reinforce the largest number of schools with the

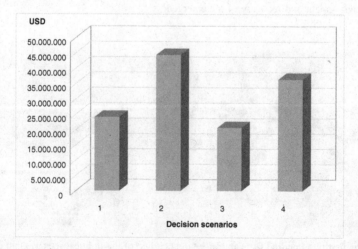

Fig. 10. Repair cost for an occasional earthquake in the main considered Decision Scenarios.

same economic investment. However, this policy may be incorrect because it could invalidate the possible resilience of the community and increase future investments. The authors are working on a broader range of scenarios (thirty-eight) and are trying to measure of these aspects in order to identify the political scenario more appropriate.

4 Conclusion

In the paper, the case of public schools in the Lima Metropolitan Area has been analyzed in order to calibrate rational procedures able to allocate the economic resources and ensure an efficient seismic risk mitigation. In order to achieve these purposes, decision makers need modern and efficient methods that allow a multidisciplinary approach with a holistic view of the problem. This approach should predict in advance the trends and changes in the urban system in respect of possible future disasters in order to plan interventions and mitigate their effects. It is very important that decision makers use effective technologies and methods to increase the availability of information and reduce the uncertainties in decision-making process. In this context, the MCDM methods and the concept of city resilience can be used and they can provide a useful support to carry out rational and multidisciplinary analysis.

The resilience of the communities has played a key role in the definition of the seismic risk mitigation strategies. Based on resilience aspects, fourteen evaluation criteria have been selected and through the assignment of their weights, thirty-eight political scenarios have been defined in order to consider the aforementioned uncertainties.

In all evaluation processes of schools with respect to each considered criterion and for the analysis and computational management of the processed data, geographic information systems have been used. The use of GIS software in all decision-making process has allowed to analyze the complex system of Lima on a smaller territorial scale. In this way, the spatial and geographical relationships of schools with the city have been more easily evaluated. The integration between GIS, MCDM methods, and resilience finds in this topic a natural application.

The proposed procedure could become a main tool for the policy makers and provide a new and rational seismic risk mitigation policy. Moreover, this procedure can be further improved. In this way, this work can be considered a work in progress.

Acknowledgments. This work has been partially completed thanks to the ELARCH project: Reference number 552129-EM-1-2014-1-IT-ERA MUNDUS-EMA21 funded with support of the European Commission. This document reflects the view only of the author, and the Commission cannot be held responsible for any use which may be made of the information contained therein.

This work has been based on the Project 70244-0034: "Evaluación Probabilística del riesgo sísmico de escuelas y hospitales de la ciudad de Lima. Componente 2: Eva-luación probabilista del riesgo sísmico de locales escolares en la ciudad de Lima". Coordinator Prof. Sandra Santa-Cruz.

References

1. Dolce, M., Kappos, A.J., Masi, A., Penelis, G., Vona, M.: Vulnerability assessment and earthquake scenarios of the building stock of Potenza (Southern Italy) using the Italian and Greek methodologies. Eng. Struct. **28**, 357–371 (2006)
2. Chiauzzi, L., Masi, A., Mucciarelli, M., Vona, M., Pacor, F., Cultrera, G., Gallovič, F., Emolo, A.: Building damage scenarios based on exploitation of Housner intensity derived from finite faults ground motion simulations. Bull. Earthq. Eng. **10**(2), 517–545 (2012)
3. Garcia-Torres, S., Kahhat, R., Santa-Cruz, S.: Methodology to characterize and quantify debris generation in residential buildings after seismic events. Resourc. Conservat. Recycl. (2016)
4. Mesta, C., Kahhat, R., Santa-Cruz, S.: Quantification of lost material stock of buildings after an earthquake. A case study of Chiclayo, Peru. In: 16th World Conference on Earthquake Engineering, 16WCEE 2017, pp. 1–12. International Association of Earthquake Engineering, Santiago (2017)
5. Vona, M., Murgante, B.: Seismic retrofitting of strategic buildings based on multi-criteria decision-making analysis. In: Furuta, H., Frangopol, D.M., Akiyama, M. (eds.) IALCCE 2014 Symposium, Tokyo, Japan, Volume: Life-Cycle of Structural Systemsi: Tsuyoshi Akiyama. Taylor and Francis Group (2014). ISBN: 978-1-138-00120-6
6. Vona, M., Harabaglia, P., Murgante B.: Thinking about resilience cities studying Italian earthquake. In: Proceedings of the Institution of Civil Engineers, Urban Design and Planning (2015). http://dx.doi.org/10.1680/udap.14.00007
7. Vona, M., Anelli, A., Mastroberti, M., Murgante, B., Santa-Cruz., S.: Prioritization Strategies to reduce the Seismic Risk of the Public and Strategic Buildings (2017)
8. INEI: PERÚ: ESTIMACIONES Y PROYECCIONES DE POBLACIÓN TOTAL POR SEXO DE LAS PRINCIPALES CIUDADES. Instituto Nacional de Estadistica e In-formatica (INEI) (in Spanish). Scribd, p. 32, March 2012. Accessed 14 Mar 2014
9. INEI (Instituto Nacional de Estadística e Informática) and MINEDU (Ministerio de Educación): Censo de Infraestructura Educativa (CIE). Lima, Peru (2013). http://www.inei.gob.pe
10. Instituto Nacional de Estadística e Informática: Una Mirada a Lima Metropolitana, Hecho el Depósito Legal en la Biblioteca Nacional del Perú N° 2014-12857. http://www.inei.gob.pe
11. Fernández de Córdova, G.: Nuevos patrones de segregación socioespacial en Lima y Callao 1990–2007. PUCP, Lima (2012)
12. Programa de Agua y Saneamiento del Banco Mundial (WSP): Perú: Gestión de Riesgo de Desastres en Empresas de Agua y Saneamiento. Tomo I: Perfil de Riesgo Catastrófico, Medidas de Mitigación y Protección Financiera. Caso Sedapal y Emapica. https://www.wsp.org/sites/wsp.org/files/publications/WSP-LAC-Peru-Gestion-De-Riesgo-De-Desastres-En-Empresas-De-Agua-Y-Saneamiento-Tomo-1.pdf
13. INDECI: Proyecto SIRAD Investigación Sobre Peligro Sísmico en el Área Metropolitana de Lima y Callao, 19, 2011. http://www.indeci.gob.pe/proyecto58530/objetos/archivos/20110606103342.pdf
14. Blondet, M., Dueñas, M., Loaiza, C., Flores, R.: Seismic vulnerability of informal construction dwellings in Lima, Peru: preliminary diagnosis. In: 13th World Conference on Earthquake Engineering, Vancouver (2004)
15. Santa-Cruz, S.: Evaluación probabilista del riesgo sísmico de escuelas y hospitales de la ciudad de Lima. Componente 2: Evaluación probabilista del riesgo sísmico de locales escolares en la ciudad de Lima. Informe Interno TAP 2. PUCP, Lima (2013)

16. Rivera, M., Velazquez, T., Morote, R.: Participación y fortalecimiento comunitario en un contexto posterremoto en Chincha, Perú. Psicoperspectivas **13**(2), 144–155 (2014)
17. Santa-Cruz, S., Palomino, J., Arana, V.: Prioritization methodology for seismic risk reduction in public schools. Study case: Lima, Peru. In: 16th World Conference on Earthquake Engineering, 16WCEE 2017, pp. 1–12. International Association of Earthquake Engineering, Santiago (2017)
18. Anelli, A., Santa-Cruz, S., Vona, M., Tarque, N., Laterza, M.: An innovative method-ology for the seismic risk mitigation on large territorial scale. In: World Engineering Confer-ence on Disaster Risk Reduction, WECDRR 2016, Lima, Peru, pp. 1–10 (2016)
19. Opricovic, S., Tzeng, G.H.: Compromise solution by MCDM methods: a comparative analysis of VIKOR and TOPSIS. Eur. J. Oper. Res. **156**, 445–455 (2004)
20. Santa-Cruz, S., Fernandez De Córdova, G., Rivera Holguin, M., Vilela, M., Arana, V., Palomino, J.: Social sustainability dimensions in the seismic risk reduction of public schools: a case study of Lima, Peru. Sustainability: Sci. Pract. Policy **12**(1), 1–13 (2016)
21. Saaty, T.L.: The Analytic Hierarchy Process. McGraw-Hill, New York (1980)
22. Saaty, T.L.: Decision Making for Leaders: The Analytic Hierarchy Process for Decision in a Complex World. RWS Publications, Pittsburgh (1999)

Workshop on Bio-inspired Computing and Applications (BIONCA 2017)

Crowd Anomaly Detection Based on Optical Flow, Artificial Bacteria Colony and Kohonen's Neural Network

Joelmir Ramos[1]([✉]), Nadia Nedjah[2], and Luiza M. Mourelle[3]

[1] Post-Graduation Program in Electronics Engineering,
State University of Rio de Janeiro, Rio de Janeiro, Brazil
joelmiramos@gmail.com
[2] Department of Electronics Engineering and Telecommunications,
State University of Rio de Janeiro, Rio de Janeiro, Brazil
nadia@eng.uerj.br
[3] Department of Systems Engineering and Computation,
State University of Rio de Janeiro, Rio de Janeiro, Brazil
ldmm@eng.uerj.br

Abstract. This paper presents a novel method for global anomaly detection in crowded scenes. The optical flow of frames is used to extract the foreground of areas with people motions in crowd. The optical flow between two frames generates one layer. The proposed method applies the metaheuristic of artificial bacteria colony as a robust algorithm to optimize the layers from optical flow. The artificial bacteria colony has the ability to adapt quickly to the most varied scenarios, extracting just relevant information from regions of interest. Moreover, the algorithm has low sensibility to noise and to sudden changes in video lighting as captured by optical flow. The bacteria population of colonies, its food storage and the colony's centroid position regarding each optical flow layer, are used as input to train a Kohonen's neural network. Once trained the network is able to detect specific events based on behavior patterns similarity, as produced by the bacteria colony during such events. Experiments are conducted on publicly available dataset. The achieved results show that the proposed method captures the dynamics of the crowd behavior successfully, revealing that the proposed scheme outperforms the available state-of-the-art algorithms for global anomaly detection.

Keywords: Crowd anomaly detection · Optical flow · Artificial bacteria colony · Kohonen's neural network

1 Introduction

Recently many research works have focused on the detection and classification of abnormalities at real-time speed in crowded scenes. Nowadays, the large amount

J. Ramos—The work of this author is supported by CAPES, the Coordination of Improvement of Higher Education Personnel of the Brazilian Federal Government.

O. Gervasi et al. (Eds.): ICCSA 2017, Part II, LNCS 10405, pp. 329–344, 2017.
DOI: 10.1007/978-3-319-62395-5_23

of public events with a high density mass of people as well as the growth number of surveillance cameras available, make the automatic detection and classification of scenes with specific events possible and necessary. An automated detection of these specific events is important because human condition, in general, does not allow for a good performance at this task manually due to its monotony. Abnormality detection in crowded videos is a challenging task for many reasons: high density of people in handled scenes, difficulty in capturing the behavior of the people in the crowd, the quality of surveillance videos available and, mainly, the computational cost required in this kind of process.

Starting from the premise that the crowd tends to behave in a stable and homogeneous way under normal conditions, an abnormality may then occur when there is a rapid and sudden variation in the motion conditions within the scene. Research works indicate that the optical flow is among the best techniques for extracting motion in scenes [1]. However, this technique is sensitive to the video lighting conditions. In addition, it requires significant computational cost. The optical flow provides the brightness variation intensity of pixels between two frames. Thereby, it is possible to generate a spatio-temporal volume representing the motion variations within the video. In order to optimize the search for areas of the video, wherein relevant activities that should be to analyzed, the application of swarm intelligence based methods is promising [2,3]. In this work, the metaheuristic inspired by real colony behavior of bacteria shows attributes that allow a precise optimization while entailing a low computational cost. Another aspect, which makes possible the application of bacteria colony optimization, is its ability to mirror the crowd behavior in the scene by the bacteria behavior. In other words, the way the artificial bacteria move has a direct relation with the way people move within a crowded scene. Therefore, understanding the motion pattern of a colony of bacteria is, somehow, equivalent to understand approximately the motion pattern of crowd.

In this work, the behavior patterns are unique for each of the studied scenarios and the classifier must be capable to recognize the abnormalities in diverse videos of the same scenario. Furthermore, the set of input data for machine learning, in this case, do not have the corresponding desired output, because these represent unpredictable movement of the crowd. Thus, this is a case of unsupervised leaning. So, only after training is completed, which is usually done using data clustering techniques, it would be possible to label manually the movement patterns that emerge. In this work, to approach this specific problem, we exploit a Kohonen's neural network for classification. This network model is chosen due to its intrinsic capability of self-organization via a competitive method. This method is capable of detecting similarities, regularities and correlations between the input patterns set by clustering them. The training set is extracted from information regarding the artificial bacteria behavior during the optimization process. After training the network, the clusters are manually labeled to distinguish between specific events.

This work presents a new method to detect abnormalities in videos of crowded scenes. It does so optimizing the optical flow of scenes using an artificial bacteria colony. Abnormalities detection is achieved by a Kohonen's neural network, trained using the data regarding the bacteria behavior during the optical flow optimization process.

First, in the Sect. 2, we describe some existing works related with human activities recognition and crowd abnormalities detection. Then, in Sect. 3, we explain the workflow of the proposed method. After that, in Sect. 4, we provide details about the methodology used to approach the abnormalities detection problem. Subsequently, in Sect. 5, we present, analyze and discuss the experimental results. Last but not least, in Sect. 6, we draw some conclusions about the work and point out some exciting directions for future works.

2 Related Works

The analysis of human behavior as captured on videos evolved gradually towards the detection and recognition of abnormal behavior and, eventually, for recognition and identification of specific events. The taxonomy for human behavior analysis, as described in [4], emphasizes the relevance of information related to movements in the video. This section presents a consolidated vision about critical aspects related to developing techniques to analyze and detect events within a given video.

A crowd abnormal behavior can be divided into two classes: either focusing on the individual [5] or on the crowd [6]. In the first case, the crowd is considered as a collection of individuals. Therefore, it is necessary to segment it and track trajectories therein. However, this approach is seriously affected by occlusions in the crowd scene. In the former case, the crowd is treated as a whole in the analysis of medium and high density scenes. Instead to tracking individuals, this approach extracts features of the crowd to represent its state.

Approaches based on supervised learning [7] use data, which are hard to be obtained, for appropriate training of support vector machine (SVMs). For automated surveillance, supervised approaches are less attractive because they usually require a new training every time an unpredictable alteration of scene occurs. Approaches based in multiple learning are techniques that are more widely used to recognize actions without supervision [8]. Unsupervised approaches offer an advantage over supervised ones in terms of feasibility, due to their ability to adapt to the most diverse situations that usually arise in applications, such as automated surveillance.

In general, the detection of anomalies in crowd videos occurs in the temporal domain, with the identification of features outside the range of normality or specific events [9]. In the literature, some of crowd abnormalities include individuals walking on a determined place or individuals running when other people around them are walking. Anomalous events are contextual and depend on other items of the scene [10]. Emergency situations, such as panic in a crowd or a threat to the people on the scene, are considered anomalies. Identifying these events is important for video surveillance applications. Spectral approaches based on clustering [11] and social force model [12] are widely used. Methods based on optical flow perform spatio-temporal volume analysis, wherein the movement is used to detect anomalous events [2].

3 Architecture of the System

In this section, we present the basic architecture of the proposed system. The Fig. 1 shows the main development stages of the proposed method to approach the problem of global abnormality detection.

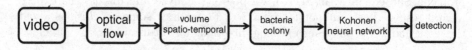

Fig. 1. Main steps of the proposed system

In the proposed method, the optical flow extraction from the input video is performed first, generating a spatio-temporal volume. All the frames of video are then converted from RGB color system to grayscale. Each layer of the optical flow, which consists of field of vectors with a magnitude variation according to the brightness patterns, is obtained considering each pair of consecutive frames of the video. So, each layer of the optical flow can be treated as a fitness function and thus optimized. Its local optima represent the regions of interest, as they entail a large amount of the scene's movements. Colonies of virtual bacteria are scattered randomly in each layer of optical flow. In each layer, the bacteria are subjected to the Darwinian natural selection, in which only the more adapted survive and reproduce. The bacteria less adapted, that are those that during foraging could not obtain enough food as represented by the objective function, are inevitably eliminated. Considering each layer of optical flow, at the end of a defined number of epochs, the colony stabilizes and the bacteria are naturally distributed on existing local optima of fitness function. The amount of bacteria in the colony (population), the food stock of all the surviving bacteria (fitness accumulated) and the midpoint module of bacteria positions (centroid) are registered for each layer. In the scenes of video, where there are large regions with little movement, the bacteria population should be dense. If there are small regions with large movement, the food stock should be bountiful. If the crowd of the video suddenly disperses, the centroid should reveal the magnitude and directions of crowd motion.

The optimization of the spatio-temporal volume of optical flow generates exit vectors containing the amount of bacteria in colonies, food stocks and centroid module. These vectors become the input set of Kohonen's neural network. The network clusters, in classes, the colonies that have similar patterns of behavior. Once trained, the network performs the detection of a specific event every time that a colony of bacteria shows the same behavior pattern. In this work, the specific event is namely crowd abnormal behavior.

4   Proposed Algorithm

In this section, we give a detailed presentation of the proposed algorithm. The proposed scheme consists of three main components: (1) optical flow, (2) artificial bacteria colony and (3) Kohonen's neural networks.

4.1   Optical Flow

Several methods use optical flow to perform event detection in frames sequence. The optical flow, as proposed in [13], estimates the motion between a pair of frames. The essence of the method is to find the estimate of the apparent motion between frames in relation to changes in the brightness patterns (pixel value of the grayscale image). The vectors of the optical flow are low-level characteristics, but interpreting them in high-level events requires a high computational cost. The data from the videos is bulky and needs to be reduced to allow real-time surveillance applications.

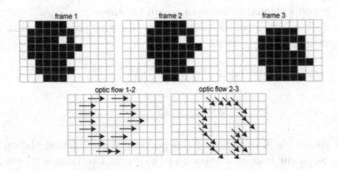

Fig. 2. Optical flow illustration

Optical flow is the distribution of the apparent velocities of brightness patterns in an image. It is a field of vectors with velocities associated to a sequence of images, as shown in Fig. 2. This arises both from the movement of the scene objects and the movement of the camera. A sequence of images can be represented by its luminance function $I = I(x, y, t)$. The luminance conservation hypothesis means that the luminance of a physical point in the image sequence does not change over a short time interval, as defined in Eq. 1:

$$I(x, y, t) \approx I(x + \Delta x, y + \Delta y, t + \Delta t). \tag{1}$$

In order to determine the optical flow, it is assumed that at each pixel of the image, the constraint, as defined in Eq. 2 applies:

$$I_x u + I_y v + I_t = 0, \tag{2}$$

wherein I_x, I_y and I_t are the partial derivatives of the image, u and v are two components of the magnitude vector of the optical flow at a pixel position. This constraint alone is not enough to compute the vector field. Since the neighboring points of a moving object have similar velocities, it is possible to assume that the vector field of the optical flow is smooth. Thus, the vector field (u, v) is found by minimizing the function of Eq. 3:

$$\int\int [(I_x u + I_y v + I_t)^2 + \alpha^2(v_x^2 + v_y^2 + u_x^2 + u_y^2)], \tag{3}$$

wherein the double integral extends to the entire image. The solution to this problem can be found, based in variational principles, by solving the following differential equations:

$$I_x^2 u + I_x I_y v = \alpha^2 \nabla^2 u - I_x I_t$$
$$I_x I_y u + I_y^2 v = \alpha^2 \nabla^2 v - I_y I_t, \tag{4}$$

wherein α represents a suitable weighting factor to compensate the error magnitude that is proportional to noise in the mesurement.

In this way, a solution can be found by an iterative procedure [13], using the following equations:

$$u^{n+1} = u^{-n} - \frac{I_x(I_x u^{-n} + I_y v^{-n} + I_t)}{\alpha + I_x^2 + I_y^2}$$
$$v^{n+1} = v^{-n} - \frac{I_y(I_x u^{-n} + I_y v^{-n} + I_t)}{\alpha + I_x^2 + I_y^2}. \tag{5}$$

Figure 3 shows an illustration of the procedure considering the optical flow between two adjacent frames. Figure 3(a)–(b) shows two images represent representing these two adjacent frames of the movement of a footstep carried out by a man. Figure 3(c) shows the calculated optical flow between these two frames. Finally, Fig. 3(d) represents a detailed view of the optical flow in which the direction field of the leg movement is shown.

(a) Frame 1   (b) Frame 2   (c) Optical flow of frames   (d) Detail of optical flow

Fig. 3. Illustration about optical flow extraction of frames

In the present work, videos containing f frames generate spatio-temporal volumes with $k = f - 1$ layers. Each layer has a vector field of magnitudes of the optical flow. The regions of interest have the largest magnitudes. Each layer is then treated as an objective function. The layers of spatio-temporal volumes, or objective functions, are optimized by artificial bacteria colony.

4.2   Artificial Bacteria Colony

Bacterial foraging optimization (BFO) is a metaheuristic inspired by the behavior of *Escherichia coli* bacteria in the search for nutrients in their environment [14]. BFO is widely applied in numerical optimizations and uses the following basic steps: chemotaxis, reproduction and elimination-dispersion.

Chemotaxis is the chemically directed movement that some living beings develop. Chemotaxis, and the chemical substances involved in it, are used by some single-celled organisms, insects, mammals and even men. It is used for various purposes, such as during the search for nutrients, in order to avoid predators, generate communication between individuals in the formation of colonies or groups, for sexual attraction, or in the territorial demarcation [15]. In addition to this broad definition, the term chemotaxis, in the scientific literature, is almost always used to refer to cell movement in response to the concentration gradient of chemicals present in the environment.

Evolved over millions of years by nature, the chemotaxis of bacteria is a highly optimized process of searching and exploring unknown environments. Due to advances in the field of computation, chemotactic strategies of bacteria and their excellent search capability can be modeled, simulated and emulated to develop nature-inspired optimization methods, which are an alternative to existing methods. In this work an algorithm based on chemotactic strategies of bacteria is developed.

In the present work, an artificial bacteria colony consists in all bacteria of a specific optical flow layer. The artificial bacteria are points in a Cartesian Plane superimposed on a layer of optical flow. The Cartesian plane has the same dimensions of the optical flow layer. Each bacterium evaluates its position through a correspondence with the position of the optical flow layer, on which it is superimposed. The value obtained in this evaluation is considered its fitness. The fitness found by each bacterium is the magnitude of the optical flow, which is at the matrix entry corresponding to the position of the bacterium in the Cartesian Plane.

The chemotaxis step simulates the movement of a *Escherichia coli* by means of swim (clockwise rotation) and tumble (counterclockwise rotation) of the flagella. A bacterium alternates between these two modes of operation throughout its life span. In the state transition, a bacterium can perform the swim movement, maintaining the same direction and tumble, moving in a random direction in the search space. Real bacteria navigate the environment in search for food, always following the direction that gives them the greatest nutritional gain.

In conjunction with chemotaxis, the Darwinian natural selection mechanism applies. Artificial bacteria use the fitness of the objective function as food to remain alive. At each epoch, in a single layer, the bacterium evaluates the fitness of its position and stores it, thus generating a food stock. For each layer, there are several values for the magnitudes of brightness variation. Therefore, bacteria that are exploring regions with low values, will have a much lower stock than those exploring regions with high values. In this way, the layer with its various regions becomes an environment that stimulates competition, so that bacteria that have a stock above a certain threshold, dynamically established for each layer, survive and reproduce. Bacteria that have a stock below this threshold are eliminated.

During the reproduction step, after chemotaxis, the most adapted bacterium is divided, via asexual reproduction, into two identical bacteria. The bacterium resulting from a reproduction is randomly positioned, but very close to the originating bacterium. Thus, at each epoch new bacteria are randomly positioned in the vicinity of regions with magnitude values well above average. Bacteria that are very distant from these regions are eliminated in the early epochs. Therefore, the outline of the colony of bacteria is shaped by the contour of the regions.

During the elimination step, bacteria that have a value in their stock lower than the dynamic threshold of the layer do not survive, as it would happen with bacteria in a real environment. During this step, mainly in the first epochs, occurs the elimination of great part of the bacteria located in regions of the layer with little or no movement. Thereafter, the reproductive process gains emphasis and there is a population explosion in regions with intense movement regarding the optical flow layer. This population explosion, whose growth is exponential, allows to reasonably delimiting the outline of the region with high magnitudes, since the bacteria that distance themselves from this region are quickly eliminated.

This work proposes the use of a modified version of the BFO algorithm to locate areas of the frames that have optical flow magnitudes above a dynamic threshold. The dynamic threshold adapts to the content of each layer. This dynamic threshold is associated with the mean and standard deviation of magnitudes of brightness variation of each optical flow layer. The processes of chemotaxis, reproduction and death of bacteria allow the used algorithm to be able to discover information about the behavior of the crowd in the sequence of frames via the analysis of the population behavior of the bacteria.

Figure 4 allows a observation of the analogy between the foraging of a real vs. virtual bacteria, because both seek the points of higher nutrient density, as the optimal ones of a multimodal function. Figure 4(d) shows a image of implementation carried out in this work, in which the bacteria are positioned in the regions that agree with the presence of intense movement. In the areas where many people congregate, the number of bacteria increases.

Figure 5 shows, in perspective, the distribution of a colony of bacteria as distributed across an optical flow layer, after few epochs. It is possible to see that the noise does not represent a good nutritional option for the bacteria, due to its relatively low value and its random and sporadic character. Bacteria always opt for more dense regions, where parts of the colony can establish themselves with stability.

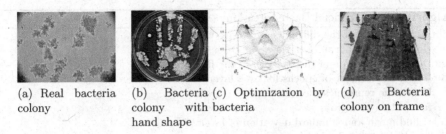

(a) Real bacteria colony

(b) Bacteria colony with bacteria hand shape

(c) Optimizarion by with bacteria

(d) Bacteria colony on frame

Fig. 4. In (a) and (b) real bacteria colony are shown. In (b) specifically, bacteria demarcate the nutritional environment in the form of a hand. In (c) an objective function is shown where bacteria seek out the local optima by chemotaxis. In (d) it is shown, on a frame, the distribution of the colony of artificial bacteria to a layer of optical flow corresponding to that frame and its consecutive one.

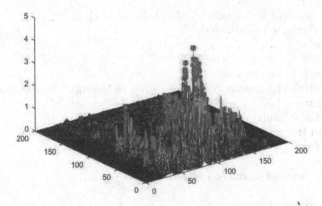

Fig. 5. The objective function is the vector field containing the magnitudes of brightness variation of an optical flow layer. Artificial bacteria are located in areas of highest magnitude.

Algorithm 1 presents the steps as executed by the optimization process inspired by the artificial bacteria colony. The spatio-temporal volume is the input of the process. The parameters of the algorithm are: the number of bacteria, initial positions of bacteria, number of epochs to be performed for each layer and reproduction condition. The reproduction condition is related to the food stock of the colony at a specific epoch. Bacteria that have a food stock above a threshold, obtained as a function of the stock of all bacteria in a given layer, reproduce asexually. The dynamic threshold of a layer is obtained as a function of its mean and standard deviation of magnitudes. Bacteria that have a fitness below the dynamic threshold of the layer are eliminated. The output is composed of vectors containing the population, the stock, and the centroid module of all layers.

Algorithm 1. Artificial Bacteria Colony – ABC

Require: volume of the difference of frames with k layers;
 1: Define the number of bacteria i;
 2: Define the position of the bacteria in the first epoch;
 3: Define the number of epochs for each layer;
 4: Define the reprodution condition;
 5: **for** each layer k **do**
 6:    find mean and standard deviation of layer k;
 7:    calculate threshold of layer (k);
 8:    **for** each epoch e **do**
 9:      **for** each bacterium i **do**
10:        calculate fitness(i) of bacterium(i);
11:        **if** fitness of bacterium$(i) \geq$ threshold(k) **then**
12:          stock$(i) :=$ stock$(i) +$ fitness(i);
13:        **end if**
14:        **if** fitness of bacterium$(i) \leq$ threshold of (k) **then**
15:          eliminate bacterium(i);
16:        **end if**
17:      **end for**
18:      **for** each bacterium i **do**
19:        **if** stock of bacterium$(i) \geq$ mean stock of layer(k) * reprodutive condition **then**
20:            bacterium(i) reproduce;
21:        **end if**
22:      **end for**
23:    **end for**
24:    calculate centroid module of colony;
25: **end for**
26: **return** population of all layers;
27: **return** stock of all layers;
28: **return** centroid module of all layers;

4.3   Kohonen's Neural Network

Kohonen's neural network or self-organizing Kohonen map (SOM) is part of a group of competitive neural networks that use competition strategies to adjust their weights during the learning process. SOMs are able to preserve the neighboring relations of input data and use unsupervised training to find similarities based only on these input patterns. The main purpose of Kohonen's neural networks is to group incoming data that are similar, thus forming classes or clusters. In Fig. 6, a Kohonen's map is depicted, where colors highlight a cluster formed after training.

Kohonen's neural networks are nonlinear projections of high dimensional spaces for a low-dimensional map M. The two-dimensional mapping, often adopted in the Kohonen's model, is a function $f : X \rightarrow M$, wherein $f : X \rightarrow \mathbb{R}^n$ and $f : M \rightarrow \mathbb{R}^2$, which assigns each element $x \in X$ a pair $(i, j) \in M$. The

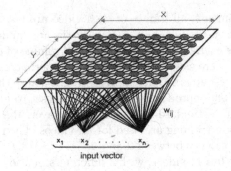

Fig. 6. Kohonen's neural network

elements $m_{i,j}$ of the map M, as well as the input data are n-dimensional vectors that adjust the values of the synaptic weights of the neural network [16].

Due to their characteristics, Kohonen's neural networks can be well applied in several areas, such as voice recognition, video behavior analysis and combinatorial optimization. The fact that similar mappings can be found in several areas of the human brain and other animals, indicates that the topology design is an important principle in signal processing systems, in the field of artificial intelligence.

In this work, after optimizing the spatio-temporal volume of the optical flow, the amount of bacteria in each colony, the food stock of all bacteria of the last epoch and the centroid module of the colonies are recorded for each layer. These values form a training set. These patterns need to be clustered according to their similarities. For this, a Kohonen's neural network is used.

The training set has three vectors: population of the colonies, food stock and centroid module. The number of neurons is chosen in order to obtain the best classification with the lowest computational cost possible. Each output unit represents a class. The number of classes is limited by the number of neurons. During training, the network determines the neuron that best responds to the input vector. The vector of weights for the winning neuron is adjusted according to the training algorithm. During the process of network self-organization, the neuron whose vector of weights is closest to the vector of the input patterns is selected as winner. The weight of the winning neuron is adjusted. Neighboring neurons are also adjusted, but their weights are updated in the inverse proportion of the distance of the winning neuron.

In this work, Kohonen's neural network cluster into classes the input data that show similarities. This allows that population, stock, and centroid values define a specific event in the scene.

5   Experiments and Results

In order to validate and evaluate the performance of the proposed method, extensive experiments were performed using public dataset UMN [17].

During the experiments, all frames of each video are resized to a resolution of 200×200 pixels. For optimization, 500 bacteria are randomly scattered in the first epoch, across each layer in the space-time volume of optical flow. In all experiments, 7 epochs were used for each layer. For experiments with more than 7 epochs, the algorithm requires too much processing, because the bacterial colony grows exponentially. The reproduction condition was set to fitness values above 3 times the value of the standard deviation of the layer. Regarding Kohonen's network, 1000 epochs of training are used for all videos. The number of neurons for all videos was set to vary between 2 and 6.

The UMN dataset has 11 videos, with 3 different scenarios. There are crowds of low and medium densities, moving indoors and outdoors. All video sequences exhibit panic situations among individuals. The video begins with individuals exhibiting normal behavior followed by sudden abnormal activity.

Figure 7 shows the sequence of steps, as followed during the experiments for each video of the dataset. Figure 7(b) depicts the foreground of the scene of Fig. 7(a), as a colormap. In Fig. 7(d), it is possible to notice that bacteria do not clump over areas with less motion than the others.

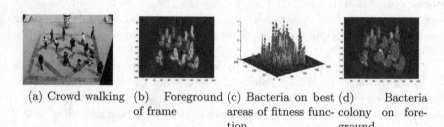

(a) Crowd walking (b)   Foreground (c) Bacteria on best (d)   Bacteria
                   of frame   areas of fitness func- colony on   fore-
                                    tion   ground

Fig. 7. Optimization steps of proposed system for UMN

Figure 8(c) shows the comparison between the detection in the ground truth and the detection by the proposed method for the UMN dataset. In Fig. 8(a), it is possible to see that the behavior of the bacteria is chaotic because it reflects the very chaotic nature of crowd behavior in UMN dataset videos. In Fig. 8(b) a two-dimensional representation of the behavior of bacteria is shown together with the position of neurons in the Kohonen network after training. In this representation some areas have a higher concentration of points. These areas represent normal behavior. Note that the larger the video, the greater the concentration of points in some areas tends to be. Areas with less density are typically areas where abnormalities are located.

Figure 9(a) shows the classification performed by the proposed method, in which only class 1 represents the normal behavior. Other classes less than 1 are other events occurring in the video. The graph with the classes, based on the ground truth of all the videos of the UMN dataset, is shown in Fig. 9(b). The graph shows that the abnormalities of the videos are accurately captured by the proposed method. Figure 10 shows the ROC curve of the classification performed

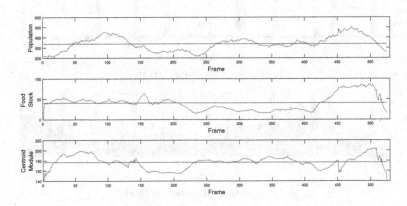

(a) Artificial bacteria colony behavior

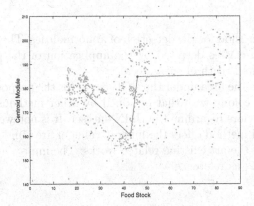

(b) Weights position of Kohonen's network

(c) Comparing abnormalities detection

Fig. 8. Detection by Kohonen network with data from bacteria colony behavior for UMN

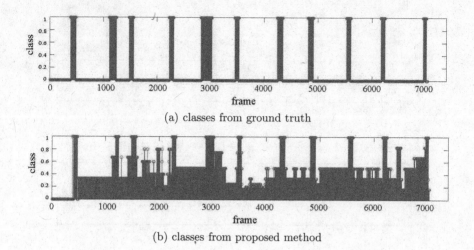

(a) classes from ground truth

(b) classes from proposed method

Fig. 9. Classes of abnormalities detection for UMN dataset

by the proposed method for the detection of abnormalities. The curve is obtained by comparing the classes detected by the application of the proposed method and from the ground truth of the videos.

Table 1 reports the area under the ROC curve for the proposed method using artificial bacteria colony with that obtained by other methods for detection of abnormality behaviors regarding the UMN dataset. It is noteworthy to point out that the proposed method offers the highest value of area under the ROC curve regarding state-of-the-art existing related works. The improvement is computed using the work [12] as reference.

Table 1. Area under ROC curve of the proposed method *vs.* other methods for the UMN dataset

Method	Area under ROC	Improvement
Optical flow [12]	0.8400	–
Social force model [12]	0.9600	14.28%
Chaotic invariant [18]	0.9900	17.85%
Sparse reconstruction [19]	0.9780	16.42%
Artificial bacteria colony	0.9929	18.02%

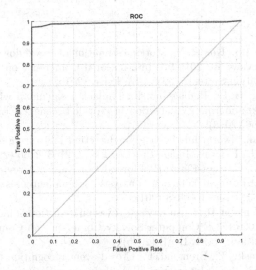

Fig. 10. The ROC curve for detection os abnormal frames in UMN dataset

6   Conclusion

This work introduces a new method to perform the detection of global abnormalities in videos of crowded scenes using optical flow, artificial bacteria colony and kohonen's neural network. This work demonstrates the ability of the method to capture the dynamics of the crowd by optimizing the optical flow of the videos. One of the main advantages of the method is that it does not need to carry out individual tracking to detect abnormalities. Another advantage of the proposed methodology is that it requires a reduced computational effort, allowing its application within real-time applications.

The experiments revealed that the colony of artificial bacteria is able to comprehend the behavior of the individuals in the videos with crowds and was robust enough to ignore the noise captured by the optical flow. The detection of abnormalities is performed by the Kohonen's neural network and the results indicate that the network is able to adapt to the most diverse scenarios quickly, accurately and with low computational cost.

Improvements of this work could be obtained by using the space-temporal volume containing magnitudes of the interaction forces, using the Social Force Model [12] instead of pure optical flow. Improvements could also be implemented in the bacterial colony algorithm to optimize spatio-temporal volumes faster and with less computational cost. These are some directions for near future works that we intend to take over to improve the achieved results.

References

1. Kajo, I., Malik, A.S., Kamel, N.: Motion estimation of crowd flow using optical flow techniques: a review. In: 2015 9th International Conference on Signal Processing and Communication Systems (ICSPCS). IEEE (2015)
2. Thida, M., Eng, H.-L., Remagnino, P.: Laplacian eigenmap with temporal constraints for local abnormality detection in crowded scenes. IEEE Trans. Cybern. **43**(6), 2147–2156 (2013)
3. Kaltsa, V., et al.: Swarm intelligence for detecting interesting events in crowded environments. IEEE Trans. Image Process. **24**(7), 2153–2166 (2015)
4. Chaaraoui, A.A., Climent-Pérez, P., Flórez-Revuelta, F.: A review on vision techniques applied to human behaviour analysis for ambient-assisted living. Expert Syst. Appl. **39**(12), 10873–10888 (2012)
5. Brostow, G.J., Cipolla, R.: Unsupervised bayesian detection of independent motion in crowds. In: 2006 IEEE Computer Society Conference on Computer Vision and Pattern Recognition, vol. 1. IEEE (2006)
6. Garate, C., Bilinsky, P., Bremond, F.: Crowd event recognition using hog tracker. In: 2009 Twelfth IEEE International Workshop on Performance Evaluation of Tracking and Surveillance (PETS-Winter). IEEE (2009)
7. Wu, X., et al.: Action recognition using multilevel features and latent structural SVM. IEEE Trans. Circuits Syst. Video Technol. **23**(8), 1422–1431 (2013)
8. Lui, Y.M.: Tangent bundles on special manifolds for action recognition. IEEE Trans. Circuits Syst. Video Technol. **22**(6), 930–942 (2012)
9. Popoola, O.P., Wang, K.: Video-based abnormal human behavior recognition - a review. IEEE Trans. Syst. Man Cybern. Part C (Appl. Rev.) **42**(6), 865–878 (2012)
10. Mahadevan, V., et al.: Anomaly detection in crowded scenes. In: 2010 IEEE Conference on Computer Vision and Pattern Recognition (CVPR). IEEE (2010)
11. Andrade, E.L., Blunsden, S., Fisher, R.B.: Modelling crowd scenes for event detection. In: 18th International Conference on Pattern Recognition, ICPR 2006, vol. 1. IEEE (2006)
12. Mehran, R., Oyama, A., Shah, M.: Abnormal crowd behavior detection using social force model. In: IEEE Conference on Computer Vision and Pattern Recognition, CVPR 2009. IEEE (2009)
13. Horn, B.K.P., Schunck, B.G.: Determining optical flow. Artif. Intell. **17**(1–3), 185–203 (1981)
14. Passino, K.M.: Biomimicry of bacterial foraging for distributed optimization and control. IEEE Control Syst. **22**(3), 52–67 (2002)
15. Murray, J.D.: Mathematical Biology I: An Introduction, Interdisciplinary Applied Mathematics, Mathematical Biology (2002)
16. Kohonen, T.: Self-organized formation of topologically correct feature maps. Biol. Cybern. **43**(1), 59–69 (1982)
17. Unusual crowd activity dataset of University of Minnesota. http://mha.cs.umn.edu/movies/crowdactivity-all.avi
18. Wu, S., Moore, B.E., Shah, M.: Chaotic invariants of lagrangian particle trajectories for anomaly detection in crowded scenes. In: 2010 IEEE Conference on Computer Vision and Pattern Recognition (CVPR). IEEE (2010)
19. Cong, Y., Yuan, J., Liu, J.: Sparse reconstruction cost for abnormal event detection. In: 2011 IEEE Conference on Computer Vision and Pattern Recognition (CVPR). IEEE (2011)

Workshop on Computational and Applied Mathematics (CAM 2017)

An Uncoupling Strategy in the Newmark Method for Dynamic Problems

Jonathan Esteban Arroyo Silva[1], Michelli Marlane Silva Loureiro[2(✉)],
Webe Joao Mansur[3], and Felipe dos Santos Loureiro[4]

[1] Postgraduate Program in Computational Modeling,
Federal University of Juiz de Fora, Juiz de Fora, Minas Gerais, Brazil
jeas560@gmail.com
[2] Department of Computer Science,
Federal University of São João del-Rei, São João del-Rei, Minas Gerais, Brazil
michelli.loureiro@ufsj.edu.br
[3] Department of Civil Engineering, COPPE,
Federal University of Rio de Janeiro, Rio de Janeiro, Brazil
webe@coc.ufrj.br
[4] Department of Thermal and Fluid Sciences,
Federal University of São João del-Rei, São João del-Rei, Minas Gerais, Brazil
felipe.loureiro@ufsj.edu.br

Abstract. When the semidiscrete formulation of the finite element method (FEM) is employed in traditional elastodynamic problems, a system of ordinary differential equations (ODEs) is obtained. The present paper focuses on the development of a numerical strategy to decouple the resulting system by means of the implicit unconditionally stable Newmark method, allowing the parts to be solved independently, and through an iterative procedure, managing to preserve the stability and accuracy properties of the original method. It is observed that only one iteration is sufficient to achieve the same level of accuracy of the solution of the fully coupled system, rendering a very efficient algorithm. The accuracy and potentialities of the proposed decoupling strategy will be studied through the solution of two 2D structural dynamic problems that present materials with functionally graded properties.

Keywords: Structural dynamics · Newmark · FEM · FGM

1 Introduction

The development of time integration schemes with distinct numerical features is still a topic of great importance and interest in many areas of the science (e.g., engineering, geophysics, etc.), once they are an efficient and accurate numerical tool when solving the system of ODEs in the time domain originated by means of semi-discretization techniques such as the FDM, FEM, Meshless, DR-BEM (Dual Reciprocity boundary element method), etc. [3,10,15,22,26,30].

© Springer International Publishing AG 2017
O. Gervasi et al. (Eds.): ICCSA 2017, Part II, LNCS 10405, pp. 347–362, 2017.
DOI: 10.1007/978-3-319-62395-5_24

Furthermore, the proper selection of time-stepping techniques that may take advantage of some characteristics of the problem under consideration, aiming at enhancing either the accuracy of the results or the computational efficiency, is an important issue to be taken into account in a simulation. For instance, using different time-step sizes into a partitioned domain (leading to a partitioned system of coupled matrix equations), one for the general part of the domain and another small time-step for the other specific part with a fine mesh, the so-called sub-cycling methods [5, 23, 28] are able to assure good results and to decrease the computational cost. Moreover, this technique allows the use of explicit-implicit, explicit-explicit or implicit-implicit coupled approaches in the different partitions according to the analyzed problem.

On the other hand, in many areas of physics and engineering, the interaction of two or more physical systems often occurs. In this way when a spatial dis-cretization technique is properly applied, a partitioned system of coupled matrix equations naturally also arises. Hence, a numerical methodology to solve the cou-pled system is required and, as a result, researches have been investigated and proposed many numerical formulations to deal with this issue. The solution of the partitioned system can be mainly accomplished through two methodologies. The first one is to consider the fully coupled system and solve it directly, whereas the other try to take advantage of the partitioned system by decoupling it into smaller systems, each one correspondent to a physical system, and then solve the smaller systems separately through an approach that correctly takes into account the interaction between systems. Uncoupling strategies, on the other hand, that allow the independent treatment of the obtained smaller systems have been becoming more attractive due to the possibility to storage separately the resulting matrices and to solve them with particular numerical procedures, taking advantage of the characteristics of each system [8, 24, 25, 27].

In this sense, this work presents an uncoupling strategy for handling elastody-namic problems. The proposed formulation decouples the vertical and horizontal displacements by means of the Newmark method, giving rise to a stable and accurate scheme. It is worth pointing out that the resulting system of coupled matrix equations has a matrix structure similar to those that appear in the prob-lems discussed above. In addition, when an iterative procedure is implemented, the accuracy is improved with almost no increase in the computational effort. The scheme proved to be robust as verified through the analysis of elastodynamic problems considering functionally graded materials [14, 20, 21, 29], that are a type of composite material in which the composition, micro-structure and properties of its constituents vary smoothly. FGMs are widely employed in contemporary science due to their distinct properties [1, 29] and, thus, the development and study of numerical methods for the solution of problems that adopt such a type of material is of great importance. For instance, this type of material is com-monly used to replace the sharp interface between two materials with a gradient transition between them (e.g., metal-ceramic), avoiding interface problems.

The present paper is organized as follows: first (Sect. 2), the governing equa-tions of elastodynamic problems are briefly presented, as well as the consid-erations for FG materials and the spatial discretization by the FEM. In the sequence, the proposed uncoupling strategy is discussed (Sects. 3 and 4). At the end of the paper (Sect. 5), numerical results are presented, illustrating the accuracy and potentialities of the proposed methodology.

2 Model Equations

Let $\Omega \subset \mathbb{R}^d$ be an open bounded domain with Lipschitz-continuous boundary $\partial \Omega$, where d is the number of spatial dimensions of the problem under consideration, and let $I = (0, T] \subset \mathbb{R}^+$ be the time domain of the analysis. The classical elastodynamic problems are modeled by the following set of equations [2, 11]:

$$\rho \ddot{u}_i - \sigma_{ij,j} = b_i \text{ in } \Omega \times I \tag{1}$$

where $u_i : \Omega \times I \to \mathbb{R}$ and $b_i : \Omega \times I \to \mathbb{R}$ stand, respectively, for the components of the displacement and the given body force per unit volume; and $\rho : \Omega \to \mathbb{R}^+$ is the mass density. Moreover, considering the partition $\partial \Omega = \Gamma = \Gamma_{D_i} \cup \Gamma_{N_i}$ with $\Gamma_{D_i} \cap \Gamma_{N_i} = \emptyset$, the common boundary conditions can be written as:

$$u_i = \bar{u}_i \text{ on } \Gamma_{D_i} \times I; \ \sigma_{ij} n_j = \bar{t}_i \text{ on } \Gamma_{N_i} \times I \tag{2}$$

where $\bar{u}_i : \Gamma_{D_i} \times I \to \mathbb{R}$ are prescribed displacements and $\bar{t}_i : \Gamma_{N_i} \times I \to \mathbb{R}$ are prescribed tractions with n_j being the unit outward normal vector components. Furthermore, the prescribed initial conditions are given by:

$$u_i = u_{0i}; \ \dot{u}_i = v_{0i} \text{ in } \Omega \text{ for } t = 0 \tag{3}$$

The constitutive relation in problems governed by linear elasticity is given by $\sigma_{ij} = C_{ijkl} \epsilon_{kl}$ in which $\epsilon_{ij} = \frac{u_{i,j} + u_{j,i}}{2}$, where $C_{ijkl} : \Omega \to \mathbb{R}$ are the material elasticity tensor components.

In this paper, numerical simulations by the FEM of two-dimensional problems (i.e., $d = 2$) considering FGM are focused. In this way, the constitutive relation, written in terms of the Lame parameters, reduces to

$$\sigma_{11} = c_1 u_{1,1} + c_2 u_{2,2}, \ \sigma_{22} = c_2 u_{1,1} + c_1 u_{2,2}, \ \sigma_{12} = c_3(u_{1,2} + u_{2,1}) \tag{4}$$

where, for plane strain problems, $c_1 = \lambda + 2\mu$, $c_2 = \lambda$ and $c_3 = \mu$, whereas for plane stress problems one can replace λ by $\bar{\lambda} = \frac{2\lambda\nu}{\lambda + 2\nu}$.

Let $\mathbf{S}$ and $\mathbf{V}$ be, respectively, the spaces of trial and test functions, the variational form of the problem under consideration can be written as: find $\mathbf{u} \in \mathbf{S}$, such that $\forall t > 0$ and $\forall \mathbf{w} \in \mathbf{V}$ [11]

$$(w_1, \rho \ddot{u}_1) + a_1(w_1, u_1) = (w_1, b_1) + (w_1, \bar{t}_1)_{\Gamma_{N_1}} \tag{5}$$

$$(w_2, \rho \ddot{u}_2) + a_2(w_2, u_2) = (w_2, b_2) + (w_2, \bar{t}_2)_{\Gamma_{N_2}} \tag{6}$$

where $(\cdot, \cdot) \equiv (\cdot, \cdot)_\Omega$ and $(\cdot, \cdot)_\Gamma$ is the classical L^2-inner product in Ω and on Γ, respectively, and the bilinear forms $a_i(\cdot, \cdot)$ are defined as:

$$a_1(w_1, u_1) = \int_{\Omega} h \left[w_{1,1} \left(c_1 u_{1,1} + c_2 u_{2,2} \right) + c_3 w_{1,2} \left(u_{1,2} + u_{2,1} \right) \right] dxdy \quad (7)$$

$$a_2(w_2, u_2) = \int_{\Omega} h \left[c_3 w_{2,1} \left(u_{1,2} + u_{2,1} \right) + w_{2,2} \left(c_2 u_{1,1} + c_1 u_{2,2} \right) \right] dxdy \quad (8)$$

where h is the thickness of the domain in the case of plane stress problems.

Then, let $\mathbf{S}^h \subset \mathbf{S}$ and $\mathbf{V}^h \subset \mathbf{V}$ be the finite element spaces consisting of piecewise polynomial functions defined as [11]:

$$\mathbf{S}^h = \left\{ \mathbf{u}^h \mid u_i^h(\cdot, t) \in H^1(\Omega), u_i^h = \bar{u}_i^h \text{ on } \Gamma_{D_i} \times I, i = 1, ..., d \right\} \quad (9)$$

$$\mathbf{V}^h = \left\{ \mathbf{w}^h \mid w_i^h \in H^1(\Omega), w_i^h = 0 \text{ on } \Gamma_{D_i}, i = 1, ..., d \right\} \quad (10)$$

where H^1 is the classical Sobolev space that denotes the space of square-integrable functions with square-integrable generalized first derivatives. By approximating the displacement components as $\mathbf{u}(\mathbf{x}, t) \approx \sum_k N_k(\mathbf{x}) \mathbf{U}_k(t)$ with $N_k(\mathbf{x})$ being the standard FEM global basis function at the kth node and $\mathbf{U}_k(t)$ the nodal displacement values, and taking into account the semi-discrete Galerkin method, the following system of ODEs arises (the reason to partition the degrees of freedom in each direction will become clear later on):

$$\begin{bmatrix} \mathbf{M}_{11} & \mathbf{0} \\ \mathbf{0} & \mathbf{M}_{22} \end{bmatrix} \left\{ \begin{matrix} \ddot{\mathbf{U}}_1 \\ \ddot{\mathbf{U}}_2 \end{matrix} \right\} + \begin{bmatrix} \mathbf{K}_{11} & \mathbf{K}_{12} \\ \mathbf{K}_{21} & \mathbf{K}_{22} \end{bmatrix} \left\{ \begin{matrix} \mathbf{U}_1 \\ \mathbf{U}_2 \end{matrix} \right\} = \left\{ \begin{matrix} \mathbf{F}_1 \\ \mathbf{F}_2 \end{matrix} \right\} \quad (11)$$

where displacement nodal vectors are given by $\mathbf{U}_i \in \mathbb{R}^{nq_i}$. The well-known mass $\mathbf{M}_{ii} \in \mathbb{R}^{nq_i \times nq_i}$ and stiffness $\mathbf{K}_{ij} \in \mathbb{R}^{nq_i \times nq_j}$ matrices, as well as load vectors $\mathbf{F}_i \in \mathbb{R}^{nq_i}$ [2,11,18], with nq_i being the number of degrees of freedom in the i-direction are defined, respectively, by

$$M_{ij}^{11} = M_{ij}^{22} = \int_{\Omega} \rho h N_i N_j dxdy \quad (12)$$

$$K_{ij}^{11} = \int_{\Omega} h \left(c_1 N_{i,1} N_{j,1} + c_3 N_{i,2} N_{j,2} \right) dxdy \quad (13)$$

$$K_{ij}^{12} = K_{ji}^{21} = \int_{\Omega} h \left(c_2 N_{i,1} N_{j,2} + c_3 N_{i,2} N_{j,1} \right) dxdy \quad (14)$$

$$K_{ij}^{22} = \int_{\Omega} h \left(c_3 N_{i,1} N_{j,1} + c_1 N_{i,2} N_{j,2} \right) dxdy \quad (15)$$

$$F_i^1 = \int_{\Omega} h N_i b_1 dxdy + \int_{\Gamma_{N_1}} h N_i \bar{t}_1 ds, \quad F_i^2 = \int_{\Omega} h N_i b_2 dxdy + \int_{\Gamma_{N_2}} h N_i \bar{t}_2 ds \quad (16)$$

As previously stated, in FGM models, the material properties are represented by continuous functions and such a characteristic needs to be correctly handled in the computer implementation to evaluate the above integrals. In this work an element level strategy was adopted, assuming that the material properties, which are governed by functions in the spatial (physical) domain, are also mapped onto the reference domain with their values being computed at the integration points in an isoparametric framework. The final task is to solve (11), commonly by numerical techniques, as discussed in Sect. 3 through a novel procedure.

3   Uncoupled Newmark Method

The present section aims at presenting an uncoupling strategy directly in the Newmark time-stepping method to carry out the time integration of (11). The Newmark method is here selected because it is widely used in dynamic analyses after a proper semi-discretization procedure [2,11]. At this point, it is important to highlight that (11) represents a system of coupled matrix equations and such a partitioned form also appears similarly in other engineering applications such as in sub-cycling techniques, dynamic fluid-structure interaction, thermo-elasticity with the non-Fourier model, etc., as already commented. Thus, we expect that the proposed formulation presented hereafter might be adopted in these more advanced engineering applications.

The key feature of the proposed formulation is the development of an algorithm to integrate the system of equations in each direction separately that correctly predicts the coupled terms and, therefore, avoiding the solution of the complete and many times very large system of equations simultaneously. To this end, the first step is to rewrite (11) as:

$$\begin{aligned} \mathbf{M}_{11}\ddot{\mathbf{U}}_1 + \mathbf{K}_{11}\mathbf{U}_1 &= \bar{\mathbf{F}}_1 \\ \mathbf{M}_{22}\ddot{\mathbf{U}}_2 + \mathbf{K}_{22}\mathbf{U}_2 &= \bar{\mathbf{F}}_2 \end{aligned} \tag{17}$$

where $\bar{\mathbf{F}}_1 = \mathbf{F}_1 - \mathbf{K}_{12}\mathbf{U}_2$ and $\bar{\mathbf{F}}_2 = \mathbf{F}_2 - \mathbf{K}_{21}\mathbf{U}_1$; the above system has a similar form of an uncoupled system in the sense that coupled terms are assumed to be handled as known vectors. The next step consists of properly applying the Newmark method. In the recurrence expressions of the Newmark method in the *a-form* implementation (see [11] for further details), the equilibrium equations are written at time t_{n+1}. In this way, we assume that for one of the directions, say for the direction 1, $\mathbf{U}_2^{n+1} = \tilde{\mathbf{U}}_2^{n+1}$, where $\tilde{\mathbf{U}}_2^{n+1}$ is the displacement predictor vector, which depends solely on the information of the previous solution at time t_n, so the first direction equation of (17) will be solved using the predicted displacement vector in the second direction. Once the displacement vector for the first equation ($\mathbf{U}_1^{n+1}$) is obtained, its value is substituted in the second equation, that can be solved as in the standard Newmark scheme, obtaining the vector $\mathbf{U}_2^{n+1}$. Then for the direction 1, one can use the obtained displacement vector $\mathbf{U}_2^{n+1}$ to recompute the displacement vector $\mathbf{U}_1^{n+1}$ initializing an iterative process, a description of the Uncoupled Newmark (named here as U-Nw) Algorithm is summarized in Table 1.

Notice that the proposed procedure provides some interesting features, namely: i) it can be programmed into existing codes in a straightforward manner; ii) if required, especially for multi-physics problems, completely different methodologies in each equation of the partitioned system (e.g., different time steps and time integration schemes) may be applied; and iii) instead of solving the fully coupled system with dimensions $(nq_1 + nq_2) \times (nq_1 + nq_2)$, one needs to solve two independent systems of dimensions $nq_1 \times nq_1$ and $nq_2 \times nq_2$; this is advantageous, mainly, if the systems present different algebraic matrix properties as in the case of multi-physics problems and if direct matrix solvers

Table 1. Step-by-step solution using the U-Nw method

A. Initial calculations:
   1. Form stiffness matrices $\mathbf{K}_{ii}$ and mass matrices $\mathbf{M}_{ii}$ (recalling that $i = 1, 2$)
   2. Initialize initial vectors $\mathbf{U}_i^0$, $\dot{\mathbf{U}}_i^0$ and $\ddot{\mathbf{U}}_i^0$
   3. Select the time-step size Δt, and set the parameters γ and β (e.g., $\gamma = 0.5$ and $\beta = 0.25$)
   4. Form effective stiffness matrices $\hat{\mathbf{K}}_{ii} = \mathbf{M}_{ii} + \Delta t^2 \beta \mathbf{K}_{ii}$ and factorize them
B. For each time step:
   1. Compute displacement and velocity predictor vectors:

$$\tilde{\mathbf{U}}_i^{n+1} = \mathbf{U}_i^n + \Delta t \dot{\mathbf{U}}_i^n + \frac{\Delta t^2}{2}(1 - 2\beta)\ddot{\mathbf{U}}_i^n; \quad \dot{\tilde{\mathbf{U}}}_i^{n+1} = \dot{\mathbf{U}}_i^n + (1 - \gamma)\Delta t \ddot{\mathbf{U}}_i^n$$

   2.1 Initialize the iteration steps (i.e., $l = 0$), and compute the acceleration vector at time t_{n+1} in one direction (named first direction) using the predictor vector $\tilde{\mathbf{U}}_{j_i}^{n+1}$ from the other direction (named second direction) by setting the index $j_i = 2\delta_{i1} + \delta_{i2}$; also compute the displacement vector in the same direction:

$$\hat{\mathbf{K}}_{ii}\ddot{\mathbf{U}}_i^{n+1,0} = \mathbf{F}_i^{n+1} - \mathbf{K}_{ii}\tilde{\mathbf{U}}_i^{n+1} - \mathbf{K}_{j_i j_i}\tilde{\mathbf{U}}_{j_i}^{n+1}; \quad \mathbf{U}_i^{n+1,0} = \tilde{\mathbf{U}}_i^{n+1} + \beta\Delta t^2\ddot{\mathbf{U}}_i^{n+1,0}$$

   2.2 Repeat the step 2.1 in the other direction using the recently found $\mathbf{U}_i^{n+1,0}$ in place of $\tilde{\mathbf{U}}_{j_i}^{n+1}$:

$$\hat{\mathbf{K}}_{ii}\ddot{\mathbf{U}}_i^{n+1,0} = \mathbf{F}_i^{n+1} - \mathbf{K}_{ii}\tilde{\mathbf{U}}_i^{n+1} - \mathbf{K}_{j_i j_i}\mathbf{U}_{j_i}^{n+1,0}; \quad \mathbf{U}_i^{n+1,0} = \tilde{\mathbf{U}}_i^{n+1} + \beta\Delta t^2\ddot{\mathbf{U}}_i^{n+1,0}$$

   3. For each *iteration step* l ($l = 1, \ldots, l_m$) repeat the step 2.2 in both directions:

$$\hat{\mathbf{K}}_{ii}\ddot{\mathbf{U}}_i^{n+1,l} = \mathbf{F}_i^{n+1} - \mathbf{K}_{ii}\tilde{\mathbf{U}}_i^{n+1} - \mathbf{K}_{j_i j_i}\mathbf{U}_{j_i}^{n+1,l-1}; \quad \mathbf{U}_i^{n+1,l} = \tilde{\mathbf{U}}_i^{n+1} + \beta\Delta t^2\ddot{\mathbf{U}}_i^{n+1,l}$$

   4. Compute velocity vectors in both directions:

$$\dot{\mathbf{U}}_i^{n+1} = \dot{\tilde{\mathbf{U}}}_i^{n+1} + \gamma\Delta t\ddot{\mathbf{U}}_i^{n+1}$$

(like SuperLU [13,16] and Pardiso [4,7,17]) are employed (sometimes a common choice in transient analysis due to their robustness [6,9,12]). This means that the factorization of the two effectiveness stiffness matrices shown in Table 1 can be performed simultaneously; and only once (i.e. one factorization) in the simulation if a constant time-step size is employed.

4   Convergence Issue

This section presents a convergence study in the time domain of the proposed uncoupled time-stepping method. Following the literature guideline, the modal technique is here selected and the convergence is proved in terms of the stability and accuracy through the analysis of the amplification matrix. To this end, the eigenproblems associated with (17) in each direction without considering the coupled terms are given by

$$\mathbf{K}_{11}\mathbf{v}_{1i} = \omega_{1i}^2\mathbf{M}_{11}\mathbf{v}_{1i}(i = 1, \ldots, nq_1); \mathbf{K}_{22}\mathbf{v}_{2j} = \omega_{2j}^2\mathbf{M}_{22}\mathbf{v}_{2j}(j = 1, \ldots, nq_2) \quad (18)$$

where ω_{1i} and ω_{2i} can be viewed as uncoupled frequencies in the sense that they do not interact with each other, whereas $\mathbf{v}_{1i}$ and $\mathbf{v}_{2j}$ are eigenvectors orthonormalized with respect to each mass matrix. Let $\mathbf{V}_1 = [\mathbf{v}_{11}, \ldots, \mathbf{v}_{1nq_1}]$ and $\mathbf{V}_2 = [\mathbf{v}_{21}, \ldots, \mathbf{v}_{2nq_2}]$ be the matrices formed by collecting the eigenvectors. This leads to the following relations, $\mathbf{V}_1^T\mathbf{M}_{11}\mathbf{V}_1 = \mathbf{I}_1$ and $\mathbf{V}_2^T\mathbf{M}_{22}\mathbf{V}_2 = \mathbf{I}_2$, in which $\mathbf{I}_1 \in {}^{nq_1 \times nq_1}$ and $\mathbf{I}_2 \in {}^{nq_2 \times nq_2}$ are identity matrices. Hence, owing to the aforementioned relations and taking the transformations $\mathbf{U}_1 = \mathbf{V}_1\mathbf{x}_1$ and $\mathbf{U}_2 = \mathbf{V}_2\mathbf{x}_2$, the original equilibrium (11) can be transformed to the modal generalized displacement equation as follows:

$$\begin{bmatrix} \mathbf{\Omega}_1^2 & \mathbf{S} \\ \mathbf{S}^T & \mathbf{\Omega}_2^2 \end{bmatrix} \begin{Bmatrix} \mathbf{x}_1 \\ \mathbf{x}_2 \end{Bmatrix} + \begin{bmatrix} \mathbf{I}_1 & \mathbf{0} \\ \mathbf{0} & \mathbf{I}_2 \end{bmatrix} \begin{Bmatrix} \ddot{\mathbf{x}}_1 \\ \ddot{\mathbf{x}}_2 \end{Bmatrix} = \begin{Bmatrix} \mathbf{r}_1 \\ \mathbf{r}_2 \end{Bmatrix} \quad (19)$$

where, $\mathbf{\Omega}_1^2 = \mathbf{V}_1^T\mathbf{K}_{11}\mathbf{V}_1 = \text{diag}\left[\omega_{1i}^2\right]$, $\mathbf{\Omega}_2^2 = \mathbf{V}_2^T\mathbf{K}_{22}\mathbf{V}_2 = \text{diag}\left[\omega_{2j}^2\right]$, $\mathbf{S} = \mathbf{V}_1^T\mathbf{K}_{12}\mathbf{V}_2$, $\mathbf{r}_1 = \mathbf{V}_1^T\mathbf{F}_1$ and $\mathbf{r}_2 = \mathbf{V}_2^T\mathbf{F}_2$.

Although matrices $\mathbf{\Omega}_1$ and $\mathbf{\Omega}_2$ are diagonal, generally $\mathbf{S}$ is not, indicating the interaction of the above system with respect to all frequencies. However, as pointed out in [19], such an interaction can be represented by just a scalar constant named φ_{ij}. In this way, the convergence study is carried out on the following 2×2 model of coupled system

$$\begin{bmatrix} \omega_{1i}^2 & \varphi_{ij} \\ \varphi_{ij} & \omega_{2j}^2 \end{bmatrix} \begin{Bmatrix} x_{1i} \\ x_{2j} \end{Bmatrix} + \begin{bmatrix} 1 & 0 \\ 0 & 1 \end{bmatrix} \begin{Bmatrix} \ddot{x}_{1i} \\ \ddot{x}_{2j} \end{Bmatrix} = \begin{Bmatrix} r_{1i} \\ r_{2j} \end{Bmatrix} \quad (20)$$

In fact, in the particular case of a regular mesh and same nodal displacement constraints in each direction, one can readily prove the above assumption. Hence, we will have $nq_1 = nq_2$ leading to $\mathbf{K}_{11} = \mathbf{K}_{22}$, $\mathbf{K}_{12} = \mathbf{K}_{21}$ and $\mathbf{K}_{12} = \alpha\mathbf{K}_{11}$ with $1/2 > \alpha \geq 1/3$, as shown in the Appendix. Hence, it is expected a similar behavior for the other cases for which the mathematical proof becomes quite cumbersome. Finally, note that if $\varphi_{ij} = 0$ there is no interaction, reducing (20) to two uncoupled scalar ODEs. Hereafter, to simplify the notation, modal indexes i and j will be dropped and once force terms do not greatly affect the convergence analysis it will no longer be considered (i.e., the influence of the so-called load operator vector is less important than that of the amplification matrix) [2]. Therefore, taking into account the aforementioned assumptions (20) reduces to

$$\begin{bmatrix} \omega^2 & \alpha\omega^2 \\ \alpha\omega^2 & \omega^2 \end{bmatrix} \begin{Bmatrix} x_1 \\ x_2 \end{Bmatrix} + \begin{bmatrix} 1 & 0 \\ 0 & 1 \end{bmatrix} \begin{Bmatrix} \ddot{x}_1 \\ \ddot{x}_2 \end{Bmatrix} = \begin{Bmatrix} 0 \\ 0 \end{Bmatrix} \quad (21)$$

After applying the proposed methodology described in Table 1 to the above system one can derive the following concisely recurrence relation: $\{\ddot{x}_1^{n+1} \; \dot{x}_1^{n+1} \; x_1^{n+1} \; \ddot{x}_2^{n+1} \; \dot{x}_2^{n+1} \; x_2^{n+1}\}^T = \mathbf{A}\{\ddot{x}_1^n \; \dot{x}_1^n \; x_1^n \; \ddot{x}_2^n \; \dot{x}_2^n \; x_2^n\}^T$, where $\mathbf{A} \in$

$\mathbb{R}^{6\times6}$ stands for the so-called amplification matrix (entries of $\mathbf{A}$ are quite long and thus not presented here). The condition to assure stability is $\rho(\mathbf{A}) \leq 1$ [2,11] and the method is said to be unconditionally stable (or A-stable), if for any choice of Δt the solution does not blow up.

On the other hand, the accuracy of a time integration method is measured by the difference between the numerical solution and the exact one. The exact solution of the system of equations given by (21) is that of a free oscillator moving at the coupled frequency $\bar{\omega} = \sqrt{1 + \alpha}\omega$. Assuming initial conditions $x_1(0) = x_1^0$, $\dot{x}_1(0) = \dot{x}_1^0$, $x_2(0) = x_1^0$, $\dot{x}_2(0) = \dot{x}_1^0$, the exact solution reads:

$$x_1^{\text{ex}} = x_2^{\text{ex}} = x_1^0 \cos(\bar{\omega}t) + \frac{\dot{x}_1^0}{\bar{\omega}} \sin(\bar{\omega}t) \tag{22}$$

The numerical dispersion is measured by the relative period error and its expression is given by $(\bar{T}^h - \bar{T})/\bar{T}$ where $\bar{T} = 2\pi/\bar{\omega}$ and $\bar{T}^h = 2\pi/\bar{\omega}^h$ are, respectively, the exact coupled natural period and its numerical counterpart.

Figure 1 displays spectral radii concerning the proposed methods, taking into account the strategy given in Table 1 without iteration (i.e., $l = 0$) named U-Nw^0 and one iteration (i.e., $l = 1$) named U-Nw^1, as well as the standard Newmark (Nw) scheme; and considering different values of the parameters β and γ. As it is possible to observe, the proposed time-stepping schemes have the same spectral radius of the Newmark method when $\gamma = 0.5$, $\beta = 0.25$ and behave in a similar manner with the other values of γ and β.

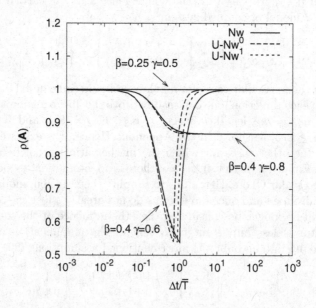

Fig. 1. Spectral radii for the Nw, U-Nw^0 and U-Nw^1 methods for different values of β and γ.

In order to better visualize this similar feature, a 3D surface plot for which $\rho(\mathbf{A}) = 1$ regarding the Nw method is displayed in Fig. 2a, whereas Fig. 2b compares the stability regions of the proposed methods to those of the Nw method at different $\Delta t/\bar{T}$ values. From these figures, it is clearly seen both stable and unstable regions as a function of the parameters β and γ (recall that stable regions means $\rho \leq 1$ and unstable $\rho > 1$). More precisely, when $\gamma < 0.5$ all methods are unstable, whereas for $\gamma \geq 0.5$ the methods can be either unconditionally stable or conditionally stable. In other words, if for some value of β and γ the line of increasing $\Delta t/\bar{T}$ crosses the surface from a stable region to an unstable one, the method is said to be conditionally stable; otherwise, it is unconditionally stable.

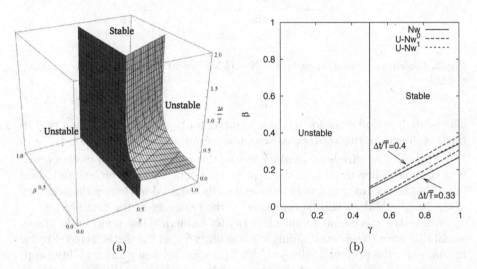

(a)   (b)

Fig. 2. Stability analysis: (a) stability regions in the Nw method; (b) 2D views for Nw, U-Nw^0 and U-Nw^1 methods at some $\Delta t/\bar{T}$ values.

Hereafter, only the so-called trapezoidal rule (i.e., $\beta = 0.25$, $\gamma = 0.5$) will be considered since it is the most accurate unconditionally stable scheme from the family of the Newmark method and, thus, widely employed.

In Fig. 3 relative period errors are plotted with the four obtained curves belonging to the Nw, U-Nw^1, U-Nw$^0_1$ and U-Nw$^0_2$ schemes, where the subscript index, for the U-Nw^0 method, means the direction for the algorithm presented in Table 1 (i.e., 1 means the first direction to be approximated whereas 2 the second one); the reason to display two curves for the U-Nw^0 method is explained in the following. It is readily seen that the curve correspondent to the proposed scheme with one iteration U-Nw^1 is coincident with the curve of the Newmark method. Although the U-Nw^1 and Nw methods present the same relative period error for both directions, this is not true for the U-Nw^0 method which, in addition to have a slightly worse accuracy, produces different curves for both directions.

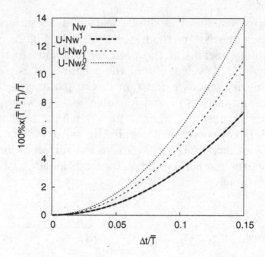

Fig. 3. Percentage period errors for the Nw, U-Nw^0 and U-Nw^1 methods for $\beta = 0.25$, $\gamma = 0.5$.

The relative period error for the second direction is slightly higher than that of the first direction; this is expected since one assumes the approximation $\mathbf{U}_{j_i}^{n+1} = \tilde{\mathbf{U}}_{j_i}^{n+1}$ for the first direction; in other words, the accuracy for each direction will be somewhat different. Hence, the application of the iterative procedure indeed improves the accuracy and only one iteration is required to recover the accuracy of the Nw method, as will be confirmed in the examples of the next section.

Notice that there is no amplitude decay (or numerical damping) in the trapezoidal rule since the spectral radius remains unity for all the methods (see Fig. 1). In summary, the proposed scheme U-Nw^1 presents accuracy and stability properties very similar to those of the standard Newmark time integration method for the fully coupled system.

5   Numerical Simulations

In order to show the potentialities in terms of stability as well as accuracy of the new methodology, a clamped rod model subject to a 45° uniform load considering homogeneous, bi-material, and functionally graded material properties is analyzed in this section. The rectangular rod is subjected to a uniform load applied at the top side while the bottom side is fixed as sketched in Fig. 4, with $f(t) = \sin(t\pi/T_p)\left(H(t) - H(t - T_p)\right)$ being the time variation of the load in which $T_p = 2.0 \times 10^{-2}$ s is its fundamental period.

The material properties for the functionally graded, bi-material and homogeneous models are presented in Table 2. A time-step size $\Delta t = 1.5 \times 10^{-3}$ s and a mesh composed of 200 quadrilateral elements are used for the proposed schemes, while a time-step size $\Delta t = 5.0 \times 10^{-4}$ s is used for the reference solution which consists of the results obtained by the Nw method using a small time-step size

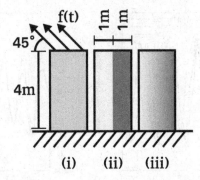

Fig. 4. Material variations: (i) homogeneous; (ii) bi-material and (iii) linear function-ally graded.

due to the lack of exact solutions. Recall that the trapezoidal rule ($\beta = 0.25$, $\gamma = 0.5$) is being employed. In the bi-material model, the material interface is placed along the height at the middle of the rod and, in the FGM model, a linear variation profile along the width with the minimum value being at the left side and the maximum one at the right side is considered as also sketched in Figs. 4(ii), (iii).

Table 2. Material properties

Material	ν	$\rho\,(\mathrm{kg/m^3})$	$E\,(\mathrm{N/m^2})$
Homo.	0.25	1	1.0×10^5
Left	0.2	0.5	5.0×10^4
Right	0.3	1.5	1.5×10^5

First we analyze the physical difference between the results furnished by the three material models. This comparison is carried out by analyzing the vertical and horizontal displacements at point $A = (1, 4)$ as displayed in Fig. 5. It is readily seen that, by adopting the functionally graded profile, we obtain an intermediate response between the bi-material and homogeneous models. In such an analysis, only the reference (converged) solutions computed by the NW method with the reference time-step size ($\Delta t = 5.0 \times 10^{-4}$ s) were employed for all models.

The next analysis consists of comparing the results obtained by the methods when only the FGM model is considered. In Figs. 6a, b, time histories for horizontal and vertical displacements at the same point $A = (1, 4)$ are depicted, taking into account the Nw, U-Nw^0 and U-Nw^1 methods. Considering the time range displayed in the figures, it can be concluded that accurate results are attained by all methods. However, as the time advances, the accumulative error due to the period error in the U-Nw^0 scheme (see Fig. 3) starts to appear in the

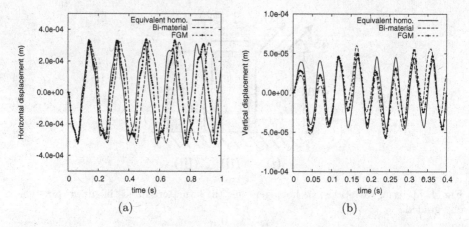

Fig. 5. Comparison of results among the Homogeneous, Bi-material and FGM models.

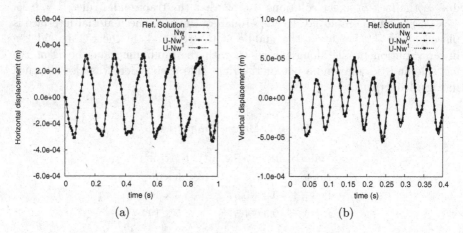

Fig. 6. Comparison of displacement-time histories at point $A = (1, 4)$ in the beginning of the time analysis.

results as clearly observed in Figs. 7a, b for a different time range far away from the initial time instant after the use of many time steps. On the other hand, the U-Nw^1 scheme stays very close to the Nw method which is in accordance with the conclusions drawn in the previous section (see also Fig. 3).

In Fig. 8 one can see the behavior of the displacement field $\|\mathbf{u}^h\|$ for all the three material models indicating more clearly the physical difference among the models (once again reference solutions by the Nw method were employed). Notice that, as shown in Fig. 5, the resulting displacement curves have different periods for all the three models. As a consequence, at the time instant in which the snapshot was taken, all of them present different phases. Furthermore, it is possible to see that, in the bi-material snapshot, the solution changes somewhat abruptly at the interface between materials, indicating a discontinuous derivative

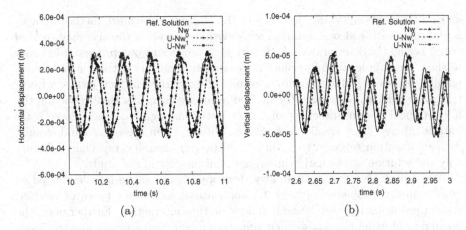

Fig. 7. Comparison of displacement-time histories at point $A = (1, 4)$ at late time instants.

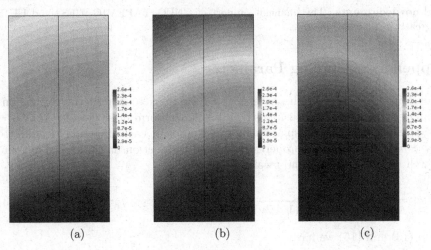

Fig. 8. Snapshots of the displacement field $\|\mathbf{u}^h\|$ at $t = 1.00\,$s: (a) Homogeneous; (b) Bi-material; and (c) FGM models.

in the results, which is not desired in a great deal of engineering applications. Notice, that this drawback does not occur with the FG material.

6   Conclusions

This paper presented the development of an uncoupling strategy directly into the Newmark method to decouple the parts of a partitioned system of ODEs, allowing the use of an unconditionally stable time-marching method while keeping its main features for treating dynamic problems. Although an iterative procedure is established, the convergence study revealed that the methodology retains the unconditional stability and accuracy of the standard Newmark method (trapezoidal rule) with just one iteration while without iteration the accuracy (period error) is slightly deteriorated. A step-by-step procedure was presented in order to show the simple computational implementation and its natural structure for parallel processing. The numerical results showed that in fact just one iteration was sufficient to achieve the desired accuracy (which is that of solving the fully coupled system directly), that does not mean significant additional computational cost since smaller systems are obtained. On the other hand, results without iteration revealed to be in a good overall accuracy, especially whether only the solution at the early time range is of interest in the study.

We expect that the methodology developed here can be later extended to more complex problems such as 3D applications and other types of coupled system problems like the dynamic fluid-structure interaction. Furthermore, the possibility of using implicit-explicit time-step integration schemes and sub-steps remains to be investigated in order to improve the presented results.

Acknowledgments. The financial support of CNPQ, FAPEMIG, UFSJ and UFJF is greatly acknowledged.

Appendix: Coupling Parameter

This appendix presents the necessary assumptions to obtain the expressions employed in Sect. 4. Thus, supposing a regular uniform mesh (leading to $N_{i,1} = N_{i,2}$) and same nodal displacement constraints in each direction (leading to $nq_1 = nq_2$) with material properties governed by functions in the spatial domain, as well as FGM models, and recalling that

$$c_1 = \lambda + 2\mu = \frac{E(1 - \nu)}{(1 + \nu)(1 - 2\nu)}, \quad c_2 = \lambda = \nu c_1, \quad c_3 = \mu = c_1 \frac{(1 - 2\nu)}{2(1 - \nu)} \qquad (23)$$

from (13) and (15) we have

$$K_{ij}^{11} = \int_{\Omega} h N_{i,1} N_{j,1}(c_1 + c_3) dxdy; \, K_{ij}^{22} = \int_{\Omega} h N_{i,1} N_{j,1}(c_3 + c_1) dxdy = K_{ij}^{11} \qquad (24)$$

Furthermore, (14) can be rewritten (after proper use of the Integral Mean Value Theorem) as follows:

$$K_{ij}^{12} = K_{ji}^{21} = \int_{\Omega} h(\nu c_1 N_{i,1} N_{j,2} + c_3 N_{i,2} N_{j,1}) dx dy \tag{25}$$

$$= \int_{\Omega} h N_{i,1} N_{j,1} (\nu c_1 + c_3) dx dy \tag{26}$$

$$= \int_{\Omega} h N_{i,1} N_{j,1} (\nu c_1 + c_3) \frac{(c_1 + c_3)}{(c_1 + c_3)} dx dy \tag{27}$$

$$= \sum_e \frac{(\nu c_1 + c_3)}{(c_1 + c_3)} \bigg|_{\xi \in \Omega_e} \int_{\Omega_e} h N_{i,1} N_{j,1} (c_1 + c_3) dx dy \tag{28}$$

$$= \sum_e \alpha_e \int_{\Omega_e} h N_{i,1} N_{j,1} (c_1 + c_3) dx dy \tag{29}$$

where $\alpha_e = \frac{(\nu c_1 + c_3)}{(c_1 + c_3)} = \frac{1 - 2\nu^2}{3 - 4\nu}$ plays a role of a parameter which varies as a function of the Poisson's ratio with $\nu \in [0, 0.5]$ in common cases. Since we are interested in the variation of such a parameter over the whole range, one can assume $K_{ij}^{12} = K_{ji}^{21} = \alpha K_{ij}^{11}$.

References

1. Arsha, A., Jayakumar, E., Rajan, T., Antony, V., Pai, B.: Design and fabrication of functionally graded in-situ aluminium composites for automotive pistons. Mater. Des. **88**, 1201–1209 (2015)
2. Bathe, K.: Finite Element Procedures. Prentice-Hall, Upper Saddle River (1996)
3. Bathe, K., Noh, G.: Insight into an implicit time integration scheme for structural dynamics. Comput. Struct. **98–99**, 1–6 (2012)
4. Belonosov, M.A., Kostov, C., Reshetova, G.V., Soloviev, S.A., Tcheverda, V.A.: Parallel numerical simulation of seismic waves propagation with Intel math kernel library. In: Manninen, P., Öster, P. (eds.) PARA 2012. LNCS, vol. 7782, pp. 153–167. Springer, Heidelberg (2013). doi:10.1007/978-3-642-36803-5_11
5. Daniel, W.: Subcycling first- and second-order generalizations of the trapezoidal rule. Int. J. Numer. Meth. Eng. **42**, 1091–1119 (1998)
6. Dongarra, J.J., Eijkhout, V.: Numerical linear algebra algorithms and software. J. Comput. Appl. Math. **123**, 489–514 (2000)
7. Fan, Z., Jia, X.: Element-free method and its efficiency improvement in seismic modelling and reverse time migration. J. Geophys. Eng. **10**, 025002 (2013)
8. Felippa, C.A., Park, K., Farhat, C.: Partitioned analysis of coupled mechanical systems. Comput. Methods Appl. Mech. Eng. **190**, 3247–3270 (2001)
9. Gao, X.W., Li, L.: A solver of linear systems of equations (REBSM) for large-scale engineering problems. Int. J. Comput. Methods **9**, 1240011 (2012)
10. Hoitink, A., Masuri, S., Zhou, X., Tamma, K.K.: Algorithms by design: Part I-on the hidden point collocation within lms methods and implications for nonlinear dynamics applications. Int. J. Comput. Methods Eng. Sci. Mech. **9**, 383–407 (2008)
11. Hughes, T.: The Finite Element Method: Linear Static and Dynamic Finite Element Analysis. Dover Publications, New York (2000)
12. Janna, C., Comerlati, A., Gambolati, G.: A comparison of projective and direct solvers for finite elements in elastostatics. Adv. Eng. Softw. **40**, 675–685 (2009)

13. Jardin, S., Breslau, J., Ferraro, N.: A high-order implicit finite element method for integrating the two-fluid magnetohydrodynamic equations in two dimensions. J. Comput. Phys. **226**, 2146–2174 (2007)

14. Kim, J.-H., Paulino, G.: Isoparametric graded finite elements for nonhomogeneous isotropic and orthotropic materials. J. Appl. Mech. Trans. **69**, 502–514 (2002). ASME

15. Kim, W., Park, S.-S., Reddy, J.N.: A cross weighted-residual time integration scheme for structural dynamics. Int. J. Struct. Stab. Dyn. **14**, 1450023 (2014)

16. Li, X.S.: An overview of SuperLU: algorithms, implementation, and user interface. TOMS **31**, 302–325 (2005)

17. Mehrabani, M., Nobari, M., Tryggvason, G.: Accelerating poisson solvers in front tracking method using parallel direct methods. Computers & Fluids. **118**, 101–113 (2015)

18. Reddy, J.N.: An Introduction to the Finite Element Method, 3rd edn. McGraw-Hill Higher Education, New York (2006)

19. Ross, M.R., Felippa, C.A., Park, K., Sprague, M.A.: Treatment of acoustic fluid-structure interaction by localized lagrange multipliers: formulation. Comput. Methods Appl. Mech. Eng. **197**, 3057–3079 (2008)

20. Santare, M., Lambros, J.: Use of graded finite elements to model the behavior of nonhomogeneous materials. J. Appl. Mech. Trans. **67**, 819–822 (2000). ASME

21. Santare, M.H., Thamburaj, P., Gazonas, G.A.: The use of graded finite elements in the study of elastic wave propagation in continuously nonhomogeneous materials. Int. J. Solids Struct. **40**, 5621–5634 (2003)

22. Shimada, M., Masuri, S., Tamma, K.K.: A novel design of an isochronous integration iIntegration framework for first/second order multidisciplinary transient systems. Int. J. Numer. Meth. Eng. **102**, 867–891 (2015)

23. Smolinski, P., Belytschko, T., Neal, M.: Multi-time-step integration using nodal partitioning. Int. J. Numer. Meth. Eng. **26**, 2 (1988)

24. Soares, D., Von Estorff, O., Mansur, W.: Iterative coupling of BEM and FEM for nonlinear dynamic analyses. Comput. Mech. **34**, 67–73 (2014)

25. Soares, D.: An optimised FEM-BEM time-domain iterative coupling algorithm for dynamic analyses. Comput. Struct. **86**, 1839–1844 (2008)

26. Soares, D.: A simple and effective new family of time marching procedures for dynamics. Comput. Methods Appl. Mech. Eng. **283**, 1138–1166 (2015)

27. Soares, D., Godinho, L.: An overview of recent advances in the iterative analysis of coupled models for wave propagation. J. Appl. Math. **2014**, 21 (2014)

28. Wu, Y., Smolinski, P.: A multi-time step integration algorithm for structural dynamics based on the modified trapezoidal rule. Comput. Methods Appl. Mech. Eng. **187**, 641–660 (2000)

29. Zhang, Z., Paulino, G.H.: Wave propagation and dynamic analysis of smoothly graded heterogeneous continua using graded finite elements. Int. J. Solid Struct. **44**, 3601–3626 (2007)

30. Zhang, Z., Yang, Z., Liu, G.: An adaptive time-stepping procedure based on the scaled boundary finite element method for elastodynamics. Int. J. Comput. Methods **9**, 1 (2012)

Continuous Extensions for Structural Runge–Kutta Methods

Alexey S. Eremin[✉] and Nikolai A. Kovrizhnykh

Saint-Petersburg State University, Saint-Petersburg, Russia
a.eremin@spbu.ru, sagoyewatha@mail.ru

Abstract. The so-called structural methods for systems of partitioned ordinary differential equations studied by Olemskoy are considered. An ODE system partitioning is based on special structure of right-hand side dependencies on the unknown functions. The methods are generalization of Runge–Kutta–Nyström methods and as the latter are more efficient than classical Runge–Kutta schemes for a wide range of systems. Polynomial interpolants for structural methods that can be used for dense output and in standard approach to solve delay differential equations are constructed. The proposed methods take fewer stages than the existing most general continuous Runge–Kutta methods. The orders of the constructed methods are checked with constant step integration of test delay differential equations. Also the global error to computational costs ratios are compared for new and known methods by solving the problems with variable time-step.

Keywords: Continuous methods · Delay differential equations · Runge–Kutta methods · Structural partitioning

1 Introduction

Despite the long history of Runge–Kutta methods they still attract great interest and people work to improve them, to make them faster and more accurate. One of the possible approaches to construction of new specialized methods is partitioning of the ordinary differential equations (ODEs) systems into two or more parts and applying different computational schemes to different parts. In many situations some partitioning is natural to the problem. Some pioneer papers were devoted to partitioning stiff problems into "stiff" and "non-stiff" components and construction of implicit-explicit methods (eg. [8]). Today multischeme methods are used to provide simplecticity [11,22], to solve partial differential equations [9], to better control fast and slow processes in large systems [19,20].

In the current paper we consider explicit Runge–Kutta type methods that are constructed with use of special structure of ODE system right-hand sides, namely the structure of their dependencies on the unknown functions. Various types of such structure are studied in Sect. 2. Such methods require fewer stages per step than classical explicit Runge–Kutta methods.

© Springer International Publishing AG 2017
O. Gervasi et al. (Eds.): ICCSA 2017, Part II, LNCS 10405, pp. 363–378, 2017.
DOI: 10.1007/978-3-319-62395-5_25

Notice that though special structures can seem unnatural and artificial, many real problems can be transformed to satisfy one of them with some reordering of variables and equations. Moreover, search for a suitable sorting order can be automatized, for instance with the algorithm for detection of such special structure presented in [15].

For many applications standard Runge–Kutta methods, which compute the approximate solution in points of some mesh, are not enough since we are interested in a continuous approximation to the solution (sometimes called *dense output* [7]). Methods that provide dense output are called *continuous*. One of the main situations when we need continuous methods is solving of delay-differential equations (DDEs) or various functional differential equations.

The so-called *standard approach* [2] to solve DDEs is to use continuous methods to approximate delayed solution in the points from between the initial point t_0 and the starting point of the current step t_m. Due to some other issues one-step methods, mainly Continuous Runge–Kutta (CRK), are most popular [2]. If the delay is non-vanishing then the step-size can always be chosen so that the delay falls before t_m and explicit continuous methods can be used (otherwise they have to be implemented as fully implicit unless specially designed to remain explicit [10]).

Consideration of continuous methods demands giving two definitions of local order: *discrete* order, the same as for usual methods, and *uniform* order of the continuous interpolant. Here we define them in the following way. Consider a problem

$$y'(t) = f(t, y(t)), \quad y(t_0) = y_0. \tag{1}$$

Definition 1. *We say that a continuous method has discrete local order p if for any problem* (1) *with sufficiently small h there exists $C > 0$ such, that the approximation $\bar{y}$ to $y(t_0 + h)$ satisfies*

$$e_d(h) = \|y(t_0 + h) - \bar{y}\| = Ch^{p+1}.$$

Definition 2. *We say that a continuous method has uniform local order q if for any problem* (1) *with sufficiently small h there exists $C > 0$ such, that the approximation $\bar{y}(t)$ provided for $t \in [t_0, t_0 + h]$ satisfies*

$$e_u(h) = \max_{t \in [t_0, t_0+h]} \|y(t) - \bar{y}(t)\| = Ch^{q+1}.$$

Discrete order obviously is not less than uniform, but the situation when $q < p$ is quite common for continuous methods (though for DDEs we need q to be not less than $p - 1$ to provide global order p [2]).

In the paper we construct continuous versions of methods that use special structure of ODE systems. In the next section we cover the background on discrete methods of the kind, and after we construct continuous methods and implement them to test problems.

2   Structural Methods

Olemskoy [16] considered systems of ODEs with special structure of right-hand side (RHS) dependencies on the unknown functions and constructed one-step Runge–Kutta type methods with fewer RHS evaluations per equation per step than it is possible for *classical* Runge–Kutta methods according to Butcher barriers [7]. Originally they were inspired by Runge–Kutta–Nyström methods for second order equations of the form

$$y''(t) = f(t, y(t)), \tag{2}$$

for which direct implementation of Runge–Kutta methods is more efficient than for equivalent systems

$$\begin{cases} y'(t) = z(t), \\ z'(t) = f(t, y(t)). \end{cases}$$

Here and in all further considered systems unknown functions can be treated as vectors. We will write all the methods below in continuous form.

2.1   Cross-Dependent Systems

Olemskoy showed that for a more general, so-called *Cross-dependent* system

$$\begin{cases} y_1'(t) = f_1(t, y_2(t)), \\ y_2'(t) = f_2(t, y_1(t)) \end{cases} \tag{3}$$

the same advantage can be obtained (here, in contrast to (2), y_1 and y_2 can be vectors of different lengths). In [12] a method of order five was constructed with only four stages, while *classical* methods require six. The method suggested by Olemskoy makes a step h from t_0 according to

$$y_1(t_0 + \theta h) \approx y_1(t_0) + h \sum_{i=1}^{s_1} b_{1i}(\theta) k_{1i},$$

$$y_2(t_0 + \theta h) \approx y_2(t_0) + h \sum_{i=1}^{s_2} b_{2i}(\theta) k_{2i} \tag{4}$$

with right-hand side functions being calculated one by one k_{11}, k_{21}, k_{12}, k_{22}, k_{13} etc. with immediate use of the values just obtained

$$\begin{aligned}
k_{11} &= f_1\big(t_0, y_2(t_0)\big), \\
k_{21} &= f_2\big(t_0 + c_{21}h, y_1(t_0) + ha_{211}k_{11}\big), \\
k_{12} &= f_1\big(t_0 + c_{12}h, y_2(t_0) + ha_{121}k_{21}\big), \\
k_{22} &= f_2\big(t_0 + c_{22}h, y_1(t_0) + ha_{221}k_{11} + ha_{222}k_{12}\big), \\
&\cdots
\end{aligned} \tag{5}$$

$$k_{1i} = f_1 \left(t_0 + c_{1i}h, \ y_2(t_0) + h \sum_{j=1}^{i-1} a_{1ij}k_{2j} \right),$$

$$k_{2i} = f_2 \left(t_0 + c_{2i}h, \ y_1(t_0) + h \sum_{j=1}^{i} a_{2ij}k_{1j} \right),$$

$$\dots$$

The method has two sets of usual Runge–Kutta coefficients. Notice that $a_{2ii} \neq 0$ for at least one i and either $s_1 = s_2$ or $s_1 = s_2 + 1$.

2.2   Two-Groups Systems

We can make a generalization of (3). For a system

$$\begin{cases} y'_{1r}(t) = f_{1r}(t, y_{11}, ..., y_{1,r-1}, y_{21}, ..., y_{2n_2}), & r = 1, ..., n_1, \\ y'_{2r}(t) = f_{2r}(t, y_{11}, ..., y_{1n_1}, y_{21}, ..., y_{2,r-1}), & r = 1, ..., n_2, \end{cases} \tag{6}$$

a similar advantage over classical methods can be obtained. Thus in [16] a third order method with two evaluations of each f_{1i} and f_{2i} is presented and in [14] a fifth order method with five stages and last stages reuse is constructed (for reuse details see Sect. 3). System (6) has two structural groups within which the right-hand sides depend only of the previous functions of the same group. We call them *Two-groups* systems, since their matrices of RHS dependencies on the unknowns have two blocks with special structure (namely, strictly lower-triangular).

In general a method for (6) looks as

$$y_{1r}(t_0 + \theta h) \approx y_{1r}(t_0) + h \sum_{i=1}^{s_1} b_{1i}(\theta)k_{1ri}, \quad r = 1, ..., n_1,$$

$$y_{2r}(t_0 + \theta h) \approx y_{2r}(t_0) + h \sum_{i=1}^{s_2} b_{2i}(\theta)k_{2ri}, \quad r = 1, ..., n_2. \tag{7}$$

Again the right-hand side functions are calculated one by one k_{111}, k_{121}, ..., k_{1n_11}, k_{211}, ..., k_{2n_21}, k_{112}, ..., k_{1n_12}, k_{212}, etc. according to

$$k_{1ri} = f_{1r} \Big(t_0 + c_{1i}h,$$

$$y_{11}(t_0) + h \sum_{j=1}^{i} a_{11ij}k_{11j}, ..., y_{1,r-1}(t_0) + h \sum_{j=1}^{i} a_{11ij}k_{1,r-1,j},$$

$$y_{21}(t_0) + h \sum_{j=1}^{i-1} a_{12ij}k_{21j}, ..., y_{2n_2}(t_0) + h \sum_{j=1}^{i-1} a_{12ij}k_{2n_2j} \Big),$$

$$k_{2ri} = f_{2r} \Big(t_0 + c_{2i}h,$$

$$y_{11}(t_0) + h \sum_{j=1}^{i} a_{21ij}k_{11j}, ..., y_{1n_1}(t_0) + h \sum_{j=1}^{i} a_{21ij}k_{1n_1j},$$

$$y_{21}(t_0) + h \sum_{j=1}^{i} a_{22ij}k_{21j}, ..., y_{2,r-1}(t_0) + h \sum_{j=1}^{i} a_{22ij}k_{2,r-1,j} \Big). \tag{8}$$

Notice, that if $n_1 = n_2 = 1$ then a two-groups system is simply cross-dependent. Still we consider (3) separately, because for it more efficient methods can be constructed (like the one of order five with four stages mentioned above).

2.3 General Group

Forms (3) and (6) demand that none of the derivatives were dependent on the function itself. However, if this is the case, another generalization of (3) can be done. All such equations are gathered in the structural group called *general* and other form the structure analogous to the second group of (6). The general group has no structural properties and all the unknown within can be considered as a single vector y_0:

$$\begin{cases} y_0'(t) = f_0(t, y_0, y_{21}, ..., y_{2n_2}), \\ y_{2r}'(t) = f_{2r}(t, y_0, y_{21}, ..., y_{2,r-1}), \quad r = 1, ..., n_2. \end{cases} \tag{9}$$

We call it a *One-group* system. It should be mentioned here that the most general structural form considered by Olemskoy [13] (called *Full*) has all three groups: general and both first and second from (6) (this explains the denotation 2 in (9)). The methods for such systems are much harder to construct and their advantage over classical methods is lower than for (6), (9) or moreover (3). In the current paper we don't consider full systems and construct continuous methods for (3), (6) and (9) only.

A method for (9) is

$$y_0(t_0 + \theta h) \approx y_0(t_0) + h \sum_{i=1}^{s_0} b_{0i}(\theta) k_{0i},$$

$$y_{2r}(t_0 + \theta h) \approx y_{2r}(t_0) + h \sum_{i=1}^{s_2} b_{2i}(\theta) k_{2ri}, \quad r = 1, ..., n_2, \tag{10}$$

$$k_{0i} = f_0 \Big(t_0 + c_{0i} h, \ y_0(t_0) + h \sum_{j=1}^{i-1} a_{00ij} k_{0j},$$

$$y_{21}(t_0) + h \sum_{j=1}^{i-1} a_{02ij} k_{21j}, ..., y_{2n_2}(t_0) + h \sum_{j=1}^{i-1} a_{02ij} k_{2n_2 j} \Big),$$

$$k_{2ri} = f_{2r} \Big(t_0 + c_{2i} h, \ y_0(t_0) + h \sum_{j=1}^{i} a_{20ij} k_{0j}, \tag{11}$$

$$y_{21}(t_0) + h \sum_{j=1}^{i} a_{22ij} k_{21j}, ..., y_{2,r-1}(t_0) + h \sum_{j=1}^{i} a_{22ij} k_{2,r-1,j} \Big).$$

As before, either $s_0 = s_2$ or $s_0 = s_2 + 1$.

We collect the coefficients of the methods into extended Butcher tables [3] of the form

$$\frac{c_1 \left| A_1 \right| b_1(\theta)}{c_2 \left| A_2 \right| b_2(\theta)}, \quad \frac{c_1 \left| A_{11} \right| A_{12} \left| b_1(\theta)}{c_2 \left| A_{21} \right| A_{22} \left| b_2(\theta)} \quad \text{and} \quad \frac{c_0 \left| A_{00} \right| A_{02} \left| b_0(\theta)}{c_2 \left| A_{20} \right| A_{02} \left| b_2(\theta)}$$

for methods (4), (5); (7), (8); and (10), (11) respectively. Here, for any possible v, $b_v(\theta) = (b_{v1}(\theta), ..., b_{vs_v}(\theta))^T$ and $c_v = (c_{v1}, ..., c_{vs_v})^T$ are vectors and $A_v = \{a_{vij}\}$ are matrices of the suitable sizes. For some methods we also include the coefficients $\hat{b}(\theta)$ of embedded error estimators (see, for instance, [4]), which we use to control automatically the integration step-size. Vectors $\hat{b}_v(\theta)$ are attached to the right of corresponding $b_v(\theta)$.

3   Continuous Structural Methods

Olemskoy [16] has constructed a method of the second order for two-groups system (6) with two stages for the first group and just one stage for the second. However it is impossible to construct a continuous extension of the second order with fewer than two stages for each group and this is not better than classical second order method. So we start from methods of order three.

The coefficients of the methods are found from the systems of algebraic equations known as *order conditions*, which exist as well for uniform orders of continuous methods [7]. Structural methods require more complicated systems of order conditions to be constructed and solved [5]. For this paper we made a straightforward combination of both. We don't present them for all methods and orders, but to get an idea, here are order conditions for the method (10), (11) of uniform order three:

$$\bullet \quad \sum_{i=1}^{s_0} b_{0i}(\theta) = \theta \qquad\qquad \circ \quad \sum_{i=1}^{s_2} b_{2i}(\theta) = \theta$$

$$\sum_{i=1}^{s_0} b_{0i}(\theta) c_{0i} = \frac{\theta^2}{2} \qquad\qquad \sum_{i=1}^{s_2} b_{2i}(\theta) c_{2i} = \frac{\theta^2}{2}$$

$$\sum_{i=1}^{s_0} b_{0i}(\theta) c_{0i}^2 = \frac{\theta^3}{3} \qquad\qquad \sum_{i=1}^{s_2} b_{2i}(\theta) c_{2i}^2 = \frac{\theta^3}{3}$$

$$\sum_{i=1}^{s_0} b_{0i}(\theta) \sum_{j=1}^{i-1} a_{00ij} c_{0j} = \frac{\theta^3}{6} \qquad\qquad \sum_{i=1}^{s_0} b_{0i}(\theta) \sum_{j=1}^{i-1} a_{02ij} c_{2j} = \frac{\theta^3}{6}$$

$$\sum_{i=1}^{s_2} b_{2i}(\theta) \sum_{j=1}^{i} a_{20ij} c_{0j} = \frac{\theta^3}{6} \qquad\qquad \sum_{i=1}^{s_2} b_{2i}(\theta) \sum_{j=1}^{i} a_{22ij} c_{2j} = \frac{\theta^3}{6}$$

These conditions can be derived by direct comparison of the Taylor series for the exact solution and the approximation under the assumptions

$$\sum_{j=1}^{i-1} a_{00ij} = \sum_{j=1}^{i-1} a_{02ij} = c_{0i}, \ i = 1, ..., s_0, \quad \sum_{j=1}^{i} a_{20ij} = \sum_{j=1}^{i} a_{22ij} = c_{2i}, \ i = 1, ..., s_2,$$

which are sometimes called *basic simplifying conditions*.

However this involves quite cumbersome algebra and would be too hard to make for high orders. Taylor series about t_0 in respect of h for the first component of the solution of (9) is

$$y_0(t_0 + \theta h) = y_0(t_0) + \theta h f_0 \big|_{t_0}$$
$$+ \frac{\theta^2 h^2}{2} \left(\frac{\partial f_0}{\partial t} \bigg|_{t_0} + \frac{\partial f_0}{\partial y_0} \bigg|_{t_0} f_0 \big|_{t_0} + \sum_{i=1}^{n_2} \frac{\partial f_0}{\partial y_{2i}} \bigg|_{t_0} f_i \big|_{t_0} \right) + O(h^3)$$

and for the approximation

$$y_0(t_0) + h \sum_{i=1}^{s_0} b_{0i}(\theta) k_{0i} = y_0(t_0) + h \sum_{i=1}^{s_0} b_{0i}(\theta) f_0 \big|_{t_0}$$

$$+ h^2 \sum_{i=1}^{s_0} b_{0i}(\theta) c_{0i} \left(\frac{\partial f_0}{\partial t} \bigg|_{t_0} + \frac{\partial f_0}{\partial y_0} \bigg|_{t_0} f_0 \big|_{t_0} + \sum_{i=1}^{n_2} \frac{\partial f_0}{\partial y_{2i}} \bigg|_{t_0} f_i \big|_{t_0} \right) + O(h^3).$$

From here the conditions $\sum_{i=1}^{s_0} b_{0i}(\theta) = \theta$ and $\sum_{i=1}^{s_0} b_{0i}(\theta) c_{0i} = \frac{\theta^2}{2}$ follow. Further expansion takes too much room and doesn't make things clearer.

However for Runge–Kutta methods the same conditions can be obtained with use of so-called *labeled trees*, the tree graphs corresponding to each condition. The number of vertices corresponds to the order provided by the condition. Different forms of trees correspond to different conditions, so we need to find all non-isomorphic trees of size $\leq p$ to write down the system of order conditions for methods of order p. In case of system (9) we need two colors of tree nodes, while it was a single color for (1). The basic simplifying conditions make the color difference of tree leaves unimportant.

The conditions correspond to the trees printed to their left. Black vertices correspond to the general group, white for the second structural group and leaves are left empty.

One can check that these are all different trees with three or less two colored vertices and one-colored leaves.

It is quite obvious that methods designed for (6) can be applied to (3), as well as methods suitable for (9) can be used to solve both (6) and (3) too. So we consider methods for narrower classes of systems only if they are more computationally efficient, i.e. demand fewer stages.

3.1 Third Order Methods

Methods with Discrete Order Three and Uniform Two. First, we construct two methods with discrete order three and continuous extension of order two only. This makes them still suitable for solving DDEs with global order three and, at the same time, easier to construct and less demanding.

In this case it is possible to make a continuous extension of one of Olemskoy's methods for two-groups systems, which has order three and two stages for both groups (cross-dependent are considered as an extreme case without anything new here). We denote it CRKOT3d2u{2} where CRKO stands for Continuous Runge–Kutta–Olemskoy, T for two-groups, 3d2u for discrete three uniform two and {2} for two stages being used. We will use the similar abbreviations later with possible variations explained where necessary. The methods coefficients are presented in the Table 1.

For obvious reasons structural methods for systems with general groups (9) can't have fewer stages for the general group than classical methods (see the system of order conditions above). So we construct a method (10), (11) with $s_0 = 3$ and $s_2 = 2$, see Table 2. We denote it CRKOG3d2u{3,2}, where G shows

Table 1. CRKOT3d2u{2} coefficients

c_{1i}	a_{11ij}		a_{12ij}	$b_{1i}(\theta)$
0	0		0	$\theta - \frac{3}{4}\theta^2$
$\frac{2}{3}$	$\frac{1}{3}$	$\frac{1}{3}$	$\frac{2}{3}$	$\frac{3}{4}\theta^2$

c_{2i}	a_{21ij}	a_{22ij}		$b_{2i}(\theta)$
$\frac{1}{3}$	$\frac{1}{3}$	$\frac{1}{3}$		$\frac{3}{2}\theta - \frac{3}{4}\theta^2$
1	0	1	1 0	$-\frac{1}{2}\theta + \frac{3}{4}\theta^2$

Table 2. CRKOG3d2u{3,2} coefficients

c_{0i}	a_{00ij}		a_{02ij}		$b_{0i}(\theta)$
0	0		0		$\theta - \frac{3}{4}\theta^2$
$\frac{1}{3}$	$\frac{1}{3}$		$\frac{1}{3}$		0
$\frac{2}{3}$	0	$\frac{2}{3}$	$\frac{1}{3}$	$\frac{1}{3}$	$\frac{3}{4}\theta^2$

c_{2i}	a_{20ij}		a_{22ij}		$b_{2i}(\theta)$
0	0		0		$\theta - \frac{3}{4}\theta^2$
$\frac{2}{3}$	0	$\frac{2}{3}$	$\frac{1}{3}$	$\frac{1}{3}$	$\frac{3}{4}\theta^2$

that the general group is considered and {3,2} stands for different number of stages for different groups.

Classical methods of the same orders can also be constructed with three stages. The coefficients of a possible variant of such method can be "extracted" from the Table 2 as matrix A_{00} and vectors $b_0(\theta)$ and c_0. So, the constructed methods have an advantage over it for structural groups.

Uniform Order Three. A classical Runge–Kutta method of uniform order three requires four stages, but it is possible to construct a discrete order three method with three stages and use the RHS computation in $y(t_0+h)$ as the fourth stage to get a continuous extension. And since we consider explicit methods, the computation made in $y(t_0+h)$ will be the first for the next step. Thus, a method is implemented with only three stages per step. This technique is called First Same as Last (FSAL) [7] or reuse. Owren and Zennaro [17] constructed a number of CRKs with reuse and local error minimization. We extend their method of order three with four stages and FSAL to systems (9).

The method from Table 3 has four stages for the general group and three for the structural group and both last computations are reused, i.e. $a_{00s_0j} = a_{20s_2j} = b_{0j}(1)$, $j = 1, ..., s_0$ and $a_{02s_0j} = a_{22s_2j} = b_{2j}(1)$, $j = 1, ..., s_2$. We also add a uniform order two embedded local error estimator. The abbreviation for the constructed method is CRKOG3(2){4F,3F}, where 3(2) stands for orders (uniform since uniform and discrete orders are equal) of the main method and the estimator, and F after the stage number shows that FSAL is implemented. The coefficients of original method by Owren and Zennaro (denoted CRK3(2){4F} here) are in $c_0|A_{00}|b_0(\theta)$ part of the Table 3. We use the same estimator $\hat{b}_0(\theta)$ for it when compare the performance of two methods.

3.2 Fourth Order Methods

Discrete Order Four Uniform Three. Classic method with such orders can be constructed with four stages. As an example see the $c_0|A_{00}|b_0(\theta)$ part

Table 3. CRKOG3(2){4F,3F} coefficients

c_{0i}	a_{00ij}			a_{02ij}			$b_{0i}(\theta)$	$\hat{b}_{0i}$
0	0			0			$\theta - \frac{65}{48}\theta^2 + \frac{41}{72}\theta^3$	$\theta - \frac{1}{2}\theta^2$
$\frac{12}{23}$	$\frac{12}{23}$			$\frac{12}{23}$			$\frac{529}{384}\theta^2 - \frac{529}{576}\theta^3$	0
$\frac{4}{5}$	$-\frac{68}{375}$	$\frac{368}{375}$		$\frac{4}{125}$	$\frac{96}{125}$		$\frac{125}{128}\theta^2 - \frac{125}{192}\theta^3$	0
1	$\frac{31}{144}$	$\frac{529}{1152}$	$\frac{125}{384}$	$\frac{1}{4}$	$\frac{3}{4}$	0	$-\theta^2 + \theta^3$	$\frac{1}{2}\theta^2$

c_{2i}	a_{20ij}			a_{22ij}			$b_{2i}(\theta)$	$\hat{b}_{2i}(\theta)$
0	0			0			$\theta - \frac{5}{4}\theta^2 + \frac{1}{2}\theta^3$	$\theta - \frac{1}{2}\theta^2$
$\frac{2}{3}$	$\frac{13}{54}$	$\frac{23}{54}$		$\frac{1}{3}$	$\frac{1}{3}$		$\frac{9}{4}\theta^2 - \frac{3}{2}\theta^3$	0
1	$\frac{31}{144}$	$\frac{529}{1152}$	$\frac{125}{384}$	$\frac{1}{4}$	$\frac{3}{4}$	0	$-\theta^2 + \theta^3$	$\frac{1}{2}\theta^2$

Table 4. CRKOX4d3u(2){4F,3} coefficients

c_{1i}	a_{1ij}			$b_{1i}(\theta)$	$\hat{b}_{1i}(\theta)$
0	0			$\theta - \frac{3}{2}\theta^2 + \frac{2}{3}\theta^3$	$\theta - \frac{3}{2}\theta^2$
$\frac{1}{2}$	$\frac{1}{2}$			$\theta^2 - \frac{2}{3}\theta^3$	θ^2
$\frac{1}{2}$	$-\frac{1}{4}$	$\frac{3}{4}$		$\theta^2 - \frac{2}{3}\theta^3$	θ^2
1	$\frac{1}{10}$	$\frac{1}{2}$	$\frac{1}{5}$	$-\frac{1}{2}\theta^2 + \frac{2}{3}\theta^3$	$-\frac{1}{2}\theta^2$

c_{2i}	a_{2ij}			$b_{2i}(\theta)$	$\hat{b}_{2i}(\theta)$
0	0			$\theta - \frac{21}{10}\theta^2 + \frac{6}{5}\theta^3$	$\theta - \frac{3}{4}\theta^2$
$\frac{1}{3}$	$\frac{2}{9}$	$\frac{1}{9}$		$\frac{5}{2}\theta^2 - 2\theta^3$	$\frac{1}{4}\theta^2$
$\frac{5}{6}$	$\frac{5}{36}$	$\frac{5}{18}$	$\frac{5}{12}$	$-\frac{2}{5}\theta^2 + \frac{4}{5}\theta^3$	$\frac{1}{2}\theta^2$

Table 5. CRKOG4d3u{4,3} coefficients

c_{0i}	a_{00ij}			a_{02ij}			$b_{0i}(\theta)$
0	0			0			$\theta - \frac{3}{2}\theta^2 + \frac{2}{3}\theta^3$
$\frac{1}{2}$	$\frac{1}{2}$			$\frac{1}{2}$			$\theta^2 - \frac{2}{3}\theta^3$
$\frac{1}{2}$	0	$\frac{1}{2}$		0	$\frac{1}{2}$		$\theta^2 - \frac{2}{3}\theta^3$
1	0	0	1	0	0	1	$-\frac{1}{2}\theta^2 + \frac{2}{3}\theta^3$

c_{2i}	a_{20ij}			a_{22ij}			$b_{2i}(\theta)$
0	0			0			$\theta - \frac{3}{2}\theta^2 + \frac{2}{3}\theta^3$
$\frac{1}{2}$	$\frac{1}{4}$	$\frac{1}{4}$		$\frac{1}{4}$	$\frac{1}{4}$		$2\theta^2 - \frac{4}{3}\theta^3$
1	0	0	1	0	0	1	$-\frac{1}{2}\theta^2 + \frac{2}{3}\theta^3$

of the Table 5, which is a continuous extension of "The Runge–Kutta" method (the one with $c_2 = c_3 = 1/2$ and $c_4 = 1$). However, it is possible to construct a method for cross-dependent systems with reuse of the fourth stage for the first equation and just three stages for the second. The Table 4 presents the coefficients of CRKOX4d3u(2){4F,3} method, where X means cross-dependent and the meaning of the other notations was explained above.

We also add an embedded error estimator. It is known that there is no order three estimator for a fourth order four-stage classical Runge–Kutta method. One of the widely used estimators for "The Runge–Kutta" method is Egorov method of order two [1]. Its continuous variant is given with $\hat{b}_1(\theta)$ part of the Table 4.

For systems (6) and (9) we were able to construct a method that gives the advantage in stage number only for the second structural group. A possible method of this kind is "The Runge–Kutta" method's extension presented in the Table 5.

Uniform Order Four. Owren and Zennaro have shown that classical CRKs of order four require six stages [17]. Still with FSAL it is possible to reduce the number of new stages per step to five. We were able to construct a five stages method with FSAL for cross-dependent systems. The free parameters were chosen to keep all coefficients a positive. We also add an embedded uniform order three local error estimator.

Table 6. CRKOX4(3){5F} coefficients

c_{1i}	a_{1ij}				$b_{1i}(\theta)$	$\hat{b}_{1i}(\theta)$
0	0				$\theta - \frac{211}{70}\theta^2 + \frac{368}{105}\theta^3 - \frac{48}{35}\theta^4$	$\theta - \frac{13}{10}\theta^2 + \frac{8}{15}\theta^3$
1	1				0	0
$\frac{7}{24}$	$\frac{7}{96}$	$\frac{7}{32}$			$\frac{540}{119}\theta^2 - \frac{936}{119}\theta^3 + \frac{432}{119}\theta^4$	0
$\frac{5}{8}$	$\frac{205}{1568}$	$\frac{135}{1568}$	$\frac{20}{49}$		$-\frac{28}{15}\theta^2 + \frac{248}{45}\theta^3 - \frac{16}{5}\theta^4$	$\frac{32}{15}\theta^2 - \frac{64}{45}\theta^3$
1	$\frac{61}{420}$	0	$\frac{1024}{1743}$	$\frac{1331}{4980}$	$\frac{35}{102}\theta^2 - \frac{176}{153}\theta^3 + \frac{16}{17}\theta^4$	$-\frac{5}{6}\theta^2 + \frac{8}{9}\theta^3$

c_{2i}	a_{2ij}				$b_{2i}(\theta)$	$\hat{b}_{2i}(\theta)$
0	0				$\theta - \frac{307}{140}\theta^2 + \frac{59}{30}\theta^3 - \frac{22}{35}\theta^4$	$\theta - \frac{23}{14}\theta^2 + \frac{16}{21}\theta^3$
$\frac{7}{36}$	$\frac{455}{2592}$	$\frac{49}{2592}$			0	0
$\frac{7}{16}$	$\frac{7}{64}$	0	$\frac{21}{64}$		$\frac{20480}{5229}\theta^2 - \frac{4096}{747}\theta^3 + \frac{11264}{5229}\theta^4$	$\frac{128}{63}\theta^2 - \frac{256}{189}\theta^3$
$\frac{10}{11}$	$\frac{1910}{9317}$	0	$\frac{750}{9317}$	$\frac{830}{1331}$	$-\frac{9317}{1660}\theta^2 + \frac{30613}{2490}\theta^3 - \frac{2662}{415}\theta^4$	0
1	$\frac{5}{42}$	0	$\frac{36}{119}$	$\frac{4}{9}$ $\quad\frac{41}{306}$	$\frac{35}{9}\theta^2 - \frac{79}{9}\theta^3 + \frac{35}{9}\theta^4$	$-\frac{7}{18}\theta^2 + \frac{16}{27}\theta^3$

In the next sections we will omit the number of stages and the order of estimator in the abbreviations of the methods, since this makes them easier to read and causes no confusion.

4   Computational Results

To test the constructed methods we choose systems of delay-differential equations which have suitable structures and don't have some issues typical of DDEs that require additional tricks to deal with, like *overlapping*, i.e. the situation when the delay falls into the step not yet computed (explicit methods for such problems can be found in [6,10]), or *discontinuity points* [2]. That's why in the following test problems the history merges smoothly into the solution and delays are bounded from below, so that the step-size can always be chosen to avoid overlapping.

We have implemented the methods as MATLAB functions. Constant time step functions just compute the solution with the given step-size (possibly making the last step smaller to hit the final time). Adaptive step-size programs use local error estimations, which are the maximal difference between the continuous approximations of two orders over the step. The step-size control algorithm uses the popular formula found, for instance, in [7]. The parameters of step control are the same as MATLAB uses [21].

Problem 1 is a cross-dependent system (3) constructed on base of the problem 1.4.8 from [18]

$$
\begin{cases}
y_1'(t) = \dfrac{1}{2}y_2(t) + \dfrac{1}{2} - \left(y_2\left(\dfrac{t-\pi}{2}\right)\right)^2, & t \geq 0, \\[2mm]
y_2'(t) = -\dfrac{1}{2}y_1(t) - \dfrac{1}{2} + \left(y_1\left(\dfrac{t}{2} - \dfrac{\pi}{4}\right)\right)^2, & t \geq 0,
\end{cases}
\tag{12}
$$

with the history for $t \leq 0$ and the exact solution both being $y_1(t) = \sin t$ and $y_2(t) = \cos t$.

Problem 2 is very similar to (12) but is a one-group system of the form (9)

$$
\begin{cases}
y_1'(t) = \dfrac{1}{2}y_2(t) + \dfrac{1}{2}\left(y_1(t)\right)^2 + \dfrac{1}{2}\left(y_2(t)\right)^2 - \left(y_2\left(\dfrac{t-\pi}{2}\right)\right)^2, & t \geq 0, \\[2mm]
y_2'(t) = -\dfrac{1}{2}y_1(t) - \dfrac{1}{2} + \left(y_1\left(\dfrac{t}{2} - \dfrac{\pi}{4}\right)\right)^2, & t \geq 0
\end{cases}
\tag{13}
$$

and has the same history and solution as (12): $y_1(t) = \sin t$ and $y_2(t) = \cos t$.

Problem 3 is another cross-dependent system (3) based of the problem 1.4.12 from [18]:

$$
\begin{cases}
y_1'(t) = -y_2(t - \pi)e^{2\sin t}, & t \geq 0, \\[2mm]
y_2'(t) = 2y_1\left(t - \dfrac{\pi}{2}\right) + y_1(t)\left(\cos^2 t - \sin t\right) - 2e^{-\cos t}, & t \geq 0.
\end{cases}
\tag{14}
$$

It has a solution $y_1 = \exp(\sin t)$ and $y_2(t) = \cos t \cdot \exp(\sin t)$ which can be used as a history for $t \leq 0$.

Problem 4 is a one-group (9) modification of (14):

$$
\begin{cases}
y_1'(t) = 2y_2(t) + y_2(t - \pi)\left(y_1(t)\right)^2, & t \geq 0, \\[2mm]
y_2'(t) = 2y_1\left(t - \dfrac{\pi}{2}\right) + y_1(t)\left(\cos^2 t - \sin t\right) - 2e^{-\cos t}, & t \geq 0.
\end{cases}
\tag{15}
$$

It has the same solution $y_1 = \exp(\sin t)$ and $y_2(t) = \cos t \cdot \exp(\sin t)$ and history for $t \leq 0$.

4.1   Declared Orders Confirmation

To check the declared orders of the methods we test their convergence as the step-size is reduced. We run methods with constant step-size and compare the exact and approximate continuous solutions by measuring their difference in multiple points within each step.

Definition 3. *We say that a continuous method has convergence order r if for any problem (1) with sufficiently small h there exists $C > 0$ such, that the continuous approximation through N steps $\bar{y}(t)$, $t \in [t_0, t_N]$ satisfies*

$$E = \max_{t \in [t_0, t_N]} \|y(t) - \bar{y}(t)\| = Ch^r.$$

It is known [2] that the convergence order r of a method for DDEs in case of smooth problems depends on its local discrete order p and uniform order q as

$$r = \max(p, q + 1).$$

This means that all methods from Sect. 3 should show the convergence order equal to their discrete local order.

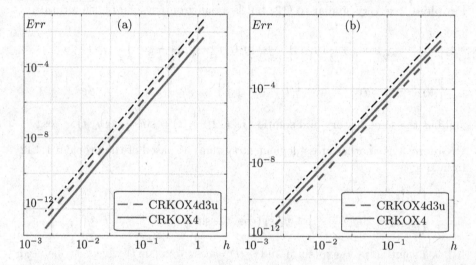

Fig. 1. Convergence orders for (a) Problem 1 and (b) Problem 3. The reference line has slope 4.

The results presented in Figs. 1 and 2 show the expected orders for five tested methods (from Tables 2, 3, 4, 5 and 6). In plots global error to the step-size ratio in double logarithmic scale is presented. We draw reference lines of the slopes four (and three were necessary) to show that the global error diminishes proportionally to fourth (or third) power of h. CRKOG4d3u for the Problem 4 (Fig. 2b) doesn't behave almost linearly as other methods do, but in average it still shows the fourth order of convergence.

4.2 Computational Cost and Global Order

We choose three methods to run variable step-size tests—those that have error estimators in the tables—CRKOG3, CRKOX4d3u and CRKOX4. We compare

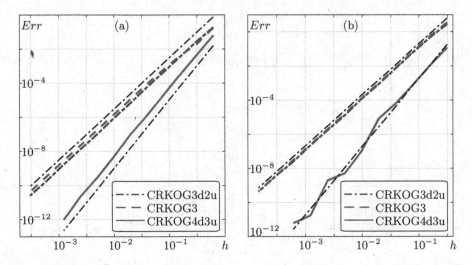

Fig. 2. Convergence orders for (a) Problem 2 and (b) Problem 4. The reference lines have slopes 3 and 4.

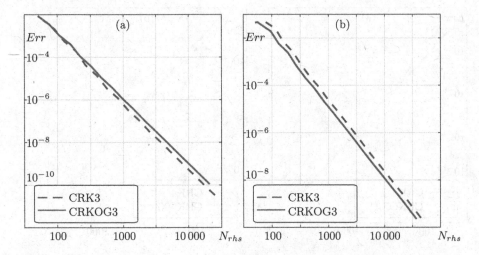

Fig. 3. Global error to right-hand side computations comparison of CRK3 and CRKOG3 with adaptive time-step for (a) Problem 2 and (b) Problem 4.

first two of them to the mentioned continuous methods they are based on (CRK3 by Owren and Zennaro and continuous variant of "The Runge–Kutta method" respectively). The last method is compared to CRK4{6F} constructed by Owren and Zennaro. We add some error estimators to CRK3 and CRK4 from [17] to make it possible to use an adaptive time-step integration.

To estimate the local error we compare two continuous solutions through the step given by the main method and the estimator in multiple points and take

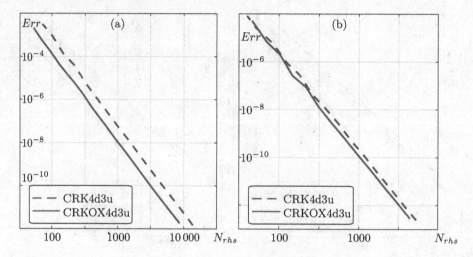

Fig. 4. Global error to right-hand side computations comparison of CRK4d3u and CRKOX4d3u with adaptive time-step for (a) Problem 1 and (b) Problem 3.

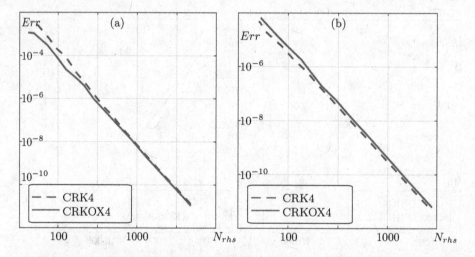

Fig. 5. Global error to right-hand side computations comparison of CRK4 and CRKOX4 with adaptive time-step for (a) Problem 1 and (b) Problem 3.

the maximal difference. We change the step-size in accordance with the formula found, for instance, in [7].

It should be noticed that in CRK3 and CRK4 free parameters were chosen to minimize the local error (i.e. the coefficients of the main terms of the local error). We didn't make the same with presented methods, since at first the possibility to construct methods with fewer stages was under consideration. It means that our methods, even with fewer right-hand side computations per step can still require more total RHS computations to get the same global error.

In Figs. 3, 4 and 5 the global error to RHS computation number ratios are plotted in double logarithmic scale.

Figure 3 shows results for CRK3 and CRKOG3. For the latter, the coefficients for general group are the same as in CRK3, and the structural group requires fewer stages. Still for Problem 2 CRK3 gives better global error to RHS computations ratio. However, for Problem 4 structural methods performs better.

CRKOX4d3u with three stages shows better results for both tests Problems 1 and 3 than its opponent that requires four RHS computations per step (Fig. 4).

The results of CRK4 and CRKOX4 (Fig. 5) are quite similar. This lets us hope, that the methods performance can be improved if its coefficients are chosen to minimize the local error.

5   Conclusion

We have constructed a number of explicit continuous Runge–Kutta type methods for systems with special structure, which require fewer stages than *classical* explicit Runge–Kutta methods. Systems with such structures are not too rare, so they seem worth studying. Comparison to known continuous Runge–Kutta methods confirms that new methods require less computations for the same accuracy at least for some problems. As it was mentioned above we didn't try to optimize coefficients of the methods from the local error minimization point of view. This should result in much better characteristics in respect of global error/RHS computations criterion. Local error estimators should also be optimized in some sense.

It might also be possible to construct continuous methods for *full* structurally partitioned systems [13] with fewer stages for structural groups than known CRKs have, though the relative advantage of such methods is expected to be less noticeable than for cross-dependent systems.

References

1. Arushanyan, O.B., Zaletkin, S.F.: Chislennoe Reshenie Obyknovennykh Differentsialnykh Uravnenii na Fortrane (Numerical Solution of Ordinary Differential Equations Using FORTRAN). Moscow State Univ., Moscow (1990). (in Russian)
2. Bellen, A., Zennaro, M.: Numerical Methods for Delay Differential Equations, 1st edn. Oxford Science Publications, Clarendon Press, Oxford (2003)
3. Butcher, J.C.: On Runge-Kutta processes of high order. J. Austral. Math. Soc. 4(2), 179–194 (1964)
4. Dormand, J.R., Prince, P.J.: A family of embedded Runge-Kutta formulae. J. Comp. Appl. Math. 6, 19–26 (1980)
5. Eremin, A.S.: Modifikatsiya teorii pomechennykh dereviev dly stukturnogo metoda integrirovaniya sistem ODU (Labelled trees theory modification for structural method of solving ODE systems). Vestn. St-Petersburg Uni. (2), 15–21 (2009). (in Russian)
6. Eremin, A.S., Olemskoy, I.V.: Functional continuous Runge-Kutta methods for special systems. In: AIP Conference Proceedings, vol. 1738, p. 100003 (2016)

7. Hairer, E., Nørsett, S.P., Wanner, G.: Solving Ordinary Differential Equations I: Nonstiff Problems, 2nd edn. Springer, Heidelberg (1993)
8. Hofer, E.: A partially implicit method for large stiff systems of ODEs with only few equations introducing small time-constants. SIAM J. Numer. Anal. **13**(5), 645–663 (1976)
9. Ketcheson, D.I., MacDonald, C., Ruuth, S.J.: Spatially partitioned embedded Runge-Kutta methods. SIAM J. Numer. Anal. **51**(5), 2887–2910 (2013)
10. Maset, S., Torelli, L., Vermiglio, R.: Runge-Kutta methods for retarded functional differential equations. Math. Model. Meth. Appl. Sci. **15**(8), 1203–1251 (2005)
11. McLachlan, R., Ryland, B., Sun, Y.: High order multisymplectic Runge-Kutta methods. SIAM J. Sci. Comput. **36**(5), A2199–A2226 (2014)
12. Olemskoy, I.V.: Fifth-order four-stage method for numerical integration of special systems. Comput. Math. Math. Phys. **42**(8), 1135–1145 (2002)
13. Olemskoy, I.V.: Structural approach to the design of explicit one-stage methods. Comput. Math. Math. Phys. **43**(7), 918–931 (2003)
14. Olemskoy, I.V.: A fifth-order five-stage embedded method of the Dormand-Prince type. Comput. Math. Math. Phys. **45**(7), 1140–1150 (2005)
15. Olemskoy, I.V.: Modifikatsiya algoritma vydeleniya strukturnykh osobennostei (Modification of structural properties detection algorithm). Vestn. St-Petersburg Uni. (2), 55–64 (2006). (in Russian)
16. Olemskoy, I.V.: Metody Integrirovaniya System Strukturno Razdelyonnykh Differentsialnykh Uravnenii (Integration of Structurally Partitioned Systems of Ordinary Differential Equations). Saint-Petersburg State Univ., Saint-Petersburg (2009). (in Russian)
17. Owren, B., Zennaro, M.: Derivation of efficient continuous explicit Runge-Kutta methods. SIAM J. Sci. Stat. Comput. **13**(6), 1488–1501 (1992)
18. Paul, C.A.H.: A test set of functional differential equations. Technical report 243, Manchester Centre for Computational Mathematics, University of Manchester, February 1994
19. Sandu, A., Günther, M.: A generalized-structure approach to additive Runge-Kutta methods. SIAM J. Numer. Anal. **53**(1), 17–42 (2015)
20. Sandu, A., Günther, M.: Multirate generalized additive Runge-Kutta methods. Numer. Math. **133**(3), 497–524 (2016)
21. Shampine, L.F., Reichelt, M.W.: The matlab ODE suite. SIAM J. Sci. Comput. **18**(1), 1–22 (1997)
22. Wang, D., Xiao, A., Li, X.: Parametric symplectic partitioned Runge-Kutta methods with energy-preserving properties for Hamiltonian systems. Comput. Phys. Comm. **184**(2), 303–310 (2013)

Polynomials over Quaternions and Coquaternions: A Unified Approach

Maria Irene Falcão[1], Fernando Miranda[1], Ricardo Severino[2],
and Maria Joana Soares[3(✉)]

[1] CMAT and Departamento de Matemática e Aplicações,
Universidade do Minho, Braga, Portugal
{mif,fmiranda}@math.uminho.pt
[2] Departamento de Matemática e Aplicações,
Universidade do Minho, Braga, Portugal
ricardo@math.uminho.pt
[3] NIPE and Departamento de Matemática e Aplicações,
Universidade do Minho, Braga, Portugal
jsoares@math.uminho.pt

Abstract. This paper aims to present, in a unified manner, results which are valid on both the algebras of quaternions and coquaternions and, simultaneously, call the attention to the main differences between these two algebras. The rings of one-sided polynomials over each of these algebras are studied and some important differences in what concerns the structure of the set of their zeros are remarked. Examples illustrating this different behavior of the zero-sets of quaternionic and coquaternionic polynomials are also presented.

Keywords: Quaternions · Coquaternions · Polynomials · Zeros

1 Introduction

Quaternions, introduced in 1843 by the Irish mathematician William Rowan Hamilton (1805–1865) as a generalization of complex numbers [11], have become a powerful tool for modeling and solving problems in classical fields of mathematics, engineering and physics [19]. One of the most widespread application of quaternions is in computer animation, where they are used to represent transformations of orientations of graphical objects [29]. The geometric and physical applications of quaternions require solving quaternion polynomial equations; see, e.g., [1]. The literature on quaternion polynomial root-finding reveals a recent growing interest on this subject; see e.g. [7,8,12,13,16,20,25,27,28].

At about the same time Hamilton discovered the non-commutative algebra of quaternions, James Cockle (1819–1895) introduced another four-dimensional hypercomplex real algebra, the algebra of real coquaternions [6]. Probably due to the fact that this is a non-division algebra, coquaternions have received much less attention than their "cousins" quaternions. There are, however, some studies

© Springer International Publishing AG 2017
O. Gervasi et al. (Eds.): ICCSA 2017, Part II, LNCS 10405, pp. 379–393, 2017.
DOI: 10.1007/978-3-319-62395-5_26

related to geometric applications of coquaternions[1] [17,23,24]. In addition, the relation between coquaternions and complexified mechanics is discussed in [5]. In what concerns coquaternionic polynomials, the bibliography is very scarce. The most relevant references on this subject are [14,15,22,26].

The main purpose of this paper is to present, in a unified manner, results which are valid in both the algebras of quaternions and coquaternions and, at the same time, point out some important differences between these algebras. We also study one-sided polynomials defined over these two algebras, emphasizing the differences that may occur in the structure of their zero-sets.

The rest of the paper is organized as follows. Section 2 contains a revision of the main definitions and results on the algebras of quaternions and coquaternions. Section 3 is dedicated to the rings of (one-sided) polynomials over these two algebras and discusses, in particular, the main differences of the structure of the sets of zeros of these polynomial rings. Finally, Sect. 4 contains carefully chosen examples illustrating some of the conclusions contained in Sect. 3.

2   The Algebras of Quaternions and Coquaternions

Let $\{1, \mathbf{i}, \mathbf{j}, \mathbf{k}\}$ be an orthonormal basis of the Euclidean vector space $\mathbb{R}^4$ with a product given according to the multiplication rules

$$\mathbf{i}^2 = \mathbf{j}^2 = \mathbf{k}^2 = -1, \quad \mathbf{ij} = -\mathbf{ji} = \mathbf{k}.$$

This non-commutative product generates the well known algebra of real quaternions, which we will denote by $\mathbb{H}$.

The algebra of real coquaternions, which we will denote by $\mathbb{H}_{\text{coq}}$, is generated by the product given according to the following rules

$$\mathbf{i}^2 = -1, \ \mathbf{j}^2 = \mathbf{k}^2 = 1, \ \mathbf{ij} = -\mathbf{ji} = \mathbf{k}.$$

In what follows, since some of the results are valid in both algebras, we will use $\mathscr{H}$ to refer to one of the algebras $\mathbb{H}$ or $\mathbb{H}_{\text{coq}}$.

We will embed the space $\mathbb{R}^4$ in $\mathscr{H}$ by identifying the element $q = (q_0, q_1, q_2, q_3)$ in $\mathbb{R}^4$ with the element $q = q_0 + q_1\mathbf{i} + q_2\mathbf{j} + q_3\mathbf{k}$ in $\mathscr{H}$. Thus, throughout the paper, we will not distinguish an element in $\mathbb{R}^4$ and the corresponding quaternion/coquaternion, unless we need to stress the context.

If $p = p_0 + p_1\mathbf{i} + p_2\mathbf{j} + p_3\mathbf{k}$ and $q = q_0 + q_1\mathbf{i} + q_2\mathbf{j} + q_3\mathbf{k}$ are two given quaternions, then

$$pq = p_0q_0 - p_1q_1 - p_2q_2 - p_3q_3 + (p_0q_1 + p_1q_0 + p_2q_3 - p_3q_2)\mathbf{i}$$
$$+ (p_0q_2 - p_1q_3 + p_2q_0 + p_3q_1)\mathbf{j} + (p_0q_3 + p_1q_2 - p_2q_1 + p_3q_0)\mathbf{k},$$

---

[1] Also known, in the literature, as *split-quaternions, para-quaternions, anti-quaternions* or *hyperbolic quaternions*.

whilst, if p and q are coquaternions, we have

$$pq = p_0q_0 - p_1q_1 + p_2q_2 + p_3q_3 + (p_0q_1 + p_1q_0 - p_2q_3 + p_3q_2)\mathbf{i}$$
$$+ (p_0q_2 - p_1q_3 + p_2q_0 + p_3q_1)\mathbf{j} + (p_0q_3 + p_1q_2 - p_2q_1 + p_3q_0)\mathbf{k}.$$

Given $q = q_0 + q_1\mathbf{i} + q_2\mathbf{j} + q_3\mathbf{k} \in \mathscr{H}$, its *conjugate* $\bar{q}$ is defined as $\bar{q} = q_0 - q_1\mathbf{i} - q_2\mathbf{j} - q_3\mathbf{k}$; the number q_0 is called the *real part* of q and denoted by $\operatorname{Re} q$ and the *vector part* of q, denoted by $\operatorname{Vec} q$, is given by $\operatorname{Vec} q = q_1\mathbf{i} + q_2\mathbf{j} + q_3\mathbf{k}$.

We will identify the set of elements in $\mathscr{H}$ whose vector part is zero with the set $\mathbb{R}$ of real numbers. We will also consider three particularly important subspaces of dimension two of $\mathscr{H}$, usually called the *canonical planes* or *cycle planes*. The first is $\{q \in \mathscr{H} : q = a + b\mathbf{i}, a, b \in \mathbb{R}\}$ which, naturally, we identify with the complex plane $\mathbb{C}$; the second, which we denote by $\mathbb{P}$ and whose elements are usually called *perplex numbers* is given by $\mathbb{P} = \{q \in \mathscr{H} : q = a + b\mathbf{j}, a, b \in \mathbb{R}\}$ and the third, denoted by $\mathbb{D}$, is the subspace of the so-called *dual numbers*, $\mathbb{D} = \{q \in \mathscr{H} : q = a + b(\mathbf{i} + \mathbf{j}), a, b \in \mathbb{R}\}$; see e.g. [3].

We will denote by $\operatorname{t}(q)$ and call *trace* of q the quantity given by

$$\operatorname{t}(q) = q + \bar{q} = 2\operatorname{Re} q \tag{1}$$

and will call *determinant* of q and denote by $\operatorname{d}(q)$ the quantity given by $\operatorname{d}(q) = q\bar{q}$. We have

$$\operatorname{d}(q) = \begin{cases} q_0^2 + q_1^2 + q_2^2 + q_3^2, & \text{if } q \in \mathbb{H}, \\ q_0^2 + q_1^2 - q_2^2 - q_3^2, & \text{if } q \in \mathbb{H}_{\text{coq}} \end{cases}. \tag{2}$$

Remark 1. Observe that, for a quaternion q, $\operatorname{d}(q)$ is always a non-negative quantity and moreover $\operatorname{d}(q) = 0$ iff $q = 0$. In this case, the square root of $\operatorname{d}(q)$ is called the *norm* of q and denoted by $|q|$.[2]

The following proposition lists some properties of quaternions/coquaternions which are easily verified; see e.g. [2,10].

Proposition 1. *Given $p, q \in \mathscr{H}$ and $\alpha \in \mathbb{R}$, we have:*

1. $q \in \mathbb{R} \iff \bar{q} = q$
2. $\bar{\bar{q}} = q$
3. $\overline{p + q} = \bar{p} + \bar{q}$
4. $\overline{pq} = \bar{q}\,\bar{p}$
5. $\operatorname{Re} q = \operatorname{Re} \bar{q}$
6. $\operatorname{Re}(pq) = \operatorname{Re}(qp)$
7. $\operatorname{d}(q) = (\operatorname{Re} q)^2 + \operatorname{d}(\operatorname{Vec} q)$
8. $\operatorname{d}(q) = \operatorname{d}(\bar{q})$
9. $\operatorname{d}(\alpha q) = \alpha^2 \operatorname{d}(q)$
10. $\operatorname{d}(pq) = \operatorname{d}(qp) = \operatorname{d}(p)\operatorname{d}(q)$
11. *q commutes with any other element in $\mathscr{H}$ if and only if $q \in \mathbb{R}$ (i.e. the center of $\mathscr{H}$ is $\mathbb{R}$):*

---

[2] This coincides with the Euclidean norm of the 4-vector (q_0, q_1, q_2, q_3).

We also have the following result, very simple to prove.

Proposition 2. *An element $q \in \mathcal{H}$ is invertible if and only if* $\mathrm{d}(q) \neq 0$. *In that case, we have* $q^{-1} = \dfrac{\bar{q}}{\mathrm{d}(q)}$.

A non-invertible element $q \in \mathcal{H}$ is also called a *singular* element. Since, as observed in Remark 1, the only quaternion with a zero determinant is $q = 0$, we immediately conclude that all non-zero quaternions are invertible, i.e. $\mathbb{H}$ is a *division algebra*. In the case of coquaternions, there are, however, non-zero singular elements, such as, for example, the coquaternion $q = 1 + \mathbf{i} + \mathbf{j} + \mathbf{k}$.

It can be shown that $q \neq 0$ is singular if and only if q is a *zero-divisor*, i.e. there exist $p, r \in \mathbb{H}_{\mathrm{coq}}$ such that $q\, p = r\, q = 0$.

Definition 1. We say that an element $q \in \mathcal{H}$ is *similar* to another element $p \in \mathcal{H}$, and write $q \sim p$, if there exists an invertible $h \in \mathcal{H}$ such that $q = h^{-1}ph$.

This is an equivalence relation in $\mathcal{H}$ (hence, we can simply say that two elements $p, q \in \mathcal{H}$ are similar), partitioning $\mathcal{H}$ in the so-called *similarity classes*. We denote by $[q]$ the similarity class containing a given element $q \in \mathcal{H}$. The next proposition shows that the equivalence classes of real elements reduce to a single element.

Proposition 3. *Let* $q \in \mathcal{H}$. *Then*

$$[q] = \{q\} \iff q \in \mathbb{R}. \tag{3}$$

Proof. If $q \in \mathbb{R}$ and $p \sim q$, then $p = h^{-1}qh = qh^{-1}h = q$. On the other hand, if $[q] = \{q\}$, we have that $h^{-1}qh = q$ or, equivalently, $qh = hq$, for all invertible h. This means, in particular, that q must commute with $\mathbf{i}, \mathbf{j}$ and $\mathbf{k}$, which can only be true if $q \in \mathbb{R}$. $\qquad\square$

In what follows, given an element $q = q_0 + q_1\mathbf{i} + q_2\mathbf{j} + q_3\mathbf{k} \in \mathcal{H}$, we will use $\mathrm{dv}(q)$ to denote the determinant of the vector part of q, i.e. $\mathrm{dv}(q) = \mathrm{d}(\mathrm{Vec}\, q)$. Note that:

- for $q \in \mathbb{H}$, $\mathrm{dv}(q) = q_1^2 + q_2^2 + q_3^2$; hence $\mathrm{dv}(q)$ is always a non-negative quantity and $\mathrm{dv}(q) = 0$ if and only if $q \in \mathbb{R}$;
- for $q \in \mathbb{H}_{\mathrm{coq}}$, $\mathrm{dv}(q) = q_1^2 - q_2^2 - q_3^2$; hence $\mathrm{dv}(q)$ can be negative, null or positive; moreover, $\mathrm{dv}(q) = 0$ does not imply $q \in \mathbb{R}$.

Definition 2. We say that two elements $p, q \in \mathcal{H}$ are *quasi-similar*, and write $p \approx q$, if and only if they satisfy the following conditions

$$\mathrm{Re}\, p = \mathrm{Re}\, q \quad \text{and} \quad \mathrm{dv}(p) = \mathrm{dv}(q). \tag{4}$$

This is an equivalence relation in $\mathcal{H}$, partitioning $\mathcal{H}$ in the so-called *quasi-similarity classes*. We denote by $[\![q]\!]$ the quasi-similarity class containing a given element $q \in \mathcal{H}$.

Remark 2. The above definition was introduced, for the case of coquaternions, by Janovská and Opfer [15], but a different notation was used.

Note that, since $d(q) = (\operatorname{Re} q)^2 + dv(q)$, the conditions (4) defining quasi-similarity are equivalent to the conditions

$$\operatorname{Re} p = \operatorname{Re} q \quad \text{and} \quad d(p) = d(q). \tag{5}$$

We will now show that the concepts of quasi-similarity and similarity coincide for any two quaternions and also for any two non-real coquaternions. Only for the case where one of the coquaternions is real and the other is non-real, the concepts of quasi-similarity and similarity do not coincide.

Theorem 1 *[4, Lemma 3]. Let $q = q_0 + q_1 i + q_2 j + q_3 k \in \mathbb{H}$. Then, q is similar to the complex number $q_0 + \sqrt{dv(q)}\,i$, i.e. $[q] = [q_0 + \sqrt{dv(q)}\,i]$.*

Corollary 1. *Two quaternions $p, q \in \mathbb{H}$ are similar if and only if they are quasi-similar.*

Proof. If p and q are similar, then it is an immediate consequence of 6. and 10. in Proposition 1 that they satisfy (5). The fact that two elements $p, q \in \mathbb{H}$ satisfying (4) are similar is an immediate consequence of the previous theorem, having in mind that similarity is an equivalence relation. □

From the previous corollary, we have, for any $q = q_0 + q_1 i + q_2 j + q_3 k \in \mathbb{H}$

$$[q] = [\![q]\!] = \{p \in \mathbb{H} : \operatorname{Re} p = \operatorname{Re} q \text{ and } dv(p) = dv(q)\}$$
$$= \{p_0 + p_1 i + p_2 j + p_3 k : p_0 = q_0 \text{ and } p_1^2 + p_2^2 + p_3^2 = dv(q)\}.$$

This shows that the similarity class of a quaternion $q = q_0 + q_1 i + q_2 j + q_3 k$ can be identified with a sphere in the hyperplane $\{(x_0, x_1, x_2, x_3) \in \mathbb{R}^4 : x_0 = q_0\}$. This sphere is centered at the point $(q_0, 0, 0, 0)$ and has radius $\sqrt{dv(q)}$, reducing to a single point when $dv(q) = 0$ i.e., when $q \in \mathbb{R}$.

Theorem 2. *Let $q = q_0 + q_1 i + q_2 j + q_3 k \in \mathbb{H}_{\mathrm{coq}}$ be a non-real coquaternion. Then:*

(i) if $dv(q) > 0$, q is similar to the complex number $q_0 + \sqrt{dv(q)}\,i$, i.e. $[q] = [q_0 + \sqrt{dv(q)}\,i]$;

(ii) if $dv(q) < 0$, q is similar to the perplex number $q_0 + \sqrt{-dv(q)}\,j$, i.e. $[q] = [q_0 + \sqrt{-dv(q)}\,j]$;

(iii) if $dv(q) = 0$, q is similar to the dual number $q_0 + i + j$, i.e. $[q] = [q_0 + i + j]$.

Proof. In each case, we will indicate how to choose an invertible $h \in \mathbb{H}_{\mathrm{coq}}$ such that $h^{-1}qh$ has the specified form.

(i) Naturally, if $q = q_0 + q_1\mathbf{i}$ with $q_1 > 0$, then $q_1 = \sqrt{\mathrm{dv}(q)}$ and there is nothing to prove. Otherwise, we can take

$$h = \begin{cases} \mathbf{j}, & \text{if } q_2^2 + q_3^2 = 0, \\ (q_1 + \sqrt{\mathrm{dv}(q)}) - q_3\mathbf{j} + q_2\mathbf{k}, & \text{if } q_2^2 + q_3^2 \neq 0. \end{cases}$$

Note that, for $h = (q_1 + \sqrt{\mathrm{dv}(q)}) - q_3\mathbf{j} + q_2\mathbf{k}$, we have

$$d(h) = (q_1 + \sqrt{\mathrm{dv}(q)})^2 - q_3^2 - q_2^2 = 2\sqrt{\mathrm{dv}(q)}\left(\sqrt{\mathrm{dv}(q)} + q_1\right).$$

Since

$$q_2^2 + q_3^2 \neq 0 \Rightarrow \mathrm{dv}(q) \neq q_1^2 \Rightarrow \sqrt{\mathrm{dv}(q)} + q_1 \neq 0,$$

the condition $q_2^2 + q_3^2 \neq 0$ guarantees that $d(h) \neq 0$, i.e. that h is an invertible element. Moreover, it is easy to verify that $qh = h(q_0 + \sqrt{\mathrm{dv}(q)}\,\mathbf{i})$.

(ii) Similarly to the previous case, if $q = q_0 + q_2\mathbf{j}$ with $q_2 > 0$, then $q_2 = \sqrt{-\mathrm{dv}(q)}$, so there is nothing to prove. Otherwise, we can consider

$$h = \begin{cases} \mathbf{i}, & \text{if } q_1^2 + q_3^2 = 0, \\ q_1 - q_3\mathbf{j} + (q_2 - \sqrt{-\mathrm{dv}(q)})\mathbf{k}, & \text{if } q_1^2 + q_3^2 \neq 0 \text{ and } q_2 \leq 0, \\ (q_2 + \sqrt{-\mathrm{dv}(q)}) + q_3\mathbf{i} + q_1\mathbf{k}, & \text{if } q_1^2 + q_3^2 \neq 0 \text{ and } q_2 > 0. \end{cases}$$

In a manner similar to the previous case, one can verify that the coquaternions h given by the above formulas are invertible and are such that $qh = h(q_0 + \sqrt{-\mathrm{dv}(q)}\,\mathbf{j})$.

(iii) We can take

$$h = \begin{cases} (1 + q_1) - q_3\mathbf{j} - (1 - q_2)\mathbf{k}, & \text{if } q_1 + q_2 \neq 0, \\ (1 + q_1)\mathbf{i} + (1 - q_1)\mathbf{j}, & \text{if } q_1 + q_2 = 0. \end{cases}$$

Once more, it is easy to verify that the coquaternions h satisfy $qh = h(1 + \mathbf{i} + \mathbf{j})$ and, under the conditions stated, are invertible.   □

Remark 3. The explicit expressions for h in the cases (i) and (ii), here included for completeness, can be seen in [17].

Corollary 2. *Two non-real coquaternions $p, q \in \mathbb{H}_{\mathrm{coq}}$ are similar if and only if they are quasi-similar.*

Proof. The proof is totally analogous to the proof of Corollary 1, now making use of the previous theorem.   □

For a given element $q = q_0 + q_1\mathbf{i} + q_2\mathbf{j} + q_3\mathbf{k} \in \mathbb{H}_{\mathrm{coq}}$, we have

$$\llbracket q \rrbracket = \{p \in \mathbb{H}_{\mathrm{coq}} : \mathrm{Re}\,(p) = \mathrm{Re}\,(q) \text{ and } \mathrm{dv}(p) = \mathrm{dv}(q)\}$$
$$= \{p_0 + p_1\mathbf{i} + p_2\mathbf{j} + p_3\mathbf{k} : p_0 = q_0 \text{ and } p_1^2 - p_2^2 - p_3^2 = \mathrm{dv}(q)\}.$$

Hence, the quasi-similarity class of q can be identified with an hyperboloid in the hyperplane $\{(x_0, x_1, x_2, x_3) \in \mathbb{R}^4 : x_0 = q_0\}$. This will be:

– an hyperboloid of two sheets, if $\mathrm{dv}(q) > 0$; in this case

$$[\![q]\!] = [q] = [q_0 + \sqrt{\mathrm{dv}(q)}\,\mathbf{i}]$$

– an hyperboloid of one sheet, if $\mathrm{dv}(q) < 0$; in this case

$$[\![q]\!] = [q] = [q_0 + \sqrt{-\mathrm{dv}(q)}\,\mathbf{j}]$$

– a degenerate hyperboloid (i.e. a cone), if $\mathrm{dv}(q) = 0$; in this case,

$$[\![q]\!] = [\![q_0]\!] \quad \text{and} \quad [q] = \begin{cases} \{q_0\}, & \text{if } q \in \mathbb{R} \\ [q_0 + \mathbf{i} + \mathbf{j}] = [\![q_0]\!] \setminus \{q_0\}, & \text{if } q \notin \mathbb{R} \end{cases};$$

see Fig. 1.

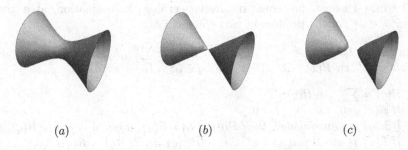

(a) $\qquad\qquad\qquad$ (b) $\qquad\qquad\qquad$ (c)

Fig. 1. Plots, in the hyperplane $\{(x_0, x_1, x_2, x_3) \in \mathbb{R}^4 : x_0 = q_0\}$, of the quasi-similarity class of a coquaternion q. (a) $\mathrm{dv}(q) < 0$; (b) $\mathrm{dv}(q) = 0$; (c) $\mathrm{dv}(q) > 0$.

3 One-Sided Polynomials

We denote by $\mathscr{H}[x]$ the set of polynomials of the form

$$P(x) = c_n x^n + c_{n-1} x^{n-1} + \cdots + c_1 x + c_0, \ c_i \in \mathscr{H}, \tag{6}$$

i.e., the set of polynomials whose coefficients are only located on the left of the variable, with the addition and multiplication of such polynomials defined as in the commutative case, where the variable is assumed to commute with the coefficients. This is a ring, referred to as the ring of (left) *one-sided, unilateral* or *simple* polynomials in $\mathscr{H}$.[3]

As usual, if $c_n \neq 0$, we will say that the *degree* of the polynomial $P(x)$ is n and refer to c_n as the leading coefficient of the polynomial. When $c_n = 1$, we say

---

[3] Right one-sided polynomials are defined in an analogous manner, by considering the coefficients on the right of the variable; all the results for left one-sided polynomials have corresponding results for right one-sided polynomials and hence we restrict our study to polynomials of the first type.

that $P(x)$ is *monic*. If the coefficients c_i in (6) are real, then we say that $P(x)$ is a *real polynomial*.

The *evaluation map* at a given element $q \in \mathscr{H}$, defined, for the polynomial $P(x)$ given by (6), by $P(q) = c_n q^n + c_{n-1} q^{n-1} + \cdots + c_1 q + c_0$, is not a homomorphism from the ring $\mathscr{H}[x]$ into $\mathscr{H}$. In fact, given two polynomials $L(x), R(x) \in \mathbb{H}[x]$, in general we do not have $(LR)(q) = L(q)R(q)$.

Remark 4. Since all the polynomials considered will be in the indeterminate x, we will usually omit the reference to this variable and write simply P when referring to an element $P(x) \in \mathscr{H}[x]$, the expression $P(q)$ being preferably reserved for the evaluation of P at a specific value $q \in \mathscr{H}$.

We say that an element $q \in \mathscr{H}$ is a *zero* (or a *root*) of a polynomial P, if $P(q) = 0$, and we use the notation $Z(P)$ to denote the *zero-set* of P, i.e. the set of all the zeros of P.

The next theorem has some results concerning the evaluation, at a given element $q \in \mathscr{H}$, of the product of two polynomials.

Theorem 3. *Let $L(x) = \sum_{i=0}^{n} a_i x^i$ and $R(x) = \sum_{j=0}^{m} b_j x^j$ be two given polynomials in $\mathscr{H}[x]$, $P(x) = L(x)R(x)$ and $q \in \mathscr{H}$. Then, we have:*

(i) $P(q) = \sum_{i=0}^{n} a_i R(q) q^i$.
(ii) *If $R(q) = 0$, then $P(q) = 0$.*
(iii) *If $R(q)$ is non-singular, then $P(q) = L(\tilde{q})R(q)$, where $\tilde{q} = R(q)q(R(q))^{-1}$.*
(iv) *If $L(x)$ is a real polynomial, then $(LR)(q) = (RL)(q) = R(q)L(q)$.*
(v) *If $q \in \mathbb{R}$, then $(LR)(q) = L(q)R(q)$.*

Proof. The proofs of (i)–(iii) are very simple adaptations of the corresponding proofs for similar results for polynomials over a division ring given in [18, Proposition (16.2)], noting that the condition "$R(q)$ is non-singular" assumed on (iii) plays the same role of the condition $R(q) \neq 0$ in the referred proposition in [18].

The results (iv)–(v) follow easily from the definition of the product of polynomials and the fact that real numbers commute with any element in $\mathscr{H}$.   □

As an immediate consequence of this theorem, we have the following result.

Corollary 3. *Let $L(x)$ and $R(x)$ be two given polynomials in $\mathscr{H}[x]$, $P(x) = (LR)(x)$ and $q \in \mathscr{H}$. If $R(q)$ is non-singular, then q is a zero of P if and only if $R(q)q(R(q))^{-1}$ is a zero of L.*

Remark 5. Since, in the case of quaternions, all non-zero elements are non-singular, the condition "$R(q)$ is non-singular" in Theorem 3-(iii) and in Corollary 3 can, in this case, be replaced by $R(q) \neq 0$.

Definition 3. Given an element $q \in \mathscr{H}$, the *characteristic polynomial* of q, denoted $\Psi_q(x)$, is the polynomial defined by

$$\Psi_q(x) = (x - q)(x - \bar{q}) = (x - \bar{q})(x - q) = x^2 - \mathrm{t}(q)x + \mathrm{d}(q). \qquad (7)$$

Note that Ψ_q is a monic quadratic real polynomial and also that we have

$$\Psi_q = \Psi_{q'} \iff [\![q]\!] = [\![q']\!] \iff q' \in [\![q]\!]. \tag{8}$$

Equation (8) shows that Ψ_q is an invariant of the quasi-similarity class of q.[4]

Remark 6. Having in mind the previous observation, it would probably be more appropriate to say that the polynomial given by (7) is the characteristic polynomial of the quasi-similarity class $[\![q]\!]$ and to denote it by $\Psi_{[\![q]\!]}$. However, for a question of simplicity, we will stick to the denomination and notation introduced in Definition 3.

Let Ψ_q be the characteristic polynomial of a given element $q \in \mathcal{H}$. The discriminant of Ψ_q is

$$\Delta = (\mathrm{t}(q))^2 - 4\,\mathrm{d}(q) = 4(\mathrm{Re}\ q)^2 - 4((\mathrm{Re}\ q)^2 + \mathrm{dv}(q)) = -4\,\mathrm{dv}(q).$$

Thus:

1. In the quaternionic case, Δ will never be positive, being zero if and only if $q \in \mathbb{R}$. Hence, the characteristic polynomial of any non-real quaternion is an irreducible (over the reals) monic quadratic polynomial. The characteristic polynomial of $q \in \mathbb{R}$ is, naturally, the polynomial $(x - q)^2$.
2. In the coquaternionic case, Δ will be negative, null or positive, depending on whether $\mathrm{dv}(q) > 0, \mathrm{dv}(q) = 0$ or $\mathrm{dv}(q) < 0$, respectively. This means that Ψ_q will be:
   - an irreducible polynomial, if $\mathrm{dv}(q) > 0$;
   - a polynomial of the form $(x - r)^2$, $r \in \mathbb{R}$, if $\mathrm{dv}(q) = 0$;
   - a polynomial of the form $(x - r_1)(x - r_2)$ with $r_1, r_2 \in \mathbb{R}, r_1 \neq r_2$, if $\mathrm{dv}(q) < 0$.

The next theorem lists some properties involving the zero-set of the characteristic polynomial of an element $q \in \mathcal{H}$.

Theorem 4. *Let Ψ_q be the characteristic polynomial of a given element $q \in \mathcal{H}$. Then:*

(i) $[\![q]\!] \subseteq Z(\Psi_q)$.
(ii) If $\mathrm{dv}_q \geq 0$, $Z(\Psi_q) = [\![q]\!]$.
(iii) If $\mathrm{dv}_q < 0$, then $\Psi_q(x) = (x - r_1)(x - r_2)$ for $r_1, r_2 \in \mathbb{R}$, $r_1 \neq r_2$ and $Z(\Psi_q) = [\![q]\!] \cup \{r_1, r_2\}$.

Proof. Let u be an arbitrary element in $[\![q]\!]$. From the relation $u = \mathrm{t}(u) - \overline{u}$ (see (1)), we obtain

$$u^2 = (\mathrm{t}(u) - \overline{u})u = \mathrm{t}(u)u - \mathrm{d}(u),$$

---

[4] Recall that, in the case of quaternions, quasi-similarity and similarity classes coincide, so, in that case, we can also say that Ψ_q is an invariant of $[q]$.

or

$$u^2 - t(u)u + d(u) = 0.$$

This shows that $\Psi_q(u) = 0$, hence establishing (i).

We now prove (ii)–(iii). Let then z be a zero of Ψ_q, i.e., let z satisfy $z^2 - t(q)z + d(q) = 0$; since z is also a zero of its characteristic polynomial (by (i)), we also have $z^2 - t(z)z + d(z) = 0$ and so we obtain $(t(z) - t(q))z + d(q) - d(z) = 0$, or $(t(z) - t(q))(\operatorname{Re} z + \operatorname{Vec} z) + d(q) - d(z) = 0$, which means that we must have

$$\begin{cases} (t(z) - t(q)) \operatorname{Vec} z = 0 \\ (t(z) - t(q)) \operatorname{Re} z + d(q) - d z = 0 \end{cases}.$$

If z is non-real, i.e. if $\operatorname{Vec} z \neq 0$, we immediately obtain $t(z) = t(q)$ and $d(z) = d(q)$, showing that $z \in [\![q]\!]$. Hence, we conclude that all non-real roots of Ψ_q belong to $[\![q]\!]$.

Let us now see what happens with the (possible) real zeros of Ψ_q. Recalling the description on the behavior of the characteristic polynomial Ψ_q done previously, we see that this type of roots only occur if $dv(q) = 0$ or $dv(q) < 0$. In the first case, we have that Ψ_q has a double real root r, but $[\![q]\!] = [\![r]\!]$, i.e., $r \in [\![q]\!]$ and so the result (ii) follows. When $dv(q) < 0$[5] we have that $\Psi_q(x) = (x - r_1)(x - r_2)$ has two real zeros r_1 and r_2 which do not belong to $[\![q]\!]$ (note that, in that case, $\operatorname{Re} q = \frac{r_1 + r_2}{2} \neq r_1, r_2$), which establishes (iii). □

The previous theorem shows us that, in what concerns the number of zeros, quaternionic and coquaternionic polynomials may behave very differently from polynomials in $\mathbb{C}[x]$, where, as is well known, a polynomial of degree n has, at most, n distinct zeros. In fact, the characteristic polynomial of an element $q \in \mathscr{H}$ (with q non-real, in the quaternionic case) gives us an example of a polynomial with an infinite number of zeros: a sphere, in the quaternionic case, and an hyperboloid or an hyperboloid and two extra zeros, in the coquaternonic case.

Although quaternionic and coquaternonic polynomials have this common feature – they both may have an infinite number of zeros – there is a very important difference between these two types of polynomials. In the case of quaternion polynomials, as proved in the pioneering paper by I. Niven [21], the Fundamental Theorem of Algebra is valid, i.e., every non-constant polynomial in $\mathbb{H}[x]$ has, at least, a zero in $\mathbb{H}$. However, this theorem is not valid in $\mathbb{H}_{coq}[x]$. In fact, as observed by Özdemir [22, Theorem 9-i.], since for a coquaternion u, we have $d(u^n) = (d(u))^n$ (see Proposition 1), any equation of the form $x^n - q = 0$, with n even and q a coquaternion with negative determinant, does not have a solution. Other examples of coquaternion polynomials with no roots can be found in [15].

We now introduce the following definition for zeros of a polynomial in $\mathscr{H}[x]$.

---

[5] Recall that this can only happen in the case of coquaternions.

Definition 4. Let $P \in \mathscr{H}[x]$ and let $z \in \mathscr{H}$ be a zero of P.

(i) If $[\![z]\!]$ contains no other zeros of P, then z is said to be an *isolated zero* of P.
(ii) If $[\![z]\!] \subseteq Z(P)$, then z is said to be:
   – a *spherical zero* of P, if $\mathscr{H} = \mathbb{H}$;
   – an *hyperboloidal zero* of P, if $\mathscr{H} = \mathbb{H}_{\mathrm{coq}}$.

Remark 7. 1. The choice of the denominations, spherical, in the case of quaternionic polynomials and hyperboloidal, in the case of coquaternionic polynomials, is a natural one, having in mind the type of sets which are the quasi-similarity classes, in each case. The term *hyperboloidal*, which we adopt here, was the choice made by Pogoruy and Ramírez-Dagnino in [26]; in [15], the authors use the term *hyperbolic* instead of hyperboloidal, which, from our point of view, does not seem so appropriate.
2. In the quaternionic case, a real zero z, is by definition, an isolated zero, since, in this case $[\![z]\!] = [z] = \{z\}$. This is not necessarily true in the coquaternionic case. Example 2, given later, will clarify this assertion.
3. When a zero z is spherical/hyperboloidal, sometimes we treat the whole class as a single zero and talk about the spherical/hyperboloidal zero $[\![z]\!]$.

Theorem 5. *Let $P \in \mathscr{H}[x]$ and let $z \in \mathscr{H}$ be a zero of P. If $[\![z]\!]$ contains another zero u of P such that $z - u$ is nonsingular, then $[\![z]\!] \subseteq Z(P)$.*

Proof. We first observe that, since the product of two polynomials in $\mathscr{H}[x]$ is defined in the usual manner, the Euclidean Division Algorithm to perform the division of two polynomials can be applied, provided that the leading coefficient of the divisor polynomial is a non-singular element in $\mathscr{H}$. We can thus divide $P(x)$ by the characteristic polynomial of z, i.e. by the quadratic monic polynomial $\Psi_z(x) = x^2 - \mathrm{t}(z)x + \mathrm{d}(z)$, obtaining

$$P(x) = Q(x)\Psi_z(x) + A + Bx \tag{9}$$

for some polynomial $Q(x)$ and values A and B which depend only on the coefficients c_i of the polynomial P and on the values $\mathrm{t}(z)$ and $\mathrm{d}(z)$. Hence, for all $w \in [\![z]\!]$, we have, $P(w) = A + Bw$, since $\Psi_z(w) = 0$ (see Theorem 4-(i)). Because, by hypothesis, z and u are both zeros of P in $[\![z]\!]$, it follows that we must have

$$A + Bz = 0 \quad \text{and} \quad A + Bu = 0, \tag{10}$$

from where we obtain $B(u - z) = 0$. The fact that $u - z$ is non-singular implies that $B = 0$ which, from (10), leads to $A = 0$ also. From (9), we then obtain $P(x) = Q(x)\Psi_z(x)$, allowing us to conclude that all the zeros of Ψ_z are zeros of P; see Theorem 3-(ii). Since $[\![z]\!] \subseteq Z(\Psi_z)$ (Theorem 4-(i)), the result follows. $\square$

Remark 8. The result of Theorem 5 for the case of quaternionic polynomials was first established, by using totally different arguments, by Gordon and Motzkin [9]. For the coquaternionic case, a more similar proof to the one presented here was given in [15].

In the case of quaternionic polynomials, the condition $u - z$ non-singular can simply be replaced by $u - z \neq 0$ i.e., by $u \neq z$. This shows that, in the quaternionic case, a zero can only be of one of two types: isolated – when its similarity class contains no other zeros – or spherical, if the class contains more than one zero. In particular, a (non-real) zero z of P is spherical if and only its conjugate $\bar{z}$ is also a zero of P. This is a very simple test one usually performs to distinguish an isolated zero from a spherical one.

In the case of coquaternionic polynomials, however, this kind of dichotomy does not hold anymore. In fact, it is possible to have a zero of a polynomial whose corresponding quasi-similarity class contains other zeros of the polynomial, but such that not all the elements in the class are zeros. Examples of such type of zeros will be given later.

Theorem 6. *Let $P \in \mathscr{H}[x]$ be a real polynomial. If z is a zero of P, then $[z] \subseteq Z(P)$.*

Proof. See e.g. the proof of [30, Lemma 2.3.] for the quaternionic case or the analogous proof in [15, Theorem 1.2], where the result is stated for the coquaternionic case.   □

Corollary 4. *Under the conditions of the previous theorem, if z is a non-real zero of P, then the zero is spherical, in the quaternionic case, and hyperboloidal, in the coquaternionic case.*

Proof. The conclusion, for the quaternionic case, is immediate, since $[z] = [\![z]\!]$. We now consider the coquaternionic case. Note that, in this case, the fact that z is non-real does not guarantee that $[\![z]\!] = [z]$ (this does not hold when $\mathrm{dv}(z) = 0$).

Let $z = z_0 + z_1 \mathbf{i} + z_2 \mathbf{j} + z_3 \mathbf{k}$. Since z is non-real, we have $z_i \neq 0$ for at least one $i \in \{1, 2, 3\}$. Assume, for example, that $z_1 \neq 0$ (the proof is analogous in other cases) and consider the coquaternion $u = z_0 - z_1 \mathbf{i} + z_2 \mathbf{j} + z_3 \mathbf{k}$. Then, $u \in [\![z]\!]$ and both u and z are non-real; this means that $u \sim z$, i.e., $u \in [z]$ and so, from the previous theorem, u is a zero of P; on the other hand, $\mathrm{d}(z - u) = \mathrm{d}(2z_1 \mathbf{i}) = 4z_1^2 \neq 0$. Theorem 5 guarantees that the zero is hyperboloidal.   □

Remark 9. The conclusion of the previous corollary is stated in [15, Example 3.6] as being an immediate consequence of Theorem 6. However, as explained before, in the coquaternionic case, the fact that $[\![z]\!] \subseteq Z(P)$ is not obvious and needs to be proved.

4   Examples

In this section, we present simple examples illustrating some of the results on the zeros of polynomials in $\mathbb{H}[x]$ and $\mathbb{H}_{\mathrm{coq}}[x]$ previously discussed.

We first summarize the main differences which can exist between quaternionic and coquaternionic polynomials, in what concerns their zeros.

1. The Fundamental Theorem of Algebra is valid in $\mathbb{H}[x]$, i.e., every non-constant quaternionic polynomial has, at least, a zero in $\mathbb{H}$, but is no longer valid in $\mathbb{H}_{\mathrm{coq}}[x]$.
2. A real zero of a quaternionic polynomial is always an isolated zero, but this is not necessarily the case for coquaternionic polynomials.
3. A zero z of a quaternionic polynomial P can be only of two types: isolated, if z is the only zero of P in $[\![z]\!]$, or spherical, if $[\![z]\!]$ contains more than one zero of P, in which case all the elements in the sphere $[\![z]\!]$ are zeros of P. In the coquaternionic case, zeros of other type may appear for a given polynomial P: zeros z such that $[\![z]\!]$ contains other zeros of P, but such that not all the class $[\![z]\!]$ is formed by zeros of P.

Example 1. As observed before, if n is even and q is a coquaternion with a negative determinant, then the polynomial $P(x) = x^n - q$ has no coquaternionic roots. Hence, for example, the polynomial $P_1(x) = x^2 - \mathbf{j}$ in $\mathbb{H}_{\mathrm{coq}}[x]$ has no zeros.

Note that if we view P_1 as a quaternionic polynomial, then P_1 has two isolated zeros: $x_1 = \frac{\sqrt{2}}{2} + \frac{\sqrt{2}}{2}\mathbf{j}$ and $x_2 = -\frac{\sqrt{2}}{2} - \frac{\sqrt{2}}{2}\mathbf{j}$.

Example 2. Consider the real polynomial

$$P_2(x) = (x-1)(x-2) = x^2 - 3x + 2.$$

The only quaternionic zeros of this polynomial are the isolated zeros $x_1 = 1$ and $x_2 = 2$.

However, if we consider P_2 as a polynomial in $\mathbb{H}_{\mathrm{coq}}[x]$, then, since $P_2(x) = \Psi_q(x)$ with q the coquaternion $q = \frac{3}{2} + \frac{1}{2}\mathbf{j}$, Theorem 4 allows us to conclude that P_2 has, apart from the two isolated zeros $x_1 = 1$ and $x_2 = 2$, an hyperboloid of zeros: the quasi-similarity class $[\![\frac{3}{2} + \frac{1}{2}\mathbf{j}]\!]$.

Example 3. Let

$$P_3(x) = (x-2)\left(x - \left(\tfrac{3}{2} + \tfrac{1}{2}\mathbf{j}\right)\right) = x^2 - \left(\tfrac{7}{2} + \tfrac{1}{2}\mathbf{j}\right)x + 3 + \mathbf{j}.$$

If we consider the polynomial P_3 as an element of $\mathbb{H}[x]$, then P_3 has only two isolated roots: $x_1 = 2$ and $x_2 = \frac{3}{2} + \frac{1}{2}\mathbf{j}$.

Considering P_3 as a polynomial in $\mathbb{H}_{\mathrm{coq}}[x]$, a simple verification shows that any coquaternion in the set

$$\mathscr{Z}_1 = \{2 + \alpha\mathbf{i} - \alpha\mathbf{k} : \alpha \in \mathbb{R}\} \subsetneqq [\![2]\!]$$

is a zero of P_3. However, not all the elements in the quasi-similarity class $[\![2]\!]$ are zeros of P_3; for example, the coquaternion $u = 2 + \mathbf{i} + \mathbf{k} \in [\![2]\!]$, but $P_3(u) = \mathbf{i} + \mathbf{k} \neq 0$. Hence, the zero $x = 2$ (and any other zero in the set $\mathscr{Z}_1$) is neither an isolated zero nor an hyperboloidal zero of P_3.

Observe that, for any two elements $z = 2 + \alpha\mathbf{i} - \alpha\mathbf{k}$ and $z' = 2 + \alpha'\mathbf{i} - \alpha'\mathbf{k}$ in $\mathscr{Z}_1$, their difference $z - z' = (\alpha - \alpha')\mathbf{i} - (\alpha - \alpha')\mathbf{k}$ is singular, so there is no contradiction with Theorem 5. One can also show that any element in the set

$$\mathscr{X}_2 = \left\{ \tfrac{3}{2} + \alpha\mathbf{i} + \tfrac{1}{2}\mathbf{j} + \alpha\mathbf{k} : \alpha \in \mathbb{R} \right\} \subsetneqq [\![\tfrac{3}{2} + \tfrac{1}{2}\mathbf{j}]\!]$$

is a zero of P_3, but not all the elements in $[\![\tfrac{3}{2} + \tfrac{1}{2}\mathbf{j}]\!]$ are zeros of P_3. For example, the coquaternion $v = \tfrac{3}{2} + \mathbf{i} + \tfrac{1}{2}\mathbf{j} - \mathbf{k} \in [\![\tfrac{3}{2} + \tfrac{1}{2}\mathbf{j}]\!]$, but $P(v) = -\mathbf{i} + \mathbf{k} \neq 0$.

Apart from confirming our statement about the existence, in the case of coquaternionic polynomials, of zeros which are neither isolated nor hyperboloidal, this example also illustrates the statement made in 2. We have a real zero, $x = 2$, which is not an isolated zero of the coquaternionic polynomial P_2, a situation that cannot occur in the case of a quaternionic polynomial.

Remark 10. We must also refer that this example contradicts the statement of [26, Theorem 2.4]: we have more than one zero in a class whose elements have a non-singular vector part — the class $[\![\tfrac{3}{2} + \tfrac{1}{2}\mathbf{j}]\!]$ — and, however, these zeros are not hyperboloidal.

Acknowledgments. Research at CMAT was financed by Portuguese Funds through FCT - Fundação para a Ciência e a Tecnologia, within the Project UID/MAT/ 00013/2013. Research at NIPE was carried out within the funding with COMPETE reference number POCI-01-0145-FEDER-006683 (UID/ECO/03182/2013), with the FCT/MEC's (Fundação para a Ciência e a Tecnologia, I.P.) financial support through national funding and by the ERDF through the Operational Programme on "Competitiveness and Internationalization - COMPETE 2020" under the PT2020 Partnership Agreement.

References

1. Adler, S.: Quaternionic Quantum Mechanics and Quantum Fields. International Series of Monographs of Physics, vol. 88. Oxford University Press, New York (1995)
2. Alagöz, Y., Oral, K.H., Yüce, S.: Split quaternion matrices. Miskolc Math. Notes **13**, 223–232 (2012)
3. Artzy, R.: Dynamics of quadratic functions in cycle planes. J. Geom. **44**, 26–32 (1992)
4. Brenner, J.L.: Matrices of quaternions. Pacific J. Math. **1**(3), 329–335 (1951)
5. Brody, D.C., Graefe, E.M.: On complexified mechanics and coquaternions. J. Phys. A Math. Theor. **44**, 1–9 (2011)
6. Cockle, J.: On systems of algebra involving more than one imaginary; and on equations of the fifth degree. Philos. Mag. **35**(3), 434–435 (1849)
7. De Leo, S., Ducati, G., Leonardi, V.: Zeros of unilateral quaternionic polynomials. Electron. J. Linear Algebra **15**, 297–313 (2006)
8. Falcão, M.I.: Newton method in the context of quaternion analysis. Appl. Math. Comput. **236**, 458–470 (2014)
9. Gordon, B., Motzkin, T.S.: On the zeros of polynomials over division rings. Trans. Am. Math. Soc. **116**, 218–226 (1965)
10. Gürlebeck, K., Sprößig, W.: Quaternionic Calculus for Engineers and Physicists. Wiley, Hoboken (1997)
11. Hamilton, W.R.: A new species of imaginaries quantities connected with a theory of quaternions. Proc. R. Ir. Acad. **2**, 424–434 (1843)

12. Janovská, D., Opfer, G.: Computing quaternionic roots in Newton's method. Electron. Trans. Numer. Anal. **26**, 82–102 (2007)
13. Janovská, D., Opfer, G.: The classification and the computation of the zeros of quaternionic, two-sided polynomials. Numer. Math. **115**(1), 81–100 (2010)
14. Janovská, D., Opfer, G.: Linear equations and the Kronecker product in coquaternions. Mitt. Math. Ges. Hamburg **33**, 181–196 (2013)
15. Janovská, D., Opfer, G.: Zeros and singular points for one-sided coquaternionic polynomials with an extension to other$\mathbb{R}^4$ algebras. Electron. Trans. Numer. Anal. **41**, 133–158 (2014)
16. Kalantari, B.: Algorithms for quaternion polynomial root-finding. J. Complex. **29**, 302–322 (2013)
17. Kula, L., Yayli, Y.: Split quaternions and rotations in semi euclidean space E_2^4. J. Korean Math. Soc. **44**, 1313–1327 (2007)
18. Lam, T.Y.: A First Course in Noncommutative Rings. Springer, Heidelberg (1991)
19. Malonek, H.R.: Quaternions in applied sciences. A historical perspective of a mathematical concept. In: 17th International Conference on the Application of Computer Science and Mathematics on Architecture and Civil Engineering, Weimar (2003)
20. Miranda, F., Falcão, M.I.: Modified quaternion Newton methods. In: Murgante, B., Misra, S., Rocha, A.M.A.C., Torre, C., Rocha, J.G., Falcão, M.I., Taniar, D., Apduhan, B.O., Gervasi, O. (eds.) ICCSA 2014. LNCS, vol. 8579, pp. 146–161. Springer, Cham (2014). doi:10.1007/978-3-319-09144-0_11
21. Niven, I.: Equations in quaternions. Amer. Math. Monthly **48**, 654–661 (1941)
22. Özdemir, M.: The roots of a split quaternion. Appl. Math. Lett. **22**(2), 258–263 (2009)
23. Özdemir, M., Ergin, A.: Some geometric applications of split quaternions. In: Proceedings 16th International Conference Jangjeon Mathematical Society, vol. 16, pp. 108–115 (2005)
24. Özdemir, M., Ergin, A.: Rotations with unit timelike quaternions in Minkowski 3-space. J. Geometry Phys. **56**(2), 322–336 (2006)
25. Pogorui, A., Shapiro, M.: On the structure of the set of zeros of quaternionic polynomials. Complex Variables Theor. Appl. **49**(6), 379–389 (2004)
26. Pogoruy, A.A., Rodríguez-Dagnino, S.: Some algebraic and analytical properties of coquaternion algebra. Adv. Appl. Clifford Algebras **20**(1), 79–84 (2010)
27. Serôdio, R., Siu, L.S.: Zeros of quaternion polynomials. Appl. Math. Letters **14**(2), 237–239 (2001)
28. Serôdio, R., Pereira, E., Vitória, J.: Computing the zeros of quaternion polynomials. Comput. Math. Appl. **42**(8–9), 1229–1237 (2001)
29. Shoemake, K.: Animating rotation with quaternion curves. SIGGRAPH Comput. Graph. **19**(3), 245–254 (1985)
30. Topuridze, N.: On roots of quaternion polynomials. J. Math. Sci. **160**, 843–855 (2009)

Mathematica Tools for Quaternionic Polynomials

M. Irene Falcão[1]([✉]), Fernando Miranda[1], Ricardo Severino[2],
and M. Joana Soares[3]

[1] CMAT and Departamento de Matemática e Aplicações,
Universidade do Minho, Braga, Portugal
{mif,fmiranda}@math.uminho.pt
[2] Departamento de Matemática e Aplicações,
Universidade do Minho, Braga, Portugal
ricardo@math.uminho.pt
[3] NIPE and Departamento de Matemática e Aplicações,
Universidade do Minho, Braga, Portugal
jsoares@math.uminho.pt

Abstract. In this paper we revisit the ring of (left) one-sided quaternionic polynomials with special focus on its zero structure. This area of research has attracted the attention of several authors and therefore it is natural to develop computational tools for working in this setting. The main contribution of this paper is a Mathematica collection of functions QPolynomial for solving polynomial problems that we frequently find in applications.

Keywords: Quaternions · Polynomial ring · Factorization · Symbolic computation

1 Introduction

In this paper we are going to study polynomials in one formal variable x whose coefficients are quaternions located on the left side of the powers of x, i.e. polynomials of the form

$$P(x) = a_n x^n + a_{n-1} x^{n-1} + \cdots + a_1 x + a_0, \quad a_k \in \mathbb{H}. \tag{1}$$

If $Q(x) = b_n x^n + b_{n-1} x^{n-1} + \cdots + b_1 x + b_0$ is another polynomial of the same kind, then one can define the operation:

$$P(x) + Q(x) = (a_n + b_n)x^n + (a_{n-1} + b_{n-1})x^{n-1} + \cdots + (a_1 + b_1)x + a_0 + b_0,$$

so that the polynomials (1) form an additive abelian group. There are several ways of defining the multiplication such that this group is made a ring. For details on the theory of non-commutative polynomials, we refer the reader to the well-known work of Ore [23]. In the classical case and in many practical applications (e.g. [11,25,26]), one assumes that the variable x commutes with

© Springer International Publishing AG 2017
O. Gervasi et al. (Eds.): ICCSA 2017, Part II, LNCS 10405, pp. 394–408, 2017.
DOI: 10.1007/978-3-319-62395-5_27

the coefficients, i.e. the multiplication of two polynomials P and Q of degrees n and m, respectively, is defined as

$$P(x)Q(x) = \sum_{k=0}^{m+n} \left(\sum_{j=0}^{k} a_j b_{k-j} \right) x^k, \tag{2}$$

with the implicit assumption that $a_k = 0$, if $k > n$ and $b_k = 0$, if $k > m$. In other words, the set of polynomials of the form (1) endowed with the aforementioned addition and multiplication is a polynomial ring over $\mathbb{H}$ and is denoted by $\mathbb{H}[x]$. In the literature, $\mathbb{H}[x]$ is usually referred to as the ring of (left) one-sided, unilateral, basic, simple or standard polynomials and the determination and classification of their zeros have attracted the attention of several authors over the years (see e.g. [1,2,6,12–14,17,18,22,27,29–31]) due to the appearance of new interesting algebraic and computational properties.

General polynomials are defined as finite sums of noncommutative monomials of type $a_0 x a_1 x \ldots x a_n$. General polynomials have been also considered in several works ([7,14,28]) specially the cases of two-sided or quadratic polynomials (e.g. [16,20,32]).

In this paper we describe a Mathematica tool for polynomial manipulation in $\mathbb{H}[x]$. The paper is organized as follows. In Sect. 2 we recall well-known concepts and review important results on quaternionic polynomials. In Sect. 3 we focus on the zeros of a polynomial and the relation to its factorization(s) in linear terms. Finally, in Sect. 4 the new Mathematica tools are presented.

2   The Ring of Quaternionic Polynomials

In 1843, Hamilton introduced numbers of the form

$$x = x_0 + \mathbf{i} x_1 + \mathbf{j} x_2 + \mathbf{k} x_3, \ x_i \in \mathbb{R}, \tag{3}$$

where $\mathbf{i}$, $\mathbf{j}$ and $\mathbf{k}$ satisfy the multiplication rules

$$\mathbf{i}^2 = \mathbf{j}^2 = \mathbf{k}^2 = -1 \text{ and } \mathbf{ij} = -\mathbf{ji} = \mathbf{k}. \tag{4}$$

This non-commutative product generates the algebra of real quaternions $\mathbb{H}$ (named after Hamilton). For a quaternion x of the form (3) we can define, in analogy with the complex case, the real part of x, $\operatorname{Re} x := x_0$, the vector part of x, $\operatorname{Vec} x := \mathbf{i} x_1 + \mathbf{j} x_2 + \mathbf{k} x_3$ and the conjugate of x, $\bar{x} := x_0 - \mathbf{i} x_1 - \mathbf{j} x_2 - \mathbf{k} x_3$. The norm of x is given by $|x| := \sqrt{x\bar{x}} = \sqrt{\bar{x}x} = \sqrt{x_0^2 + x_1^2 + x_2^2 + x_3^2}$. It immediately follows that each non-zero $x \in \mathbb{H}$ has an inverse given by $x^{-1} = \frac{\bar{x}}{|x|^2}$ and therefore $\mathbb{H}$ is a non-commutative division ring or a skew field.

Let us review now some well-known facts on quaternionic polynomials and fix some notation. In what follows, q denotes a quaternion, P is a left one-sided polynomial and $\mathbf{Z}_P$ denotes the zero-set of P.

Definition 1 (*Similarity*). Two quaternions q and q' are called *similar*, $q \sim q'$, if there exists $h \neq 0$ such that $q' = hqh^{-1}$. Similarity is an equivalence relation and the *similarity class* of q, denoted by $[q]$, is the set $[q] = \{q' \in \mathbb{H} : q \sim q'\}$.

It can easily be shown (see, e.g. [3]) that

$$[q] = \{q' \in \mathbb{H} : \operatorname{Re} q = \operatorname{Re} q' \text{ and } |q| = |q'|\}.$$

Example 1. The similarity class of $1 + \mathbf{i} + \mathbf{j} + \mathbf{k}$ is the three-dimensional sphere in the hyperplane $\{(1, x, y, z) \in \mathbb{R}^4\}$, with center $(1, 0, 0, 0)$ and radius $\sqrt{3}$, i.e.

$$[1 + \mathbf{i} + \mathbf{j} + \mathbf{k}] = \{1 + \mathbf{i}x + \mathbf{j}y + \mathbf{k}z \in \mathbb{H} : x^2 + y^2 + z^2 = 3\}.$$

Definition 2 (*Special polynomials*)

1. The *conjugate* of P is the polynomial

$$\overline{P}(x) := \overline{a}_n x^n + \overline{a}_{n-1} x^{n-1} + \cdots + \overline{a}_1 x + \overline{a}_0.$$

2. The *characteristic polynomial* of q is the polynomial

$$\Psi_q(x) := (x - q)(x - \overline{q}) = x^2 - 2\operatorname{Re}(q)\, x + |q|^2. \tag{5}$$

3. The *companion polynomial* of P is the real polynomial

$$\mathcal{C}_P(x) := P(x)\overline{P}(x) = \overline{P}(x)P(x). \tag{6}$$

Remark 1. The characteristic polynomial of q depends only on the real part and norm of q and, if $q \in \mathbb{H} \setminus \mathbb{R}$, then Ψ_q is an irreducible quadratic trinomial from the ring of polynomials $\mathbb{R}[x]$.

Remark 2. In [17], Janovská and Opfer had called the polynomial

$$\sum_{k=0}^{2n} \left(\sum_{j=\max\{0, k-n\}}^{\min\{k, n\}} \overline{a}_j a_{k-j} \right) x^k$$

the companion polynomial of the quaternionic polynomial P, without mentioning that this is nothing else than the usual product of $\overline{P}$ and P in $\mathbb{H}[x]$ (cf. (2)). As pointed out by these authors, the idea of constructing such a polynomial came from the work of Niven [22], where the construction of the polynomial is described as follows: *By replacing the quaternions a_k by their conjugates $\overline{a}_k$ we obtain a polynomial $\overline{P}$. We multiply this on the right by $P(x)$, and allow x to be commutative with the coefficients. Thus we obtain a polynomial with coefficients in $\mathbb{R}$.*

In the literature one can find different designations to the polynomial (6), namely basic polynomial, semi-norm or symmetrization of a polynomial. We prefer the designation companion polynomial due to its connection (not recognized explicitly in [17]) to the characteristic polynomial of the companion matrix which was already apparent from the work of [29] (see also [4]).

Example 2. For $P(x) = x^2 + (2 - \mathbf{i})x + \mathbf{j} + \mathbf{k}$ and $q = \mathbf{i} + \mathbf{j}$ we have

$$\Psi_q(x) = x^2 + 2,$$
$$\mathcal{C}_P(x) = \big(x^2 + (2 - \mathbf{i})x + \mathbf{j} + \mathbf{k}\big)\big(x^2 + (2 + \mathbf{i})x - \mathbf{j} - \mathbf{k}\big) = x^4 + 4x^3 + 5x^2 + 2.$$

Definition 3 (*Evaluation map and zeros*)

1. The *evaluation* of P at q is defined as $P(q) := a_n q^n + a_{n-1} q^{n-1} + \cdots + a_1 q + a_0$.
2. A quaternion q is a *zero* (or a *root*) of P, if $P(q) = 0$.
3. A zero q is called an *isolated zero* of P, if $[q]$ contains no other zeros of P.
4. A zero q is called a *spherical zero* of P, if q is not an isolated zero; $[q]$ is referred to as a sphere of zeros.[1]

Remark 3. We recall here the well-known fact that the evaluation map (at a quaternion q) is not an algebra homomorphism. In fact, if $P(x) = L(x)R(x)$, then (see e.g. [19])

$$P(q) = \begin{cases} 0, & \text{if } R(q) = 0, \\ L(\tilde{q})\, R(q), & \text{if } R(q) \neq 0, \end{cases} \qquad \text{where} \quad \tilde{q} = R(q)\, q\, (R(q))^{-1}. \quad (7)$$

In particular, if q is a zero of P which is not a zero of R, then $\tilde{q}$ is a zero of L. Next example illustrates this remark.

Example 3. Consider the polynomial P and the quaternion q of last example. If $Q(x) = x - q$, then we have

$$P(q) = -1 + 2\mathbf{i} + 3\mathbf{j} \qquad \text{and} \qquad Q(q) = 0.$$

On the other hand, since

$$(QP)(x) = x^3 + (2 - 2\mathbf{i} - \mathbf{j})x^2 + (-1 - 2\mathbf{i} - \mathbf{j})x + 1 - \mathbf{i} + \mathbf{j} - \mathbf{k},$$

we remark that

$$(QP)(q) = -2\mathbf{k} \neq 0 = Q(q)P(q).$$

3 Zeros and Factor-Terms in $\mathbb{H}[x]$

We list now the main results concerning the zero-structure and the factorization of polynomials in $\mathbb{H}[x]$. For the proofs and other details we refer the reader to [14, 19, 23].

We recall first that it can be proved (see e.g. [15]) that $\mathbb{H}[x]$ is a principal ideal domain and therefore also left and right division algorithms can be defined.

Theorem 1 (Euclidean division). *If $A(x)$ and $B(x)$ are polynomials in $\mathbb{H}[x]$ (with $\deg B \leq \deg A$ and $B \neq 0$), then there exist unique $Q(x)$, $Q'(x)$, $R(x)$ and $R'(x)$ such that*

$$A(x) = Q(x)B(x) + R(x) \tag{8}$$

and

$$A(x) = B(x)Q'(x) + R'(x) \tag{9}$$

with $\deg R < \deg B$ and $\deg R' < \deg B$.

---

[1] In such case, it can be proved that all quaternions in $[q]$ are in fact zeros of P and therefore the choice of the term spherical to designate this type of zeros is natural.

If $R(x) = 0$ (resp. $R'(x) = 0$) in (8) (resp. (9)), then $B(x)$ is called a right (resp. left) divisor of $A(x)$. The greatest common (right and left) divisor polynomial of two polynomials can be computed using the Euclidean algorithm, by a basic procedure similar to the one used in the complex setting. In this context we also refer to [5] where issues of division within $\mathbb{H}[x]$ are addressed. In what follows, if nothing is specified, right divisors are always assumed.

Concerning the zero structure, the theory of quaternionic polynomials is very different from that of complex polynomials. Nevertheless, in 1941, Niven [22] proved the Fundamental Theorem of Algebra for quaternionic polynomials, establishing that a non-constant polynomial P in $\mathbb{H}[x]$ has a zero in $\mathbb{H}$. As in the classical case, it is also possible to write P as a product of linear factors; however the link between zeros and factors is not straightforward.

Theorem 2 (Factorization into linear terms [14,19]). *Any monic polynomial P of degree $n \in \mathbb{N}$ in $\mathbb{H}[x]$ admits a factorization into linear factors, i.e. there exist $x_1, \ldots, x_n \in \mathbb{H}$, such that*

$$P(x) = (x - x_n)(x - x_{n-1}) \cdots (x - x_1). \tag{10}$$

Definition 4 (*Chain*). In a factorization of P of the form (10), the quaternions $x_1, \ldots, x_n$ will be called *factor-terms* of P and the n-uple $(x_1, \ldots, x_n)$ will be called a *factor-terms chain associated with P* or simply a *chain of P*.

Theorem 3 ([19,30]). *Let $(x_1, \ldots, x_n)$ be a chain of a polynomial P. Then every zero of P is similar to some factor-term x_k in the chain and reciprocally every factor-term x_k is similar to some zero of P.*

Remark 4. If $(y_1, \ldots, y_n)$ is another chain of P, then there exists a permutation π of $(1, \ldots, n)$ and $h_1, \ldots, h_n \in \mathbb{H}$, such that $y_{\pi(i)} = h_i x_i h_i^{-1}$; $i = 1, \ldots n$, i.e. $[y_{\pi(i)}] = [x_i]$. The explicit expression of h_i will be given later on. In such cases we write $(x_1, \ldots, x_n) \sim (y_1, \ldots, y_n)$ and say that the chains $(x_1, \ldots, x_n)$ and $(y_1, \ldots, y_n)$ are *equivalent* or *similar*.

The explicit relation between factor-terms and zeros of a quaternionic polynomial is addressed in the following results. The first result is useful if one knows a factorization of the polynomial; for example in [10] a numerical method for computing a Weierstrass factorization of a quaternionic polynomial is proposed and the non-spherical zeros are obtained via (11). The second result plays an important role in the construction of polynomials with prescribed zeros.

Theorem 4 (Zeros from factors [19]). *Consider a chain $(x_1, \ldots, x_n)$ of a polynomial P. If the similarity classes $[x_k]$ are distinct, then P has exactly n zeros ζ_k which are related to the factor-terms x_k as follows:*

$$\zeta_k = \overline{\mathcal{P}}_k(x_k)\, x_k \left(\overline{\mathcal{P}}_k(x_k) \right)^{-1}; \quad k = 1, \ldots, n, \tag{11}$$

where

$$\mathcal{P}_k(x) := \begin{cases} 1, & \text{if } k = 1, \\ (x - x_{k-1}) \ldots (x - x_1), & \text{otherwise.} \end{cases} \tag{12}$$

Theorem 5 (Factors from zeros [1]**).** *If $\zeta_1, \ldots, \zeta_n$ are quaternions such that the similarity classes $[\zeta_k]$ are distinct, then there is a unique polynomial P of degree n with zeros $\zeta_1, \ldots, \zeta_n$, which can be constructed from the chain $(x_1, \ldots, x_n)$, where*

$$x_k = \mathcal{P}_k(\zeta_k)\, \zeta_k \, (\mathcal{P}_k(\zeta_k))^{-1}; \quad k = 1, \ldots, n \tag{13}$$

and $\mathcal{P}_k$ is the polynomial (12).

In [1,13] more general forms of Theorems 4 and 5 can be found, where the notion of multiplicity of a zero plays a fundamental role. The following result motivates the definition of multiplicity we are going to adopt (in this context we refer also to [1,2,9,24]).

Theorem 6 ([1,24]). *Let P be a quaternionic polynomial with degree n. Then $x_1 \in \mathbb{H} \setminus \mathbb{R}$ is the unique zero of P if and only if P admits a unique chain $(x_1, \ldots, x_n)$ which has the property*[2]

$$x_l \in [x_1] \quad and \quad x_l \neq \overline{x}_{l-1}, \tag{14}$$

for all $l = 2, \ldots, n$.

Definition 5 (*Multiplicity*). The *multiplicity of a zero q of P*, $m_P(q)$, is defined as the maximum degree of the right factors of P having q as their unique zero. The *multiplicity of a sphere of zeros $[q]$ of P*, $m_P([q])$, is the largest $k \in \mathbb{N}_0$ for which Ψ_q^k divides P.

Example 4. The polynomial $P(x) = (x - \mathbf{k})(x - \mathbf{j})(x - \mathbf{i})$ has $\mathbf{i}$ as its unique zero of multiplicity 3. Concerning the polynomial $Q(x) = (x + \mathbf{i})(x - \mathbf{i})(x - \mathbf{i})$ we observe that $m_Q(\mathbf{i}) = 2$ while $m_Q([\mathbf{i}]) = 1$. In [9] such a zero has been called a *mixed zero*.[3]

The final results of this section are presented/rewritten by taking into account the purpose of the present paper.

Theorem 7. *Consider a chain $(x_1, \ldots, x_{l-1}, x_l, \ldots, x_n)$ of a polynomial P.*

1. If $h := \overline{x}_l - x_{l-1} \neq 0$ then

$$(x_1, \ldots, h^{-1}x_l h, h^{-1}x_{l-1}h, \ldots, x_n) \sim (x_1, \ldots, x_{l-1}, x_l, \ldots, x_n). \tag{15}$$

2. If the chain $(x_1, \ldots, x_n)$ has the property (14) *and Q is a polynomial of degree m in $\mathbb{H}[x]$ such that $y_1 \in \mathbb{H} \setminus \mathbb{R}$ is its unique zero and $y_1 \notin [x_1]$, then the polynomial of degree $n + m$, QP, has only two zeros, namely x_1 and $\overline{P}(y_1)y_1(\overline{P}(y_1))^{-1}$.*

---

[2] In [2], such a chain was called a *spherical chain*.
[3] More precisely, q is a mixed zero of P if $m_P([q]) > 0$ and $m_P(q) > m_P(q')$, for all $q' \in [q]$.

Proof. The first result follows by simple manipulation (see also [30] for a different, but equivalent result).

By Theorem 6, the polynomial Q can be written as

$$Q(x) = (x - y_m)\ldots(x - y_2)(x - y_1), \quad y_l \in [y_1], \quad y_l \neq \overline{y}_{l-1}, \quad l = 2,\ldots, m.$$

and therefore

$$T(x) := Q(x)P(x) = (x - y_m)\ldots(x - y_2)(x - y_1)(x - x_n)\ldots(x - x_2)(x - x_1).$$

It is clear that x_1 is a zero of T with multiplicity n. Multiplying both sides of last expression by $\overline{P}(x)$ and recalling (5) and (14) we obtain

$$\begin{aligned}T(x)\overline{P}(x) &= (x - y_m)\ldots(x - y_2)(x - y_1)\Psi_{x_1}^n(x) \\ &= \Psi_{x_1}^n(x)(x - y_m)\ldots(x - y_2)(x - y_1).\end{aligned}$$

Since $\Psi_{x_1}(q) \neq 0$ if $q \notin [x_1]$, we conclude that y_1 is the only zero of $T(x)\overline{P}(x)$ which is not in $[x_1]$. From $\overline{P}(y_1) \neq 0$ and (7), it follows at once that the only zero of T apart from x_1 is $\overline{P}(y_1)y_1(\overline{P}(y_1)^{-1})$.   □

The repeated use of Theorem 7 allows to classify (and identify) the zeros of a polynomial from one of its factorization.

Example 5. Consider the polynomial P associated with the chain

$$(5\mathbf{j} - 5\mathbf{k}, 5\mathbf{j}, -7\mathbf{j} + \mathbf{k}, 1 + \mathbf{i}, -3\mathbf{i} - 4\mathbf{j}, 3\mathbf{j} - 4\mathbf{k}).$$

The repeated use of (15) allows to state that this chain is equivalent to the following ones:

$$(5\mathbf{j} - 5\mathbf{k}, -5\mathbf{j} + 5\mathbf{k}, 3\mathbf{j} - 4\mathbf{k}, 1 + \mathbf{i}, -3\mathbf{i} - 4\mathbf{j}, 3\mathbf{j} - 4\mathbf{k}).$$

$$(5\mathbf{j} - 5\mathbf{k}, -5\mathbf{j} + 5\mathbf{k}, 1 - \tfrac{23}{27}\mathbf{i} - \tfrac{2}{27}\mathbf{j} - \tfrac{14}{27}\mathbf{k}, \tfrac{50}{27}\mathbf{i} + \tfrac{83}{27}\mathbf{j} - \tfrac{94}{27}\mathbf{k}, -3\mathbf{i} - 4\mathbf{j}, 3\mathbf{j} - 4\mathbf{k}).$$

Therefore

- $5\mathbf{j} - 5\mathbf{k}$ is a spherical zero of P with multiplicity 1;

- $1 - \tfrac{23}{27}\mathbf{i} - \tfrac{2}{27}\mathbf{j} - \tfrac{14}{27}\mathbf{k}$ is an isolated zero of P with multiplicity 1;

- $3\mathbf{j} - 4\mathbf{k}$ is an isolated zero of P with multiplicity 3.

Finally, one can construct a polynomial with assigned zeros, by the repeated use of the following result.

Theorem 8. *A polynomial having ζ_1 and ζ_2 as its isolated zeros of multiplicity n and m, respectively, and a sphere of zeros $[\zeta_s]$ with multiplicity k, can be constructed through the chain*

$$(\underbrace{\zeta_1,\ldots,\zeta_1}_{n}, \underbrace{\tilde{\zeta}_2,\ldots,\tilde{\zeta}_2}_{m}, \underbrace{\zeta_s,\overline{\zeta}_s,\ldots,\zeta_s,\overline{\zeta}_s}_{2k}), \tag{16}$$

where $\tilde{\zeta}_2 = \mathcal{Q}(\zeta_2)\,\zeta_2\,(\mathcal{Q}(\zeta_2))^{-1}$ and $\mathcal{Q}(x) = (x - \zeta_1)^n$.

Proof. The use of Theorem 2.8 in [13], adapted to the case of left one-sided polynomials, allows the construction of a family of polynomials with the assigned zeros, namely $P = \Psi_{\zeta_s}^k(x)(x - y_m)\ldots(x - y_1)(x - x_n)\ldots(x - x_1)$, where

- $x_1 = \zeta_1$ and x_l are chosen in $[x_1]$, so that $x_l \neq \overline{x}_{l-1}$, $l = 2,\ldots,n$. In particular, one can take $x_n = \cdots = x_1 = \zeta_1$ in line with the classical definition of multiplicity;
- $\tilde{\zeta}_2 = y_1 = \mathcal{P}_{n+1}(\zeta_2)\,\zeta_2\,(\mathcal{P}_{n+1}(\zeta_2))^{-1}$, where $\mathcal{P}_{n+1}$ is given by (12) (cf. (13));
- y_k, $k = 2,\ldots,m$ are chosen in $[y_1]$, so that $y_k \neq \overline{y}_{k-1}$. The particular choice[4] $y_m = y_{m-1} = \cdots = y_1 = \tilde{\zeta}_2$ leads to the final result.   □

Example 6. An example of a polynomial P which has $x_1 = \mathbf{i}$ as a zero of multiplicity 3, $x_2 = -1 + \mathbf{j} + \mathbf{k}$ as a zero of multiplicity 2 and $[2 + \mathbf{i}]$ as a sphere of zeros with multiplicity 2 is:

$$P(x) = \Psi_{2+\mathbf{i}}^2(x - \tilde{x}_2)^2(x - x_1)^3,$$

where $\tilde{x}_2 = \mathcal{P}_4(x_2)x_2(\mathcal{P}_4(x_2))^{-1}$ and $\mathcal{P}_4(x) = (x - x_1)^3$, i.e.

$$P(x) = (x - 2 - \mathbf{i})^2(x - 2 + \mathbf{i})^2(x + 1 + \tfrac{7}{5}\mathbf{i} + \tfrac{1}{5}\mathbf{j})^2(x - \mathbf{i})^3. \tag{17}$$

It follows immediately from the above considerations that the polynomial

$$Q(x) = (x - 2 - \mathbf{i})^2(x - 2 + \mathbf{i})^2(x + 1 - \mathbf{j} - \mathbf{k})(x + 1 + \tfrac{7}{5}\mathbf{i} + \tfrac{1}{5}\mathbf{j})(x - \mathbf{k})(x - \mathbf{j})(x - \mathbf{i})$$

solves the same problem.

4   QPolynomial

Some years ago, two of the authors of this paper introduced some new tools on the Mathematica standard package Quaternions for implementing Hamilton's Quaternion Algebra [8]. Later on, a new add-on QuaternionAnalysis [21] was developed to provide also tools for manipulating regular quaternion valued functions. This package can be downloaded at the Wolfram Library archive and full description of the QuaternionAnalysis functions as well as illustrative examples can be found in the tutorial included in [21].

QPolynomial is a collection of Mathematica functions, depending on the package QuaternionAnalysis, for treating usual problems in $\mathbb{H}[x]$: evaluation, euclidean division, greatest common divisor polynomial, construction of a polynomial with prescribed zeros, etc. In this context, a quaternionic polynomial is an object of the form Polynomial$[a_n, a_{n-1}, \ldots, a_1, a_0]$ accordingly to (1). For such objects, rules as Plus, NonCommutativeMultiply, Power and functions as Conjugate, Eval, CharacteristicPolynomial, CompanionPolynomial, etc., are defined. We summarize in Table 1 the most important functions included in QPolynomial. Auxiliary functions which come with its own usefulness are also listed in the same table. The use of these functions is illustrated by several examples.

---

[4] In [2] the same problem is also addressed and one can find a particular (different) choice of $\tilde{x}_l$ ($l = 2,\ldots,n$) and y_l ($l = 2,\ldots,m$).

Table 1. QPolynomial functions

Algebraic operations on polynomials	
Conjugate[P]	Gives the conjugate of P
CharacteristicPolynomial[q]	Gives the characteristic polynomial of q
Eval[P][q]	Gives $P(q)$
CompanionPolynomial[P]	Gives the companion polynomial of P
PolynomialDivisionL[P_1,P_2]	Gives a list with the quotient and remainder $\{Q,R\}$ of left division i.e. $P_1 = P_2Q+R$
PolynomialDivisionR[P_1,P_2]	Gives a list with the quotient and remainder $\{Q,R\}$ of right division i.e. $P_1 = QP_2+R$
PolynomialGCDL[P_1,...,P_n]	Gives the greatest common left divisor of $P_1,\ldots,P_n$
PolynomialGCDR[P_1,...,P_n]	Gives the greatest common right divisor of $P_1,\ldots,P_n$
ZerosFromChain[$\{x_1,\ldots,x_n\}$]	Gives the (isolated) zeros from a chain of a polynomial
ChainFromZeros[$\{r_1,\ldots,r_n\}$]	Gives a chain of a polynomial with prescribed (simple) zeros
ChainFromZeros[$\{\{r_1,m_1\},\ldots,\{r_p,m_p\}\}$, $\{\{s_1,n_1\},\ldots,\{s_q,n_q\}\}$]	Gives a polynomial with isolated zeros r_i and spherical zeros s_j, with multiplicities m_i and n_j, respectively

Auxiliary functions	
PSimplify[P]	Gives P with simplified coefficients
PNormalize[P]	Gives the normalized (monic) polynomial
SimilarQ[q_1,q_2]	Gives True if $[q_1] = [q_2]$
SimilarChainQ[c_1,c_2]	Gives True if the chains c_1 and c_2 are similar
SphericalQ[P, q]	Gives True if q is a spherical zero of P
FactorShift[$\{q_1,q_2\}$]	Gives a list $\{\tilde{q}_2,\tilde{q}_1\}$ such that $(x-q_1)(x-q_2)=(x-\tilde{q}_2)(x-\tilde{q}_1)$
PolynomialFromChain[$x_1,\ldots,x_n$]	Gives the Polynomial object corresponding to the polynomial $(x-x_n)\ldots(x-x_1)$

Example 7. Rules on the object Polynomial

```
In[1]:= P1 = Polynomial[1,Quaternion[2,-1,0,0]];
        P2 = Polynomial[Quaternion[0,0,0,1],Quaternion[0,0,1,1]];
        α = Quaternion[0,1,1,0];
In[4]:= P1 + P2
```
Out[4]= Polynomial[Quaternion[1, 0, 0, 1], Quaternion[2, −1, 1, 1]]

```
In[5]:= α ** P1
```
Out[5]= Polynomial[Quaternion[0, 1, 1, 0], Quaternion[1, 2, 2, 1]]

```
In[6]:= P1 ** P2
```
Out[6]= Polynomial[Quaternion[0, 0, 0, 1], Quaternion[0, 0, 2, 3],

Quaternion[0, 0, 3, 1]]

Example 8. Polynomial Functions (cf. Example 2)

```
In[1]:= P = Polynomial[1,Quaternion[2,-1,0,0],Quaternion[0,0,1,1]];
        q = Quaternion[0,1,1,0];
In[3]:= CharacteristicPolynomial[q]
```
Out[3]= Polynomial[1, 0, 2]

```
In[4]:= CompanionPolynomial[P]
```
Out[4]= Polynomial[1, 4, 5, 0, 2]

```
In[5]:= P ** Conjugate[P] // PSimplify
```
Out[5]= Polynomial[1, 4, 5, 0, 2]

Example 9. Evaluation and zeros (cf. Example 3)

```
In[6]:= Q = Polynomial[1,-q]
```
Out[6]= Polynomial[1, Quaternion[0, −1, −1, 0]]

```
In[7]:= Eval[P][q]
```
Out[7]= Quaternion[−1, 2, 3, 0]

```
In[8]:= Eval[Q][q]
```
Out[8]= 0

```
In[9]:= Eval[Q ** P][q]
```
Out[9]= Quaternion[0, 0, 0, −2]

Next example addresses issues of division in $\mathbb{H}[x]$. Both left and right Euclidean division are implemented in QPolynomial (see Theorem 1).

Example 10. Euclidean division and GCD

```
In[1]:= P1=Polynomial[1,Quaternion[1,0,2,1],-2,Quaternion[1,1,1,0]];
        P2=Polynomial[1,1,Quaternion[1,0,-1,1]];
In[3]:= {quotientR,remainderR} = PolynomialDivisionR[P1,P2]
```
Out[3]= {Polynomial[1, Quaternion[0, 0, 2, 1]],

   Polynomial[Quaternion[−3, 0, −1, −2], Quaternion[0, −2, −1, −1]]}

```
In[4]:= P1 - quotientR ** P2 - remainderR // PSimplify
```
Out[4]= 0

```
In[5]:= P1 - P2 ** quotientR - remainderR // PSimplify
```
Out[5]= Polynomial[Quaternion[0, −6, 0, 0]]

```
In[6]:= {quotientL,remainderL} = PolynomialDivisionL[P1,P2]
```
Out[6]= {Polynomial[1, Quaternion[0, 0, 2, 1]],
        Polynomial[Quaternion[−3, 0, −1, −2], Quaternion[0, 4, −1, −1]]}

```
In[7]:= P1 - quotientL ** P2 - remainderL // PSimplify
```
Out[7]= Polynomial[Quaternion[0, 6, 0, 0]]

```
In[8]:= P1 - P2 ** quotientL - remainderL // PSimplify
```
Out[8]= 0

In the quaternion case, greatest common left and right divisors are computed by a procedure similar to the classical Euclidean algorithm via PolynomialGCDL and PolynomialGCDR, respectively.

```
In[1]:= P=Polynomial[1,Quaternion[1,0,1,0]];
        Q=Polynomial[1,Quaternion[2,1,0,1]];
        R=Polynomial[1,Quaternion[-1,1,0,1]];
        S=Polynomial[1,Quaternion[0,0,1,0]];
In[5]:= PolynomialGCDR[P**Q**R,S**Q**R] // PSimplify
```
Out[5]= Polynomial[1, Quaternion[1, 2, 0, 2], Quaternion[−4, 1, 0, 1]]

```
In[6]:= PSimplify[PolynomialGCDL[P**Q**R,S**Q**R]]
```
Out[6]= 1

```
In[7]:= PolynomialGCDR[Q**P**R,S**Q**R]
```
Out[7]= Polynomial[1, Quaternion[−1, 1, 0, 1]]

```
In[8]:= PolynomialGCDL[Q**P**R,S**Q**R] // PSimplify
```
Out[8]= 1

For complex arguments, PolynomialDivisionL and PolynomialDivisionR coincide with the PolynomialQuotientRemainder built-in function and the same is true for PolynomialGCDL, PolynomialGCDR and PolynomialGCD.

```
In[1]:= P1=Polynomial[1,1]**Polynomial[1,2]**Polynomial[1,3];
        P2=Polynomial[1,-1]**Polynomial[1,2];
```

```
In[3]:= {quotientR,remainderR}=PolynomialDivisionR[P1,P2]
```
Out[3]= $\{\text{Polynomial}[1,5], \text{Polynomial}[8,16]\}$

```
In[4]:= {quotientL,remainderL}=PolynomialDivisionL[P1,P2]
```
Out[4]= $\{\text{Polynomial}[1,5], \text{Polynomial}[8,16]\}$

```
In[5]:= PolynomialQuotientRemainder[(x+1)(x+2)(x+3),(x-1)(x+2),x]
```
Out[5]= $\{x + 5, 8x + 16\}$

```
In[6]:= PolynomialGCDR[P1,P2] // PSimplify
```
Out[6]= $\text{Polynomial}[1,2]$

```
In[7]:= PolynomialGCDL[P1,P2] // PSimplify
```
Out[7]= $\text{Polynomial}[1,2]$

```
In[8]:= PolynomialGCD[(x+1)(x+2)(x+3),(x-1)(x+2)]
```
Out[8]= $x + 2$

The functions ZerosFromChain and ChainFromZeros are based on the use of Theorems 4 and 5, respectively.

Example 11. Factors from zeros and zeros from factors

```
In[1]:= l ={Quaternion[1,1,0,0],Quaternion[2,0,1,0],Quaternion[3,0,0,1]};
In[2]:= r = ZerosFromChain[l]
```
Out[2]= $\left\{\text{Quaternion}[1,1,0,0], \text{Quaternion}[2,\frac{2}{3},\frac{1}{3},\frac{2}{3}], \text{Quaternion}[3,\frac{9}{11},\frac{2}{11},\frac{6}{11}]\right\}$

```
In[3]:= P=PolynomialFromChain[l]
```
Out[3]= $\text{Polynomial}[1,\text{Quaternion}[-6,-1,-1,-1],\text{Quaternion}[11,4,5,2],$
$\qquad \text{Quaternion}[-7,-5,-5,1]]$

```
In[4]:= Table[Eval[P][r[[i]]],{i,1,3}]
```
Out[5]= $\{0,0,0\}$

```
In[5]:= c=ChainFromZeros[RotateLeft[r]]
```
Out[6]= $\left\{\text{Quaternion}[2,\frac{2}{3},\frac{1}{3},\frac{2}{3}], \text{Quaternion}[3,\frac{13}{21},-\frac{4}{21},\frac{16}{21}],\right.$

$\qquad\qquad \left.\text{Quaternion}[1,-\frac{2}{7},\frac{6}{7},-\frac{3}{7}]\right\}$

```
In[6]:= SimilarChainQ[l,c]
```
Out[8]= True

Observe that ZerosFromChain/ChainFromZeros can not be used when some of the arguments are similar. This problem can be overcome in the case of Chain-FromZeros by using an alternative argument syntax (see Example 13).

In[1]:= l={Quaternion[1,1,0,1],Quaternion[0,1,0,0],Quaternion[0,1,0,0]};

In[2]:= ZerosFromChain[l]

Out[2]= ZerosFromChain :: arg :arguments in the same similarity class.

In[3]:= ChainFromZeros[l]

Out[3]= ChainFromZeros :: arg :arguments in the same similarity class.
                     Use an alternative syntax.

Next example shows how to use the function FactorShift for successive applications of (15).

Example 12. Identifying the zeros from a given chain (cf. Example 5)

In[1]:= x1=Quaternion[0,0,5,-5];x2=Quaternion[0,0,5,0];
       x3=Quaternion[0,0,-7,1];x4=Quaternion[1,1,0,0];
       x5=Quaternion[0,-3,-4,0];x6=Quaternion[0,0,3,-4];

In[4]:= {y2,y3}=FactorShift[{x2,x3}];

In[5]:= TraditionalForm/@(c1={x1,y2,y3,x4,x5,x6})

Out[5]= $\{5j - 5k, 5k - 5j, 3j - 4k, i + 1, -3i - 4j, 3j - 4k\}$

In[6]:= {z3,z4}=FactorShift[{y3,x4}];

In[7]:= TraditionalForm/@(c2={x1,y2,z3,z4,x5,x6})

Out[7]= $\left\{5j - 5k, 5k - 5j, -\frac{23i}{27} - \frac{2j}{27} - \frac{14k}{27} + 1, \frac{50i}{27} + \frac{83j}{27} - \frac{94k}{27}, -3i - 4j, 3j - 4k\right\}$

In[8]:= (P=PolynomialFromChain[c1])==PolynomialFromChain[c2]

Out[8]= True

In[9]:= Eval[P]/@{x1,-x1,z3,y3}

Out[9]= $\{0, 0, 0, 0\}$

Our last example illustrates the use of the function ChainFromZeros which allows the construction of a polynomial with assigned zeros and corresponding multiplicities. It is based on the repeated use of Theorem 8.

Example 13. Polynomial with prescribed zeros (cf. Example 6)

In[1]:= r1=Quaternion[0,1,0,0];m1=3;
       r2=Quaternion[-1,0,1,1];m2=2;
       s1=Quaternion[2,1,0,0];mS1=3;

In[4]:= c=ChainFromZeros[{{r1, m1},{r2, m2}},{{s1, mS1}}]

Out[4]= {Quaternion[0, 1, 0, 0], Quaternion[0, 1, 0, 0], Quaternion[0, 1, 0, 0],
     Quaternion$[-1, -\frac{7}{5}, -\frac{1}{5}, 0]$, Quaternion$[-1, -\frac{7}{5}, -\frac{1}{5}, 0]$,
     Quaternion[2, 1, 0, 0], Quaternion[2, −1, 0, 0], Quaternion[2, 1, 0, 0],
     Quaternion[2, −1, 0, 0), Quaternion[2, 1, 0, 0], Quaternion[2, −1, 0, 0]}

In[5]:= P=PolynomialFromChain[c];

In[6]:= Eval[P]/@{r1,r2,s1,Conjugate[s1]}

Out[9]= {0, 0, 0, 0}

Acknowledgments. Research at CMAT was financed by Portuguese funds through Fundação para a Ciência e a Tecnologia, within the Project UID/MAT/00013/2013. Research at NIPE was carried out within the funding with COMPETE reference number POCI-01-0145-FEDER-006683 (UID/ECO/ 03182/2013), with the FCT/MEC's (Fundação para a Ciência e a Tecnologia, I.P.) financial support through national funding and by the ERDF through the Operational Programme on "Competitiveness and Internationalization - COMPETE 2020" under the PT2020 Partnership Agreement.

References

1. Beck, B.: Sur les équations polynomiales dans les quaternions. Enseign. Math. **25**, 193–201 (1979)
2. Bolotnikov, V.: Zeros and factorizations of quaternion polynomials: the algorithmic approach. arXiv:1505.03573 (2015)
3. Brenner, J.L.: Matrices of quaternions. Pacific J. Math. **1**(3), 329–335 (1951)
4. Chapman, A., Machen, C.: Standard polynomial equations over division algebras. Adv. Appl. Clifford Algebras **27**, 1065–1072 (2016). doi:10.1007/s00006-016-0740-4
5. Damiano, A., Gentili, G., Struppa, D.: Computations in the ring of quaternionic polynomials. J. Symbolic Comput. **45**(1), 38–45 (2010)
6. De Leo, S., Ducati, G., Leonardi, V.: Zeros of unilateral quaternionic polynomials. Electron. J. Linear Algebra **15**, 297–313 (2006)
7. Eilenberg, S., Niven, I.: The "fundamental theorem of algebra" for quaternions. Bull. Am. Math. Soc. **50**, 246–248 (1944)
8. Falcão, M.I., Miranda, F.: Quaternions: a Mathematica package for quaternionic analysis. In: Murgante, B., Gervasi, O., Iglesias, A., Taniar, D., Apduhan, B.O. (eds.) ICCSA 2011. LNCS, vol. 6784, pp. 200–214. Springer, Heidelberg (2011). doi:10.1007/978-3-642-21931-3_17
9. Falcão, M.I., Miranda, F., Severino, R., Soares, M.J.: Quaternionic polynomials with multiple zeros: a numerical point of view. In: AIP Conference Proceedings, vol. 1798, no. 1, p. 020099 (2017)
10. Falcão, M.I., Miranda, F., Severino, R., Soares, M.J.: Weierstrass method for quaternionic polynomial root-finding. arXiv:1702.04935 (2017)
11. Farouki, R.T., Gentili, G., Giannelli, C., Sestini, A., Stoppato, C.: A comprehensive characterization of the set of polynomial curves with rational rotation-minimizing frames. Adv. Comput. Math. **43**(1), 1–24 (2017)
12. Gentili, G., Stoppato, C.: Zeros of regular functions and polynomials of a quaternionic variable. Mich. Math. J. **56**(3), 655–667 (2008)
13. Gentili, G., Struppa, D.C.: On the multiplicity of zeroes of polynomials with quaternionic coefficients. Milan J. Math. **76**, 15–25 (2008)

14. Gordon, B., Motzkin, T.: On the zeros of polynomials over division rings I. Trans. Am. Math. Soc. **116**, 218–226 (1965)
15. Jacobson, N.: The Theory of Rings. Mathematical Surveys and Monographs. American Mathematical Society, New York (1943)
16. Janovská, D., Opfer, G.: The classification and the computation of the zeros of quaternionic, two-sided polynomials. Numer. Math. **115**(1), 81–100 (2010)
17. Janovská, D., Opfer, G.: A note on the computation of all zeros of simple quaternionic polynomials. SIAM J. Numer. Anal. **48**(1), 244–256 (2010)
18. Kalantari, B.: Algorithms for quaternion polynomial root-finding. J. Complex. **29**(3–4), 302–322 (2013)
19. Lam, T.Y.: A First Course in Noncommutative Rings. Graduate Texts in Mathematics. Springer, New York (1991)
20. Lianggui, F., Kaiming, Z.: Classifying zeros of two-sided quaternionic polynomials and computing zeros of two-sided polynomials with complex coefficients. Pacific J. Math **262**(2), 317–337 (2013)
21. Miranda, F., Falcão, M.I.: Modified quaternion newton methods. In: Murgante, B., Misra, S., Rocha, A.M.A.C., Torre, C., Rocha, J.G., Falcão, M.I., Taniar, D., Apduhan, B.O., Gervasi, O. (eds.) ICCSA 2014. LNCS, vol. 8579, pp. 146–161. Springer, Cham (2014). doi:10.1007/978-3-319-09144-0_11
22. Niven, I.: Equations in quaternions. Am. Math. Monthly **48**, 654–661 (1941)
23. Ore, O.: Theory of non-commutative polynomials. Ann. Math. **34**(3), 480–508 (1933)
24. Pereira, R.: Quaternionic Polynomials and Behavioral Systems. Ph. D. Thesis, Universidade de Aveiro (2006)
25. Pereira, R., Rocha, P.: On the determinant of quaternionic polynomial matrices and its application to system stability. Math. Methods Appl. Sci. **31**(1), 99–122 (2008)
26. Pereira, R., Rocha, P., Vettori, P.: Algebraic tools for the study of quaternionic behavioral systems. Linear Algebra Appl. **400**(1–3), 121–140 (2005)
27. Pogorui, A., Shapiro, M.: On the structure of the set of zeros of quaternionic polynomials. Complex Variables Theor. Appl. **49**(6), 379–389 (2004)
28. Pumplün, S., Walcher, S.: On the zeros of polynomials over quaternions. Commun. Algebra **30**(8), 4007–4018 (2002)
29. Serôdio, R., Pereira, E., Vitória, J.: Computing the zeros of quaternion polynomials. Comput. Math. Appl. **42**(8–9), 1229–1237 (2001)
30. Serôdio, R., Siu, L.S.: Zeros of quaternion polynomials. Appl. Math. Lett. **14**(2), 237–239 (2001)
31. Topuridze, N.: On the roots of polynomials over division algebras. Georgian Math. J. **10**(4), 745–762 (2003)
32. Xu, W., Feng, L., Yao, B.: Zeros of two-sided quadratic quaternion polynomials. Adv. Appl. Clifford Algebras **24**(3), 883–902 (2014)

Shifted Generalized Pascal Matrices in the Context of Clifford Algebra-Valued Polynomial Sequences

Isabel Cação[1], Helmuth R. Malonek[1], and Graça Tomaz[1,2(✉)]

[1] Centro de Investigação e Desenvolvimento em Matemática e Aplicações, Universidade de Aveiro, 3810-193 Aveiro, Portugal
{isabel.cacao,hrmalon}@ua.pt
[2] Unidade de Investigação para o Desenvolvimento do Interior, Instituto Politécnico da Guarda, 6300-559 Guarda, Portugal
gtomaz@ipg.pt

Abstract. The paper shows the role of shifted generalized Pascal matrices in a matrix representation of hypercomplex orthogonal Appell systems. It extends results obtained in previous works in the context of Appell sequences whose first term is a real constant to sequences whose initial term is a suitable chosen polynomial of n variables.

Keywords: Shifted generalized Pascal matrix · Generalized Appell polynomials · Matrix representation

1 Introduction

The role of the so-called *creation matrix* H defined by

$$(H)_{il} = \begin{cases} i, & i = l + 1 \\ 0, & i \neq l + 1, \end{cases} \qquad i, l = 0, 1, \ldots, m,$$

has been highlighted in [2] as the main tool for the matrix representation of Appell polynomial sequences of one real variable and its extension to the representation of Sheffer sequences (cf. [1]). The importance of the creation matrix was also confirmed in [3] where the authors developed the matrix representation of homogeneous Appell polynomials that are solutions of a generalized Cauchy-Riemann system in Euclidean spaces of arbitrary dimensions $n > 2$. In that work the first term of the considered sequence is a real constant. The consideration of a suitable polynomial in spaces of dimension less than n as first term leads to more general Appell polynomials (see [16]). Those polynomials also appear in the approach used to construct Gelfand-Tsetlin bases related to branching techniques ([7,15]). The construction process gives relevance to polynomials obtained by shifting coefficients of the sequence constructed in [13] and reveals their fundamental importance (cf. [10]). The present paper focuses mostly on the representation of those polynomials and stresses that besides H, also the *shift matrix* defined by

© Springer International Publishing AG 2017
O. Gervasi et al. (Eds.): ICCSA 2017, Part II, LNCS 10405, pp. 409–421, 2017.
DOI: 10.1007/978-3-319-62395-5_28

$$(J)_{il} = \begin{cases} 1, & i = l+1 \\ 0, & i \neq l+1, \end{cases} \qquad i,l = 0,\ldots,m, \tag{1}$$

plays a significant role. We notice that the pre-multiplication of J by a matrix A results in a matrix whose entries are those of A shifted downward by one position, with zeros in the first row. Also, the post-multiplication causes in A a left shift by one position, with zeros appearing in the last column.

The paper is organized as follows. In Sect. 2 we recall some notions useful in the sequel, namely basic concepts on Clifford analysis and the definitions of generalized and shifted Pascal matrices. In Sect. 3 a matrix approach to the general orthogonal Appell sequences is proposed and some examples are given. A block matrix representation for general systems of orthogonal Appell polynomials is deduced in Sect. 4. We finalize the paper with some conclusions in Sect. 5.

2   Basic Notions

Let $\{e_1, e_2, \ldots, e_n\}$ be an orthonormal basis of the Euclidean vector space $\mathbb{R}^n$ endowed with a non-commutative product according to the multiplication rules

$$e_i e_j + e_j e_i = -2\delta_{ij}, \quad i,j = 1,2,\ldots,n,$$

where δ_{ij} is the Kronecker symbol. A basis for the associative 2^n−dimensional real Clifford algebra $\mathcal{C}\ell_{0,n}$ is the set $\{e_A : A \subseteq \{1,\ldots,n\}\}$ with

$$e_A = e_{h_1} e_{h_2} \ldots e_{h_r}, \quad 1 \leq h_1 < \cdots < h_r \leq n, \quad e_\emptyset = e_0 = 1.$$

In general, the vector space $\mathbb{R}^{n+1}$ is embedded in $\mathcal{C}\ell_{0,n}$ by identifying the element $(x_0, x_1, \ldots, x_n) \in \mathbb{R}^{n+1}$ with the element

$$x = x_0 + \sum_{k=1}^{n} e_k x_k = x_0 + \underline{x} \in \mathcal{A}_n := \mathrm{span}_{\mathbb{R}}\{1, e_1, \ldots, e_n\} \subset \mathcal{C}\ell_{0,n}.$$

The conjugate $\bar{x}$ and the norm $|x|$ of x are given by $\bar{x} = x_0 - \underline{x}$ and $|x| = (x\bar{x})^{1/2} = (\bar{x}x)^{1/2} = \left(\sum_{k=0}^{n} x_k^2\right)^{1/2}$, respectively.

The generalized Cauchy-Riemann operator in $\mathbb{R}^{n+1}$ is defined by

$$\overline{\partial} := \frac{1}{2}(\partial_0 + \partial_{\underline{x}}),$$

with $\partial_0 := \frac{\partial}{\partial x_0}$ and $\partial_{\underline{x}} := \sum_{k=1}^{n} e_k \frac{\partial}{\partial x_k}$. The conjugate generalized Cauchy-Riemann operator, also called the hypercomplex differential operator, is denoted by

$$\partial := \frac{1}{2}(\partial_0 - \partial_{\underline{x}}).$$

We consider $\mathcal{C}\ell_{0,n}$−valued functions defined in an open subset $\Omega \subseteq \mathbb{R}^{n+1} \cong \mathcal{A}_n$, i.e., functions of the form $f(z) = \sum_A f_A(z) e_A$ with $f_A(z)$ real valued.

A function f is called *left (right) monogenic* in Ω if it is a solution of the differential equation $\overline{\partial} f = 0$ $(f\overline{\partial} = 0)$. From now on we will only use left monogenic functions which we refer to as monogenic functions (right monogenic functions are treated analogously).

The hypercomplex differentiability as generalization of complex differentiability has to be understood in the following way (see [14]): a function f defined in an open domain $\Omega \subseteq \mathbb{R}^{n+1}$ is hypercomplex differentiable supposing there exists in each point of Ω a uniquely defined areolar derivative f'. Then f is real differentiable and $f' := \partial f$. Furthermore, f is hypercomplex differentiable in Ω if and only if f is monogenic. In addition, the monogenicity of f implies that $f' = \partial_0 f = -\partial_{\underline{x}} f$.

Further on we will consider hypercomplex polynomials orthogonal with respect to the Clifford algebra-valued inner product

$$(f,g)_{\mathcal{C}\ell_{0,n}} = \int_{B^{n+1}} \overline{f} g \, d\lambda^{n+1}, \tag{2}$$

where λ^{n+1} is the Lebesgue measure in $\mathbb{R}^{n+1}$ and $\overline{f}$ the conjugate of $f \in \mathcal{C}\ell_{0,n}$.

With regard to the use of matrices in the context of this work, we remember that the well-known Pascal matrix P, considered as a lower triangular matrix, contains the binomial coefficients as its non-zero entries, i.e.,

$$(P)_{il} = \begin{cases} \binom{i}{l}, & i \geq l \\ 0, & i < l, \end{cases} \quad i,l = 0,\ldots,m.$$

Such matrix is an essential tool in many applications, particularly in matrix theory and combinatorics. Since all Appell polynomials satisfy a binomial theorem (cf. [12]), it becomes clear that the Pascal matrix plays a central role in the matrix representation of such polynomials. In order to achieve a matrix representation of general systems of orthogonal Appell polynomials we introduce the following matrix:

Definition 1. *The matrix* $S_r(x), r \in \mathbb{N}_0$, *with entries given by*

$$(S_r(x))_{il} = \begin{cases} \binom{i+r}{l+r} x^{i-l}, & i \geq l \\ 0, & i < l, \end{cases} \quad i,l = 0,\ldots,m \tag{3}$$

is called shifted generalized Pascal matrix.

It is easy to see that the matrix (3) can be written as an exponential matrix expressed jointly by the creation matrix H and the shift matrix J:

$$S_r(x) = e^{(H+rJ)x}. \tag{4}$$

Since $S_0(1) = e^H = P$ is the Pascal matrix of order $m+1$ and $S_0(x) = e^{Hx} = P(x)$ the generalized Pascal matrix of the same order, Definition 1 includes both special cases (see [5]). But for variable r, and values of x equal 0 or 1, it is evident that $S_r(0)$ and $S_r(1)$ coincide with the identity matrix I_{m+1} and the shifted Pascal matrix S_r, respectively (see [4]).

Considering the diagonal matrix Λ_r whose non-zero entries are

$$\binom{r+l}{r}, \quad l = 0, \ldots, m, \tag{5}$$

the following relation between $S_r(x)$ and $P(x)$ can be easily established (cf. [5]):

$$\Lambda_r^{-1} S_r(x) \Lambda_r = P(x). \tag{6}$$

Of course, formula (6) shows that $P(x)$ and $S_r(x)$ are similar matrices for given values of m and r as the chosen shift parameter. Moreover, since $P^{-1}(x) = P(-x)$ (cf. [11]), from (6) it follows immediately that the inverse of the shifted generalized Pascal matrix is given by

$$S_r^{-1}(x) = S_r(-x). \tag{7}$$

Remark 1. All the matrices introduced above could be defined as infinite dimensional square matrices in correspondence with infinite sequences of Appell polynomials. However, we restrict ourselves to the representation of vectors of polynomials up to a certain degree and, consequently, we only need to consider square matrices of order $m + 1$.

3   Matrix Approach to Orthogonal Hypercomplex Appell Sequences

Appell's sequence definition, derived from the original work of P. Appell [6], has been adapted to the Clifford analysis framework in the following way (see [13]):

Definition 2. *A sequence of homogeneous $\mathcal{C}\ell_{0,n}$-valued polynomials $\{\phi_k^{(n)}(x)\}_{k \geq 0}$ in the variable $x \in \mathcal{A}_n$ is called an hypercomplex Appell sequence if, for each $k \geq 0$, the following conditions are verified:*

(i) $\phi_k^{(n)}(x)$ has exact degree k;
(ii) $\bar{\partial}\, \phi_k^{(n)}(x) = 0$ for all $x \in \mathcal{A}_n$;
(iii) $\partial\, \phi_k^{(n)}(x) = k\phi_{k-1}^{(n)}(x), \quad k \geq 1$.

The approach developed in [3] on the matrix representation of hypercomplex Appell sequences shows that, like in the case of real Appell sequences ([2]), the creation matrix H is the main tool for carrying out that representation.

In fact, the n-dimensional homogeneous polynomials

$$\phi_k^{(n)}(x) = \sum_{j=0}^{k} \binom{k}{j} c_j(n)\underline{x}^j x_0^{k-j}, \quad k = 0, 1, \ldots, \tag{8}$$

whose coefficients $c_j(n)$ verify

$$c_0(n) \neq 0, \quad c_{2k}(n) = c_{2k-1}(n) = \frac{\left(\frac{1}{2}\right)_k}{\left(\frac{n}{2}\right)_k} c_0(n), \qquad n > 1; \, k = 1, 2, \ldots \quad (9)$$

$((.)_r$ denotes the Pochhammer symbol defined by $(a)_r = \frac{\Gamma(a+r)}{\Gamma(a)}, \, r \geq 0)$, fulfil all the conditions of the Definition 2, i.e., they form a hypercomplex Appell sequence.

Introducing the vector $\boldsymbol{\phi}^{(n)}(x) = [\phi_0^{(n)}(x) \ \phi_1^{(n)}(x) \ \cdots \ \phi_m^{(n)}(x)]^T$, as consequence of the condition *(iii)* of the referred definition, we obtain the differential equation

$$\partial_0 \boldsymbol{\phi}^{(n)}(x) = H \boldsymbol{\phi}^{(n)}(x)$$

which together with the vector of initial values on the hyperplane $x_0 = 0$, $\boldsymbol{\phi}^{(n)}(0, \underline{x}) = [c_0 \underline{x}^0 \ c_1 \underline{x} \cdots c_m \underline{x}^m]^T$, has the solution

$$\boldsymbol{\phi}^{(n)}(x) = e^{Hx_0} \boldsymbol{\phi}^{(n)}(0, \underline{x}) \qquad (10)$$
$$= P(x_0) \boldsymbol{\phi}^{(n)}(0, \underline{x}).$$

Furthermore, introducing the diagonal matrix

$$D_{c(n)} = \mathrm{diag}[c_0(n) \ c_1(n) \cdots c_m(n)], \qquad (11)$$

and the vector $\xi(\underline{x}) = [1 \ \underline{x} \cdots \underline{x}^m]^T$, the vector of Appell polynomials (10) becomes

$$\boldsymbol{\phi}^{(n)}(x) = e^{Hx_0} D_{c(n)} \xi(\underline{x}). \qquad (12)$$

Extensions of hypercomplex Appell polynomial sequences can be achieved by considering their first terms, not only a real or complex constant, but as an arbitrary chosen monogenic polynomial $Q_j(\underline{x})$, of a fixed degree $j = 0, 1, \ldots$. Since these polynomials are independent of x_0, their hypercomplex derivative is the constant zero, i.e., they behave under differentiation like an ordinary constant number. For that reason they are usually called generalized constants or monogenic constants.

With that starting point, the reached sequences include all orthogonal polynomial sequences obtained in the framework of the Gelfand-Tsetlin branching approach ([7,15]). The referred orthogonal polynomials are constructed as follows.

Considering, for each integer k and each $j = 0, \ldots, k$, the homogeneous polynomial $\tilde{X}_{n+1,j}^{(k)}(x)$ of degree k and index j resulting from the product of a, in general, non-monogenic polynomial of the form

$$X_{n+1,j}^{(k-j)}(x) = \sum_{s=0}^{k-j} \binom{k}{j+s} d_{j,s}(n) \, \underline{x}^s \, x_0^{k-j-s} \qquad (13)$$

of degree $k - j$, by a homogeneous $\mathcal{C}\ell_{0,n}$-valued monogenic polynomial $Q_j(\underline{x})$ of degree j, we obtain (see [10])

$$
\begin{aligned}
\tilde{X}_{n+1,j}^{(k)}(x) &:= X_{n+1,j}^{(k-j)}(x)Q_j(\underline{x}) \\
&= \sum_{s=0}^{k-j} \binom{k}{j+s} d_{j,s}(n)\, \underline{x}^s\, x_0^{k-j-s}\, Q_j(\underline{x}), \quad x \in \mathcal{A}_n,
\end{aligned}
\tag{14}
$$

where $d_{j,s}(n)$ are real constants.

The monogenicity of $\tilde{X}_{n+1,j}^{(k)}(x)$ depends on the values of the coefficients $d_{j,s}(n)$. In order to obtain such values, we use the following Lemma.

Lemma 1. *For each fixed j $(j = 0,1,\ldots)$, let $Q_j(\underline{x})$ be a homogeneous $\mathcal{C}\ell_{0,n}$-valued monogenic polynomial of degree j, defined in $\mathbb{R}^n$. Then*

$$
\partial_{\underline{x}}(\underline{x}^s Q_j(\underline{x})) = \begin{cases} -s\underline{x}^{s-1}Q_j(\underline{x}), & s \text{ even} \\ -(n + 2j + s - 1)\underline{x}^{s-1}Q_j(\underline{x}), & s \text{ odd}, \end{cases} \quad s=0,1,2,\ldots.
$$

Proof. For $s = 0$ the result is immediate due to the monogenicity of $Q_j(\underline{x})$. For $s = 1$, by applying the *generalized Leibniz rule* [8] to the product $\underline{x}\, Q_j(\underline{x})$ with respect to the operator ∂, one has

$$
-\partial_{\underline{x}}(\underline{x}\, Q_j(\underline{x})) = nQ_j(\underline{x}) + 2\mathbb{E}_n Q_j(\underline{x}) = (n + 2j)Q_j(\underline{x}),
$$

where $\mathbb{E}_n = \sum_{k=1}^{n} x_k \dfrac{\partial}{\partial x_k}$ is the Euler operator in $\mathbb{R}^n$. Repeatedly applying the mentioned rule, the result follows by induction on s. $\qquad\square$

Theorem 1. *A polynomial of the form (14) is monogenic if and only if the coefficients $d_{j,s}(n)$ verify the following relations:*

$$
d_{j,s}(n) = \begin{cases} \dfrac{j+s}{n+2j+s-1}d_{j,s-1}(n), & s \text{ odd} \\[2mm] \dfrac{j+s}{s}d_{j,s-1}(n), & s \text{ even}, \end{cases}
\tag{15}
$$

with $d_{j,0}(n) \neq 0$.

Proof. Making use of the Lemma 1, the action of the operator $\bar{\partial} = \partial_0 + \partial_{\underline{x}}$ on $\tilde{X}_{n+1,j}^{(k)}(x)$ results in

$$
\begin{aligned}
\bar{\partial}\tilde{X}_{n+1,j}^{(k)}(x) = & \sum_{s=0}^{k-j-1} \binom{k}{j+s} d_{j,s}(n)\,(k-j-s)x_0^{k-j-s-1}\, \underline{x}^s\, Q_j(\underline{x}) \\
& - \sum_{s=1}^{\lfloor \frac{k-j}{2} \rfloor} \binom{k}{j+2s} d_{j,2s}(n)\, x_0^{k-j-2s}(2s)\, \underline{x}^{2s-1}\, Q_j(\underline{x}) \\
& - \sum_{s=0}^{\lfloor \frac{k-j-1}{2} \rfloor} \binom{k}{j+2s+1} d_{j,2s+1}(n)\, x_0^{k-j-2s-1}(n+2j+2s)\, \underline{x}^{2s}\, Q_j(\underline{x}),
\end{aligned}
$$

where $\lfloor . \rfloor$ is the floor function.

Splitting the odd and the even powers of $\underline{x}$ on the first sum, and gathering the same powers of x_0 and $\underline{x}$, $\bar{\partial}\tilde{X}_{n+1,j}^{(k)}(x)$ becomes

$$
\bar{\partial}\tilde{X}_{n+1,j}^{(k)}(x) = \sum_{s=0}^{\lfloor \frac{k-j-1}{2} \rfloor} \left[\binom{k}{j+2s}(k-j-2s)d_{j,2s}(n) \right.
$$
$$
\left. - \binom{k}{j+2s+1}(n+2j+2s)d_{j,2s+1}(n) \right] x_0^{k-j-2s-1}\underline{x}^{2s}\,Q_j(\underline{x})
$$
$$
+ \sum_{s=0}^{\lfloor \frac{k-j}{2}-1 \rfloor} \left[\binom{k}{j+2s+1}(k-j-2s-1)d_{j,2s+1}(n) \right.
$$
$$
\left. - \binom{k}{j+2s+2}(2s+2)d_{j,2s+2}(n) \right] x_0^{k-j-2s-2}\underline{x}^{2s+1}\,Q_j(\underline{x}).
$$

The polynomial $\tilde{X}_{n+1,j}^{(k)}(x)$ is monogenic if and only if $\bar{\partial}\tilde{X}_{n+1,j}^{(k)}(x) = 0$, i.e.,

$$
\binom{k}{j+2s}(k-j-2s)d_{j,2s}(n) - \binom{k}{j+2s+1}(n+2j+2s)d_{j,2s+1}(n) = 0 \quad (16)
$$

and

$$
\binom{k}{j+2s+1}(k-j-2s-1)d_{j,2s+1}(n) - \binom{k}{j+2s+2}(2s+2)d_{j,2s+2}(n) = 0. \quad (17)
$$

Simplifying (16) and (17) we get, respectively,

$$
(j+2s+1)d_{j,2s}(n) = (n+2j+2s)d_{j,2s+1}(n),
$$

and

$$
(j+2s+2)d_{j,2s+1}(n) = (2s+2)d_{j,2s+2}(n),
$$

and the result follows. $\qquad\square$

Notice that, from (15) we can also obtain the recurrence relations, for each fixed $j \geq 0$:

$$
d_{j,2s}(n) = \frac{(j+2s)!(n+2j-2)!!}{j!(2s)!!(n+2j+2s-2)!!}d_{j,0}(n); \quad (18)
$$
$$
d_{j,2s-1}(n) = \frac{(j+2s-1)!(n+2j-2)!!}{j!(2s-2)!!(n+2j+2s-2)!!}d_{j,0}(n), \quad s = 1,2,\dots. \quad (19)
$$

Corollary 1. *For each fixed $j \in \mathbb{N}_0$, the sequence $\{\tilde{X}_{n+1,j}^{(k)}(x)\}_{k \geq j}$ formed by the polynomials (14) whose coefficients verify (15) is an Appell sequence.*

Proof. From the guarantee of monogenecity of the polynomials (14) given by the Theorem 1, and by applying ∂_0 to those polynomials it follows,

$$\partial_0 \tilde{X}_{n+1,j}^{(j)}(x) = 0,$$
$$\partial_0 \tilde{X}_{n+1,j}^{(k)}(x) = k\tilde{X}_{n+1,j}^{(k-1)}(x), \quad k > j,$$

and the proof is complete. □

The property mentioned in the previous corollary was already used in [9] and [10].

Remark 2. The system $\{\tilde{X}_{n+1,j}^{(k)}(x); j = 0, \ldots, k\}_{k \geq 0}$, besides being formed by Appell polynomials, is orthogonal with respect to the inner product (2) (see [9] and [10], for details).

According to the previous corollary, the sequence $\{\tilde{X}_{n+1,j}^{(r+j)}(x)\}_{r \geq 0} \equiv \{\tilde{X}_{n+1,j}^{(k)}(x)\}_{k \geq j}$ verifies the Appell property

$$\partial_0 \tilde{X}_{n+1,j}^{(r+j)}(x) = \begin{cases} 0, & r = 0 \\ (r+j)\tilde{X}_{n+1,j}^{(r+j-1)}(x), & r > 0. \end{cases} \tag{20}$$

Noting that the initial values of $\tilde{X}_{n+1,j}^{(r+j)}(x)$ on the hyperplane $x_0 = 0$ are

$$\tilde{X}_{n+1,j}^{(r+j)}(0, \underline{x}) = d_{j,r}(n)\underline{x}^r Q_j(\underline{x}), \quad r \geq 0,$$

and denoting the vector

$$\tilde{\boldsymbol{X}}_j(x) = [\tilde{X}_{n+1,j}^{(j)}(x) \ \tilde{X}_{n+1,j}^{(j+1)}(x) \cdots \tilde{X}_{n+1,j}^{(j+m)}(x)]^T, \tag{21}$$

the corresponding vector of the initial values is

$$\tilde{\boldsymbol{X}}_j(0, \underline{x}) = \mathcal{D}_j \, \xi_j(\underline{x}), \tag{22}$$

where

$$\mathcal{D}_j = \text{diag}[d_{j,0}(n) \ d_{j,1}(n) \cdots d_{j,m}(n)], \tag{23}$$

and

$$\xi_j(\underline{x}) = [Q_j(\underline{x}) \ \underline{x} \, Q_j(\underline{x}) \ \cdots \ \underline{x}^m \, Q_j(\underline{x})]^T. \tag{24}$$

Now, we can establish the following result.

Theorem 2. *For each fixed j, the vector $\tilde{\boldsymbol{X}}_j(x)$ can be decomposed in the form*

$$\tilde{\boldsymbol{X}}_j(x) = S_j(x_0) \, \mathcal{D}_j \, \xi_j(\underline{x}). \tag{25}$$

Proof. According to the property (20), $\tilde{\boldsymbol{X}}_j(x)$ verifies the vector differential equation

$$\partial_0 \tilde{\boldsymbol{X}}_j(x) = (H + jJ)\tilde{\boldsymbol{X}}_j(x)$$

whose general solution is

$$\tilde{\boldsymbol{X}}_j(x) = e^{(H+jJ)x_0} \tilde{\boldsymbol{X}}_j(0, \underline{x}).$$

From (4) and (22) the result follows straightforwardly. □

Remark 3. We notice that, the matrix representation of the vector $\boldsymbol{X}_j(x) = [X_{n+1,j}^{(0)}(x)\ X_{n+1,j}^{(1)}(x)\ \cdots\ X_{n+1,j}^{(m)}(x)]^T$ of the polynomials (13) is similar to (25); the difference consists in the vector $\xi_j(\underline{x})$ which in this case corresponds to $\xi_0(\underline{x}) \equiv \xi(\underline{x}) = [1\ \underline{x}\ \cdots\ \underline{x}^m]^T$, i.e.,

$$\boldsymbol{X}_j(x) = S_j(x_0)\,\mathcal{D}_j\,\xi(\underline{x}). \tag{26}$$

Moreover, when $j = 0$, if $d_{0,0}(n) = c_0(n)$, it follows that $\mathcal{D}_0 \equiv D_{c(n)}$ and $\tilde{\boldsymbol{X}}_0(x) \equiv \boldsymbol{X}_0(x) \equiv \phi^{(n)}(x)$.

Example 1. In order to display an example for $n = 2$, we consider the monogenic polynomials $Q_j(\underline{x})$ coincident with the polynomials $\zeta^j := (x_1 - e_1 e_2 x_2)^j = (-e_1\underline{x})^j$ used in [15] for the construction of an orthogonal basis for the space of homogeneous monogenic polynomials with values in $\mathcal{Cl}_{0,n}$. In this case, by means of (25) and choosing $d_{j,0}(n) = 1$, we obtain the first polynomials for $j = 1$ and $j = 2$:

$$\tilde{\boldsymbol{X}}_1(x) = S_1(x_0)\,\mathcal{D}_1\,\xi_1(\underline{x})$$

$$\Leftrightarrow
\begin{bmatrix}
\tilde{X}_{3,1}^{(1)}(x) \\
\tilde{X}_{3,1}^{(2)}(x) \\
\tilde{X}_{3,1}^{(3)}(x) \\
\tilde{X}_{3,1}^{(4)}(x)
\end{bmatrix}
=
\begin{bmatrix}
1 & 0 & 0 & 0 \\
2x_0 & 1 & 0 & 0 \\
3x_0^2 & 3x_0 & 1 & 0 \\
4x_0^3 & 6x_0^2 & 4x_0 & 1
\end{bmatrix}
\begin{bmatrix}
1 & 0 & 0 & 0 \\
0 & \frac{1}{2} & 0 & 0 \\
0 & 0 & \frac{3}{4} & 0 \\
0 & 0 & 0 & \frac{1}{2}
\end{bmatrix}
\begin{bmatrix}
\zeta \\
\underline{x}\zeta \\
\underline{x}^2\zeta \\
\underline{x}^3\zeta
\end{bmatrix}$$

$$=
\begin{bmatrix}
\zeta \\
2x_0\zeta + \frac{1}{2}\underline{x}\zeta \\
3x_0^2\zeta + \frac{3}{2}x_0\underline{x}\zeta + \frac{3}{4}\underline{x}^2\zeta \\
4x_0^3\zeta + 3x_0^2\underline{x}\zeta + 3x_0\underline{x}^2\zeta + \frac{1}{2}\underline{x}^3\zeta
\end{bmatrix},$$

$$\tilde{\boldsymbol{X}}_2(x) = S_2(x_0)\,\mathcal{D}_2\,\xi_2(\underline{x})$$

$$\Leftrightarrow
\begin{bmatrix}
\tilde{X}_{3,2}^{(2)}(x) \\
\tilde{X}_{3,2}^{(3)}(x) \\
\tilde{X}_{3,2}^{(4)}(x) \\
\tilde{X}_{3,2}^{(5)}(x)
\end{bmatrix}
=
\begin{bmatrix}
1 & 0 & 0 & 0 \\
3x_0 & 1 & 0 & 0 \\
6x_0^2 & 4x_0 & 1 & 0 \\
10x_0^3 & 10x_0^2 & 5x_0 & 1
\end{bmatrix}
\begin{bmatrix}
1 & 0 & 0 & 0 \\
0 & \frac{1}{2} & 0 & 0 \\
0 & 0 & 1 & 0 \\
0 & 0 & 0 & \frac{5}{8}
\end{bmatrix}
\begin{bmatrix}
\zeta^2 \\
\underline{x}\zeta^2 \\
\underline{x}^2\zeta^2 \\
\underline{x}^3\zeta^2
\end{bmatrix}$$

$$=
\begin{bmatrix}
\zeta^2 \\
3x_0\zeta^2 + \frac{1}{2}\underline{x}\zeta^2 \\
6x_0^2\zeta^2 + 2x_0\underline{x}\zeta^2 + \underline{x}^2\zeta^2 \\
10x_0^3\zeta^2 + 5x_0^2\underline{x}\zeta^2 + 5x_0\underline{x}^2\zeta^2 + \frac{5}{8}\underline{x}^3\zeta^2
\end{bmatrix}.$$

Theorem 3. *Let* $\phi^{(n+2j)}(x) = [\phi_0^{(n+2j)}(x)\ \phi_1^{(n+2j)}(x)\cdots\phi_m^{(n+2j)}(x)]^T$ *be the vector of polynomials (8) on the parameter* $n + 2j$, *for each fixed* j. *Then*

$$\boldsymbol{X}_j(x) = \Lambda_j\,\phi^{(n+2j)}(x).$$

Proof. Rewriting coefficients (9) in the form

$$c_{2s} = c_{2s-1} = \frac{(2s-1)!!(n-2)!!}{(n+2s-2)!!} c_0(n), \quad s = 1, 2, \ldots,$$

and comparing with (18) and (19), we have

$$d_{j,s}(n) = \binom{j+s}{j} c_s(n+2j), \quad s = 0, 1, \ldots.$$

By considering the matrix Λ_j as defined in (5), the diagonal matrix (11) on the parameter $n + 2j$, and (23), the matrix form of that equality is

$$\mathcal{D}_j = \Lambda_j D_{c(n+2j)}.$$

Then, taking into account (6), (12), and (26) we have

$$\boldsymbol{X}_j(x) = S_j(x_0)\,\mathcal{D}_j\,\xi(\underline{x}) = \Lambda_j P(x_0)\Lambda_j^{-1}\Lambda_j D_{c(n+2j)}\,\xi(\underline{x}) = \Lambda_j \phi^{(n+2j)}(x)$$

and the proof is complete.   □

Remark 4. We observe that the result stated in Theorem 3 converts in matrix form the property established in Theorem 4.1 of [10].

Since the matrix $S_j(x_0)\mathcal{D}_j$ of (26) is non-singular, powers of $\underline{x}$ may be expressed in terms of the polynomials $X_{n+1,j}^{(k-j)}(x)$. More concretely, we get the following theorem.

Theorem 4. *Let $\xi(\underline{x})$ the vector of powers of $\underline{x}$. Then*

$$\xi(\underline{x}) = \mathcal{D}_j^{-1} S_j(-x_0)\,\boldsymbol{X}_j(x). \tag{27}$$

Proof. Considering (7) together with (26) the proof is immediate.   □

Noting that the matrix $\mathcal{D}_j^{-1} S_j(-x_0)$ is defined by

$$(\mathcal{D}_j^{-1} S_j(-x_0))_{il} = \begin{cases} \binom{i+j}{l+j} \frac{(-x_0)^{i-l}}{d_{j,i}(n)}, & i \geq l \\ 0, & i < l, \end{cases} \quad i, l = 0, \ldots, m,$$

(27) is equivalent to

$$\sum_{l=0}^{r} \binom{r+j}{l+j} \frac{(-x_0)^{r-l}}{d_{j,r}(n)}\, X_{n+1,j}^{(l)}(x) = \underline{x}^r, \quad r = 0, 1, \ldots, m.$$

As consequence, the recurrence relation

$$X_{n+1,j}^{(r)}(x) = d_{j,r}(n)\underline{x}^r - \sum_{l=0}^{r-1} \binom{r+j}{l+j}(-x_0)^{r-l}\, X_{n+1,j}^{(l)}(x), \quad r = 0, 1, \ldots$$

for the polynomials (13) is obtained.

4   Block Matrix Representation of a Global Recursive Construction Scheme

Based on the previous section, it is possible to obtain a block matrix representation for the whole orthogonal system $\{X^{(k-j)}_{n+1,j}(x)Q_j(\underline{x}); \; j = 0,\ldots,k\}_{k\geq 0}$.

Setting $k = 0,\ldots,m$, the truncated system consists of the non-vanishing entries of the lower triangular matrix $\tilde{\mathcal{X}}$ defined by

$$(\tilde{\mathcal{X}})_{kj} = \begin{cases} X^{(k-j)}_{n+1,j}(x)Q_j(\underline{x}), & k \geq j \\ 0, & k < j, \end{cases} \quad k,j = 0,\ldots,m.$$

It means that its r-th column, $0 \leq r \leq m$, is

$$(\tilde{\mathcal{X}})_r = J^r \, \tilde{\boldsymbol{X}}_r(x),$$

where $\tilde{\boldsymbol{X}}_r(x) = [\tilde{X}^{(r)}_{n+1,r}(x) \; \tilde{X}^{(r+1)}_{n+1,r}(x) \cdots \tilde{X}^{(r+m)}_{n+1,r}(x)]^T$ as defined in (21) and

$$(J^r)_{kj} = \begin{cases} 1 & k = j + r \\ 0, & k \neq j + r, \end{cases} \quad k,j = 0,\ldots,m, \tag{28}$$

are the powers of the shift matrix (1).

Theorem 5. *Let* $\mathcal{D} = \mathrm{diag}[\mathcal{D}_0 \; \mathcal{D}_1 \cdots \mathcal{D}_m]$, $\mathcal{J} = [J^0 \; J^1 \cdots J^m]^T$, *and* $\mathcal{E} = \mathrm{diag}[\xi_0(\underline{x}) \; \xi_1(\underline{x}) \cdots \xi_m(\underline{x})]^T$ *be the block matrices where* $\mathcal{D}_r$, $\xi_r(\underline{x})$, J^r, $r = 0,\ldots,m$, *are defined by (23), (24), and (28), respectively. Considering the generalized Pascal matrix* $P(x_0)$, *the matrix* $\tilde{\mathcal{X}}$ *can be factorized in the form*

$$\tilde{\mathcal{X}} = P(x_0) \, \mathcal{J}^T \, \mathcal{D}\,\mathcal{E}.$$

To prove this theorem we need the following lemma.

Lemma 2. *For every* $r = 0,\ldots,m$,

$$P(x) \, J^r = J^r \, S_r(x).$$

Proof. Recalling that

$$(P(x))_{il} = \begin{cases} \binom{i}{l}x^{i-l} & i \geq l \\ 0, & i < l, \end{cases} \quad i,l = 0,\ldots,m,$$

and by using (3) and (28), the product of matrices leads to

$$(P(x) \, J^r)_{il} = (J^r \, S_r(x))_{il} = \begin{cases} \binom{i}{l+r}x^{i-l-r} & i \geq l + r \\ 0, & i < l + r, \end{cases} \quad i,l = 0,\ldots,m,$$

which proves the assertion. □

Proof (of Theorem 5). The r-th column of $P(x_0) \, \mathcal{J}^T \, \mathcal{D}\,\mathcal{E}$ is

$$(P(x_0) \, \mathcal{J}^T \, \mathcal{D}\,\mathcal{E})_r = P(x_0) \, J^r \, \mathcal{D}_r \, \xi_r(\underline{x}).$$

By Lemma 2 and (25) follows that

$$(P(x_0) \, \mathcal{J}^T \, \mathcal{D}\,\mathcal{E})_r = J^r \, S_r(x_0) \, \mathcal{D}_r \, \xi_r(\underline{x})$$
$$= J^r \, \tilde{\boldsymbol{X}}_r(x),$$

which coincides with $(\tilde{\mathcal{X}})_r$. Thus, the proof is concluded. □

5   Conclusions

In [15] orthogonal bases for the space $\mathcal{M}_k(\mathbb{R}^{n+1}, \mathcal{Cl}_{0,n})$, of $\mathcal{Cl}_{0,n}$-valued homogeneous polynomials of degree k, were constructed resulting in polynomials

$$f_{k,\mu} = X_{n+1,k_n}^{(k-k_n)} X_{n,k_{n-1}}^{(k_n-k_{n-1})} \cdots X_{3,k_2}^{(k_3-k_2)} \zeta^{k_2}, \tag{29}$$

where $\zeta = x_1 - x_2 e_1 e_2$ and $\mu = (k_{n+1}, k_n, \ldots, k_3, k_2)$ is an arbitrary sequence of integers such that $k = k_{n+1} \geq k_n \geq \cdots \geq k_3 \geq k_2 \geq 0$.

We stress that (26) gives a matrix representation for each building block used in the construction of the polynomial basis (see (29)). In the particular case $n = 2$, for each $k \geq 0$, the set $\{X_{3,j}^{(k-j)}(x)\zeta^j; \ j = 0, \ldots, k\}$ is itself an orthogonal basis for the space $\mathcal{M}_k(\mathbb{R}^3, \mathcal{Cl}_{0,2})$ (cf. [7]). In other words, in terms of the general constructive scheme in the block matrix representation of Theorem 5, every k-th row contains $k+1$ entries which form a basis of the space $\mathcal{M}_k(\mathbb{R}^3, \mathcal{Cl}_{0,2})$.

Finally we emphasize that in this paper the key point was the role of both, the creation and the shift matrices. In fact, we showed the relevance of those matrices in the representation of general hypercomplex sequences unifying and generalizing the approach already considered in [2,3].

Acknowledgments. This work was supported in part by the Portuguese Foundation for Science and Technology ("FCT-Fundação para a Ciência e Tecnologia"), through CIDMA-Center for Research and Development in Mathematics and Applications, within project UID/MAT/04106/2013.

References

1. Aceto, L., Cação, I.: A matrix approach to Sheffer polynomials. J. Math. Anal. Appl. **446**, 87–100 (2017)
2. Aceto, L., Malonek, H.R., Tomaz, G.: A unified matrix approach to the representation of Appell polynomials. Integral Transforms Spec. Funct. **26**, 426–441 (2015)
3. Aceto, L., Malonek, H.R., Tomaz, G.: Matrix approach to hypercomplex Appell polynomials. Appl. Num. Math. **116**, 2–9 (2017)
4. Aceto, L., Trigiante, D.: The matrices of Pascal and other greats. Amer. Math. Monthly **108**, 232–245 (2001)
5. Aceto, L., Trigiante, D.: Special polynomials as continuous dynamical systems. In: Cialdea, A., Dattoli, G., He, M.X., Shrivastava, H.M. (eds.) Lecture Notes of Seminario Interdisciplinare di Matematica, vol. 9, pp. 33–40 (2010)
6. Appell, P.: Sur une classe de polynômes. Ann. Sci. École Norm. Sup. **9**(2), 119–144 (1880)
7. Bock, S., Gürlebeck, K., Lávička, R., Souček, V.: Gelfand-Tsetlin bases for spherical monogenics in dimension 3. Rev. Mat. Iberoam. **28**(4), 1165–1192 (2012)
8. Cação, I., Falcão, M.I., Malonek, H.R.: Laguerre derivative and monogenic Laguerre polynomials: an operational approach. Math. Comput. Modelling **53**, 1084–1094 (2011)
9. Cação, I., Falcão, M.I., Malonek, H.R.: A matrix recurrence for systems of Clifford algebra-valued orthogonal polynomials. Adv. Appl. Clifford Algebras **24**, 981–994 (2014)

10. Cação, I., Falcão, M.I., Malonek, H.R.: Three-term recurrence relations for systems of Clifford algebra-valued orthogonal polynomials. Adv. Appl. Clifford Algebras **27**, 71–85 (2017)

11. Call, G.S., Velleman, D.J.: Pascal's matrices. Am. Math. Monthly **100**(4), 372–376 (1993)

12. Carlson, B.C.: Polynomials satisfying a binomial theorem. J. Math. Anal. Appl. **32**, 543–558 (1970)

13. Falcão, M.I., Malonek, H.R.: Generalized exponentials through Appell sets in $\mathbb{R}^{n+1}$ and Bessel functions. In: Simos, T.E., Psihoyios, G., Tsitouras, C. (eds.) AIP Conference Proceedings, vol. 936, pp. 738–741 (2007)

14. Gürlebeck, K., Malonek, H.R.: A hypercomplex derivative of monogenic functions in $\mathbb{R}^{m+1}$ and its applications. Complex Variables **39**, 199–228 (1999)

15. Lávička, R.: Complete orthogonal Appell systems for spherical monogenics. Complex Anal. Oper. Theory **6**, 477–489 (2012)

16. Peña Peña, D.: Shifted Appell sequences in Clifford analysis. Results Math. **63**, 1145–1157 (2013)

A Trade-off Analysis of the Parallel Hybrid SPIKE Preconditioner in a Unique Multi-core Computer

Leonardo Muniz de Lima[1], Lucia Catabriga[1,2(✉)], Maria Cristina Rangel[2],
and Maria Claudia Silva Boeres[2]

[1] High Performance Computing Lab, Federal University of Espírito Santo,
Vitória, ES, Brazil
lmuniz@ifes.edu.br
[2] Optimization Lab, Federal University of Espírito Santo, Vitória, ES, Brazil
{luciac,crangel,boeres}@inf.ufes.br

Abstract. In this paper we apply the parallel hybrid SPIKE algorithm as a preconditioner for a nonstationary iterative method to solve large sparse linear systems. In order to obtain a good preconditioner, we employ several strategies solving combinatorial problems such as matching, reordering, partitioning, and quadratic knapsack. Our SPIKE implementation combines MPI and OpenMP paradigms in a unique multi-core computer. The computational experiments show the influence of each strategy evaluating the number of iterations and CPU time of the iterative solver in a set of large systems from miscellaneous application areas. The experiments suggest that the SPIKE preconditioner is very advantageous when a suitable set of parameters is chosen. The choice of the number of MPI ranks and OpenMP threads is not an easy task, because the SPIKE algorithm can increase the number of iterations when the number of MPI ranks grows. Moreover, the increase in the number of threads does not ensure a better performance.

Keywords: Parallel hybrid SPIKE preconditioner · Combinatorial strategies · Nonstationary iterative method

1 Introduction

In the recent years, iterative methods to solve large sparse linear systems have become very usual in several areas of scientific computing. Saad [17] highlights memory consumption and facilities for an efficient high performance computing implementation as the main motivations to use iterative instead of direct solvers. The convergence of linear systems is based on eigenstructure of their matrices and depends on a good preconditioner. The Conjugate Gradients method [7] for symmetric positive-definite matrices and the Generalized Minimal Residual (GMRES) method [16] for nonsymmetric matrices are the most widely used methods to solve large sparse linear systems.

O. Gervasi et al. (Eds.): ICCSA 2017, Part II, LNCS 10405, pp. 422–437, 2017.
DOI: 10.1007/978-3-319-62395-5_29

The GMRES convergence has attracted a lot of attention in the last few years [21]. Presently, the absence of any reliable way to guarantee (or bound) the number of iterations needed to achieve convergence means that there is not good a priori way to identify the desired qualities of a preconditioner. Thus, heuristics can play an important role in view of that difficulties found in the theoretical analysis of the GMRES-like methods.

Benzi [2] and Wathen [21] present excellent surveys about preconditioners for large linear systems. Two main classes of algebraic preconditioners deserve enhance: incomplete LU factorization (ILU) and sparse approximate inverses. Despite their popularity, the traditional ILU preconditioner have their limitations. Those include potential instabilities, difficulty of parallelization, and lack of algorithmic scalability. Sparse approximate inverses, in turn, are widely applicable, but usually they do not generate spectrally equivalent preconditioners for elliptic Partial Differential Equations (PDEs).

Another option to accelerate the convergence of iterative methods lies in adjustments of hybrid solvers such as SPIKE [14,15], that is more used as a preconditioners. They are named hybrid once are able to take advantage of robustness of direct methods and lower computational cost of the iterative methods. The PSPIKE [18] is a variant of SPIKE where an iterative solver is used with the direct solver PARDISO [19], that solves the systems on each node based on a shared-memory parallelization, whereas the communication between the nodes is established by the distributed parallelization of the preconditioned iterative solver.

Variants of SPIKE algorithm as a preconditioner or as a solver, have been used in different situations. Li et al. [9] purposed a Graphics Processing Unit (GPU) version of SPIKE preconditioner which is two or three times faster than the Math Kernel Library (MKL)[1] solver, when applied in large dense banded matrices. The SPIKE solver demonstrated to be suitable and quite powerful in the solution of diagonally dominant tridiagonal linear systems as shown by [8, 11,20].

The SPIKE preconditioner effectiveness depends on some combinatorial strategies [18]. For example, a banded structure of the input matrix is needful, then several reordering techniques are applied to the original matrix to achieve all entries of the matrix being covered, ideally either by the diagonal blocks. If all entries are confined within the blocks, a nonstationary iterative solver could converge in a few iterations. However, in practical applications this situation does not occur very often; thus, a realistic goal would be to find a preconditioner such that possibly all heavy-weighted entries are included in the block structures and some small-weighted entries are not.

In this work, we present a trade-off analysis among several parameters on the SPIKE preconditioner performance in the solution of large sparse linear systems. We evaluate the use of the combinatorial strategies, the influence of the coupling blocks size and the behavior of this preconditioner when we combine different numbers of MPI ranks and threads in a unique multi-core computer.

---

[1] https://software.intel.com/en-us/intel-mkl

The remainder of this work is organized as follows. Section 2 addresses a brief discussion about preconditioning and some details about SPIKE as a preconditioner. In Sect. 3, combinatorial strategies applied to the preconditioning are presented. Section 4 shows the numerical experiments analyzing how combinatorial strategies, coupling blocks size and the combinations of different numbers of MPI ranks and threads in a unique multi-core computer affect the SPIKE preconditioner performance. Finally, Sect. 5 concludes this paper.

2   SPIKE Preconditioner

A good preconditioner for a linear system $Ax = b$ is an appropriate matrix M such that $M^{-1}A$ is well conditioned [17], so that the preconditioned linear system $M^{-1}Ax = M^{-1}b$ is solved more easily than $Ax = b$. Because that, the best choice for M would be A, but the calculation A^{-1} could be more difficult than solving the initial system $Ax = b$. The SPIKE preconditioner [15] is able to provide a matrix M which is a high quality approximation of A and can be computed in parallel. SPIKE was conceived as a hybrid parallel solver for narrow-banded linear systems. However, if linear systems are diagonally dominant, the truncated SPIKE version can be used as a preconditioner for external iterative schemes. The SPIKE preconditioner is defined from three stages. Firstly, the narrow-banded matrix M is obtained, named preprocessing stage, followed by the factorization of the matrix M and, finally, the postprocessing, where the preconditioning is actually performed.

Combinatorial strategies are used in the first stage to transform the original sparse general matrix A into a narrow-banded matrix. The bandwidth bw of a matrix A of order n is defined as $bw(A) = \max\limits_{i,j:a_{ij}\neq 0} |i - j|$. Then, A is considered a narrow-band matrix if $bw(A) \ll n$. The narrow-banded matrix can be divided according to the appropriate number of processors, as we can see in Fig. 1 – an example with 4 partitions.

The preconditioning matrix M is formed by diagonal blocks A_i $(i = 1, 2, ..., p)$ and coupling blocks matrices B_i $(i = 1.2, ..., p-1)$ and C_i $(i = 2, ..., p)$, where p is the number of partitions. A_i has dimension $n_i \times n_i$ in each partition i and B_i and C_i dimension $k \times k$, which value is chosen conveniently [12]. To finalize the first stage, the preconditioting matrix M is obtained discarding the coefficients outside the coupling blocks B_i and C_i.

The second stage begins when the narrow-banded matrix is factored in DS, where D is diagonal block matrix and S is named Spike matrix. Figure 2 illustrates a factorization divided in four partitions. D is equal A without coupling blocks. S has diagonal blocks I_i given by correspondent identity matrix and dense matrices V_i $(i = 1, 2, ..., p-1)$ and W_i $(i = 2, ..., p)$ which dimension are $n_i \times k$. When blocks A_i are diagonally dominant, dense matrices V_i and W_i can be calculated using LU/UL factorizations [15]. The bottom of the Spike V_i (denominated $V_i^{(b)}$) can be computed using only the bottom $k \times k$ blocks of L and U. Similarly, the top of the Spike W_i (denominated $W_i^{(t)}$) may be obtained

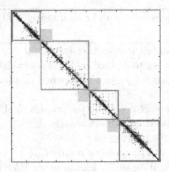

Fig. 1. Matrix A partitioning: example with 4 partitions.

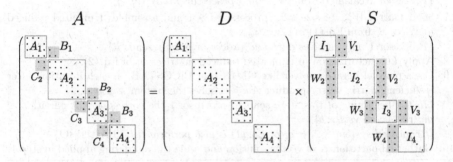

Fig. 2. Matrix A factorization: example with 4 partitions.

if it performs the UL-factorization. Thus, after some communications, a good approximation of S, named $\tilde{S}$, is built and the preconditioning matrix $M = D\tilde{S}$ is defined. Then a parallel LU-factorization is applied to each partition of $\tilde{S}$.

In the last stage, called postprocessing, the preconditioning is performed, whenever the iterative method calls for a matrix vector product like $\tilde{A}u = z$ and $\tilde{A} = M^{-1}A$. Obviously, the M^{-1} matrix is not calculated, but a parallel solution of the linear system $M\tilde{x} = z$ is done by two steps:

$$\tilde{D}_i \tilde{g}_i = z_i \tag{1}$$

$$\tilde{S}_i \tilde{x}_i = \begin{bmatrix} I & V_i^{(b)} \\ W_{i+1}^{(t)} & I \end{bmatrix} \begin{bmatrix} x_i^{(b)} \\ x_{i+1}^{(t)} \end{bmatrix} = \begin{bmatrix} g_i^{(b)} \\ g_{i+1}^{(t)} \end{bmatrix} = \tilde{g}_i \tag{2}$$

The application of SPIKE preconditioner to solve the system $Ax = b$ is presented in Algorithm 1.

Algorithm 1. Solve a linear system $Ax = b$ using SPIKE preconditioner for p processors

---

Require: Given matrix A stored in compact format.

1: Apply reordering techniques to reduce matrix bandwidth of A.
   {*Improvements in preconditioning can be reached if other combinatorial techniques are utilized in matrix A (see [18])*}

2: Employ on the right side vector b same techniques applied to matrix A in step 1.

3: Partition A and obtain preconditioning matrix M.

4: Send M_i, B_i and C_i blocks to processor i.
   {*processor 1 has only B_1 and processor p only C_p*}

5: Make LU and UL factorization of M_i blocks using PARDISO software in parallel.

6: Solve with PARDISO: $L_i U_i [V_i^{(b)}, W_i^{(t)}] = [\left(\begin{smallmatrix} 0 \\ B_i \end{smallmatrix} \right), \left(\begin{smallmatrix} C_i \\ 0 \end{smallmatrix} \right)]$.
   {*Processor 1 calculates only $V_1^{(b)}$ and processor p only $W_p^{(t)}$*}.

7: Send matrix $W_i^{(t)}$ of size k to processor $i - 1$ and assembles truncated reduced matrices $\tilde{S}_i$ from Eq. (1) of size $2k$.
   {*Processor 1 only receives data and processor p, only sends it*}.

8: Apply LU factorization in truncated reduced matrices $\tilde{S}_i$ of Eq. (2).

9: Use a parallel iterative solver like GMRES or BiCGSTAB. In each iteration, after a efficient matrix-vector product $Au = z$, solve the system $M\tilde{x} = z$.
   {*In the beggining of this step, send to processor i the correspondent partition of matrix A and vector b*}
   {*Truncated system $\tilde{S}_i \tilde{x}_i = g_i$ of Eq. (1) is also performed by PARDISO*}

10: Gather all partition x_i of vector solution and undo the reordering applied in step 1.

---

3 Combinatorial Strategies

As already pointed out in Sect. 2, the first step of the SPIKE algorithm is concerned with the computation of a narrow-banded matrix, permuted from the original matrix A by reordering procedures, aiming to have ideally all its entries covered by the diagonal blocks A_i or by the coupling blocks B_i and C_i [18]. A matrix organized with diagonal and coupling blocks can be denoted as a matrix with a SPIKE structure (Figs. 1 and 2). As in real applications, this ideal matrix format is hard to achieve, a good approximation should be a banded matrix with its heavy-weighted entries confined into the block matrix structures. This matrix can be obtained if, after applying the reordering procedures, we manage to: (i) find the diagonal blocks; (ii) guarantee their non-singularity (i.e., they must admit an inverse); (iii) organize the original matrix into diagonal and coupling blocks, moving the most significant entries into the coupling blocks. These three goals can be respectively formalized as graph partitioning, graph matching, and quadratic knapsack problem [18]. Graph algorithms seem to be very suitable tools to transform the original matrix if they are designed to perform search procedures and solve these known combinatorial problems.

The application of the combinatorial strategies to the original matrix A provides the definition of the SPIKE preconditioner matrix M. For this, the matrix A is multiplied by scaling factors and permutation matrices (reordering procedures) on the left and right. After that, row permutations are applied to move large entries onto the diagonal (graph matching problem) and techniques for solving the quadratic knapsack problem are applied in order to maximize the entries weights in the coupling blocks. Finally, the matrix partitioning is performed for parallelization purposes. The algorithms implemented to accomplish all these matrix operations are briefly described below.

In the factorization stage of the SPIKE algorithm – Algorithm 1, line 5 – the diagonal blocks A_i must be solved by direct method using LU /UL factorizations, which can be applied only if these blocks are non-singular. Every preconditioning matrix M can be associated to a graph G by means of the adjacency information of every matrix row. Thus, the non-singularity of M can be achieved if a perfect matching [18] is found in G, in a way that it is able to move the most significant elements to the main diagonal. A perfect matching in G is a set of independent edges that are incident to all vertices of G. For more details on graph concepts, see [5].

The algorithm used in this work to obtain a perfect matching in G is called *weighted bipartite matching* and is available in HSL_MC64[2] [6]. This algorithm returns row permutations that maximize the smallest element on the diagonal, maximize the sum of the diagonal entries and also finds scaling factors that may be used to scale the original matrix. Therefore, the nonzero diagonal entries of the permuted and scaled matrix are equal to one in absolute value and all the off-diagonal entries are less than or equal to one, also in absolute value.

The task of the matrix reordering is performed in this work by three different algorithms: the Reverse Cuthill-Mckee (RCM) [3], the spectral algorithm [1], and weighted spectral ordering (WSO) [12]. All of them try to put almost all entries onto the SPIKE structure and leave only few elements uncovered.

Given the original matrix A seem as the adjacency matrix of the associated graph G, the RCM algorithm visits all vertices of G, according to some criterion. The order of the vertices visited corresponds to a permutation of rows and columns. The algorithm uses the inverse order of the visits performed in order to reduce the bandwidth and the envelope of the matrix A [10]. The Spectral and WSO algorithms consider the concept of the algebraic connectivity, also known by Fiedler vector of the unweighted or weighted Laplacian matrix of the original matrix A. These algorithms compute a symmetric permutation that reduces the profile and wavefront of A by using a multilevel algorithm [12]. Both algorithms codes are available in HSL_MC73[3]. The WSO algorithm attempts to move the most significant coefficients as close as possible to the main diagonal, ensuring a more effective SPIKE preconditioner.

For an efficient parallelization of the SPIKE algorithm, the partitioning step needs to balance the processor loads during each phase of Algorithm 1 and should reduce the communication cost. In this work we consider the well known *chains-on-chains* partitioning problem to model this step. This problem is solved in polynomial time considering the *MinMax* algorithm [13] to find the exact solution.

---

[2] http://www.hsl.rl.ac.uk/catalogue/mc64.html.
[3] http://www.hsl.rl.ac.uk/catalogue/hsl_mc73.html.

Considering the blocks $M_1, \ldots, M_p$ of the banded matrix (see Fig. 2), it is important to have $k \times k$ coupling matrices B_i and C_{i+1}, for $i = 1, \ldots, p-1$, with many of their entries as close as possible to the diagonal blocks. To achieve this, we consider this step modeled by the quadratic knapsack problem and solved it by the *DeMin* heuristic, as described in [18].

4 Numerical Experiments

This section analyzes the SPIKE preconditioner performance when a unique multi-core computer is used, considering different parameters, which are: the coupling blocks size (k), the number of MPI ranks, and the number of OpenMP threads.

The tests are performed for 10 sparse matrices from the University of Florida matrix collection [4]. Table 1 shows the main characteristics of these matrices, i.e., matrix name, dimension (n), number of nonzero (nnz), the ratio of the nonzero by the dimension (nnz/n), the initial bandwidth (bw), and the application area. We highlight that the sparsity of all matrices is greater than 99%. Our experiments were run in a multi-core computer, that has 32 GB of RAM, 20 M of L3 Cache and two Intel Xeon ES-2630 Octa Core, each one been a hyper-thread with two threads, totalizing 32.

All codes are developed in C language and compiled with Intel version 2015 Update 3 optimized with Library Math Kernel Library MKL version 11.2 Update 3 and MPI version 5.0 Update 3. The SPIKE preconditioner steps involving calculations using direct methods that are solved by pardiso500-INTEL1301-X86-64 library of PARDISO software. The matrix operations as reordering, scaling, and matching are performed using the HSL[4] Mathematical Software Library.

Table 1. Features of the tested matrices

Matrix	n	nnz	nnz/n	Initial bw	Application area
dw8192	8,192	41,746	5.10	4,160	Dielectric waveguide
H20	67,024	2,216,736	33.07	14,976	Quantum chemistry
rail_79841	79,841	553,921	6.94	79,811	Model reduction
F1	343,791	26,837,113	78.06	343,754	Structural problem
CoupCons3D	416,800	22,322,336	53.56	83,363	Structural problem
largebasis	440,020	5,560,100	12.64	400,019	Optimization problem
parabolic_fem	525,825	3,674,625	6.99	525,820	Comp. fluid dynamics
atmosmodj	1,270,432	8,814,880	6.94	21,904	Atmospheric models
G3_circuit	1,585,478	7,660,826	4.83	947,128	Circuit simulation
rajat31	4,690,002	20,316,253	4.33	4,688,751	Circuit simulation

---

[4] http://www.hsl.rl.ac.uk/.

Section 4.1 presents an overview of the combinatorial strategies influence, that is, the solver behavior when is applied or not matching+scaling algorithms (ms), reordering schemes (rs), the quadratic knapsack problem heuristic (qkp), and also if the preconditioning is considered or not. In Sect. 4.2 the coupling blocks size, i.e., the k value (see Fig. 2, for details) is analyzed. Section 4.3 shows the parallel SPIKE preconditioner performance when different numbers of MPI ranks and OpenMP threads are considered.

4.1 · The Combinatorial Strategies

Firstly, we evaluate the influence of the combinatorial strategies in the SPIKE preconditioner for all matrices described in Table 1, that is, it is discussed how scaling+matching algorithms, reordering schemes, and the quadratic knapsack problem heuristic can affect the number of iterations and CPU time of the GMRES solver. We consider coupling blocks with size $k = 50$, 2 OpenMP threads in PARDISO and take into account 8 MPI ranks. The GMRES method consider 30 vectors to restart and tolerance equal to 10^{-8}. The CPU time is measured considering over the average of five executions.

Tables 2 and 3 show, respectively, the best and the worst choices for CPU time of each matrix from Table 1 considering all possible strategies combinations. The worst choice is defined among the possibilities that the preconditioning can offer at least a convergent solution. Both tables have the same template: the first column contains the matrices names; in the second column (strategies), there are three symbols per matrix indicating if a specific combinatorial strategy – scaling+matching (sm), reordering scheme (rs), and quadratic knapsack problem (qkp) – is used or not, respectively represented by symbols ● and ○. If the reordering schemes are on, the symbol ● is replaced by ● (rcm), ● (spectral), or ● (wso); the final bw are presented in the third column and the last four columns bring the GMRES number of iterations and CPU time, respectively without or with preconditioner. The symbol † means that the GMRES method not converged after 30,000 iterations. The results without preconditioner in both tables (Tables 2 and 3) are identical, since for them it is not applied any combinatorial strategy.

Generally, the SPIKE preconditioner is very advantageous when the best set of parameters are chosen. Furthermore, preconditioning is very needful to reach the solver convergence in almost all matrices tested. In our set of matrices, only for H20 matrix the CPU time became worse when SPIKE is used (Table 2), once the number of iterations decreases, but the CPU time increases, since the CPU time for the combinatorial strategies is meaningful.

Combinatorial strategies can affect positively or negatively the SPIKE preconditioner convergence. The rs scheme is the more notable strategy. The wso, as expected (see [12]), presented the best choices for five matrices. The spectral was better in three situations (matrices rail_79841, CoupCons3D, and rajat31) because the wso do not reduce the bandwidth enough for these matrices. Evaluating the number of iterations, rcm is the best choice for matrix H20, because no other reordering decreases the bandwidth size for this matrix, although the CPU time increases. For `atmosmodj` matrix, the preconditioner reduces the number of iterations and CPU time, even when no combinatorial scheme is considered (best choice). It means that this matrix already has a suitable structure for the SPIKE preconditioner. The sm and qkp strategies introduce secondaries effects. The sm were indispensable in `rail_79841` and `rajat31` matrices and the qkp in `dw8192`, `H2O` and `G3_circuit`. We mean an indispensable strategy if the best result is generated when it is chosen and gets the worst result, otherwise.

Table 2. Influence of the combinatorial strategies - best choices

Strategies and reorderings			Without preconditioner		With preconditioner	
Matrix	Strategies	Final bw	Iterations	CPU time	Iterations	CPU time
dw8192	○ ● ●	255	†	†	86	0.06
H2O	○ ● ●	7335	2641	3.38	117	15.65
rail_79841	● ● ●	952	†	†	1330	4.87
F1	● ● ○	21552	†	†	1032	1043.80
CoupCons3D	○ ● ○	18043	†	†	21	40.59
largebasis	● ● ○	479	†	†	16	6.17
parabolic_fem	○ ● ●	1195	†	†	381	40.65
atmosmodj	○ ○ ○	21904	1563	34.24	27	27.44
G3_circuit	● ● ●	12134	†	†	253	95.14
rajat31	● ● ●	19350	†	†	13	45.71

Strategy (sm,rs,qkp): ● used ○ unused
Reordering (rs): ● rcm ● spectral ● wso

Traditionally, SPIKE preprocessing is inherently sequential. Although, there are some parallel combinatorial implementation applied in general purpose, those implementations are a hard task [18]. Table 4 presents the CPU time to execute each strategy and the overall time to solve the linear system considering the best choices presented in Table 2. In some cases, the matrices preprocessing take a considerable portion of the total runtime to solve the linear system using the SPIKE preconditioner, as for `dw8192`, `CoupCons3D`, `largebasis`, and `rajat31` matrices. However, the strategies are crucial to get the convergence, as one can see in Tables 2 and 3.

4.2   Coupling Block Size (k)

In this section we show how the coupling block size – B_i and C_i matrices dimension, Fig. 2 – can affect the preconditioner performance. Table 5 shows the number of iterations and CPU time (in seconds) taking into account coupling blocks with size $k = 2, 50, 150, 350$ for all matrices considering the same strategies and parameters of the Table 2.

Table 3. Influence of the combinatorial strategies - worst choices

Strategies and reorderings			Without preconditioner		With preconditioner	
Matrix	Strategies	Final bw	Iterations	CPU time	Iterations	CPU time
dw8192	○ ● ○	255	†	†	90	0.13
H2O	● ● ○	7335	2641	3.38	175	21.91
rail_79841	○ ● ○	952	†	†	6508	23.71
F1	● ● ●	21552	†	†	8750	8014.25
CoupCons3D	● ● ●	18043	†	†	41	58.16
largebasis	● ● ○	487	†	†	37	15.51
parabolic_fem	○ ● ●	514	†	†	639	65.35
atmosmodj	● ● ●	7772	1563	34.24	162	120.66
G3_circuit	● ● ○	5068	†	†	668	216.51
rajat31	○ ● ●	7486	†	†	138	261.21

Strategy (sm,rs,qkp): ● used ○ unused
(rs): ● rcm ● spectral ● wso

Table 4. CPU time of combinatorial strategies - best choices

Matrix	sm	rs	qkp	Strategies	Solver	Total	Strategies (%)
dw8192	0.00	0.06	-	0.06	0.06	0.12	50.00
H2O	-	0.14	0.34	0.48	15.17	15.65	3.07
rail_79841	0.02	0.51	0.08	0.61	5.74	6.35	9.61
F1	0.91	9.59	-	10.50	1043.80	1054.30	1.00
CoupCons3D	-	13.97	-	13.97	26.62	40.59	34.42
largebasis	0.31	2.34	-	2.65	3.52	6.17	42.95
parabolic_fem	-	2.81	0.61	3.42	37.23	40.65	8.41
atmosmodj	-	-	-	-	27.44	27.44	00.00
G3_circuit	0.38	8.07	1.16	9.61	85.53	95.14	10.10
rajat31	-	14.58	5.71	20.29	25.42	45.71	44.39

The coupling block size has a strong influence at the number of float point operations in the preprocessing stage. Generally, when the coupling block size increases, the number of iterations can decrease, but it is possible that the CPU time increases meaningly, as for matrices dw8192, CoupCons3D, largebasis, atmosmodj, G3_circuit, and rajat31. For half of the matrices $k = 50$ got the best or reasonable SPIKE performance.

4.3   Number of MPI Ranks and OpenMP Threads

This section intends to show how the choice of the CPU core number used in MPI ranks or OpenMP threads can influence the SPIKE preconditioner convergence.

Table 5. Influence of coupling blocks size

k	iter	time	iter	time	iter	time	iter	time	iter	time
	dw8192		H2O		rail_79841		F1		CoupCons3D	
2	235	0.18	131	17.01	1656	5.65	**787**	**778.55**	**21**	**24.17**
50	**90**	**0.06**	**117**	**15.17**	3352	11.57	1032	1043.80	21	24.83
150	47	0.25	116	20.69	†	†	2008	2005.37	21	33.03
350	47	0.77	130	20.91	**27**	**1.53**	5445	2402.32	21	34.09
	largebasis		parabolic_fem		atmosmodj		G3_circuit		rajat31	
2	37	9.89	441	42.64	**27**	**27.20**	410	133.84	38	87.90
50	**16**	**3.52**	381	37.23	27	27.44	**13**	**85.53**	**13**	**25.42**
150	16	9.02	399	42.21	26	59.59	170	113.81	15	75.13
350	16	10.07	**292**	**34.36**	26	62.96	166	127.90	16	163.69

Table 6 shows the SPIKE preconditioner scalability when it is applied for the matrices in Table 1 considering only one OpenMP thread to each MPI rank for 2, 4, 8, 16, and 32 ranks. Table 6 also outlines, for each number of ranks, the amount of iterations and the CPU time, considering the best choices for the combinatorial strategies and coupling block size, according to Sects. 4.1 and 4.2. The symbols † and ⊘ indicate respectively that the SPIKE preconditioner do not converge and that is not possible to divide the matrix in agreement with MPI ranks because bandwidth is not sufficiently small. As we can see, the bold settings highlight the shorter processing times for each matrix. Increasing the number of MPI ranks does not mean improving the preconditioner performance because the number of iterations can also increase dramatically. For example, when we divide the matrix rail_79841 in 2 partitions the number of iterations is 2. However, when MPI ranks is 32 the number of iterations is 1041. On the other hand, as greater the number of MPI ranks, smaller are the A_i blocks size and consequently, the time to perform the LU and UL factorizations. Thus, the SPIKE preconditioner runtime depends on factorizations plus GMRES iterations.

A detailed investigation considering the 32 available threads combining MPI and OpenMP contexts over the matrix set is presented in Fig. 3. Whenever it is possible, MPI ranks were divided in 2, 4, 8, 16, and 32 partitions and the number of OpenMP threads inside each MPI rank was 1, 2, 4, 8, or 16. In the graphs of Fig. 3, the x-axis is the number of cores and the number of OpenMP threads, while the y-axis is the CPU time, in seconds, it takes to solve the linear system. For each set of colored bars it is shown the CPU time to perform the solution of the linear system – colored for different number of OpenMP threads according to the captions – and the time to perform the combinatorial strategies colored in gray. The best combinatorial strategies and the coupling block size discussed, respectively, in Sects. 4.1 and 4.2 are kept again.

A simplistic analysis would indicate that the scalability of the SPIKE algorithm would only be maintained if the partitions were large enough to com-

Table 6. Scalability with 1 OpenMP thread

# MPI ranks	2		4		8		16		32	
Matrix	iter	time	iter	time	iter	time	iter	time	iter	time
dw8192	12	0.11	25	0.10	**89**	**0.09**	†	†	†	†
H20	55	141.18	95	52.32	**117**	**17.65**	⊘	⊘	⊘	⊘
rail_79841	**2**	**1.37**	13	1.54	27	1.72	343	7.77	1041	45.73
F1	507	678.28	543	453.93	787	378.74	**808**	**276.73**	⊘	⊘
CoupCons3D	16	82.41	17	32.57	21	17.74	**32**	**13.40**	⊘	⊘
largebasis	1	5.26	6	4.36	**16**	**3.57**	37	4.68	80	8.27
parabolic_fem	221	48.88	**262**	**34.43**	382	37.80	616	53.05	833	65.12
atmosmodj	15	480.23	21	121.28	27	32.82	38	17.99	**57**	**15.84**
G3_circuit	85	78.80	**95**	**52.50**	253	86.96	290	84.78	429	110.95
rajat31	5	49.59	7	34.53	**13**	**25.56**	25	28.46	52	44.38

pensate the time lost in the communications of the MPI routines. In addition, the chains-on-chains algorithm presents an excellent result in load balancing. However, when we analyze the case atmosmodj, there is an excellent scalability and we could deduce that matrices with nnz greater than or equal to matrix atmosmodj would have similar scalability. Unfortunately, it is not the case when we compare the scalability of the matrices G3_circuit and rajat31. For details see Fig. 3.

Tuning and Analysis Utilities (TAU)[5] enables us to identify that SPIKE preconditioner scalability not only depends on MPI operations but also on other important operations of the Algorithm 1, which are the LU and UL factorizations in line 5, the calculations of $V_i^{(b)}$'s and $W_i^{(t)}$'s in line 6 and the solutions of the triangular linear systems in line 9. These operations are respectively denoted by SPIKE_factorization, SPIKE_top_bottom_tips, and SPIKE_back_forward_substitution. It is important to note that, the routines SPIKE_factorization, SPIKE_top_bottom_tips are performed just once and SPIKE_back_forward_substitution is executed according to the number of GMRES iterations.

Matrices rail_79841 and atmosmodj have an opposite behavior which can be explained when the main parts of the execution is analyzed. Figure 4 shows the CPU time in seconds for MPI operations, SPIKE_factorization, SPIKE_top_bottom_tips, and SPIKE_back_forward_substitution considering 1 OpenMP thread and 2 up to 32 MPI ranks. In general, this figure illustrates why scalabilities presented in Fig. 3 are very different. In matrix rail_79841 the run-time of MPI operations and SPIKE_back_forward_substittion increases dramatically when the number of MPI ranks grows up. It occurs because those two parts depend directly on the number of GMRES iterations. Once the

---

[5] https://www.cs.uoregon.edu/research/tau/home.php.

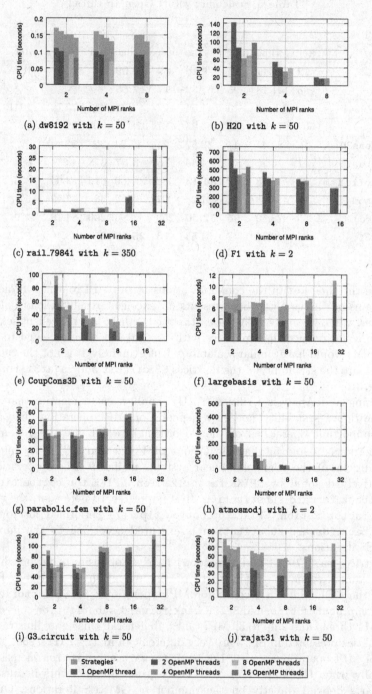

(a) dw8192 with $k = 50$

(b) H20 with $k = 50$

(c) rail_79841 with $k = 350$

(d) F1 with $k = 2$

(e) CoupCons3D with $k = 50$

(f) largebasis with $k = 50$

(g) parabolic_fem with $k = 50$

(h) atmosmodj with $k = 2$

(i) G3_circuit with $k = 50$

(j) rajat31 with $k = 50$

| Strategies | 2 OpenMP threads | 8 OpenMP threads |
| 1 OpenMP thread | 4 OpenMP threads | 16 OpenMP threads |

Fig. 3. Scalability - MPI ranks versus OpenMP threads. (Color figure online)

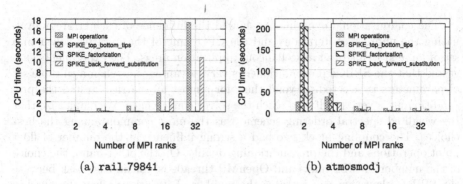

Fig. 4. Analysis of main routines: matrices `rail_79841` and `atmosmodj`.

matrix `rail_79841` is extremely ill-conditioned, the preconditioner becomes sensible to MPI rank divisions (see Table 6). Moreover, `SPIKE_factorization` and `SPIKE_top_bottom_tips` – that are performed just once – predominate in matrix `atmosmodj` runtime and the number of GMRES iteration does not increase significantly when the number of ranks increases leading to a good scalability.

Generally, when the CPU time of the `SPIKE_factorization` and `SPIKE_top_bottom_tips` are preponderant and the number of iterations do not increase exceedingly, the use of more OpenMP threads per MPI rank can be very helpfull. Actually, we are facing a trade-off problem. Since our hardware has only 32 threads available, it is possible to use them in divisions of MPI ranks or OpenMP threads. The increasing number of MPI ranks implies in reducing of the A_i's blocks size, which can be advantageous. However, depending on the matrix conditioning, that division can increase significantly the number of iterations. On the other hand, the increasing number of OpenMP threads can enable runtime reduction once it acts directly on the SPIKE routines. In short, while the number of iterations do not increase overly it is worth continuing to divide into MPI ranks. Besides, if the scalability is lost before reach the maximum number of threads available, idle threads could be used as OpenMP threads. The experiments show that on average the best results are obtained considering the 32 available threads divided into 8 MPI ranks and 4 OpenMP threads per rank.

5   Conclusions

In this paper we applied the parallel hybrid SPIKE algorithm as a preconditioner for the GMRES nonstationary iterative method to solve large sparse linear systems combining MPI and OpenMP paradigms. Combinatorial strategies such as matching, reordering, partitioning, and quadratic knapsack problems are employed in order to obtain a good preconditioner. A trade-off analysis on the SPIKE preconditioner performance was conducted evaluating the use of the combinatorial strategies, the influence of the coupling blocks size, and the behavior of the SPIKE preconditioner when we combine different numbers of MPI ranks and OpenMP threads in a unique multi-core computer.

The computational experiments showed the influence of each strategy evaluating the number of iterations and the CPU time of the iterative solver in a set of large systems from miscellaneous application areas. The experiments suggested that the SPIKE preconditioner is very advantageous when a suitable set of parameters is selected. Beyond that, for almost all matrices tested the preconditioning is very needful to reach the solver convergence. We emphasize that the weighted spectral ordering scheme was the most common among the best choices. The coupling block size had a strong influence at the number of float point operations and the preconditioning quality. On the other hand, the choice of the number of MPI ranks and OpenMP threads is not an easy task because the SPIKE algorithm can increase the number of iterations when the number of MPI ranks grows. Moreover, the increase in the number of threads does not ensure a better performance.

Acknowledgments. This work has been supported in part by CNPq, CAPES, and FAPES.

A   Available source codes

All codes are developed in C language and compiled with Intel version 2015 Update 3 optimized with Library Math Kernel Library MKL version 11.2 Update 3 and MPI version 5.0 Update 3, available at https://github.com/leomunizlima/SPIKE. The SPIKE preconditioner steps involving calculations using direct methods that are solved by pardiso500-INTEL1301-X86-64 library of PARDISO software. The matrix operations as reordering, scaling and matching are performed using the HSL Mathematical Software Library.

References

1. Barnard, S.T., Pothen, A., Simon, H.: A spectral algorithm for envelope reduction of sparse matrices. Numer. Linear Algebra Appl. **2**(4), 317–334 (1995)
2. Benzi, M.: Preconditioning techniques for large linear systems: a survey. J. Comput. Phys. **182**(2), 418–477 (2002)
3. Cuthill, E., McKee, J.: Reducing the bandwidth of sparse symmetric matrices. In: Proceedings of the 1969 24th National Conference, pp. 157–172. ACM, New York (1969). http://doi.acm.org/10.1145/800195.805928
4. Davis, T.A., Hu, Y.: The university of Florida sparse matrix collection. ACM Trans. Math. Softw. **38**(1), 1:1–1:25 (2011). http://www.cise.ufl.edu/research/sparse/matrices
5. Diestel, R.: Graph Theory. Graduate Texts in Mathematics. Springer, Heidelberg (2006)
6. Duff, I.S., Koster, J.: On algorithms for permuting large entries to the diagonal of a sparse matrix. SIAM J. Matrix Anal. Appl. **22**(4), 973–996 (2001)
7. Hestenes, M.R., Stiefel, E.: Methods of conjugate gradients for solving linear systems, vol. 49. NBS (1952)

8. Kouris, A., Sobczyk, A., Venetis, I., Gallopoulos, E., Sameh, A.: Revisiting the SPIKE-based framework for GPU banded solvers: a givens rotation approach for tridiagonal systems in CUDA (2014)

9. Li, A., Deshmukh, O., Serban, R., Negrut, D.: A comparison of the performance of SPIKE: GPU and intel's math kernel library (MKL) for solving dense banded linear systems (2014)

10. Liu, W.H., Sherman, A.H.: Comparative analysis of the Cuthill-mckee and the reverse Cuthill-mckee ordering algorithms for sparse matrices. SIAM J. Numer. Anal. **13**(2), 198–213 (1976)

11. Macintosh, H.J., Warne, D.J., Kelson, N.A., Banks, J.E., Farrell, T.W.: Implementation of parallel tridiagonal solvers for a heterogeneous computing environment. ANZIAM J. **56**, 446–462 (2016)

12. Manguoglu, M., Koyutürk, M., Sameh, A.H., Grama, A.: Weighted matrix ordering and parallel banded preconditioners for iterative linear system solvers. SIAM J. Sci. Comput. **32**(3), 1201–1216 (2010)

13. Manne, F., Sørevik, T.: Optimal partitioning of sequences. J. Algorithms **19**(2), 235–249 (1995). doi:10.1006/jagm.1995.1035

14. Polizzi, E., Sameh, A.: SPIKE: a parallel environment for solving banded linear systems. Comput. Fluids **36**(1), 113–120 (2007)

15. Polizzi, E., Sameh, A.H.: A parallel hybrid banded system solver: the SPIKE algorithm. Parallel Comput. **32**(2), 177–194 (2006)

16. Saad, Y., Schultz, M.H.: GMRES: a generalized minimal residual algorithm for solving nonsymmetric linear systems. SIAM J. Sci. Stat. Comput. **7**(3), 856–869 (1986)

17. Saad, Y.: Iterative Methods for Sparse Linear Systems. Siam, Philadelphia (2003)

18. Sathe, M., Schenk, O., Uçar, B., Sameh, A.: A scalable hybrid linear solver based on combinatorial algorithms. In: Naumann, U., Schenk, O. (eds.) Combinatorial Scientific Computing, pp. 95–128. Taylor & Francis, Chapman-Hall/CRC Computational Science, Boca Raton (2012)

19. Schenk, O., Gärtner, K.: On fast factorization pivoting methods for sparse symmetric indefinite systems. Electron. Trans. Numerical Anal. **23**(1), 158–179 (2006)

20. Situ, Y., Martha, C.S., Louis, M.E., Li, Z., Sameh, A.H., Blaisdell, G.A., Lyrintzis, A.S.: Petascale large eddy simulation of jet engine noise based on the truncated SPIKE algorithm. Parallel Comput. **40**(9), 496–511 (2014)

21. Wathen, A.: Preconditioning. Acta Numerica **24**, 329–376 (2015)

Workshop on Computer Aided Modeling, Simulation, and Analysis (CAMSA 2017)

Solution of the Inverse Bioheat Transfer Problem for the Detection of Tumors by Genetic Algorithms

Antonio Marcio Gonçalo Filho[1], Lucas Lagoa Nogueira[1],
Joao Victor Caetano Silveira[1], Michelli Marlane Silva Loureiro[1(✉)],
and Felipe dos Santos Loureiro[2]

[1] Department of Computer Science, Federal University of São João del-Rei,
São João del-Rei, Minas Gerais, Brazil
antoniomarciofilho@gmail.com, lucaslagoanogueira@hotmail.com,
caetanosjoao@gmail.com, michelli.loureiro@ufsj.edu.br
[2] Department of Thermal and Fluid Sciences, Federal University of São João del-Rei,
São João del-Rei, Minas Gerais, Brazil
felipe.loureiro@ufsj.edu.br

Abstract. The problem of determining the size and location of a tumor situated underneath the skin by means of an inverse bioheat transfer analysis is considered in this article. The problem is posed by minimizing an error norm that considers the temperature information at the skin surface. Since Genetic Algorithms (GA's) are powerful and versatile tools, a GA of steady–state type is implemented to solve the optimization problem originated from the inverse analysis. A finite element program based on the discretization of Pennes' bioheat equation coupled with the Gmsh software is also developed to solve the set of direct problems required by the GA. A 2D numerical model is analyzed in order to demonstrate the effectiveness and robustness of the proposed approach.

Keywords: Genetic Algorithm · Pennes' equation · FEM · Inverse problem

1 Introduction

Millions of people over the years are affected by cancer and this is the second leading cause of death in the world. Because of that several research fronts are being developed to detect and treat cancer. Among these research fronts we can mention computational and mathematical modeling that aims at identifying the physical parameters, location and size of the tumor through inverse methodologies. As an inverse problem can be cast as an optimization one, either deterministic (e.g., conjugated gradient, Levenberg Marquardt, Broyden Fletcher Goldfarb Shanno) or stochastic (e.g., simulated annealing, evolutionary strategies, ant colony) algorithms are being applied by researchers. One of the optimization techniques that deserves to stand out among evolutionary strategies is the Genetic Algorithm (GA) that is based on Darwin's theory.

© Springer International Publishing AG 2017
O. Gervasi et al. (Eds.): ICCSA 2017, Part II, LNCS 10405, pp. 441–452, 2017.
DOI: 10.1007/978-3-319-62395-5_30

Partrige and Wrobel [1] applied a binary GA in conjunction with the dual reciprocity boundary element method (DR–BEM) in the search of tumors, considering the skin temperature information on the inverse process. Das and Mishra [2] proposed a newly curve fitting technique for the estimation of the size and location of breast tumors, while a GA and a control volume based finite difference method (FDM) were employed by the same authors to estimate the blood perfusion parameter, as well as geometry characteristics of tumors [3]. Souza et al. [4] performed a comparative study on different deterministic and heuristic algorithms for the estimation of the blood perfusion parameter in which the direct problems were solved by the integral transform technique. Concerning the bioheat transfer mathematical model, the well–known Pennes' equation was employed in all the aforementioned cited papers. In fact, among different mathematical models of bioheat transfer [5], Pennes' equation [6] is still widely employed due to its simplicity and overall satisfactory representation of the thermal process in the biological tissue [7–10].

Differently from other numerical methods such as the BEM and the FDM, the finite element method (FEM) requires a more involved remeshing procedure to deal with the change of the geometry originated from an inverse methodology. Therefore, a great effort in terms of computer programing is needed, especially for 3D simulations. Hence, in this paper, we present a new front of tumor detection embedded in a healthy tissue, using the genetic algorithm, the FEM and the Gmsh software to automatically perform the meshes in the FEM program. The paper is organized into the following sections. In Sect. 2, a description of the inverse bioheat transfer problem to detect the location and the size of the tumor is described. In Sect. 3, the GA and FEM procedures to solve the inverse problem are discussed. In Sect. 4, the results are presented and explained. Finally, in Sect. 5, conclusions are drawn as well as future works that can be explored.

2   Problem Statement

The problem to be analyzed consists of estimating the location and size of a tumor inside a healthy tissue by taking into account the tissue temperature raise along with the skin surface. To this end, let $\Omega_H, \Omega_{T_u} \subset \mathbb{R}^2$ be the healthy tissue and tumor domains with $\Omega_{T_u}(\mathbf{x}_r, r) = \{\mathbf{x}, \mathbf{x}_r \in \mathbb{R}^2, r \in \mathbb{R} : \|\mathbf{x} - \mathbf{x}_r\| \leq r\}$ being the tumor represented by a circle of radius r centered at point $\mathbf{x}_r$ such that $\Omega = \Omega_H \cup \Omega_{T_u}$ and $\Omega_H \cap \Omega_{T_u} = \emptyset$. The inverse bioheat transfer problem can be stated as an optimization problem as follows: find $\mathbf{x}_r$ and r (i.e., $\Omega_{T_u}(\mathbf{x}_r, r)$) that minimize the following least–squares norm of the temperature difference:

$$J(\mathbf{x}_r, r) = \frac{1}{2} \sum_{i \in \eta_{\Gamma_S}} \left(\hat{T}_i - T_i(\mathbf{x}_r, r) \right)^2 \tag{1}$$

where η_{Γ_S} is the set of nodes over the skin surface Γ_S, $\hat{T}_i \equiv \hat{T}(\mathbf{x}_i)$ stands for the measured temperature at the coordinate of the ith node over the skin surface available from a high precision temperature sensor, while $T_i(\mathbf{x}_r, r) \equiv T(\mathbf{x}_i; \mathbf{x}_r, r)$ is the computed temperature at the ith node by using estimated values of the unknown parameters $\mathbf{x}_r$ and r. The computed temperatures are obtained from the solution of the direct problem which is mathematically modeled by the bioheat Pennes' equation defined as [5,6]:

$$\nabla \cdot k\nabla T + \omega_b \rho_b c_b (T_a - T) + Q_m = 0, \text{ in } \Omega \qquad (2)$$

where $T : \bar{\Omega} \to \mathbb{R}^+$ denotes the tissue temperature field, $k : \bar{\Omega} \to \mathbb{R}^+$ stands for the thermal conductivity, $\rho_b, c_b : \Omega \to \mathbb{R}^+$ stand for the density and specific heat of the blood, respectively and $\omega_b : \Omega \to \mathbb{R}^+$ and $T_a : \Omega \to \mathbb{R}^+$ are, respectively, the blood perfusion rate and the arterial temperature. The metabolic heat generation is denoted by $Q_m : \Omega \to \mathbb{R}^+$. In addition to the above equation, boundary conditions on $\Gamma \subset \mathbb{R}^2$ for the closed domain $\bar{\Omega} = \Omega \cup \Gamma$ need also to be imposed in order to have a well–posed problem as will be described later on. It is worth mentioning that the effects of the heat exchange between the blood flow through the capillaries and the tissue are taking into account in the second term of Eq. (2).

3   Inverse Problem Solution

In this section the genetic algorithm and the FEM that are implemented in a computer program are briefly described. Firstly, some concepts of the GA and its characteristics such as encoding, fitness function, selection scheme, genetic operators, among others are presented. Finally, the standard FEM Galerkin method and an approach to cope with the mesh adaptivity for each candidate solution generated by the GA are discussed.

3.1   Genetic Algorithm

Genetic Algorithms (GA's) can be defined as search procedures based on genetics and natural selection of species [11]. As in the natural environment, in a GA, there is a group of candidate solutions, also known as individuals, that compete with each other to ensure their own survival and ensure that their characteristics are passed on through the new generation [11].

Basically, a genetic algorithm presents five fundamental aspects when used to solve an optimization problem, namely [11]: (i) a genetic coding of solutions to the problem; (ii) a procedure to create an initial population of solutions; (iii) an evaluation function that returns the fitness of each candidate solution; (iv) genetic operators that manipulate the parents coding during the reproduction process, giving rise to new candidate solutions; and (v) parameter values to be used in the algorithm, i.e., crossover and mutation probabilities.

The coding is a way of representing possible solutions to the problem, since GA's do not work on the candidate solutions, but on their representations. In this work, we have chosen to use the real coding in which the representation is carried out directly on vectors of real values.

The initialization of the population has the role of determining the process of creating candidate solutions for the first cycle of the algorithm. Typically, the elements constituting the initial population are chosen randomly within a search space defined for each experiment.

The fitness function is used by the GA to determine the quality of an individual as a solution to the problem in question. In general, the fitness function, for an unconstrained optimization problem, is the same as the objective function.

The selection process is responsible for identifying candidate solutions (individuals) that will serve as parents during the reproduction. This simulates the mechanism of natural selection, in which more capable parents generate more children. In this paper, we prefer to use the roulette and tournament selection methods. In the roulette, each individual of the population is represented in a roulette–wheel proportionally to its fitness index. On the other hand, in the tournament, some individuals are chosen randomly in the population and the best of them is the chosen one.

In relation to genetic operators, we choose to use crossover and mutation. The crossover operator has the purpose of choosing parts of "parent genetic material" to produce the "child" chromosome. There are several ways of performing the crossover, here the crossover of an applied point with probability equal to 0.8 is selected. The mutation operator is used after the application of the recombination operator in order to introduce diversity into the newly generated individuals. This is applied with a probability equal to 0.01. In this work, a random mutation is employed [12] in which an element is randomly chosen in the chromosome and its value is also determined randomly in a defined interval.

In order to solve the optimization problem under consideration, i.e., find $\mathbf{x}_r$ and r such that:

$$\min_{[\mathbf{x}_r, r] \in \Omega} J(\mathbf{x}_r, r) \tag{3}$$

in which $J(\mathbf{x}_r, r)$ stands hereafter for the so–called objective function, an Octave program based on real–coding steady–state GA (SSGA) with both roulette and tournament selections, as displayed in Algorithm 1, has been developed.

Algorithm 1. SSGA pseudocode

1: Initialize population P
2: Compute fitness of population P
3: Order population P according to fitness
4: **repeat**
5:     Select parent(s)
6:     Apply Genetic Operator
7:     Generate offspring
8:     Evaluate offspring
9:     Select offspring **x** to survive
10:     **if** (offspring x is better then the worst in the population P) **then**
11:     worst is removed
12:     offspring x is inserted in the P According to the "ranking"
13:     **endif**
14: **until** satisfy stopping criterion

3.2   Finite Element Analysis and Mesh Adaptivity

Since the analytical solution of Eq. (2) is not feasible for a general heterogeneous medium, a numerical solution based on the FEM [13] is employed here to solve Pennes' equation. Let $\mathcal{S} = \{T \in \mathcal{H}^1(\Omega) \,|\, T = \bar{T} \text{ on } \Gamma_D\}$ and $\mathcal{V} = \{v \in \mathcal{H}^1(\Omega) \,|\, v = 0 \text{ on } \Gamma_D\}$ be, respectively, the spaces of admissible and test functions in which $\bar{T}$ is the Dirichlet boundary condition imposed on the boundary portion Γ_D. The FEM is based on the weak form of Eq. (2) which can be stated as: find $T \in \mathcal{S}$ such that

$$\int_{\Omega} \left(\nabla v \cdot k \nabla T - v \omega_b \rho_b c_b (T_a - T) - v Q_m \right) d\Omega - \int_{\Gamma_N} v k \nabla T \cdot \mathbf{n} \, d\Gamma = 0, \forall v \in \mathcal{V} \tag{4}$$

In the above expression Γ_N is the boundary portion of Neumann type with **n** being the outward normal vector. To discretize the weak form, the closed domain $\bar{\Omega}$ is first divided into nonoverlapping finite elements $\bar{\Omega}_e$ taking into consideration the finite–dimensional spaces $\mathcal{S}^h \subset \mathcal{S}$ and $\mathcal{V}^h \subset \mathcal{V}$ usually represented by piecewise polynomial functions. Thus, considering the approximation $T(\mathbf{x}) \approx T^h(\mathbf{x}) = \sum_{i=1}^{n} N_i(\mathbf{x}) T_i$ with n being the number of nodes in the mesh and after some proper algebraic manipulations, the following system of linear equations arises:

$$(\mathbf{K} + \mathbf{M})\mathbf{T} = \mathbf{F} \tag{5}$$

where $\mathbf{K} \in \mathbb{R}^{N \times N}$ is the so–called conductivity matrix, $\mathbf{M} \in \mathbb{R}^{N \times N}$ is the perfusion matrix (similar to the so–called mass matrix), $\mathbf{T} \in \mathbb{R}^N$ is the temperature nodal vector and $\mathbf{F} \in \mathbb{R}^N$ is the known vector with N being the total number of unknown degrees of freedom.

A computer FEM program based also upon the Octave environment has been developed in association with Gmsh, which is an open–source mesh generator software written in C^{++} [14]. Gmsh is employed to create both the geometry and the mesh of the tissue domain. While Octave performs most of the calculations due to FEM analysis, Gmsh is capable of generating and displaying the whole created model. Gmsh, described as "a three–dimensional finite element mesh generator", allows the user to generate geometries by simple programming specific input '.geo' files and passing it to the program, generating an output '.msh' file. The input file must contain the geometry information, type of element and its size, refinement around the nodes, besides which nodes are to be connected by lines or other entities, like physical groups that will be worked on. The application also permits refinements to specific entities, enabling the user a great flexibility in adopting different levels of refinements over the regions. Conversely, the generated output file contains nodes and connectivities that are grouped into elements, the latter being grouped into two physical groups, one representing the healthy tissue and another for the tumor.

Figure 1 presents distinct meshes generated by Gmsh adopting linear triangle conforming elements for two possible candidate solutions obtained randomly during the evolutionary process of the GA. The model includes a rectangle representing the healthy tissue portion, and a circle representing the tumor. The term

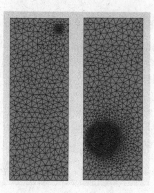

Fig. 1. Mesh adaptivity.

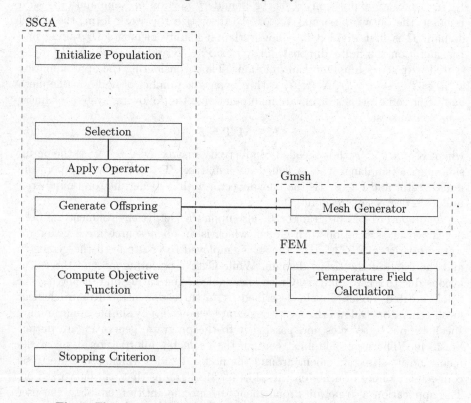

Fig. 2. Flowchart of the main steps required by the inverse analysis.

mesh adaptivity should be here understand as the process of dynamic remeshing the whole new domain generated during the evolutionary process. From this figure, it becomes clear that Gmsh readily allows the user to explore the flexibility offered by FEM, which is the use of nonuniform and unstructured meshes (a characteristic not presented in other methods such as the standard FDM).

In fact, the small tumor depicted in the figure is correctly represented by adopting a refinement differently from that of the healthy tissue and, thus, optimizing the analysis.

Finally, Fig. 2 shows the main steps required in the whole inverse analysis by joining the SSGA with the FEM as well as the Gmsh programs. Notice that the choice of the GA to perform the inverse analysis stems mainly from the generality with respect to the objective function and from the relatively straightforward implementation of the FEM into the GA program. Indeed, the GA just requires the evaluation of the objective function with no derivative being computed.

4   Numerical Results and Discussion

In order to validate the methodology presented above as well as the developed code, the bioheat transfer effect of a circular tumor embedded in a square healthy tissue is analyzed. A sketch of the model in focus is depicted in Fig. 3. The tumor has a domain $\Omega_{T_u} = \left\{ x, y \in \mathbb{R} : (x - 0.01)^2 + y^2 \leq 2.5 \times 10^{-5} \right\}$ while that of the healthy tissue is defined as $\Omega_H = (]0, 0.03[\times]-0.04, 0.04[) \setminus \Omega_{T_u}$. Figure 3 also depicts the boundary labels for which $\Gamma = \Gamma_S \cup \Gamma_N \cup \Gamma_D$ such that $\Gamma_S \cap \Gamma_N = \Gamma_S \cap \Gamma_D = \Gamma_N \cap \Gamma_D = \emptyset$. The boundary conditions $\nabla T \cdot \mathbf{n} = 0$ are applied on Γ_N as well as on the skin surface Γ_S, whereas a value of $T = 37\,^{\circ}\mathrm{C}$ is prescribed on Γ_D in order to represent the body core temperature. Notice that the skin surface is assumed to be insulated and, thus, there is no heat exchange with the surround environment. The material properties of both healthy and tumor tissues are presented in Table 1; besides, the arterial temperature is assumed to be $T_a = 37\,^{\circ}\mathrm{C}$. It is worth to note that due to a more vascularized blood vessel network, the tumor possesses a high value for the perfusion rate as well as for the metabolic heat generation.

In this problem the design variables are the radius and center coordinates of the tumor. Although the skin temperature may be obtained experimentally from a practical application point of view to perform the inverse analysis, here a synthetic tissue model (as described above) with *a priori* known design variables is adopted and a reference solution is computed by adopting a fine mesh. To solve the inverse problem a SSGA is employed in which the population size and generation number are equal to 100 and 250, respectively with the latter

Table 1. Model parameters

Properties	Healthy tissue	Tumor
k [W/m°C]	0.5	0.55
c_b [J/kg°C]	3800	4200
ρ_b [kg/m³]	1000	1000
w_b [1/s]	0.0005	0.02
Q_m [W/m³]	430	4600

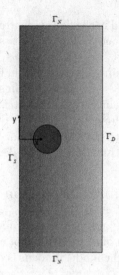

Fig. 3. Sketch of the model with boundary labels.

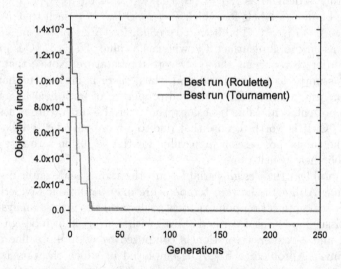

Fig. 4. Objective function evolution.

being used as a stopping criterion. Note that the search space is limited by the predefined domain Ω; besides, this type of constraint is trivially enforced in the SSGA and does not require further consideration here.

We perform 30 independent runs to analyze the SSGA behavior. The simulation has been performed in an $I7$, 4.0 GHz, 32 GB computer and only a few hours are required to execute all the runs. With the results obtained from these 30 runs, a statistical study is carried out and the results are presented in Table 2. Analyzing Table 2 it is possible to observe that the standard deviation for

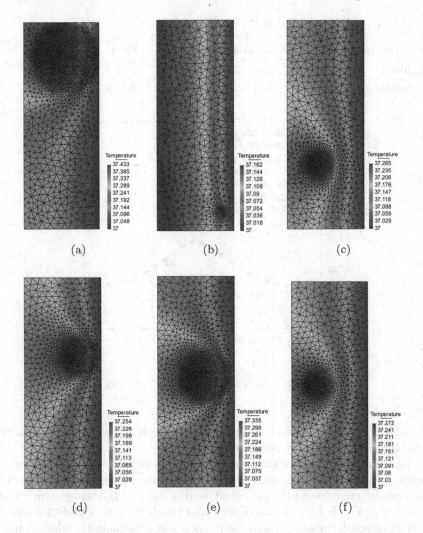

Fig. 5. Temperature field of some candidate solutions during the SSGA process.

tournament selection method is slightly higher compared to the standard deviation of the roulette selection method (i.e., 2.22×10^{-5} vs. 1.52×10^{-5}). In contrast, the tournament selection method yields the smallest value for the objective function when compared to the roulette selection method (i.e., 1.85×10^{-7} vs. 2.48×10^{-7}). Such a difference occurs due to randomness in the choice of the individual for reproduction when the tournament selection method is employed.

Figure 4 shows the objective function evolution for the SSGA for the problem analyzed here, describing the convergence of the best run until reach the best value obtained from both the tournament and roulette selection schemes

Table 2. Statistics of 30 runs

Selection type	Best	Median	Mean	St. dev.	Worst
Roulette	2.48×10^{-7}	2.06×10^{-6}	8.48×10^{-6}	1.52×10^{-5}	6.79×10^{-5}
Tournament	1.85×10^{-7}	2.44×10^{-7}	9.05×10^{-6}	2.22×10^{-5}	9.12×10^{-5}

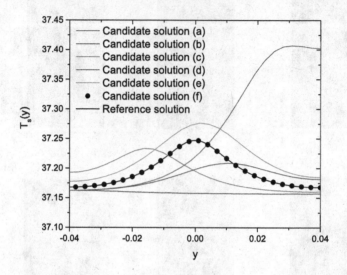

Fig. 6. Skin surface temperature.

(see Table 2). It can be observed that a monotonically convergence is obtained during the generations as expected by a SSGA.

To analyze the change of the tumor size and its location as well as the mesh adaptivity, a set of six candidate solutions with decreasing objective function values are selected during the optimization process, and the temperature fields are plotted in Fig. 5. First, it can be seen that Gmsh provides a perfect means of representing small tumors without the requirement of refining the whole domain and, thus, resulting in an efficient FEM analysis with respect to the computational time and memory storage savings without degrading the overall accuracy. Finally, the SSGA converges to the correct design variables once the temperature field shown in Fig. 5(f) is the converged solution corresponding to the best candidate solution. To confirm this fact, temperature distributions over the skin surface (Γ_s) for all the candidate solutions displayed in Fig. 5 as well as the reference solution are plotted in Fig. 6. From this figure, it becomes clear that the skin temperature of the candidate solution(f) is almost equal to that of the reference solution, indicating the good prediction accuracy of the SSGA. This is in accordance with Fig. 4 in which the objective function value gradually decreases to a value around 10^{-7} as mentioned previously.

Concerning the physical aspects of the simulated model, Fig. 7 clearly shows a temperature increase inside the tumor when compared to temperature of the

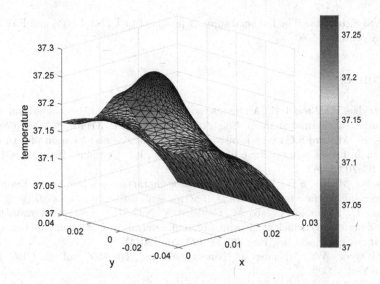

Fig. 7. Temperature field over the whole tissue domain.

healthy tissue which is caused by a high value of the metabolic heat genera-tion in the tumor. It is important to stress that the presented bioheat transfer model results are in accordance with the results found in the literature [15,16], validating the computer program developed for this paper.

5   Conclusions and Future Works

A real–coding steady-state GA (SSGA) based program focused on tumor search with the assistance of an open source mesh generator Gmsh was developed. By taking into account the simulated skin surface temperature, it was possi-ble to estimate the location and size of the tumor through an inverse analysis. The many forward (direct) problems based on Pennes' equation required by the inverse analysis were efficiently solved by considering a FEM program. The final developed SSGA–FEM Octave program linked with the Gmsh reveals to be quite effective and robust in the sense that the many geometries and meshes were successfully generated for the forward problems. In fact, the FEM can be a very powerful tool in an inverse methodology to simulate complex problems with geometry changes once an efficient unstructured mesh generator is pro-vided. Hence, this paper shows that computer simulation modeling could be viable for research on non–invasive medical assistance in an attempting to aid the treatment of tumors. Thus, the results presented here are very encourage and, as future works, we plan to continue the development of the computer pro-gram by inserting more design variables, nonlinear bioheat transfer models and 3D analyses with real data. Furthermore, comparison with other optimization techniques and parallel computing may also be performed.

Acknowledgments. The financial support provided by UFSJ, CNPq and FAPEMIG is greatly acknowledged.

References

1. Partridge, P., Wrobel, L.: An inverse geometry problem for the localisation of skin tumours by thermal analysis. Eng. Anal. Bound. Elem. **31**(10), 803–811 (2007)
2. Das, K., Mishra, S.C.: Non-invasive estimation of size and location of a tumor in a human breast using a curve fitting technique. Int. Commun. Heat Mass Transf. **56**, 63–70 (2014)
3. Das, K., Mishra, S.C.: Estimation of tumor characteristics in a breast tissue with known skin surface temperature. J. Therm. Biol. **38**(6), 311–317 (2013)
4. Souza, C., Souza, M., Colaço, M., Caldeira, A., Neto, F.S.: Inverse determination of blood perfusion coefficient by using different deterministic and heuristic techniques. J. Braz. Soc. Mech. Sci. Eng. **36**(1), 193–206 (2014)
5. Minkowycz, W.: Advances in Numerical Heat Transfer, vol. 3. CRC Press, New York (2009)
6. Pennes, H.H.: Analysis of tissue and arterial blood temperatures in the resting human forearm. J. Appl. Physiol. **1**(2), 93–122 (1948)
7. Loureiro, F., Mansur, W., Wrobel, L., Silva, J.: The explicit greens approach with stability enhancement for solving the bioheat transfer equation. Int. J. Heat Mass Transf. **76**, 393–404 (2014)
8. Reis, R.F., Loureiro, F., Lobosco, M.: 3D numerical simulations on GPUs of hyperthermia with nanoparticles by a nonlinear bioheat model. J. Comput. Appl. Math. **295**, 35–47 (2016)
9. Iljaž, J., Škerget, L.: Blood perfusion estimation in heterogeneous tissue using bem based algorithm. Eng. Anal. Bound. Elem. **39**, 75–87 (2014)
10. Hossain, S., Mohammadi, F.A.: Tumor parameter estimation considering the body geometry by thermography. Comput. Biol. Med. **76**, 80–93 (2016)
11. Goldberg, D.E., Holland, J.H.: Genetic algorithms and machine learning. Mach. Learn. **3**(2), 95–99 (1988)
12. Michalewicz, Z., Hartley, S.J.: Genetic algorithms + data structures = evolution programs. Math. Intell. **18**(3), 71 (1996)
13. Hughes, T.J.: The Finite Element Method: Linear Static and Dynamic Finite Element Analysis, vol. 4. Courier Corporation, New York (2012)
14. Geuzaine, C., Remacle, J.-F.: Gmsh: a 3-D finite element mesh generator with built-in pre-and post-processing facilities. Int. J. Num. Methods Eng. **79**(11), 1309–1331 (2009)
15. Liu, J., Xu, L.X.: Boundary information based diagnostics on the thermal states of biological bodies. Int. J. Heat Mass Transf. **43**(16), 2827–2839 (2000)
16. Cao, L., Qin, Q.-H., Zhao, N.: An RBF-MFS model for analysing thermal behaviour of skin tissues. Int. J. Heat Mass Transf. **53**(7), 1298–1307 (2010)

Workshop on Computational Geometry and Security Applications (CGSA 2017)

Nearest Neighbour Graph and Locally Minimal Triangulation

Ivana Kolingerová[1,3]([✉]), Andrej Ferko[2], Tomáš Vomáčka[1], and Martin Maňák[3]

[1] Department of Computer Science, Faculty of Applied Sciences,
University of West Bohemia, Pilsen, Czech Republic
{kolinger,tvomacka}@kiv.zcu.cz
[2] Department of Algebra, Geometry and Didactics of Mathematics,
Faculty of Mathematics, Physics and Informatics,
Comenius University, Bratislava, Slovakia
Andrej.Ferko@fmph.uniba.sk
[3] New Technologies for the Information Society, Faculty of Applied Sciences,
University of West Bohemia, Pilsen, Czech Republic
manak@ntis.zcu.cz

Abstract. Nearest neighbour graph (NNG) is a useful tool namely for collision detection tests. It is well known that NNG, when considered as an undirected graph, is a subgraph of Delaunay triangulation (DT) and this relation can be used for efficient NNG computation. This paper concentrates on relation of NNG to the locally minimal triangulation (LMT) and shows that, although NNG can be proved not to be a LMT subgraph, in most cases LMT contains all or nearly all NNG edges. This fact can also be used for NNG computation, namely in kinetic problems, because LMT computation is easier.

Keywords: Nearest Neighbour Graph · Locally Minimal Triangulation · Delaunay triangulation · Kinetic problem

1 Introduction

Nearest neighbour graph is one of the best known concepts in computational geometry, with applications such as collision detection. Although it is a directed graph by definition, it is often used as undirected. Then a relation between the nearest neighbour graph and Delaunay triangulation, the former being a subgraph of the latter, holds and brings an easy way to compute the nearest neighbour graph edges. A relation to minimum weight triangulation has also been examined. However, there is no research devoted to the relation of the nearest neighbour graph to locally minimal triangulation. This research direction has seemed useless as locally minimal triangulation is rarely used in real applications – although its definition is simpler than that of Delaunay triangulation, the difference is not so substantial to justify the replacement of the Delaunay triangulation, with all its known and proved qualities and efficient algorithms to compute it.

© Springer International Publishing AG 2017
O. Gervasi et al. (Eds.): ICCSA 2017, Part II, LNCS 10405, pp. 455–464, 2017.
DOI: 10.1007/978-3-319-62395-5_31

The situation is different for kinetic data where the points or vertices move. The importance of a simpler condition for the edges or triangles to be tested in computation grows when the triangulation is to be recomputed again and again which is the case for kinetic points. This is a good reason to inspect whether there is a relation between the locally minimal triangulation and the nearest neighbour graph as, if so, the computation of the nearest neighbour graph edges could be simplified.

This paper shows that although generally the nearest neighbour graph is not a subgraph of the locally minimal triangulation, all or nearly all nearest neighbour graph edges can be found in the locally minimal triangulation and thus the replacement of the Delaunay triangulation in kinetic problems, such as a collision application, is possible.

Section 2 contains preliminaries for the static data, Sect. 3 concentrates on kinetic triangulations. Section 4 explains the relation between NNG and LMT. Section 5 presents experiments and results, Sect. 6 concludes the paper.

2   Preliminaries for Static Data

Let us survey definitions of the geometrical graphs used in the paper.

Definition 1 (Triangulation). *The triangulation $T(P)$ of a set of points P in the plane, $P = \{p_0, \ldots, p_{n-1}\}$, $n > 2$, is a maximum set of edges E such that*

- *edges from E intersect only in the points from P and*
- *edges from E subdivide the convex hull of P into triangles.*

The Delaunay triangulation definition is oriented rather to triangles than to edges:

Definition 2 (Delaunay triangulation). *The triangulation $DT(P)$ of a set of points P in the plane, $P = \{p_0, p_1, \ldots, p_{n-1}\}$, $n > 2$, is a Delaunay triangulation of P if and only if the circumcircle of any triangle of $DT(P)$ does not contain a point of P in its interior.*

For our purposes it is useful to have also a condition of emptiness of the circumcircle formed for the edges of DT as for the purpose of nearest neighbour edges search it is more proper to see the triangulations as graphs than as sets of triangles. Thus we include also the edge emptiness property:

Property 1 (Edge emptiness). An edge is in the Delaunay triangulation if and only if the edge has an empty circumcircle [7].

Delaunay triangulation has several important subgraphs – Euclidean minimum spanning tree (EMST), relative neighbourhood graph (RNG), Gabriel graph (GG), with EMST $\subseteq$ RNG $\subseteq$ GG $\subseteq$ DT [15]. These graphs are useful, first of all, in networks optimization and pattern recognition – they help to find structure among the points.

These graphs are non-directed graphs. If we introduce a direction of the edges, we can define another subgraph of DT, a directed subgraph, called the nearest neighbour graph [14]:

Definition 3 (Nearest neighbour graph – $\overline{\text{NNG}}$). *Let P be a set of points in the plane, $P = \{p_0, p_1, \ldots, p_{n-1}\}$, $n > 1$. $\overline{\text{NNG}}$ is a directed graph with P being its vertex set and with a directed edge from p_i to p_j whenever p_j is the nearest neighbour of p_i (i.e., the distance from p_i to p_j is no larger than from p_i to any other point from P) [15].*

$\overline{\text{NNG}}$ is often replaced by an undirected graph underlying $\overline{\text{NNG}}$, let us denote it NNG. There is a relation between NNG and EMST as follows: NNG $\subseteq$ EMST which implies also the relationship NNG $\subseteq$ DT.

Now let us proceed to other triangulations used in this paper:

Definition 4 (Locally minimal triangulation – LMT). *The triangulation $LMT(P)$ of a set of points P in the plane, $P = \{p_0, p_1, \ldots, p_{n-1}\}$, $n > 2$, is a locally minimal triangulation of P if and only if every edge $p_i p_j$ shared by two triangles $p_i p_j p_k$, $p_i p_j p_l$ forming a convex quadrilateral is not longer than the diagonal $p_k p_l$ [9].*

It should be pointed out that LMT is not unique on the given point set P - more triangulations with locally minimal edges can be constructed. LMT is a well-known term in relation to the minimum weight triangulation. Usability of some of its edges to the minimum weight triangulation construction have been examined [1, 4, 5, 8, 9].

Definition 5 (Greedy triangulation – GT). *The triangulation $GT(P)$ of a set of points P in the plane, $P = \{p_0, p_1, \ldots, p_{n-1}\}$, $n > 2$, is a greedy triangulation of P if and only if it consists of the shortest possible mutually non-intersecting edges.*

Existing research results and namely practical experiments show that different types of triangulations are for general data sets not so much mutually different as could be expected. Cho in [6] shows that at least 40% of DT edges are common with the edges of GT for points uniformly distributed in a region. According to his experiments, even about 90% of edges are common between the two triangulations. Kim et al. in [13] found experimentally that more than 90% of the edges of a triangulated terrain model belong to DT and they utilize this fact in a compression.

Many algorithms exist to compute the DT triangulation - local improvement by edge flipping, divide & conquer, incremental insertion, incremental construction, sweeping and high-dimensional embedding. [16] provided a comparison of sequential DT algorithms; although nowadays a bit dated, still this study may provide useful general ideas about fundamental algorithmic strategies utilized to construct DT. [17] provides a set of benchmarks to check the correctness of the DT implementations.

LMT algorithms have not been so deeply researched, however, either the incremental insertion or the local improvement by edge flipping are the easiest choice as the algorithm is nearly the same as for DT, the only difference being a different test on the validity of a triangulation edge. The non-uniqueness of LMT does not bring any problems - the algorithms converge to one of the triangulations satisfying the LMT definition.

3   Preliminaries for Kinetic Data

As shown in [2, 12], we can say that spatial subdivision data structures are defined by functions of the input data called predicates which determine the resulting topology of the data structure depending on the input data. Each geometric structure constructed over a finite set of primitives may be proved valid by checking a finite number of predicates of these primitives. These checks are called certificates. In DT, the certificates are represented by the incircle test function which is a determinant formulation of the empty circumcircle test (1).

$$p_{DT} = det \begin{bmatrix} x_i & y_i & x_i^2 + y_i^2 & 1 \\ x_j & y_j & x_j^2 + y_j^2 & 1 \\ x_k & y_k & x_k^2 + y_k^2 & 1 \\ x_l & y_l & x_l^2 + y_l^2 & 1 \end{bmatrix} \tag{1}$$

In LMT the certificates are comparisons of lengths of the two possible edges - diagonals of the convex quadrilateral formed by two neighbouring triangles (2).

$$p_{LMT} = (x_i - x_j)^2 + (y_i - y_j)^2 - (x_k - x_l)^2 - (y_k - y_l)^2 \tag{2}$$

where p_i, p_j, p_k and p_l are points from P.

As we are considering the kinetic counterparts of these data structures, it is necessary to note that the coordinates of the source data are in fact functions of time. They are most commonly in the form of polynomials, but they may be any functions as long as it is possible to solve them to find their roots with a necessary precision.

In order to maintain a kinetic data structure, one has to maintain a priority queue of the currently active predicates sorted by the increasing value of the nearest next root (greater than the current time value). Each time the value of current time reaches the value determined by the first certificate function in the queue, at least one of the certificates becomes invalid and the topology of the kinetic data structure needs to be changed. This change depends on the type of the data structure - in the case of kinetic Delaunay triangulation (KDT) and kinetic locally minimal triangulation (KLMT) they are represented by an edge swap in the triangle pair associated with the predicate (and certificate). Also, usually there are new certificates that need to be placed into the queue. Swap conditions for moving circles and line segments can be found in [10] and for spheres in [11].

From the facts stated above it is obvious that in order to maintain a kinetic data structure one has to solve a large number of algebraic equations. Let us now have a set of points $P = \{p_0, p_1, \ldots, p_{n-1}\}$, $n > 2$, KDT(P) and KLMT(P). Let also $p_i = (x_i(t), y_i(t))$, where $x_i(t)$ and $y_i(t)$ are polynomial functions of degree $R \geq 1$.

Lemma 1. *The computation of certificates for KDT(P) is at least as computationally complex as for KLMT(P) constructed over the same set of kinetic points moving along polynomial trajectories of degree up to R.*

Proof. If the two data structures are constructed over a set of static points ($R = 0$, thus voiding their kinetic property), their kinetic maintenance is no longer necessary which applies to both of the structures. In every other case we can see from (1) that in order to compute the predicates for KDT(P), we need to search periodically for the roots of polynomials of the degree up to $4R$ and from (2) we can see that in case of KLMT(P) we only need to solve polynomials of degree up to $2R$.

From the practical point of view, it may be worth mentioning that the polynomials $x_i(t)$ and $y_i(t)$ are often of degree one (meaning that the points move along linear trajectories) which results in solving quadratic equations for KLMT and fourth-degree polynomials for KDT. Note that solving fourth-degree polynomials usually cannot be done analytically which further increases the overall complexity of maintaining the KDT, especially when compared to KLMT.

Having all necessary information, we can proceed to the relation between NNG and LMT, the core topic of this paper.

4   Relation of NNG and LMT

The expectation that NNG edges are not only a subset of DT but also of LMT is supported by the fact that LMT contains "short edges". Unfortunately, the situation is not so simple and, generally, NNG is not contained in LMT.

Lemma 2. *Let us have a LMT on the set of points P. Then NNG of the given set of points can contain edges which do not belong to LMT.*

Proof. To prove the lemma it is enough to find a data set P, which will produce an edge in NNG that will not belong to LMT. Such a data set can be found even on four points: $P = \{p_0, p_1, p_2, p_3\}$. If the line segments p_0p_1 and p_2p_3 intersect in exactly one point $\notin P$ and if $|p_2p_3| < |p_0p_1| < min(|p_0p_2|, |p_0p_3|)$ then the points of P constitute a convex quadrilateral with diagonals p_0p_1 and p_2p_3, the edge from p_0 to p_1 is in NNG but not in LMT since p_2p_3 is shorter than p_0p_1.

Figure 1 illustrates the proof of the lemma. Points p_0 and p_1 can be chosen arbitrarily, then the point p_2 is chosen from the first highlighted region and after that, the point p_3 is chosen from the second highlighted region. The circles have centres in p_0, p_1, p_2 and they all have the same radius $|p_0p_1|$.

From this example it is easy to see that NNG is generally not included in LMT. But due to the fact that triangulations used in real applications have a considerable common subset of edges, it can be expected that many edges of NNG can be produced from LMT. In real applications, e.g., for collision tests in virtual reality, where savings in computation can be substantial, some small omissions in NNG can be tolerated in favour of increased efficacy of computation. Thus in the next section we will concentrate on the expected behaviour of the DT, LMT and NNG in experiments for various data. Our effort is to show that most of the NNG edges are mostly present in LMT in spite of the fact that one-hundred-percent-inclusion cannot be guaranteed.

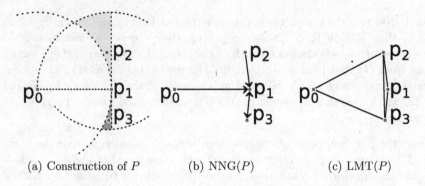

(a) Construction of P (b) NNG(P)   (c) LMT(P)

Fig. 1. A set $P = \{p_0, p_1, p_2, p_3\}$ of points for which NNG(P) $\not\subseteq$ LMT(P)

5   Experiments and Results

Testing data sets were both real terrain data (kindly provided by T. Bayer [3]) and artificially generated data sets with randomly generated points distributed uniformly, in clusters, with Gaussian distribution and on an arc, see Fig. 2. The data sets did not contain any duplicated points. The cluster data are represented by 10 uniformly generated points used as a centre of a Gaussian distribution. The artificial data sets were from 100 up to 10000 points, the real data from about 11000 up to about 150000. Examples of real data sets are depicted in Fig. 3. As there can be more LMT on one data set, each test was repeated for several LMT versions but the differences in results were negligible.

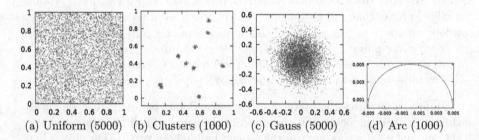

(a) Uniform (5000)   (b) Clusters (1000)   (c) Gauss (5000)   (d) Arc (1000)

Fig. 2. Examples of artificial data sets used in experiments

5.1   Similarity Between DT and LMT

Similarity S of two triangulations is given by (3).

$$S = \frac{ne_{LD}}{ne_L} 100 \ [\%] \tag{3}$$

where ne_L is the total number of edges of LMT, ne_{LD} the number of common edges of LMT and DT.

(a) Data ID 9   (b) Data ID 10

Fig. 3. Examples of real data sets

Table 1. Similarity between LMT and DT computed for real data as the percentage $S\%$ of LMT edges that are also present in DT. Columns ne_L contain the total number of LMT edges, ne_{LD} the number of LMT edges in DT.

Data ID	Vertices nv	Edges ne_L	ne_{LD}	$S\%$	Data ID	Vertices nv	Edges ne_L	ne_{LD}	$S\%$
1	11495	34457	30835	89.49	9	78659	235946	210409	89.18
2	18552	55623	50174	90.20	10	79903	239678	215309	89.83
3	21407	63502	57609	90.72	11	92845	275664	250548	90.89
4	25392	74194	66486	89.61	12	98616	295807	266949	90.24
5	26207	78588	75216	95.71	13	100632	301863	272425	90.25
6	30264	86372	77524	89.76	14	110369	326589	289296	88.58
7	49391	148146	132322	89.32	15	141045	423083	375461	88.74
8	74110	221423	196823	88.89	16	148559	445641	398941	89.52

Table 1 shows results of DT and LMT comparison for real data.

From Table 1 it can be seen that similarity of both types of triangulations is relatively high for real data, DT and LMT have about 90% edges in common. No dependence of results on the size of data set was detected. These results are in correspondence with the previous observation in [6,13].

Table 2 shows similarity for artificially generated data. Gauss data indicate a bit lower similarity but still about 85%. Uniform data provided similar percentage, but the decrease of S with growing triangulation size. Cluster data presented very low similarity for two largest data sets. Arc data have only about 59% similarity.

Although the cases with high similarity, corresponding to the previous results from [6,13], were prevailing in the tests, during experiments we came also across the data sets which had a substantially lower similarity of LMT and DT, such as cluster data sets in the rows 6 and 7 of Table 2. These were the cases of data sets which contained many configurations with concentric points. Let us consider a small data set of several points sitting on a circle empty from other points, then DT of this point group is ambiguous - there are more than one DT. The simplest example is a group of four points in a square position, for which any of both square diagonals (but only one of them) is present in DT. In such a case even two DTs of the same set would have lower similarity than 100%. If many such cases are present in the testing data sets, similarity of two triangulations is decreased.

Table 2. Similarity between LMT and DT computed for sets of random points (various distributions) as the percentage $S\%$ of LMT edges that are also present in DT. Columns ne_L contain the total number of LMT edges, ne_{LD} the number of LMT edges in DT.

Vertices	Gauss			Uniform			Clusters			Arc		
nv	ne_L	ne_{LD}	$S\%$	ne_L	ne_{LD}	$S\%$	ne_L	ne_{LD}	$S\%$	ne_L	ne_{LD}	$S\%$
100	285	252	88	286	262	92	280	241	86	102	64	63
500	1486	1294	87	1479	1307	88	1481	1305	88	243	144	59
1000	2984	2633	88	2982	2641	89	2987	2635	88	335	183	55
5000	14980	13191	88	14966	13063	87	14980	13185	88	1019	361	35
10000	29965	25738	86	29967	26281	88	29978	26386	88	25594	19188	75
50000	149620	118979	80	149870	127395	85	149952	46377	31	47876	31903	67
100000	298596	231843	78	299613	228589	76	299917	51960	17	63695	38101	60

5.2   NNG Edges Inclusion in LMT

Next we tested how many NNG edges (computed from DT) can be found in LMT, see Tables 3 and 4. The tests were done on condition that each vertex has only one nearest neighbour – i.e., for each vertex, only one outgoing edge was examined. Such a condition would be improper for degenerated cases, such as the points sitting on a regular grid but such data are usually not subject to DT as they can be triangulated directly.

It can be seen that all or nearly all NNG edges can be found in LMT, even for the triangulations which showed a lower mutual similarity in the previous group of tests.

Table 3. The percentage of NNG edges found also in LMT for real data sets

Data ID	Vertices nv	Edges $ne[\%]$	Data ID	Vertices nv	Edges $ne[\%]$
1	11495	99.93	9	78659	99.95
2	18552	99.91	10	79903	92.94
3	21407	98.42	11	92845	98.92
4	25392	96.79	12	98616	99.94
5	26207	99.99	13	100632	99.95
6	30264	93.37	14	110369	98.16
7	49391	99.90	15	141045	99.96
8	74110	99.43	16	148559	99.96

Table 4. The percentage of NNG edges (*ne*) found also in LMT for various artificial data sets.

Vertices	Gauss	Uniform	Clusters	Arc
nv	*ne* [%]	*ne* [%]	*ne* [%]	*ne* [%]
100	100	100	100	100
500	99.80	100	99.60	100
1000	99.90	100	100	100
5000	99.94	100	99.98	100
10000	99.99	100	99.97	100
50000	99.98	100	99.89	100
100000	99.98	100	99.79	100

6   Conclusion

From these experiments a conclusion can be drawn that for practical applications, LMT can serve as a good source of NNG edges. It may be interesting namely for applications with kinetic data, such as the collision detection or collision avoidance, where kinetic triangulations are needed, as even the small savings achieved by a simpler predicate/certificate computation in the kinetic LMT can be important due to a high number of their recomputations.

In future work it would be interesting and useful to examine the 3D versions of the triangulations and their relation to NNG where even higher impact on collision detection tests can be expected.

Acknowledgements. This work was supported by the Ministry of Education, Youth and Sports of the Czech Republic, the project SGS-2016-013 Advanced Graphical and Computing Systems and the project PUNTIS (LO1506) under the program NPU I. We would like to thank to T. Bayer from the Charles University in Prague, Czech Republic for supplying us the real terrain data for the experiments.

References

1. Aichholzer, O., Aurenhammer, F., Cheng, S.W., Katoh, N., Rote, G., Taschwer, M., Xu, Y.F.: Triangulations intersect nicely. Discrete Comput. Geom. **16**(4), 339–359 (1996)
2. Basch, J., Guibas, L.J., Hershberger, J.: Data structures for mobile data. J. Algorithms **31**(1), 1–28 (1999)
3. Bayer, T.: Department of Applied Geoinformatics and Cartography, Faculty of Science, Charles University, Prague, Czech Republic. https://web.natur.cuni.cz/~bayertom. Accessed 15 May 2017
4. Beirouti, R., Snoeyink, J.: Implementations of the LMT heuristic for minimum weight triangulation. In: Proceedings of the Fourteenth Annual Symposium on Computational Geometry, SCG 1998, pp. 96–105. ACM, New York (1998)

5. Bose, P., Devroye, L., Evans, W.: Diamonds are not a minimum weight triangulation's best friend. Int. J. Comput. Geom. Appl. **12**(06), 445–453 (2002)
6. Cho, H.G.: On the expected number of common edges in Delaunay and greedy triangulation. J. WSCG **5**(1–3), 50–59 (1997)
7. Dey, T.K.: Curve and Surface Reconstruction: Algorithms with Mathematical Analysis (Cambridge Monographs on Applied and Computational Mathematics). Cambridge University Press, New York (2006)
8. Dickerson, M.T., Keil, J.M., Montague, M.H.: A large subgraph of the minimum weight triangulation. Discrete Comput. Geom. **18**(3), 289–304 (1997)
9. Dickerson, M.T., Montague, M.H.: A (usually?) connected subgraph of the minimum weight triangulation. In: Proceedings of the Twelfth Annual Symposium on Computational Geometry, SCG 1996, pp. 204–213. ACM, New York (1996)
10. Gavrilova, M., Rokne, J.: Swap conditions for dynamic Voronoi diagrams for circles and line segments. Comput. Aided Geom. Des. **16**(2), 89–106 (1999)
11. Gavrilova, M., Rokne, J.: Updating the topology of the dynamic Voronoi diagram for spheres in euclidean d-dimensional space. Comput. Aided Geom. Des. **20**(4), 231–242 (2003)
12. Guibas, L., Russel, D.: An empirical comparison of techniques for updating Delaunay triangulations. In: Proceedings of the Twentieth Annual Symposium on Computational Geometry, SCG 2004, pp. 170–179. ACM, New York (2004)
13. Kim, Y.S., Park, D.G., Jung, H.Y., Cho, H.G., Dong, J.J., Ku, K.J.: An improved TIN compression using Delaunay triangulation. In: Proceedings of Seventh Pacific Conference on Computer Graphics and Applications (Cat. No. PR00293), pp. 118–125 (1999)
14. Okabe, A., Boots, B., Sugihara, K., Chiu, S.N.: Spatial Tessellations: Concepts and Applications of Voronoi Diagrams. Probability and Statistics, 2nd edn. Wiley, New York (2000)
15. Preparata, F.P., Shamos, M.: Computational Geometry: An Introduction. Springer, New York (1985)
16. Su, P., Drysdale, R.L.S.: A comparison of sequential Delaunay triangulation algorithms. Comput. Geom. **7**(5), 361–385 (1997)
17. Špelič, D., Novak, F., Žalik, B.: Delaunay triangulation benchmarks. J. Electr. Eng. **59**(1), 49–52 (2008)

A Clustering Approach to Path Planning for Groups

Jakub Szkandera[1,2(✉)], Ondřej Kaas[1,2], and Ivana Kolingerová[1,2]

[1] Department of Computer Science and Engineering,
Faculty of Applied Sciences, University of West Bohemia,
Univerzitni 8, 30614 Plzen, Czech Republic
{szkander,kaas,kolinger}@kiv.zcu.cz
[2] New Technologies for the Information Society,
Univerzitni 8, 30614 Plzen, Czech Republic

Abstract. The paper introduces a new method of planning paths for crowds in dynamic environment represented by a graph of vertices and edges, where the edge weight as well as the graph topology may change, but the method is also applicable to environment with a different representation. The utilization of clusterization enables the method to use the computed path for a group of agents. In this way a speed-up and memory savings are achieved at a cost of some path suboptimality. The experiments showed good behaviour of the method as to the speed-up and relative error.

Keywords: Path planning · Agent based model · Graph representation · Clustering

1 Introduction

The path planning or more specifically the task of finding a path between two or more spots in some environment is an important research problem, useful in many different applications, e.g., in robotics, molecular biology or simulations of crowds. A graph of vertices and edges often describes the environment, which can be static or dynamic. In the dynamic environment, the environment properties or its topology may change over time. The static environment does not allow anything of that. The paths are planned for entities, which are often called agents.

While the path planning for one agent in the static environment can be considered as a solved problem, the situation is different for more agents or even crowds in a dynamic environment: a repeated recomputation of path for many changes and many agents may be too slow. For a huge amount of agents the optimality of the paths is less important than the speed of computation as the application for this version of path planning problem usually rather needs the crowd to look well and realistic than to compute and use optimal paths. Therefore, there is a research space for algorithms, sacrificing optimality to speed.

© Springer International Publishing AG 2017
O. Gervasi et al. (Eds.): ICCSA 2017, Part II, LNCS 10405, pp. 465–479, 2017.
DOI: 10.1007/978-3-319-62395-5_32

We proposed an algorithm that takes advantage of path similarity of some agents [42]. The agents are grouped according to their initial and target position and the path is found only once for the whole group. The method brings a speed-up and memory savings but there is still space for an improvement. In this paper we proposed to incorporate a more sophisticated approach of groups formation based on clustering. In this way it is possible to increase the speed-up without fatal consequences on the relative error of the produced paths.

This paper is organized as follows. Section 2 outlines existing path planning methods, which are suitable for the simulation of crowds, and clustering methods. Section 3 describes the proposed solution. Section 4 contains the results of experiments performed on real environment data and a comparison to classic path planning approach. Section 5 concludes the paper.

2   Related Work

2.1   Path Planning

The models for the movement simulation and planning of crowds can be divided into two main categories – the agent-based model and the continuum model.

The more natural for human beings is the first mentioned model where a path is planned for each agent separately. Every agent can have his or her individual requirements and the biggest advantage is that the agent-based model respects them. On the other hand the path planning for this model may be quite challenging in terms of time and memory complexity. Moreover, with an increasing number of agents this approach stops to be a real time and may become unsuitable.

Agent-based methods can be divided into local and global methods. The global methods which help to locate possible evacuation-critical spots in buildings and simulate an emergency scenario belong to fire-escape algorithms [10,30]. Algorithms [4,34] similar to the fire-escape algorithms have been proposed but with a focus on the bottlenecks. Also human behaviour have been examined by many interesting proposed approaches [15,31].

The graph representation of the environment is often used for the global agent-based path planning. The most widespread A* algorithm [17], which uses a heuristics to speed up the planning, is proposed for a static environment as well as the basic algorithms – Breadth First Search (BFS), Depth First Search (DFS), Dijkstra's algorithm or Floyd-Warshall algorithm [12,45], which rank among all-pair algorithms. The minimal cost path for all pairs of vertices is found in the memory complexity $O(n^2)$ and time complexity $O(n^3)$ in the worst case for a graph with n vertices. In the case of dynamic graphs, where weights of vertices and edges may change over time, the D* algorithm [38] and its improvement D* Focused [39] are more suitable. The D* Lite [22], which modifies the backwards algorithm LPA* [23], may yield even better results than D* Focused. The memory complexity $O(k(n+m))$ for k agents, n vertices and m edges can be easily reached by these algorithms because each agent needs its own graph ranking. Hierarchical Annotated A* [16] creates a hierarchical graph to find an almost optimal path and Anytime D* [24] finds a sub-optimal path in a limited time. The path planning for a moving target solves Moving Target D*-Lite [41] or Generalized Fringe-Retrieving A* [40].

Local methods are less diverse because they are mostly focused on the collision detection between agents themselves or between agents and obstacles. Methods of collision avoidance have been developed including grid-based rules [25] or Bayesian decision processes [28]. The parallel computation of the collision avoidance has also been proposed [14]. The smoothing of the planned path for autonomous vehicles [7] or robots [44] belongs into another bigger group of local approaches.

The disadvantages of the agent-based model solves the continuum path planning model which is suitable for dense crowds. The Navier-Stokes equations [13] can describe the movement in a dense crowd which is similar to the fluid flow [8] as well as the car traffic [5]. A continuum model that introduces an evolving dynamic potential function and describes a crowd as a continuous density field was proposed by Hughes [18]. These partial differential equations describe the continuous density field. This field can be changed into a particle description of the crowd [43], which reproduces emerging phenomena of real crowds. The smooth movement of the agents in a complex environments can be achieved with the sophisticated continuum model [20] that was expanded from [43]. Moreover, the continuum model problems can be solved in parallel [27]. Although the continuum model is suitable for the dense crowds, the disadvantage is that the fluid simulation is ruled by the laws of physics. Therefore individual requirements of crowd members are neglected.

2.2 Creating Groups

The clustering algorithms group similar elements (called *clients*) together [1,19] into groups (called *clusters*) represented by one centroid (called a *cluster center* or a *facility*). The nature of the elements are based on the application and in a general case they could be any objects, which can be described by a characteristic N-dimensional vector. Depending on the area of application, a transformation into the N-dimensional space may be needed to express elements. For example, for processing a point in a N-dimensional space its N spatial coordinates are used. A similarity of elements is measured by a distance function. The function can be modified based on application requirements and basically might not fulfill the properties of the metrics such as the non-negativity, symmetry or the triangle inequality. However, the Euclidean distance is used most often.

The literature shows several clustering methods such as single-link [33,37], complete-link [21], sweep-line [46], k-means [26] and facility location [6].

Now we will define the clustering task more precisely. As we have chosen the facility location algorithm for our purpose, we will present the definition suitable for this type of algorithm.

A given data set of N input data points $x_1, x_2, \ldots, x_N$ is subdivided into k disjoint subsets $F_i, i = 1, \ldots, k$ each containing n_i data points, $0 < n_i < N$ by minimizing the following mean-square-error (MSE) clustering cost:

$$J_{MSE} = \sum_{i=1}^{k} \sum_{x_j \in F_i} \|x_j - f_i\|^2 \tag{1}$$

where x_j is a vector representing the j-th data point in the cluster F_i and f_i is the cluster centre of the cluster F_i. The J_{MSE} function in Eq. (1) represents the distance between the data point x_j and the cluster center f_i.

As we can see, the number of clusters has to be determined before clustering, which is not suitable in all scenarios. Moreover, if k equals to N, J_{MSE} loses its measuring property and the result of clustering is not correct. The facility location algorithm keeps a reasonable amount of clusters by minimization of the following clustering cost:

$$J_{FL} = \sum_{f_i \in F} fc + \sum_{j \in C} c_{f,j} \tag{2}$$

where $c_{f,j}$ is the distance between a data point $j \in C$ and its facility $f_i \in F$. The set C contains all data points. To open a new cluster center, a cost fc must be paid, this way a quantity of cluster centers can be controlled.

The clustering task is an NP-hard problem, so the most algorithms produce only approximate results or have some restrictions. On the other hand, in many scenarios the approximate solution suffices.

Clustering methods are used in technical as well as non-technical disciplines, such as data analysis [2,9], data mining [11], image segmentation [19], pattern recognition [3], information retrieval [32].

There are several ways how to categorize the clustering algorithms [35]. One of possible subdivisions is into *partitional* and *hierarchical* methods. The partitional ones divide the data into an exact number of clusters (partitions), the hierarchical ones create a hierarchy of small clusters grouped into larger clusters forming a tree structure. Another possible subdivision is into *agglomerative* and *divisive* (or *partitional*) approaches. The agglomerative ones start with each element in a single cluster. The final result is formed by successively merging these clusters according to a similarity measure until a stopping condition is met. The divisive approaches start with one large cluster which contains all the data. By repeatedly splitting clusters according to a dissimilarity measure smaller clusters are created. Both agglomerative and divisive algorithms stop when there is a predetermined number of clusters or when existing clusters are homogeneous enough so that no further iteration is needed.

The clustering is called hard if each element is assigned into exactly one cluster. Fuzzy clustering determines for each element a degree of membership in several clusters.

Clustering algorithms can be *deterministic* or *stochastic*. Stochastic techniques usually use randomized algorithms which are more suitable for processing large amounts of data due to their smaller time complexity.

3 Proposed Solution

This section describes the proposed path planning approach for many agents in an environment represented by an undirected graph $G = (V, E)$, where V is a finite, non-empty set of vertices, and E is a set of unordered pair of vertices $(E \subset \binom{V}{2})$ called edges. First, we will summarize the idea of the groups path planning from the submitted paper [42] and then we propose a better and faster solution with the use of clusters.

Let $P = \{p_1, p_2, \ldots, p_c\}$ be a set of agents. Each $p_i \in P$ needs to individually rate vertices of an undirected graph. The graph represents the dynamic environment the dynamics of which can be interpreted as a set of pairs $D = \{(d_1, t_1), (d_2, t_2), \ldots, (d_r, t_r)\}$, where d_i is a graph change and t_i is a simulation time. Each vertex of the graph describes a point in $\mathbb{R}^2$. Moreover, every agent p_i has a starting vertex with the position $s(p_i) \in \mathbb{R}^2$ and a goal vertex with $e(p_i) \in \mathbb{R}^2$.

3.1 Summarized Group Approach

Let $g \subseteq P$ be a group of agents. For each group we find a leader $p_m \in g$ who will be followed by others. Figure 1 illustrates the idea. First, any standard path planning algorithm is used to compute the path of the leader from the start $s(p_m)$ to the end $e(p_m)$. After that, the path of every agent $p_i \in g \setminus \{p_m\}$ is planned: First from $s(p_i)$ to $s(p_m)$, then the path of the leader p_m is reused, and finally the part from $e(p_m)$ to $e(p_i)$ is computed.

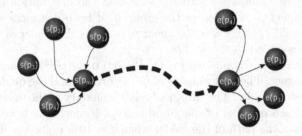

Fig. 1. The idea of path planning for a group of agents. The path of each agent p_i starts in $s(p_i)$ and ends in $e(p_i)$, p_m is the leader. The others can reuse leader's path.

In some cases the path requires further optimization, which will be discussed next, because the path cost is not necessarily optimal. For example, the nearest vertex of the leader's path for p_i may be other vertex than $s(p_m)$, then it has no sense to go to $s(p_m)$ and back again. Therefore, the path planning is modified in the following way to handle such situations. When the agent p_i reaches the path of the leader p_m, the path planning to the start position $s(p_m)$ is stopped and the essential part of the path of the leader p_m is used instead (Fig. 2b). Once the agent p_i reaches the destination position $e(p_m)$, path planning algorithm from the position $e(p_m)$ to $e(p_i)$ should be started. However, Fig. 2c illustrates that a situation similar to Fig. 2a may happen. The change of the path planning direction transforms this problem to the already solved problem (Fig. 2b).

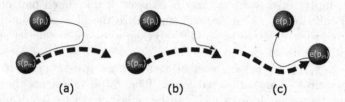

(a)   (b)   (c)

Fig. 2. Group path planning problems and solutions (a) Problem near the start position, (b) Solution for the case (a) (c) Problem near the destination.

The criteria for creating groups of agents with similar start positions and destinations and choosing a leader are as follows. A list P of agents and a grouping threshold τ (will be discussed later) are taken as the input of the group approach and a set $Gr = \{g_1, g_2, \ldots, g_q\}$ of non-empty distinct groups of agents is produced. A path planning strategy is also needed because the computation of the path of an agent from the group g_j depends on the path of its leader. Any classic path planning algorithms, e.g., A*, D* or D* Lite, may be used.

The groups are created as follows: The group algorithm iterates over all agents and tries to add them to all groups created so far. The first agent added to the group is always the leader p_m of the group. For every agent p_i its positions $s(p_i)$ and $e(p_i)$ are compared with the start and the destination of the leader $p_m \in g_j$. The agent p_i is added to the group g_j if both distances are less than the error threshold τ. Otherwise the agent p_i is set as the leader of a new group and the group is added to the set Gr.

The worst-case time complexity is $O(|P||Gr|)$ or $O(|P|^2)$ if each group contains only one agent. The performance of the algorithm and the quality of results depend on the threshold τ. The threshold τ is a boundary of the maximal allowed relative error. Each member of the group (except leader) may have its path 1.1 times longer than the path of the leader when $\tau = 10\%$ is chosen. The higher τ the faster the path computation is. On the other hand with the growing threshold τ the path inaccuracy grows.

3.2 Clustering Approach

The output groups of the group approach are sensitive to the order of the input data because the algorithm has a greedy character - the first possible solution is accepted and never reconsidered. An additional optimization of the found groups would improve the final paths but deteriorate the complexity which is already $O(n^2)$. What is more, although the group approach speeds up the path computation, there is still room for acceleration. Therefore, we incorporate to the group creation a clustering by a non-modified local search algorithm [29, 36] with relevant parameters setup discussed in Sect. 4. The algorithm implicitly optimizes the agent groups and their found paths. Moreover, the clustering even speeds up the computation because the clustering can be done in $O(n \log n)$.

For clustering purposes the agent p in the input set $P = \{p_1, p_2, \ldots, p_N\}$ is described as the following 4-dimensional vector:

$$v_i = \big(s(p_i), e(p_i)\big)^T, v_i \in \mathbb{R}^4 \tag{3}$$

where $s(p_i) \in \mathbb{R}^2$ is the starting and $e(p_i) \in \mathbb{R}^2$ the destination position of the agent.

The main idea of the proposed improvement is to incorporate Eq. (2) as a heuristic to avoid checking so many possible cases compared to the original group approach at a price of a lowered accuracy of the clustering result.

Let us describe the used *local search* clustering algorithm steps in detail. At first, a coarse initial solution is generated. A cluster center is always created at the first point and further points are then taken in the random order. A point v_i is connected to the closest already open cluster center based on the measured distance d between point the v_j and the open cluster center. A new cluster center is opened at the point v_j with probability d/fc (or one if $d > fc$). This initial coarse solution is improved by $O(n \log n)$ iterations of the following local search step. The explanation for $O(n \log n)$ steps of iterations is given in [36].

A single local search step can be described as a random selection of a point $v_i \in C \bigcup F$ (it does not matter whether it is a cluster center or not) and it is computed whether the agent p_i can improve the current solution (if v_i is not already a cluster center, the facility cost would have to be paid for its opening). Some clients (points) may be closer to the investigated (new) cluster center f_{v_i} than to their current facility. All such clients can be re-assigned to f_{v_i}, it decreases the connection cost. If these changes result in some cluster center having only a few clients remaining, the cluster center could be closed and its facility cost spared.

A possible improvement of the current solution by declaring the point v_i a new cluster center f_{v_i} and reassigning all near clients from their cluster centers to f_{v_i} is determined by a gain function according to the following relation:

$$gain(v_i) = -fc + \sum_{c_i \subseteq C} ds_i + \sum_{f_j \subseteq F} cs_j \tag{4}$$

where fc is the facility cost, or zero if the cluster center f_{v_i} is already open, ds_i is the distance spared by reassigning the client c_i from its current cluster center to the cluster center candidate f_{v_i} and cs_j (close spare) is the facility cost minus expenses for reassigning all remaining clients from their current cluster center f_j to f_{v_i}. If the current cluster center f_j lies closer to v_i than f_{v_i} then $ds_i < 0$ and ds_i needs to be set to 0. Again, if $cs_j < 0$ (cluster center f_j has enough clients, so no spare can be achieved by closing the cluster center and reassigning all their clients to the new cluster center f_{v_i}) then cs_j is set to 0. If $gain(v_i) > 0$, the cluster center at point v_i is opened (if not already open) and reassignments and closures are performed.

The algorithm of the group approach with the facility location clustering is summed up as Algorithm 1. When the clusters, which represent the groups of agents, are computed (Algorithm 1, line 1) from the input set of agents P, the path of each agent is planned by any standard path planning strategy, e.g., Dijkstra, A* or D*. First the path of the leader p_m of the group g_i is found (Algorithm 1, line 4). Then two paths are computed each other member of the group g_i - first between vertices $s(p_j)$ and $s(p_m)$ and second between $e(p_j)$ and $e(p_m)$ (Algorithm 1, lines 6–7). Finally, this paths and path of the leader are joined (Algorithm 1, lines 8). This process goes in cycle for each group $g_i \in Gr$. The procedure $cluster_agents$ generates first a coarse solution of agent groups, then it repeatedly tries to improve it by reassignment of a randomly chosen point to another group. If the new connection improves the overall clustering cost J_{FL}, then it is preserved and another randomly chosen point is investigated. The improvement phase is repeated $O(n \log n)$ times.

Algorithm 1. The clustering path planning approach

Data: A list P of agents, a path planning strategy find_path(...)
Result: The list P of agents with computed paths
Algorithm cluster_approach
1      $Gr \leftarrow$ cluster_agents(P); // groups of agents
2   **foreach** $g_i \in Gr$ **do**
3          $p_m \leftarrow$ the first member of g_j; // the leader of the group
4   p_m.path $\leftarrow$ find_path($s(p_m)$, $e(p_m)$);
5   **foreach** $p_j \in \{g_i \setminus \{p_m\}\}$ **do**
6   path_start $\leftarrow$ find_path($s(p_i)$, $s(p_m)$);
7   path_end $\leftarrow$ find_path($e(p_i)$, $e(p_m)$);
8   p_j.path $\leftarrow$ Join_paths($path\_start$, p_m.path, $path\_end$)
          end
      end
      return P;
  Procedure cluster_agents(*the list of agents P*)
1   Generate a coarse solution of groups.
2   **repeat** $O(n \log n)$ times
3   Pick $v_i \in P$ at random;
4   **if** $gain(v_i) > 0$ **then**
5   Perform reassignments and closures.
      end
      return groups;

---

4   Experiments and Results

The proposed method was implemented in C# and all experiments were performed on a computer with the CPU Intel®Core™ i7-950 (8 MB Cache,

3.07 GHz) and 12 GB 668 MHz RAM. The proposed solution was tested on two types of the testing data. First type ('unsuitable data') contains agents generated at random positions and the second type ('suitable data') contains groups of agents (many agents with similar paths). The environment is represented by real data – the Open Street Map of the City of Pilsen, Czech Republic.

We were unable to determine the computational time and space generally because they highly depend on the distribution of the agents and the chosen path planning algorithm. For instance the A* algorithm as the path planning method will not save computational space because it does not need to store a unique graph rating for each agent. However, dynamic algorithms such as D* Lite algorithm, as mentioned in Sect. 2, are able to save up to $O(kn)$ space for k similar paths and n vertices. Therefore, the experiments focus on the computational time and especially on the path correctness of the proposed solution.

4.1   Computational Time

The computational time of the clustering method without the path planning is shown in Fig. 3, where the time dependencies on the facility cost fc are depicted. The higher value of fc produces bigger clusters and speeds up the computation of the clusters. The reasonable facility cost is for $fc > 0.1$, where the computational time becomes relatively very low (Fig. 3) and with the growing fc the time still descents.

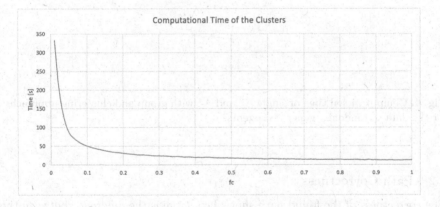

Fig. 3. Computational time of the clustering method for 100 k agents

Figure 4 shows the computational time of the standard A* algorithm, the A* with the group approach and with the proposed clustering approach (Algorithm 1). The experiments were made for $100k$ randomly generated agents and the measured computational time of both approaches includes the groups creation and path planning strategy. The group approach depicted in Fig. 4 uses $\tau = 10\%$ and the clustering approach is showed with the facility cost $fc = 0.5$.

The proposed clustering approach is the fastest. Small fluctuations on the time curve are caused by the random character of the clustering algorithm. The curve of the clustering approach is $O(n \log n)$, the group approach curve is $O(n^2)$ and the A* curve is $O(n)$. Although the A* is in the graph the slowest, obviously it thanks to its better algorithmic complexity for some number of agents overruns both the group and the clustering approaches. For the given size of the environment it will be $\approx 300\,k$ agents for the group approach and over millions of agents for the clustering approach. This number is relatively high and so it is not such a big problem as it might seem.

However, the data of randomly generated agents are the worst possible for the proposed approach. The clustering approach is most suitable for the groups of agents where the clustering approach is a clear choice because it is able to reuse many paths and save a huge amount of the computational time (Fig. 5). The same computational time of the A* algorithm and the clustering approach with A* algorithm is for billions of agents for the data of the groups of agents.

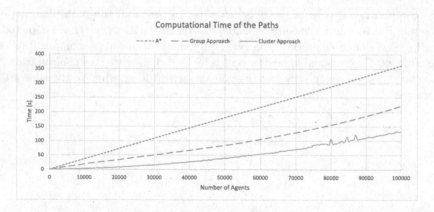

Fig. 4. Computational time of single A* and A* with group and clustering approaches for the data of randomly generated agents.

4.2   Path Correctness

The correctness of the found path shows how large is the difference between the minimal path and the path found by the proposed solution. The path correctness is measured by a relative error (5).

$$\delta = \frac{length(group.path) - length(minimal.path)}{length(minimal.path)} \tag{5}$$

where *group.path* is the path found with the group or the clustering approach including existing path planning method and *minimal.path* is the path found with the same path planning algorithm without the proposed solution. Note that $\delta \geq 0$.

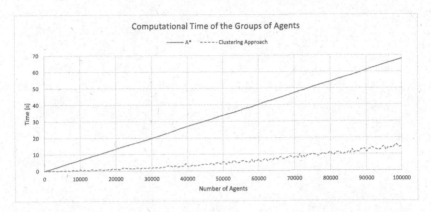

Fig. 5. Computational time of single A* and A* with the clustering approach for the data of the groups of agents.

The average relative error dependency on the chosen facility cost is shown in Fig. 6 for two above mentioned types of the testing data. The upper curve, which reaches the relative error up to 15%, represents $100k$ randomly generated agents positions and the lower curve represents data with generated groups of agents (a lot of similar paths). The average relative error of the groups does not exceed 2%. This is an expected result because this type of data are most suitable for the proposed clustering approach. Moreover, the result of the worst case (randomly positioned agents) is acceptable and for the facility cost up to the value 0.1 is comparatively very good. It can be also seen that the relative error does not change too much if $fc > 0.5$. It means that if a higher relative error is acceptable, number of clusters can be kept relatively small. The parameter fc can be understood as percentage expression number of clusters, where approximately $100(1 - fc)$ per cent of agents from the input data become cluster centers, see Subsect. 3.2.

Figure 7 shows the dependency of the relative error on the relative number of the created groups and the difference between the original group approach (without clustering) and the group approach with clustering (Algorithm 1). The x-axis describes the relative number of groups $\lambda = \frac{|Gr|}{|P|}$, where $|Gr|$ is the number of created groups and $|P|$ the number of tested agents. The experiments have been performed on the randomly generated data of 100 agents. The higher the number of groups is, the fewer agents each group contains. For example, when the proposed solution creates $100\,k$ groups (clusters) from $100\,k$ agents, the average relative error will be zero because every agent is a leader of a group. The other extreme is only one group (a cluster) created from $100\,k$ agents. In that case, relative error will be enormous because everyone has to follow the leader. The average relative error of both approaches grows with the decreasing number of groups, as can be expected. Although both approaches are almost the same for a high number of groups, the clustering approach gives better results for a smaller number of groups, which contain a higher number of agents. The average relative

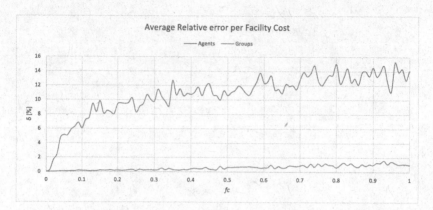

Fig. 6. The average relative error δ for $100\,k$ generated agents, either random or in groups, in dependence on facility cost fc.

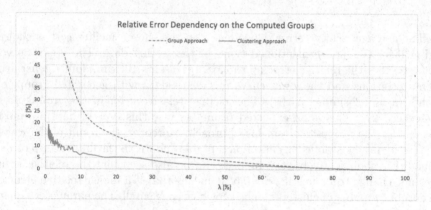

Fig. 7. The average relative error δ of the group and clustering approaches in dependence on the relative number of groups λ.

error of the clustering approach oscillates for the small number of groups because it contains randomized operations which are more visible for greater groups.

In the graphs, attention was drawn to the average relative error. The maximal relative error is influenced by the setting of the parameters - how big attraction of the group leaders for other agents is allowed. If big, then even more distant agents are pushed to group with a leader which may result in a high suboptimality of the path for the given agent. Such a path then has a high relative error and so contributes to a high maximal error. Fortunately, maximal errors concern individuals while the approach is aimed at groups or even crowds, so one strongly non-optimal path is not important in the intended applications and will be amortized by good behavior of the whole crowd.

5 Conclusion

In this paper we introduced a path planning approach for many agents in an environment represented by a graph of vertices and edges. The approach can be also used for other types of environment, e.g., grids or polygonal meshes. Our approach creates groups of agents based on their start and target positions and optimizes the size of the created groups by clustering. Any standard graph-based path planning algorithm, e.g., A*, D* or D* Lite is then used to plan the paths. The substantial part of the computed path of group leaders is shared with the rest of the group. In this way, computation time and memory can be saved at a price of non-optimality of paths. The proposed approach was tested with the use of the A* algorithm but it could be easily extended to employ other path planning methods. The running time and the influence of the clustering on the quality of the paths has been measured. The proposed approach proved to be suitable for the data with a significant number of potentially similar paths.

Acknowledgement. This work was supported by the Ministry of Education, Youth and Sports of the Czech Republic, project SGS-2016-013 Advanced Graphical and Computing Systems, and Czech Science Foundation, project 17-07690S.

References

1. Jain, A.K., Dubes, R.C.: Algorithms for Clustering Data. Prentice-Hall Inc., Upper Saddle River (1988)
2. Ball, G.H., Hall, D.J.: Isodata, a novel method of data analysis and pattern classification. Technical report, DTIC Document (1965)
3. Baraldi, A., Blonda, P.: A survey of fuzzy clustering algorithms for pattern recognition. I. IEEE Trans. Syst. Man Cybern. Part B (Cybern.) **29**(6), 778–785 (1999)
4. Bonabeau, E.: Agent-based modeling: methods and techniques for simulating human systems. Proc. Nat. Acad. Sci. **99**(suppl 3), 7280–7287 (2002)
5. Bretti, G., Natalini, R., Piccoli, B.: A fluid-dynamic traffic model on road networks. Arch. Comput. Methods Eng. **14**(2), 139–172 (2007)
6. Charikar, M., Guha, S.: Improved combinatorial algorithms for the facility location and k-median problems. In: 40th Annual Symposium on Foundations of Computer Science, 1999, pp. 378–388. IEEE (1999)
7. Chu, K., Lee, M., Sunwoo, M.: Local path planning for off-road autonomous driving with avoidance of static obstacles. IEEE Trans. Intell. Transp. Syst. **13**(4), 1599–1616 (2012)
8. Darken, C.J., Burgess, R.G.: Realistic human path planning using fluid simulation (2004)
9. Dubes, R., Jain, A.K.: Clustering methodologies in exploratory data analysis. Adv. Comput. **19**, 113–228 (1980)
10. Fang, Z., Zong, X., Li, Q., Li, Q., Xiong, S.: Hierarchical multi-objective evacuation routing in stadium using ant colony optimization approach. J. Transport Geogr. **19**(3), 443–451 (2011)
11. Fayyad, U., Piatetsky-Shapiro, G., Smyth, P.: From data mining to knowledge discovery in databases. AI Mag. **17**(3), 37 (1996)
12. Floyd, R.W.: Algorithm 97: shortest path. Commun. ACM **5**(6), 345 (1962)

13. Glowinski, R., Ciarlet, P., Lions, J.: Handbook of numerical analysis: Numerical methods for fluids (2003)

14. Guy, S.J., Chhugani, J., Kim, C., Satish, N., Lin, M., Manocha, D., Dubey, P.: Clearpath: highly parallel collision avoidance for multi-agent simulation. In: Proceedings of the 2009 ACM SIGGRAPH/Eurographics Symposium on Computer Animation, pp. 177–187. ACM (2009)

15. Guy, S.J., Kim, S., Lin, M.C., Manocha, D.: Simulating heterogeneous crowd behaviors using personality trait theory. In: Proceedings of the 2011 ACM SIGGRAPH/Eurographics Symposium on Computer Animation, pp. 43–52. ACM (2011)

16. Harabor, D., Botea, A.: Hierarchical path planning for multi-size agents in heterogeneous environments. In: 2008 IEEE Symposium On Computational Intelligence and Games, pp. 258–265. IEEE (2008)

17. Hart, P.E., Nilsson, N.J., Raphael, B.: A formal basis for the heuristic determination of minimum cost paths. IEEE Trans. Syst. Sci. Cybern. **4**(2), 100–107 (1968)

18. Hughes, R.L.: A continuum theory for the flow of pedestrians. Transp. Res. Part B Methodological **36**(6), 507–535 (2002)

19. Jain, A.K., Murty, M.N., Flynn, P.J.: Data clustering: a review. ACM Comput. Surv. (CSUR) **31**(3), 264–323 (1999)

20. Jiang, H., Xu, W., Mao, T., Li, C., Xia, S., Wang, Z.: Continuum crowd simulation in complex environments. Comput. Graph. **34**(5), 537–544 (2010)

21. King, B.: Step-wise clustering procedures. J. Am. Stat. Assoc. **62**(317), 86–101 (1967)

22. Koenig, S., Likhachev, M.: D* lite. In: AAAI/IAAI, pp. 476–483 (2002)

23. Koenig, S., Likhachev, M., Furcy, D.: Lifelong planning A*. Artif. Intell. **155**(1), 93–146 (2004)

24. Likhachev, M., Ferguson, D.I., Gordon, G.J., Stentz, A., Thrun, S.: Anytime dynamic A*: an anytime, replanning algorithm. In: ICAPS, pp. 262–271 (2005)

25. Loscos, C., Marchal, D., Meyer, A.: Intuitive crowd behavior in dense urban environments using local laws. In: Theory and Practice of Computer Graphics, 2003 Proceedings, pp. 122–129. IEEE (2003)

26. MacQueen, J., et al.: Some methods for classification and analysis of multivariate observations. In: Proceedings of the Fifth Berkeley Symposium on Mathematical Statistics and Probability, Oakland, CA, USA, vol. 1, pp. 281–297 (1967)

27. Mao, T., Jiang, H., Li, J., Zhang, Y., Xia, S., Wang, Z.: Parallelizing continuum crowds. In: Proceedings of the 17th ACM Symposium on Virtual Reality Software and Technology, pp. 231–234. ACM (2010)

28. Metoyer, R.A., Hodgins, J.K.: Reactive pedestrian path following from examples. Vis. Comput. **20**(10), 635–649 (2004)

29. Meyerson, A.: Online facility location. In: 42nd IEEE Symposium on Foundations of Computer Science, 2001 Proceedings, pp. 426–431. IEEE (2001)

30. Okazaki, S., Matsushita, S.: A study of simulation model for pedestrian movement with evacuation and queuing. In: International Conference on Engineering for Crowd Safety, vol. 271 (1993)

31. Pellegrini, S., Ess, A., Schindler, K., Van Gool, L.: You'll never walk alone: modeling social behavior for multi-target tracking. In: 2009 IEEE 12th International Conference on Computer Vision, pp. 261–268. IEEE (2009)

32. Rasmussen, E.M.: Clustering algorithms. Inf. Retrieval Data Struct. Algorithms **419**, 442 (1992)

33. Rohlf, F.J.: 12 Single-link clustering algorithms. In: Krishnaiah, P.R., Kanal, L.N. (eds.) Handbook of Statistics, vol. 2, pp. 267–284. Amsterdam, North-Holland (1982)

34. Singh, S., Kapadia, M., Hewlett, B., Reinman, G., Faloutsos, P.: A modular framework for adaptive agent-based steering. In: Symposium on Interactive 3D Graphics and Games, pp. 141–150. ACM (2011)

35. Skála, J.: Algorithms for manipulation with large geometric and graphic data. Technical report, Technical Report DCSE/TR-2009-02, Department of Computer Science and Engineering, University of West Bohemia (2009)

36. Skála, J., Kolingerová, I.: Accelerating the local search algorithm for the facility location. In: Proceedings of the 12th WSEAS International Conference on Mathematical and Computational Methods in Science and Engineering, pp. 98–103. World Scientific and Engineering Academy and Society (WSEAS) (2010)

37. Sokal, R.: The principles of numerical taxonomy: twenty-five years later. In: Goodfellow, M., Jones, D., Priest, F.G. (eds.) Computer-Assisted Bacterial Systematics, vol. 15, p. 1. Academic Press, New York (1985)

38. Stentz, A.: Optimal and efficient path planning for partially-known environments. In: 1994 IEEE International Conference on Robotics and Automation, 1994 Proceedings, pp. 3310–3317. IEEE (1994)

39. Stentz, A., et al.: The focussed D* algorithm for real-time replanning. In: IJCAI, vol. 95, pp. 1652–1659 (1995)

40. Sun, X., Yeoh, W., Koenig, S.: Generalized fringe-retrieving A*: faster moving target search on state lattices. In: Proceedings of the 9th International Conference on Autonomous Agents and Multiagent Systems: Volume 1-Volume 1, pp. 1081–1088. International Foundation for Autonomous Agents and Multiagent Systems (2010)

41. Sun, X., Yeoh, W., Koenig, S.: Moving target D* lite. In: Proceedings of the 9th International Conference on Autonomous Agents and Multiagent Systems: Volume 1-Volume 1, pp. 67–74. International Foundation for Autonomous Agents and Multiagent Systems (2010)

42. Szkandera, J., Kolingerová, I., Maňák, M.: Path planning for groups on graphs. Procedia Comput. Sci. 108, 2338–2342 (2017)

43. Treuille, A., Cooper, S., Popović, Z.: Continuum crowds. ACMTrans. Graph. (TOG) 25, 1160–1168 (2006). ACM

44. Vadakkepat, P., Tan, K.C., Ming-Liang, W.: Evolutionary artificial potential fields and their application in real time robot path planning. In: Proceedings of the 2000 Congress on Evolutionary Computation, vol. 1, pp. 256–263. IEEE (2000)

45. Warshall, S.: A theorem on boolean matrices. J. ACM 9(1), 11–12 (1962)

46. Žalik, K.R., Žalik, B.: A sweep-line algorithm for spatial clustering. Adv. Eng. Softw. 40(6), 445–451 (2009)

The Refutation of Amdahl's Law and Its Variants

Ferenc Dévai[1,2]([envelope]) [iD]

[1] London South Bank University, London, UK
fdevai@acm.org
[2] Hungarian Academy of Sciences, Budapest, Hungary

Abstract. Amdahl's law, imposing a restriction on the speedup achievable by a multiple number of processors, based on the concept of sequential and parallelizable fractions of computations, has been used to justify, among others, asymmetric chip multiprocessor architectures and concerns of "dark silicon". This paper demonstrates flaws in Amdahl's law that (i) in theory no inherently sequential fractions of computations exists (ii) sequential fractions appearing in practice are inherently different from parallelizable fractions and therefore usually have different growth rates and that (iii) the time requirement of sequential fractions can be proportional to the number of processors. However, mathematical analyses are also provided to demonstrate that sequential fractions have negligible effect on speedup if the growth rate of the parallelizable fraction is higher than that of the sequential fraction. Examples from computational geometry are given that Amdahl's law and its variants fail to represent limits to parallel computation. In particular, Gustafson's law, claimed to be a refutation of Amdahl's law by some authors, is shown to contradict established theoretical results. We can conclude that no simple formula or law governing concurrency exists.

1 Introduction

The computing literature about parallel and distributed computing can roughly be divided in two categories: publications relying on Amdahl's law [2] and its variants, as well as publications about designing and implementing multiprocessor algorithms. Amdahl's law and its variants, including Gustafson's law [18], provide little help with issues the second group of publications are dealing with, therefore hardly ever mentioned by that group of publications. One of the few exceptions is Preparata [34] asking in an 1995 invited presentation if Amdahl's law should be repealed. However, at the time of writing, according to Google Scholar, Amdahl's paper [2] has been cited by 4,621, and Gustafson's paper [18] by 1,662 publications, including several published in 2017.

Both Amdahl's law and Gustafson's law attempt to predict the speedup of computation time achievable by a multiple number of processors. Amdahl [2] places a pessimistic upper bound on the multiprocessor approach by concluding that the achievable speedup is a constant of five to seven, even if the number of processors approaches infinity.

© Springer International Publishing AG 2017
O. Gervasi et al. (Eds.): ICCSA 2017, Part II, LNCS 10405, pp. 480–493, 2017.
DOI: 10.1007/978-3-319-62395-5_33

On the other hand, Gustafson's law is overly optimistic. Gustafson [18] states that the achievable speedup is proportional to the number of processors. Having two drastically different versions of a scientific law, up for choice by the advocates, is awkward. For example, if a re-evaluated version of Ohm's law predicted more electric current than voltage divided by resistance, it would not be credible. Similarly, one of Amdahl's law and Gustafson's law cannot be correct.

This paper demonstrates that both Amdahl's law and its variants, including Gustafson's law, are fundamentally wrong. Some results from the second group of publications will be used, but interestingly, none of the publications used ever mention Amdahl's law and its variants.

In Sect. 2 Amdahl's paper [2] is discussed in more detail. Section 3 reviews some of the most frequently cited and most recent publications advocating Amdahl's law and its variants. Section 4 summarises arguments that no substantial impediments to parallel computation exist and reviews earlier attempts to refute Amdahl's law. Section 5 provides refutations of Amdahl's law and its variants, including Gustafson's law. Section 6 reviews the possibilities for the development of fast parallel algorithms and the known limits to parallel computation. Section 7 discusses the contributions of this paper. Finally Sect. 8 concludes that no simple formula or law governing concurrency exists and that experimental results should be interpreted with theoretical results in mind.

2 Amdahl's Law

Amdahl's paper [2] published in 1967, criticizes *"prophets"* voicing *"the contention that the organization of a single computer has reached its limits and that truly significant advances can be made only by interconnection of a multiplicity of computers"*. Amdahl argues that *"the fraction of the computational load ... associated with data management housekeeping ... accounts for 40% of the executed instructions"*.

He goes on that this fraction *"might be reduced by a factor of two"* but it is highly unlikely that *"it could be reduced by a factor of three"* and that *"this overhead appears to be sequential so that it is unlikely to be amenable to parallel processing techniques"*. He concludes that this overhead places *"an upper limit on throughput of five to seven times the sequential processing rate"*.

Amdahl aims to provide an upper limit on speedup and therefore assumes that, apart from the sequential fraction, the remaining computations are perfectly parallelizable. Let t_1 be the time taken by one processor solving a computational problem and t_p be the time taken by p processors solving the same problem. Finally let us denote the supposed inherently sequential fraction of instructions by f. Then $t_p = t_1(f + (1 - f)/p)$ and the speedup obtainable by p processors can be expressed as

$$\frac{t_1}{t_p} = \frac{1}{f + (1 - f)/p}. \tag{1}$$

Indeed if we substitute $f = 0.4/2$ and $f = 0.4/3$, and assume that p approaches infinity, we get 5 and 7.5 respectively. Formula (1) is referred to as Amdahl's law by his followers (though Amdahl never presented the formula).

3   Followers

Some authors, e.g., Denning and Lewis [8] and Sun and Chen [39], claim that Gustafson's law [18] is a refutation of Amdahl's law. Using high-performance computing facilities, such as a 1024-processor hypercube, it is the parallel time t_p on p processors what the user can readily observe. Gustafson, in an attempt to interpret experimental results, argues that a single processor solving the same problem would spend ft_p time on the sequential, and $(1 - f)pt_p$ time on the parallelizable part of the problem, therefore $t_1 = ft_p + (1 - f)pt_p$, from which the speedup achievable by p processors is

$$\frac{t_1}{t_p} = f + (1 - f)p, \tag{2}$$

where f is the same "inherently sequential" fraction of instructions as in the case of Amdahl's law. As Gustafson's law is based on the same concepts as the bases of Amdahl's law, it is a variant, rather than a refutation of Amdahl's law. Indeed, the title of Gustafson's paper [18] is *"Reevaluating Amdahl's Law"*. However, Amdahl's law is still more widely accepted than Gustafson's law.

Gustafson does not take sequential input-output requirements proportional to input and output sizes into account. Patterson et al. [32], in an attempt to defeat Amdahl's law, propose disk arrays to reduce input-output requirements.

In order to mitigate the effects of Amdahl's law, Annavaram et al. [3] make a case for varying the amount of energy expended to instructions according to the amount of available parallelism. Power consumption is the product of energy per instruction (EPI) and instructions per second (IPS). Annavaram et al. [3] propose that during phases of limited parallelism (low IPS) a chip multiprocessor should spend more EPI, and during phases of higher parallelism (high IPS) it should spend less EPI.

Borkar [5], without explicitly citing Amdahl's paper, presents (1) and argues that, due to the limitation by Amdahl's law, it will be difficult to exploit the performance enhancements of many-core architectures. However, he adds that multiple applications running simultaneously mitigate the effect of Amdahl's law and therefore could benefit from many-core architectures.

Many authors [21, 24, 38, 42] propose asymmetric (or heterogeneous) chip multiprocessor architectures consisting of at least one large, high-performance core and several small, low-performance cores. Serial program portions would execute on a large core to reduce the performance impact of the serial bottleneck imposed by Amdahl's law. Some of these proposals are summarised below.

Kumar et al. [24], relying on Annavaram et al. [3], claim that asymmetric chip multiprocessor architectures present unique opportunities for mitigating the adverse effect of Amdahl's law on the achievable speedup.

Hill and Marty [21] argue that robust, general-purpose multicore designs should operate under Amdahl's more pessimistic, rather than Gustafson's optimistic assumptions. They assume that architects have techniques for using the resources of multiple cores to create a single core with greater sequential performance. However, Hill and Marty [21] conclude that the future of scalable multicore processors is questionable.

Woo and Lee [42] attempt to extend Amdahl's law in order to take power and energy into account, independently of Annavaram et al. [3]. They suggest many small, energy-efficient cores integrated with a large powerful core.

Suleman et al. [38] propose a technique, called Accelerated Critical Sections (ACS), leveraging the high-performance core(s) of asymmetric chip multiprocessors to accelerate the execution of critical sections. When a small core encounters a critical section, it requests a large one to execute that critical section. The large core acquires the lock, executes the critical section and notifies the requesting small core when the critical section is complete.

As Suleman et al. [38] properly note, ACS has disadvantages. To execute critical sections, the large core may require some private data from the small core e.g. input parameters on a stack. Such data is transferred from the cache of the small core via the regular cache coherence mechanism. These transfers increase cache misses.

Executing critical sections exclusively on a large core can have another disadvantage. Paraphrasing Suleman et al. [38], multi-threaded applications often try to improve concurrency by using data synchronization at a fine granularity, i.e., having multiple critical sections, each guarding a disjoint set of the shared data (e.g., a separate lock for each element of an array). Then executing all critical sections on the large core can lead to false serialization of different, disjoint critical sections that otherwise could have been executed in parallel.

Sun and Chen [39] study symmetric multicore architectures. As noted above, they accept Gustafson's law and criticize Hill and Marty's pessimistic view [21]. They conclude that asymmetric multicore architectures are much more complex and therefore are worth exploring only if their symmetric counterparts cannot deliver satisfactory performance.

Eyerman and Eeckhout [14] propose to augment Amdahl's law with the notion of critical sections. They claim that it leads to a new fundamental law, asserting that parallel performance is not only limited by the sequential fraction (according to Amdahl's law) but also limited by critical sections.

Juurlink and Meenderinck [22] consider Amdahl's law harmful. In an attempt to reconcile the differences between Amdahl's law and Gustafson's law, they offer a new law assuming that the parallelizable work grows as a *sublinear* function of p, the number of processors. In fact, the non-parallelizable work can grow as a *linear* function of p, as we demonstrate in Sect. 5.

Esmaeilzadeh et al. [13] predict that as the number of cores increases, power constraints may prevent powering up all the cores at full speed. According to their models — based on Amdahl's law — the fraction of chips remaining "dark" could be as much as 50% within three process generations.

Yavits et al. [43] consider asymmetric chip multiprocessor architectures, with a general-purpose core responsible for executing the sequential fraction and a number of other cores for executing the parallelizable fraction of the code. This arrangement incurs data exchange between the sequential and the parallel cores and also a data exchange among the parallel cores. Based on these observations, the authors suggest a modification of Amdahl's law, offering a pessimistic conclusion that a highly parallelizable yet highly synchronization- and connectivity-intensive workload might be more efficiently processed by a sequential core rather than a parallel multicore.

Morad et al. [30] propose a shared memory to speed up communication between parallel and sequential cores. Yavits et al. [44] study the thermal effects of three-dimensional integration on the performance and scalability of chip multiprocessors from the perspective of Amdahl's law.

Mittal [29] provides a comprehensive survey on asymmetric multicore processors, mainly justified by Amdahl's law, emphasising the challenges these architectures face, including the optimization of both sequential and parallel performance and thread migration overheads. The latter can take millions of cycles, based on independent experimental results taken from the literature.

Patterson and Hennessy [31] and Hennessy and Patterson [19, pp. 46–47] generalize Amdahl's law for any performance enhancement by defining speedup as the execution time for a particular task without using the enhancement divided by the execution time for the same task using the enhancement. If the enhancement cannot affect, say, one third of the execution time of the task, then the speedup can only be at most threefold. This is hardly more than triviality. For dismissing Gustafson's law, Hennessy and Patterson [19] merely refer to Amdahl's paper [2] without explicitly citing Gustafson's paper [18].

Shavit [37] presents Amdahl's law (without explicitly citing Amdahl's paper) and suggests that it is worthwhile to invest efforts to derive as much parallelism as possible from parts of programs dealing with inter-thread interaction and coordination in order to get highly parallel, concurrent data structures. This is an important research direction, regardless of the validity of Amdahl's law.

The textbook by Herlihy and Shavit gives an example of a ten-processor machine and a program with a 90% parallelizable fraction [20, p. 14]. According to Amdahl's law (1) the speedup is merely 5.26. Then the authors state that the major focus of their book is to help understanding the tools and techniques that allow software developers effectively preparing the parts of the program code dealing with coordination and synchronization. This approach makes the book useful, irrespective of the validity of Amdahl's law.

4   Related Work

In 1975 Valiant [40] proved that, using p processors doing binary comparisons, speedup proportional to $p/\log \log p$ can be achieved for problems of finding the maximum, sorting, and merging a pair of sorted lists, assuming $p \leq n$, where n the size of the input set. Then in 1990 Valiant [41] argued that *"no substantial impediments to general-purpose parallel computation ... have been uncovered"* without the refutation of Amdahl's law or any reference to its refutation.

In 1995 Preparata [34] stated that research on algorithms has shown that most problems are parallelizable and that, in a realistic computational model fully accounting for the finiteness of the speed of light, uniprocessors incur a slowdown not only due to loss of parallelism but also due to loss of locality.

Preparata [34] assumed that Amdahl's law may only apply to P-complete problems. In Sect. 6 we demonstrate that even some P-complete problems are parallelizable, such that the product of the parallel running time and the number of processors used matches the sequential lower bound for the problem, and therefore break Amdahl's law.

Although Amdahl's law has already been classified as a "folk theorem" [1, 23] and claimed to be refuted for a special case of *superlinear speedup*, when the amount of input data keeps increasing during computation [1, 26] it is still widely accepted. Superlinear speedup is nicely illustrated by Luccio and Pagli's [26] snow-shoveler metaphor: In a winter morning a man shovels snow from his driveway, while the snow falls. On a neighbouring driveway three men do the same job. As a smaller amount of snow falls while they work, they clean the driveway in less than one third of their neighbour's time.

In 2007 Paul and Meyer [33] stated that Amdahl's law is *"one of the few, fundamental laws in computing"*, but went on that it can fail when applied to single-chip heterogeneous (asymmetric) multiprocessor designs. However, as we have already seen in Sect. 3, even after Paul and Meyer's publication [33] much research continued on the mitigation of the adverse effect of Amdahl's law on asymmetric chip multiprocessors.

Some of these researchers [14, 37, 38, 43] observed that, in practice, parallel performance is not only limited by a sequential fraction, as Amdahl's law claims, but also limited by communication, synchronization and concurrent objects. In Sect. 5 we demonstrate that the time requirement of these can be proportional to the number of processors.

Amdahl's and Gustafson's followers never gave examples of "inherently sequential" computations and not even specific examples when sequential computations appear in concurrent systems in practice. The theory community never provided refutations, merely ignores Amdahl's law and its variants.

In the next section we demonstrate that inherently sequential computations do not exist in theory and that Gustafson's law contradicts established theoretical results. We also demonstrate, using three practical examples of sequential operations (such as reading input, sequential concurrent objects [12] and Lamport's bakery algorithm [25]) that Amdahl's law is broken in practice.

5   Refutation

One of the most spectacular success stories of the computing industry is the parallelization of the hidden-surface problem by using graphics processing units

(GPUs). The hidden-surface image of 10 random triangles in three-dimensional space is given in Fig. 1 and the hidden-surface problem is stated as follows.

Given a set S of pairwise disjoint, opaque and planar simple polygons possibly with holes and with a total of n edges in three-dimensional space; find each region ρ of each polygon in S, such that all points of ρ are visible from a viewpoint u, where $u = (0, 0, \infty)$.

Fig. 1. A hidden-surface image

The hidden-surface algorithms with the best possible worst-case time take time proportional to n^2 [9, 28]. The input set S is usually too large to fit in main memory, therefore we need to read the relevant parts from a disk or a solid-state drive. Ignoring disk seek time, reading the data sequentially takes time proportional to n. Although the output can be as large as $O(n^2)$, it is perfectly parallelizable; each processor writes its result into a memory block, called the frame buffer. Then the running time for a single processor is

$$t_1(n) = an + bn^2$$

for some positive constants a and b.

For some small n, we can observe sequential reading times of e.g., one third or one fourth of $t_1(n)$. Substituting $f = 1/3$ or $1/4$ into (1) in Sect. 2, we would get upper limits of 3 or 4 for speedup, while GPUs successfully use hundreds or thousands of processors.

To resolve the contradiction, we should go back to the original definition of speedup: $t_1(n)$ divided by $t_p(n)$. According to Amdahl (Sect. 2) for the development of an upper bound we assume that, apart from the sequential fraction, the remaining computations are perfectly parallelizable. Therefore

$$t_p(n) = an + \frac{bn^2}{p},$$

from which it follows

$$\frac{t_1(n)}{t_p(n)} = \frac{a + bn}{a + bn/p}.$$

If p approaches infinity, we get the upper limit

$$\frac{t_1(n)}{t_p(n)} = 1 + \frac{b}{a}n \qquad (3)$$

that grows linearly with n, rather than *"an upper limit"* of *"five to seven"*.

However, it is a more realistic assumption that p is proportional to n. If we substitute $n = cp$, for some positive constant c, we get

$$\frac{t_1(n)}{t_p(n)} = \frac{a + bcp}{a + bc},$$

a speedup that grows linearly with p.

Even though Amdahl's law is overly pessimistic, it does not take into account that increasing the number of processors can have adverse effect on speedup. Ellen, Hendler and Shavit [12,16] show that performing p operations, each by a different process on objects with nonblocking linearizable implementations, takes p times the time of a single operation. Also the implementation of Lamport's bakery algorithm [25] for p processes takes time proportional to p.

Assuming that the parallelizable fraction has a quadratic growth rate as before and that the number of processes is the same as the number of processors, using the notation as above we have

$$\frac{t_1(n)}{t_p(n)} = \frac{ap + bn^2}{ap + bn^2/p}.$$

Substituting $n = cp$, we get

$$\frac{t_1(n)}{t_p(n)} = \frac{a + bc^2p}{a + bc^2},$$

a speedup that still grows linearly with p.

It should be noted that the above does not suggest that any problem with a linear sequential fraction and a quadratic parallelizable fraction would have a speedup proportional to the number of processors. It merely indicates that the speedup is heavily dependent on the growth rates of the fractions of the code executed sequentially and in parallel. If the growth rate of the fraction executed sequentially is the same or higher than that of the fraction executed in parallel, there would indeed be a small upper limit for speedup.

To examine if this is a possibility, we need to see if any fraction of instructions executed sequentially is indeed *"unlikely to be amenable to parallel processing techniques*". A result by Dymond and Tompa suggests that this is not the case. Dymond and Tompa [11] proved that every deterministic Turing machine running in time t can be simulated by a concurrent-read, exclusive-write (CREW) parallel random-access machine (PRAM) in time $O(t^{1/2})$.

The linear tapes of the Turing machine forbid random access of individual tape cells. Dymond and Tompa did not yet clarify if the more flexible storage structure of the PRAM, the parallelism, or the combination of both that realizes such a spectacular speedup. Mak [27] has demonstrated that parallelism alone suffices to achieve an almost quadratic speedup.

From here it follows that no inherently sequential computations exist in theory. All the above demonstrate that Amdahl's law is fundamentally wrong.

Gustafson's law contradicts not only the results by Dymond and Tompa and Mak, but also other established theoretical results known before Gustafson's publication [18] in 1988, as already noted [10] and shown below.

The sequential running time for finding the maximum of n integers, $t_1(n) \leq cn$, accounting for $n - 1$ comparisons and, in the worst case, $n - 1$ assignments, where c is a positive constant. Based on Gustafson's law (2) the time to find the maximum of n integers by using p processors would be

$$t_p(n) \leq \frac{cn}{f + (1 - f)p},$$

which is bounded by a constant for any $f < 1$ and any p proportional to n, and approaches 0 for any fixed n if p approaches infinity.

However, in 1982 Cook and Dwork [7] provided an $\Omega(\log n)$ lower bound for finding the maximum of n integers allowing infinitely many processors of any PRAM without simultaneous writes.

In 1985 Fich et al. [15] proved an $\Omega(\log \log n)$ lower bound for the same problem for $p < \binom{n}{2}$ under the priority model, where the processor with the highest priority is allowed to write in case of a write conflict. The priority model is the strongest PRAM model allowing simultaneous writes. (For $p = n^2$ there exists an $O(1)$-time algorithm, but it does not justify Gustafson's law.)

6   Fast Algorithms and Limits to Parallel Computation

In complexity theory a problem is said to be in the class P if it can be solved in sequential time $n^{O(1)}$, where n is the problem size. Similarly, a problem is said to be in the class NC if it can be solved in parallel time $(\log n)^{O(1)}$ using $n^{O(1)}$ processors, or in other words, if it can be solved in polylogarithmic time by using a polynomial number of processors.

Problems in P that are unlikely to be in NC are called *P-complete* problems and sometimes are also referred to as "inherently sequential", e.g., by Reif [35], who provided evidence that depth-first search is hard to parallelize. As we have already seen in Sect. 5, no inherently sequential problems exist, therefore referring to P-complete problems as such is inappropriate.

Problems in the class NC are the ones having fast parallel algorithms. Another important class of algorithms, called *cost-optimal*, or *work-optimal* algorithms, is the group algorithms with the product of the parallel running time and the number of processors used matching the sequential lower time bound for the problem. These algorithms with the parallel running time

$$t_p(n) = \frac{\Theta(t_1(n))}{p},$$

where n is the problem size, p is the number of processors and $t_1(n)$ is the running time of the best possible sequential algorithm for the problem, clearly break Amdahl's law.

There exist cost-optimal algorithms even for P-complete problems. For example, Castanho et al. [6] devise cost-optimal algorithms for computational-geometry problems, such as the convex-layers and the envelope-layers problem.

On the other hand, Atallah et al. [4] prove that, unless $P = NC$, it is impossible to solve a number of two-dimensional geometric problems in polylogarithmic time by using a polynomial number of processors.

Greenlaw et al. [17] provide a good discussion of P-completeness theory and limits to parallel computation in general. However, these have nothing to do with Amdahl's law and its variants, which are not even mentioned in the book.

7 Discussion

Probably one of the most important contributions of this paper is the demonstration that sequential fractions of computations have negligible effect on speedup if the growth rate of the parallelizable fraction is higher than that of the sequential fraction. The successful application of GPUs for general-purpose computations seems to justify that this is very often the case in practice.

Sequential computations, e.g., sequential concurrent objects [12] and Lamport's bakery algorithm [25], may appear in concurrent systems with the result that the amount of non-parallelizable work grows with the number of processors. However, these still have negligible effect on speedup if the growth rate of the parallelizable work is higher than the growth rate of the number of processors.

On the theoretical level, many researchers (including the author of this paper) published parallel algorithms ignoring sequential fractions, such as reading input, and therefore ignoring Amdahl's law. This concern was one of the motivations of this research. However, as long as the growth rate of the parallelizable fractions is higher, ignoring sequential fractions (including the ones needed, e.g., for synchronisation) is just the same as using the "big Oh" (Θ and Ω) notation for analysis of algorithms.

An unexpected result of this research is that, very likely, unnecessary attention has been paid to the presumed adverse effect of Amdahl's law, in order to justify asymmetric chip multiprocessor architectures, as summarised in Sect. 3. In addition to the disadvantages already mentioned in Sect. 3, asymmetric chip multiprocessors require more complicated system software. As Sun and Chen [39] properly conclude (though based on the wrong assumption of Gustafson's law) asymmetric multicore architectures would only be worth exploring if their symmetric counterparts could not deliver satisfactory performance.

Let us suppose that an^q and bn^r, respectively, are the time requirements of the sequential and the parallelizable fractions of an application, where a, b, q and r are positive constants and n is the problem size. Assuming a perfectly parallelizable fraction, the speedup achievable by p processors is

$$\frac{t_1(n)}{t_p(n)} = \frac{an^q + bn^r}{an^q + bn^r/p}.$$

If p approaches infinity, the upper limit for speedup is

$$\frac{t_1(n)}{t_p(n)} = 1 + \frac{b}{a}n^{r-q}. \tag{4}$$

For $r - q = 1$, this is the same as (3) in Sect. 5. If $r - q = 0$, we get the same upper limit as given by (1) in Sect. 2 (since $\frac{a+b}{a} = \frac{1}{f}$).

However, as we have already seen, inherently sequential fractions do not exist in theory, and sequential fractions in practice are inherently different from parallelizable fractions, therefore $r-q = 0$ is hardly ever the case. Another reason why we should resist the temptation of proposing a generalisation of Amdahl's law is that (4) cannot take into account that in practice the time requirement of sequential fractions can be proportional to the number of processors.

While teaching concurrency, it would be tempting to rely on rules of thumb or simple scientific laws, such as (4). However, for proving upper and lower bounds [7,15,36] more sophisticated arguments are needed than the ones Amdahl's law and its variants are based on.

One of the most important open problems of computer science is to decide if $P = NC$. The results by Dymond and Tompa and Mak do not imply that $P = NC$, as Dymond et al. require exponential numbers of processors, but fair enough for the refutation of Amdahl's law that assumes an infinite number of processors. Therefore the question "Should Amdahl's law be repealed?" in the title of Preparata's 1995 invited talk [34] should be answered affirmatively.

8   Conclusions

We have seen that ignoring the difference of the growth rates of the numbers of operations is one of the reasons why Amdahl's law and its variants fail in practice. We have demonstrated that inherently sequential computations, the cornerstone of Amdahl's law and its variants, do not exist in theory. Some authors observed that in practice parallel performance is limited by communication, synchronization and concurrent objects. We have shown that the time requirement of these can be proportional to the number of processors. However, we have demonstrated that such sequential fractions have negligible effect on speedup if the growth rate of the parallel fraction is higher than that of the sequential fraction.

We have also demonstrated that Gustafson's law and other variants of Amdahl's law contradict established theoretical results. We can conclude that no simple formula or law governing concurrency exists. Experimental results should not be interpreted by rules of thumb, but theoretical results in mind.

Acknowledgements. The author thanks three anonymous reviewers for their support and constructive criticism that helped to improve the presentation of the paper.

References

1. Akl, S.G., Cosnard, M., Ferreira, A.G.: Data-movement-intensive problems: two folk theorems in parallel computation revisited. Theor. Comput. Sci. **95**(2), 323–337 (1992). doi:10.1016/0304-3975(92)90271-G
2. Amdahl, G.M.: Validity of the single processor approach to achieving large scale computing capabilities. In: Proceedings of Spring Joint Computer Conference, pp. 483–485. ACM, New York (1967). doi:10.1145/1465482.1465560
3. Annavaram, M., Grochowski, E., Shen, J.: Mitigating Amdahl's law through EPI throttling. SIGARCH Comput. Archit. News **33**(2), 298–309 (2005). http://doi.acm.org/10.1145/1080695.1069995
4. Atallah, M.J., Callahan, P.B., Goodrich, M.T.: P-complete geometric problems. Int. J. Comput. Geom. Appl. **03**(04), 443–462 (1993). doi:10.1142/S0218195993000282
5. Borkar, S.: Thousand core chips: a technology perspective. In: Proceedings of 44th Annual Design Automation Conference, DAC 2007, pp. 746–749. ACM, New York (2007). http://doi.acm.org/10.1145/1278480.1278667
6. Castanho, C.D., Chen, W., Wada, K., Fujiwara, A.: Parallelizability of some P-complete geometric problems in the EREW-PRAM. In: Wang, J. (ed.) COCOON 2001. LNCS, vol. 2108, pp. 59–63. Springer, Heidelberg (2001). doi:10.1007/3-540-44679-6_7
7. Cook, S.A., Dwork, C.: Bounds on the time for parallel RAM's to compute simple functions. In: Proceedings of 14th Annual ACM Symposium on Theory of Computing, STOC 1982, pp. 231–233. ACM, New York (1982). doi:10.1137/0215006
8. Denning, P.J., Lewis, T.G.: Exponential laws of computing growth. Commun. ACM **60**(1), 54–65 (2017). http://doi.acm.org/10.1145/2976758
9. Dévai, F.: An optimal hidden-surface algorithm and its parallelization. In: Murgante, B., Gervasi, O., Iglesias, A., Taniar, D., Apduhan, B.O. (eds.) ICCSA 2011. LNCS, vol. 6784, pp. 17–29. Springer, Heidelberg (2011). doi:10.1007/978-3-642-21931-3_2
10. Dévai, F.: Gustafson's law contradicts theory results. (Letter to the Editor). Commun. ACM **60**(4), 8–9 (2017). http://doi.acm.org/10.1145/3056859
11. Dymond, P.W., Tompa, M.: Speedups of deterministic machines by synchronous parallel machines. J. Comput. Syst. Sci. **30**(2), 149–161 (1985). doi:10.1145/800061.808763
12. Ellen, F., Hendler, D., Shavit, N.: On the inherent sequentiality of concurrent objects. SIAM J. Comput. **41**(3), 519–536 (2012). doi:10.1137/08072646X
13. Esmaeilzadeh, H., et al.: Power challenges may end the multicore era. Commun. ACM **56**(2), 93–102 (2013). http://doi.acm.org/10.1145/2408776.2408797
14. Eyerman, S., Eeckhout, L.: Modeling critical sections in Amdahl's law and its implications for multicore design. SIGARCH Comput. Archit. News **38**(3), 362–370 (2010). http://doi.acm.org/10.1145/1816038.1816011

15. Fich, F.E., Meyer auf der Heide, F., Ragde, P., Wigderson, A.: One, two, three ... infinity: lower bounds for parallel computation. In: Proceedings of 17th Annual ACM Symposium on Theory of Computing, STOC 1985, pp. 48–58. ACM, New York (1985). http://doi.acm.org/10.1145/22145.22151

16. Fich, F.E., Hendler, D., Shavit, N.: Linear lower bounds on real-world implementations of concurrent objects. In: Proceedings of 46th Annual IEEE Symposium on Foundations of Computer Science, FOCS 2005, pp. 165–173 (2005). http://dx.doi.org/10.1109/SFCS.2005.47

17. Greenlaw, R., Hoover, H.J., Ruzzo, W.L.: Limits to Parallel Computation: P-Completeness Theory. Oxford University Press Inc., New York (1995)

18. Gustafson, J.L.: Reevaluating Amdahl's law. Commun. ACM **31**(5), 532–533 (1988). doi:10.1145/42411.42415

19. Hennessy, J.L., Patterson, D.A.: Computer Architecture: A Quantitative Approach, 5th edn. Morgan Kaufmann Publishers Inc., San Francisco (2011)

20. Herlihy, M., Shavit, N.: The Art of Multiprocessor Programming, Revised Reprint, 1st edn. Morgan Kaufmann Publishers Inc., San Francisco (2012)

21. Hill, M.D., Marty, M.R.: Amdahl's law in the multicore era. Computer **41**(7), 33–38 (2008). doi:10.1109/MC.2008.209

22. Juurlink, B., Meenderinck, C.H.: Amdahl's law for predicting the future of multicores considered harmful. SIGARCH Comput. Archit. News **40**(2), 1–9 (2012). doi:10.1145/2234336.2234338

23. Kuck, D.J.: A survey of parallel machine organization and programming. ACM Comput. Surv. **9**(1), 29–59 (1977). http://doi.acm.org/10.1145/356683.356686

24. Kumar, R., Tullsen, D.M., Jouppi, N.P., Ranganathan, P.: Heterogeneous chip multiprocessors. Computer **38**(11), 32–38 (2005). doi:10.1109/MC.2005.379

25. Lamport, L.: A new solution of Dijkstra's concurrent programming problem. Commun. ACM **17**(8), 453–455 (1974). http://doi.acm.org/10.1145/361082.361093

26. Luccio, F., Pagli, L.: The p-shovelers problem: (computing with time-varying data). SIGACT News **23**, 72–75 (1992). doi:10.1145/130956.130960

27. Mak, L.: Parallelism always helps. SIAM J. Comput. **26**(1), 153–172 (1997). doi:10.1137/S0097539794265402

28. McKenna, M.: Worst-case optimal hidden-surface removal. ACM Trans. Graph. **6**, 19–28 (1987). doi:10.1145/27625.27627

29. Mittal, S.: A survey of techniques for architecting and managing asymmetric multicore processors. ACM Comput. Surv. **48**(3), 45:1–45:38 (1987). http://doi.acm.org/10.1145/2856125

30. Morad, A., Yavits, L., Kvatinsky, S., Ginosar, R.: Resistive GP-SIMD processing-in-memory. ACM Trans. Archit. Code Optim. **12**(4), 57:1–57:22 (2016). http://doi.acm.org/10.1145/2845084

31. Patterson, D., Hennessy, J.: Computer Organization and Design: The Hardware/Software Interface. The Morgan Kaufmann Series in Computer Architecture and Design, ARM® edn. Elsevier Science, Amsterdam (2016)

32. Patterson, D.A., et al.: A case for redundant arrays of inexpensive disks (RAID). SIGMOD Rec. **17**(3), 109–116 (1988). doi:10.1145/971701.50214

33. Paul, J.M., Meyer, B.H.: Amdahl's law revisited for single chip systems. Int. J. Parallel Program. **35**(2), 101–123 (2007). http://dx.doi.org/10.1007/s10766-006-0028-8

34. Preparata, F.P.: Should Amdahl's law be repealed? (abstract). In: Staples, J., Eades, P., Katoh, N., Moffat, A. (eds.) ISAAC 1995. LNCS, vol. 1004, p. 311. Springer, Heidelberg (1995). doi:10.1007/BFb0015436

35. Reif, J.H.: Depth-first search is inherently sequential. Inf. Process. Lett. **20**(5), 229–234 (1985). doi:10.1016/0020-0190(85)90024-9

36. Roughgarden, T., Vassilvitskii, S., Wang, J.R.: Shuffles and circuits: (on lower bounds for modern parallel computation). In: Proceedings of 28th ACM Symposium on Parallelism in Algorithms and Architectures, SPAA 2016, pp. 1–12. ACM, New York (2016). http://doi.acm.org/10.1145/2935764.2935799

37. Shavit, N.: Data structures in the multicore age. Commun. ACM **54**, 76–84 (2011). http://doi.acm.org/10.1145/1897852.1897873

38. Suleman, M.A., Mutlu, O., Qureshi, M.K., Patt, Y.N.: Accelerating critical section execution with asymmetric multi-core architectures. SIGPLAN Not. **44**(3), 253–264 (2009). http://doi.acm.org/10.1145/1508284.1508274

39. Sun, X.H., Chen, Y.: Reevaluating Amdahl's law in the multicore era. J. Parallel Distrib. Comput. **70**(2), 183–188 (2010). doi:10.1016/j.jpdc.2009.05.002

40. Valiant, L.G.: Parallelism in comparison problems. SIAM J. Comput. **4**(3), 348–355 (1975). http://dx.doi.org/10.1137/0204030

41. Valiant, L.G.: A bridging model for parallel computation. Commun. ACM **33**(8), 103–111 (1990). http://doi.acm.org/10.1145/79173.79181

42. Woo, D.H., Lee, H.H.: Extending Amdahl's law for energy-efficient computing in the many-core era. Computer **41**(12), 24–31 (2008). doi:10.1109/MC.2008.494

43. Yavits, L., Morad, A., Ginosar, R.: The effect of communication and synchronization on Amdahl's law in multicore systems. Parallel Comput. **40**(1), 1–16 (2014). http://dx.doi.org/10.1016/j.parco.2013.11.001

44. Yavits, L., Morad, A., Ginosar, R.: The effect of temperature on Amdahl law in 3D multicore era. IEEE Trans. Comput. **65**(6), 2010–2013 (2016). http://dx.doi.org/10.1109/TC.2015.2458865

A Distance Matrix Completion Approach to 1-Round Algorithms for Point Placement in the Plane

Md. Zamilur Rahman[1], Asish Mukhopadhyay[1(✉)], Yash Pal Aneja[2], and Cory Jean[1]

[1] School of Computer Science, University of Windsor,
Windsor, ON N9B 3P4, Canada
asish.mukerji@gmail.com
[2] Odette School of Business, University of Windsor,
Windsor, ON N9B 3P4, Canada

Abstract. In this paper we propose a 1-round algorithm for point placement in the plane in an adversarial model. The distance query graph presented to the adversary is chordal. The remaining distances are uniquely determined using a distance matrix completion algorithm for chordal graphs, based on a result by Bakonyi and Johnson [4].

Keywords: Distance geometry · Point placement · Distance matrix completion · Embed algorithm · Eigenvalue decomposition

1 Introduction

The problem of locating n distinct points on a line, up to translation and reflection, in an adversarial setting has been extensively studied [1,7–9]. The best known 2-rounds algorithm that makes $9n/7$ queries and has a query lower bound of $9n/8$ queries is due to Alam and Mukhopadhyay [2]. In this paper we propose a 1-round algorithm for the same problem in the plane. To the best of our knowledge there is no prior work extant on this problem. A practical motivation for this study is the extensively researched and closely related sensor network localization problem [3,5].

2 Preliminaries

Let $D = [d_{ij}]$ be an $n \times n$ symmetric matrix, whose diagonal entries are 0 and the off-diagonal entries are positive. It is said to be an Euclidean distance matrix if there exists points $p_1, p_2, \ldots, p_n$ in some k-dimensional Euclidean space such that $d_{ij} = d(p_i, p_j)^2$, where $d(p_i, p_j)$ is the Euclidean distance between the points p_i and p_j. A set of necessary and sufficient conditions for this was given by Schoenberg [16], as well as Young and Householder [17]. A partial distance matrix is one in which some entries are missing.

© Springer International Publishing AG 2017
O. Gervasi et al. (Eds.): ICCSA 2017, Part II, LNCS 10405, pp. 494–508, 2017.
DOI: 10.1007/978-3-319-62395-5_34

A *graph* G consists of a finite set of vertices (also called nodes) $\{v_1, v_2, \ldots, v_n\}$ and a set of edges $\{\{v_i, v_j\}, i \neq j\}$ joining some pairs of vertices. A standard description is $G = (V, E)$, where V is the set of vertices set and E is the set of edges. A *path* in G is a sequence of vertices $v_i, v_{i+1}, \ldots, v_k$, where $\{v_j, v_{j+1}\}$ for $j = i, i+1, \ldots, k-1$, is an edge of G. A *cycle* is a closed path. The *size* of a cycle is the number of edges in it. A *chord* of a cycle is an edge joining two non-consecutive vertices. A graph G is said to be *chordal* if it has no chordless cycles of size 4 or more.

The *distance graph* of an $n \times n$ distance matrix, is a graph on n vertices with an edge connecting two vertices v_i and v_j if there is a non-zero entry in i-th row and j-th column of the distance matrix.

The *neighbourhood* $N(v)$ of a vertex v of G consists of those vertices in G that are adjacent to v. A vertex v is said to be *simplicial* if $N(v)$ is a clique, that is, a complete subgraph on $N(v)$. A *simplicial ordering* of the vertices of G is a map $\alpha : V \to \{1, 2, \ldots, n\}$ such that v_i is simplicial in the induced graph on the vertex set $\{v_i, v_{i+1}, \ldots, v_n\}$. The following result is well-known.

Theorem 1 [11]. *A graph $G = (V, E)$ is chordal iff there exists a simplicial ordering of its vertices.*

3   Point Placement on a Line: A Quick Review

To provide a context and motivation for the results of this paper, we provide a quick review of the main ideas underlying point placement algorithms for points on a line, with reference to a state-of-the-art algorithm [2].

Let $P = \{p_1, p_2, \ldots, p_n\}$ be n distinct points on a line. A distance graph on n vertices (corresponding to the n points in P) has edges joining pairs of points whose distances on the line are sought of an adversary. An assignment of lengths to the edges of this graph by an adversary is assumed to be valid if there exists a linear layout consistent with these lengths. The distance graph is said to be *line-rigid* if a consistent layout exists for all valid, adversarial assignments of lengths. All the distance graphs shown in Fig. 1 are line-rigid. However, a 4-cycle is not line-rigid as there exists an assignment of lengths that makes it a parallelogram, whose vertices have two distinct linear layouts.

A prototypical 1-round algorithm uses the line-rigid 3-cycle (or triangle) graph as the core structure and constructs the following distance graph on n points (see Fig. 2). As the figure shows, the graph has $n - 2$ triangles, hanging from a common strut. The number of distance queries made is $2n - 3$.

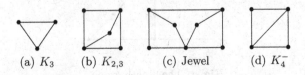

(a) K_3 (b) $K_{2,3}$ (c) Jewel   (d) K_4^-

Fig. 1. Some examples of line rigid graphs.

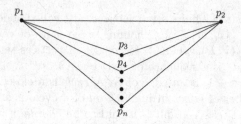

Fig. 2. Distance graph for a 1-round algorithm

A prototypical 2-rounds algorithm uses the 4-cycle (or triangle) graph as the core structure and constructs the following distance graph on n points (see Fig. 3). As the figure shows, the graph has b and $b+2$ edges, hanging from the left and right end-points respectively of a fixed edge. The explanation is that a 4-cycle is not line-rigid and the rigidity condition that at least one pair of opposite edges are not equal can be satisfied over two rounds. The discrepancy of 2 in the number of edges hanging from the two end-points allows us to pair edges which are not equal and thus meet the line rigidity condition for a 4-cycle. The number of distance queries made is $3n/2-2$. Thus by increasing the number of rounds and constructing a more complex query graph we reduce the number of distance queries by a constant factor.

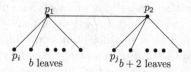

Fig. 3. Distance graph for a 2-rounds algorithm

The best known 2-rounds algorithm to-date [2], builds a distance query graph using the 3-path graph of Fig. 4. Its query complexity is $9n/7+O(1)$. This comes at the expense of 55 rigidity conditions that must be satisfied over two rounds.

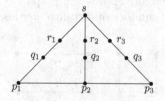

Fig. 4. The 3-path graph

The main tool for obtaining these rigidity conditions is the concept of a layer graph, introduced in [7]. A layer graph is an orthogonal re-drawing (if possible) of the distance query graph that must satisfy the following conditions:

P1. Each edge e of G is parallel to one of the two orthogonal directions $\mathbf{x}$ and $\mathbf{y}$.
P2. The length of an edge e is the distance between the corresponding points on L.
P3. Not all edges are along the same direction (thus a layer graph has a two-dimensional extent).
P4. When the layer graph is folded onto a line, by a rotation either to the left or to the right about an edge of the layer graph lying on this line, no two vertices coincide.

Chin et al. [7] showed that a given distance query graph is not line rigid iff it has a layer graph drawing. The different rigidity conditions are derived from a painstaking enumeration of all possible layer graph drawings of the given distance query graph. This is a challenging task.

Our experience with the implementation of 2-round algorithms (see [14]) has shown that the rigidity conditions are easy to verify when exact arithmetic is used; indeed, we simulated an adversary by generating layouts with integral coordinates. However if pairwise distances are not integral, the rounding errors introduced in finite-precision calculations can make checking the rigidity conditions difficult. This is an unavoidable issue for point-placement in the plane.

Another difficulty of generalizing this approach to two and higher dimensions is that of obtaining a suitable generalization of the layer graph concept and the associated theorem. This motivates the approach taken in this paper. The advantage of this approach is that it is susceptible to generalization to higher dimensions.

4 Overview of Our Results

We first discuss a reductionist approach to this problem: reducing point placement in the plane to point placement on a line. We consider the case when the points lie on a circle, using stereographic projection to reduce this to a 1-dimensional point placement problem. For points lying on an integer grid, we reduce the problem to two 1-dimensional point placement problems.

The algorithms for point placement on a line, requires testing a very large number of constraints involving edge lengths of a distance graph. Our experiments have shown that these work well when the points on a line have integral coordinates. To circumvent this problem we consider a matrix distance completion approach, when the distance graph is chordal. In our adversarial setting, we seek the lengths of this chordal graph from an adversary (An adversary can be thought of as a source of correct distance measurements).

Once the adversary has returned edge lengths for the chordal distance graph, we solve a matrix distance completion problem. Bakonyi and Johnson [4] showed that if the distance graph corresponding to a partial distance matrix is chordal, there exists a completion of this partial distance matrix. Precisely, they proved the following result.

Theorem 2 [4]. *Every partial distance matrix in R^k, the graph, G, of whose specified entries is chordal, admits a completion to a distance matrix in R^k.*

Finally, we compute the planar coordinates of the vertices of this complete distance graph, using an algorithm based on a result of Young and Householder [17].

5   Point Placement in the Plane

When the points $p_1, p_2, \ldots, p_n$ lie on an integer grid, we can solve the problem by solving two 1-dimensional point placement problems by projecting them on the x and y-axes. We assume that no two points lie on the same vertical or horizontal line of the grid (see Fig. 5).

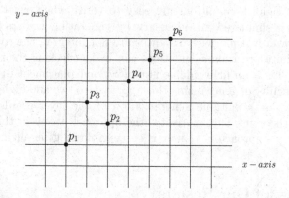

Fig. 5. Points on a two-dimensional integer grid

When the points lie on a circle, we can solve the problem by stereographic projection of the points on a line and then applying a 1-dimensional point location algorithm to the projected points (see Fig. 6).

When the distance query graph is complete, we can compute the locations of the points using an algorithm, based on the following result due to Young and Householder [17].

Theorem 3 [17]. *A necessary and sufficient condition for a set of numbers $d_{ij} = d_{ji}$ to be the mutual distances of a real set of points in Euclidean space is that the matrix $B = [d_{1i}^2 + d_{1j}^2 - d_{ij}^2]$ be positive semi-definite; and in this case the set of points is unique apart from a Euclidean transformation.*

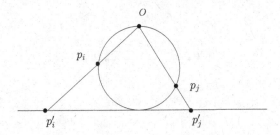

Fig. 6. Stereographic projection of points on a circle

In this case, there exists an orthogonal matrix σ such that

$$B = \sigma L^2 \sigma^t$$

where $L^2 = [\lambda_1^2, \lambda_2^2, \ldots, \lambda_r^2, 0, \ldots, 0]$, σ^t is the transpose of σ, and r is the embedding dimension. Thus, we have

$$B = (\sigma L)(\sigma L)^t$$

Since $B = AA^t$, where the rows of the matrix A are the coordinates of the points $p_1, p_2, \ldots, p_n$ in some r dimensional Euclidean space, the coordinates of the points are determined by solving the system of linear equations.

$$A = \sigma L$$

When the (distance) graph of the partial distance matrix is chordal, we use a distance matrix completion algorithm, the major components of which are discussed below.

5.1   Computing a Simplicial Ordering, α, of G

A simplicial ordering can be found by a breadth-first search of G, combined with a lexicographic labeling of its vertices. The following well-known LEX-BFS algorithm is due to Rose et al. [15].

Algorithm 1. Simplicial Ordering

1: Assign the empty label list, (), to each vertex in V
2: **for** $i = n$ downto 1 **do**
3:     Pick a vertex $v \in V$ with the lexicographically largest label list
4:     Set $\alpha(v) = i$
5:     For each unnumbered vertex w adjacent to v, add i to the label list of w
6: **end for**
7: return α

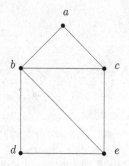

Fig. 7. A chordal graph on five vertices

For the chordal graph of Fig. 7, a simplicial ordering of is: $\alpha(a) = 5, \alpha(b) = 4, \alpha(c) = 3, \alpha(e) = 2$ and $\alpha(d) = 1$, obtained from the following label lists of the vertices over five steps.

	a	b	c	d	e
Step 0	()	()	()	()	()
Step 1	()	(5)	(5)	()	()
Step 2	()	(5)	(5,4)	(4)	(4)
Step 3	()	(5)	(5,4)	(4)	(4,3)
Step 4	()	(5)	(5,4)	(4,2)	(4,3)

5.2   Computing a Chordal Graph Sequence

An algorithm for generating the sequence of chordal graphs depends on the following results, proved in [12].

Theorem 4 [12]. *G has no minimal cycles of length exactly 4 if and only if the following holds: For any pair of vertices u and v with $u \neq v$, $\{u, v\} \notin E$, the graph $G + \{u, v\}$ has a unique maximal clique which contains both u and v. (That is: if C and C' are both cliques in $G + \{u, v\}$ which contain u and v, then so is $C \cup C'$)*

In particular, Theorem 4 holds for chordal graphs. The next theorem suggests an iterative algorithm for solving the distance matrix completion problem.

Theorem 5 [12]. *Let $G = (V, E)$ be chordal. Then there exists a sequence of chordal graphs $G_i = (V, E_i)$, $i = 0, 1, \ldots, k$, such that $G = G_0, G_1, G_2, \ldots, G_k$ is the complete graph and G_i is obtained by adding to G_{i-1} an edge $\{u, v\}$ as in Theorem 4.*

Such an edge $\{u, v\}$ is selected using the following scheme described in [12].

Assume that a simplicial ordering α of the vertices of the input chordal graph G is available. Let v_k be the vertex $\alpha^{-1}(k)$. Set $k_i = max\{k | (v_k, v_m) \notin E_i$ for some $m\}$ and $r_i = max\{r | (v_r, v_{k_i}) \notin E_i\}$. Then the edge to be added is $\{u, v\} = \{u_{k_i}, v_{r_i}\}$. In the next section we discuss an algorithm for selecting a maximal clique, containing this edge.

5.3 Computing a Maximal Clique Containing a Given Edge

An interesting algorithm due to Bron-Kerbosch [6] computes all maximal cliques, from which we can select the maximal clique that contains this edge.

The Bron-Kerbosch algorithm is a recursive backtracking algorithm, and a version based on choosing a pivot can be described thus.

Algorithm 2. Computing Cliques

1: BronKerbosch2(R, P, X):
2: **if** (P and X are both empty) **then**
3:     report R as a maximal clique
4: **end if**
5: choose a pivot vertex $u \in P \cup X$
6: **for** each vertex $v \in P \setminus N(u)$ **do**
7:     BronKerbosch2($R \cup \{v\}, P \cap N(v), X \cap N(v)$)
8:     $P := P \setminus \{v\}$
9:     $X := X \cup \{v\}$
10: **end for**

The above algorithm maintains three sets R, P and X, reporting the set R as the vertices of a maximum clique when at any level of the recursive calls, the sets P and X become empty.

We have implemented a simple algorithm that starts with the edge of interest and grows this into a maximal clique. In greater details, start with a clique containing two vertices of the given edge, and grow the current clique one vertex at a time by looping through the graph's remaining vertices. For each vertex v examined, add v to the clique if it is adjacent to every vertex that is already in the clique; otherwise, discard v.

5.4 Distance Matrix Completion of a Clique

The distance matrix of a clique with the distance of one edge missing can be formulated as the problem of completing a partial distance matrix with one missing entry. The following lemma proposes a solution to this problem.

Theorem 6 [4]. *The partial distance matrix*

$$\begin{pmatrix} 0 & D_{12} & x \\ D_{12}^t & D_{22} & D_{23} \\ x & D_{23}^t & 0 \end{pmatrix}$$

admits at least one completion to a distance matrix F. Moreover, if

$$\begin{pmatrix} 0 & D_{12} \\ D_{12}^t & D_{22} \end{pmatrix}$$

and

$$\begin{pmatrix} D_{22} & D_{23} \\ D_{23}^t & 0 \end{pmatrix}$$

are distance matrices with embedding dimensions p and q then x can be chosen so that the embedding dimension of F is $s = max\{p, q\}$.

This is equivalent to finding completions of the partial distance matrix:

$$\begin{pmatrix} 0 & 1 & 1 & e^t & 1 \\ 1 & 0 & d_{12} & \overline{D}_{13} & d_{14} \\ 1 & d_{12} & 0 & \overline{D}_{23} & x \\ e & \overline{D}_{13}^t & \overline{D}_{23}^t & \overline{D}_{33} & \overline{D}_{34} \\ 1 & d_{14} & x & \overline{D}_{34}^t & 0 \end{pmatrix}$$

to a matrix in which the Schur complement

$$\begin{pmatrix} a & B & x - d_{12} - d_{14} \\ B^t & C & D \\ x - d_{12} - d_{14} & D^t & f \end{pmatrix}$$

of the upper left 2×2 prinicipal submatrix

$$\begin{pmatrix} 0 & 1 \\ 1 & 0 \end{pmatrix}$$

has a positive semidefinite completion of rank s. This provides a solution for x that follows from the following result.

Theorem 7 [10]. *Let*

$$R = \begin{pmatrix} a & B & x \\ B^t & C & D \\ x & D^t & f \end{pmatrix}$$

be a real partial positive semidefinite matrix in which $rank \begin{pmatrix} a & B \\ B^t & C \end{pmatrix} = p$ *and* $rank \begin{pmatrix} C & D \\ D^t & f \end{pmatrix} = q$. *Then there is real positive semidefinite completion F of R such that the rank of F is* $max\{p, q\}$. *This completion is unique iff rank* $C = p$ *or rank* $C = q$.

5.5   Experimental Results

We have implemented the above algorithm in Python on a laptop with an AMD x4 955 processor with 6 GB RAM, running under Windows 10. The software includes a module for the generation of chordal graphs to be used as input. The chordal graph generation uses one of two algorithms proposed by Markenzon et al. [13]. A formal description of this algorithm is given below.

Algorithm 3. ChordalGraphGeneration(G)

1: Generate a binary tree, G
2: Pick two random vertices u and v from G.
3: **if** there is an edge exists between these two vertices (u and v) **then**
4:     go back to step 2.
5: **else**
6:     find the neighbors of u and v.
7: **end if**
8: **if** there is no common neightbors between u and v **then**
9:     go back to step 2.
10: **else**
11:     choose one vertex x from the common neighbor of u and v
12: **end if**
13: Perform BFS from u to v on the graph consisting of all the adjacent vertices of x minus the intersection of u and v's neighbors
14: **if** BFS finds a path from u to v **then**
15:     go back to step 2.
16: **else**
17:     insert an edge between u and v
18: **end if**

The result of an experiment is described below for a manually constructed chordal graph. The following partial distance matrix, where the off-diagonal 0's represent unknown distances,

$$\begin{pmatrix} 0 & 9 & 0 & 0 & 0 & 0 & 5 & 20 \\ 9 & 0 & 2 & 25 & 40 & 34 & 20 & 17 \\ 0 & 2 & 0 & 17 & 0 & 0 & 0 & 13 \\ 0 & 25 & 17 & 0 & 5 & 0 & 0 & 2 \\ 0 & 40 & 0 & 5 & 0 & 2 & 0 & 5 \\ 0 & 34 & 0 & 0 & 2 & 0 & 10 & 5 \\ 5 & 20 & 0 & 0 & 0 & 10 & 0 & 13 \\ 20 & 17 & 13 & 2 & 5 & 5 & 13 & 0 \end{pmatrix}$$

is obtained from the distances returned by the adversary based on the layout of a chordal graph, G_1, shown in Fig. 8.

The matrix distance completion algorithm outputs the following matrix.

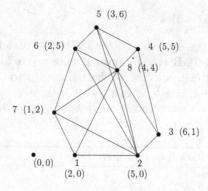

Fig. 8. A chordal graph on 8 vertices

$$\begin{pmatrix}
0 & 9 & 17.0 & 34.0 & 37.0 & 25.0 & 5 & 20 \\
9 & 0 & 2 & 25 & 40 & 34 & 20 & 17 \\
17.0 & 2 & 0 & 17 & 34.0 & 32.0 & 26.0 & 13 \\
34.0 & 25 & 17 & 0 & 5 & 9.0 & 25.0 & 2 \\
37.0 & 40 & 34.0 & 5 & 0 & 2 & 20.0 & 5 \\
25.0 & 34 & 32.0 & 9.0 & 2 & 0 & 10 & 5 \\
5 & 20 & 26.0 & 25.0 & 20.0 & 10 & 0 & 13 \\
20 & 17 & 13 & 2 & 5 & 5 & 13 & 0
\end{pmatrix}$$

where the computed entries are shown with a decimal point, followed by a single 0. The correctness of the computed entries can be checked against Fig. 8.

We ran the above complete distance matrix through our implementation of the Young-Householder algorithm. A plot of the output is shown in Fig. 9.

As can be seen that, apart from scale and orientation and missing edges, it is the same as the plot of the original graph shown in Fig. 8. More examples are shown in an appendix.

6   Computational Complexity

The computational complexity of the algorithm can be parametrized with respect to three different measures: (a) number of rounds, which is 1 in our case; (b) query complexity, which is the number of distance queries posed to the adversary and is a function of the number of rounds; (c) the time complexity of the algorithm.

The query complexity is the number of edges in the initial chordal graph. We have tried to make it as sparse as possible. It is always a tree on n vertices, with a few more edges added, to meet the requirements of the distance matrix completion algorithm. Thus the query complexity is $O(n)$. The time complexity of the algorithm is dominated by the number of times we have to perform distance

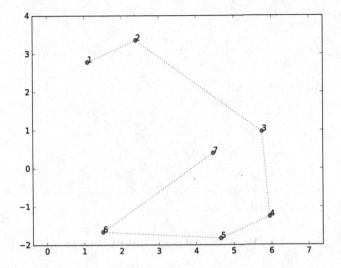

Fig. 9. A plot of the output of the Young-Householder algorithm

matrix completion of a clique. This is $O(n^2 f(n))$, as we go from a chordal graph which is nearly a tree to a complete (chordal) graph. This explains the n^2 term. The factor $f(n)$ is the complexity of the distance matrix completion algorithm. A loose upper bound is $O(n^3)$. Thus the time complexity of our algorithm is in $O(n^5)$.

7   Conclusion

In this paper we have proposed a 1-round algorithm for point placement in the plane in an adversarial setting, taking advantage of an existing infrastructure for completing partial distance matrices whose distance graphs are chordal. The locations of the points in the plane are recovered from the complete distance matrix, using an algorithm based on a result in [17].

Much more work remains to be done. Of particular interest are the extensions of the algorithm to other classes of graphs than chordal graphs and the design of 2-round algorithms.

A   Appendix

Below we show two more examples where the input chordal graphs are automatically generated by our software. Also shown are the incomplete distance matrices and the coordinates of the points obtained from their corresponding matrix completions (Tables 1, 2, 3 and 4).

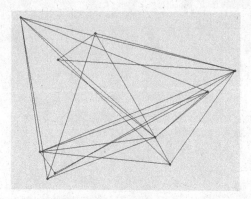

Fig. 10. A chordal graph with 10 vertices

Table 1. Distance matrix corresponding to the chordal graph of Fig. 10

	p_1	p_2	p_3	p_4	p_5	p_6	p_7	p_8	p_9	p_{10}
p_1	0	5513	5570	7069	0	0	0	0	1017	0
p_2	5513	0	157	4964	2740	0	0	7933	2180	0
p_3	5570	157	0	0	0	0	0	0	0	0
p_4	7069	4964	0	0	9376	7177	0	8125	6760	0
p_5	0	2740	0	9376	0	0	0	3077	232	0
p_6	0	0	0	7177	0	0	6137	8874	2333	0
p_7	0	0	0	0	0	6137	0	3497	3492	421
p_8	0	7933	0	8125	3077	8874	3497	0	2249	5066
p_9	1017	2180	0	6760	232	2333	3492	2249	0	3121
p_{10}	0	0	0	0	0	0	421	5066	3121	0

Table 2. x and y coordinates of points obtained from the matrix completion of the chordal graph of Fig. 10

	p_1	p_2	p_3	p_4	p_5	p_6	p_7	p_8	p_9	p_{10}
x	84	17	23	9	69	20	39	95	63	24
y	54	22	11	92	16	8	84	65	30	70

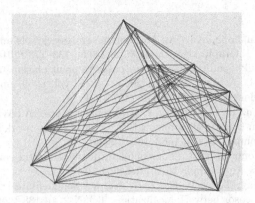

Fig. 11. A chordal graph with 15 vertices

Table 3. Distance matrix corresponding to the chordal graph of Fig. 11

	p_1	p_2	p_3	p_4	p_5	p_6	p_7	p_8	p_9	p_{10}	p_{11}	p_{12}	p_{13}	p_{14}	p_{15}
p_1	0	17	160	122	37	85	221	425	200	232	125	373	197	265	274
p_2	17	0	89	49	4	26	136	306	109	149	72	0	234	0	265
p_3	160	89	0	26	61	45	5	65	8	8	5	0	221	225	146
p_4	122	49	26	0	25	5	45	145	18	58	37	113	325	353	272
p_5	37	4	61	25	0	10	100	250	73	113	52	0	250	0	257
p_6	85	26	45	5	10	0	74	200	0	89	0	0	0	0	293
p_7	221	136	5	45	100	74	0	34	0	1	16	0	0	0	149
p_8	425	306	65	145	250	200	34	0	61	29	90	2	360	320	193
p_9	200	109	8	18	73	0	0	61	0	16	25	0	313	317	218
p_{10}	232	149	8	58	113	89	1	29	16	0	17	17	233	221	130
p_{11}	125	72	5	37	52	0	16	90	25	17	0	68	162	170	109
p_{12}	373	0	0	113	0	0	0	2	0	17	68	0	338	306	185
p_{13}	197	234	221	325	250	0	0	360	313	233	162	338	0	8	37
p_{14}	265	0	225	353	0	0	0	320	317	221	170	306	8	0	17
p_{15}	274	265	146	272	257	293	149	193	218	130	109	185	37	17	0

Table 4. x and y coordinates of points obtained from the matrix completion of the chordal graph of Fig. 11

	p_1	p_2	p_3	p_4	p_5	p_6	p_7	p_8	p_9	p_{10}	p_{11}	p_{12}	p_{13}	p_{14}	p_{15}
x	18	19	14	19	19	20	13	10	16	12	13	11	4	2	7
y	0	4	12	11	6	9	14	19	14	14	10	18	1	3	7

References

1. Alam, M.S., Mukhopadhyay, A.: More on generalized jewels and the point placement problem. J. Graph Algorithms Appl. **18**(1), 133–173 (2014)
2. Alam, M.S., Mukhopadhyay, A.: Three paths to point placement. In: Ganguly, S., Krishnamurti, R. (eds.) CALDAM 2015. LNCS, vol. 8959, pp. 33–44. Springer, Cham (2015). doi:10.1007/978-3-319-14974-5_4
3. Aspnes, J., Eren, T., Goldenberg, D.K., Stephen Morse, A., Whiteley, W., Yang, Y.R., Anderson, B.D.O., Belhumeur, P.N.: A theory of network localization. IEEE Trans. Mob. Comput. **5**(12), 1663–1678 (2006)
4. Bakonyi, M., Johnson, C.R.: The euclidian distance matrix completion problem. SIAM J. Matrix Anal. Appl. **16**(2), 646–654 (1995)
5. Biswas, P., Liang, T.-C., Wang, T.-C., Ye, Y.: Semidefinite programming based algorithms for sensor network localization. TOSN **2**(2), 188–220 (2006)
6. Bron, C., Kerbosch, J.: Algorithm 457: finding all cliques of an undirected graph. Commun. ACM **16**(9), 575–577 (1973)
7. Chin, F.Y.L., Leung, H.C.M., Sung, W.K., Yiu, S.M.: The point placement problem on a line – improved bounds for pairwise distance queries. In: Giancarlo, R., Hannenhalli, S. (eds.) WABI 2007. LNCS, vol. 4645, pp. 372–382. Springer, Heidelberg (2007). doi:10.1007/978-3-540-74126-8_35
8. Damaschke, P.: Point placement on the line by distance data. Discrete Appl. Math. **127**(1), 53–62 (2003)
9. Damaschke, P.: Randomized vs. deterministic distance query strategies for point location on the line. Discrete Appl. Math. **154**(3), 478–484 (2006)
10. Ellis, R.L., Lay, D.: Rank-preserving extensions of band matrices. Linear Multilinear Algebra **26**, 147–179 (1990)
11. Golumbic, M.C.: Algorithmic Graph Theory and Perfect Graphs (Annals of Discrete Mathematics, vol. 57). North-Holland Publishing Co., Amsterdam (2004)
12. Grone, R., Johnson, C.R., Sa, E.M., Wolkowicz, H.: Positive definite completions of partial hermitian matrices. Linear Algebra Appl. **58**, 109–124 (1984)
13. Markenzon, L., Vernet, O., Araujo, L.H.: Two methods for the generation of chordal graphs. Annals OR **157**(1), 47–60 (2008)
14. Mukhopadhyay, A., Sarker, P.K., Kannan, K.K.V.: Point placement algorithms: an experimental study. Int. J. Exp. Algorithms **6**(1), 1–13 (2016)
15. Rose, D.J., Tarjan, R.E., Lueker, G.S.: Algorithmic aspects of vertex elimination on graphs. SIAM J. Comput. **5**(2), 266–283 (1976)
16. Schoenberg, I.J.: Remarks to Maurice Frechet's article "sur la definition axiomatique d'une classe d'espace distancis vectoriellement applicable sur l'espace de hilbert". Ann. Math. **36**(3), 724–731 (1935)
17. Young, G., Householder, A.S.: Discussion of a set of points in terms of their mutual distances. Psychometrika **3**(1), 19–22 (1938)

Computing the Triangle Maximizing the Length of Its Smallest Side Inside a Convex Polygon

Sanjib Sadhu[1]([⊠]), Sasanka Roy[2], Soumen Nandi[2], Subhas C. Nandy[2], and Suchismita Roy[1]

[1] Department of CSE, National Institute of Technology Durgapur, Durgapur, India
sanjibsadhu411@gmail.com
[2] Indian Statistical Institute, Kolkata, India

Abstract. Given a convex polygon with n vertices, we study the problem of identifying a triangle with its smallest side as large as possible among all the triangles that can be drawn inside the polygon. We show that at least one of the vertices of such a triangle must be common with a vertex of the polygon. Next we propose an $O(n^2 \log n)$ time algorithm to compute such a triangle inside the given convex polygon.

Keywords: Computational geometry · Algorithms · Properties of isosceles and equilateral triangles · Optimal inclusion problem

1 Introduction

Optimizing a property of any geometric object inside another given geometric object has drawn a lot of interests to the researchers not only due to its theoretical interest, but also for its practical applications e.g. the problem of computing a triangle [3,8], rectangle [4] or a parallelogram [10] of largest area, or largest line segment inside a polygon [9]. Another well known area of research is computing the inner approximation of the polygon P, which is the largest convex body of desired shape inside P [9]. In this paper, our main objective is to compute a triangle $\triangle$ inside a given convex polygon P so that the smallest side of $\triangle$ is maximum over the smallest side of all possible triangles inside P. This problem arose while we were working on designing a streaming algorithm for the 2-center problem of a convex polygon [11]. Here we exploited two properties: (i) the radii of the two smallest congruent disks covering a convex polygon P must cover a triangle lying inside it, and (ii) if we cover a triangle by two congruent disks, one of the disks must cover the smallest side of that triangle. Thus, the lower bound of the diameter of the disks for the 2-center problem of P must be the smallest side of a triangle inside the polygon. In order to have a better approximation factor for the 2-center problem of a convex polygon, it would be better to have a larger value of the lower bound of the problem. The problem of computing such a triangle inside P with its smallest side largest is itself an interesting optimization problem, and may find some other important applications.

© Springer International Publishing AG 2017
O. Gervasi et al. (Eds.): ICCSA 2017, Part II, LNCS 10405, pp. 509–524, 2017.
DOI: 10.1007/978-3-319-62395-5_35

1.1   Related Work

There exists many results on optimal inclusion problem. The longest line segment inside a simple polygon was first solved by Chazelle and Sharir [5] in $O(n^{1.99})$ time. Later, Hall-Holt et al. [9] proposed a $\frac{1}{2}$-approximation algorithm in $O(n \log n)$ time and a PTAS with time complexity $O(n \log^2 n)$ for this problem. Aggarwal et al. [1] showed that the largest area convex k-gon inside a convex n-gon can be found in $O(kn + n \log n)$ time. DePano et al. [7] showed that $O(n^3)$ time is required to compute maximum area inscribed equilateral triangles or squares within a simple polygon[1]. An $O(n \log n)$ time constant factor approximation algorithm for computing a largest area triangle or a convex polygon inside a simple n-gon was proposed by Hall-Holt et al. [9]. Recently, Jin and Matulef [10] proposed an $O(n^2)$ time algorithm to compute a maximum area parallelogram inside a convex n-gon. Alt et al. [2] computed largest axis-parallel rectangle inside a convex n-gon in $O(\log n)$ time. Later Karen et al. [6] solved the same problem for any simple n-gon in $O(n \log^2 n)$ time. In 2016, Cabello et al. [4] proposed an $O(n^3)$ time algorithm to solve the maximum area and maximum perimeter rectangle (of any arbitrary orientation) contained inside a convex n-gon.

1.2   Our Result

We show that one vertex of the triangle, whose smallest side is longest among all possible triangles inside a given convex n-gon P, must coincide with a vertex of P. We also show that if such a triangle is isosceles whose two equal sides are the smallest, then two of its vertices will coincide with two vertices of P. Finally we propose an $O(n^2 \log n)$ time algorithm for computing such a triangle in a given polygon P. To the best of our knowledge, our work is the first study on this problem, i.e. computing a triangle maximizing the length of its smallest side inside P. We also compute the largest equilateral triangle inside P in $O(n^2)$ time.

1.3   Notations and Terminologies

Throughout the paper we use the following notations. The line segment joining any two points p and q is denoted by $\overline{pq}$ and its length will be denoted by $|pq|$. A triangle is represented by $\triangle$. The smallest side of a triangle $\triangle$ and its corresponding length are denoted by $\delta(\triangle)$ and $|\delta(\triangle)|$ respectively. A parallelogram is denoted by $\Diamond$. The boundary of a polygon P is denoted by $boundary(P)$.

---

[1] We could not access the paper [7] in the Internet and in the library of many reputed universities and institutions. Even we requested one of the authors of the paper for a copy and he could not give it.

2   Problem Formulation and Properties

A triangle Δ is said to be **inside** a convex polygon P, if $P \cap \Delta = \Delta$, i.e. no vertices of Δ lie outside *boundary(P)*. The edges of the polygon P on which the vertices of a triangle Δ lie are said to be **supporting edges** of the triangle Δ.

Problem definition: Given a convex polygon P, compute a triangle Δ inside P that maximizes the length of its smallest side. Note that, such a triangle inside P may be scalene or isosceles or equilateral.

Observation 1. *If a vertex of a triangle Δ_1 lies properly inside P, then there exists another triangle Δ_2 whose all vertices lie on the* boundary(P) *so that* $|\delta(\Delta_1)| \leqslant |\delta(\Delta_2)|$ *(see Fig. 1).*

Observation 2. *For every scalene triangle Δ_1, there exists an isosceles triangle Δ_2 so that* $|\delta(\Delta_1)| = |\delta(\Delta_2)|$ *and Δ_2 lies inside Δ_1 (Fig. 2).*

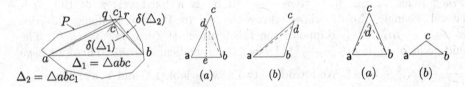

Fig. 1. Observation 1   **Fig. 2.** Observation 2   **Fig. 3.** Observation 3

Observation 2 says that, in order to compute the required triangle as per problem definition, we consider all possible isosceles triangles (instead of scalene triangle) inside P. Now, there are two types of isosceles triangles in terms of the number of the smallest side: (i) **Type I:** Δ has a unique smallest side as shown in Fig. 3(a) and (ii) **Type II:** Δ has two smallest sides, say $|ac|$ and $|bc|$ as shown in Fig. 3(b).

Observation 3. *The* **Type I** *isosceles triangle Δ must have an equilateral triangle inside it whose smallest side is the smallest side of Δ (see Fig. 3(a)).*

Observation 3 says that, instead of considering **Type-I** isosceles triangles inside convex polygon P, we may consider all possible equilateral triangles inside P. Thus, we need to consider all possible equilateral triangles and all **Type-II** isosceles triangles inside P.

2.1   Properties of Equilateral Triangles Inside a Convex Polygon P

Lemma 1. *If one vertex of an equilateral triangle Δ is kept fixed while another vertex of it moves along an edge of the polygon P, then the third vertex of Δ must move along a line segment.*

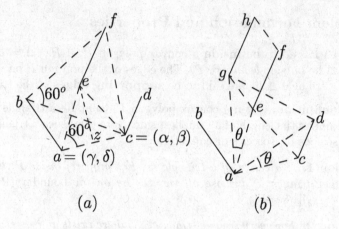

Fig. 4. (a) Proof of Lemma 1, and (b) Proof of Corollary 2.

Proof. Refer to Fig. 4(a). Here $c = (\alpha, \beta)$ is a fixed vertex of the equilateral triangle $\triangle ace$, whose vertex $a = (\gamma, \delta)$ moves along an edge $\overline{ab}$: $y = m_1 x + c_1$. Suppose, the third vertex of $\triangle ace$ be at $e(h, k)$. The mid-point of $\overline{ac}$ is $z(\frac{\alpha+\gamma}{2}, \frac{\beta+\delta}{2})$. Draw $\overline{ez}$ perpendicular to $\overline{ac}$. Thus, we have $\frac{k-\left(\frac{\beta+\delta}{2}\right)}{h-\left(\frac{\alpha+\gamma}{2}\right)} \times \frac{\delta-\beta}{\gamma-\alpha} = -1$. We introduce two new variables X and Y as follows

$$X = h - \left(\frac{\alpha+\gamma}{2}\right) \quad and \quad Y = k - \left(\frac{\beta+\delta}{2}\right), \tag{1}$$

and we have $\frac{Y}{X} = -\left(\frac{\gamma-\alpha}{\delta-\beta}\right)$. In $\triangle ezc$ and $\triangle ace$ (equilateral), we have $|ez|^2 + |zc|^2 = |ec|^2$, and $|ec| = |ac| = 2|zc|$. Thus, $|ez|^2 = 3|zc|^2$ which gives $\{h - (\frac{\alpha+\gamma}{2})\}^2 + \{k - \left(\frac{\beta+\delta}{2}\right)\}^2 = 3\{\left(\frac{\alpha-\gamma}{2}\right)^2 + \left(\frac{\beta-\delta}{2}\right)^2\}$. We can write this as $X^2 + Y^2 = \frac{3}{4}\{(\alpha-\gamma)^2 + (\beta-\delta)^2\}$. Writing Y in terms of X, we have $X = \pm\frac{\sqrt{3}}{2}(\beta-\delta)$, i.e. $h = (\frac{\alpha+\gamma}{2}) \pm \frac{\sqrt{3}}{2}(\beta-\delta)$. The dual sign of h is due to the fact that two different equilateral triangles $\triangle ace$ and $\triangle ace'$ are possible at the two opposite sides of $\overline{ac}$. Without loss of generality, we take $X = \frac{\sqrt{3}}{2}(\beta-\delta)$. This gives $Y = -\frac{\sqrt{3}}{2}(\alpha-\gamma)$. Hence from Eq. 1, $h = \frac{\sqrt{3}}{2}(\beta-\delta) + \frac{\alpha+\gamma}{2}$ and $k = -\frac{\sqrt{3}}{2}(\alpha-\gamma) + \frac{\beta+\delta}{2}$. Now $a(\gamma, \delta)$ lies on $\overline{ab}$: $y = m_1 x + c_1$. So, $\delta = m_1\gamma + c_1$. Substituting δ, we get $h = \frac{\gamma}{2}(1 - \sqrt{3}m_1) + \frac{1}{2}(\alpha + \sqrt{3}(\beta - c_1)) = K_1\gamma + K_2$, where $K_1 = \frac{1-\sqrt{3}m_1}{2}$ and $K_2 = \frac{1}{2}(\alpha + \sqrt{3}(\beta - c_1))$ are two constants. Hence,

$$\gamma = \frac{h}{K_1} - \frac{K_2}{K_1} \tag{2}$$

Equation 1 gives $k = \frac{1}{2}(\beta + \delta) - \frac{\sqrt{3}}{2}(\alpha - \gamma) = \frac{1}{2}(\beta + m_1\gamma + c_1) - \frac{\sqrt{3}}{2}(\alpha-\gamma) = \frac{\gamma}{2}(m_1 + \sqrt{3}) + \frac{1}{2}(\beta + c_1 - \sqrt{3}\alpha)$. Using Eq. 2, we have $k = h\left(\frac{m_1+\sqrt{3}}{1-\sqrt{3}m_1}\right) - \mathcal{K}$

where $\mathcal{K} = \frac{K_2}{2K_1}(m_1 + \sqrt{3}) - \frac{1}{2}(\beta + c_1 - \sqrt{3}\alpha)$ is a constant. Therefore, the locus of the vertex $e(h, k)$ which is a straight line $\overline{ef}$ as shown in Fig. 4, is given by

$$y = \left(\frac{m_1 + \sqrt{3}}{1 - \sqrt{3}m_1}\right) x - \mathcal{K} \tag{3}$$

When vertex a moves to b, the vertex e of the equilateral triangle $\triangle ace$ moves to f along the line segment $\overline{ef}$ and the equilateral triangle $\triangle bcf$ is obtained. $\square$

Corollary 1. *When a vertex c of an equilateral triangle $\triangle ace$ is fixed and its other vertex a moves along an edge $\overline{ab}$, then the length of $\overline{ef}$, along which the the vertex e moves to keep $\triangle ace$ always equilateral, is equal to $|ab|$.*

Proof. Refer to Fig. 4(a). Here $\angle ecb = 60° - \angle bca$ since $\triangle eca$ is equilateral. Also $\angle ecf = 60° - \angle ecb = \angle bca$, because $\triangle bcf$ is equilateral. From $\triangle abc$, we get $|ab|^2 = |ac|^2 + |bc|^2 + 2|ac|.|bc| \cos \angle bca$. Similarly from $\triangle cef$ we have $|ef|^2 = |ce|^2 + |cf|^2 + 2|ce|.|cf| \cos \angle ecf = |ac|^2 + |bc|^2 + 2|ac|.|bc| \cos \angle bca$, because $|ce| = |ac|$, $|cf| = |bc|$ and $\angle ecf = \angle bca$. Therefore, $|ef| = |ab|$. $\square$

It needs to be mentioned that if vertex a of the equilateral triangle $\triangle ace$ is kept fixed and c moves along the edge $\overline{cd}$, then the third vertex e moves along $\overline{eg}$ and $|eg| = |cd|$ (see Corollary 1).

Corollary 2. *If for an equilateral triangle $\triangle ace$, the vertex a moves along $\overline{ab}$, c moves along $\overline{cd}$, then the locus of the third vertex e will lie in the closed region bounded by the parallelogram $\Diamond efhg$.*

Proof. Refer to Fig. 4(b). When the vertex c of $\triangle ace$ is kept fixed at a point c on $\overline{cd}$ and vertex a moves along $\overline{ab}$, the slope of $\overline{ef}$ along which the third vertex e moves, depends only on the slope of $\overline{ab}$ (see slope of Eq. 3). If c changes its position to c' on $\overline{cd}$, the slope of the new line segment on which the third vertex e moves will not be changed, rather it will be shifted parallely (see the intercept term of Eq. 3). Similarly, to keep the triangle $\triangle ace$ equilateral, if vertex a changes its position to a' on $\overline{ab}$, and c moves along the edge $\overline{cd}$, then the line segment $\overline{eg}$, on which third vertex e moves, will be shifted parallely to itself. Hence the vertex e always lies inside $\Diamond efhg$. $\square$

Now we will investigate the properties of largest equilateral triangle supported by any three edges, say $\overline{ab}$, $\overline{cd}$ and $\overline{k_1k_2}$, of a convex polygon P. We will refer to the vertices of equilateral triangle lying on edges $\overline{cd}$ and $\overline{ab}$ as first vertex and second vertex respectively. As stated in Corollary 2, while the first and second vertices of an equilateral triangle $\triangle$ move along the edges $\overline{ab}$ and $\overline{cd}$ of P, its third vertex must trace out parallel line segments within the parallelogram $\Diamond efhg$, as shown in Fig. 5(a), (b). Since $\triangle$ is supported by the three edge $\overline{ab}$, $\overline{cd}$ and $\overline{k_1k_2}$ of P, the edge $\overline{k_1k_2}$ must intersect the *boundary* $(\Diamond efhg)$ once or twice; or $\overline{k_1k_2}$ will lie completely inside the *boundary* $(\Diamond efhg)$. At first we assume that the two end points k_1 and k_2 will lie inside the parallelogram $\Diamond efhg$ (see

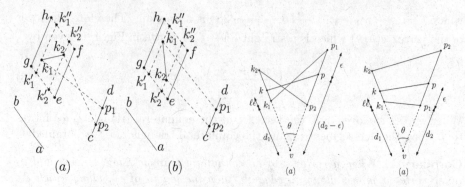

Fig. 5. Edge $\overline{k_1k_2}$ of P lies inside $\Diamond efhg$. Quadrilateral $k_1k_2p_2p_1$ is (a) simple (b) non-simple

Fig. 6. Proof of Lemma 2 where $\overline{k_1k_2}$ and $\overline{p_1p_2}$ are not parallel to each other

Fig. 5(a), (b)). Now the end points k_1 and k_2 lie on two different parallel line segments $\overline{k_1'k_1''}$ and $\overline{k_2'k_2''}$ which are the loci of third vertex of the triangle while the first vertex is at p_1 and p_2 respectively. In Fig. 5(a), (b), $|ek_1'| = |cp_1|$ and $|ek_2'| = |cp_2|$. If first vertex of $\triangle$ is at p_1 or p_2, then the third vertex of it will be at k_1 and k_2 respectively; and the intermediate points of $\overline{p_1p_2}$ are mapped proportionately on $\overline{k_1k_2}^2$. Thus, the first vertex of $\triangle$ lies on $\overline{p_1p_2}$ otherwise the equilateral triangle will not be supported by the edge $\overline{k_1k_2}$. Without loss of generality we take $|p_1p_2| = 1$ and $|k_1k_2| = \ell$ where $\ell > $ or $=$ or < 1. Now, assume that $|p_1p| = \epsilon$, $0 < \epsilon \leqslant 1$. This gives $|k_1k| = \ell\epsilon$. Now consider the quadrilateral $k_1k_2p_2p_1$. Depending on whether the two sides $\overline{k_1p_1}$ and $\overline{k_2p_2}$ intersect or not, the quadrilateral $k_1k_2p_2p_1$ will be either a non-simple (self intersecting) or a simple quadrilateral as shown in Fig. 5(a) and (b) respectively.

Lemma 2. *For both non-simple and simple quadrilateral $k_1k_2p_2p_1$, among all possible pairs (k, p) of first and third vertices ($k \in \overline{k_1k_2}$ and $p \in \overline{p_1p_2}$), $|kp|$ will be maximum when $(k, p) = (k_1, p_1)$ or $(k, p) = (k_2, p_2)$.*

Proof. Refer to Fig. 6(a), (b) for non-simple quadrilateral $k_1k_2p_2p_1$ and simple quadrilateral $k_1k_2p_2p_1$ respectively. Assume that both $\overline{k_1k_2}$ are $\overline{p_1p_2}$ are smaller than a line segment $\overline{kp}$ where $k \in \overline{k_1k_2}$, $p \in \overline{p_1p_2}$; and $|p_1p| = \epsilon$, $|k_1k| = \ell\epsilon$. Depending on the slope of $\overline{p_1p_2}$ and $\overline{k_1k_2}$ we consider following two cases:

Case (i) $\overline{p_1p_2}$ and $\overline{k_1k_2}$ are not parallel: Refer to Fig. 6(a), (b), where the extended $\overline{k_1k_2}$ and $\overline{p_1p_2}$ meet at a point v. Let $\overline{k_1k_2}$ and $\overline{p_1p_2}$ make an angle θ. Let us assume $|vk_1| = d_1$ and $|vp_1| = d_2$. Now, the following two types of quadrilateral are generated.

Non-simple quadrilateral $k_1k_2p_2p_1$ (see Fig. 6(a)): From $\triangle vkp$, we have $|kp|^2 = (d_1 + \ell\epsilon)^2 + (d_2 - \epsilon)^2 + 2|d_1 + \ell\epsilon| . |d_2 - \epsilon| \cos\theta$. Therefore, $|kp|^2 = C_1\epsilon^2 +$

---
[2] i.e., if the first vertex be $p \in \overline{p_1p_2}$ then the corresponding third vertex $k \in \overline{k_1k_2}$ such that $|k_1k| = \frac{|k_1k_2|}{|p_1p_2|}|p_1p|$.

$C_2\epsilon + C_3$, where C_1, C_2 and C_3 are constants[3]. Take $|kp|^2 = \mathcal{Y}_1(\epsilon) = C_1\epsilon^2 + C_2\epsilon + C_3$, which is a parabola, and it attains minimum at a particular value $\epsilon = \epsilon_o$, say. Since $|kp| > 0$, $|kp| = \sqrt{\mathcal{Y}_1}$ is also minimized at $\epsilon = \epsilon_o$ and increases in both sides of ϵ_o monotonically.

Simple quadrilateral $k_1 k_2 p_2 p_1$ (see Fig. 6(b)): From $\triangle vkp$, we have $|kp|^2 = (d_1 + \ell\epsilon)^2 + (d_2 + \epsilon)^2 + 2|d_1 + \ell\epsilon|.|d_2 + \epsilon|\cos\theta$. Therefore, $|kp|^2 = C_4\epsilon^2 + C_5\epsilon + C_6$, where C_4, C_5 and C_6 are constants. Here also, the function $|kp|^2 = \mathcal{Y}_2(\epsilon) = C_4\epsilon^2 + C_5\epsilon + C_6$ is a parabola and arguing as earlier, $|kp|$ is minimized at a particular value ϵ_o, and it increases monotonically in both the sides of ϵ_o.

Case (ii): $\overline{p_1 p_2}$ and $\overline{k_1 k_2}$ are parallel to each other: Without loss of generality, let us assume $|p_1 p_2| \geqslant |k_1 k_2|$.

Quadrilateral $k_1 k_2 p_2 p_1$ is non-simple. Refer to Fig. 7(a). Assume that slope of $\overline{k_1 p_1}$ and $\overline{k_2 p_2}$ are positive and negative respectively. Let the slope of $\overline{kp}$ be positive and z be the projection of p on $\overline{k_1 k_2}$ (extend, if required). Therefore, we have $|zp| \leqslant |kp| < |k_1 p| < |k_1 p_1|$.

Quadrilateral $k_1 k_2 p_2 p_1$ is simple. Refer to Fig. 7(b), (c). Since $|p_1 p_2| \geqslant |k_1 k_2|$, we have $|p_1 p| > |k_1 k|$ and $|p_2 p| > |k_2 k|$. Let z be the projection of k on $\overline{p_1 p_2}$. Now, if z lies on $\overline{p_1 p}$, draw a line segment $\overline{k_2 p_3'}$ parallel to $\overline{kp}$, where $p_3' \in \overline{p_2 p}$ as shown in Fig. 7(b). The point p_2 will lie above p_3' since $|p_2 p| > |k_2 p|$. Therefore, $|k_2 p_2| > |k_2 p_3'| = |kp|$. However, if z lies outside $\overline{p_1 p}$ (as shown in Fig. 7(c)), then draw a line segment $\overline{k_1 p_3'}$ parallel to $\overline{kp}$, where p_3' lies above p_1 because $|p_1 p| > |k_1 k|$. Hence, $|k_1 p_1| > |k_1 p_3'| = |kp|$. $\qquad\square$

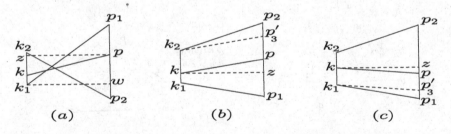

Fig. 7. Proof of Lemma 2 where $\overline{k_1 k_2}$ is parallel to $\overline{p_1 p_2}$: (a) $k_1 k_2 p_2 p_1$ is non-simple quadrilateral (b), (c) $k_1 k_2 p_2 p_1$ is simple quadrilateral

Therefore, Lemma 2 shows that the length of the sides of the equilateral triangle $\triangle$ becomes maximum when its third vertex is either at k_1 or k_2. This proves that the largest equilateral triangle supported by the three edges must coincide with an end point of the supporting edges. Similarly, we can prove that if the $\overline{k_1 k_2}$ intersect the *boundary ($\Diamond efhg$)*, then the length of the sides of an equilateral triangle will be maximum when one of its vertices coincides with

---

[3] Because they depend on constants d_1, d_2 and ℓ.

any one of the intersection points of $\overline{k_1 k_2}$ and the boundary($\Diamond efhg$). Now, the boundary($\Diamond efhg$) is locus of the third vertex of an equilateral $\triangle$ whose one of the other two vertices lies at one end point of $\overline{ab}$ or $\overline{cd}$. Thus we have the following lemma

Lemma 3. *Among all possible equilateral triangles supported by three edges of a convex polygon P, the largest one has at least one vertex coinciding with an end point of its supporting edge.*

Lemma 4. *If $\Diamond efhg$ lies inside boundary(P), then one vertex of the largest equilateral triangle supported by the edges $\overline{ab}$ and $\overline{cd}$ must coincide with one of the corners of $\Diamond efhg$.*

Proof. Refer to Fig. 8. Assume that $\Diamond efhg$ lies inside boundary(P). Let $\triangle stu$ be the largest equilateral triangle supported by the edges $\overline{ab}$ and $\overline{cd}$ of P, where u, the third vertex of $\triangle stu$, lies inside the boundary($\Diamond efhg$). Let u lie on $\overline{u'u''}$, where $u' \in \overline{fh}$ and $u'' \in \overline{eg}$, and $\overline{u'u''}$ is parallel to $\overline{ef}$. Thus, in $\triangle su'u''$, $|su| < \max(|su'|, |su''|)$. Assume that $|su| < |su'|$ and the corresponding equilateral triangle is $\triangle sbu'$ (since $\overline{fh}$ is the locus of third vertex of the triangle whose one vertex is at b). Similarly we can show that $|bu'|$ is maximized either at $u' = f$ or $u' = h$. Suppose this is maximized at h and the corresponding equilateral triangle becomes $\triangle bhd$, since third vertex lies on $\overline{gh}$ while first vertex is at d. Thus, $|\delta(\triangle bhd)| > |\delta(\triangle stu)|$ which contradicts our assumption that $\triangle stu$ was largest. $\qquad\square$

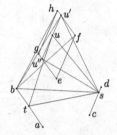

Fig. 8. Proof of Lemma 4

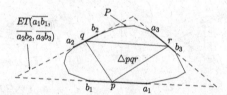

Fig. 9. Extended triangle for three edges $\overline{a_1 b_1}$, $\overline{a_2 b_2}$, and $\overline{a_3 b_3}$

Since the pair (first vertex, second vertex) of equilateral triangles are (a, c), (b, c), (a, d) and (b, d) while the corresponding third vertex of it is at e, f, g and h respectively, we get the following theorem using Lemmas 3 and 4.

Theorem 1. *One of the vertices of the largest equilateral triangle inside a given convex polygon P must coincide with a vertex of P.*

2.2   Properties of Isosceles Triangles Inside a Convex Polygon P

Before studying any special properties of isosceles triangle we will explore some properties that are common to all possible triangles inside a convex polygon.

Observation 4. *If three vertices of a triangle $\triangle$ are rotated by an angle θ about any arbitrary point v, then the resulting triangle $\triangle'$ will be congruent to the $\triangle$.*

Definition 1. *If the three non-parallel edges $\overline{a_1b_1}$, $\overline{a_2b_2}$ and $\overline{a_3b_3}$ of a convex polygon P are extended, they form a triangle termed as **extended triangle**, which will be denoted by $ET(\overline{a_1b_1},\ \overline{a_2b_2},\ \overline{a_3b_3})$ (see Fig. 9). In case, the two considered edges of P are parallel, this triangle $ET(\overline{a_1b_1},\ \overline{a_2b_2},\ \overline{a_3b_3})$ degenerates to an open region enclosed by three intersecting supporting lines two of which are parallel.*

Now refer to Fig. 10(a). We consider a triangle $\triangle pqr$ whose vertices p, q and r are supported by three edges $\overline{a_1b_1}$, $\overline{a_2b_2}$ and $\overline{a_3b_3}$ respectively, of the convex polygon P. The perpendiculars at the points p, q and r on the supporting edges of the three vertices p, q and r are denoted by ℓ_1, ℓ_2 and ℓ_3 respectively, and the points of their pairwise intersection are u, v and w respectively, as shown in Fig. 10(a). Here two cases can arise: (i) at least one of u, v and w lies inside $ET(\overline{a_1b_1},\ \overline{a_2b_2},\ \overline{a_3b_3})$, and (ii) none of the u, v and w lies inside $ET(\overline{a_1b_1},\ \overline{a_2b_2},\ \overline{a_3b_3})$.

Lemma 5. *If any two perpendiculars among ℓ_1, ℓ_2 and ℓ_3 intersect inside the extended triangle $ET(\overline{a_1b_1},\ \overline{a_2b_2},\ \overline{a_3b_3})$, then a vertex of $\triangle pqr$ whose smallest side is maximum, must coincide with an end point of its supporting edge which is also a vertex of P.*

Proof. Refer to Fig. 10(a). We assume that the smallest side of $\triangle pqr$, supported by the three edges $\overline{a_1b_1}$, $\overline{a_2b_2}$, $\overline{a_3b_3}$ of polygon P, is the largest among all possible triangles supported by $\overline{a_1b_1}$, $\overline{a_2b_2}$, $\overline{a_3b_3}$. Furthermore, we assume that no vertex of $\triangle pqr$ lies at the end point of its supporting edge. In this case at least two perpendiculars intersect inside $ET(\overline{a_1b_1},\ \overline{a_2b_2},\ \overline{a_3b_3})$ at point, say v. If all the three points of intersection u, v and w lies inside $ET(\overline{a_1b_1},\ \overline{a_2b_2},\ \overline{a_3b_3})$, choose any one of them, say v as shown in Fig. 10(b). Since $\overline{vq}$ and $\overline{vr}$ are perpendiculars to the supporting edges $\overline{a_2b_2}$ and $\overline{a_3b_3}$ respectively, if q and r are rotated in any direction about the point v, then $\overline{qv}$ and $\overline{rv}$ will not intersect their corresponding supporting edges. However, since $\overline{vp}$ is not perpendicular to its supporting edge $(\overline{a_1b_1})$, any one of $\angle a_1pv$ or $\angle b_1pv$ is obtuse. Thus, if we rotate $\triangle pqr$ by a very small amount ϵ in anticlockwise or clockwise direction depending on whether $\angle a_1pv$ or $\angle b_1pv$ is obtuse or not, then $\triangle pqr$ remains inside the polygon P. The amount ϵ can be determined as follows. Let ϵ_1, ϵ_2 and ϵ_3 be the angles of rotation of $\overline{vp}$, $\overline{vq}$ and $\overline{vr}$ respectively, about the point v so that they hit the boundary(P). Take $\epsilon < \min(\epsilon_1, \epsilon_2, \epsilon_3)$. Since the triangle $\triangle$ remains inside P after rotation, there must exist another triangle $\triangle_1$ supported by the same three edges so that $|\delta(\triangle_1)| > |\delta(\triangle)|$ which contradicts our assumption. Therefore, the triangle $\triangle pqr$ whose smallest side is largest must coincide with an end point of its supporting edge which is also a vertex of polygon P. $\square$

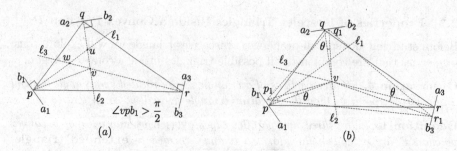

Fig. 10. Proof of Lemma 5(a) At least two of the perpendiculars ℓ_1, ℓ_2 and ℓ_3 intersect inside the $ET(\overline{a_1b_1},\ \overline{a_2b_2},\ \overline{a_3b_3})$ (b) $\triangle pqr$ after rotation lies inside the polygon P

Lemma 6. *If all the perpendiculars ℓ_1, ℓ_2 and ℓ_3 intersect outside the extended triangle $ET(\overline{a_1b_1},\ \overline{a_2b_2},\ \overline{a_3b_3})$ at three different points, then $\delta(\triangle pqr)$ will be largest if a vertex of the $\triangle pqr$ coincides with an end point of its supporting edge.*

Proof. Refer to Fig. 11(a), where $\triangle pqr$ is supported by the three edges $\overline{a_1b_1}$, $\overline{a_2b_2}$, $\overline{a_3b_3}$. We assume that $\delta(\triangle pqr)$ is largest among all possible triangles supported by these edges and no vertices of $\triangle pqr$ coincide with the end point of their corresponding supporting edges. The perpendiculars ℓ_1, ℓ_2 and ℓ_3 intersect outside $ET(\overline{a_1b_1},\ \overline{a_2b_2},\ \overline{a_3b_3})$ at u, v *and* w behind[4] the edge $\overline{a_1b_1}$ as shown in Fig. 11(a). Consider the two supporting edges $\overline{a_2b_2}$ and $\overline{a_3b_3}$ and the point of intersection v between the corresponding perpendiculars ℓ_2 and ℓ_3. As argued earlier, $\overline{vq}$ and $\overline{vr}$ are perpendiculars to $\overline{a_2b_2}$ and $\overline{a_3b_3}$ respectively, the line segment $\overline{vq}$ and $\overline{vr}$ after rotation in any direction about the point v do not intersect with $\overline{a_2b_2}$ and $\overline{a_3b_3}$ respectively. However, $\overline{vp}$ is not perpendicular to the $\overline{a_1b_1}$, the supporting edge of p. Thus, $\overline{vp}$ will make an acute angle with one end point (a_1 as shown in Fig. 11(a)) of $\overline{a_1b_1}$. Now, if $\overline{vp}$ is rotated towards point a_1, then the point p_1, the rotated position of p will lie inside $ET(\overline{a_1b_1},\ \overline{a_2b_2},\ \overline{a_3b_3})$. There-fore, if we rotate all the vertices of the triangle $\triangle pqr$ clockwise (if $\angle vpa_1 < 90°$) about the point v by an angle ϵ (as explained in proof of the Lemma 5), then we obtain a congruent triangle $\triangle p_1q_1r_1$ that completely lies inside the *boundary(P)* (see Fig. 11(b)). Thus, we have another triangle $\triangle_1$ supported by the same three edges so that $|\delta(\triangle_1)| > |\delta(\triangle)|$ which contradicts our assumption. Therefore, the $\triangle pqr$ whose smallest side is largest must coincide with an end point of a supporting edge which is also a vertex of polygon P.   □

However, if all the three perpendiculars ℓ_1, ℓ_2 and ℓ_3 intersect at the same point which is outside the extended triangle $ET(\overline{a_1b_1},\ \overline{a_2b_2},\ \overline{a_3b_3})$ behind any supporting edge say $\overline{a_1b_1}$, then rotation is not possible, because one of the ver-tex of the triangle (vertex p in Fig. 11(b)) comes outside the polygon after the rotation. Hence, this type of degeneracies should be handled in different way.

---

[4] 'Behind' the edge $\overline{a_1b_1}$ in the sense that the three points of intersection (u, v *and* w) and the polygon P lies at the two opposite sides of the edge $\overline{a_1b_1}$.

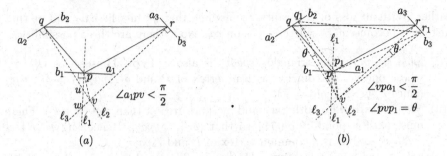

Fig. 11. (a) The three perpendiculars ℓ_1, ℓ_2 and ℓ_3 intersect outside the $ET(\overline{a_1b_1},\ \overline{a_2b_2},\ \overline{a_3b_3})$ (b) $\triangle pqr$ after rotation lies inside the polygon P

Thus we are concerned with only **Type II** isosceles triangle, where we have two smallest sides in the concerned triangle.

Now there are also two types of vertices in an isosceles triangle Δ - (i) one vertex of it is of type *red* with two equal adjacent sides and (ii) the other two vertices are of type *blue* with two adjacent sides of unequal length.

Lemma 7. *If the perpendiculars ℓ_1, ℓ_2 and ℓ_3 on the supporting edges of a Type II isosceles triangle $\triangle pqr$ intersect at common point v which is outside the $ET(\overline{a_1b_1},\ \overline{a_2b_2},\ \overline{a_3b_3})$, then $\delta(\triangle pqr)$ is largest, provided one of the vertices of $\triangle pqr$ coincides with an end point of its supporting edge.*

Proof. Depending on the type of vertex of the isosceles triangle $\triangle pqr$, we consider following:
Case (i): The three perpendiculars intersect at the same point v behind the supporting edge $\overline{a_1b_1}$ of 'red' type vertex p of the $\triangle pqr$ where $|pq| = |pr|$, as shown in Fig. 12(a).

Translate $\overline{qr}$ parallely towards p so that its end points q and r always slide on $\overline{a_2b_2}$ and $\overline{a_3b_3}$ respectively, until $\overline{qr}$ hits the vertex a_2 or b_3, whichever occurs

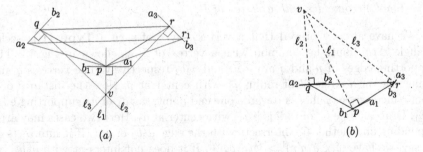

Fig. 12. (a) Three perpendiculars intersect behind the supporting edge of distinguished vertex p, and (b) three perpendiculars intersect behind the supporting edge of vertex q. (Color figure online)

earlier. Without loss of generality, we assume that $\overline{qr}$ hits first a_2 and at that instant it becomes $\overline{a_2r_1}$ where r_1 lies on $\overline{rb_3}$. Now, there are three possibilities:

(i) $|pa_2| = |pr_1|$: The triangle $\triangle pa_2r_1$ is also a **Type II** isosceles triangle whose one vertex coincide with a vertex of P and $\delta(\triangle pa_2r_1) > \delta(\triangle pqr)$ (since $\angle pqr < \angle vqb_2 = \frac{\pi}{2}$).

(ii) $|pa_2| < |pr_1|$: Here both $|pa_2|$ and $|pr_1|$ are greater than $|pq|$ ($= |pr|$). There must exist a point r' on $\overline{rr_1}$ so that $|pr'| = |pa_2|$. Hence, $\delta(\triangle pa_2r') > \delta(\triangle pqr)$ and a_2 is a common vertex of P and $\triangle pa_2r'$.

(iii) $|pa_2| > |pr_1|$: consider **Type II** $\triangle pa_2b_3$. Three sub-cases are there as follows:

  (a) $|pb_3| = |pa_2|$, we obtain an isosceles triangle $\triangle pa_2b_3$ with $\delta(\triangle pa_2b_3) > \delta(\triangle pqr)$.

  (b) $|pb_3| > |pa_2|$. Since $|pa_2| > |pr_1|$, there must be a point r'' on $\overline{r_1b_3}$ so that $|pr''| = |pa_2|$ and we consider the **Type II** isosceles $\triangle pa_2r''$ with $\delta(\triangle pa_2r'') > \delta(\triangle pqr)$ and one vertex of it is common with that of P.

  (c) $|pb_3| < |pa_2|$. Since $|pb_3| > |pr_1| > |pq|$, we have $|pa_2| > |pb_3| > |pq|$. There must be a point q' on $\overline{a_2q}$ so that $|pq'| = |pb_3|$ and we consider the **Type II** isosceles triangle $\triangle pq'b_3$ where $\delta(\triangle pq'b_3) > \delta(\triangle pqr)$.

Case (ii): The three perpendiculars intersect at the same point v behind the supporting edge of a *blue* type vertex of $\triangle pqr$.

Without loss of generality, let the supporting edge be $\overline{a_2b_2}$, behind which the perpendiculars ℓ_1, ℓ_2 and ℓ_3 intersect at v (see Fig. 12(b)). Now, vertex q lies to the left of both ℓ_1 and ℓ_3. Hence, $\angle qpa_1 > 90°$ and $\angle qra_3 > 90°$. Let ϵ_1 and ϵ_2 be the angles of anticlockwise rotation of $\overline{qp}$ and $\overline{qr}$ about q so that they hits boundary(P). Now, if the $\triangle pqr$ is rotated anticlockwise about the vertex q by a angle $\epsilon < \min(\epsilon_1, \epsilon_2)$, then the vertices p and r will lie inside the polygon P. This proves that there exists an isosceles triangle $\triangle_1$ so that $|\delta(\triangle_1)| > |\delta(\triangle)|$. Hence one vertex of the isosceles triangle must coincide with a vertex of P. $\square$

Therefore Lemmas 5, 6 and 7 give the following theorem.

Theorem 2. *One of the vertices of a* Type II *isosceles triangle $\triangle$ inside P with largest $|\delta(\triangle)|$ coincides with a vertex of P.*

We have already proved that a vertex, say r of such a **Type II** isosceles triangle $\triangle pqr$ must be common with a vertex of P as shown in Fig. 13. The supporting edges of p and q are $\overline{a_1b_1}$ and $\overline{a_2b_2}$ respectively. The vertices q and r must lie on a circle C of radius $\overline{pr}$ with center at p. Now the distance of p from r increases as p moves towards one end point, say a_1 of its supporting edge $\overline{a_1b_1}$. Draw a circle C' of radius $|ra_1|$ with center at a_1. Here two cases may arise depending on whether C' intersects with the edge $\overline{a_2b_2}$ or not. If it intersects at q', say, we have $|\delta(\triangle a_1q'r)| > |\delta(\triangle pqr)|$. If it does not intersect with $\overline{a_2b_2}$, we have a point p_1 on $\overline{a_1p}$ such that $\triangle b_2p_1r$ is isosceles and $|\delta(\triangle b_2p_1r)| > |\delta(\triangle pqr)|$.

Therefore, we see that one of the two vertices (p and q) of the **Type II** isosceles triangle $\triangle pqr$ supported by the three edges, must coincide with an

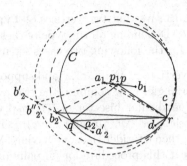

Fig. 13. Vertex p or q of **Type II** isosceles triangle $\triangle pqr$ will be a vertex of P

end-point of supporting edge ($\overline{a_2b_2}$ or $\overline{a_1b_1}$) so that $|\delta(\triangle pqr)|$ is maximized. Now any end point of a supporting edge is also a vertex of P and hence, the following theorem is obtained.

Theorem 3. *Two vertices of the* **Type II** *isosceles triangle whose smallest side is largest among all possible* **Type II** *isosceles triangles lying inside a given convex polygon P, must coincide with two vertices of P.*

3   Algorithm

We compute a triangle $\triangle$ inside a given convex n-gon P so that $\delta(\triangle)$ becomes largest. The algorithm works in two phases. The first phase and second phase computes an equilateral triangle $\triangle_1$ with maximized $\delta(\triangle_1)$ and a **Type II** isosceles triangle $\triangle_2$ with maximized $\delta(\triangle_2)$, respectively. Finally we choose a triangle $\triangle \in \{\triangle_1, \triangle_2\}$ with $\max(\delta(\triangle_1), \delta(\triangle_2))$. The phases are described below:

First Phase: Fix a vertex v_i and an edge $e_j = \overline{v_j v_{(j+1)\%n}}$ of P. By Lemma 1, compute the locus $\overline{st}$ of the third vertex, when one vertex of the equilateral triangle is fixed at v_i and the other vertex moves on e_j. If $\overline{st}$ intersects P (assume s lies inside boundary(P)), then compute the point of intersection q between $\overline{st}$ and boundary(P). Now $\max(|v_i s|, |v_i q|)$ gives the length of the sides of the equilateral triangle. If $\overline{st}$ lies completely outside P, no such triangle is possible. On the other hand, if the $\overline{st}$ lies completely inside the *boundary(P)*, then the side length of the largest equilateral triangle will be $\max(|v_i s|, |v_i t|)$. Now, keeping the vertex v_i fixed, choose the clockwise next edge of e_j as the new e_j. As e_j rotates clockwise, so does the third vertex to keep the triangle equilateral. Hence the new point of intersection q will appear only after old q (and before v_i) in clockwise direction and we keep computing the largest triangles for all possible e_j of P, while v_i is fixed. The computation of all such q corresponding to all possible e_j takes $O(n)$ time for a fixed v_i. The maximum of all such equilateral triangles give the largest equilateral triangle inside P with one of its vertex on a vertex v_i of P. Now repeat the above procedure for each vertex v_i of the n-gon P. The entire process will take $O(n^2)$ time.

Second Phase: Theorem 3 says that two vertices of **Type II** isosceles triangle Δ coincide with that of P. Depending on the type of these vertices, we compute largest isosceles triangle for the following two cases separately:

Case (i): Consider two vertices v_i and v_j of P and suppose they are of *red* type in a **Type II** isosceles triangle. Now compute the two points of intersection p'_j and p''_j between the perpendicular bisector of $\overline{v_i v_j}$ and the *boundary(P)*. This generates two isosceles triangles $\triangle v_i v_j p'_j$ and $\triangle v_i v_j p''_j$. If any or both of these triangles are of **Type II**, then consider that one having larger $|\delta(\triangle)|$. This takes $O(\log n)$ time. Now, repeat this procedure for all pairs of vertices (v_i, v_j) in P. This takes overall $O(n^2 \log n)$ time.

Case (ii): Choose a vertex v_i on P. Let v_j be another vertex of P. We will construct a **Type II** isosceles triangle using two vertices v_i and v_j so that v_i is of type *'blue'* and v_j is of type *'red'*. Now choose a point v in boundary(P) such that $\triangle v_i v_j v$ is isosceles and v is also of type *'blue'*. We may have many choices of such a v, but we will choose one which appears first in the clockwise traversal of boundary(P) from v_j to v_i. If it is of **Type II**, then compute $\delta(\triangle v_i v_j v)$. By fixing v_i, we choose the clockwise next vertex of v_j as the new v_j and compute v as explained earlier. It needs to be mentioned that the new v will appear after the old v in clockwise direction. Thus for each v_i, we need to consider all the vertices of P as v_i. Thus the whole process needs $O(n^2)$ time.

Finally we consider the triangle $\triangle$ returned by Phase I and Phase II for which $\delta(\triangle)$ is larger. Thus we have the following result.

Theorem 4. *Given a convex polygon P, (a) the largest equilateral triangle inside P can be determined in $O(n^2)$ time, and (b) the triangle $\triangle$ inside P with maximum $\delta(\triangle)$ can be obtained in $O(n^2 \log n)$ time.*

4   Conclusion and Future Work

In this paper, we have explored the various properties of an inscribed triangle inside a convex polygon P so that the smallest side of this triangle becomes maximum. For such a triangle, at least one vertex of it will coincide with a vertex of P. This helps to solve the problem in $O(n^2 \log n)$ time which can be further improved to $O(n^2)$, if the Case (i) of second phase of the algorithm in Sect. 3 can be improved to $O(n^2)$. In other words, the challenge here is to compute the isosceles triangle whose two blue vertices coincide with any two vertices of polygon and its smallest side becomes maximized, in $O(n^2)$ time.

Acknowledgement. We thank Prof. Joseph O'Rourke for valuable suggestion on the Proof of Lemma 2.

Appendix

Proof of Observation 1

Suppose the vertex c of a triangle $\triangle_1 = \triangle abc$ lies properly inside the polygon P (see Fig. 14). Draw an arc $\overparen{cq}$ with a as center and with $|ac|$ as radius which intersect the boundary(P) at a point q. Similarly, draw another arc $\overparen{cr}$ with b as the center and $|bc|$ as the radius, which intersect the boundary(P) at the point r. Take any point c_1 on the portion qr of the *boundary(P)*. The triangle $\triangle_2 = \triangle abc_1$ has $|ac_1| > |ac|$ and $|bc_1| > |bc|$. If the smallest side of the $\triangle_1$ is either $\overline{ac}$ or $\overline{bc}$, then the smallest side of $\triangle_2$ will be larger than that of $\triangle_1$, otherwise (i.e. when the smallest side of $\triangle_1$ is $\overline{ab}$) it remains same in $\triangle_2$. Thus, the result follows.　□

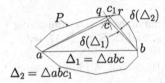
$\triangle_2 = \triangle abc_1$

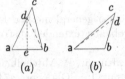

Fig. 14. Observation 1　　　　**Fig. 15.** Observation 2　　　　**Fig. 16.** Observation 4

Proof of Observation 2

Take any scalene triangle $\triangle_1 = \triangle abc$ whose smallest side $\delta(\triangle abc)$ is $\overline{ab}$ (see in Fig. 15(a), (b)) and consider the angles adjacent to $\overline{ab}$. If both the angles $\angle abc$ and $\angle bac$ are at least $60°$ (see Fig. 15(a)), then draw a perpendicular bisector $\overline{de}$ of the edge $\overline{ab}$ which intersects $\overline{ac}$ at a point d. Now, $\triangle abd\ (= \triangle_2)$ is an isosceles triangle, satisfying $|ad| = |bd| \geqslant |ab|$. However, if one of the aforesaid angles, say $\angle bac < 60°$ (see Fig. 15(b)), we take a point $d \in \overline{bc}$ so that $|bd| = |ab|$. Now connect the points a and d to obtain an isosceles triangle $\triangle_2 = \triangle abd$ which lies inside $\triangle_1 = \triangle abc$. Thus in both cases, we have $|\delta(\triangle_2)| = |\delta(\triangle_1)|$.　□

Proof of Observation 4

Refer to Fig. 16. The three vertices p, q and r of the triangle $\triangle pqr$ is rotated about the point v by an angle θ and we obtain a new triangle $\triangle p_1q_1r_1$. Here, $\angle qvr = \angle q_1vr_1$. Now from $\triangle qvr$, we have $|qr|^2 = |qv|^2 + |rv|^2 + 2|qv|.|rv| \cos \angle qvr$. Similarly, from $\triangle q_1vr_1$, we get $|q_1r_1|^2 = |q_1v|^2 + |r_1v|^2 + 2|q_1v|.|r_1v| \cos \angle q_1vr_1$. Since $|qv| = |q_1v|$ and $|rv| = |r_1v|$, we have $|qr| = |q_1r_1|$. Similarly we can prove that $|pq| = |p_1q_1|$ and $|rp| = |r_1p_1|$. Therefore, $\triangle pqr$ is congruent to $\triangle p_1q_1r_1$.　□

References

1. Aggarwal, A., Klawe, M.M., Moran, S., Shor, P.W., Wilber, R.E.: Geometric applications of a matrix-searching algorithm. Algorithmica **2**, 195–208 (1987)
2. Alt, H., Hsu, D., Snoeyink, J.: Computing the largest inscribed isothetic rectangle. In: CCCG, pp. 67–72 (1995)
3. Boyce, J.E., Dobkin, D.P., Drysdale III, R.L.S., Guibas, L.J.: Finding extremal polygons. SIAM J. Comput. **14**(1), 134–147 (1985)
4. Cabello, S., Cheong, O., Knauer, C., Schlipf, L.: Finding largest rectangles in convex polygons. Comput. Geom. **51**, 67–74 (2016)
5. Chazelle, B., Sharir, M.: An algorithm for generalized point location and its applications. J. Symb. Comput. **10**(3/4), 281–309 (1990)
6. Daniels, K.L., Milenkovic, V.J., Roth, D.: Finding the largest area axis-parallel rectangle in a polygon. Comput. Geom. Theor. Appl. **7**, 125–148 (1997)
7. DePano, A., Ke, Y., O'Rourke, J.: Finding largest inscribed equilateral triangles and squares. In: 25th Allerton Conference on Communication, Control, and Computing, pp. 869–878 (1987)
8. Dobkin, D.P., Snyder, L.: On a general method for maximizing and minimizing among certain geometric problems (extended abstract). In: FOCS, pp. 9–17. IEEE Computer Society (1979)
9. Hall-Holt, O.A., Katz, M.J., Kumar, P., Mitchell, J.S.B., Sityon, A.: Finding large sticks and potatoes in polygons. In: SODA, pp. 474–483 (2006)
10. Jin, K., Matulef, K.: Finding the maximum area parallelogram in a convex polygon. In: CCCG (2011)
11. Sadhu, S., Roy, S., Nandi, S., Maheshwari, A., Nandy, S.C.: Approximation algorithms for the two-center problem of convex polygon. CoRR abs/1512.02356 (2015)

Workshop on Central Italy 2016 Earthquake: Computational Tools and Data Analysis for the Emergency Response, the Community Support and the Reconstruction Planning (CIEQ 17)

Collecting and Managing Building Data to Perform Seismic Risk Assessment – Palestine Case Study

Antonella Di Meo[✉], Marta Faravelli, Diego Polli, Marco Denari,
Alessio Cantoni, and Barbara Borzi

Eucentre - European Centre for Training and Research
in Earthquake Engineering, Pavia, Italy
{antonella.dimeo,marta.faravelli,diego.polli,
marco.denari,alessio.cantoni,
barbara.borzi}@eucentre.it

Abstract. This paper describes a web interface with GIS functionality (Web-GIS) that Eucentre (EUropean CENtre for Training and Research in Earthquake engineering) developed for SASPARM 2.0 (Support Action for Strengthening PAlestine capabilities for seismic Risk Mitigation) project [1].

The SASPARM 2.0 WebGIS is a simple and intuitive platform intended for people with different backgrounds, such as citizens, students, practitioners, governmental and non-governmental institutions. The final aim of the implemented WebGIS application is to calculate the seismic risk of residential buildings. Nablus has been taken as case study to demonstrate, implement, and calibrate project actions. To calculate the seismic risk, residential buildings data are collected. Such activity can be conducted by both practitioners and citizens who compile two standard forms that differ from each other only in terms of detail. The survey forms can be compiled directly on WebGIS, at the dedicated tabs, or through two mobile apps designed for the purpose. All filled forms are shown on the homepage map of the WebGIS platform. Starting from the collected data, the seismic risk of each single building is evaluated by combining the hazard and the vulnerability with its exposure. In particular, the seismic demand to which each building is subjected to is defined from the hazard curve. In the specific case study of Nablus, the hazard curve is obtained by referring to "West Bank and Gaza Strip: Seismic Hazard Map Distribution". The structural vulnerability, instead, is quantified through fragility curves calculated with the mechanical method SP-BELA (Simplified Pushover-Based Earthquake Loss Assessment), modified to represent the building environment of Nablus.

Keywords: Vulnerability assessment · Seismic risk · WebGIS platform · Mobile app

© Springer International Publishing AG 2017
O. Gervasi et al. (Eds.): ICCSA 2017, Part II, LNCS 10405, pp. 527–542, 2017.
DOI: 10.1007/978-3-319-62395-5_36

1   Introduction

The seismic risk is defined by the combination of three different factors:

- hazard, i.e. the probability of occurrence of an earthquake exceeding a certain threshold of intensity magnitude (i.e. PGA) in a given area and in a certain interval of time;
- exposure, i.e. the importance of the object exposed to the risk;
- vulnerability, i.e. the level of damage that a structure can suffer when subjected to an earthquake of a certain level of intensity.

It is evident that to reduce the seismic risk it is necessary to intervene on the vulnerability. This awareness underlined the importance of a new Palestinian design code, introduced at the end of SASPARM (Support Action for Strengthening Palestinian-administrated Areas capabilities for seismic Risk Mitigation) project [2] which is a project financed in October 2012 within a FP7 framework with the aim to spread the concept of seismic risk in Palestine. The second aspect is the need for an assessment of the seismic vulnerability of buildings. This activity has been conducted in SASPARM 2.0 project [1], founded in a DG ECHO framework, where a methodology of four different steps has been carried out.

The first step consisted in the identification of the most common structural types in Palestine. From the survey activities, it has been evident that Palestinian buildings can be classified in four main categories: i.e. reinforced concrete frame buildings, reinforced concrete soft storey buildings (*pilotis*), reinforced concrete shear wall buildings and masonry buildings. Reinforced concrete buildings generally exceed ten floors and have various uses. On the contrary, masonry structures have not more than four floors and are used predominantly as residential buildings.

After the definition of the structural type, survey forms have been identified to collect geometrical and structural data of buildings (second step). In particular, two different forms are designed: one for citizens and a more detailed one for practitioners. The survey forms can be compiled on WebGIS, at the tabs *Building form – Practitioners* and *Building form – Citizens*, or through two mobile apps. The latter were specifically created for the Android operating system version 4.0 or higher and allow to upload data also when internet is not available. All survey data are then uploaded and managed on a WebGIS platform that is the main deliverable of SASPARM 2.0 project [1].

The WebGIS platform collects all compiled forms and, once selected a building, allows to check the corresponding form with all the data, thanks to which vulnerability curves can be calculated (third step). Fragility curves are defined with SP-BELA (Simplified Pushover-Based Earthquake Loss Assessment) method, that is an analytical procedure initially implemented to assess the seismic vulnerability of Italian buildings [3, 4]. In the specific case of Palestine, SP-BELA has been modified to best perform the building environment of those territories, herein represented by the case study of Nablus city.

The last step of the methodology to estimate the seismic vulnerability of Palestinian buildings regards the definition of a criteria that connects each form to a set of fragility curves. The identified criteria allows the association of each set of fragility curves to a building in relation to its structural type and its number of floors.

Once defined the structural vulnerability, seismic risk is calculated by combining the vulnerability to the hazard and the exposure. The WebGIS platform herein presented allows to calculate the seismic risk as defined above and proposes also different retrofit measures for the reduction of it [5, 6].

2 Data Collection

The final result of SASPARM 2.0 project [1] is a WebGIS platform that collects the structural data of the buildings in the city of Nablus. The platform is available to all users, i.e. citizens, students, practitioners, and governmental/non-governmental institutions, and allows the management of data for the definition of vulnerability and seismic risk of each selected building.

Structural data can be collected by compiling two forms, one for citizens and one for practitioners, different from each other only in terms of detail. The compilation of the forms can occur directly on the WebGIS platform or through two apps for smartphones and tablets with Android operating system.

Some information of the forms are mandatory. After clicking on the button *Save* to save the compiled form, the system will check whether the compiler missed some obligatory data. In particular, the message *"Form has some errors or missing fields"* will appear and lacking fields will be notified in red color. To further reduce the number of mistakes, others checks are performed. For instance, if geographical coordinates are wrong (e.g., coordinates results outside the city of Nablus) or the year of construction is bigger than the renovation one, the WebGIS will notify them and will not allow the compiler to proceed until everything will be corrected.

Once filled in properly, all forms are available in the WebGIS homepage at the tab *Map*. The marker corresponding to each form can have two different colors, green or blue, depending on whether the survey has been carried out by practitioners or by citizens. The WebGIS homepage is shown in Fig. 1.

Fig. 1. Tab *Map* - Screenshot of the initial view of the WebGIS to select surveyed buildings. (Color figure online)

By clicking on the marker of the building, in relation to the compiler, the tab *Building form - Practitioners* or *Building form - Citizens* will be activated and all buildings data can be examined.

The buildings detected with the forms described herein are considered as structural units of ordinary construction type (such as masonry, reinforced concrete frame or walls) used for housing and/or services. Therefore, these forms are not suitable for monumental or specialized types (such as industrial depots, sport constructions, theaters, churches etc.).

2.1   Building Form – Citizens

The tab *Building form - Citizens* reflects the survey paper form used by citizens, downloadable in *pdf* format by pressing the button *Help*.

By clicking the button *New Form*, an empty form generates in the WebGIS platform. Once opened, the page can be deleted or closed by using respectively the buttons *Delete Form* and *Close Form*.

At the beginning, the form presents a section dedicated to the compiler information, such as name and level of education (*Education level/Faculty/Department*), as shown in Fig. 2. In this section the compilation date of the form, the name of the compiler and the educational level are mandatory.

Fig. 2.   Initial part of *Building form-Citizens*.

Then, five sections follow, namely respectively *Identification of the Building*, *Description of the Building*, *Structural Data*, *Notes,* and *Photos*, in which the compiler defines the geometric and structural characteristics of the building. Each section contributes to collect the right information to make a quick assessment of seismic performance of the building.

In the section *Identification of the Building*, the citizen enters the data to locate the building, including the geographic coordinates expressed in decimal degrees according

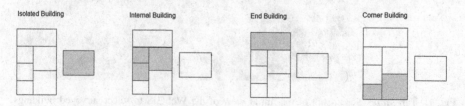

Fig. 3.   Description of the different type of building locations.

to WGS84 (World Geodetic System 1984) and detected at the entrance of the structure. Both municipality and geographical coordinates are important to georeference the building so that these data are obligatory.

Since closely related to the seismic response of the building, the definition of the building location is also necessary: in particular, the compiler has to specify if the building is an isolated, inner, end, or corner structure, as indicated in Fig. 3. Figure 4 shows the section *Identification of the Building*.

Fig. 4. Section 1 – *Identification of the Building*.

The section *Description of the Building* contains all the information about the number of floors, the age, the type of use and exposure, and property. The total number of floors of a building has to be marked in *N° Total of floors with the basement*, counting from the foundation, including basements and attic floor, only if accessible. The basements, whose number is marked in *N° Basement*, are all that floors whose height above ground (i.e. the average height above ground in the case of buildings located on a slope) is less than 1/2 of the total height of the floor.

To have an idea about the methods of construction and seismic behavior of the building, the compiler has to indicate the year of construction of the detected building. He can make only two different choices: one is for the period in which the building has been built and the other one is for the possible age in which the building might have had a significant restructuring from a structural point of view. The reference time period for the year of construction is divided into intervals and ranges from <1919 to ≥ 2002.

Regarding the use, it will concern all types of use that may coexist in the building with the related number of units. Finally, the compiler has to specify the percentage of use, the type of property (i.e. public or private) and the number of people that occupy the building. The latter, together with the number of floors and basements are mandatory data.

The section *Description of the Building* is shown Fig. 5.

In addition to the number of floors and the use of the building, to calculate the vulnerability it is important to specify the type of structure.

In *Structural Data* (see Fig. 6), in fact, the compiler has to choose between masonry and reinforced concrete, being the two materials mainly used for the realization of vertical structures of Palestinian buildings.

If the buildings are in reinforced concrete, the total or partial absence of cladding has to be specified, together with the corresponding floors. The absence of cladding, in

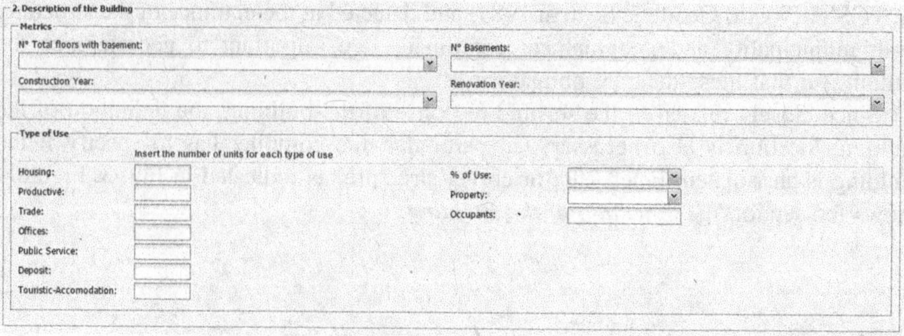

Fig. 5. Section 2 – *Description of the Building.*

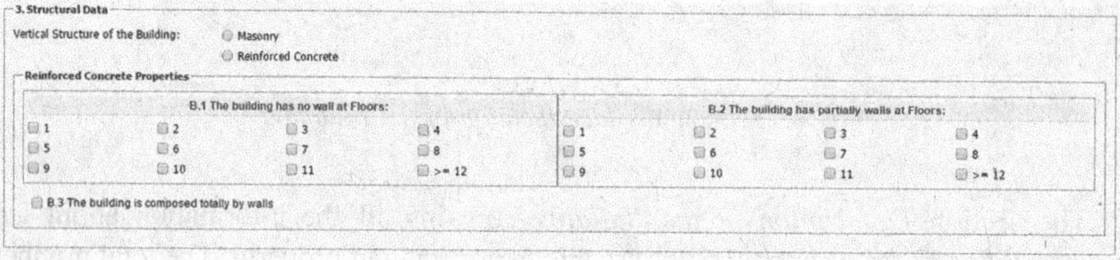

Fig. 6. Section 3 – *Structural Data.*

fact, favors the occurrence of soft story mechanism in case of medium or high-intensity ground shaking, thus increasing the vulnerability of the building. Since citizens could not identify the vertical structure of the surveyed building, they can avoid to insert this information. In this case, the fragility curves (see Sect. 3.2) are associated to the building only through the number of floors. In particular, if the building has until 4 storeys, it is considered a masonry building otherwise a reinforced concrete one.

Fig. 7. Sections 4 and 5 – *Notes* and *Photos.*

The tab *Building form – Citizen* ends with *Notes* and *Photos* (see Fig. 7) where the compiler can write notes and/or information that cannot be caught in the sections of the form as well as attach photos of the reference building. As shown in Fig. 7, photos can be uploaded or deleted by clicking on *Upload Photo* and *Delete Photo*.

2.2   Building Form – Practitioners

The tab *Building form – Practitioners* in the WebGIS platform is filled in by practitioners. This form is more detailed than the one for citizens. The first difference is in the section *Description of the building* where, in addition to the information required to citizens described above, the compiler has to input the height and floor area, both obtained as the average of heights and surfaces of all detected floors of the building. Figure 8 shows the structure of the section *Description of the Building*.

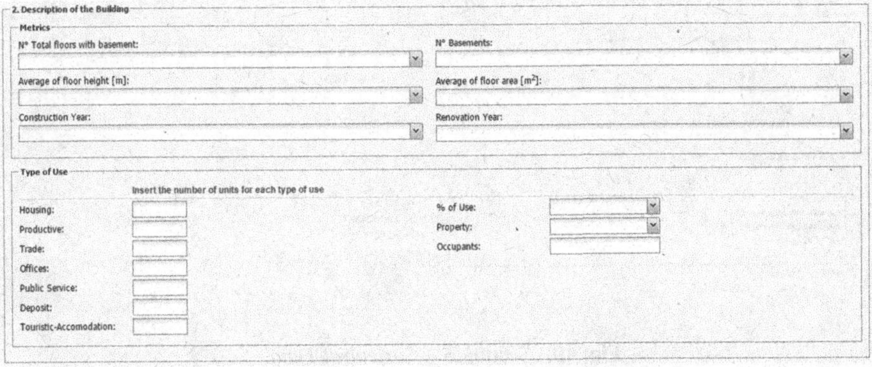

Fig. 8. Section 2 – *Description of the Building.*

Besides the types of vertical structure already proposed in the tab *Building form - Citizens*, the section *Structural Data* includes also reinforced concrete shear wall buildings. This further classification is relevant because reinforced concrete shear wall buildings have a seismic behavior and a level of vulnerability different from reinforced concrete frame buildings ones. Unlike citizens, practitioners have to indicate if the surveyed building has a masonry vertical structure or a reinforce concrete one.

The section *Structural Data* contains also the part *Horizontal Structure and Roof* where there is a list of horizontal structures, for both interfloor and roof slab, mainly used in Palestinian residential buildings (see Fig. 9). *Horizontal Structure and Roof* provides the possible presence of cantilever structures (e.g., balconies): the latter, in fact, are an important cause to the increase of the detected buildings vulnerability especially for the very common practice to have cladding walls all around cantilever structures. Both horizontal structure and the roof type are mandatory data. Since the type of slab could be a difficult data to gather, the option *Non identified* is automatically selected.

Figure 10 displays the structure of the section *Structural Data*.

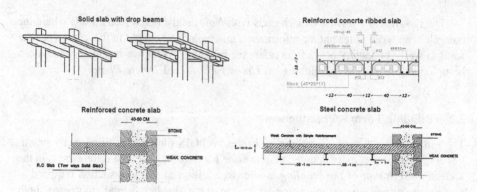

Fig. 9. Types of slabs mainly used in Palestinian residential buildings.

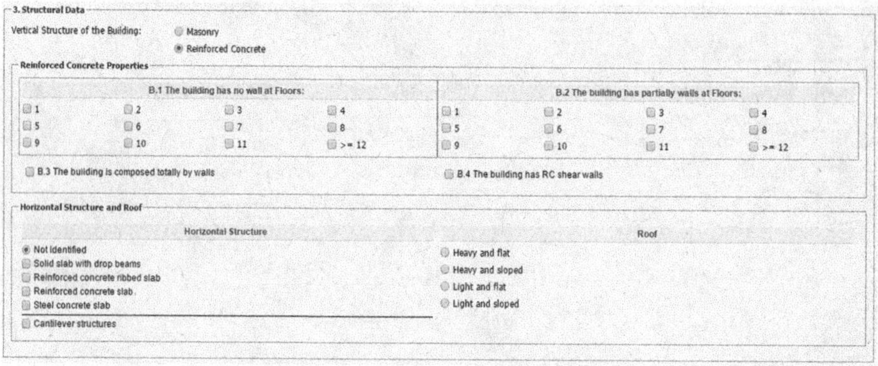

Fig. 10. Section 3 – *Structural Data.*

Practitioners have to compile also other two obligatory sections, i.e. *Regularity* and *Geomorphological Data*. The latter regard respectively the regularity of the building (both in plan and in elevation), the morphology of the site and the category of the soil foundation on which the building was constructed. The criteria for the definition of the structural regularity and the type of soil foundation are described in Eurocode 8 [7]. Figure 11 shows sections *Regularity* and *Geomorphological Data.*

As for the form of citizens, also the one targeted to practitioners ends with the sections *Notes* and *Photos* to give more detailed information on the detected building.

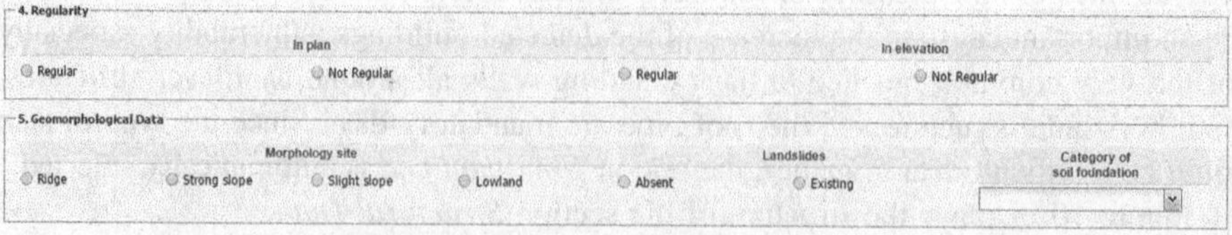

Fig. 11. Sections 4 and 5 – *Regularity* and *Geomorphological Data.*

2.3 Apps for Compiling the e-Forms Through Smart Phones and Tablets

To make more efficient and faster the collection of survey data, two apps for compiling the e-forms through smartphones and tablets have been developed. These apps run for Android operating system version 4.0 or higher. The apps become useful in case of internet absence: all forms, in fact, can be loaded directly on the smartphone or tablet for offline use. The compiled data will be sent to the web portal once internet becomes available again.

Figure 12 shows the home page of the apps for: (a) citizens and (b) practitioners. The user has to compile all the sections of the form as the ones reported in Fig. 13 for practitioners. The user can save, upload or delete the form at any time only by clicking the button in the upper right, as shown in Fig. 14. In the same list shown in Fig. 14, the button *Send* can be selected once completed the form.

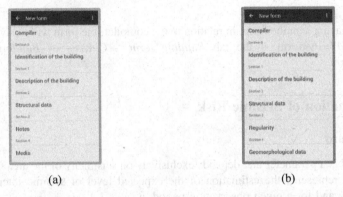

(a)   (b)

Fig. 12. Home Page of the SASPARM 2.0 apps for: (a) citizens and (b) practitioners.

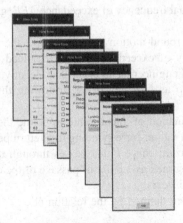

Fig. 13. Sections to insert data in the SASPARM 2.0 app for practitioners.

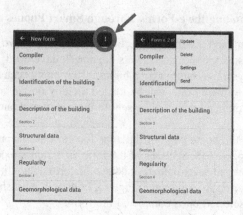

Fig. 14. Operations on the app.

After having sent the form, in relation to the compiler, the form will be available on the WebGIS Platform at the tab *Building form – Citizens* or *Building form – Practitioners*.

3 Estimation of Seismic Risk

3.1 Hazard

The hazard is a parameter that depends exclusively on seismicity of the area. Precisely, the hazard represents the estimation of the expected level of seismic intensity in a certain area and for a given observation period.

The definition of seismic hazard occurs through the hazard curve. In particular, the curve relates the severity of shaking, in this case defined by the peak ground acceleration *PGA*, with the annual frequency of exceedance *AFE*, given by the inverse of the return period T_r.

The logarithm of a ground-motion parameter and the logarithm of the corresponding annual frequency of exceedance can be assumed to be linearly-related, at least for return periods of engineering interest. The negative gradient of the log–log hazard curve is referred to as *k* in this paper, following the definition in Part 1 of Eurocode 8 [7].

Thanks to the approximation of linear trend, to define the hazard curve it is sufficient to determine the *PGA* value, corresponding to a return period T_r, and the negative gradient of the log–log hazard curve *k* that passes through the reference point. Conventionally, the latter is assumed as a point of passage of the curve that corresponds to the return period T_r of 475 years.

The hazard curve is then defined by the relation (1):

$$AFE = AFE_{475} \left(\frac{S \cdot PGA_{475}}{PGA} \right)^k \tag{1}$$

where AFE_{475} e PGA_{475} are the annual frequency of exceedance and the peak ground acceleration corresponding to the return period T_r of 475 years while S is the soil factor. Once defined the hazard curve, it is possible to calculate the seismic demand related to PGA for any return period T_r.

For the case study of Nablus, we considered the "West Bank and Gaza Strip: Seismic Hazard Distribution Map" shown in Fig. 15 that displays the PGA corresponding to a return period T_r of 475 years. Since the gradient k of hazard curve is unknown, the value of k is assumed equal to 3 as suggested in the Eurocode 8 [7].

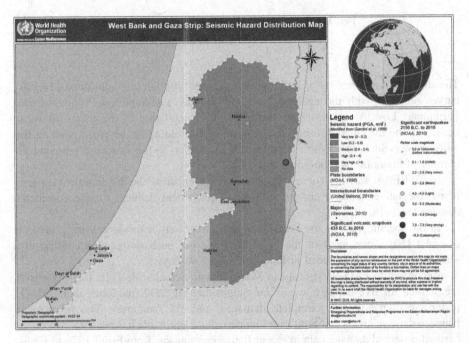

Fig. 15. Hazard Map (West Bank and Gaza Strip).

In summary, the parameters that allow to calculate the hazard curve in the Eq. (1) are:

- AFE_{475} = 1/475;
- PGA_{475} = 0.24 g;
- k = 3;
- S = 1 (soil A), 1.2 (soil B or C), 1.4 (soil D or E) from Eurocode 8 [7].

Figure 16 shows tab *Hazard* corresponding to the selected building and it displays the hazard curve for a soil "C". Since the type of the soil can be specified only in the form filled in by practitioners, the hazard curve related to the forms compiled by citizens is determined by considering the soil type "B".

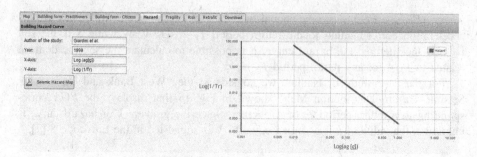

Fig. 16. Tab *Hazard* – Hazard curve for soil "C".

3.2 Fragility

The structural vulnerability of buildings is evaluated through the definition of fragility curves obtained with the analytical method SP-BELA (Simplified Pushover-Based Earthquake Loss Assessment). In particular, the compilation of the forms allows to collect all the information useful to associate each building to the corresponding fragility curves. Such association is performed in relation to the structural type and the number of floors of each building.

In SP-BELA, fragility curves are obtained by comparing the displacement capacity of representative building classes with the displacement demand for the considered damage levels [8, 9].

In the specific case of the Palestinian structures, only a soft storey mechanism is assumed [10], since most of the buildings have not been designed according to a seismic regulation, introduced in Palestine only nowadays.

According to EMS-98 (European Macroseismic Scale) scale [11], fragility curves are defined for five damage levels: *D1* (slight damage), *D2* (moderate damage), *D3* (extensive damage), *D4* (complete damage), and *D5* (collapse). The relationship between damage levels and limit states numerically identifiable has been defined on the basis of observed damage data in recent Italian earthquakes from Friuli 1976 to Emilia 2012, since specific data for Palestine are not available.

Given a peak ground acceleration *PGA* value equal to 0.24 g (assumed equivalent to a return period T_r of 475 years), Table 1 shows the probability of exceeding of each damage level for 7-storeys reinforced concrete frame buildings and for 3-storeys masonry ones. The corresponding fragility curves, instead, are respectively displayed in Figs. 17 and 18 where the tab *Fragility* of the WebGIS platform is illustrated.

Table 1. Probability of exceeding each damage level calculated by applying SP-BELA and HAZUS [12] for a PGA = 0.24 g (T_r = 475 years).

	DL1	DL2	DL3	DL4	DL5
7-storeys concrete frame buildings	98.30	77.61	71.12	55.36	14.85
3-storeys masonry buildings	100	99.9	99.8	97.5	55.7

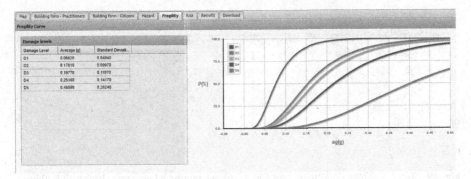

Fig. 17. Tab *Fragility* – Fragility curves for 7-storeys reinforced concrete frame buildings.

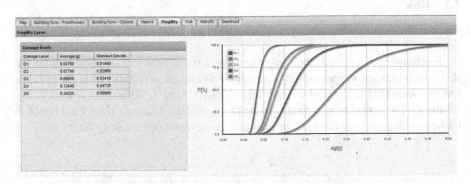

Fig. 18. Tab *Fragility* – Fragility curves for 3-storeys masonry buildings.

The definition of fragility curves for reinforced concrete shear wall buildings has not been tackled trough mechanical methodology since it is not possible to identify a prototype building which can be representative of the whole building stock. As a matter of fact, the shear wall layout does not have a standard which could be assumed as representative.

To face this problem, fragility curves of reinforced concrete shear wall buildings have been obtained from the fragility curves of reinforced concrete frame buildings with the same number of storeys. The assumption herein considered is that the presence of shear walls reduces the average value of the fragility curves calculated for reinforced concrete frame buildings whereas the coefficient of variation remains constant.

To calibrate the correction factor that allows to determine the fragility curves of reinforced concrete shear wall buildings, a careful bibliographic study has been undertaken. Specifically, the vulnerability study presented in HAZUS [12] has been selected: HAZUS manual, in fact, proposes fragility curves for different structural types. By comparing HAZUS fragility curves of frame buildings with masonry infill walls with HAZUS fragility curves of shear wall buildings, a correction factor equal to 1.3 has been defined.

Figure 19 shows the fragility curves calculated for 10-storeys reinforced concrete shear wall buildings located in Nablus.

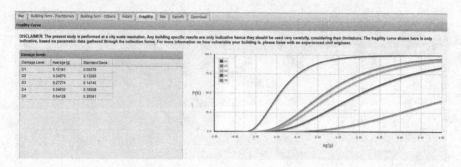

Fig. 19. Tab *Fragility* - Fragility curves for 10-storeys reinforced concrete shear wall buildings.

3.3   Risk

The combination of vulnerability, exposure and hazard allows to calculate the seismic risk.

Precisely, once selected a building, the seismic risk related to it is reported at the tab *Risk* which contains the probability to reach the five levels of damage from *D1* to *D5* of EMS98 scale [11] in three observation time frames T_d of 1, 10 and 50 years.

Figure 20 shows the risk values related to a selected 8-storeys reinforced concrete frame building, not regular both in plan and in elevation and recently built.

| Map | Building form - Practitioners | Building form - Citizens | Hazard | Fragility | Risk | Retrofit | Download |

DISCLAIMER: The present study is performed at a city scale resolution. Any building specific results are only indicative hence they should be used very carefully, considering their limitations. The results in terms of seismic risk shown here are only indicative, based on parametric data gathered through the collection forms. For more information on the seismic response of your building, please liaise with an experienced civil engineer.

Risk (%)

Time Window (Years)	D1	D2	D3	D4	D5
1	19.3	3.7	2.8	1.4	0.2
10	60.9	21.9	17.8	10.6	2.2
50	87.2	50.8	44.5	31.0	6.8

Fig. 20. Tab *Risk* - Risk values related to a selected 8-storeys reinforced concrete frame building, not regular both in plan and in elevation and recently built.

To reduce the seismic vulnerability and consequently the risk, different structural interventions are suggested in the tab *Retrofit* of the WebGIS [5, 6]. As shown in Fig. 21, the tab *Retrofit* is organized according to the different types of buildings defects, mainly due to a non-seismic design. Structural defects induce the building to have a non-ductile behavior and to be more predisposed to the damage, even in case of low or medium intensity ground shaking. As a consequence, the proposed rehabilitation measures vary from case to case and depend on the type of building as well as on

| Map | Building form - Practitioners | Building form - Citizens | Hazard | Fragility | Risk | Retrofit | Download |

DISCLAIMER: The present study is performed at a city scale resolution. Any building specific results are only indicative and should be treated very carefully. The retrofitting measures proposed below are only general recommendations, based on parametric data gathered through the collection forms. The rehabilitation measures are proposed according to their applicability, from the most popular technique to the more advanced, that requires experience, training and familiarity. For more information on how to retrofit your building and how to select the appropriate technique, please contact an experienced civil engineer.

Table for Seismic deficiencies and potential rehabilitation techniques

Torsion

Elevation Not Regular

Diaphragm

Fig. 21. Tab *Retrofit.*

its geometrical and structural characteristics that can be identified through the completion of the forms.

4   Conclusion

This paper describes the WebGIS realized for SASPARM 2.0 project [1]. It is structured in a very simple and intuitive way and allows free access to all users, i.e. citizens, practitioners, students, and governmental and non-governmental institutions. In particular, each user can upload or consult all information relevant to the definition of the vulnerability and the seismic risk of residential buildings in the city of Nablus.

This information regards the geometrical and the structural data of buildings which are surveyed through the use of two forms, one for citizens and one for practitioners, designed for this purpose. Citizens and practitioners can compile the forms either directly on the WebGIS platform or by using two mobile apps that were specifically created for Android operating system.

By using the survey data and SP-BELA method, in relation to the structural type and the number of floors, fragility curves are defined for five different levels of damage.

The structural vulnerability is then combined to the local seismic hazard, thus obtaining the seismic risk for three observation time windows.

Finally, possible retrofit measures are also suggested to reduce the seismic risk of each selected building.

Although it started as an instrument to be used on the Palestinian territory, the WebGIS of SASPARM 2.0 is a very useful tool that could be used also in others countries, such as Italy, to collect all vulnerability data of residential buildings that are still not available on a national scale.

Acknowledgements. This research has been conducted within the project "ECHO/SUB/2014/694399 SASPARM 2.0 Support Action for Strengthening Palestine capabilities for seismic risk mitigation", a project co-financed by DG-ECHO - Humanitarian Aid and Civil Protection. We thank R. Monteiro, P. Ceresa, V. Cerchiello and I. Grigoratos (IUSS, Pavia) for the WebGIS section to retrofit and Prof. J. Dabbeek and his staff (An-Najah National University, Nablus) for structural information on Palestinian buildings.

References

1. SASPARM 2.0 Project: Support Action for Strengthening Palestine's capabilities for seismic Risk Mitigation, DG-ECHO 2014, ECHO/SUB/2014/694399. SASPARM 2.0. http://www.sasparm2.com
2. SASPARM Project: Support Action for Strengthening Palestinian-administrated Areas capabilities for Seismic Risk Mitigation, FP7-INCO, ID. 295122. SASPARM. http://www.sasparm.ps/en/
3. Rasulo, A., Testa, C., Borzi, B.: Seismic risk analysis at urban scale in Italy. In: Gervasi, O., et al. (eds.) ICCSA 2015. LNCS, vol. 9157, pp. 403–414. Springer, Cham (2015). doi:10.1007/978-3-319-21470-2_29

4. Rasulo, A., Fortuna, M.A., Borzi, B.: A seismic risk model for Italy. In: Gervasi, O., et al. (eds.) ICCSA 2016. LNCS, vol. 9788, pp. 198–213. Springer, Cham (2016). doi:10.1007/978-3-319-42111-7_16

5. Federal Emergency Management Agency (FEMA): FEMA 547 - Techniques for the seismic rehabilitation of existing buildings, Washington D.C., United States (2006)

6. Deliverable D.C.1: Report on the identification of retrofit measures. SASPARM 2.0 Project, DG-ECHO 2014, ECHO/SUB/2014/694399 (2016)

7. EUROCODE 8, prEN 1998-1: Design of structures for earthquake resistance – Part 1: General rules, seismic actions and rules for buildings. European Committee for Standardization (CEN), Brussels, Belgium (2004)

8. Borzi, B., Pinho, R., Crowley, H.: Simplified pushover-based vulnerability analysis for large scale assessment of RC buildings. Eng. Struct. 30(3), 804–820 (2008)

9. Borzi, B., Crowley, H., Pinho, R.: Simplified pushover-based earthquake loss assessment (SP-BELA) method for masonry buildings. Int. J. Architectural Herit. 2(4), 353–376 (2008)

10. Borzi, B., Di Meo, A., Faravelli, M., Ceresa, P., Monteiro, R., Dabbeek, J.: Definition of fragility curves for frame buildings in Palestine. In: 1st International Conference on Natural Hazards and Infrastructure, Chania, Greece (2016)

11. Grünthal, G. (ed.): European Macroseismic Scale 1998 (EMS-98). In: European Seismological Commission, sub commission on Engineering Seismology, Working Group Macroseismic Scales, vol. 15. Conseil de l'Europe, Cahiers du Centre Européen de Géodynamique et de Séismologie, Luxembourg (1998)

12. Federal Emergency Management Agency (FEMA): HAZUS 99 - Earthquake Loss Estimation Methodology. Technical Manual, Washington D.C., United States (1999)

Workshop on Computational and Applied Statistics (CAS 2017)

Receiver Operating Characteristic (ROC) Packages Comparison in R

Daniela Ferreira da Cunha[1] and Ana Cristina Braga[2(✉)]

[1] Departamento de Engenharia Geotécnica,
Instituto Superior de Engenharia do Porto, Porto, Portugal
daniela.cunha@outlook.com
[2] ALGORITMI Centre, University of Minho, 4710-057 Braga, Portugal
acb@dps.uminho.pt

Abstract. The Receiver Operating Characteristic (ROC) curve analysis and the resulting plot can be used as a tool to select optimal models of possibility and to discard those of inferior quality from the cost of context (or class distribution). Presently, this type of analysis is used in a variety of fields from the medical community, bioinformatics, military and finance. There is a variety of software packages available for ROC analysis, and this analysis will focus on those specific of R and open source. The chosen packages were: ROCR, Verification, caTools, Comp2ROC, and Epi available on CRAN, and the ROC library from Bioconductor. This work intends to make a comparative analysis of the main characteristics of these R packages.

Keywords: R · ROC curves · Comp2ROC · ROCR · pROC · caTools · Verification · Bioconductor

1 Introduction

Receiver Operating Characteristic (ROC) curve analysis was originally developed for the detection of radar signals during World War II [1], and the resulting plot allowed the evaluation of the discriminatory performance of a discrete or continuous variable, which represented the classifier. Its usefulness was later recognized by other fields, and soon the methodology was introduced and adapted to psychology (to study the perceptual detection of stimuli), and to medicine (radiology) [2–4].

Currently, ROC curves are widely applied in many domains of knowledge, including but not restricted, to the medical decision-making community, as well as, bioinformatics (pattern classification, microarray analysis, genome annotation), evaluating biomarker performances, or comparing scoring methods, atmospheric sciences or finances [4,5,7–9]. ROC analysis has also been gradually used in data mining and machine learning (evaluating and comparing algorithms, and economics [4,8,9].

Many software packages are available for ROC analysis; this work will focus only on *open access* R packages. After a first search the chosen packages were:

© Springer International Publishing AG 2017
O. Gervasi et al. (Eds.): ICCSA 2017, Part II, LNCS 10405, pp. 545–559, 2017.
DOI: 10.1007/978-3-319-62395-5_37

pROC, ROCR, Verification, caTools, Comp2ROC, Bioconductor ROC library and *Epi*. After an initial individual assessment, their main features will be compared. As seen before, the ROC curves are by themselves useful in assessing the performance of a test, but these tools can provide several different measurements. Our goal will be to determine which package can be chosen to perform ROC analysis and what kind of measures it offers. For this purpose, the common functions will also be analysed and the same analysis made.

2 State of the Art

ROC is an analytic assessment tool that provides a measurement used to establish some condition of interest. This measurement might be a rating along a discrete scale or a value along a continuous scale. Those can come from a variety of sources as long as its measurements or interpretations are ranked in magnitude, for example: the objective measurements of a test, objective evaluation of image features or subjective diagnostic interpretations [7,10].

This lead to a binary classification system, where the measurements produce two discrete results, and will be label as positive (P) when above (or identified) or as negative (N) when below (or rejected) the threshold. However, in this case there will be four different possible outcomes, as each of these labels can also be classified as true or false.

So, if the condition of interest is correctly identified, it is classified as a True Positive (TP); if the condition of interest is incorrectly identified, it is classified as a False Positive (FP). On the contrary, if the condition of interest is correctly rejected, it is classified as a True Negative (TN); and if the condition of interest is incorrectly rejected, it is classified as a False Negative (FN). This can be represented by a confusion matrix as illustrated in Table 1.

Table 1. Confusion matrix for classification

		True condition	
		Positive	*Negative*
Predicted condition	*Positive*	True Positive	False positive (Type I error)
	Negative	False Negative (Type II error)	True Negative

The metrics that will allow a more in depth analysis and comparisons are:

- The True Positive Fraction or Sensitivity is the proportion of those with the condition of interest that are correctly identified as such.

$$Sensitivity = TPF = \frac{TP}{TP + FN}$$

$$FNF = \frac{FN}{TP + FN} = 1 - Sensitivity$$

- The specificity is the proportion of those without the condition of interest that are correctly identified as such (TNF), this is the complement of the False Positive Fraction.

$$Specificity = TNF = \frac{TN}{TN + FP}$$

$$FPF = \frac{FP}{TN + FP} = 1 - Specificity$$

- Prevalence is the overall proportion of those with the condition of interest (it does not depend on the cut-off value).

$$Prevalence = \frac{TP + FN}{TP + FN + FP + TN}$$

- Precision will be the proportion of those with the condition of interest and of those that are identified as such.

$$Precision = \frac{TP}{TP + FP}$$

- Accuracy is used to describe the closeness of a measurement to the true value; therefore, it is the proportion of true results among the total number of cases examined.

$$Accuracy = \frac{TP + TN}{TP + FN + FP + TN}$$

$$ACC = (Sensitivity)(Prevalence) + (Specificity)(1 - Prevalence)$$

The ROC curve is used to evaluate the performance of a test, since it is used to plot the compromise between True Positive Fraction (TPF or Sensitivity) and False Positive Fraction (FPF or 1-Specificity) for a threshold decision value [11].

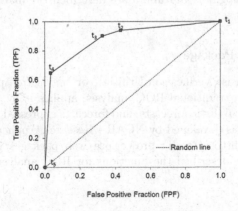

Fig. 1. Empirical ROC curve tresholds (adapted from [1])

In Fig. 1 each point on the curve represents the trade off between sensitivity (TPF) and 1-specificity (FPF), any increase in sensitivity will be accompanied by a decrease in specificity. The closer the curve follows the upper-left border of the ROC space, the more accurate the test, and therefore the area under the curve (AUC) is a measure of accuracy.

3   Materials and Methods

Currently there are several tools for ROC curve analysis. This software development was driven by the necessity to study new data in different areas of knowledge. In this study only those in R (packages) will be explored. The following assignment will proceed with a comparative analysis of the main characteristics of the chosen R packages[1].

The software used the R version 3.3.2 (https://www.r-project.org/). R is a programming language and software environment for statistical computing and graphics supported by the R Foundation for Statistical Computing [17].

3.1   PROC Package

The *pROC* package full title is "Display and Analyze ROC Curves" and is accessible at http://expasy.org/tools/pROC/ under the GNU General Public License. This package was designed in order to simplify ROC curve analysis and to apply suitable statistical tests for their comparison; it's main feature is the comparison between ROC curves [12]. In Table 2 are described the functions associated with this package.

3.2   ROCR Package

This package was created for evaluating and visualizing the performance of scoring classifiers [5,6] and is accessible though the CRAN repository or at http://rocr.bioinf.mpi-sb.mpg.de/. In Table 3 are described the main functions associated with this package.

3.3   Verification Package

The "Weather Forecast Verification Utilities" or *verification* package, while not directly aimed for a traditional ROC analysis, applies it by verifying discrete, continuous and probabilistic forecasts, and forecasts expressed as parametric distributions [13]. It was developed by NCAR - Research Applications Laboratory and is available at https://cran.r-project.org/web/packages/verification/index.html. In Table 4 are described the functions for ROC analysis associated with this package.

---

[1] As a rule, each of these packages brings specific dataset upon installation

Table 2. Functions provided in pROC package [12]

Function	Description
roc	Main function of the package. Builds a ROC curve and returns a "roc" object, a list of class "roc"
are.paired	Determines if two ROC curves are paired
auc	Computes the numeric value of area under the ROC curve (AUC) with the trapezoidal rule
ci	Computes the confidence interval (CI) of a ROC curve
ci.auc	Computes the CI of an area under the curve (AUC)
ci.coords	Computes the CI of the coordinates of a ROC curves with the coords function
ci.se	Computes the CI of the sensitivity at the given specificity points
ci.sp	Computes the CI of the specificity at the given sensitivity points
ci.thresholds	Computes the CI of the sensitivity and specificity of the thresholds given in argument
coords	Returns the coordinates of the ROC curve at the specified point
cov.roc	Computes the covariance between the AUC of two correlated (or paired) ROC curves
groupGeneric	Redefines base groupGeneric functions to handle auc and ci objects
has.partial.auc	Determines if the ROC curve has a partial AUC
lines.roc	Adds a ROC line to a ROC curve
multiclass.roc	Builds multiple ROC curve to compute the multi-class AUC
plot.ci	Adds CI to a ROC curve plot
plot.roc	Plots a ROC curve
power.roc.test	Computes sample size, power, significance level or minimum AUC for ROC curves
print	Prints a ROC curve, AUC or CI object and returns it invisibly
roc.test	Compares the AUC or partial AUC of two correlated (or paired) or uncorrelated (unpaired) ROC curves
smooth	Smoothes a ROC curve of numeric predictor
var	Computes the variance of the AUC of a ROC curve

Table 3. Functions provided in ROCR package [5]

Function	Description
performance	Performs all kinds of predictor evaluations
plot-methods	Plot all objects of class performance
prediction	Transforms the input data (vector, matrix, data frame, list form) into a standardized format

3.4 Comp2ROC Package

The *Comp2ROC* package, "Compare two ROC Curves that Intersect", allows the comparison of two diagnostic systems that cross each other [14]. Like the others,

Table 4. Functions provided in the verification package [13]

Function	Description
roc.area	Calculates the area underneath a ROC curve
roc.plot	Creates ROC plots for one or more models
verify	Calculates a range of verification statistics and skill scores, it also creates a verify class object

Table 5. Functions provided in the Comp2ROC package [14]

Function	Description
comp.roc.curves	Calculates by bootstrapping the real distribution for the entire length set
comp.roc.delong	Allows to calculate the areas under the curve for each curve and some statistical measure
curvesegslope	Allows to calculate the ROC curve segments slope through the points that are given by parameter
curvesegsloperef	Allows to calculate the segments slope that connect the ROC curve segments with the reference point (1, 0)
areatriangles	Allows to calculate the triangles area formed with two points that were next to each other and the reference point
diffareatriangles	Allows to calculate the difference between triangles areas formed by the same sampling lines in two different ROC curves
linedistance	Allows to calculate the intersection points between the ROC curve and the sampling lines
roc.curves.boot	Controls the whole package
roc.curves.plot	Allows to plot the two roc curves in comparison
rocboot.summary	Allows to see the information obtained through function roc.curve.boot
rocsampling	Allows to calculate some statistical measures like extension and location
rocsampling.summary	Allows to see with a simple interface the results obtained in rocsampling

Table 6. Functions provided in the caTools package

Function	Description
colAUC	Calculates Area Under the ROC Curve (AUC) for every column of a matrix. Also, can be used to plot the ROC curves

it is also available through the CRAN repository on https://cran.r-project.org/web/packages/Comp2ROC/index.html. In Table 5 are described the main functions associated with this package.

3.5   caTools Package

This package is available at https://cran.r-project.org/web/packages/caTools/index.html, and it has several basic utility functions, including the ROC and AUC [18]. In Table 6 are described the roc functions associated with this package.

3.6   Bioconductor ROC Library

The *ROC* library is a collection of R classes and functions related to ROC curves. These functions are targeted at the use of ROC analysis with DNA microarrays [15]. This library is accessible at http://bioconductor.org/packages/ROC/. In Table 7 are described the roc functions associated with this package.

Table 7. Functions provided in Bioconductor ROC library [15]

Function	Description
rocc	Representation of a ROC curve
AUC	Various functionals of ROC curves
plot-methods	Plot method for ROC curves
rocc-class	Object representing ROC curve
rocdemo.sca	Demonstrates "rocc" class construction using a scalar marker and simple functional rule
trapezint	Trapezoidal rule for AUC

3.7   Epi Package

The *Epi* package is an R-package for epidemiological analysis and it is also available from CRAN too at https://cran.r-project.org/web/packages/Epi/index.html [16]. This package has only one function for ROC analysis, called ROC. This function computes sensitivity, specificity and positive and negative predictive values for a test based on dichotomizing along the variable test, for prediction of state [16].

3.8   Dataset Characterization

For illustration purposes, the packages were tested with a dataset consisting of three fictitious indicators (Perfect, Reasonable, Useless) in ordinal scale of measurement for the assessment of a result (presence/absence). A higher value of the indicators tends to have a positive value of the result (presence of attribute).

4   Results

In the following section the selected ROC packages will be analysed in terms of their main features, specifically in terms of precision measures, plots, curve comparison, using the chosen dataset.

Table 8. Comparison of AUC for the Reasonable Test

Package	AUC
pROC	0.9006
ROCR	0.900641
verification	0.900641
caTools	0.900641
Epi	0.901
Comp2ROC	0.900641
ROC (Bioconductor)	0.9038462

All the chosen packages had the ability to produce a plot for the visualization of the ROC curves. These plots were built using the dataset provided, in which the 'Reasonable Test' was chosen as the indicator.

The resulting plots are replicated in the Appendix, Fig. 2a for the *ROCR* package, Fig. 3a for the Verification package, Fig. 4a for the *Comp2ROC* package, Fig. 5 for the *caTools*, Fig. 6 for the *Epi* package, Fig. 7 for the ROC library of Bioconductor, and Fig. 8a for the *pROC* package.

One of the measures that is possible to obtain from these packages is the Area Under the Curve (AUC), which is given by all the analysed packages, as shown in Table 8, and its possible to observe that the values are similar.

The *ROCR* package also had the ability to plot the Area Under the Curve, Fig. 2b, and to show multiple ROC curves in the same plot, Fig. 2c. Other features of this package was the possibility of retrieving relevant performance measures like accuracy, error rate, TPF, FPF, TNF, FNF, sensitivity, specificity, recall and precision (among others).

The *verification* package is capable of creating the plot, Fig. 3a, and of some basic calculations, Fig. 3b. The *caTools*, likewise, only produces a fast calculation of AUC and the plot when specifically requested.

The ROC function in the *verification* package provides the plot, Fig. 6, and it is possible to select a report of the model summary. The AUC values can also be given, as well as the sensitivity (Sens), specificity (Spec) and predictive values (PV) at the optimal response cut-points (Ir.eta - 'optimal cutpoint' (i.e. where sens+spec is maximal)).

The *Comp2ROC* package has a very specific objective, it is able to compare two sets of data, plot and analyse their characteristics, see Fig. 4. This package is able to compare two indexes, paired or not paired, in this example the Reasonable and the Useless Test, are paired, and plot empirical ROC curves from this analysis. At any moment was observed an intersection between the indexes, as seen in Fig. 4a, from the implied calculations, it is possible to obtain a second plot were it is expressed the statistic "area difference between the curves" and its respective confidence intervals (CI), see Fig. 4b. It is also possible to obtain an individual analysis of the indexes, acquiring the AUC value, standard error, confidence interval (with minimum and maximum limit). It also estimates the coefficient of correlation between areas and statistical tests of the differences between the areas. This is possible to export directly from R to an external document, Fig. 4d.

In regard to the *ROC* package of the Bioconductor, it plots the ROC curve, Fig. 7, and provides the AUC if requested.

Concerning the *pROC* package, it also provides the ROC curve plot (Fig. 8a), the AUC ROC plot and its value (Fig. 8c), and it compares two curves and their data (Fig. 8e). It also provides a series of performance measures (Fig. 8b), both in relation to only the original curve analysed and to the correlation of the two curves (Fig. 8f). If we consider the confidence intervals of the AUC ROC, it also provides a visual display, in the CI of area (light blue in the plot) and its thresholds (Fig. 8d).

5   Discussion

Let consider the Table 9, where is given a brief analysis of the main common features of the packages.

Table 9. Features of the R packages for analysis

Package name	pROC	ROCR	Verification	caTools	Epi	Comp2ROC	ROC (Bio-conductor)
Available on CRAN	Yes	Yes	Yes	Yes	Yes	Yes	No
Plot curve	Yes	Yes	Yes[a]	Yes[a]	Yes[a]	Yes[a]	Yes
Smoothing	Yes	No	Yes	No	No	No	No
AUC	Yes	Yes	Yes	Yes	Yes	Yes	Yes
Plot AUC	Yes	Yes	No	No	No	No	No
Confidence intervals	Yes	Partial	Partial	No	No	Yes	No
Performance measures	Yes	Yes	No	No	Yes	Yes	No
Comparing curves	Yes[b]	Yes	No	No	No	Yes	No

[a]These packages import the plot function from the ROCR package
[b]Only graphically

All of the packages, are available in CRAN, except the ROC package that is available in an independent repository from CRAN, Bioconductor, specially designed for biological packages.

The ROCR package produces the plot, see Fig. 2a, that is the standard reference for many of the other packages, like the *verification* (Fig. 3a), the *Comp2ROC* (Fig. 4a), the *caTools* (Fig. 5) and the *Epi* (Fig. 6) packages [13,14,16,18]. This is the case when it was not necessary to create a new code for this particular function, because the main focus of these packages was not to do a ROC analysis from zero. So these packages 'call' one function that already exists and is recognized as producing satisfactory results and it is used to complement their main purpose. Two of the packages (*pROC* and *verification*) performed a smoothing step where the ROC curve was adjusted for the existing values. In both this can be performed selecting TRUE or FALSE operators.

It was possible to calculate some performance measures with a few of the packages, however, its form of calculation was not always visible and require additional commands or editing of the functions already in use.

It was not expected that the *verification*, *caTools* and *Epi* packages had the ROC curve functionality. In terms of usability, they were not first packages chosen to use to perform this type of analysis, unless the said package was the base of the work and it would complement the other functionalities. So, in that regard, it was not very intuitive. However, the functions to obtain the results were relatively simple to understand and the tutorial provided the necessary guidance [13, 16, 18].

In regard to the determination of the AUC value, every package perform this calculus, and as seen in Table 8, the results are very similar across the different packages, and the differences observed can be attributed to the type of algorithm used. In complement, some of these packages also plotted the AUC curve, which provided a visual reference in the original plot.

The *ROCR* is the simplest package, having just three main functions that can be adjusted to obtain lots of operations that it is able to perform [5]. It is also the more widely recognised and used as source of the plot function by other packages. The *pROC* package is one of the most complete analysed packages and it is capable of performing all the analyses required [4]. The ROC function in Bioconductor did what it was supposed to do, but it had inherent issues related to the Bioconductor installation, which sometimes causes unrelated errors to the functions in use. The *caTools* is a useful and quick tool that performs very well what is asked of it. In the same way, the ROC function encoded in the *Epi* package appears to be well adjusted for the type of analysis that this package is used for, including the performance measures that can be provided next to the plot and which can facilitate the use of the image for other purposes. The same can be said for the *verification* package, because the function of interest is incorporated seamlessly with the rest of the functions, and can be used independently. The *Comp2ROC* package was a good surprise among these packages, because it provides a special view of a comparison of two curves with all the related statistical information.

6   Final Remarks

As is well known, R is a widely applied free software in the areas of statistics and bioinformatics, among others.

All packages analysed in this work, except one, as well as all dependencies, are found in CRAN. The *ROC* is in a specific repository, the Bioconductor which is also free.

On this work we explored each of the packages, pointing to their versatility for the analysis through ROC curves. Taking into account the characteristics inherent to this analysis, we can say that:

- all packages have the ability to trace the empirical ROC curve. If we want an adjustment to the ROC curve, only *pROC* and *ROCR* are able to do this;
- all of them estimate the curve precision index (AUC) in the same way;
- *pROC*, *ROCR* and *Comp2ROC* perform comparison of indicators based on the AUC index, with *Comp2ROC* being the only one that does it for independent and paired samples, and also in situations where the two curves intersect.

Acknowledgments. This work was supported by FCT - (Fundação para a Ciência e Tecnologia) within the Project Scope: UID/CEC/00319/2013.

Appendix

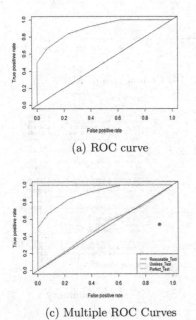

(a) ROC curve

(b) AUC ROC

(c) Multiple ROC Curves

Fig. 2. ROC package (blue: Reasonable, red: Useless, green: Perfect) (Color figure online)

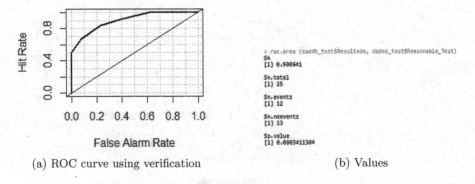

(a) ROC curve using verification

(b) Values

Fig. 3. ROC curve with verification package

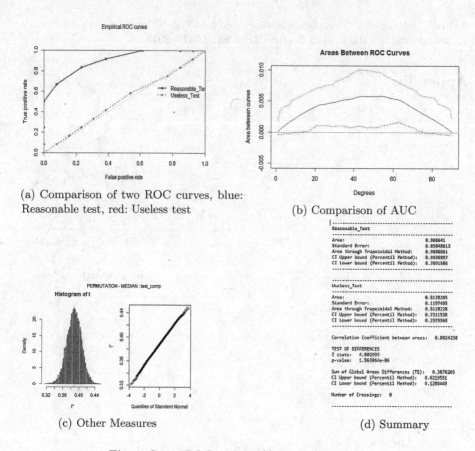

(a) Comparison of two ROC curves, blue:
Reasonable test, red: Useless test

(b) Comparison of AUC

(c) Other Measures

(d) Summary

Fig. 4. Comp2ROC package (Color figure online)

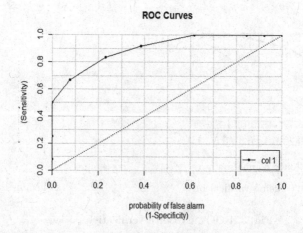

Fig. 5. ROC curve with caTools package for the Reasonable Test

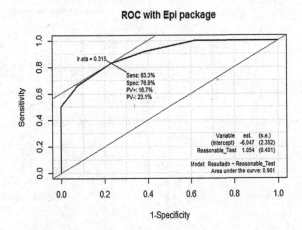

Fig. 6. ROC curve with Epi package for the Reasonable Test. The Area Under the Curve (AUC) values are given, as well as the sensitivity (Sens), specificity (Spec) and predictive values (PV) at the optimal response cut-points (Ir.eta).

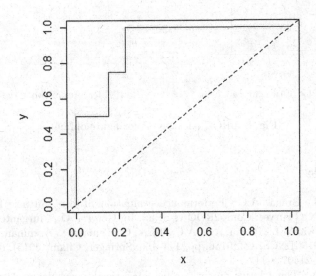

Fig. 7. ROC curve with Bioconductor for the Reasonable Test (blue) (Color figure online)

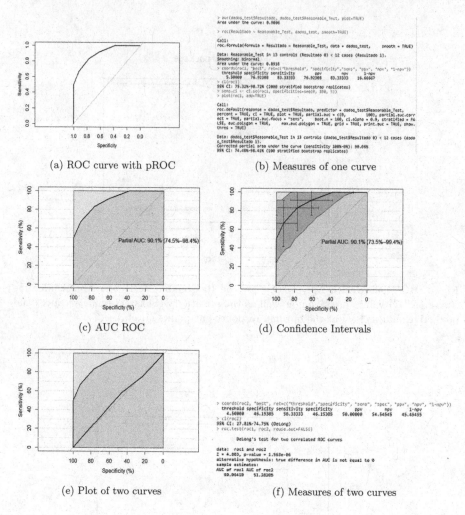

(a) ROC curve with pROC

(b) Measures of one curve

(c) AUC ROC

(d) Confidence Intervals

(e) Plot of two curves

(f) Measures of two curves

Fig. 8. pROC package (Color figure online)

References

1. Coelho, S., Braga, A.C.: Performance evaluation of two software for analysis through ROC curves: Comp2ROC vs SPSS. In: Gervasi, O., Murgante, B., Misra, S., Gavrilova, M.L., Rocha, A.M.A.C., Torre, C., Taniar, D., Apduhan, B.O. (eds.) ICCSA 2015. LNCS, vol. 9156, pp. 144–156. Springer, Cham (2015). doi:10.1007/978-3-319-21407-8_11

2. Swets, J.A.: The relative operating characteristic in psychology: a technique for isolating effects of response bias finds wide use in the study of perception and cognition. Science **182**, 990–1000 (1973)

3. Metz, C.E.: Some practical issues of experimental design and data analysis in radiological ROC studies. Invest. Radiol. **24**(3), 234–45 (1989)

4. Robin, X., Turck, N., Hainard, A., Tiberti, N., Lisacek, F., Sanchez, J.C., Müller, M.: pROC: an open-source package for R and S+ to analyze and compare ROC curves. BMC Bioinform. **12**, 77 (2011)
5. Sing, T., Sander, O., Beerenwinkel, N., Lengauer, T.: ROCR: visualizing classifier performance in R. Bioinformatics **21**, 3940–3941 (2005)
6. Sing, T., Sander, O., Beerenwinkel, N., Lengauer, T.: ROCR: visualizing the performance of scoring classifiers. Version 1.0-7 (2015). https://cran.r-project.org/web/packages/ROCR/index.html
7. Lasko, T., Bhagwat, J.G., Zou, K.H., Ohno-Machado, L.: Evaluation, receiver operating characteristic. Test Accuracy J. Biomed. Inform. **38**, 404–415 (2005)
8. Fawcett, T.: ROC graphs: notes and practical considerations for data mining researchers, pp. 1–27. In: HP Inven (2003)
9. Gonçalves, L., Subtil, A., Oliveira, M.R., Bermudez, P.Z.: ROC curve estimation: an overview. REVSTAT Stat. J. **1**, 1–20 (2014)
10. Obuchowski, N.A.: Receiver operating characteristic curves and their use in radiology. Radiology **229**, 3–8 (2003)
11. Braga, A.C., Costa, L., Oliveira, P.: ROC Curves in medical decision. In: 46th Scientific Meeting of the Italian Statistical Society (SIS), Rome, Italy, 20–22 June 2012
12. Robin, X., Turck, N., Hainard, A., Tiberti, N., Lisacek, F., Sanchez, C., Müller, M.: Package pROC: display and analyze ROC curves version. Version 1.7.2 (2015). http://web.expasy.org/pROC/files/pROC_1.7.2_R_manual.pdf
13. NCAR: verification: weather forecast verification utilities. Version 1.42 (2015). https://cran.r-project.org/web/packages/verification/index.html
14. Braga, A., Carvalho, S., Santiago, A.M.: Comp2ROC: compare two ROC curves that intersect. Version 1.1.4 (2016). https://cran.r-project.org/web/packages/Comp.2ROC/Comp.2ROC.pdf
15. Carey, V., Redestig, H.: Utilities for ROC, with uarray focus. Version 1.48.0 (2016). https://www.bioconductor.org/packages/release/bioc/manuals/ROC/man/ROC.pdf
16. Carstensen, B., Plummer, M., Laara, E., Hills, M.: Package 'Epi': a package for statistical analysis in epidemiology. Version 2.0 (2016). https://cran.r-project.org/web/packages/Epi/index.html
17. Hornik, K.: R-FAQ (2016). https://CRAN.R-project.org/doc/FAQ/R-FAQ.html
18. Tuszynski, J.: Package 'caTools': tools- moving window statistics, GIF, Base64, ROC AUC, etc. Version 1.17.1 (2015). https://cran.r-project.org/web/packages/caTools/index.html

A Neural Network Model for Team Viability

Isabel Dórdio Dimas[1,2], Humberto Rocha[3,4(✉)], Teresa Rebelo[5,6],
and Paulo Renato Lourenço[5,6]

[1] ESTGA, Universidade de Aveiro, 3750-127 Águeda, Portugal
idimas@ua.pt
[2] GOVCOPP, Universidade de Aveiro, 3810-193 Aveiro, Portugal
[3] FEUC, CeBER, Universidade de Coimbra, 3004-512 Coimbra, Portugal
hrocha@mat.uc.pt
[4] INESC-Coimbra, 3030-290 Coimbra, Portugal
[5] IPCDVS, Universidade de Coimbra, 3001-802 Coimbra, Portugal
[6] FPCEUC, Universidade de Coimbra, 3000-115 Coimbra, Portugal
{terebelo,prenato}@fpce.uc.pt

Abstract. Team effectiveness has been the focus of numerous studies
since teams play an increasingly decisive role in modern organizations.
In the present paper, our attention is centered on team viability, which
is one dimension of team effectiveness. Given the challenges that actual
teams face today, exploring the conditions and processes that enhance
the capacity of teams to adapt and continue to work together is a fundamental research path to pursue. In this study, team psychological capital
and team learning were considered as antecedents of team viability. The
relationships that team psychological capital and team learning establish with team viability were explored as accurately as possible. Typically,
these relationships are assumed to be linear as multivariate linear models
are often used. However, these linear models fail to explain possible nonlinear relations between variables, expected to exist in dynamic systems
as teams. Adopting computational modeling strategies in the context
of organizational psychology has become more common. In this paper,
radial basis function models and neural networks were used to study the
complex relationships between team psychological capital, team learning
and team viability.

Keywords: Team viability · Radial basis functions · Neural networks

1 Introduction

Teams are, nowadays, a fundamental cornerstone of organizations. Since teams are
created with the aim of generating value, understanding the conditions that contribute to team performance has been one of the major focus of research (e.g., [16]).
However, although performance is a fundamental team effectiveness dimension,
effectiveness is much more than performance. In fact, the literature is consensual
about the need to consider different criteria to assess effectiveness [17,21]. We will
focus our attention on team viability, which is in line with one of the dimensions

© Springer International Publishing AG 2017
O. Gervasi et al. (Eds.): ICCSA 2017, Part II, LNCS 10405, pp. 560–573, 2017.
DOI: 10.1007/978-3-319-62395-5_38

of the Hackman's three dimensional effectiveness approach [17,18]. Team viability can be defined as the team's capacity to adapt to internal and external changes as well as the extent to which team members are able to continue to work together in the future [39]. Given the challenges that actual teams face today, such as working with new technology or the need to constantly adapt to market fluctuations and changes, exploring the conditions and processes that enhance the capacity of the team to adapt and continue to work together is a fundamental path to pursue. In this study, team psychological capital and team learning will be considered as antecedents of team viability.

The term psychological capital can be defined as a positive, individual psychological state, or, in other words, a personal characteristic that can be measured and developed [22,24]. Luthans and colleagues [22,24], using a number of key criteria, identified four main psychological resources that formed the higher-order concept of psychological capital: self-efficacy, hope, optimism, and resilience. Self-efficacy, based on Bandura's social cognitive theory [4], refers to the individual belief in his or her ability to successfully execute a specific task [38]. Hope is characterized by two dimensions: will power, i.e., the drive and determination to attain a goal; way power thinking, i.e., the ability to plan alternative ways for attaining a desired goal [23]. Optimism is defined as the individual's expectancy of positive outcomes and integrates both realism and flexibility [41]. Finally, resilience is the ability to withstand and recover from challenges, stressful events or any other threat to well-being [43].

The relationship between individual-level psychological capital and effectiveness, namely, performance, has been investigated and established by a large number of studies [2,41]. Results are, however, less consistent when we consider some psychological resources separately, namely self-efficacy. For instance, concerning the relationship between self-efficacy and performance, whereas some studies, focusing on a socio-cognitive theory of self-regulation [4], found a positive relationship between self-efficacy and performance, others report that self-efficacy might lead to overconfidence, increasing, in consequence, the chance of committing errors during the tasks and affecting performance negatively [42].

The research developed until this moment almost never focused psychological capital as a team level phenomenon [30]. Teams have a major influence on the perceptions, decisions, beliefs and emotions of individuals and as a result of the interaction of knowledge between team members, shared mental models are created [26], which might lead to the development of a collective psychological capital [20]. Studies that focused psychological capital as a collective phenomenon, although scarce, point to a positive influence of this collective psychological state on team outcomes (e.g., [9,20,27]).

Concerning team learning, this construct can be conceptualized as both a process and an outcome [12]: the former concerns the group member' behaviors through the interaction processes, and the latter is the manifested outcome or result that emerges as a collective property of the team. In the present research our focus will be on team learning behaviors.

In line with Edmondson [14], team learning behaviors can be conceived as a continuous process of reflection and action, characterized by five fundamental behaviors: *seeking feedback*, both internally and externally, in order to measure group's effectiveness and to investigate possible improvements; *exploring* through the sharing of knowledge and perspectives, as well as through constructively managing different opinions; *experimenting* collectively new ways of achieving goals; *reflecting* on past achievements and on future aims and goals; and *discussing errors* and unexpected outcomes collectively and exploring ways to prevent them.

Previous research presented team learning as a crucial process of adaptation of teams to their environment and highlighted its importance to team effectiveness (e.g., [8,14,15]). Nevertheless, some studies suggest that team learning may act as a double-edge sword. Bunderson and Sutcliffe [6], for instance, found that too much emphasis on learning can compromise efficiency because detract the team from results, and this is particular salient for teams that have been performing well. In line with these findings, we have found, previously, increasing trends on team effectiveness up to a certain threshold of team learning, followed by a deflation for the highest values of this variable [13]. Taken together, these results highlight that more team learning is not always better and claim for more studies that support the non-linear relationship between team learning and team outcomes.

Recently, the importance of adopting computational modeling strategies in the context of organizational psychology was stressed by Cortina et al. [11]. As outlined by Hanges et al. [19], the adoption of more complex designs, such as radial basis functions neural networks, are useful for modeling nonlinear behavior as produced by dynamic systems, such as, teams. In this paper, radial basis functions (RBF) regression and RBF neural networks are used to study the complex relationships between team psychological capital, team learning and team viability. RBF regression has been successfully applied in different contexts, including aeronautics [32,33] or radiotherapy [35,36]. RBF models proved to mimic well unknown responses providing reliable surrogates that can be used either for prediction or to extract relationships between variables. On the other hand, neural networks have been widely used for deep learning with large data sets [37]. Incorporating RBF in the neural network learning process might enhance the extraction of nonlinear relationships between explanatory variables and response variable(s).

2 Materials and Methods

2.1 Sample

A quantitative study with a cross-sectional design was conducted, between November 2016 and January 2017, in which we surveyed teams from the Portuguese organizational setting. In order to be selected for the present study, teams must meet the following criteria: teams must consist of at least three members (1), who are perceived by themselves and others as a team (2), and who interact regularly and interdependently to accomplish a common goal (3) [10]. Different kind of measures were administered to team members and their respective leaders. Team members were surveyed about team psychological capital. Team leaders were surveyed about team learning and team viability. The use of different data sources contribute both to obtain a broader diagnose about the team and to prevent the problems related with common method variance. Two different strategies were used in data collection. In the majority of the organizations, data collection occurred in the organization facilities, with the physical presence of trained research assistants. When this strategy was not possible to implement, the questionnaires were filled in online via an electronic platform, with the link being provided to the participants. In both cases, the participation in the study was voluntary and was clarified on the front page of the survey that only aggregated data would be reported and that all identifying information would be removed. Besides, informed consent was obtained from all participants.

Surveys were administered to 452 members and to 104 leaders of 104 workgroups from 66 Portuguese organizations. After eliminating from the sample the teams where less than 50% of members answered the questionnaire and also the questionnaires where more than 10% of the answers were missing [5], the sample remained with 82 teams (353 members and 82 leaders) from 57 organizations, the majority from services sector (73%). Team size ranged from three to 18 members, with an average of six members (SD = 3.55).

2.2 Measures

To measure **team psychological capital** we adapted to the Portuguese language the scale *Psychological Capital Questionnaire* (PCQ) developed by Luthans et al. [25]. All items were reworded to reflect the group, rather than the individual, as the referent. This scale is composed of 24 items that assess the four dimensions of psychological capital (six items per dimension) that are measured on a 6-point Likert type scale from 1 (strongly disagree) to 6 (strongly agree). Sample items for self-efficacy, hope, resilience and optimism are "We feel confident analyzing a long-term problem to find a solution", "At the present time, we are energetically pursuing our work goals", "We usually take stressful things at work in stride" and "I always look on the bright side of things regarding my job", respectively. An Exploratory Factorial Analysis (EFA) was conducted, returning a five-factor solution that explained 62.76% of the total variance. Since the fifth factor had no theoretical support because it was only composed of reverse items from different dimensions, these items were sequentially eliminated. Besides, items with loadings below .50, were also sequentially eliminated from the solution. The final solution retained 18 items organized in four factors that jointly explained 65.83%. The Cronbach alpha for efficacy was .90, for hope was .85, for resilience was .75 and for optimism was .80. Since psychological capital has been consistently analyzed as a second-order factor (e.g., [9]) and also because the four psychological capital dimensions were highly correlated (bi-variate correlations varied from .54 to .73) we decided to conduct a Confirmatory Factor Analysis (CFA) with the retained four-factor structure obtained from EFA and with psychological capital as a second order factor. The fit indices obtained were acceptable [$\chi^2(131) = 302.97$, $p < .001$, $\chi^2/\text{gl.} = 2.31$, CFI $= .95$, RMSEA $= .06$] and, in consequence, we will consider, in the following analyses, the global score obtained from the average score of the four dimensions. A sample item is "Team members go out and get all the information they possibly can from others such as customers, or other parts of the organization".

To measure **team learning** we adapted to the Portuguese language the scale developed by [14]. This scale is constituted by seven items that are measured on a 5-point scale from 1 (almost never happens) to 5 (almost always happens). An EFA was conducted that returned a two-factor solution. Since the original scale is constituted only by one dimension, we decided to drop off, sequentially the items from the second dimension (items 2, 4 and 6). After eliminating items 2 and 4, a one-factor solution was obtained. However, the communalities and loadings of items 1 and 6 were low. As the referred items were both related to obtaining feedback and were the reverse of each other, and since it was important to have at least one item related to feedback, we decided to eliminate item 1 and to maintain item 6. Besides, this solution presented a higher reliability. The final solution was then composed of four items (3, 5, 6 and 7) that explained 53.3% of variance and presented a Cronbach alpha of .69.

To measure **team viability** we used the scale developed by Aubé and Rou- sseau [3], which was previously adapted to the Portuguese language by Albuquerque [1]. The scale is constituted by four items that are measured on a 5-point scale from 1 (almost doesn't apply) to 5 (almost totally applies). Since the Cronbach alpha for this scale was .68, and since the exclusion of item 3, which was the item with the content less related to the definition of the construct, increased the alpha to .73, the referred item was dropped from the scale. A sample item is "The members of this team could work a long time together".

2.3   Radial Basis Functions Models

Radial basis functions can furnish response surfaces able to explore/explain the nonlinear relationships between different input or explanatory variables and output or response variable(s). Furthermore, RBFs are often used to predict the response of a variable given the value of explanatory variables. For any given set of data points, a RBF model (surface) can be calculated even for poorly distributed data points in a high dimensional space. However, the RBF surface landscape, i.e. the relationship between and beyond the data points, depends on the choice of the basis function. Some RBFs can provide desirable trends while other may exhibit undesirable trends. Numerical selection of the most appropriate RBF for the given data set is advisable instead of an *a priori* choice based on the literature [34]. A brief description of RBF model calculation is provided next.

RBF Interpolation Problems

Let $y(\mathbf{x})$ denote the response for a given data point $\mathbf{x}$ (of n components) such that the value of y is only known at a finite set of N input data points $\mathbf{x}^1, \ldots, \mathbf{x}^N$, i.e., only $y(\mathbf{x}^k)$ $(k = 1, \ldots, N)$ are known. A RBF interpolation model $h(\mathbf{x})$ can be generically represented as

$$h(\mathbf{x}) = \sum_{j=1}^{N} \alpha_j \varphi(\|\mathbf{x} - \mathbf{x}^j\|), \tag{1}$$

where $\varphi(x)$ is the selected RBF, α_j are the coefficients determined by the interpolation equations $h(\mathbf{x}^k) = y(\mathbf{x}^k)$ $(k = 1, \ldots, N)$, $\|\mathbf{x} - \mathbf{x}^j\|$ corresponds to the parameterized distance between $\mathbf{x}$ and $\mathbf{x}^j$, $\|\mathbf{x} - \mathbf{x}^j\| = \sqrt{\sum_{i=1}^{n} |\theta_i|(x_i - x_i^j)^2}$, and $\theta_1, \ldots, \theta_n$ are scalars [34]. Coefficients $\alpha_1, \ldots, \alpha_N$ in Eq. (1) are computed for fixed parameters θ_i using the interpolation equations of the following linear system:

$$\sum_{j=1}^{N} \alpha_j \varphi(\|\mathbf{x}^k - \mathbf{x}^j\|) = y(\mathbf{x}^k), \quad \text{for } k = 1, \ldots, N. \tag{2}$$

Multiquadric, $\varphi(x) = \sqrt{1 + x^2}$, thin plate spline, $\varphi(x) = x^2 \ln x$, cubic spline, $\varphi(x) = x^3$, and Gaussian, $\varphi(x) = \exp(-x^2)$, are examples of RBFs that are commonly used to model linear, almost quadratic and cubic growth rates, as well as exponential decay of the response, respectively [31].

Calculation of the RBF model $h(\mathbf{x})$ in Eq. (1) requires the selection of a RBF $\varphi(x)$ and the choice of model parameters $\theta_1, \ldots, \theta_n$. While selection of the most appropriate RBF for the given data set can be done iteratively by testing the different possible choices of $\varphi(x)$, there is an infinite number of possible choices for $\theta_1, \ldots, \theta_n$. For different fixed sets of model parameters $\theta_1, \ldots, \theta_n$, distinct models with different behaviors between data points are calculated for a given selection of $\varphi(x)$. Cross-validation (CV) can be used for model parameter tuning leading to models with enhanced prediction capability [40]. Furthermore, the most appropriate basis function $\varphi(x)$ can be numerically computed using prediction accuracy (CV error) as main criterion. The leave-one-out CV procedure can be used in model parameter tuning for RBF interpolation [34]:

Algorithm 1 (Leave-one-out cross-validation for RBF interpolation)

1. Fix a set of model parameters $\theta_1, \ldots, \theta_n$.
2. For $j = 1, \ldots, N$, construct the RBF model $h_{-j}(\mathbf{x})$ of the data points $(\mathbf{x}^k, y(\mathbf{x}^k))$ for $1 \leq k \leq N, k \neq j$.
3. Set prediction error as the following CV root mean square error:

$$E^{CV}(\theta_1, \ldots, \theta_n) = \sqrt{\frac{1}{N} \sum_{j=1}^{N} (h_{-j}(\mathbf{x}^j) - y(\mathbf{x}^j))^2}. \tag{3}$$

The goal of model parameter tuning by CV is to find $\theta_1, \ldots, \theta_n$ that minimize the CV error, $E^{CV}(\theta_1, \ldots, \theta_n)$, so that the interpolation model has the highest prediction accuracy when CV error is the measure. Using different θ_i allows the model parameter tuning to scale each variable x_i based on its significance in modeling the variance in the response, thus, has the benefit of implicit variable screening built in the model parameter tuning.

2.4 RBF Neural Networks

Neural Networks (NN) is a class of bio-inspired computer algorithms that attempt to mimic the human brain thinking process. This class of artificial intelligence methods have been widely used in machine learning, data mining or statistics. The main feature of NN is the ability to extract trends from large amounts of data whose patterns are difficult to perceive by simple inspection or to detect by other methods. Unlike computer algorithms that follow a set of instructions, NN learn by mean of examples that will largely determine the quality of the results obtained.

The human nervous system is composed by particular cells called neurons that send signals rapidly through their myelinated projections (axons) with the goal of inhibiting or exciting the neighbor neuron(s) or cell(s). The junctions between neurons are called synapses and the network of connected neurons is responsible for the human body perception of the world stimulus and its feedback. Similarly, NN algorithms consist of a large number of connected units (neurons), typically organized in several layers, with the signal passing from the input neuron layer to the output neuron layer in a feed-forward process. The synapses between neurons are weighted connections and the learning process can be simply described as finding the weights that enable the NN to best capture the trends buried in the data set. Typically, several computational stages are required for an accurate learning process, where each stage updates the network weights. There are many different forms of network architectures, signal propagation or weights transformation. For a detailed description of different types of NN, Schmidhuber provides an overview of deep learning using NN [37]. Here, RBF are used for a non-linear update of the weights of a multi-layer feed-forward NN.

3 Results

Since the unit of analysis in the present study was the group, and team psychological capital was obtained from team members, it was necessary to aggregate efficacy, hope, resilience and optimism to the team level (the remaining variables were already at the group level). To justify aggregation, the Average Deviation Index (AD_M Index) developed by Burke, Finkelstein, and Dusig [7] was calculated. The average AD_M values obtained for efficacy, hope, resilience and optimism were 0.43, 0.41, 0.45 and 0.42, respectively. Since all the values were below the upper-limit criterion of 1.01, team members' scores were aggregated, with confidence, to the team level.

Table 1 displays the (Pearson) correlation analysis performed to assure that team psychological capital and team learning are correlated with team viability. Significant and positive correlations were found between the independent

Table 1. Correlation analysis.

	Viability	Learning	Psychological capital
Viability	1	.590**	.499**
Learning		1	.395**
Psychological capital			1

Note: **p < .01.

variables, psychological capital and learning, and team viability. Significant and positive correlation was also found between psychological capital and learning. Despite that, the inclusion of both variables as predictors of team viability is licit since in social sciences correlations of that magnitude are considered of medium-size [5].

Table 2. Optimal CV errors for the data set.

Multiquadric CV error	Thin plate CV error	Cubic CV error	Gaussian CV error
3.84	0.64	1.38	3.11

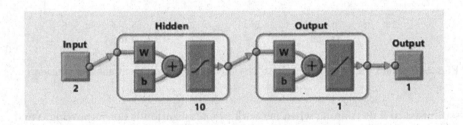

Fig. 1. Neural network architecture.

Optimal RBF model parameters $\theta_1, \ldots, \theta_n$ of Eq. 3 were computed by minimizing the CV error using a MATLAB implementation (*fminsearch*) of a derivative-free optimization algorithm called Nelder-Mead [29]. The optimal CV error obtained for the different basis functions tested was used as proxy of their prediction ability [34]. Optimal CV errors for the RBF interpolation models obtained by using different basis functions are displayed in Table 2. Thin Plate RBF was selected as basis function since the corresponding RBF model presented the lowest CV error.

MATLAB Neural Network Toolbox [28] was used to compute the RBF NN model. The data set was randomly divided into three subsets: training data set – used by the network to adjust its weights during learning/fitting, validation data

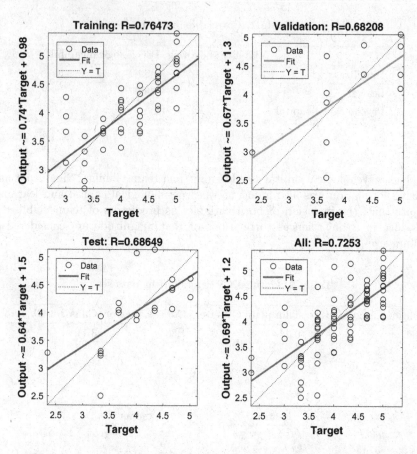

Fig. 2. R-values obtained by the RBF NN model for the different subsets considered.

set – used to halt training when network's generalization is not improving – and testing data set – to provide an independent measure of the network's prediction ability. The training data set considered 65% of the entire data set (53 data points – teams), the validation data set 10% (8) and the training data set 25% (21). The network architecture considered two inputs (psychological capital and learning), one output (team viability) and 10 neurons in the fitting network's hidden layer as illustrated in Fig. 1. The network was trained/fitted using RBF regression. R-values obtained for the three data subsets and the entire data set are displayed in Fig. 2. The predicting ability inherent to the R-value (and consequently R^2-value) obtained for the training set is quite good. Since RBF regression is an interpolation method, the R^2-value for the data set where the model is fit will be exactly one. Thus, in order to compare the results with the RBF NN model, a Thin Plate RFB model was fit using the data set composed by the training set and the validation set of the RBF NN model. Then, that Thin Plate RBF model was used to predict team viability for the remaining 21

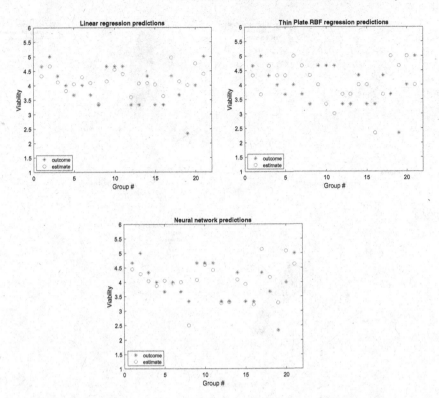

Fig. 3. Prediction of linear, Thin Plate RBF and RBF NN models.

data points corresponding to the testing set of the RBF NN model. Figure 3 display the prediction of team viability for 21 data points using the RBF NN model and the Thin Plate RBF model. In order to benchmark these results, the following multiple linear regression model, obtained using SPSS, was also fitted and tested considering the exact same fitting and testing data sets:

$$\text{Team viability} = 0.01 + 0.315 \times (\text{psychological capital}) + 0.465 \times (\text{learning}).$$

R^2-values obtained by the linear model, the Thin Plate RBF model and the RBF NN model for the testing data set were .39, .31 and .47, respectively.

Figure 4 display the relationships between both explanatory variables and team viability, captured by the different models. Data points of the training set and the testing set were added to the plots for a better perception of the scatter in the data.

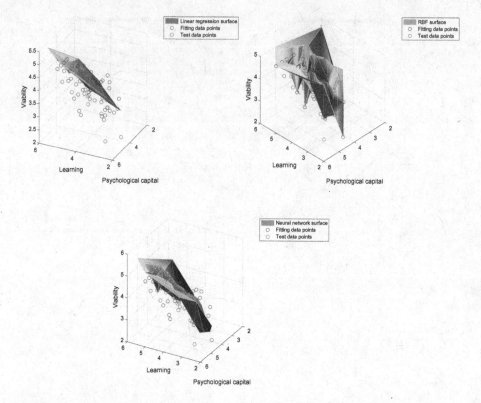

Fig. 4. Three-dimensional surface of linear, Thin Plate RBF and RBF neural networks models.

4   Discussion and Conclusions

It is possible to find many different models that present similar responses for a given data set. This is also valid for models whose response is 100% accurate for all data points (interpolation models) as RBF models. However, model's behavior between data points, i.e. trends or relationships between variables, may be completely different for distinct models. A simple way to find the model that present the most reliable trends is to verify which model present the most accurate predictions for new data points.

Prediction results obtained by RBF NN model clearly outperform linear and Thin Plate RBF model's results. That is clear by simple inspection of Fig. 3 or by comparing the R^2-values obtained for the testing data set. Thus, the trends or relationships between team psychological capital, team learning and team viability are more reliable for the RBF NN model. These trends can be easily inspected in Fig. 4. These plots, obtained by MATLAB, are actually dynamic 3D surfaces whose inspection from different angles enable a better understanding of the nonlinear trends. For the RBF NN response surface, an increase in either team psychological capital or team learning leads to an increase in team viability up to a certain threshold where team viability cease to increase. As expected, the linear response is a plane that presents similar trends except that they are linear and thus team viablity always increase for an increase of either team psychological capital or team learning. Finally, although the Thin Plate RBF response surface presents a nonlinear trend similar to the RBF NN response surface, it also presents many peaks (sharp oscilations). This is related to the fact of RBF models being interpolation models and this particular data set clearly do no favor interpolation type of methods.

The results presented revealed nonlinear patterns between the predictor variables, team psychological capital and team learning, and the criterion variable, team viability. An increasing trend up to a certain threshold is obtained, followed by a deflation for the highest values of the predictors. Thus, our results, which extend the conclusions of previous studies to different samples and variables [13], clearly highlight that more is not always better. Concerning team learning, the present findings are in line with those obtained by Bunderson and Sutcliffe [6] and also Dimas et al. [13], and add to the growing body of knowledge that considers the negative effect on effectiveness of an excessive focus on learning. Hence, the involvement of the team in learning behaviors such as exploring new ways of performing the tasks, experimenting alternatives or discussing errors, is positive for team results but when all the resources and energy of the team is focused on that behaviors, goals achievement might suffer threatening the viability of the team. As for team psychological capital, our results extends the findings obtained, at the individual level, by Vancouver et al. [42], and also the remarks made by Veraharen regarding the construct of self-efficacy [44], to the team level and to the broader concept of psychological capital. Our findings highlighted that when the levels of team psychological capital are too high, team viability might suffer, probably because the team became overconfident, neglecting some important aspects of the task and committing more errors, undermining, in consequence, the future of the team.

Acknowledgements. This work was supported by the Fundação para a Ciência e a Tecnologia (FCT) under project grants UID/MULTI/00308/2013 and POCI-01-0145-FEDER-008540.

References

1. Albuquerque, L.B.: Team resilience and team effectiveness: adaptation of measuring instruments. Master thesis, Faculdade de Psicologia e de Ciências da Educação da Universidade de Coimbra, Coimbra (2016)
2. Avey, J.B., Luthans, F., Youssef, C.M.: The additive value of positive psychological capital in predicting work attitudes and behaviors. J. Manage. **36**, 430–452 (2010)
3. Aubé, C., Rousseau, V.: Team goal commitment and team effectiveness: the role of task interdependence and supportive behaviors. Group. Dyn. Theor. Res. **9**, 189–204 (2005)
4. Bandura, A.: Social Learning Theory, Revised edn. Prentice Hall, Upper Saddle River (1977)

5. Bryman, A., Cramer, D.: Quantitative Data Analysis for Social Scientists, rev edn. Routledge, Florence (1994)
6. Bunderson, J.S., Sutcliffe, K.M.: Management team learning orientation and business unit performance. J. Appl. Psychol. **88**, 552–560 (2003)
7. Burke, M.J., Finkelstein, L.M., Dusig, M.S.: On average deviation indices for estimating interrater agreement. Organ. Res. Methods **2**, 49–68 (1999)
8. Chan, C., Pearson, C., Entrekin, L.: Examining the effects of internal and external team learning on team performance. Team Perform. Manage. **9**, 174–181 (2003)
9. Clapp-Smith, R., Vogelgesang, G.R., Avey, J.B.: Authentic leadership and positive psychological capital: the mediating role of trust at the group level of analysis. J. Leadersh. Organ. Stud. **15**, 227–240 (2008)
10. Cohen, S.G., Bailey, D.E.: What makes teams work: group effectiveness research from the shop floor to the executive suite. J. Manage. **23**, 239–290 (1997)
11. Cortina, J.M., Aguinis, H., DeShon, R.P.: Twilight of dawn or of evening? A century of research methods. J. Appl. Psychol. **102**, 274–290 (2017)
12. Decuyper, S., Dochy, F., Bossche, P.V.: Grasping the dynamic complexity of team learning: an integrative model for effective team learning in organizations. Educ. Res. Rev. **5**, 111–133 (2010)
13. Dimas, I.D., Rocha, H., Rebelo, T., Lourenço, P.R.: A nonlinear multicriteria model for team effectiveness. In: Gervasi, O., et al. (eds.) ICCSA 2016. LNCS, vol. 9789, pp. 595–609. Springer, Cham (2016). doi:10.1007/978-3-319-42089-9_42
14. Edmondson, A.C.: Psychological safety and learning behavior in work teams. Admin. Sci. Q. **44**, 350–383 (1999)
15. Flood, P., Maccurtain, S., West, M.: Effective Top Management Teams: An International Perspective. Blackhall Publishing, Dublin (2001)
16. Gully, S.M., Incalcaterra, K.A., Joshi, A., Beauien, M.J.: A meta-analysis of team-efficacy, potency, and performance: interdependence and level of analysis as moderators of observed relationships. J. Appl. Psychol. **87**, 819–832 (2002)
17. Hackman, J.R.: The design of work teams. In: Lorsch, J. (ed.) Handbook of Organizational Behavior, pp. 315–342. Prentice Hall, Englewood Cliffs (1987)
18. Hackman, J.R.: From causes to conditions in group research. J. Organ. Behav. **33**, 428–444 (2012)
19. Hanges, P.J., Lord, R.G., Godfrey, E.G., Raver, J.L.: Modeling nonlinear relationships: neural networks and catastrophe analysis. In: Rogelberg, S.G. (ed.) Handbook of Research Methods in Industrial and Organizational Psychology, pp. 431–455. Blackwell Publishing, Maiden (2004)
20. Heled, E., Somech, A., Waters, L.: Psychological capital as a team phenomenon: mediating the relationship between learning climate and outcomes at the individual and team levels. J. Posit. Psychol. **9760**, 1–12 (2015)
21. Kozlowski, S.W.J., Ilgen, D.R.: Enhancing the effectiveness of work groups and teams. PSPI **7**, 77–124 (2006)
22. Luthans, F., Avey, J.B., Avolio, B.J., Norman, S.M., Combs, G.M.: Psychological capital development: toward a micro-intervention. J. Organ. Behav. **27**, 387–393 (2006)
23. Luthans, F., Norman, S.M., Avolio, B.J., Avey, J.B.: The mediating role of psychological capital in the supportive organizational climate-employee performance relationship. J. Organ. Behav. **29**, 219–238 (2008)
24. Luthans, F., Youssef, C.M.: Human, social, and now positive psychological capital management. Organ. Dyn. **33**, 143–160 (2004)

25. Luthans, F., Avolio, B.J., Avey, J.B., Norman, S.M.: Positive psychological capital: measurement and relationship with performance and satisfaction. Pers. Psychol. **60**, 541–572 (2007)
26. Mathieu, J.E., Heffner, T.S., Goodwin, G.F., Salas, E., Cannon-Bowers, J.A.: The influence of shared mental models on team process and performance. J. Appl. Psychol. **85**, 273–283 (2000)
27. Mathe-Soulek, K., Scott-Halsell, S., Kim, S., Krawczyk, M.: Psychological capital in the quick service restaurant industry: a study of unit-level performance. J. Hosp. Tour. Res. 24 (2014)
28. MATLAB 2016a. The MathWorks Inc., Natick (2016)
29. Nelder, J., Mead, R.: A simplex method for function minimization. Comput. J. **7**, 308–313 (1965)
30. Newman, A., Ucbasaran, D., Zhu, F., Hirst, G.: Psychological capital: a review and synthesis. J. Organ. Behav. **35**, s1 (2014)
31. Powell, M.: Radial basis function methods for interpolation to functions of many variables. HERMIS Int. J. Comput. Math. Appl. **3**, 1–23 (2002)
32. Rocha, H., Li, W., Hahn, A.: Principal component regression for fitting wing weight data of subsonic transports. J. Aircr. **43**, 1925–1936 (2006)
33. Rocha, H.: Model parameter tuning by cross validation and global optimization: application to the wing weight fitting problem. Struct. Multidiscip. Optim. **37**, 197–202 (2008)
34. Rocha, H.: On the selection of the most adequate radial basis function. Appl. Math. Model. **33**, 1573–1583 (2009)
35. Rocha, H., Dias, J.M., Ferreira, B.C., Lopes, M.C.: Selection of intensity modulated radiation therapy treatment beam directions using radial basis functions within a pattern search methods framework. J. Global Optim. **57**, 1065–1089 (2013)
36. Rocha, H., Dias, J.M., Ferreira, B.C., Lopes, M.C.: Beam angle optimization for intensity-modulated radiation therapy using a guided pattern search method. Phys. Med. Biol. **58**, 2939 (2013)
37. Schmidhuber, J.: Deep learning in neural networks: an overview. Neural Netw. **61**, 85–117 (2015)
38. Stajkovic, A.D., Luthans, F.: Self-efficacy and work-related performance: a meta analysis. Psychol. Bull. **44**, 580–590 (1998)
39. Sundstrom, E., De Meuse, K., Futrell, D.: Work teams: applications and effectiveness. Am. Psychol. **45**, 120–133 (1990)
40. Tu, J.: Cross-validated multivariate metamodeling methods for physics-based computer simulations. In: Proceedings of the IMAC-XXI (2003)
41. Youssef, C.M., Luthans, F.: Positive organizational behavior in the workplace: the impact of hope, optimism, and resilience. J. Manage. **33**, 774–800 (2007)
42. Vancouver, J.B., Thompson, C.M., Williams, A.A.: The changing signs in the relationships among self-efficacy, personal goals, and performance. J. Appl. Psychol. **8**, 605–620 (2001)
43. West, B.J., Patera, J.L., Carsten, M.K.: Team level positivity: investigating positive psychological capacities and team level outcomes. J. Organ. Behav. **30**, 249–267 (2009)
44. Verhaeren, T.: Is a strong sense of self-efficacy always beneficial? BUT Philolos. Soc. Sci. **5**, 193–200 (2012)

Arrow Plot for Selecting Genes in a Microarray Experiment: An Explorative Study

Catarina Lemos[1], Gustavo Soutinho[2], and Ana Cristina Braga[2(✉)]

[1] Departamento de Informática, Escola de Engenharia,
Universidade do Minho, Braga, Portugal
cifmlemos@hotmail.com
[2] ALGORITMI Centre, University of Minho, 4710-057 Braga, Portugal
b6675@algoritmi.uminho.pt, acb@dps.uminho.pt

Abstract. Genetic expression analysis is essential for the identification of gene functions and take even more importance when they are directly related with diseases. For the performing of a large-scale study of changes in gene expression it is necessary to find a method to do it with precision and accuracy. Thus, the analysis by the microarray technology is an important tool in the diagnosis of diseases.

An important role of the analysis of microarray data involves the determination of which genes could be differentially expressed (DE) across two or more kind of tissue samples.

The traditional methods to detect DE genes are generally based on simple measures of distances and could failed in this classification.

In this work it is explored a new tool proposed by Silva-Fortes [21] that overcome this difficulty, the arrow plot. This tool is also compared with other methods mostly used to this purpose.

The arrow plot is a graphical tool based on two measures of distances between two probability density functions: the overlapping coefficient (OVL) between two densities and the area under the receiver operating characteristic (ROC) curve (AUC), for each gene of a microarray experience.

For illustrative purpose we will use a dataset of pancreatic adenocarcinoma. All computation will be done in R software.

Keywords: Microarrays · Genes · Area under the ROC curve · Overlapping Coefficient · Arrow plot

1 Introduction

The fast evolution of the cell and molecular biology in the past few decades, associated to the evolution of bioinformatics, has proportioned a big advance in the genomic sequencing and in obtaining informations about mechanisms of regulation, cell functions and differences in many types of tissues.

The microarray technology became a ruling tool when associated to genomics and cell mechanisms [3]. It is a technique with big influence in the analysis of

© Springer International Publishing AG 2017
O. Gervasi et al. (Eds.): ICCSA 2017, Part II, LNCS 10405, pp. 574–585, 2017.
DOI: 10.1007/978-3-319-62395-5_39

the genetic composition of a certain organism, but also in the identification of genes which contain a variation on their expression levels when they are under a certain condition. Microarray experiences allow one to simultaneously evaluate thousands of potential biomarkers that distinguish different tissue types [3].

According [9] we can resume the chain of events between biological sample and final outcome in a several experimental and computational steps:

- Sample preparation;
- Oligonucleotide probe design;
- Oligonucleotide synthesis;
- Slide preparation and spotting;
- Hybridization;
- Washing;
- Image analysis;
- Statistical data analysis.

In each step of this sequence, errors will occur and consequently will affect considerably the accuracy of the final results. Indeed data generated from microarrays usually show a large number of false positives (i.e., genes which are not DE but appear as being so) [9].

Statistics is an essential science for the evaluation and interpretation of collected data in biological experiences that allows the analysis and validation of genetic expression data.

Throw statistical analysis it is possible to select DE (differentially expressed) genes in tissues with or without pathology, for example carcinoma. The receiver operating characteristic (ROC) methodology could be used to select differential expressed genes, because it discriminates between two mutual exclusive conditions, for instance, genes differentially expressed and genes non differentially expressed.

A very common use of ROC methodology is the identification of genes related to a certain kind of cancer [8], it compares the expression levels between them, always taking into account those who belong to the experimental group and those who belong to the control group. As the use of this methodology for itself was not enough to take all the conclusions needed, another tool, the *arrow plot*, proposed by Silva-Fortes [21], was used in this work. This method is based on two estimates, the AUC (area under the ROC curve) and the OVL (overlapping coefficient for densities), both estimated using a non parametric approach [16].

According the ROC methodology, genes with up regulation, also called positive regulation, have values of AUC near 1 and genes with down regulation, also called negative regulation, have values of AUC below 0.5, the genes with DE will have a AUC around 0.5 (Fig. 1). So the estimated value for AUC, alone is not enough to select the "special genes" (Fig. 1 C and D) since they are not distinguished from genes not DE (Fig. 1 B), as the AUC values in both cases are close to 0.5.

In this paper the designation of "special genes" refers to the group of differentially expressed genes that are not usually identified by the statistical methods

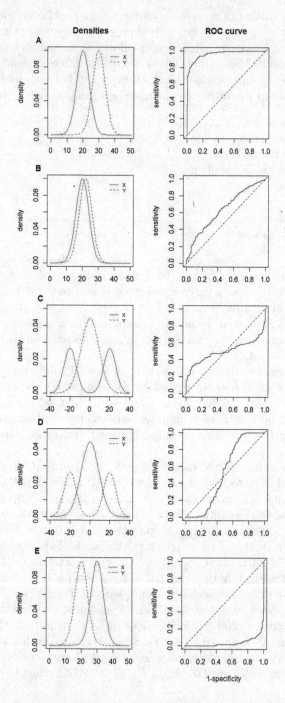

Fig. 1. Relationship between densities and ROC curves. (Source: Adapted from [21])

commonly used in the identification of genes with differential expression and which can provide useful information about the functions of the cells.

In Fig. 1 are represented the probability density functions of gene expression values of two variables and their corresponding empirical ROC curves, where Y represents the random variable of the expression values under the experimental condition and X represents the random variable of the expression values for the control group. The same classification rule was considered for all ROC plots, i.e., high values of the decision variable correspond to positive regulation. Density plots were obtained using density estimation from two samples of size 200 simulated from normal distributions:

A: $X \sim N(20,4), Y \sim N(30,4)$;
B: $X \sim N(20,4), Y \sim N(22,4)$;
C: $X \sim 0.4N(-20,6) + 0.4N(20,6), Y \sim N(0,9)$;
D: $X \sim N(0,9), Y \sim 0.4N(-20,6) + 0.4N(20,6)$;
E: $X \sim N(30,4), Y \sim N(20,4)$.

Analysing both OVL and AUC values obtained from all genes allows to distinguish these cases through a simple graphical tool, which expresses in the axis of the ordinates the AUC value and in the abscissa axis the value of OVL. This graph is called *arrow plot*. In the case of the differentially expressed genes the OVL is supposed to be a low value (less than 0.5) [21].

The main goal of this study is to determine which genes have characteristics of differential expression in different experimental conditions. In this way, the arrow plot tool was used to identify, not only positively and negatively regulated genes, but also differentially expressed genes in a control and in a experimental group [10].

Another situation, that is important take into account, in the DE genes selection, is the analysis of the bimodality and multimodality, since the distinction between the group of patients with high or low genetic expression is more easily to conduct when we compared genes with unimodal expression. Genes with a bimodal or a multimodal distribution usually have significant roles in what comes to differentiation and cellular signaling, that may also be indicative of the presence of unknown subclasses. This means, that there are two separate peaks in the distribution, one peak related to a subclass with a low expression level and the other peak related to the subclass with a high expression level. The next step in this process is to identify those subclasses where it would be possible to know some biological mechanisms under a certain pathological condition [20]. The algorithm proposed by Silva-Fortes in [21] make this analysis.

Another objective of this work is to show the versatility of using the arrow plot tool in the identification of DE genes and to compare it with existing traditional methods such as FC (Fold Change), AD (Average Difference), WAD (Weighted Average Difference), RP (Rank Products), ibmT (Intensity-based moderated t-statistic), modT (Moderated t-statistic), samT (Significance Analysis of Microarrays based on t-statistic), Welch-t and methods based on AUC measurement, such as samROC (Significance Analysis of Microarrays based on

ROC methodology), kerAUC and empAUC. For this purpose, a set of data was used on the collection of microarrays in tissue from individuals with pancreatic cancer.

Pancreatic cancer is the malignant tumor of the pancreas and may also arise with the name of pancreatic carcinoma. It is one of the most lethal cancers, being the fifth most common cause of cancer deaths in the world and the fourth leading cause of death in the European Union.

2   Materials and Methods

There are numerous methods to predict differential expression, and in all of them, one way to do this is to calculate the statistics for each gene and then order the given values for each gene according to the calculated values.

In this section we will present a brief description of the used methods to select genes with DE.

2.1   Methods for the Selection of Differentially Expressed Genes

Fold Change (FC). The FC performs a biological interpretation of the data through a determination of arbitrary cut-off points, which are generally values greater than 2. Thus, for values above 1, the gene has a positive regulation, for values below 1, the gene has a negative regulation. If the value is zero, the genes are not differentially expressed [7]. It is usually the most commonly used. However, this can lead to errors since considers a constant variability in all the genes selected.

Average Difference (AD). According to this method, genes are ordered taking into account the AD value. High values usually have a great significance in terms of differential expression. However, there may be some genes classified in the first positions that have low expression values. This happens due to this measure does not take into account the variability of the data [11].

Weighted Average Difference (WAD). This method is the product of the combination of the AD method with a significant weight of mean signal intensity, the higher the WAD value, the more relevant the gene is considered [11].

Significance Analysis of Microarrays (SAM). The SAM method is based on t-statistic procedure. This measure takes into account fluctuations that occur in the samples, related to the specificity of each gene. This method consists of three phases, the first one for the calculation of a t-statistic, the second one is based on a permutation test and the third based on the control of "false positives" [22].

Moderated t-statistic (modT) and Intensity-Based Moderated t-statistic (ibmT). ModT and imT are two methods based on t-statistics, both take into account the variance between genes. The modT has utility when the experiment addresses the analysis of two or one channel microarrays. The ibmT is a modified version of modT, when the relationship between gene variance and signal intensity is weak or even non-existent, the modT method is applied instead of ibmT.

Rank Products (RP). It is a method that makes possible to identify genes with positive and negative regulation. It is more advisable when there is a reduced number of arrays. RP is a test based on non-parametric procedure and only require few assumption about the data. Unlike the AD method, it does not depend on an estimate of the gene-specific measurement variance and is, therefore, particularly powerful when only a small number of replicates are available [2].

Significance Analysis of Microarrays Based on ROC Methodology (samROC). The samROC methodology minimizes the number of false negatives and false positives, which had been classified as DE. It is based on the SAM method and the ROC methodology.

Empirical AUC (empAUC) and Kernel AUC (kerAUC). The area under the ROC curve (AUC) is the most used method in the ROC methodology to evaluate the discriminative power of a classification model. It is a non-parametric estimate that measures the distance between two variables of distribution present in the two classes.

Thus, two methods are proposed to estimate this parameter: the empirical AUC, that could be computed using the Wicoxon-Mann-Whitney approximation, that is a more optimistic method having the tendency to overestimate the AUC; the kernel method, that is a more robust method.

Generally the number of genes with biological interest does not have a great difference when comparing these two methods. On the other hand, in what comes to the selection of genes with the positive and negative regulation, the empirical AUC selected a higher number of genes with these characteristics. Because, these last genes do not make much difference to the selection of DE genes, we used the AUC estimated by the kernel method in arrow plot procedure.

Arrow Plot. In order to determine the genes which present differential expression characteristics when inserted in different samples submitted to different experimental conditions we used the most recent graphical tool, the *arrow plot*. With this technology it was possible to simultaneously analyse two non-parametric measures, the AUC and the OVL. In addition to identifying positive and negative regulated genes, this technique also identifies all the differentially expressed genes present in all the different groups [20]. Up-regulated genes will have an AUC near 1 and an OVL below 0.5; down-regulated genes will have an AUC near 0 and the same OVL as last one referred; the differentially expressed "special genes" will have an AUC between 0.4 and 0.6 and the OVL again below 0.5 [21].

The cut-off values were arbitrary chosen, however, if these are wrongly selected, can lead to misinterpretation errors [20].

2.2 Data Set

For the illustration of the arrow plot methodology was used a microarray experience for the identification of the gene expression differences between normal and pancreatic cancer samples.

GEO (Gene Expression Omnibus) database was selected, which is mostly used for the information of genetic expression from NCBI (National Center Biotechnology Information), the microarrays data were produced by the *Affymetrix*. The set chosen contained 54675 genes in 52 arrays, of which 36 were composed of tumoral samples and 16 of normal samples. From the 36 tumoral samples, 16 had tumor and normality data, while the other 20 had just tumoral data (available on https://www.ncbi.nlm.nih.gov/geo/query/acc. cgi?acc=GSE16515).

For the procedure was used an R code available in [20], and carefully adapted to this new data set.

3   Results and Discussion

To perform the analysis of DE genes using the arrow plot technique it was previously performed the evaluation of data quality and preprocessing through the background correction, normalization, PM correction and summarization.

The selection of DE genes was made by the simultaneous analysis of AUC and OVL.

The cut-off point considered for OVL was less than or equal to 0.4. For the AUC, genes with positive and negative regulation were defined, respectively, as having a value greater than 0.9 and less than 0.1. To select the "special genes" with DE, an AUC value between 0.4 and 0.6 was selected.

To compute the AUC and OVL estimates we use the algorithm proposed by Silva-Fortes [21]. Using a computer with an Intel Core i7-3667U 2.0 GHz processor and 8.00 GB of RAM the time spent for the estimations of kernel AUC and OVL was 2 h and 41 min.

The first arrow plot produced is in Fig. 2.

From 54675 genes, only 170 were differentially expressed, 10 genes with a possible biological interest ("special genes"), 153 genes with positive regulation and 7 genes with negative regulation (Table 1 and Fig. 3).

With kerAUC > 0.9 and OVL < 0.4 was considered to select up-regulated genes, corresponding to red dots on the plot. To select down-regulated genes an kerAUC < 0.1 and OVL < 0.4 was considered, corresponding to blue dots on

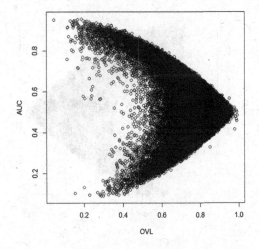

Fig. 2. Arrow plot with the pancreatic cancer data.

Table 1. Number of selected DE genes.

Genes	Condition	Total
With biological interest	$0.4 < $ kerAUC $ < 0.6$ and OVL < 0.4	10
Positive regulation	kerAUC > 0.9 and OVL < 0.4	153
Negative regulation	kerAUC < 0.1 and OVL < 0.4	7

Table 2. AUC and OVL values and bimodality group identification of the 7 selected special genes.

ID of gene	Name of gene	kerAUC	OVL	Group
214218_s_at	XIST	0.578	0.195	Both
224588_at	XIST	0.565	0.284	Control
224589_at	XIST	0.563	0.280	Experimental
224590_at	XIST	0.565	0.235	Control
226448_at	MIR1182	0.539	0.393	Experimental
227671_at	XIST	0.568	0.203	Experimental
241943_at	C18orf61	0.421	0.398	Control

the plot. To select special genes an OVL < 0.4 and $0.4 < $ kerAUC $ < 0.6$ was considered, corresponding to orange dots.

After analysing the bimodality, it was possible to verify that only 7 of the 10 initially selected really had biological interest by presenting bimodality in at least one of the groups, 3 genes had bimodality in the experimental group, another 3 had bimodality in the control group and only 1 gene had bimodality in both groups (Table 2).

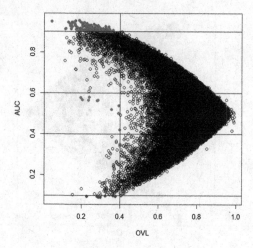

Fig. 3. Arrow plot of pancreatic cancer data. Points in red represent the genes with positive regulation (up-regulation), the blue dots represent the down-regulated genes, the green dots represent the candidate genes to have biological interest. (Color figure online)

Falling back on the Table 2, it is possible to identify the functions of the genes selected with biological interest. Five of the 7 genes selected are *XIST's* (X-inactive specific transcripts) which means that are RNA genes on the X chromosome of the placental humans that act as a major effector of the X inactivation process. The XIST gene has been shown to interact with *BRCA1* which is a human tumor suppressor gene and normally expressed in the cells of breast and other tissue.

Other gene that revealed biological interest was the *MIR1182* (microRNA 1182) and is involved in post-transcriptional regulation of gene expression by affecting both the stability and translation of mRNAs.

The last gene identified was the *C18orf61* which is a non-coding RNA gene, meaning that it is not translated into a protein so the protein synthesis does not occur. With the lack of protein synthesis, the organism has much more difficulties to react positively to chemotherapy.

In the Fig. 4, it is possible to see the *arrow plot* with the results mentioned above, where red dots on the plot represent positive regulated genes; blue dots represent negative regulated genes; orange dots represent genes with biological interest with bimodality in the experimental group; cyan dots represent bimodality in the control group; the green dot represent bimodality in both experimental and control groups and the black dots represent the 3 genes who possible had biological interest but after the bimodality analysis have not revealed bimodality in any of the groups.

Eleven feature selection methods mentioned above were applied to the full dataset. We assessed the overlap between gene lists produced by different feature selection methods and ranked lists of differentially expressed genes were produced. We examined the top 100 mostly highly ranked genes, and for all

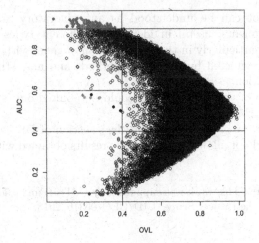

Fig. 4. Arrow plot of pancreatic cancer data with the multiple groups and levels expression. (Color figure online)

methods the 7 special genes selected by our methodology are missed. It should be noted that of 100 DE genes selected by the arrow plot method, the following results were obtained for each method: kerAUC 100; empAUC 93; RP 32; modT 84; ibmT 84; samT 84; Welch-T 78; samROC 85; AD 24; WAD 18 and FC 25.

4   Conclusions

Currently in the oncological research, microarray technology has been widely used. One of the main applications is the identification of genes associated with the development of a certain type of cancer. The ROC methodology revealed itself insufficient to identify the genes with biological interest and, consequently, was necessary to call upon the *arrow plot*.

The arrow plot is presented as a complementary tool to the ROC methodology in the analysis of DE genes as it is obtained by the plotting of the AUC against the OVL. It is a tool for microarray experiments, useful in the identification of different kinds of DE genes, and it is important in the identification of genes which are not detected by usual methods but can bring great biological information. This methodology allows us to see which genes have a special role in a particular pathological condition.

The bimodality analysis showed up to be an essential tool to distinguish the non DE genes who also presented an AUC around 0.5 and low values to the OVL.

This graph should be seen as an exploratory statistical tool and not as an inference tool. According Silva-Fortes [21], the purpose of the arrow plot is the visual identification of genes that may play a special role. No other method is able to achieve this goal.

The arrow plot can be understood as an exploratory graphical tool for microarray experiments, useful in identifying different types of differentially expressed genes, particularly in the identification of genes with a special behaviour that are not detected by the usual methods and may still bring relevant biological information.

One limitation of this method that could be pointed out is that the cut-off values used were arbitrary chosen.

Given the scarcity of information on the selection of DE genes for the data used, this work did not allow to compare the results obtained with those of other authors.

Acknowledgments. This work was supported by FCT - (Fundação para a Ciência e Tecnologia) within the Project Scope: UID/CEC/00319/2013.

References

1. Behzadi, P., Behzadi, E., Ranjbar, R.: Microarray data analysis. Albanian Med. J. **4**, 84–90 (2014)
2. Breitling, R., Armengaud, P., Amtmann, A., Herzyk, P.: Rank products: a simple, yet powerful, new method to detect differentially regulated genes in replicated microarray experiments. FEBS Lett. **573**, 83–92 (2004)
3. Bumgarner, R.: DNA microarrays: types, applications and their future. Curr. Protoc. Mol. Biol. **6137**(206), 1–17 (2013)
4. Clemons, T.E., Bradley, E.L.: Nonparametric measure of the overlapping coefficient. Comput. Stat. Data Anal. **34**(1), 51–61 (2000)
5. Cleves, M.A.: Comparative assessment of three common algorithms for estimating the variance of the area under the nonparametric receiver operating characteristic curve. Stata J. **2**(3), 280–289 (2002)
6. Cooper, M.G.: The Cell: A Molecular Approach. Sinauer Associates, Sunderland (2011)
7. Dalman, M.R., Deeter, A., Nimishakavi, G., Duan, Z.-H.: Fold change and p-value cutoffs significantly alter microarray interpretations. BMC Bioinform. **13**(2), 1–4 (2012)
8. Florkowski, C.M.: Sensitivity, specificity, receiver-operating characteristic (ROC) curves and likelihood ratios: communicating the performance of diagnostic tests. Clin. Biochem. Rev. **29**(Suppl 1), S83–S87 (2008)
9. Hariharan, R.: The analysis of microarray data. Pharmacogenomics **4**(4), 477–497 (2003)
10. Jeffery, I.B., Higgins, D.G., Culhane, A.C.: Comparison and evaluation of methods for generating differentially expressed gene lists from microarray data. BMC Bioinform. **7**, 359 (2006)
11. Kadota, K., Nakai, Y., Shimizu, K.: A weighted average difference method for detecting differentially expressed genes from microarray data. Algorithms Mol. Biol. AMB **3**, 8 (2008)
12. Larson, B.: Meet the Overlapping Coefficient: A Measure for Elevator Speeches, June 2014
13. Metz, C.E.: Basic principles of ROC analysis. Semin. Nucl. Med. **VIII**(4), 283–298 (1978)

14. National Center for Biotechnology Information: http://www.ncbi.nlm.nih.gov
15. National Human Genome Research Institute: https://www.genome.gov/10000533/dna-microarray-technology/
16. Park, S.H., Goo, J.M., Jo, C.-H.: Receiver operating characteristic (ROC) curve: practical review for radiologists. Korean J. Radiol. **5**(1), 11 (2004)
17. Parodi, S., Izzotti, A., Muselli, M.: Re: the central role of receiver operating characteristic (ROC) curves in evaluating tests for the early detection of cancer. J. Nat. Cancer Inst. **97**(7), 511–515 (2003)
18. Parodi, S., Pistoia, V., Muselli, M.: Not proper ROC curves as new tool for the analysis of differentially expressed genes in microarray experiments. BMC Bioinform. **9**, 410 (2008)
19. Pepe, M.S., Longton, G., Anderson, G.L., Schummer, M.: Selecting differentially expressed genes from microarray experiments. Biometrics **59**, 133–142 (2003)
20. Silva-Fortes, C.: Aplicação da metodologia ROC na análise de dados de microarrays. Ph.D. thesis, Faculdade de Ciéncias de Lisboa (2012)
21. Silva-Fortes, C., Turkman, M.A.A., Sousa, L.: Arrow plot: a new graphical tool for selecting up and down regulated genes and genes differentially expressed on sample subgroups. BMC Bioinform. **13**, 147 (2012)
22. Tusher, V.G., Tibshirani, R., Chu, G.: Significance analysis of microarrays applied to the ionizing radiation response. Proc. Nat. Acad. Sci. Unit. States Am. **98**, 5116–5121 (2001)

Applying Spatial Copula Additive Regression to Breast Cancer Screening Data

Elisa Duarte[1]([⊠]), Bruno de Sousa[2], Carmen Cadarso-Suárez[1],
Jenifer Espasandín-Domínguez[1], Oscar Lado-Baleato[1], Giampiero Marra[3],
Rosalba Radice[4], and Vítor Rodrigues[5]

[1] Unit of Biostatistics, Department of Statistics, Mathematical Analysis, and
Optimization, School of Medicine, University of Santiago de Compostela,
Santiago de Compostela, Spain
duarte.elisa@gmail.com
[2] Faculty of Psychology and Education Sciences, CINEICC,
University of Coimbra, Coimbra, Portugal
[3] Department of Statistical Science, University College London,
Gower Street, London WC1E 6BT, UK
[4] Department of Economics, Mathematics and Statistics, Birkbeck,
University of London, Malet Street, London WC1E 7HX, UK
[5] Faculty of Medicine, University of Coimbra, Coimbra, Portugal

Abstract. Breast cancer is associated with several risk factors.
Although genetics is an important breast cancer risk factor, environmental and sociodemographic characteristics, that may differ across populations, are also factors to be taken into account when studying the disease. These factors, apart from having a role as direct agents in the risk of the disease, can also influence other variables that act as risk factors. The age at menarche and the reproductive lifespan are considered by the literature as breast cancer risk factors so that, there are several studies whose aim is to analyze the trend of age at menarche and menopause along generations. Also, it is believed that these two moments in a woman's life can be affected by environmental, social status, and lifestyles of women. Using the information of 278,282 registries of women which entered in the breast cancer screening program in Central Portugal, we developed a bivariate copula model to quantify the effect a woman's year of birth in the association between age at menarche and a woman's reproductive lifespan, in addition to explore any possible effect of the geographic location in these variables and their association. For this analysis we employ Copula Generalized Additive Models for Location, Scale and Shape (CGAMLSS) models and the inference was carried out using the R package SemiParBIVProbit.

1 Introduction

The age at menarche and age at menopause are well known breast cancer risk factors, since these moments set a woman's reproductive lifespan, during which the woman is exposed to endogenous hormones responsible to ensure the regular

© Springer International Publishing AG 2017
O. Gervasi et al. (Eds.): ICCSA 2017, Part II, LNCS 10405, pp. 586–599, 2017.
DOI: 10.1007/978-3-319-62395-5_40

functioning of her reproductive system. There are several factors that affect the beginning and the end of a woman's reproductive lifespan. A downward trend in the age at menarche has been highlighted in recent researches [15,18]. The results of a study conducted by Nichols et al. [11], shows an upward trend in the number of a woman's reproductive years.

The natural menopause is defined as a complex bio-social and bio-cultural phenomenon [4]. Other studies, such as the one presented in [21], analyse the association between age at menarche and socio-economic characteristics showing that the environmental conditions may have influence in the onset of a woman's reproductive lifespan. Therefore, it should be of the utmost importance to explore how individual characteristics such as age at menarche and a woman's reproductive lifespan can be a reflection of environment factors. In addition, rather than analyse the effect of a woman's cohort and of the environment in the age at menarche and reproductive lifespan as two separated response variables, quantify these effects in the association between them is a major topic. Thus, to address this problem methodological approaches are needed, enabling a joint analysis of the age of menarche and reproductive lifespan in a regression context that allows to specify the parameters of the marginal distributions as functions of additive predictors including several types of covariates as non-linear and spatial effects.

The newest methodological developments to approach multivariate response distributions are based on Copula theory [7,16,17]. In the current work we employ Copula Generalized Additive Models for Location, Scale and Shape [9,10] to build a spatial model, in order to study the effect of a woman's year of birth and place of residence on relevant parameters of the joint distribution of age at menarche and a woman's reproductive lifespan. The copula approach allows one to model the marginal distributions and the dependence structure of a multivariate distribution using a functional association that binds them together and which is not influenced by marginal behaviour. The choice of the CGAMLSS is based on the fact that such models extend the scope of univariate GAMLSS by binding two equations with binary, discrete or continuous responses. The equations can be flexibly specified using smoothers with single or multiples penalties, thus allowing for several types of covariate effects. The copula dependence parameter can also be specified as a function of flexible covariate effects. All the model's parameters are estimated simultaneously. The inference is carried out using the R package SemiParBIVProbit [8].

2   Breast Cancer Screening Data

This study is based on data provided by the Central Regional Nucleus of the Portuguese Cancer League (LPCC-NRC), sponsored by the Breast Cancer Screening Program (BCSP) in 78 municipalities located in central Portugal. Figure 1 shows the map of Portugal, with the blue regions representing the municipalities under study. The database consists of 278,282 women who were registered for the BCSP in central Portugal between 1990 and 2010.

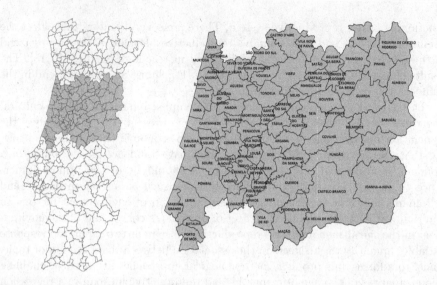

Fig. 1. The map of Portugal, with the blue regions representing the municipalities under study. (Color figure online)

Women considered in this study have a screening age between 45 and 69, with 76% (212,517) of them reaching menopause. Since we are dealing with the reproductive lifespan cycle of a woman, only the post-menopausal women were considered in the study.

The variables involved in this study are: age of menarche, a woman's reproductive lifespan cycle (calculated by subtracting the age of menarche from the age of menopause), year of birth, and the code of the municipality where a woman resides. Table 1 shows a summary description of these variables.

Table 1. Statistics of the variables in the study.

Variable	Mean	Standard Deviation (SD)	Min-Max
Birth year	1946	9.8	1920–1965
Age of menarche	13.3	8.0	8–18
Reproductive life span	34.9	5.5	3–50

3 Bivariate Copula Regression Models (CGAMLSS)

The main goal of this study is to apply the Bivariate Copula Additive Models for Location, Scale and Shape in order to explain the dependence structure of a bivariate response consisting of age at menarche and a woman's reproductive lifespan regressing the complete distributional of the response on the year of birth and a woman's place of residence. The Bivariate Copula Additive Models for Location, Scale and Shape extends the use of GAMLSS [14] models to situations in which two responses are modeled simultaneously conditional on some covariates [9]. Using additive predictors, casting several types of covariates such as nonlinear effects of continuous covariates, random effects, interactions or spatial dependence, the approach allows to model a bivariate response through of a copula function. Besides that, the regression is not restrict to the response expectation, being able to be extended to other distributional parameters.

One of the strengths of the copula approach is the possibility of the marginal distribution be of different families, providing different types of response distributions (continuous, discrete, and mixed discrete continuous) as opposed to the classical statistical bivariate response models, that assume each marginal response as Gaussian. A complete description of the CGAMLSS theory can be found in the recent work of Marra and Radice [9,10].

In the present application of the CGAMLSS, it is considered two bivariate continuous responses, Y_1 and Y_2, representing, respectively, the age at menarche and the reproductive lifespan and covariate information (year of birth and a woman's place of residence) collected in the generic vector z_i. The joint cumulative distribution function (cdf) of Y_1 and Y_2 can be expressed in terms of the marginal cdfs of Y_1 and Y_2 and a copula function C that binds them together [9] as follows:

$$F(y_1, y_2 | \vartheta) = C\left(F_1(y_1 | \mu_1, \sigma_1, \nu_1), F_2(y_2 | \mu_2, \sigma_2, \nu_2); \zeta, \theta\right), \tag{1}$$

where $\vartheta = (\mu_1, \sigma_1, \nu_1, \mu_2, \sigma_2, \nu_2, \zeta, \theta)^T$, $F_1(y_1 | \mu_1, \sigma_1, \nu_1)$ and $F_2(y_2 | \mu_2, \sigma_2, \nu_2)$ are the marginal cdfs of Y_1 and Y_2 taking values in $(0, 1)$, μ_m, σ_m, ν_m, for $m = 1, 2$ are the marginal distribution parameters. $C(\cdot, \cdot)$ is a uniquely defined two-place copula function which does not depend on the marginals, and θ is an association copula parameter measuring the dependence between the two random variables [16,17] and ζ represents in this case the number of degrees of freedom of the Student-t copula (which only appears in C and ϑ when such copula is employed). A summary of the copulas currently implemented in the package SemiParBIVProbit are reported in Table 2.

By considering a suitable additive predictors $\eta's$ for all parameters of the bivariate response distribution defined above for an observation i, the predictor could be written as:

$$\eta_i = \beta_0 + \sum_{k=1}^{K} s_k(\mathbf{z}_{ki}), \ i = 1, \dots, n, \dots \tag{2}$$

where $\beta_0 \in \mathbb{R}$ is an overall intercept, and the function s_k represent the different covariate effects (as binary, categorial, continuous and spatial variables). The K functions s are chosen according the type of covariate considered ($\mathbf{z}_{ki}$).

As defined in Generalize Additive Models (GAM [20]), each function s_k can be approximated as a linear combination of J_k basis functions $b_{kj_k}(\mathbf{z}_{ki})$ and regression coefficients $\beta_{kj_k} \in \mathbb{R}$, i.e.

Table 2. Some classic copulae functions, with corresponding parameter range of association parameter θ and relation between Kendall's τ and θ. $\Phi_2(\cdot,\cdot;\theta)$ denotes the cdf of a standard bivariate normal distribution with correlation coefficient θ, and $\Phi(\cdot)$ the cdf of a univariate standard normal distribution. $D_1(\theta) = \frac{1}{\theta}\int_0^\theta \frac{t}{\exp(t)-1}dt$ is the Debye function and $D_2(\theta) = \int_0^1 t\log(t)(1-t)^{\frac{2(1-\theta)}{\theta}}dt$. Rotated versions of the Clayton, Gumbel and Joe copulae can also be obtained following [1].

Copula	$C(u,v;\zeta,\theta)$	Ranges of θ and ζ	Transformation	Kendall's τ
AMH	$\frac{uv}{1-\theta(1-u)(1-v)}$	$\theta \in [-1,1]$	$\tanh^{-1}(\theta)$	$-\frac{2}{3\theta^2}\{\theta+(1-\theta)^2\log(1-\theta)\}+1$
Clayton	$(u^{-\theta}+v^{-\theta}-1)^{-1/\theta}$	$\theta \in (0,\infty)$	$\log(\theta-\epsilon)$	$\frac{\theta}{\theta+2}$
FGM	$uv\{1+\theta(1-u)(1-v)\}$	$\theta \in [-1,1]$	$\tanh^{-1}(\theta)$	$\frac{2}{9}\theta$
Frank	$-\theta^{-1}\log\{1+(e^{-\theta u}-1)(e^{-\theta v}-1)/(e^{-\theta}-1)\}$	$\theta \in \mathbf{R}\setminus\{0\}$	–	$1-\frac{4}{\theta}[1-D_1(\theta)]$
Gaussian	$\Phi_2(\Phi^{-1}(u),\Phi^{-1}(v);\theta)$	$\theta \in [-1,1]$	$\tanh^{-1}(\theta)$	$\frac{2}{\pi}\arcsin(\theta)$
Gumbel	$\exp\left[-\{(-\log u)^\theta+(-\log v)^\theta\}^{1/\theta}\right]$	$\theta \in [1,\infty)$	$\log(\theta-1)$	$1-\frac{1}{\theta}$
Joe	$1-\{(1-u)^\theta+(1-v)^\theta-(1-u)^\theta(1-v)^\theta\}^{1/\theta}$	$\theta \in (1,\infty)$	$\log(\theta-1-\epsilon)$	$1+\frac{4}{\theta^2}D_2(\theta)$
Student-t	$t_{2,\zeta}(t_\zeta^{-1}(u),t_\zeta^{-1}(v);\zeta,\theta)$	$\theta \in [-1,1]$, $\zeta \in (2,\infty)$	$\tanh^{-1}(\theta)$, $\log(\zeta-2-\epsilon)$	$\frac{2}{\pi}\arcsin(\theta)$

$$\sum_{j_k=1}^{J_k} \beta_{kj_k} b_{kj_k}(\mathbf{z}_{ki}) \tag{3}$$

Equation (3) implies that the vector of evaluations $\{s_k(z_{k1}),...,s_k(z_{kn})\}^T$ can be written as $\mathbf{Z}_k\boldsymbol{\beta}_k$ with $\boldsymbol{\beta}_k = (\beta_{k1},...,\beta_{kJ_k})^T$ and the design matrix $\mathbf{Z}_k[i,j_k] = b_{kj_k}(\mathbf{z}_{ki})$. This allows the predictor in Eq. (2) to be written as:

$$\boldsymbol{\eta} = \beta_0\mathbf{1}_n + \mathbf{Z}_1\boldsymbol{\beta}_1 + ... + \mathbf{Z}_k\boldsymbol{\beta}_k \tag{4}$$

where $\mathbf{1}_n$ is an n-dimensional vector made up of ones.

Equation (4) can be written in matrix notation as $\boldsymbol{\eta} = \mathbf{Z}\boldsymbol{\beta}$ where $\mathbf{Z} = (\mathbf{1}_n, \mathbf{Z}_1,...,\mathbf{Z}_K)$ and $\boldsymbol{\beta} = (\beta_0, \boldsymbol{\beta}_1^T,...,\boldsymbol{\beta}_K^T)^T$. Each $\boldsymbol{\beta}_k$ has an associated quadratic penalty $\lambda_k\boldsymbol{\beta}_k^T\mathbf{D}_k\boldsymbol{\beta}_k$ whose role is to enforce specific properties on the kth function, such as smoothness. It is important to note that $\mathbf{D}_k$ only depends on the choice of basis functions. The smoothing parameter $\lambda_k \in [0,\infty)$ controls the trade-off between fit and smoothness, and plays a crucial role in determining the shape of $\hat{s}_k(\mathbf{z}_{ki})$. The overall penalty can be defined as $\boldsymbol{\beta}^T\mathbf{D}\boldsymbol{\beta}$, where $\mathbf{D} = \text{diag}(0, \lambda_1\mathbf{D}_1,...,\lambda_K\mathbf{D}_K)$.

To model spatial information, [10] proposed the use of a Markov random field smoother, that is useful in our application where we have the spatial information split up in discrete contiguous geographic units. In this case, Eq. (4) becomes $\mathbf{z}_{ki}^T\boldsymbol{\beta}_k$ where $\boldsymbol{\beta}_k = (\beta_{k1},...,\beta_{kR})^T$ represents the vector of spatial effects,

R denotes the total number of regions z_{ki}. Thus, the design matrix linking an observation i with the corresponding spatial effect is defined as:

$$Z_k[i,r] = \begin{cases} 1 \text{ if the observation belongs to region r} \\ 0 \text{ otherwise} \end{cases}$$

where $r = 1, \ldots, R$. The smoothing penalty D_λ associated with the Markov random field is constructed based on the neighborhood structure of the geographic units:

$$D_k[r,q] = \begin{cases} -1 \text{ if } r \neq q \wedge \text{r and q are adjacent neighbors} \\ 0 \quad \text{if } r \neq q \wedge \text{r and q are not adjacent neighbors} \\ N_r \text{ if } r = q \end{cases}$$

where r and q are two regions and N_r the total number of regions.

The non-linear effects of continuous covariates, are modeled using the regression spline approach popularized by [3], with a second order penalties.

3.1 Inference in the CGAMLSS Models

The inference is based on penalised maximum likelihood estimation. First, it is considered the log-likelihood function for a copula model with two continuous margins [7]:

$$l(\delta) = \sum_{i=1}^{n} log \left\{ C(F_{1i}(y_{1i}|\mu_{1i}, \sigma_{1i}, \nu_{1i}), F_2(y_{2i}|\mu_{2i}, \sigma_{2i}, \nu_{2i}); \zeta_i, \theta_i) \right\}$$

$$+ \sum_{i=1}^{n} \sum_{m=1}^{2} log \left\{ f_m(y_{mi} \mid \mu_{mi}, \sigma_{mi}, \nu_{mi})) \right\}$$

where parameter δ is defined as $(\beta_{\mu 1}^T, \beta_{\mu 2}^T, \beta_{\sigma 1}^T, \beta_{\sigma 2}^T, \beta_{\nu 1}^T, \beta_{\nu 2}^T, \beta_{\zeta}^T, \beta_{\theta}^T)^T$.

The use of a classic unpenalized optimization algorithm is likely to result unduly wiggly estimates, therefore [9] proposed a penalised maximum likelihood estimation of the form:

$$l_p(\delta) = l(\delta) - \frac{1}{2}\delta^T S_\lambda \delta \tag{5}$$

where

$$S_\lambda = \text{diag}(\lambda_{\mu 1}D_{\mu 1}, \lambda_{\mu 2}D_{\mu 2}, \lambda_{\sigma 1}D_{\sigma 1}, \lambda_{\sigma 2}D_{\sigma 2}, \lambda_{\nu 1}D_{\nu 1}, \lambda_{\nu 2}D_{\nu 2}, \lambda_{\zeta}D_{\zeta}, \lambda_{\theta}D_{\theta})$$

with each smoothing parameters related to the corresponding D component and the overall λ is defined as $(\lambda_1, \ldots, \lambda_K)^T$.

In [9] a two-step algorithm using a trust region algorithm with integrated multiple smoothing parameter selection was proposed to maximize the Eq. (5):

1. At a iteration index a, holding $\boldsymbol{\lambda}$ fixed at a vector of values and for a given $\boldsymbol{\delta}^{[a]}$ they maximize Eq. (5) using a Trust region algorithm, as follows:

$$\min_{\boldsymbol{p}} \breve{l}_p \overset{\text{def}}{=} -\{l_p(\boldsymbol{\delta}^{[a]}) + \boldsymbol{p}^T \boldsymbol{g}_p^{[a]} + \frac{1}{2}\boldsymbol{p}^T \boldsymbol{H}_p^{[a]} \boldsymbol{p}\} \quad \text{so that} \quad \|\boldsymbol{p}\| \leq \Delta^{[a]}$$

$$\boldsymbol{\delta}^{[a]} = \arg\min_{\boldsymbol{p}} \ \breve{l}_p(\boldsymbol{\delta}^{[a]}) + \boldsymbol{\delta}^{[a]} \tag{6}$$

where $\boldsymbol{g}_p^{[a]}$ and the $\boldsymbol{H}_p^{[a]}$ are the gradient vector and Hessian matrix of the log-likelihood function penalized by the $\boldsymbol{S}_{\hat{\lambda}}$ matrix (vector $\mathbf{g}(\boldsymbol{\delta}^{[a]})$ consists of $\mathbf{g}_{\mu_1}(\boldsymbol{\delta}^{[a]}) = \partial \ell(\boldsymbol{\delta})/\partial \beta_{\mu_1}|_{\beta_{\mu_1} = \beta_{\mu_1}^{[a]}}, \dots, \mathbf{g}_{\theta}(\boldsymbol{\delta}^{[a]}) = \partial \ell(\boldsymbol{\delta})/\partial \beta_{\theta}|_{\beta_{\theta} = \beta_{\theta}^{[a]}}$, the Hessian matrix has elements $\boldsymbol{H}(\boldsymbol{\delta}^{[a]})_{o,h} = \partial^2 \ell(\boldsymbol{\delta})/\partial \beta_o \partial \beta_h |_{\beta_o = \beta_o^{[a]}, \beta_h = \beta_h^{[a]}}$ where $o, h = \mu_1, \mu_2, \sigma_1, \sigma_2, \nu_1, \nu_2, \zeta, \theta)$, $\|\cdot\|$ the euclidean norm and $\Delta^{[a]}$ is the radius of the trust region algorithm of each iteration. Trust region algorithms are generally more stable and faster than the line search methods such as Newton-Raphson, particularly for functions that are, for example, non-concave and/or exhibit regions that are close to flat [12].

2. Holding the model's parameter vector value fixed at $\boldsymbol{\delta}^{[a+1]}$, solve the problem:

$$\boldsymbol{\lambda}^{[a+1]} = \arg\min_{\boldsymbol{\lambda}} \ \|\mathbf{M}^{[a+1]} - \mathbf{A}^{[a+1]}\mathbf{M}^{[a+1]}\|^2 - \check{n} + 2\text{tr}(\mathbf{A}^{[a+1]}) \tag{7}$$

where $\mathbf{M}^{[a+1]} = \sqrt{-\boldsymbol{H}(\boldsymbol{\delta}^{[a+1]})}\boldsymbol{\delta}^{[a+1]} + \sqrt{-\boldsymbol{H}(\boldsymbol{\delta}^{[a+1]})}^{-1}\mathbf{g}(\boldsymbol{\delta}^{[a+1]})$,
$\mathbf{A}^{[a+1]} = \sqrt{-\boldsymbol{H}(\boldsymbol{\delta}^{[a+1]})}\left(-\boldsymbol{H}(\boldsymbol{\delta}^{[a+1]}) + \mathbf{S}\right)^{-1}\sqrt{-\boldsymbol{H}(\boldsymbol{\delta}^{[a+1]})}$, $\text{tr}(\mathbf{A}^{[a+1]})$ is the number of effective degrees of freedom (edf) of the penalized model and $\check{n} = 5n$ (if we employ a two parameter distribution and a gaussian copula).
This expression is equivalent to the Un-Biased Risk Estimator (UBRE) [10, 20], solved with the methodology proposed in [19].

3.2   Model Building

The adequate selection of the distributions to use for the margins of the bivariate response and the copula function that best modelize the structure of dependence between this margins is a dificulty in the formulation of these types of models that the researcher needs to solve.

In our study, from the continuous distribution families available in the R package `SemiParBIVProbit` [8], a Log-normal distribution for the age at menarche and a Gumbel to the reproductive lifespan of the woman were chosen. This choice was based on the AIC (Akaike information criterion) and on the BIC (Bayesian information criterion) following [9]. Both distributions are defined by two parameters: a location parameter μ and a scale parameter σ^2.

$$\begin{cases} \eta_i^{\mu_1} = \beta_{0_i}^{\mu_1} + s_i^{\mu_1}(Year\ of\ birth) + s_i^{\mu_1}(Municipality) \\ \eta_i^{\sigma_1^2} = \beta_{0_i}^{\sigma_1^2} + s_i^{\sigma_1^2}(Year\ of\ birth) + s_i^{\sigma_1^2}(Municipality) \\ \eta_i^{\mu_2} = \beta_{0_i}^{\mu_2} + s_i^{\mu_2}(Year\ of\ birth) + s_i^{\mu_2}(Municipality) \\ \eta_i^{\sigma_2^2} = \beta_{0_i}^{\sigma_2^2} + s_i^{\sigma_2^2}(Year\ of\ birth) + s_i^{\sigma_2^2}(Municipality) \\ \eta_i^{\theta} = \beta_{0_i}^{\theta} + s_i^{\theta}(Year\ of\ birth) + s_i^{\theta}(Municipality) \end{cases} \quad (8)$$

The two first equations refer to the location and scale parameter of the age at menarche, the next two refer to the location and scale parameter of the reproductive lifespan of women and the last one refers to the association between both variables. All parameters were modeled using an additive composition of intercept β_0 representing the overall level of the predictor, and smooth effects $s_i(z)$ reflecting non-linear effects of the continuous covariate (Year of Bith) and the spatial information (Municipality).

Model (8) can be easily computed in R [13] by using the copulaReg() function in the R package SemiParBIVProbit:

```
> mu1 = Menarche ~ s(Year of Birth)+s(Municipality,bs="mrf",
  xt=map)
> mu2 = Life.span ~ s(Year of Birth)+s(Municipality,bs="mrf",
  xt=map)
> sd1 = ~s(Year of Birth)+s(Municipality,bs="mrf",xt=map)
> sd2 = ~s(Year of Birth)+s(Municipality,bs="mrf",xt=map)
```

Table 3. Comparison of AIC, BIC and Run-time values under different copula assumptions. Run-Time is the time that the model required to reach the optimal estimation of the regression parameters in a computer with a Intel(R) Core(TM) i5-4570s CPU 2.90 GHz with windows 7 Professional.

Family	AIC	BIC	Run.time
Gaussian	2100208	2103537	24' 15"
Gumbel (90)	2101223	2101223	34' 28"
Frank	2101914	2105217	28' 17"
Clayton (180)	2102480	2105722	52' 05"
Gumbel (270)	2105144	2108527	54' 46"
Joe (90)	2105342	2108590	43' 04"
Clayton (270)	2108901	2112265	29' 32"
Joe (270)	2111979	2115350	33' 17"
Clayton	2123078	2125939	49' 33"
Joe	2123078	2125939	44' 13"
Joe (180)	2123078	2125939	46' 39"
Gumbel	2123078	2125939	50' 27"
Gumbel (180)	2123078	2125939	52' 00"

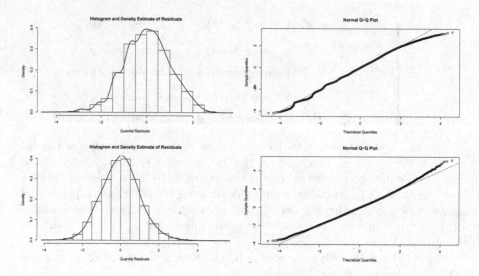

Fig. 2. Quantile residuals of the selected model

```
> theta = ~s(Year of Birth)+s(Municipality,bs="mrf",xt=map)
> f = list(mu1,mu2,sd1,sd2,theta)
> model = copulaReg(f,margin=c("LN","GU"),data)
```

where s are smooths functions of Year of bith and the Municipality, the former was modeled using penalized low rank regression splines with default second order penalties and the latter using a Markov random field smoother.

Concerning the choice of the copula function in terms of AIC, BIC and runtime the gaussian seems to be the best one, see Table 3.

Marginal residual plots and QQ-plots of normalized quantile residuals obtained from the fit model are shown in Fig. 2 we see plots that look normally distributed as expected for a good model fit.

4   Results

The estimates of the smooth effect of the year of birth with the associated 95% point-wise intervals, and the spatial effects on the parameters of the marginal distributions of the variables age at menarche and a woman's reproductive lifespan, are presented in Figs. 3, 4, 5, 6 and 7. The left side of Fig. 3 shows a clear decreasing effect of the year of birth on the expectation of the age at menarche. Regarding to the effect on the expectation of a woman's reproductive lifespan, Fig. 4 shows an increasing effect along the cohorts before 1952, followed by a sharp decrease. The spatial effects presented in right side of the same figures, show the inland regions of central Portugal associated with lower ages at menarche and higher reproductive lifespans.

The effects of the year of birth on the variance of the marginal distributions of the age at menarche and reproductive lifespan, are shown in the left side of

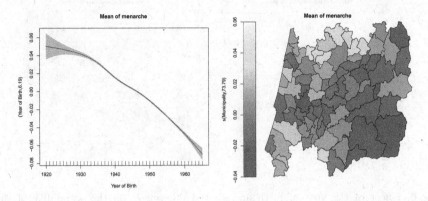

Fig. 3. Effect of the year of birth and spatial differences on the age at menarche.

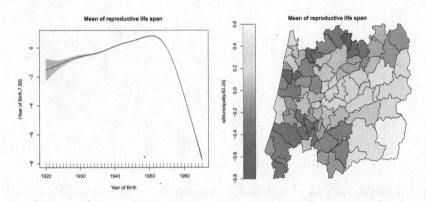

Fig. 4. Effect of the year of birth and spatial differences on the reproductive lifespan.

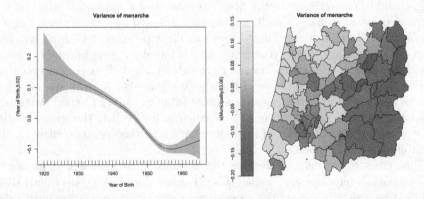

Fig. 5. Effect of the year of birth and spatial differences on the variability of the age at menarche.

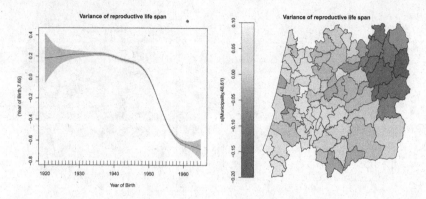

Fig. 6. Effect of the year of birth and spatial differences on the variability of the reproductive lifespan.

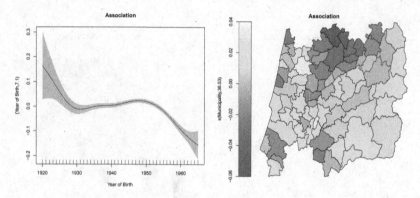

Fig. 7. Effect of the year of birth and spatial differences on the association between the menarche and reproductive lifespan.

Figs. 5 and 6, respectively. In the first, it is observed a downward effect on the variance of age at menarche until the year 1955 followed by a slight upward effect. The second, shows a steady effect of year of birth prior to 1938, from which it starts decreasing up to the year 1949, and followed by a significantly decline. Looking to the spatial effects on the right side of Fig. 5, it is clear a east-west increasing effect of the variance of the age at menarche. A weak spatial effect of a woman's reproductive lifespan variance in almost all of Central Portugal's municipalities is depicted in Fig. 6. Nevertheless, notice that the municipalities located in the northeast part of the region have a marked negative effect on the variability of the lifespan.

The effect of the year of birth on the association between age at menarche and a woman's reproductive lifespan (Fig. 7) shows a decreasing effect until 1930, followed by a slow increase from a negative to a positive association until 1950, decreasing afterwards. Regarding the spatial effects (right side of Fig. 7), only a couple of municipalities in the north part of the region show a negative effect in this association.

5   Discussion

This study was conducted in order to apply the CGAMLSS to quantify the effect of a woman's cohort and of the environment (represented by a woman's place of residence) in the association between age at menarche and a woman's reproductive lifespan and the effects of this covariates in the location and scale parameters of the marginal distributions. This approach allows to assess easily a suitable marginal distributions for the response variables, and find the best fitted copula additive model.

The package SemiParBIVProbit [8] has shown a good performance in the fit of models with big datasets such as the breast cancer screening, including complex covariates in the predictors equations, as the spatial effects. To our knowledge, this is the first time that the CGAMLSS is applied to a database with a considerable number of records, with the selected copula function converging in 24' 15' in a Intel(R) Core(TM) i5-4570s CPU 2.90 GHz with operating system Windows 7 Professional.

The results achieved suggested that earlier menarche is associated with younger women and an increasing effect of the year of birth in a woman's reproductive lifespan, showing that the pattern reported in the literature [11,15,18] is also observed for Portuguese women of the central region. Although, the increasing effect in in a woman's reproductive lifespan is followed by a sharp decrease for women born after 1952. This drop is justified by the fact that women born after 1952 are those cases who already reported a menopause, despite of their young age. The decreasing effect of year of birth in the variability of the age at menarche may be explained by the fact that that moment is easily remembered by younger women. Since the west region in central Portugal is in general wealthier and more economically developed region in comparison to interior part, the expectation of an increasing effect on a woman's reproductive lifespan and a decreasing one on age at menarche in the east-west direction was not verified. These may be explained by factors that were not taken into account in the model, leading to the conclusion that a woman's place of residence is not the only factor that may affect women's individual characteristics, such as age at menarche and a woman's reproductive lifespan cycle. This issue should be considered in future analysis involving spatial effects. Nevertheless, the CGAMLSS were found to be an important tool when modeling epidemiological data where a spatial covariate should be considered in the joint analysis of two variables of interest.

Duarte et al. [2] made a similar approach to this problem carrying out a joint analysis of the age of menarche and reproductive lifespan using a Bayesian structured additive distributional regression for multivariate responses [5], where the spatial effect was evaluated at each cohort through the introduction of varying coefficients. Since CGAMLSS allows the inclusion of varying coefficients in the predictor equations, in future research it will be of interest, assess the results achieved with the two methodologies and also compare their performance with such big datasets.

Acknowledgments. This work was financed by Spanish Ministry of Science and Innovation grant MTM2015-69068-REDT, and the projects MTM2014-52975-C2-1-R cofinanced by the Ministry of Economy and Competitiveness (SPAIN) and the European Regional Development Fund (FEDER).

References

1. Brechmann, E.C., Schepsmeier, U.: Modeling dependence with C-and D-vine copulas: the R-package CDVine. J. Stat. Softw. **52**(3), 1–27 (2013)
2. Duarte E., de Sousa B., Cadarso-Suárez C., Klein N., Kneib T., Rodrigues V.: Studying the relationship between a woman's reproductive lifespan and age at menarche using a Bayesian multivariate structured additive distributional regression model. Biometrical J. (in press)
3. Eilers, P.H., Marx, B.D.: Flexible smoothing with B-splines and penalties. Stat. Sci. **11**(2), 89–102 (1996)
4. Kaczmarek, M.: The timing of natural menopause in Poland and associated factors. Maturitas **57**, 139–153 (2007)
5. Klein, N., Kneib, T., Klasen, S., Lang, S.: Bayesian structured additive distributional regression for multivariate responses. J. Roy. Stat. Soc.: Ser. C (Appl. Stat.) **64**, 1024–1052 (2015). doi:10.1214/15-AOAS823
6. Klein, N., Kneib, T.: Simultaneous inference in structured additive conditional copula regression models: a unifying Bayesian approach. Stat. Comput. **26**, 841–860 (2016)
7. Kolev, N., Paiva, D.: Copula-Based Regression Models. Department of Statistics, University of São Paulo, nkolev@ime.usp.br (2007)
8. Marra, G., Radice, R., Marra, G.: Package SemiParBIVProbit (2016)
9. Marra, G., Radice, R.: A Bivariate Copula Additive Model for Location, Scale and Shape. Cornell University Library (2017). arxiv:1605.07521 [stat.ME]
10. Marra G., Radice R., Barnighausen T., Wood S.N., McGovern M.E.: A simultaneous equation approach to estimating hiv prevalence with non-ignorable missing responses. J. Am. Stat. Assoc. (in press)
11. Nichols, H.B., Trentham-Dietz, A., Hampton, J.M., Titus-Ernstoff, L., Egan, K.M., Willett, W.C., Newcomb, P.A.: From menarche to menopause: trends among US women Born from 1912 to 1969. Am. J. Epidemiol. **164**(10), 1003–1011 (2006). doi:10.1093/aje/kwj282
12. Nocedal, J., Wright, S.J.: Numerical Optimization. Springer-Verlag, New York (2006)
13. R Core Team. R: A language and environment for statistical computing. R Foundation for Statistical Computing, Vienna, Austria (2016). https://www.R-project.org/
14. Rigby, A., Stasinopoulos, D.M.: Generalized additive models for location, scale and shape (with discussion). Appl. Stat. **54**, 507–554 (2005)
15. Rigon, F., Bianchin, L., Bernasconi, S., Bona, G., Bozzola, M., Buzi, F., et al.: Update on age at menarche in Italy: toward the leveling off of the secular trend. J. Adolesc. Health **46**, 238–244 (2010)

16. Sklar, A.: Fonctions de répartition é n dimensions et leurs marges. Publications de l'Institut de Statistique de l'Université de Paris **8**, 229–231 (1959)
17. Sklar, A.: Random variables, joint distributions, and copulas. Kybernetica **9**, 449–460 (1973)
18. Talma, H., Schönbeck, Y., van Dommelen, P., Bakker, B., van Buuren, S., HiraSing, R.A.: Trends in menarcheal age between 1955 and 2009 in the Netherlands. PLoS ONE **8**(4), e60056 (2013). doi:10.1371/journal.pone.0060056
19. Wood, S.N.: Stable and efficient multiple smoothing parameter estimation for generalized additive models. J. Am. Stat. Assoc. **99**(467), 673–686 (2004)
20. Wood, S.: Generalized Additive Models: An Introduction with R. CRC Press, Boca Raton (2006)
21. Wronka, I., Pawlińska-Chmara, R.: Menarcheal age and socio-economic factors in Poland. Ann. Hum. Biol. **32**(5), 630–638 (2005). doi:10.1080/03014460500204478

A Simple Graphical Way of Evaluating Coverage and Directional Non-coverages

Adelaide Freitas[1(✉)], Sara Escudeiro[2], and Vera Afreixo[1]

[1] University of Aveiro, Aveiro, Portugal
{adelaide,vera}@ua.pt
[2] Polytechnic of Coimbra (ESAC), Coimbra, Portugal
sarae@esac.pt

Abstract. Evaluation of the coverage probability and, more recently, of the intervalar location of confidence intervals, is a useful procedure if exact and asymptotic methods for constructing confidence intervals are used for some populacional parameter. In this paper, a simple graphical procedure is presented to execute this kind of evaluation in confidence methods for linear combinations of k independent binomial proportions. Our proposal is based on the representation of the mesial and distal non-coverage probabilities on a plane. We carry out a simulation study to show how this graphical representation can be interpreted and used as a basis for the evaluation of intervalar location of confidence interval methods.

Keywords: Coverage probability · Mesial and distal non-coverage probabilities

1 Introduction

Inference involving more than one population parameter is very common in Statistics. For instance, the effect of the interaction between the presence and absence of two treatments A and B can be established in terms of a linear combination of four independent binomial proportions, $p_1 - p_2 - p_3 + p_4$, where each p_i, $i = 1, 2, 3, 4$, denotes the unknown population proportion in one of four possible groups. In order to analyze the existence of interaction between treatments A and B, the following statistical test could be carried out:

$$H_0 : p_1 - p_2 - p_3 + p_4 = 0 \quad vs \quad H_1 : p_1 - p_2 - p_3 + p_4 \neq 0. \quad (1)$$

Moreover, due to the dual relationship between statistical tests and confidence intervals (CIs), this testing problem can also be addressed in terms of the CI for $p_1 - p_2 - p_3 + p_4$.

In this paper, we deal with the evaluation of the performance of asymptotic methods used to construct two-sided CIs involving two or more population parameters. In particular, we focus our study on proportions. Asymptotic methods are generally preferred because they are computationally simpler and faster than

© Springer International Publishing AG 2017
O. Gervasi et al. (Eds.): ICCSA 2017, Part II, LNCS 10405, pp. 600–610, 2017.
DOI: 10.1007/978-3-319-62395-5_41

exact ones. Several approximate methods have been proposed in the literature for constructing confidence intervals (CIs) for the difference of two independent binomial proportions [1–6]. However, few authors have been discussing approximate methods for obtaining CIs for any linear combination of two [7,8] and more than two [9–13] independent binomial populations. Within the context of investigating the properties of each of the different approaches to construct CIs, the performance of each method is commonly evaluated through simulation studies. Such evaluations are usually based on the exact coverage probabilities of each method. More recently [13,15,16], the expected interval location of the CIs, which is based on the mesial and distal non-coverage probabilities, has also been considered as an important performance measure. In the present work we discuss a graphical representation of the two directional non-coverage probabilities, aimed at facilitating the characterization of interval location.

The paper is organized as follows. In Sect. 2, a brief overview of four variants of the classic Wald CI is provided. In Sect. 3, evaluative indexes related to the directional non-coverages are highlighted. In Sect. 4 we propose a graphical technique to evaluate the CIs location. Section 5 shows some examples of the application of this graphical technique. Finally, we summarize our findings in Sect. 6.

2   Classic and Adjusted Wald CIs for Linear Combinations of Proportions

Due to the dual relationship between statistical tests and CIs, the most common approach to obtain large-sample interval estimates for a combination $L = \sum_{i=1}^{k} \beta_i p_i$ of $k \geq 1$ binomial proportions $p_1, p_2, \ldots, p_k$ from independent binomial populations $X_1, X_2, \ldots, X_k$ with $n_1, n_2, \ldots, n_k$ trials, respectively, and weights given by k fixed constants $\beta_1, \beta_2, \ldots, \beta_k \neq 0$, consists in inverting the standard two-sided Wald test $H_0 : L = \lambda_0$, where λ_0 is any real constant admissible for $\sum_{i=1}^{k} \beta_i p_i$, meaning that λ_0 should belong to the support scale $\left[\sum_{\beta_i < 0} \beta_i \; ; \; \sum_{\beta_i > 0} \beta_i \right]$. The general formula of the classic Wald CI is

$$\hat{L} \mp z_{\alpha/2} \sqrt{\hat{v}(\hat{L})} ,$$

where $z_{\alpha/2}$ is the $\alpha/2$ upper quantile of the standard normal distribution and $\hat{L}$ and $\hat{v}(\hat{L})$ represent an estimate of L and the variance of estimator $\hat{L}$, respectively. When variance is estimated using the maximum likelihood estimate (MLE) of each p_i, $i = 1, 2, \ldots, k$, the classic version of the Wald CI, known for its poor coverage properties (e.g. [4,11,14]), is obtained. When variance is estimated using a shrinkage estimator for each p_i given by

$$\frac{X_i + h_i}{n_i + 2h_i}$$

for some $h_i > 0$, $i = 1, \ldots, k$, that is, by adding h_i successes and h_i failures to the original data, a modified version of the Wald CI, the so-called adjusted Wald

CI is obtained. Depending on the particular h_i chosen, different variants of the adjusted Wald CI can be established. A list of h_i values herein considered and the names given to their corresponding variants is presented below:

- $h_i = 0$ (variant-0, classic version)
- $h_i = \frac{2}{k}$ (variant-1, [10])
- $h_i = \frac{z_{\alpha/2}^2}{2k}$ (variant-2)
- $h_i = \frac{z_{\alpha/2}^2}{2}\left(1_{\mathcal{A}_i}(x_i) + \frac{1}{k}\right)$ (variant-3, [11])
- $h_i = \frac{z_{\alpha/2}^2}{2}\left(1_{\mathcal{A}_i}(x_i) + \frac{\beta_i^2/n_i}{\sum_{i=1}^k \frac{\beta_i^2}{n_i}}\right)$ (variant-4, [13])

where $1_{\mathcal{A}_i}(\cdot)$ is the indicator function of

$$\mathcal{A}_i = \left\{x_i \in \{0, n_i\} : (n_i - 2x_i)(\hat{L} - \lambda_0)\beta_i < 0\right\}.$$

It is obvious that variant-2 is equal to variant-3 when $0 < x_i < n_i$ for all i, variant-1 is approximately equal to variant-2 when $\alpha = 5\%$, and variant-3 is equal to variant-4 when $\frac{\beta_i^2}{n_i}$ is a constant for all i. The adjusted Wald CIs have better performance than the classic Wald CI.

3 Coverage and Directional Non-coverages

In order to assess and compare the performance of methods to construct CIs for any linear combination $L = \sum_{i=1}^k \beta_i p_i$, evaluations of the exact coverage probabilities and locations (characterized by its mesial and distal non-coverage probabilities) can be performed.

Given the weights $(\beta_1, \beta_2, \ldots, \beta_k)$ and a set of k independent binomials with parameters $(n_1, p_1), (n_2, p_2), \ldots, (n_k, p_k)$, the exact coverage probability (R) can be computed as

$$R(L) = \sum_{x_1=0}^{n_1} \sum_{x_2=0}^{n_2} \cdots \sum_{x_k=0}^{n_k} \prod_{i=1}^k \binom{n_i}{x_i} p_i^{x_i}(1-p_i)^{n_i-x_i} 1_{[l(\boldsymbol{x}), u(\boldsymbol{x})]}(L)$$

where $[l(\boldsymbol{x}), u(\boldsymbol{x})]$ is the CI obtained from the observation $\boldsymbol{x} = (x_1, x_2, \ldots, x_k)$ for the linear combination $L = \sum_{i=1}^k \beta_i p_i$. To examine the interval location, we considered the procedure suggested by [13] for the linear combination. Concretely, for each CI of a linear combination, we analyse the existence of equilibrium between the directions of the mesial non-coverage probability (MNR) and distal non-coverage probability (DNR). These directions indicate whether the CIs are located too distally or too mesially from the midpoint c of the support scale relatively to the true value L ($c = \sum_{i=1}^k \beta_i/2$). The MNR and DNR are defined as

$$MNR(L) = \sum_{x_1=0}^{n_1} \sum_{x_2=0}^{n_2} \cdots \sum_{x_k=0}^{n_k} \prod_{i=1}^k \binom{n_i}{x_i} p_i^{x_i}(1-p_i)^{n_i-x_i} 1_{\mathcal{M}}(\boldsymbol{x}),$$

with $\mathcal{M} = \{\boldsymbol{x} : (L \leq c \wedge u(\boldsymbol{x}) < L) \vee (L \geq c \wedge l(\boldsymbol{x}) > L)\}$, and

$$\text{DNR}(L) = \sum_{x_1=0}^{n_1} \sum_{x_2=0}^{n_2} \cdots \sum_{x_k=0}^{n_k} \prod_{i=1}^{k} \binom{n_i}{x_i} p_i^{x_i} (1-p_i)^{n_i-x_i} \, \mathbf{1}_{\mathcal{D}}(\boldsymbol{x}),$$

with $\mathcal{D} = \{\boldsymbol{x} : (L < c \wedge l(\boldsymbol{x}) > L) \vee (L > c \wedge u(\boldsymbol{x}) < L)\}$.

The non-coverage is said to be mesial iff the interval is located too distally to include the true parameter value L (see Fig. 1(a)) and the non-coverage is said to be distal iff the interval is located too mesially to include the true parameter value L (see Fig. 1(b)).

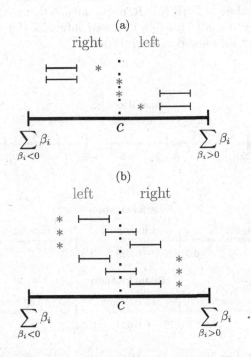

Fig. 1. Examples of CIs located too distally (a) and too mesially (b) relatively to the true value L (identified by an asterisk), on the support scale of the linear combination. The midpoint of the support scale is denoted by c.

According to [15], the interval location of the CIs can be characterized by the ratio

$$Q = \frac{\text{MNR}}{1-\text{R}} = \frac{\text{MNR}}{\text{MNR}+\text{DNR}}$$

This ratio expresses the balance condition between MNR and DNR. Based on a partition of the range of values of Q (see Fig. 2, on the top), [15] and, more recently, [13] established a classification criterion for the location of CIs for linear combinations of independent binomial proportions. Concretely, values of

Q between 0.375 and 0.625 correspond to satisfactorily located CIs, less than 0.375 to CIs located too mesially to include the true value of L, and greater than 0.625 to CIs located too distally to include L. Hence, when (see Fig. 2(a)),

- MNR and DNR are balanced, CIs are satisfactorily located;
- MNR predominates, CIs are too or much too distally located; and,
- DNR predominates, CIs are too or much too mesially located.

Furthermore, for situations where it is more adequate to evaluate the two directional non-coverages, MNR and DNR, (e.g., for CIs constructed when extremal observation exist), values of DNR (MNR, resp.) between $\alpha \times 0.375$ and $\alpha \times 0.625$ correspond to CI methods which yield intervals with a satisfactory mesial (distal) location and values of DNR (MNR, resp.) outside that range will correspond to non-satisfactory mesially (distally) located intervals (for the nominal level $\alpha = 0.05$, see Fig. 2(b); more details in [13]).

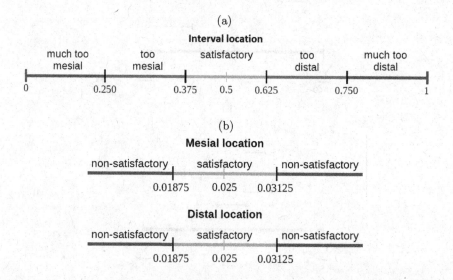

Fig. 2. (a) Interpretation of interval location in terms of Q index. (b) Interpretation of the mesial and distal locations in terms of DNR and MNR. For a 95%-confidence level, the CIs for a linear combination are expected to produce values of DRN and MNR between $(1 - R) \times 0.375 = 0.05 \times 0.375 = 0.01875$ and $(1 - R) \times 0.625 = 0.05 \times 0.625 = 0.03125$.

4   Graphical Representation

In the evaluation of the performance of CI methods for a linear combination of $k \geq 2$ binomial proportions, $L = \sum_{i=1}^{k} \beta_i p_i$, over the whole or part of the parameter space, summary statistics of the exact or simulated coverage probabilities (e.g., [10,13]) or two-dimensional plots of the coverage probabilities, by holding one proportion p_i and fixing the other ones (e.g., [4] for the difference of two proportions), are usually given. In addition to frequently not being able to draw the main conclusions in terms of the key parameter L, these types of summarizing procedures can also be misleading [17].

In order to assess the performance of CI methods for a linear combination L, for any trial setting $n_1, n_2, \ldots, n_k$ and weights $\beta_1, \beta_2, \ldots, \beta_k$, we propose a different approach which consists of using a simple two-dimensional plot of DNR versus MNR. In this plot, all pairs $(\text{MNR}(L), \text{DNR}(L))$, computed via exact or Monte Carlo methods for different possible values of L, are represented by points. By inspecting the location of these points, we can visually check for the existence of desirable properties concerning R, MNR and DNR. Concretely, the following properties could be easily analyzed:

(i) The values of R should be close to the nominal level of $100(1 - \alpha)\%$.
    Since $1 - R = \text{DNR} + \text{MNR}$, then the closer the value of DNR + MNR is to α the better the performance of the CI in terms of coverage. Hence, the line DNR + MNR $= \alpha$ and two parallel lines, for instance DNR + MNR $= \alpha + \varepsilon$ and DNR + MNR $= \alpha - \varepsilon$, for some tolerance $\varepsilon > 0$, allow to examining whether the coverage probabilities produced by the CI method under analyzed, for different values of the parameter L, are sufficiently close to the expected nominal level. Graphically, the points should be between the two border parallel lines;

(ii) When R values are below the nominal level, the CI method is classified as liberal. When R values are above the nominal level, the CI method is conservative.
    Hence, graphically, having points above (below, respectively) the reference line DNR + MNR $= \alpha$ means that the method produces more liberal (conservative, resp.) CIs;

(iii) The quantity MNR $-$ DNR provides information about the balance between MNR and DNR. If MNR $-$ DNR is close to 0, the value of $Q = \frac{\text{MNR}}{\text{MNR}+\text{DNR}}$ will be around 0.5, which is interpreted as satisfactorily located CIs. Values of Q between the bounding references 0.375 and 0.625 will correspond to CI methods which yield intervals with a satisfactory location. Values of Q outside that range will correspond to non-satisfactory located intervals: less than 0.375 to intervals located too mesially to include the true value of L, and greater than 0.625 to intervals located too distally to include L [13]. Points between the two lines DNR $= 0.600$ MNR and DNR $= 1.667$ MNR will correspond to satisfactorily located CIs, since

$$0.375 \le Q \le 0.625 \iff 0.375 \le \frac{\text{MNR}}{\text{MNR} + \text{DNR}} \le 0.625 \iff$$

$$\frac{1 - 0.625}{0.625}\text{MNR} \le \text{DNR} \le \frac{1 - 0.375}{0.375}\text{MNR} \iff$$

$$0.600\text{MNR} \le \text{DNR} \le 1.667\text{MNR}.$$

(iv) For situations in which extremal observations exist, it is convenient to apply CI methods for which it is expected that MNR $= (1 - R)/2 = \alpha/2$ and

DNR $= (1 - R)/2 = \alpha/2$. Values of DNR (MNR, resp.) between $\alpha \times 0.375$ and $\alpha \times 0.625$ will correspond to CI methods which yield intervals with a satisfactory mesial (distal, resp.) location. Values of DNR (MNR, resp.) outside that range will correspond to non-satisfactory mesially (distally, resp.) located intervals.

5   Examples

To illustrate the usefulness of the proposed graphical technique, we carried out a simulation study to show the advantages of using it to evaluate the coverage probabilities and interval location of the five variants of the adjusted Wald CIs for two types of linear combinations of $k = 3$ independent binomial proportions: one with balanced weights, $\frac{1}{3}p_1 + \frac{1}{3}p_2 + \frac{1}{3}p_3$ (Scenario 1) and the other with unbalanced weights, $\frac{1}{3}p_1 + \frac{1}{2}p_2 + 3p_3$ (Scenario 2). Similar results were observed for others unbalanced and *quasi*-balanced configurations.

Using the Monte Carlo method, estimates of R, MNR, DNR and Q were obtained using 100 sets of 3-samples simulated from binomial distributions, $Bin(n_i, p_i)$, $i = 1, 2, 3$, where each p_i is randomly generated from $U[0, 1]$. Four configurations with different sample sizes (n_1, n_2, n_3) were considered: (10, 10, 10), (30, 10, 10), (30, 20, 10) and (30, 30, 30). The quantities R, MNR, DNR and Q were computed for each parameter setting. Only the results for 95% will be herein discussed. All simulations were carried out using R software [18].

For each Wald CI variant, average values of R, MNR, DNR and Q (R_{mean}, MNR_{mean}, DNR_{mean} and Q_{mean}) were calculated for each parameter setting (Tables 1 and 2). Based on these averaged quantities, we concluded the following:

R_{mean}: Wald CI variants-1, 2 yielded the best performances in terms of averaged coverage probabilities for both balanced and unbalanced weights;

Q_{mean}: Wald variants-1, 2 were also the best to produce CIs with satisfactory locations for most scenarios; Wald variant-4 was found to be the worst in terms of absence of equilibrium between the mesial and distal non-coverage probabilities, particularly in unbalanced scenarios;

MNR_{mean} and DNR_{mean}: Wald variant-4 was the best to produce CIs with mesially satisfactory location and a very low probability of being distally located, for almost all scenarios. This result is consistent with the capability of this variant to handle extremal observations (i.e., when there are estimates of p_i equal to 0 or 1, for some i).

For each scenario and each variant of the Wald method, all the pairs of estimated values (MNR,DNR) and the centroid (MNR_{mean}, DNR_{mean}) were displayed in a plot (see Fig. 3 for variants-0, 4). The centroids correspond to values shown in Tables 1 and 2. Based on these plots, we concluded:

R: All the classic Wald CIs (variant-0) constructed were very conservative (points above the line with negative slope). Some Wald variant-4 CIs exhibited good performance in terms of coverage (points close to the line with negative slope);

Table 1. Results of averaged coverage and directional non-coverages for the five variants of the Wald CI, when $k = 3$, for Scenarios 1 and 2. Confidence level $1 - \alpha = 95\%$.

$(\beta_1, \beta_2, \beta_3)$	$n_1/n_2/n_3$	Variants					Variants					Variants				
		R_{mean} (%)					MNR_{mean} (%)					DNR_{mean} (%)				
		0	1	2	3	4	0	1	2	3	4	0	1	2	3	4
$\left(\frac{1}{3}, \frac{1}{3}, \frac{1}{3}\right)$	10/10/10	91.6	95.6	95.5	97.0	97.0	5.18	1.42	1.49	0.57	0.57	3.22	2.98	3.01	2.45	2.45
	30/30/30	94.0	95.2	95.2	95.6	95.6	3.47	1.95	1.98	1.62	1.62	2.53	2.85	2.82	2.74	2.74
	30/10/10	91.6	95.5	95.4	96.7	96.7	5.15	1.61	1.69	0.82	0.77	3.25	2.89	2.91	2.45	2.52
	30/20/10	92.2	95.4	95.3	96.4	96.3	4.75	1.74	1.82	1.06	1.03	3.05	2.86	2.88	2.53	2.65
$\left(\frac{1}{3}, \frac{1}{2}, 3\right)$	10/10/10	86.1	95.3	95.2	96.8	96.7	14.24	2.59	2.70	1.16	0.33	2.20	2.12	2.10	2.07	2.90
	30/30/30	91.1	95.0	94.9	95.5	96.0	7.16	3.04	3.14	2.53	1.18	1.74	1.97	1.96	1.92	2.85
	30/10/10	83.4	95.2	95.1	96.7	96.7	14.38	2.65	2.77	1.54	0.33	2.22	2.15	2.13	1.20	2.95
	30/20/10	82.4	95.1	95.0	96.6	96.7	15.45	2.74	2.86	1.25	0.34	2.15	2.16	2.14	2.12	3.00

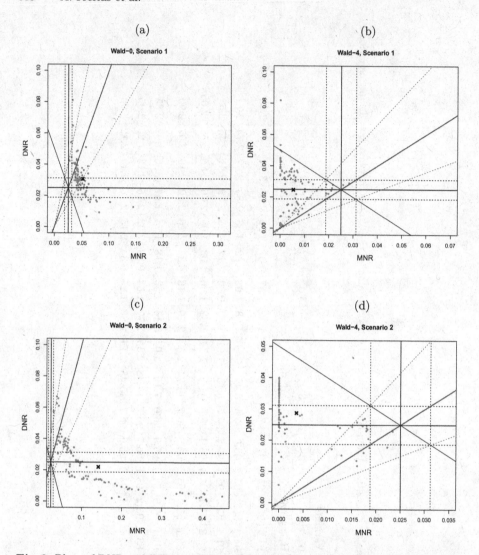

Fig. 3. Plots of DNR vs MNR for variant-0 and variant-4 of the adjusted Wald CI, for the linear combinations $\frac{1}{3}p_1 + \frac{1}{3}p_2 + \frac{1}{3}p_3$ (Scenario 1, (a) and (b)) and $\frac{1}{3}p_1 + \frac{1}{2}p_2 + 3p_3$ (Scenario 2, (c) and (d)), in both cases for the sample sizes $(10, 10, 10)$. The centroid is marked by a cross. The solid lines are the reference lines associated to R (lines with negative slope), Q (lines with positive slope), MNR (vertical lines) and DNR (horizontal lines). The dashed lines bound the region around the corresponding reference lines where the interval location is classified as satisfactory in terms of Q, DNR and DNR.

Q: Only Wald variant-0 tends to produce more satisfactorily located CIs when the weights are balanced (points between the two border lines with positive slopes);

Table 2. Results of averaged index Q for the five variants of the Wald CI, when $k = 3$, for Scenarios 1 and 2. Confidence level $1 - \alpha = 95\%$.

$(\beta_1, \beta_2, \beta_3)$	$n_1/n_2/n_3$	Variants				
		0	1	2	3	4
		Q_{mean}				
$\left(\dfrac{1}{3}, \dfrac{1}{3}, \dfrac{1}{3}\right)$	10/10/10	0.617	0.322	0.332	0.188	0.188
	30/30/30	0.579	0.406	0.413	0.371	0.371
	30/10/10	0.613	0.358	0.367	0.250	0.234
	30/20/10	0.609	0.379	0.387	0.296	0.281
$\left(\dfrac{1}{3}, \dfrac{1}{2}, 3\right)$	10/10/10	0.792	0.550	0.562	0.359	0.101
	30/30/30	0.805	0.607	0.615	0.568	0.293
	30/10/10	0.866	0.553	0.566	0.364	0.102
	30/20/10	0.878	0.559	0.572	0.372	0.102

MNR and DNR: Although the location of the centroids indicates that the averaged values of DNR have a satisfactory level for all situations represented in Fig. 3, there are some CIs produced by variant-0 and variant-4 of the adjusted Wald method that fall outside the region corresponding to a mesially satisfactory location. Therefore, one should be aware that the averaged values of DNR and MNR may not always be reliable indicators of interval location.

6   Conclusions

Using the shrinkage estimator $(X_i + h_i)/(n_i + 2h_i), h_i > 0$, in the estimation of the proportion p_i, several types of adjustments of the Wald CI for a linear combination of k independent binomial proportions can be constructed. To fully characterize the location of these different variants of the adjusted Wald CI and determine how satisfactory is their mesial (distal) location, particularly in situations in which extremal observations can exist, the mean of evaluation measures like the Q index, the MNR and the DNR may not be sufficient. In fact, these summarized statistics do not describe how well the values of R, MNR and DNR are distributed for different values of the key parameter L.

Our examples show that a graphical representation of the DNR vs MNR is easier to interpret, facilitating the evaluation of the location of the CIs. Moreover, this type of representation draws the attention to the need of analyzing the dispersion of the values of MNR and DNR in such evaluations.

Acknowledgments. This work was partially supported by Portuguese funds through the CIDMA - Center for Research and Development in Mathematics and Applications, and the Portuguese Foundation for Science and Technology (FCT – Fundação para a Ciência e a Tecnologia), within project UID/MAT/04106/2013.

References

1. Anbar, D.: On estimating the difference between two proportions, with special reference to clinical trials. Biometrics **39**, 257–262 (1983)
2. Mee, R.: Confidence bounds for the difference between two proportions. Biometrics **40**, 1175–1176 (1984)
3. Newcombe, R.: Interval estimation for the difference between independent proportions: comparison of eleven methods. Stat. Med. **17**, 873–890 (1998)
4. Agresti, A., Caffo, B.: Simple and effective confidence intervals for proportions and differences of proportions result from adding two successes and two failures. Am. Stat. **54**, 280–288 (2000)
5. Brown, L., Li, X.: Confidence intervals for two sample binomial distribution. J. Stat. Plan. Infer. **130**, 359–375 (2005)
6. Fagerland, M., Lydersen, S., Laake, P.: Recommended confidence intervals for two independent binomial proportions. Stat. Methods Med. Res., 1–31 (2011)
7. Decrouez, G., Robison, A.: Confidence intervals for the weighted sum of two independent binomial proportions. Aust. N. Z. Stat. **54**, 281–299 (2012)
8. Martín Andrés, A., Álvarez Hernández, M.: Optimal method for realizing two-sided inferences about a linear combination of two proportions. Commun. Stat. Simul. Comput. **42**, 327–343 (2013). doi:10.1080/03610918.2011.650263
9. Tebbs, J., Roths, S.: New large-sample confidence intervals for a linear combination of binomial proportions. J. Stat. Plan. Infer. **138**, 1884–1893 (2008)
10. Price, R., Bonett, D.: An improved confidence interval for a linear function of binomial proportions. Comput. Stat. Data Anal. **45**, 449–456 (2004)
11. Martín Andrés, A., Álvarez Hernández, M., Herranz Tejedor, I.: Inferences about a linear combination of proportions. Stat. Methods Med. Res. **20**, 369–387 (2012). doi:10.1177/0962280209347953. Erratum in Stat. Methods Med. Res. **21**(4), 427–428 (2012). doi:10.1177/0962280211423597
12. Martín Andrés, A., Herranz Tejedor, I., Álvarez Hernández, M.: The optimal method to make inferences about a linear combination of proportions. J. Stat. Comput. Simul. **82**, 123–135 (2012). doi:10.1080/00949655.2010.530601
13. Escudeiro, S., Freitas, A., Afreixo, A.: Approximate confidence intervals for a linear combination of binomial proportions: a new variant. Commun. Stat. Simul. Comput. (2016). doi:10.1080/03610918.2016.1241408
14. Agresti, A., Coull, B.: Approximate is better than exact for interval estimation of binomial proportions. Am. Stat. **52**, 119–126 (1998)
15. Newcombe, R.: Measures of location for confidence intervals for proportions. Commun. Stat. Theory Methods **40**, 1743–1767 (2011)
16. Newcombe, R.: Confidence Intervals for Proportions and Related Measures of Effect Size. CRC Press, Boca Raton (2013)
17. Laud, P., Daneb, A.: Confidence intervals for the difference between independent binomial proportions: comparison using a graphical approach and moving averages. Pharmaceut. Statist. **13**, 294–308 (2014)
18. R Core Team: R: A Language and Environment for Statistical Computing. R Foundation for Statistical Computing (2016)

Household Packaging Waste Management

João A. Ferreira[1,2], Manuel C. Figueiredo[1,2(✉)],
and José A. Oliveira[1,2]

[1] ALGORITMI Research Centre, University of Minho,
Campus de Azurém, 4800-058 Guimarães, Portugal
mcf@dps.uminho.pt
[2] Department of Production and Systems, University of Minho,
Campus de Gualtar, 4710-057 Braga, Portugal

Abstract. Household packaging waste (HPW) has an important environmental impact and economic relevance. Thus there are networks of collection points (named *"ecopontos"* in Portugal) where HPW may be deposited for collection by waste management companies.

In order to optimize HPW logistics, accurate estimates of the waste generation rates are needed to calculate the number of collections required for each *ecoponto* in a given period of time.

The most important factors to estimate HPW generation rates are linked to the characteristics of the population and the social and economic activities around each *ecoponto* location.

We developed multiple linear regression models and artificial neural networks models to forecast the number of collections per year required for each location. For operational short term planning purposes, these forecasts need to be adjusted for seasonality in order to determine the required number of collections for the relevant planning period. In this paper we describe the methodology used to obtain these forecasts.

Keywords: Forecasting · Household packaging waste · Waste collection · Recycling · Seasonality

1 Introduction

Recycling of waste materials became a very important issue for society, as the environment benefits greatly from any advances made in direction of a cleaner future. Process collecting for recycling involves teams of workers and vehicles. One of the main problems lies in finding optimal collection routes, where a set of collection points is targeted, and each point is given a priority level. This problem can be described as the Vehicle Routing Problem (VRP). However more flexibility is needed when it comes to choose only a part of the collection points to be visited, instead of the whole set. Thus, a more fitting description of the selective waste collection process may be the Team Orienteering Problem (TOP).

In this context, the TOP can be described as the problem of designing the routes to be assigned to a fleet of vehicles that perform the collection of different types of waste stored along a network of collection points. Each one of these collection points contains

© Springer International Publishing AG 2017
O. Gervasi et al. (Eds.): ICCSA 2017, Part II, LNCS 10405, pp. 611–620, 2017.
DOI: 10.1007/978-3-319-62395-5_42

a certain amount of waste that is directly linked to the respective priority level. The collection routes have maximum durations or distances, and consequently, the selection of collection points to be visited by the vehicles is made by balancing their priorities and their contributions for the route duration or distance. The objective is to maximize the total amount of waste collected by all routes while respecting the time or distance constraints.

Aside from the routing problem, there are other issues related to the process of waste collection for recycling, especially when dealing with real scenarios and the activity of real waste collection companies. One of these issues is the estimate of waste material quantities generated over time at each collection point in a given collection network, which enables the estimation of a waste generation rate (WGR) for each collection point. Estimating WGRs is crucial for designing collection networks since collection points are located according to these estimates. Moreover, forecasts for waste generated along a network are used to design collection routes as each collection point is given a priority according to their WGR, which translates into the frequency of service by collection vehicles.

Considering the goals Portugal has to fulfil for the recycling and recovery of HPW, there is a permanent need for increased efficiency in waste collection. Thus, one important task is the development of models to forecast the quantities of waste generated at collection points.

The paper is structured in 6 sections. Section 2 presents the problem. A literature review is presented in Sect. 3. The forecasting models developed are described in Sect. 4. Seasonal adjustments to the base forecasts are described in Sect. 5. Finally, in Sect. 6, the results and main conclusions of this study are presented.

2 The Problem

The main objective is to solve a real-world problem faced by an intermunicipal waste management company, in six municipalities: Braga, Vieira do Minho, Vila Verde, Povoa do Lanhoso, Amares and Terras de Bouro, in Northern of Portugal (Fig. 1).

Fig. 1. Collection area of HPW

The problem focus is on HPW collection, and a major step to achieve a good performance is to obtain accurate forecasts for waste generation rates at the waste collection points in order to determine the frequency of collection.

In Portugal, household packaging waste (HPW) is disposed of in collection points called *ecopontos*. Each *ecoponto* can have three types of containers, identified with different colors: glass (green), paper/cardboard (blue), and plastic/metal (yellow), as can be seen in Fig. 2.

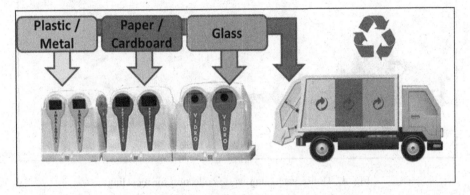

Fig. 2. *Ecopontos* illustration (Color figure online)

The company currently operates a network of more than 1,200 *ecopontos* located across. These six municipalities are characterized by a mix of urban and rural areas, which require different strategies concerning waste management.

The company vehicles do not visit all *ecopontos* every workday. It is necessary to select a subset of *ecopontos* to visit, according to their estimated fill rate, each time the route planning is done. Thus, given a planning horizon, for example a week, the company must decide which *ecopontos* must be visited (because they are near full) and which can be skipped (because they are near empty) by the collection fleet in order to design efficient routes for the selective collection of HPW. Thus the priority level of an *ecoponto* is highly related to the estimated amount of waste it holds during the route planning phase, Fig. 3.

Fig. 3. Estimated amount of waste held

Taking into account its priority level, an *ecoponto* may or may not be selected to be visited during the established planning horizon, Fig. 4.

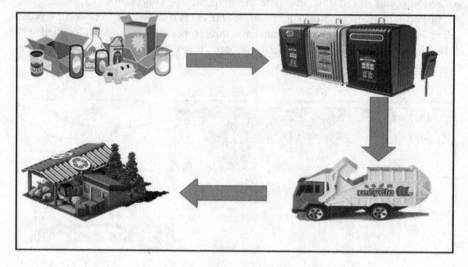

Fig. 4. House packaging waste collection for recycling

Thus, to improve route planning, reliable forecasts of the amount of waste generated at each *ecoponto* are necessary. In a previous phase of this study [5], significant factors for HPW generation rates were identified in order to develop forecasting models to predict the number of times each *ecoponto* should be visited during a certain period of time. However, due to a variety of seasonal factors, such as holidays, periodic events, weather conditions, etc., waste generation rates need to be seasonally adjusted. For example, the number of collection visits required in this study increases significantly during summer months. In this paper we present a method for making these seasonal adjustments that allow a significant reduction of the number of unnecessary visits to "nearly empty" *ecopontos* and therefore saving resources and reducing the logistics costs of the operation, Fig. 5.

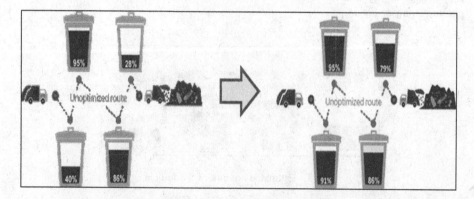

Fig. 5. Optimizing collection routes

3 Literature Review

Forecasting the generation rates of municipal solid waste (MSW) is important for both, the development of waste management infrastructures, and the implementation of the logistics processes for waste collection.

In a review, Beigl et al. [2] concluded that MSW generation is best predicted by time series analysis and regression methods. According to Denafas et al. [4] seasonal factors may have a strong impact on MSW generation and should be taken into account in the decision-making processes concerning waste management. Concerning waste generation contributing factors, Wang and Nie [11] identified the growth of urban population and gross domestic product (GDP) as the most important ones. Grossman et al. [6] considered factors such as: increase of population, income level and housing type. Later studies pointed out that waste generation can be related to production and consumption levels [3, 9]. More detailed analyses showed that the growth of the urban population had a greater impact than GDP on the total amount of MSW produced. Also, with factors like increasing income, MSW seems to change more in composition rather than in total amount generated. Other factors that may influence the generation and composition of waste are climate, living habits, level of education, religious and cultural beliefs, and social and public attitudes [1, 8, 10].

Frequently, time series forecasting models may be a good choice to estimate MSW generation when there is access to a substantial amount of historical data. Based on the comparison of several forecasting methods, Beigl et al. [2] imply that a forecasting tool based on socio-economic variables was more suitable than single time series analyses. In most cases, the application of modelling methods such as regression analyses, and group comparisons, seems to be the best option when the goal is to test the relationship between the level of affluence and the generation of total MSW or a material-related fraction, and to identify significant effects of waste management activities on recycling quotas. The application of time series analyses and input-output analyses is advantageous when there is a need for special information (i.e., assessment of seasonal effects for short-term forecasts).

After this review, it became clear that most of the previous research in this area focused on a different level of analysis that did not match our purpose of predicting waste generation at each collection point in order to determine when it needs to be emptied.

4 Forecasting Models

The available data consisted on all the waste collections performed by the company in all the six municipalities it operates from 2013. It contained monthly records showing how many times each *ecoponto* in the network was emptied during each month of the year.

Our aim was to forecast the number of times each *ecoponto* needs to be emptied each year. Therefore, this number of collections per year (and per *ecoponto*), hereafter referred as CPY, was set as the dependent variable considered in the forecasting models developed. In the next subsections, a brief description of the developed models is given, followed by a detailed description of the seasonal adjustments made to estimate the required number of collections per month.

The forecasting models for CPY use data from waste collection records. The factors used to estimate waste generation rates, were determined using several sources for demographic information and socioeconomic indicators. In Table 1, a list with all explaining factors for HPW generation is presented.

Table 1. Factors used to estimate waste generation rates.

Factor	Description	Acronym
1	Number of Ecopontos in the area	NE
2	Population Density in the area	PD
3	Number of Inhabitants per Ecoponto	NRE
4	Ecoponto Density	ED
5	Ecoponto Type (street level or underground) – qualitative	ET
6	Ecoponto Position (closed or open area) – qualitative	EP
7	Ecoponto Capacity	EC
8	Number of Ecopontos within a 300 ms radius	NE300
9	Demographic Factor (household density around each ecoponto in a 300 m radius) - qualitative	DF
10	SocioEconomic Factor (based on the number of schools, businesses, local attractions, leisure and sports infrastructures, restaurants, etc.)	SEF

The data used consisted of information on these factors for all the *ecopontos* (185) from two municipalities: Amares and Vila Verde.

We used two different methods to develop the forecasting models: Multiple Regression (MR) and Artificial Neural Networks (ANN). MR models were developed using *Forecast Pro* software and ANN models were developed using *Encog* [7]. Regression models are widely known and used for forecasting purposes, and ANN are attaining more recently some relevance in this field, with promising results being reported in the last years.

The best performing forecasting models for the collection of cardboard at the 185 *ecopontos* sample are described in Eq. 1 (regression model) and Table 2 (ANN model).

$$CPYi = 18.209DF - 0.006EC + 8.314ED + 21.141ET + 0.035NRE - 0.027PD + \in \quad (1)$$

The performance of these models at forecasting the actual number of collections per year (Observed collections) in a test sample of 50 *ecopontos* can be seen in Fig. 6.

The mean absolute deviation (MAD) of forecasting errors for both models is shown in Table 3.

Table 2. Artificial neural network model.

Network type	Multilayer perceptron
Network structure	3 layers (4-4-1)
Learning algorithm	Levenberg-Marquardt
Activation function	Sigmoid
Training, test, validation	125, 10, 50
Training epochs	1000
Inputs	DF, ET, NE, SEF
Output	CPY

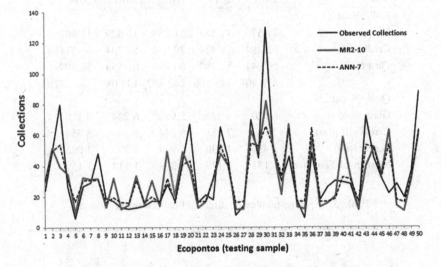

Fig. 6. Forecast results for the best ANN and regression models.

Table 3. Mean absolute deviation of forecasting errors.

Multiple regression	Artificial neural network
10.44	9.14

5   Seasonal Adjustments

The models described in Sect. 4 are used to estimate the number of collections per year required for each *ecoponto*. However, as previously stated, there are seasonal variations on the waste generation rates (WGRs). Therefore there is a need to adjust the number of collections to these variations in order to forecast the number of collections required for each month of the year. For example, in summer months, there are usually higher WGRs in this region due to emigrants returning to their home country and tourism activities resulting in significant population increases. There are also seasonal changes in social habits. For example, there is a higher consumption of beverages in plastic, metal and glass containers. Many other factors may also have contributed for seasonal variations in WGRs throughout the year.

Table 4 shows general performance indicators of the company operation from 2010 to 2014. In order to describe the seasonal adjustment procedures we will use data concerning cardboard collection in a sample of 185 *ecopontos* from two municipalities, Amares and Vila Verde. Table 5 shows the average monthly number of effective cardboard collections done for the 185 *ecopontos* in the sample.

Table 4. Company indicators

Year	2010	2011	2012	2013	2014
Collections					
Glass	11 570	11 555	11 238	11 829	11 887
Cardboard	62 294	72 431	70 234	69 704	65 571
Plastic	56 044	59 180	60 880	61 556	58 899
	129 908	143 166	142 352	143 089	136 357
Quantity coll.					
Glass (ton)	6 320	6 282	5 968	6 224	6 170
Cardboard (ton)	7 137	7 151	6 244	6 245	6 354
Plastic (ton)	1 434	1 730	1 909	1 992	2 095
Number of *Ecopontos*	1 131	1 170	1 208	1 114	1 159

Table 5. Average number of cardboard collections per month.

Month	January	February	March	April	May	June	July	August	September	October	November	December
Number of Collections	754	719	617	783	783	677	909	828	868	769	775	793

The analysis of the data in Table 4 shows that there is not any significant trend in the number of collections from 2010 to 2014. Thus to determine the seasonal factors we may simply divide the average number of collections for each particular month by the monthly collections average. Table 6 shows the seasonal factors for each month of the year.

Once determined the seasonal adjustment factors, the forecasts of total collections for each month can be corrected using the corresponding seasonal adjustment factor. For example, if the forecast of CPY_i for a certain *ecoponto* i is 120, then the average number of collections per month will be 10. To correct forecasts for example for the

Table 6. Seasonal factors

Month	January	February	March	April	May	June	July	August	September	October	November	December
Seasonal Factor	0.976	0.930	0.798	1.013	1.013	0.876	1.176	1.071	1.123	0.995	1.003	1.026

months of March and July, the factors 0.798 and 1.176 are used. Thus the revised forecast for required collections for *ecoponto* i would be 8 collections for March and 12 collections for July. Since the company operates approximately 25 days per month (there is no collection on Sundays), in operational terms this means that *ecoponto* i should be emptied every 3 days in March and every 2 days in July. The use of these revised forecasts at the route planning phase makes the collection routes much more efficient by avoiding visits to "nearly empty" *ecopontos*.

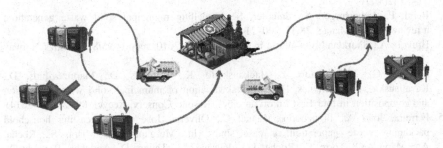

Fig. 7. House packaging waste collection for recycling

6   Discussion and Conclusions

The use of the seasonal adjustments resulted in an important reduction of the errors at forecasting the number of effective collections required per month. The impact of the revised forecast values resulted in an estimated 10% decrease in the number of visits to *ecopontos* not resulting in any collection. For the company involved in this study this means avoiding hundreds of unnecessary visits to *ecopontos* per year, Fig. 7.

In this study we presented a real-world problem faced by a company that collects House Packaging Waste (HPW) deposited in a network of waste collection points (*ecopontos*).

The main problem faced by the company was the high number of visits made by the collection teams to *ecopontos* that are "empty". This represented an important waste of resources that translated in thousands of kilometers traveled and hours spent by the collection teams that do not resulted in any collection.

Our main goal was to improve the forecasts for the number of monthly waste collections required. We achieved this by correcting the base forecasts for the number of waste collections per year for each *ecoponto* (CPYi), developed with Regression and Artificial Neural Networks models, using seasonal factors for each month of the year.

These seasonal factors where determined using actual data from effective collections. The results achieved show that these seasonal adjustments reduce the number of unnecessary visits to *ecopontos,* thus leading to important savings in environmental and logistics costs.

Acknowledgments. This research has been partially supported by COMPETE: POCI-01-0145-FEDER-007043 and FCT – *Fundação para a Ciência e Tecnologia* within the Project Scope: UID/CEC/00319/2013.

References

1. Bandara, J., Hettiaratchi, J., Wirasinghe, S., Pilapiiya, S.: Relation of waste generation and composition to socio-economic factors: a case study. Environ. Monit. Assess. **135**(1–3), 31–39 (2007)
2. Beigl, P., Lebersorger, S., Salhofer, S.: Modelling municipal solid waste generation: a review. Waste Manag. **28**(1), 200–214 (2008)
3. Bruvoll, A., Spurkland, G.: Waste in Norway up to 2010, reports 95/8. Statistics Norway (1995)
4. Denafas, G., Vitkauskaitơ, L., Jankauskaitơ, K., Staniulis, D., Martuzeviþius, D., Kavaliauskas, A., Tumynas, D.: Seasonal variation of municipal solid waste generation and composition in four East European cities. Resour. Conserv. Recycl. **89**, 22–30 (2014)
5. Ferreira, João A., Figueiredo, Manuel C., Oliveira, José A.: Forecastihg household packaging waste generation: a case study. In: Murgante, B., Misra, S., Rocha, Ana Maria A.C., Torre, C., Rocha, J.G., Falcão, M.I., Taniar, D., Apduhan, Bernady O., Gervasi, O. (eds.) ICCSA 2014. LNCS, vol. 8581, pp. 523–538. Springer, Cham (2014). doi:10.1007/978-3-319-09150-1_38
6. Grossman, D., Hudson, J.F., Mark, D.H.: Waste generation methods for solid waste collection. J. Environ. Eng. **6**, 1219–1230 (1974). ASCE
7. Heaton, J.: Encog: library of interchangeable machine learning models for Java and C#. J. Mach. Learn. Res. **16**, 1243–1247 (2015)
8. Marquez, M.Y., Ojeda, S., Hidalgo, H.: Identification of behavior patterns in household solid waste generation in Mexicali's city: study case. Resour. Conserv. Recycl. **52**(11), 1299–1306 (2008)
9. Nagelhout, D., Joosten, M., Wierenga, K.: Future waste disposal in the Netherlands. Resour. Conserv. Recycl. **4**(4), 283–295 (1990)
10. Oribe-Garcia, I., Kamara-Esteban, O., Martin, C., Macarulla-Arenaza, A., Alonso-Vicario, A.: Identification of influencing municipal characteristics regarding household waste generation and their forecasting ability in Biscay. Waste Manag. **39**, 26–34 (2015)
11. Wang, H.T., Nie, Y.F.: Municipal solid waste characteristics and management in China. J. Air Waste Manag. Assoc. **51**(2), 250–263 (2001)

Assessing the Predictive Performance of Survival Models with Longitudinal Data

Ipek Guler[1]([✉]), Christel Faes[2], Francisco Gude[3], and Carmen Cadarso-Suárez[1]

[1] Center for Research in Molecular Medicine and Chronic Diseases (CiMUS),
University of Santiago de Compostela,
15782 Santiago de Compostela, A Coruña, Spain
ipek.guler@usc.es
[2] Interuniversity Institute for Biostatistics and Statistical Bioinformatics,
Universiteit Hasselt, Hasselt, Belgium
[3] Clinical Epidemiology Unit, Hospital Clinico Santiago de Compostela,
A Coruña, Spain

Abstract. In many follow-up studies, different types of outcomes are collected including longitudinal measurements and time-to-event outcomes. Commonly, it is of interest to study the association between them. Different regression modelling techniques have been proposed in the literature to study such association. In this paper, we will focus on two of them: two-stage models and joint modelling framework and compare them in terms of the predictive capacity of the time-to-event model. Our interest is twofold: discussing the occurrence of bias estimations and predictive capacity of the models in short-term and long-term mortality in our particular case study, Cardiac Resynchronization Therapy (CRT).

1 Introduction

In most of the biomedical research studies, the patients are followed up during a period and produce different types of outcomes. The clinicians are often interested in time to recovery, recurrence of a disease or mortality in terms of baseline covariates and repeated measurements that are taken during the research study. The objective is often to identify prognostic factors that can be used to guide clinical management of patients. Besides, the estimation of the risk for death, the predictive accuracy for individual patients is also important for the this type of studies. Accuracy summaries such as sensitivity and specificity are well established for simple binary (disease) variables with either discrete or continuous marker measurements. The most popular method for sensitivity and specificity analysis can be the ROC curves [4].

The ROC curves methodology is used to evaluate the predictive capacity of survival models with different types of biomarkers. However, many follow-up studies produce repeated measurements which are related to risk for death and the interest often lies on the predictive accuracy of the joint model including the longitudinal biomarker and the survival process.

© Springer International Publishing AG 2017
O. Gervasi et al. (Eds.): ICCSA 2017, Part II, LNCS 10405, pp. 621–631, 2017.
DOI: 10.1007/978-3-319-62395-5_43

Depending on the research question, different regression modelling techniques exist in the literature for this kind of joint models. For instance, one of the initial approaches to study longitudinal and time-to-event data is the two-stage models [9] where both processes are modeled separately. Furthermore, the joint modelling approaches were developed to feature the association between the longitudinal and time-to-event process.

The aim of this paper is to compare two possible regression techniques to study the longitudinal biomarker and patients'mortality in a Cardiac Resynchronization Therapy (CRT) study. Although there are already existing comparisons of naive two-stage modelling with joint models in the literature (see for example: Wu et al. 2012), most of them are based on the comparison of the coefficient estimations. However, many clinician's interest is on the predictions of the time-to-event model, especially on risk for death. Thus, our interest of the model comparison is twofold: discussing the occurrence of bias estimations and predictive capacity of the models in short-term (2, 4, 5 years) and long-term (9, 12 years) mortality in our particular case study, CRT.

We will use time-dependent ROC curves [4] to assess the predictive capacity of the models using new features of joint modelling approaches and compare the two-stage modelling approaches and joint models in the context of discrimination. The outline of this paper is as follows. In the first section we will describe our motivating database, the statistical methods are then given in the following section. In Sect. 4 we give a brief information about the model comparison methods for both two-stage approach and joint modelling. We end with the results and a discussion sections. Software information and the code can be found in software section.

2   CRT Study

Cardiac Resynchronization Therapy (CRT) includes 328 patients with the average age of 70.19 (sd $= 9.48$) and with heart failure which received a cardiac resynchronization therapy device. There were 23 % females and 77 % males. Various measurements are recorded by an echography in each visit to track the heart failure of the patients such as the ejection fraction (EF) which is an important measurement to diagnose the heart failure. As we can observe in Fig. 1, the EF measurements exhibit different trends between patients who were observed the event and who were not. Therefore, we need to model the longitudinal outcome and survival together to study the association between them.

3   Statistical Models

There are various regression techniques in the literature to study the association between a longitudinal biomarker and survival data. We will illustrate two different techniques to explore the association between the EF measurements and the mortality. The first model will be the two-stage modelling approach where the longitudinal and survival processes are modeled by a linear mixed effects model

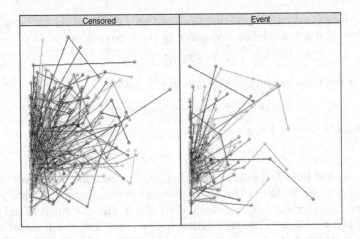

Fig. 1. Subject specific profiles of EF measurements for censored and failed patients.

[6] and a Cox proportional hazard model [3] respectively. The two-stage approach is a naive technique for longitudinal and survival data, modelling independently the two different processes. The joint modelling techniques differentiate from these models by a full joint likelihood approach where the conditional distributions of both models are calculated.

We will illustrate these different techniques on CRT database and discuss the advantages and disadvantages of each them in this section.

3.1   The Two-Stage Approach [9]

The two-stage approach presented by [9] is described in two steps as follows,

(i) In the first stage a linear mixed effects model is fitted to estimate the true longitudinal process [6]

$$\mathrm{EF}_{ij} = \beta_0 + \beta_1 age + \beta_2 time_{ij} + \beta_3 time_{ij}^2$$
$$+ u_{0i} + u_{1i} t_{ij} + \epsilon_i(t_{ij}) \tag{1}$$

where EF_{ij} is the EF measured on subject $i = 1, ..., n$ at time point t_{ij}, with $j = 1, ..., m_i$, β_0, β_1, β_2, β_3 are the fixed effects coefficients, u_{0i} is the random intercept and u_{1i} is the random slope parameter.

It is assumed that

$$\begin{pmatrix} u_{0i} \\ u_{1i} \end{pmatrix} \sim N(\begin{pmatrix} 0 \\ 0 \end{pmatrix}, \Sigma)$$

with

$$\Sigma = \begin{pmatrix} \sigma_{u_0}^2 & \sigma_{u_0}\sigma_{u_1}\rho_{12} \\ \sigma_{u_0}\sigma_{u_1}\rho_{12} & \sigma_{u_1}^2 \end{pmatrix}$$

(ii) In the second stage the estimations at time t of Model (1) are incorporated into the survival sub-model as covariates in the following manner;

$$t_i \sim Weibull(p, \mu_i(t))$$

where t_i is the time-to-event for subject i, $p > 0$ and,

$$\log(\mu_i(t)) = \gamma_1 age + \gamma_2 gender + \omega_i(t)$$

The hazard at time t is given as,

$$\lambda_i(t) = pt^{p-1} \exp(\gamma_1 age + \gamma_2 gender + \omega_i(t)) \tag{2}$$

where γ_1, γ_2 are the fixed effects corresponding to baseline covariates: age and gender, $\omega_i(t)$ is a function of the association structure between the longitudinal and survival data including the same random effects with the longitudinal process and $\omega_i(t)$ is a function of the association between the longitudinal and survival processes. In the following, we present two different possibilities of this function.

– Parameterization 1: *The random effects predictions at time t* [10]
  The first proposal takes the random time trend into account in the time to event model. In this model, the assocation structure $\omega_i(t)$ is defined as a sum of intercept and slope of trajectory as shown below,

$$\omega_i(t) = U_{0i} + U_{1i}t_{ij}$$

Thus the survival sub-model becomes,

$$\lambda(t) = \lambda_0(t) exp(\beta X + \alpha(U_{0i} + U_{1i}t_{ij}))$$

in which α represent the association between the longitudinal biomarker and the risk for death at time t with a unit change in the marker corresponding to a $exp(\alpha)$ fold change in the risk for death.

– Parameterization 2 *The true unobserved (current) value at time t* [8]
  The second parameterization includes the true value of the longitudinal biomarker at time t into the survival model. In this case, the association structure is defined via:

$$\omega_i(t) = \beta_0 + \beta_1 age + \beta_2 time_{ij} + \beta_3 time_{ij}^2$$

The survival sub-model becomes,

$$\lambda(t) = \lambda_0(t) exp(\beta X + \alpha(\beta_0 + \beta_1 t_{ij} + U_{0i} + U_{1i}t_{ij}))$$

in which α represent the association between the longitudinal biomarker and the risk for death at time t taking into account the true value of the longitudinal biomarker both with fixed and random effects predictions.

The main advantages of two-stage models include fast computing and existing software. However they may lead biased inference because of estimating parameters in the first stage based only observed covariate data. That is, the trajectories of the longitudinal data who experience an event may be very different from those who do not.

3.2 The Joint Modelling Approach

Joint modelling for longitudinal and survival data (JMLS) are based on a full joint distribution of both processes. There are different modelling strategies for the mentioned joint distribution. For instance, the shared random effects framework is based on the simultaneous estimation of both longitudinal and survival through an incorporation of shared random effect which underlines the conditional distributions [10]. We will focus on this framework for the joint modelling approach.

Let the Y be the longitudinal process, T be the survival processes and U be a latent random effect, the joint distribution in a shared random effects framework is as follows,

– *Shared Random Effect Models:* In this framework the latent random effect underlines both longitudinal and survival processes.

$$f(Y, T, U) = f(U)f(Y|U)f(T|U)$$

In JMLS, the longitudinal response is modeled by a linear mixed model (see Model 1) and the survival process is usually modeled by a Cox proportional hazard model [3] including estimations of Model (1) (see Model (2)). In the joint modelling approach we will also use the different parameterizations presented in Sect. 3.1.

4 Model Comparisons

We will use different criterias for model comparison such as log-likelihood and Akaike Information Criterian (AIC) [1]. The log-likelihood and AIC criteria can be useful when the focus is on the overall predictive performance of the joint model, for both longitudinal and survival processes at the same time. However, often the interest is on the survival process and its predictive capacity, particularly how good the longitudinal marker predict the survival.

Due to current trends in medical practice towards to personalized medicine, we use the sensitivity and specificity analysis for chosen patient to explore the predictive performances of the mentioned models by using ROC curves. As a representative example, we choose a specific patient who had a higher EF values and a decreasing levels after a certain time point. Figure 2 shows the EF measurements of this patient over time.

The standard approach proposed by [4] using linear predictors of the survival model. To compute the time-dependent ROC curves for two-stage models we will use the standard approach proposed by [4] using the linear predictors of the survival sub-model. However, in the joint modelling concept, the prediction of the survival sub-model has a dynamic nature that is, as the follow-up study continues additional biomarkers are recorded thus their predictions can be updated. To assess the predictive performances of these models, [8] presented discrimination measures based on Receiver Operating Characteristic (ROC) methodology.

In the following, we present the calculation of ROC curves for each different approach.

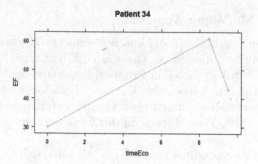

Fig. 2. EF measurements for patient 34.

4.1   Predictive Accuracy of the Two-Stage Modelling

To compute the ROC curves and area under the curve (AUC) two-stage approaches, we based on the idea of [4] using the survival predictions of a specific patient, 34, at time t,

The main idea [4]:

Incident sensitivity/Dynamic specifity (I/D): Each individual is a "control" until time $t < T_i$, and a "case" when $t = T_i$.

$$\text{AUC}(t) = \int_0^1 \text{ROC}_i^{I/D}(p)dp$$

where TP is the True Positive rate (Sensitivity) measures the cases which are correctly identified having the condition and FP is the False Positive rate (Specificity) measures the cases who are correctly identified not having the condition.

4.2   Dynamic Predictions and Predictive Accuracy of Joint Models

The new features of joint modelling approaches include ROC curves analysis with dynamic predictions of the model [8]. In this concept of modelling, the predictions at time t are dynamically changed by a the new longitudinal measure taken at time t. This allows to update the prediction when we have new information recorded for the patient. Thus, the conditional probability is of primary interest, described as,

$$\pi_i(u|t) = P(T_i^* \geq u|T_i^* > t, Y_i(t), \omega_i, D_n),$$

where u is the followed-up time ($u > t$), D_n denotes the sample on which joint model was fitted and ω_i is the baseline covariates. [8] uses a Bayesian formulation of the problem and Monte Carlo estimates of $\pi_i(u/t)$.

   Figure 3 represents the dynamic predictions of the patient 34 for joint model with parameterization 1, for each time of the repeated measurements are taken. We can observe in this figure, that the EF measurements at time 8.5 increase

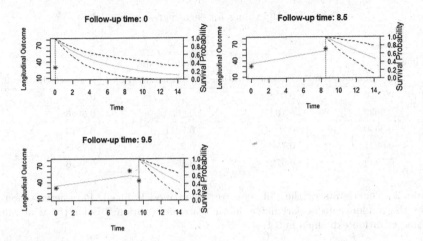

Fig. 3. Dynamic predictions for the specific patient, 34, for the baseline EF measurement only, for the EF measured at time 8.5 and time 9.5 respectively.

the survival probability and the last EF measurement decreases the survival probability.

The ROC curves and AUC values are calculated within this dynamic nature of the survival predictions using the idea of [4].

$$TP_t^I I(c) : P(y_i > c | T_i* = t)$$

$$1 - FP_t^D(c) : P(y_i <= c | T_i* > t)$$

5   Results

The results obtained for each regression model, given in Table 1, show a statistically significant association between the EF measurements and survival process with different coefficients.

According to these results the patients with lower EF tend to have a worse survival. In case of using the true value parameterization, the association coefficients of two-stage approach and joint models are similar. We should note that, the random effects parameterization has slightly different estimations in survival, due to not including the longitudinal fixed effects estimations in the survival model.

Table 2 represents the AUC values for each regression model for specific time points: 2, 4, 5, 9, 12 years. The joint modelling approaches and two-stage models with parameterization 1 show closer AUC values at time points 2 and 4. As the time increases, joint model approaches show better discrimination according to time dependent AUC values on both parameterizations.

Table 1. AUC values for each regression model.

AUC values				
Time	Two-stage		Joint model	
	Parameterization 1	Parameterization 2	Parameterization 1	Parameterization 2
2	0.6550	0.6921	0.6650	0.6575
4	0.6622	0.7027	0.6930	0.6926
5	0.6550	0.7043	0.7211	0.7113
9	0.6821	0.7366	0.7795	0.7898
12	0.5982	0.6725	0.8611	0.8626

Table 2. The results of the different regression models fitted. Parameterization 1 is for the random effects parameterization, Paramaterization 2 is for the true value parameterization explained in 3.1.

	Two-stage		Joint model	
	Parameterization 1	Parameterization 2	Parameterization 1	Parameterization 2
	Coef. (SE)	Coef. (SE)	Coef. (SE)	Coef. (SE)
Longitudinal process				
Intercept	24.33(3.28)	24.33(3.28)	24.18(3.04)	24.30(3.30)
Age	0.08(0.04)	0.08(0.04)	0.09(0.04)	0.08(0.04)
Time	4.99(0.43)	4.99(0.43)	4.38(0.47)	5.00(0.43)
$Time^2$	−0.37(0.06)	−0.37(0.06)	−0.34(0.06)	−0.41(0.06)
Loglikelihood	−3390.608	−3390.608	-	-
Event process				
Intercept	6.66(0.93)	5.44(0.93)	5.46(0.82)	4.61(0.91)
Age	−0.05(0.01)	−0.06(0.01)	−0.04(0.01)	−0.04(0.01)
Gender	−0.61(0.24)	−0.51(0.24)	−0.61(0.23))	−0.49(0.21)
EF (α)	0.01(0.009)	0.05(0.01))	0.01(0.009)	0.03(0.01)
Loglikelihood	−383.2	−369.4	-	-
Joint loglikelihood	−3773.808	−3760.008	−3772.85	−3768.188

6   Discussion

This study shows a comparison of two different regression techniques to study longitudinal and survival data in terms of the predictive capacity of the survival model. Depending on the research question, the interest is often on the predictions of the risk for death in survival studies. Thus, this model comparison showed us important results in term of predictions.

The joint modelling approaches are the most appropriate techniques to study the association between a longitudinal biomarker and the survival process properly. However, in some cases, the joint modelling approaches are difficult to implement. Computational problems arise when the number of longitudinal biomarkers are getting higher. Two-stage modelling is easy to implement and existing software allows different extensions for both longitudinal model and survival model.

The study showed that in terms of early-term mortality, the predictive capacity of both two-stage and joint models are similar. Thus, the two-stage modelling approaches can be used instead of joint models to study longitudinal and survival data when the interest is not especially on the long-term mortality predictions. This allow us to use two-stage models in different extensions such as to study multivariate longitudinal and survival data, avoiding computational problems on full joint likelihood calculation.

7   Software

In this section we will illustrate fitting described regression techniques with two different software: R, version 2.12.0 (R Development Core Team, Vienna, Austria) and SAS (Institute Inc., Cary, NC, USA), version 9.2.

- Fitting two stage and joint model with SAS : We will illustrate the SAS code to fit the two-stage models and joint models with parameterization 1 and 2. We use NLMIXED procedure to fit both models in SAS. The details are given inside the code below.

```
/* We first read our database in long format */
/* and save it as database.long*/

proc nlmixed data=database.long;

/* We need to introduce initial parameters */
/* for procedure nlmixed, these parameters */
/* are coming from a simple linear mixed model for EF measurements*/

parameters beta1=24.33 beta2=4.99 beta3=-0.37
beta4=0.08 bs0=6.66 bs1=-0.05 bs2=-0.61;

/* For the parameterization 1*/
b=u1+u2*ttot;

/*For the parameterization 2, true value at time t */
b=(beta1 +beta2*tsurv+ beta3*tsurv*tsurv +beta4*age+ u1+u2*tsurv);;

/* Because of having a long format of the */
/* database with repeated observations, we need to */
/* indicate the last variable whether is equal to 1, */
/* it indicates the unique survival times. */

if last=1 then do;
linpsurv= bs0+bs1*age+bs2*(gender-1)+w1*b;
alpha=exp(-linpsurv);
G_t=exp(-alpha*tsurv);
g=alpha*G_t;
llsurv=(exitus=1)*log(g)+(exitus=0)*log(G_t);
end; else llsurv=0;
```

```
mean = (beta1+u1) + (beta2+u2)*timeEco + beta3*timeEco*timeEco
+beta4*age;
resid = (EF-mean);
if (abs(resid) > 1.3E100) or (s2 < 1e-12) then do;
lllong = -1e20;
end; else do;
lllong = -0.5*(1.837876 + resid**2 / s2 + log(s2));
end;

var1=tau11*tau11;
cov12=tau11*tau12;
var2=tau12*tau12+tau22*tau22;

/* Depending on this following comment on likelihood calculation,
the model fits separately or jointly the longitudinal and survival model.
 In case of the sum of them we have a joint model estimations*/

model last ~ general(lllong+llsurv);
random u1 u2 ~normal([0,0],[var1,cov12,var2]) subject=id;
run;
```

- Fitting two stage and joint model with R : We will illustrate the R code to fit
the two-stage models and joint models with parameterization 1 and 2. The
following packages are used to fit the models: nlme, survival, JM [7].

```
>library(nlme)
>library(survival)

>lmeFit <- lme(EF~ timeEco+I(timeEco^2)+age,
random = ~ timeEco| id, data = data.long,
na.action=na.omit)

#For the parameterization 1#
>ran=random.effects(lmeFit)
>wt=ran[,1]+ran[,2]*data$tsurv

#For the parameterization 2#

>wt=lmeFit$coefficients$fixed[1]+
>lmeFit$coefficients$fixed[2]*data$tsurv+
>lmeFit$coefficients$fixed[3]*(data$tsurv)^2+
>lmeFit$coefficients$fixed[4]*data$age+
ran[,1]+ran[,2]*data$tsurv

>library(survival)

>fitSurv <- coxph(Surv(tsurv, exitus) ~age+gender+wt,
data = data.id)
```

```
#For joint model we will use JM package (Rizopoulos, 2014)

>library(JM)
>lmeFit <- lme(EF~ timeEco+I(timeEco^2)+age,
random = ~ timeEco| id, data = data.long,
na.action=na.omit)

>fitSurv <- coxph(Surv(tsurv, exitus) ~age+gender,
data = data, x = TRUE)

>jointFit <- jointModel(lmeFit, fitSurv, timeVar ="timeEco")
```

References

1. Akaike, H.: A new look at the statistical model identification. IEEE Trans. Autom. Control **19**(6), 716–723 (1974)
2. Anderson, P.K., Gill, R.D.: Cox's regression model for counting process: a large sample study. Ann. Stat. **10**, 1100–1120 (1982)
3. Cox, D.: Regression models and life-tables (with discussion). J. Roy. Stat. Soc. **34**, 187–220 (1972)
4. Heagerty, P.J., Zheng, Y.: Survival model predictive accuracy and ROC curves. Biometrics **61**(1), 92–105 (2005)
5. Williamson, P., Kolamunnage-Dona, R., Henderson, R.: joineR: joint modelling of repeated measurements and time-to-event data (2012)
6. Pinheiro, J., Bates, D.: Mixed Effects Models in S and S-plus. Springer, New York (2000)
7. Rizopoulos, D.: JM: an R package for the joint modelling of longitudinal and time-to-event data. J. Stat. Softw. **35**(9), 1–33 (2010)
8. Rizopoulos, D.: Joint Models for Longitudinal and Time-to-Event Data. With Applications in R. Chapman Hall/CRC Biostatistics Series, Boca Raton (2012)
9. Self, S., Pawitan, Y.: Modelling a marker of disease progression and ofset of disease. In: Jewell, N.P., Dietz, K. Farewell, V.T. (eds.) AIDS Epidemiology: Methodological Issues. Birkhauser, Boston (1992)
10. Wulfsohn, M.S., Tsiatis, A.A.: A joint model for survival and longitudinal data measured with error. Biometrics **53**, 330–339 (1997)

Workshop on Computational Methods for Business Analytics (CMBA 2017)

Sub-hour Unit Commitment MILP Model with Benchmark Problem Instances

Paula Carroll[1,2]([✉]), Damian Flynn[2], Bernard Fortz[3,4], and Alex Melhorn[2]

[1] Centre for Business Analytics, University College Dublin, Dublin, Ireland
paula.carroll@ucd.ie
[2] Electricity Research Centre, University College Dublin, Dublin, Ireland
[3] Département d'informatique, Univeristé Libre de Bruxelles, Bruxelles, Belgium
[4] INOCS, INRIA Lille Nord-Europe, Lille, France

Abstract. Power systems are operated to deliver electricity at minimum cost while adhering to operational and technical constraints. The introduction of smart grid technologies and renewable energy sources offers new challenges and opportunities for the efficient and reliable management of the grid. In this paper we focus on a Mixed Integer Programming sub-hour Unit Commitment model. We present analysis of computational results from a large set of problem instances based on the Irish system and show that problem instances with higher variability in net demand (after the integration of renewables) are more challenging to solve.

1 Integrating Renewable Energy Sources (RESs)

The Irish Government is aiming for 40% of electricity to be generated from renewable energy sources (RESs) by 2020 in response to an EU directive. Ireland is rich in wind resources but integrating wind energy creates new operational and planning challenges for Transmission System Operators (TSOs).

In a deregulated market the TSO plays a central role in determining which generating units should be committed to meet estimated demand. The classical Unit Commitment (UC) problem determines which generators to start up (shut down) on a day-ahead scheduling basis. The demand to be met by traditional thermal generators can be estimated as the forecast load. This gross load can be offset by the power available from RESs and demand response. This approach can lead to net demand load patterns that are quite different to the typical diurnal electricity demand pattern. An example of the typical diurnal demand pattern is shown in Fig. 1a. The more variable net load instances, such as that in Fig. 1b, prove more computationally challenging than instances exhibiting the traditional diurnal pattern.

The need for more detailed UC models is addressed in [23]. Operation of the system at sub-hourly levels offers increased flexibility [15,24], but leads to computational challenges for mixed integer linear programming (MILP) models. Using MILP models to solve continuous time problems leads to issues of discretisation. We need to adjust the models to cater for the finer time step granularity.

O. Gervasi et al. (Eds.): ICCSA 2017, Part II, LNCS 10405, pp. 635–651, 2017.
DOI: 10.1007/978-3-319-62395-5_44

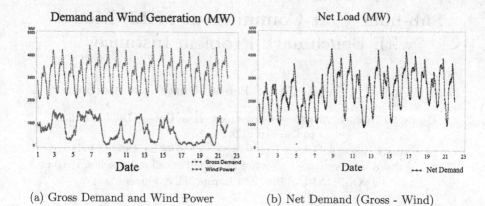

(a) Gross Demand and Wind Power   (b) Net Demand (Gross - Wind)

Fig. 1. Comparison of gross and net demand instances, Ireland January 2014.

These challenges provide new opportunities for the business analytics and optimisation communities to design efficient solution approaches.

The contributions of this paper are a detailed sub-hourly UC MILP model, and a set of insights gained from computational experiments on realistic UC test instances based on the Irish system. Our analysis gives some insight into what makes a UC instance more difficult to solve.

2   The Unit Commitment Problem

We focus on the Thermal Unit Commitment problem. i.e., conventional thermal generating units where fuel is converted to produce electric power. The UC problem can be stated as follows:

Instance:

- a set of thermal generating units (GU) G and their operating characteristics,
- a set of load demands D and required reserves R per time step of a planning horizon of K periods. In our case, a set of wind energy levels are also given.

Problem Statement:

Determine the minimum cost GU dispatch schedule that meets forecast demand and satisfies the GU operating characteristics.

We consider a planning horizon of a single day. In practice UC MILP models are solved by the TSO for the day-ahead market with subsequent in-day updates and real time adjustments. Estimated demand must be met. Hydro, interconnect, renewable energy and demand side interventions can be reflected as a simple reduction in demand. In this paper focusing on the integration of wind energy, the amount of demand D to be met by thermal generation can be reduced by the amount of wind power available.

Reserve power in the system is specified in case of failures or outages. The higher the reserve value, the better the operational security, but at a higher cost. In much of the literature a simple 10% reserve rule is suggested, which means that demand is effectively inflated by 10%. Blackouts in recent years have increased the focus on the design and operation of secure grids [4]. So called "$n - 1$" constraints ensure the reserve must be equal to or greater than the largest generator. In the case of systems with significant utilisation of RESs, the variable nature of the source leads to additional focus on forecasting techniques to quantify reserve requirements and to reduce the supply side forecast error. See for example [10].

In addition, there is increased interest in algorithmic techniques that address the data uncertainty in energy problems. Stochastic programming and robust optimisation techniques have been applied to various aspects of power grid operations. See for example [2,18,25].

The focus of this paper is on a more detailed unit commitment model which is of practical interest as electricity markets migrate to sub-hourly operation to facilitate the integration of renewable energy sources, demand response and smart grid initiatives.

Each GU G, indexed by g, has a set of operating characteristics specified as:

$\overline{P}_g$ Maximum output (MW)

$\underline{P}_g$ Minimum output (MW)

UT_g Minimum time that a unit must stay online (up) once it has been switched online

DT_g Minimum time that a unit must stay offline (down) once it has been switched offline

IS_g Initial state, number of time steps unit on (off) line at $k = 0$, the time step just prior to the start of planning horizon

a_g, b_g, c_g Coefficients of quadratic power production cost function

hc_g, cc_g Hot (cold) start cost coefficients

t_g^{cold} Number of time steps for unit to cool fully (after min down time)

GU power production costs are described by quadratic functions. MILP approaches can be used in conjunction with piecewise approximations to solve UC instances. There are many additional GU parameters that may be considered, such as:

ramp-up/down (RU_g, RD_g) **limits:** rate of power change when running

start-up/shut-down (SU_g, SD_g) **limits:** rate of power change at start up/shut down

shut-down costs, C_g^d**:** the cost of lost fuel

Further variants of the basic UC problem may also include power output at start of planning horizon, final state requirements, operational requirements, maintenance schedules, cycling constraints, fuel usage constraints, plant crew considerations, emission constraints (CO_2, NO_x, and SO_x), reserve constraints to ensure security (Simple rule (x%), Contingency or Control) and network or transmission constraints.

A review of UC solution approaches is given in [19]. Approaches include dynamic programming (DP), Lagrangian relaxation (LR), MILP, simulated annealing (SA), expert systems and artificial neural networks, fuzzy systems, genetic algorithms (GA), evolutionary programming (EP), ant colony heuristics, particle swarm optimisation and hybrid approaches. In many cases the test systems are not fully described making reproduction and comparison of the empirical results difficult. Table 1 gives a summary of highly cited UC solution approaches.

Table 1. UC solution approaches

Reference	Year	Approach	Test data
[14]	1996	GA	10 base units, no ramping rates or shutdown costs
[5]	2006	MILP	10–100 units based on [14]
[27]	1988	LR	100 units, details not available
[13]	1999	EP	10–100 units based on [14]
[17]	1983	LR	172 units, details not available
[7]	2000	LR, GA	10–100 units based on [14]
[21]	1996	SP	not specified
[20]	1987	DP	not specified
[26]	1990	SA	10 and 100 units, details not available
[9]	1978	SP	5 units, details given
[8]	1983	MILP	not specified
[25]	2009	MILP	45 unit test system, details not available
[22]	2006	PSO	10 units from [14]

3   Unit Commitment (Sub-hour) MILP Model

A UC model with ramping constraints is described in [5]. This paper presents a sub-hour variant of the UC MILP model with ramping constraints. We include the following ideas:

1. A set of real variables are introduced to simplify the implementation of hot and cold start costs;
2. A set of start up and shut down variables are introduced to allow a slow unit to start up or shut down over a number of time steps. This is important in sub-hour models with finer time step granularity;
3. The ramp (start/shutdown) constraints are adapted for slow units and sub-hourly models;

Let k index the time steps in K, giving demand and reserve per time step D_k and R_k. $k = 0$ is the time step immediately prior to the start of the planning horizon.

Let $c_{g,k}^p, c_{g,k}^u, c_{g,k}^d \in \mathbb{R}^+$ be sets of decision variables representing the power production, start-up and shut-down costs respectively. Let $p_{g,k}$ and $\overline{p}_{g,k} \in \mathbb{R}^+$ be the power output and power availability variables. $v_{g,k}$ are binary variables set to 1 if unit g is on, zero otherwise.

With these sets of variables, a UC model can be formulated as follows:

$$\min \quad \sum_{k \in K} \sum_{g \in G} c_{g,k}^p + c_{g,k}^u + c_{g,k}^d \tag{1}$$

$$\text{s.t.} \quad \sum_{g \in G} p_{g,k} = D_k \qquad\qquad \forall k \in K \tag{2}$$

$$\sum_{g \in G} \overline{p}_{g,k} \geq D_k + R_k \qquad\qquad \forall k \in K \tag{3}$$

$$\underline{P}_g \cdot v_{g,k} \leq p_{g,k} \leq \overline{p}_{g,k} \qquad\qquad \forall g \in G, k \in K \tag{4}$$

$$\overline{p}_{g,k} \leq \overline{P}_g \cdot v_{g,k} \qquad\qquad \forall g \in G, k \in K \tag{5}$$

$$c_{g,k}^p = a_g + b_g \cdot p_{g,k} + c_g \cdot p_{g,k}^2 \qquad\qquad \forall g \in G, k \in K \tag{6}$$

Objective: The objective is to minimise operating costs over a planning horizon, usually a (rolling) daily horizon. Traditionally power production, start-up and shut-down costs of each generator are included. Work is ongoing on how to best capture the cost of RESs, reserve, cycling and emissions in the objective function. Note that the objective function in this model does not explicitly charge for reserve, i.e., only p appears in the objective, not $\overline{p}$.

Production Constraints: The primary constraint is the production constraint (2). Total production of the units at a given time must equal the demand. An approach to integrating RESs, is to reduce the demand target by the amount of renewable capacity predicted to be available to give *net* demand.

Reserve constraints (3) ensure that the maximum production available meets the additional reserve target. Constraints (4) and (5) ensure the production of an individual unit lies between its minimum and maximum output when online.

Production Cost: The quadratic power production cost in (6) is usually approximated by a piecewise linear approach [1,5,12,23]. We use a delta-approach to a piecewise linear approximation of $c_{g,k}^p$ by L line segments, $\delta_{g,k,l}, l = 1, \ldots, L$ are the variables, giving:

$$c_{g,k}^p = A_g \cdot v_{g,k} + \sum_{l \in L} F_{l,g} \cdot \delta_{l,g,k} \qquad\qquad \forall g \in G, k \in K \tag{7}$$

$$p_{g,k} = \underline{P}_g \cdot v_{g,k} + \sum_{l \in L} \delta_{l,g,k} \qquad\qquad \forall g \in G, k \in K \tag{8}$$

$$\delta_{1,g,k} \leq T_{1,g} - \underline{P}_g \qquad\qquad \forall g \in G, k \in K \tag{9}$$

$$\delta_{l,g,k} \leq T_{l,g} - T_{l-1,g} \qquad\qquad \forall g \in G, k \in K, l = 2, \ldots, L \tag{10}$$

where A_g is the no-load cost given by: $A_g = a_g + b_g \cdot \underline{P}_g + c_g \cdot \underline{P}_g^{\,2} \quad \forall g \in G$, $F_{l,g}$ is the slope of line segment $l \in L$, and $T_{l,g}$ are the breakpoints of the power intervals from $\underline{P}_g$ to $\overline{P}_g$. Note in this implementation $T_{L,g} = \overline{P}_g \quad \forall g \in G$.

Minimum Up and Down Times: Unit g is required to stay on initially for at least $G_g := \min(|K|, \max((UT_j - IS_g)v_{g,0}, 0))$ steps once turned on. $v_{g,0}$ indicates the initial status of unit g. Variables $v_{g,k}$ can be fixed for the required number of steps for units that must be kept on initially.

During the operating horizon the minimum up constraints are given by:

$$\sum_{n=k}^{k+UT_g-1} v_{g,n} \geq UT_g(v_{g,k} - v_{g,k-1}) \quad \forall g \in G, k = G_g+1, \ldots, |K|-UT_g+1 \tag{11}$$

Minimum up constraints for the final steps of the horizon are:

$$\sum_{n=k}^{|K|} (v_{g,n} - (v_{g,k} - v_{g,k-1})) \geq 0 \quad \forall g \in G, k = |K| - UT_g + 2, \ldots, |K| \tag{12}$$

Similarly, (13) and (14) enforce a minimum down time of L_g time steps when unit g is switched off where $L_g = \min(|K|, \min(DT_g + IS_g)(1 - v_{g,0}), 0))$.

$$\sum_{n=k}^{k+DT_g-1} (1 - v_{g,n}) \geq DT_g(v_{g,k-1} - v_{g,k}) \qquad \forall g \in G, k = L_g + 1, \ldots, |K| - DT_g + 1 \tag{13}$$

$$\sum_{n=k}^{|K|} (1 - v_{g,n} - (v_{g,k-1} - v_{g,k})) \geq 0 \qquad \forall g \in G, k = |K| - DT_g + 2, \ldots, |K| \tag{14}$$

Minimum up and down constraints (11) and (13) can be strengthened by disaggregation, but our experience shows that it decreases the performance of the solver as it considerably increases the size of the model.

3.1   Slow and Sub-hour Ramping

UC models such as the one used in [5] allow a unit to turn on if it can ramp up to $\underline{P}_g$ in a single time step. Likewise, a unit can only be turned off if it can do so in a single step. This results in unrealistic solutions that reflect the problem instance initial status. A unit with slow start-up or shut-down rates may need to be turned on (off) and allowed to start-up (shut-down) over a series of steps. This issue becomes a particular concern when operating at a sub-hourly resolution when even fast units may require a number of sub-hourly time steps to reach operating power limits.

A BigM approach is used in [5] to model the ramping constraints. As an alternative approach to the ramping constraints in [5], we introduce additional binary variables u and $w \in \{0,1\}$ a unit can be allowed to turn on (shut down) over a series of steps before (after) the unit is in the synchronised production state as shown in Fig. 2. Power generated during the start up and shut down phases can be used to satisfy demand but stable system operation is only ensured when the production power for a unit that is on is maintained within its generation limit bounds, $\underline{P}_{g,k}$ and $\overline{P}_{g,k}$. The constraints below ensure that a unit is started up (shut down) as quickly as possible. The costs of power produced during the starting up and shut down phases are captured in the start up and shut down fixed costs respectively.

The minimum number of time steps required for a unit to start up is calculated as $SUT_g = \lfloor \underline{P}_g / SU_g \rfloor$. Similarly, the minimum number of time steps for a unit to shut down is $SDT_g = \lfloor \underline{P}_g / SD_g \rfloor$. This allows a unit to be brought just below $\underline{P}_g$ at time step k by starting up at the maximum startup rate. It is then ready to breach $\underline{P}_g$ at or below the ramp up rate in step $k+1$.

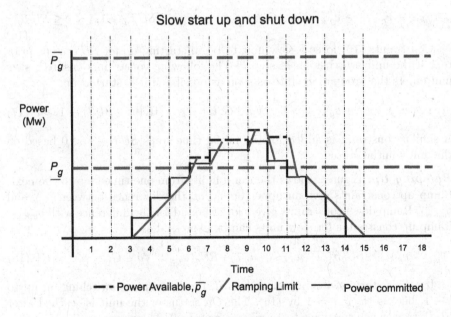

Fig. 2. Slow/sub-hour ramping restrictions.

Likewise, a unit can be brought from just above $\underline{P}_g$ at time $k-1$ to below $\underline{P}_g$ at k and shut down in SDT_g steps at the maximum shut down rate. The available power during normal operations is still restricted to within the unit's limits. However using this approach, a small amount of additional power is available during a start up or shut down. Constraints (2) can be modified to:

$$\sum_{g \in G} p_{g,k} + SU_g \left(\sum_{l=\min(1,k-SUT_g+1)}^{k} u_{g,k} \right) + SD_g \left(\sum_{l=k}^{\min(k+SDT_g-1,|K|)} w_{g,k} \right) = D_k \quad \forall k = 1 \ldots |K|$$

(2a)

The amount of power during start up or shut down is bounded by the minimum power threshold $\underline{P}_g$. If a unit is on at time k, the power available from start up completion at $k-1$ or start of shut down at $k+1$ is bounded by $\underline{P}_g$.

$$SU_g \left(\sum_{l=k-SUT_g}^{k-1} u_{g,l} \right) + SD_g \left(\sum_{l=k+1}^{k+SDT_g} w_{g,l} \right) \leq \underline{P}_g v_{g,k} \quad \forall g \in G, k = 1 \ldots |K|$$

(15)

Fast units can start or shut down in a single step. In such cases u or w variables are not required. The u and v variables for slow units are linked as follows:

$$u_{g,k-l} + v_{g,k-1} \geq v_{g,k} \quad \forall g \in G, k = 1, \ldots, |K| \| k - SUT_g \geq 1, 1 \leq l \leq SUT_g$$

(16)

Constraints (16) force a slow unit to begin starting up for SUT steps prior to k if the unit switches on in step k. The following constraints ensure a slow unit enters the production state as soon as possible if it is started up:

$$v_{g,k} \geq u_{g,k-SUT} - u_{g,k-SUT-1} \quad \forall g \in G, k = 1, \ldots, |K| \| k - SUT_g - 1 \geq 1 \quad (17)$$

A similar constraint is added for the initial time steps $SUT - 1 < 0$ based on the unit's initial status.

Ramping Up: Figure 2 shows the ramp limits from one time step to the next. Ramp up constraints give an upper bound on the difference between $\overline{p}_{g,k}$ and $p_{g,k-1}$. Ramp down constraints give a lower bound for the difference with $p_{g,k+1}$. Ramp up constraints for slow units can be expressed as:

$$\overline{p}_{g,k} \leq p_{g,k-1} + SU_g SUT_g (u_{g,k-1} - u_{g,k}) + RU_g (v_{g,k}) \quad \forall g \in G, k = 2 \ldots |K| \quad (18)$$

If a unit starts switching on at $k - SUT_g$, it continues switching on up to $k - 1$, because $u_{g,k-1} = 1$ by (16). The On status of the unit is reached after SUT_g steps when $v_{g,k}$ goes to 1 and $u_{g,k}$ goes to 0. The power available in step k is the SU_g term of (18) plus some ramp up in step k (bounded by RU_g).

If the unit is already on and was not started in step $k-1$ ($u_{g,k-1} = 0$), then the u terms are zero and the power available in k can increase from the power output in the previous step plus an amount up to the ramp-up rate RU_g. This is captured in the RU_g term of (18).

The equivalent ramp up constraint for a fast start unit is:

$$\overline{p}_{g,k} \leq p_{g,k-1} + SU_g (v_{g,k} - v_{g,k-1}) + RU_g v_{g,k-1} \quad \forall g \in G, k = 2 \ldots |K| \quad (19)$$

The power available in the first time step needs to be handled separately and is dependent on the initial conditions of the system.

Constraints (19) ensure that the system is capable of ramping up to $\overline{p}_{g,k}$ from $p_{g,k-1}$ in a single time step. In general, the system deploys an amount of power equal to $p_{g,k}$ but must be capable of delivering $\overline{p}_{g,k}$ in the event of a failure in the system or significant deviation from forecast load values during operation.

Ramping Down: A similar approach is taken to the shutting down of slow units. The v and w variables are linked as follows:

$$w_{g,k+l}+v_{g,k+1} \geq v_{g,k} \quad \forall g \in G, k = 1,\ldots,|K|-1|k+SDT_g \leq |K|, 1 \leq l \leq SDT_g \tag{20}$$

These constraints are valid for $k+SDT_g \leq |K|$ as the model is only concerned with the planning horizon of K steps. They force a unit to shut down over SDT_g steps if a slow unit switches off ($v_{g,k} = 1$ and $v_{g,k+1} = 0$).

Ramp down constraints for slow units can then be expressed as:

$$\overline{p}_{g,k-1} \leq p_{g,k} + SD_g SDT_g(w_{g,k} - w_{g,k-1}) + RD_g v_{g,k-1} \quad \forall g \in G, k = 2,\ldots,|K| \tag{21}$$

The available power in time step $k - 1$ less the shut down or ramp down rate cannot exceed the output power $p_{g,k}$ at the next step k. The slow unit shuts down over SDT_g steps after k.

Ramp down constraints for fast units are:

$$\overline{p}_{g,k-1} \leq p_{g,k} + SD_g(v_{g,k-1} - v_{g,k}) + RD_g v_{g,k} \quad \forall g \in G, k = 2,\ldots,|K| \tag{22}$$

A slow unit can only be in any one state (starting up, in production or shutting down) at any time k so the following are valid:

$$u_{g,k} + v_{g,k} + w_{g,k} \leq 1 \quad \forall g \in G, k \in K \tag{23}$$

Power Output Ramping: In addition to the constraints on $\overline{p}_{g,k}$, the difference between the power output in step k, $p_{g,k}$ and the power output in its neighbouring time step must also be bounded.

$$p_{g,k-1} \leq p_{g,k} + SD_g(v_{g,k-1} - v_{g,k}) + RD_g v_{g,k} \forall k = 2,\ldots,|K| \tag{24}$$

A similar constraint adapted for slow units is:

$$p_{g,k-1} \leq p_{g,k} + SD_g SDT_g(w_{g,k} - w_{g,k-1}) + RD_g v_{g,k-1} \quad \forall g \in G, k = 2,\ldots,|K| \tag{25}$$

3.2   Start Up and Shut Down

A set of real start up and shutdown variables (bounded by 1) $cstart_{g,k}$, $hstart_{g,k}$ were introduced to simplify implementation of hot and cold start costs, they indicate a unit is cold-started/hot-started in time k. As noted in [12], such real variables are helpful in implementing some of the inequalities in UC MIP models and do not substantially impact computational performance.

Start-up Cost: It takes time for a unit to warm up before it can be synchronised to the system. It takes further time to reach minimum production limit, $\underline{P}_g$. During these times the units incur fuel costs. The longer the unit has been offline, the colder it will be and the more fuel it will require starting up.

In most MILP formulations, the start-up exponential costs are approximated by a simple step function. It is assumed that the start up cost is triggered when any slow unit begins to start up or any fast unit switches state from off to on. For any fast unit ($SUT_g = 1$) the following is valid:

$$cstart_{g,k} \geq v_{g,k} - \sum_{n=1|n<k}^{t_g^{Cold}+DT_g} v_{g,k-n} \quad \forall g \in G, k \in K \tag{26}$$

In the case of slow units, $SUT_g > 1$, a unit that begins to start up in time step k will be fully synchronised and ready to deliver power in time $k + SUT_g$. The start up cost is assumed to be incurred at time $k + SUT_g$:

$$cstart_{g,k+SUT_g} \geq u_{g,k} - \sum_{n=1|n<k}^{t_g^{Cold}+DT_g} u_{g,k-n} \quad \forall g \in G, k \in K \tag{27}$$

In both cases the summation term is only valid for $k > n$. Conditions for the periods before the planning horizon are handled separately.

The unit is hot-started if online in k, offline in $k-1$ and not cold started:

$$hstart_{g,k} \geq v_{g,k} - v_{g,k-1} - cstart_{g,k} \quad \forall g \in G, k = 2,\ldots,|K| \tag{28}$$

Start-up costs can be captured as:

$$c_{g,k}^u \geq cc_g(cstart_{g,k}) + hc_g(hstart_{g,k}) \quad \forall g \in G, k \in K \tag{29}$$

Start-up costs occur only in time-step k if the generator was offline in time-step $k-1$ and then either hot or cold started in k, so the following are also valid for each GU and time step:

$$v_{g,k-1} + hstart_{g,k} + cstart_{g,k} \leq 1 \quad \forall g \in G, k = 2,\ldots,|K| \tag{30}$$

$$hstart_{g,k} + ctart_{g,k} \leq v_{g,k} \quad \forall g \in G, k \in K \tag{31}$$

Shut-down Cost: A traditional thermal generator must lower its output and then de-synchronise from the system before shutting off. During this time, units are using fuel and generating power between minimum generation and zero, therefore incurring a cost.

$$c_{g,k}^d \geq C_g^d(v_{g,k-1} - v_{g,k}) \quad \forall g \in G, k = 2,\ldots,|K| \tag{32}$$

4 Methodology

The MILP model is not full dimensional so that a polyhedral analysis is difficult. Strong formulations can be identified by empirical testing on meaningful test instances. The UC MILP model described in Sect. 3 was implemented in C and solved using XpressMP 7.7 on a Dell 64 bit Windows 8 machine with Intel i5 3.2 GHz processor and 8 GB of Ram. The implementation was first tested and verified on the Kazarlis 10 unit system [14].

The MILP model was then tested on UC instances based on the 54 unit Irish system with demand and wind power data for 2014 at a 15 time step. The year was solved on a rolling basis. Each 24 h period was solved, the system settings at the end of the day were used as the initial conditions for the following day. We tested (1) the gross demand and (2) the net demand (the gross demand offset by the available wind). The demand and wind power data for Ireland in 2014 were extracted from [11]. For the purposes of testing, actual wind power (which may have been curtailed at certain times) was used rather than forecasted wind.

The Irish generation system data is derived from the Single Energy Market data available from Ireland's Commission for Energy Regulation, [6]. Figure 3 shows the fuel mix used is 2014. Many units have low min up/down times which provides flexibility to integrate wind power. The Irish system has approximately 8,500 MW of conventional power, some pumped storage, no nuclear units, approximately 3,000 MW wind capacity and two HVDC interconnectors to the UK. Hydro-units and pumped storage were removed for the purposes of testing. Initial states were based on the GUs most likely to be online, [6]. Reserve was assumed to be 10% of demand with more realistic rules to be tested later.

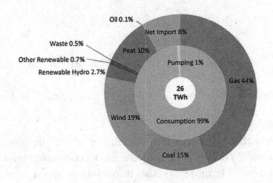

Fig. 3. Fuel mix used in Ireland to satisfy demand in 2014

5 Results and Analysis

Missing values of the load data were imputed during pre-processing. Initial analysis of the 2014 load data revealed seasonal, trend and diurnal patterns. The profile of the diurnal pattern is similar across weekdays with only slight weekend

variation at weekends as shown in the relative daily demand profiles in Fig. 4a.
The amplitude of the profile differs by season with stronger evidence of distance
between the seasonal profiles in Fig. 4b which was confirmed using a Euclidean
measure.

We also tested fitting an ARIMA model to describe the load data. The data
were smoothed and adjusted by the soil temperature at Dublin Airport, the seasonal trend was removed. The ARIMA model could be used in future stochastic
programming implementations.

There was evident variability in the wind power data. Our interest in the
load and wind data in this paper is to compare the gross load and net load after
integrating the wind power. Several similarity or distance measures can be used
to quantify the similarity/difference between time series data. We calculated the
absolute difference between the two series as a simple measure and found an
average of 577 MW. Figure 5 shows the distributions of the gross and net loads
for comparison.

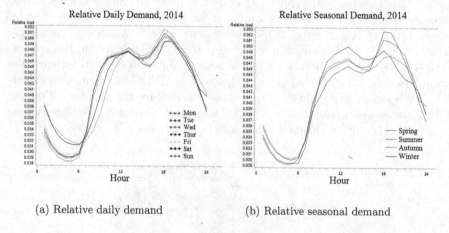

(a) Relative daily demand   (b) Relative seasonal demand

Fig. 4. Comparison of daily and seasonal load profiles, Ireland 2014

Next we discuss the MILP results. A number of UC instances based on the
test system in [14] were tested. It is not possible to compare these results directly
to those published in [14] or [5] as the information given there is incomplete
e.g., we have no knowledge of the system initial power levels, ramping rates or
shutdown costs. These have a significant bearing on the feasibility of problem
instances and solution times. The number G_g of steps a unit must stay on initially
depends on the initial state of the system. This has particular significance for
the practical application of UC MILP models which TSOs use to manage the
electricity system on a rolling daily basis. Very small changes to the initial system
commitment can make the same load profile become a difficult problem instance.

The MILP model was then tested on the 54 unit Irish test system at a
15 min time resolution. Results below show the impact of the problem instance

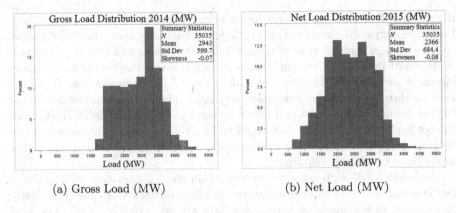

(a) Gross Load (MW)   (b) Net Load (MW)

Fig. 5. Distributions of gross and net load (MW)

variability on solution times. The average solution times for the gross load values i.e., disregarding the wind power, was 70 s with a standard deviation of 10 s. All instances solved at the root node. The minimum load value of 1,665 MW occurs on a summer night while the maximum demand of 4,614 MW occurs on a winter evening. This is consistent with traditional demand profiles in a temperate climate like Ireland. Figure 6 shows distributions of MILP run times for the gross and net instances.

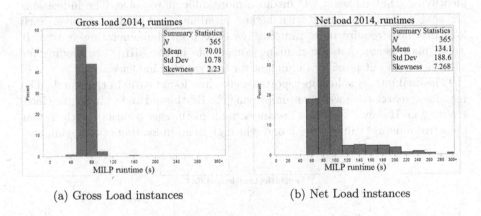

(a) Gross Load instances   (b) Net Load instances

Fig. 6. MILP runtime distributions

The net load instances proved more challenging. In these instances the gross demand is offset by the wind power to give a net load instance. The minimum net load value of 642 MW occurs on a winter morning while the maximum net load of 4,487 MW occurs on a winter evening. The depth of the annual net load low gives some indication of the challenges in managing systems with significant

RESs. Only five thermal generating units are required to meet this demand. Using the simple 10% reserve rule highlights the security issue that would arise.

The average solution time of the net load instances was 134 s with a standard deviation of 187s. There was an average of 72.3 nodes in the tree although many instances solved quickly at the root node. Figure 7 shows an example of the net load for three consecutive days in January 2014. The instances on either side of 23rd of January solve in 76 and 83 s respectively at the root node. In contrast the net load of 23rd solves in just over 1,024 s after exploring 3,698 nodes. It can be observed that the more challenging instance has a deeper trough and higher peak. More ramping is required.

A simple MLR regression model to explain the solution time of a problem instance was tested. The load instances are effectively a set of time series data. Approaches to summarising time series data are described in [3,16]. The load data can be represented as a set of summary statistics such as measures of central tendency and variation. In the case of load data, areas such as night time valley, morning peak, evening peak may also be useful in summarising the load instance. In all, 21 possible explanatory variables were identified. A stepwise backward elimination approach was used to identify which were statistically significant in the regression model. The final regression model used only the average down ramp, variance and standard deviation of the load instance (R^2 value of 0.42). A GLM model with an interaction term between the down ramp and variance improved the fit of the final model slightly. While this is not a particularly good model to predict the runtime for this UC model, it is useful for our purpose of identifying what makes a UC instance more difficult to solve. The indications are that problem instances with higher variability are harder to solve. Such load instances require more ramping response from the generation system. It would make sense that the ramping constraints in the MILP are binding for such instances but possibly redundant for less variable instances.

The final integer solutions reported were often found early in the search. The remaining search time was spent improving the Bestbound and reducing the integrality gap. However for some instances, multiple integer solutions with similar objective function values were found which gives an indication of the symmetry

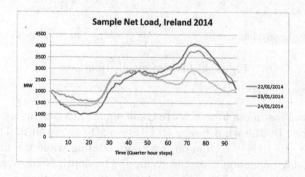

Fig. 7. January 2014

problems that can arise. MILP techniques are considered exact approaches. However commercial solvers employ a number of heuristics and cut strategies to improve performance. This suggests that the solver plays a significant role in improving solution times. The nature and details of the techniques used by the solvers are not generally publicly available so the solver can be treated as a black-box, with each of the control parameters analogous to a treatment effect with possible interactions. The MILP model was tested in a Design of Experiments framework to evaluate which solver control parameters settings might be beneficial for more challenging UC instances. The control parameters considered were presolve, cutstrategy and heurstrategy. The default control parameter settings were most effective in general.

6 Conclusion

This paper presents a UC sub-hourly MILP model and demonstrates the model's performance variability on a large set of test instances based on the Irish electricity system. Problem instances where the demand is offset by the available wind power are more challenging to solve and involve more ramping of the generation system. Solutions tend to shut down in the last time step(s) as the approach does not look forward beyond the 24 h horizon. The power available during start-up and shut-down of slow units is not directly costed in the objective function but is included in the startup and shut down costs. In the case of the Irish system, the shut down costs are zero, so units can be shut down freely. This overly flexible approach to operating the grid may not be desirable. A symmetry problem was noted in the MILP solutions which additional reserve and cycling constraints may reduce. Switching to sub-hourly time steps allows more flexibility in grid operations. A sub-hourly approach not only requires changes to traditional UC constraints, but may also require new constraints to better approximate the future grid desired operations.

Acknowledgements. The work of Bernard Fortz is supported by the Interuniversity Attraction Poles Programme P7/36 "COMEX" initiated by the Belgian Science Policy Office. Alex Melhorn is funded by the Science Foundation Ireland Sustainable Electrical Energy Systems Strategic Research Cluster grant (SFI/09/SRC/E1780).

References

1. Arroyo, J., Conejo, A.: Optimal response of a thermal unit to an electricity spot market. IEEE Trans. Power Syst. **15**(3), 1098–1104 (2000)
2. Bertsimas, D., Litvinov, E., Sun, X., Zhao, J., Zheng, T.: Adaptive robust optimization for the security constrained unit commitment problem. IEEE Trans. Power Syst. **28**(1), 52–63 (2013)
3. Bickel, P.J., Lehmann, E.L.: Descriptive statistics for nonparametric models I. Introduction. Ann. Stat. **3**(5), 1038–1044 (1975)
4. Bienstock, D., Mattia, S.: Using mixed-integer programming to solve power grid blackout problems. Discrete Optim. **4**(1), 115–141 (2007)

5. Carrion, M., Arroyo, J.: A computationally efficient mixed-integer linear formulation for the thermal unit commitment problem. IEEE Trans. Power Syst. **21**(3), 1371–1378 (2006)
6. CER: Validated 2011–12 sem generator data parameters. Technical report, Commission for Energy Regulation (2011). http://www.allislandproject.org/en/market_decision_documents.aspx?page=4&article=151a9561-cef9-47f2-9f48-21f6c62cef34. Accessed Nov 2013
7. Cheng, C.P., Liu, C.W., Liu, C.C.: Unit commitment by Lagrangian relaxation and genetic algorithms. IEEE Trans. Power Syst. **15**(2), 707–714 (2000)
8. Cohen, A.I., Yoshimura, M.: A branch-and-bound algorithm for unit commitment. Power Eng. Rev. IEEE PER **3**(2), 34–35 (1983)
9. Dillon, T., Edwin, K.W., Kochs, H.D., Taud, R.J.: Integer programming approach to the problem of optimal unit commitment with probabilistic reserve determination. IEEE Trans. Power Appar. Syst. PAS **97**(6), 2154–2166 (1978)
10. Doherty, R., O'Malley, M.: A new approach to quantify reserve demand in systems with significant installed wind capacity. IEEE Trans. Power Syst. **20**(2), 587–595 (2005)
11. Eirgrid: System performance data (2016). http://smartgriddashboard.eirgrid.com. Accessed Mar 2017
12. Hedman, K., O'Neill, R., Oren, S.: Analyzing valid inequalities of the generation unit commitment problem. In: Power Systems Conference and Exposition, PSCE 2009, pp. 1–6. IEEE/PES (2009)
13. Juste, K.A., Kita, H., Tanaka, E., Hasegawa, J.: An evolutionary programming solution to the unit commitment problem. IEEE Trans. Power Syst. **14**(4), 1452–1459 (1999)
14. Kazarlis, S., Bakirtzis, A., Petridis, V.: A genetic algorithm solution to the unit commitment problem. IEEE Trans. Power Syst. **11**(1), 83–92 (1996)
15. Kiviluoma, J., Meibom, P., Tuohy, A., Troy, N., Milligan, M., Lange, B., Gibescu, M., O'Malley, M.: Short-term energy balancing with increasing levels of wind energy. IEEE Trans. Sustain. Energy **3**(4), 769–776 (2012)
16. McLoughlin, F., Duffy, A., Conlon, M.: Characterising domestic electricity consumption patterns by dwelling and occupant socio-economic variables: an Irish case study. Energy Build. **48**, 240–248 (2012)
17. Merlin, A., Sandrin, P.: A new method for unit commitment at electricite de France. IEEE Trans. Power Appar. Syst. PAS **102**(5), 1218–1225 (1983)
18. Papavasiliou, A., Oren, S.S.: Multiarea stochastic unit commitment for high wind penetration in a transmission constrained network. Oper. Res. **61**(3), 578–592 (2013)
19. Sheble, G., Fahd, G.: Unit commitment literature synopsis. IEEE Trans. Power Syst. **9**(1), 128–135 (1994)
20. Snyder, W.L., Powell, H., Rayburn, J.C.: Dynamic programming approach to unit commitment. IEEE Trans. Power Syst. **2**(2), 339–348 (1987)
21. Takriti, S., Birge, J., Long, E.: A stochastic model for the unit commitment problem. IEEE Trans. Power Syst. **11**(3), 1497–1508 (1996)
22. Ting, T., Rao, M.V.C., Loo, C.: A novel approach for unit commitment problem via an effective hybrid particle swarm optimization. IEEE Trans. Power Syst. **21**(1), 411–418 (2006)
23. Troy, N., Flynn, D., Milligan, M., O'Malley, M.: Unit commitment with dynamic cycling costs. IEEE Trans. Power Syst. **27**(4), 2196–2205 (2012)

24. Troy, N., Flynn, D., O'Malley, M.: The importance of sub-hourly modeling with a high penetration of wind generation. In: 2012 IEEE Power and Energy Society General Meeting, pp. 1–6, July 2012

25. Tuohy, A., Meibom, P., Denny, E., O'Malley, M.: Unit commitment for systems with significant wind penetration. IEEE Trans. Power Syst. **24**(2), 592–601 (2009)

26. Zhuang, F., Galiana, F.: Unit commitment by simulated annealing. IEEE Trans. Power Syst. **5**(1), 311–318 (1990)

27. Zhuang, F., Galiana, F.: A more rigorous and practical unit commitment by lagrangian relaxation. IEEE Trans. Power Syst. **3**(2), 763–773 (1988)

A Hybrid Evolutionary Approach for Solving the Traveling Thief Problem

Mahdi Moeini$^{(\boxtimes)}$, Daniel Schermer, and Oliver Wendt

BISOR, Technical University of Kaiserslautern,
Postfach 3049, Erwin-Schrödinger-Str., 67653 Kaiserslautern, Germany
{mahdi.moeini,daniel.schermer,wendt}@wiwi.uni-kl.de

Abstract. In this paper, we are interested in a recent variant of the
Traveling Salesman Problem (TSP) in which the Knapsack Problem
(KP) is integrated. In the literature, this problem is called *Traveling
Thief Problem (TTP)*. Interested by the inherent computational challenge of the new problem, we investigate a hybrid evolutionary approach
for solving the TTP. In order to evaluate the efficiency of the proposed
approach, we carried out some numerical experiments on benchmark
instances. The results show the efficiency of the introduced method in
solving medium sized instances. In particular, the algorithm improves
some best known solutions of the literature. In the case of large instances,
our approach gives competitive results that need to be improved in order
to obtain better results.

Keywords: Traveling Salesman Problem · Knapsack Problem · Genetic
Algorithms · Local Search · Variable Neighborhood Search

1 Introduction

During the last decades, there was a particular attention in studying combinatorial optimization problems [7–9,11]. Many of these optimization problems
are known to be NP-hard; consequently, it is very difficult to solve their large
instances within a reasonable computation time. Due to importance of this class
of problems, many algorithms have been developed for solving them as well as
their numerous variants [7–9].

The Traveling Salesman Problem (TSP) is one of the well-known NP-hard
combinatorial optimization problems. A recent variant of TSP is obtained by
integrating the Knapsack Problem (KP), that is a popular optimization problem
as well, into the TSP [2]. This problem is known under the name *Traveling Thief
Problem (TTP)* and, its restricted variant (in which, the tour is fixed) is called
Packing While Traveling (PWT) Problem [17,18]. In the TTP, it is assumed that
a certain number of items are given. These items are scattered among a given set
of cities. The mission of a vehicle (or a voyager), known also as a thief, consists
of visiting all cities (each of them exactly once) and picking up items in order to
put them in the vehicle (or in his rented knapsack) and finally, returning to the

© Springer International Publishing AG 2017
O. Gervasi et al. (Eds.): ICCSA 2017, Part II, LNCS 10405, pp. 652–668, 2017.
DOI: 10.1007/978-3-319-62395-5_45

starting city. The limited capacity of the vehicle/knapsack as well as its weight have influence on the packing plan of the thief, as the speed of the thief decreases by higher weight. Hence, the objective consists in finding the best Hamiltonian path as well as the best packing plan in order to maximize the total benefit of the thief [5,6,12,14,17,18,20,21].

Several practical problems have common points with the TTP. The main class of problems that have similarities with the TTP are TSP under KP constraints (excluding the interdependencies between TSP and KP) and also the vehicle routing problems. In the latter case, the less speed of the thief for higher weights can be interpreted as the higher fuel consumption of vehicles [5].

1.1 Literature Review

No matter the potential practical applications of the TTP, an interesting aspect of the TTP relies on the fact that it is a challenging combinatorial optimization problem [12,17,18,21]. The interdependency of two components of the TTP, i.e., TSP and KP, makes the problem particularly difficult to solve [14,21]. Due to this complexity, since the introduction of the problem, several research works have been dedicated to solving the TTP. The algorithms that have been introduced for solving the problem are mainly evolutionary approaches as well as heuristics and metaheuristics [5,6,12,14,17,18,20,21].

In [12], Mei et al. proposed a Memetic Algorithm with Two-stage Local Search (MATLS) for solving large instances of the TTP. In this context, Mei et al. analyze some heuristics for solving the TTP and use complexity reduction strategies. In [3], Bonyadi et al. proposed a heuristic, named Density-based Heuristic (DH), and a socially inspired method, called CoSolver. According to their results, CoSolver outperforms DH. Faulkner et al. investigated the performance of a variety of packing strategies that were ran on top of TSP solutions, provided by the well-known Chained Lin-Kernighan heuristic [6]. El Yafrani et al. were interested in studying the performance of population-based approaches versus single-solution heuristics [5]. According to the reported results, the population based method performs better than their proposed single solution heuristic (both methods are proposed by El Yafrani et al.). Lourenço et al. proposed an evolutionary algorithm by addressing, simultaneously, both components of the TTP [15]. Chand et al. studied a variant of the TTP where several thieves are considered. They claim that this variant is more realistic than the TTP [4]. Mei et al. studied the interdependence between sub-problems of the TTP and showed that the objective function of the TTP is not additively separable and proposed two meta-heuristics. In [20], Wagner used a swarm intelligence approach for solving the TTP and claims that the approach outperforms some of the existing methods. Recently, Wagner et al. carried out an extensive computational study on 21 known algorithms for the TTP and compared their efficiency [21].

To the best of our knowledge, there is neither an exact nor an approximation (with performance guarantee) algorithm for solving the TTP.

For the restricted variant of the TTP, i.e., the Packing While Traveling (PWT) problem [17,18], Polyakovskiy et al. propose mathematical programming approaches, such as linear approximation and lower/upper bounds, for solving the PWT problem.

Although we believe that *Packing While Traveling* is a more appropriate term than the *Traveling Thief Problem*, from both practical and scientific point of view, we will use the latter term throughout this paper in order to respect the original name of the problem.

1.2 Our Contributions

The contributions of this paper are twofold: First, we propose a hybrid evolutionary approach for solving the TTP problem. This method is based on a combination of several methods, namely Genetic Algorithms (GA), Local Search (LS), and Variable Neighborhood Search (VNS). The GA and LS have already been used for solving the TTP [15, 21]; however, to the extent of our knowledge, using VNS for solving this problem is a new contribution. VNS is a general framework that is based on using various neighborhood structures, an improvement procedure, and random perturbation (called *shake*). This approach has been used (successfully) for solving a large variety of combinatorial optimization problems (see e.g., [7–9] and references therein). In order to assess the performance of the approach in solving the TTP, we carried out a series of experiments by using benchmark instances. Our observations confirm the positive impact of the hybrid approach in improving the quality of the best known solutions. Second, we investigate the importance of the knapsack *used capacity* in solving the TTP. This investigation leads to interesting observations that can be useful for future research works on the TTP.

This paper is organized as follows: In Sect. 2, we first introduce the main notation and basic concepts that we need in this paper. Then, we present the mathematical model as well as an intuitive example of the TTP. In Sect. 3, we explain our proposed algorithms for solving the TTP. Afterwards, in Sect. 4, we examine the performance of our approach through numerical results. We sum up this paper with some concluding remarks, in Sect. 5.

2 Packing While Traveling Problem

2.1 Notation and Basic Definitions

Suppose that $G = (V, E)$ is a complete graph with a set of n vertices V and a set of edges E. Let the set of vertices be $V = \{V_1, \ldots, V_n\}$ and d_{ij} be the length of the edge connecting vertices V_i and V_j (for $i, j \in \{1, \ldots, n\}$).

A (TSP) tour π is a sequence of vertices such that it visits each vertex of G exactly once and then returns to the starting vertex. The *Traveling Salesman Problem (TSP)* asks for a tour π while minimizing the distance (or time) to complete the tour. Let $I = (P, W)$ be a list of $|I|$ items. Each item $I_i \in I$ with $i = \{1, \ldots, |I|\}$ has a profit $p_i \in P$ and a weight $w_i \in W$. Let $C \in \mathbb{R}_{\geq 0}$ be the maximum capacity of a given knapsack. The objective of the *Knapsack Problem*

(KP) consists in maximizing the overall profit of the picked items such that we may not exceed the capacity C of the knapsack.

We ask a voyager (thief) to find a TSP tour π, as well as a valid *packing plan* Y (serves to pick up items, distributed among vertices, and put them in his knapsack) in order to maximize the total profit of the picked items while minimizing the total distance (or time) needed to complete the tour. Furthermore, the total weight of the picked items must respect the capacity of the knapsack. This problem, that combines TSP and KP, is called the *Traveling Thief Problem* (TTP). In TTP, it is not mandatory to pick an item at each vertex. This might be strange from practical point of view; however, we would like to respect the original assumptions of this problem (see also [5,6,12,14,20,21]) and focus on solving the existing description of the TTP.

2.2 Mathematical Description of the Problem

Assume that each vertex $V_i \in V$ (for $i \neq 1$) has an equal number of $m \in \mathbb{Z}$ available items, except the origin (starting point) V_1 that holds no item. Hence, $n = |G| = |V|$ and $m = |I|/(n-1)$. Let $d_{i,i+1}$ be the distance connecting the i-th to the $(i+1)$-th vertex of the tour π. Suppose that p_{ij} and w_{ij} are the profit and weight of item j at vertex i of the tour, respectively. We define $R \in \mathbb{R}_{\geq 0}$ as the *renting ratio*, that is a penalty based on the total travel time.

For $i \in \{2, \ldots, n\}$ and $j \in \{1, \ldots, m\}$, define y_{ij} as binary decision variable, that is equal to 1, if the j-th item is picked at the i-th vertex and equal to 0, otherwise. We refer to the set of all non-zero y_{ij}, noted by Y, as the *packing plan* of the thief.

The thief has a knapsack with maximum available capacity C. He can travel at a maximum velocity v_{max} if his knapsack is empty and at a minimum velocity v_{min} if his knapsack is full. At each vertex i of the tour π, the weight W_i (of the knapsack) is the sum of the weights of all items picked up at the i-th vertex and its predecessors in π (according to Y). At a given vertex i in the tour π, the thief travels at a velocity v_i. We assume that the relation between the current weight W_i and the current velocity v_i is linearly proportional [10], as follows:

$$v_i = v_{max} - \frac{v_{max} - v_{min}}{C} \cdot W_i \quad : \quad \forall i. \tag{1}$$

Furthermore, the time required for the transition from vertex i to vertex $i+1$ is equal to $\frac{d_{i,i+1}}{v_i}$. The objective of the TTP problem consists in finding a TSP tour π and a packing plan Y in order to maximize the total gain of the travel, given by the following equation [6,12,14,16,20,21]:

$$\max Z(\pi, Y) = \left(\sum_{i=2}^{n} \sum_{j=1}^{m} p_{ij} \cdot y_{ij}\right) - R \cdot \left(\frac{d_{n,1}}{v_n} + \sum_{i=1}^{n-1} \frac{d_{i,i+1}}{v_i}\right), \tag{2}$$

where,

- the first term defines the *revenues*, i.e., the profit based on the packing plan and the associated profit of each item,

– the second term defines the *costs*, i.e., the price that the thief must pay based on the renting ratio R and the total time required to complete the tour.

Finally, the packing plan must be valid, i.e., it must respect the capacity constraint of the knapsack:

$$W_n = \sum_{i=2}^{n} \sum_{j=1}^{m} y_{ij} \cdot w_{ij} \leq C \qquad (3)$$

The objective function gives us an insight into the complexity of the TTP problem and highlights why, for solving the TTP problem, we can not solve the TSP and the KP independently from each other. Indeed, for a given packing plan Y, if we change the tour π, then the profit gained by the picked items will not change. However, the price that the thief must pay will change, as it is not just a function of π but rather a function π and Y. Similarly, for a given tour π, if we change the packing plan Y the velocities v_i at each vertex i might change and thus the travel time be impacted. Therefore, we need to always keep in mind the dependence of each sub-problem to the other one.

2.3   An Example

The complete and symmetric graph $G = (V, E)$ as well as a set of items $I = (P, W)$, as shown in Fig. 1, are given. We consider a maximum velocity $v_{max} = 1.0$, a minimum velocity $v_{min} = 0.1$, a renting ratio $R = 1.0$, and a capacity of $C = 10$. The packing plan Y_{ij} (that shows if item j is picked at vertex i) is given by $Y_{i1} = 1$ and $Y_{i2} = 0$, for all $i \neq 1$. Suppose that a tour π is the following sequence of vertices $\pi = \{V_1, V_2, V_3, V_4\}$. The tour π and the packing plan Y are valid. The total profit of the packing plan Y is 160 and the total penalty, based on the packing plan Y_{ij} and tour π, is equal to 81.81. Hence, $Z(\pi, Y) = 160 - 81.81 = 78.19$.

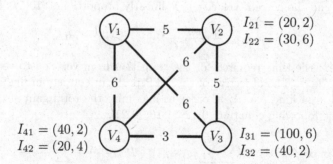

Fig. 1. An example of a TTP with a graph $G = (V, E)$ and a set of items $I = (P, W)$.

3   An Evolutionary Algorithm for Solving the TTP

In order to solve the TTP, we propose a hybrid Evolutionary Algorithm (EA). The main components of our algorithm are based on Genetic Algorithm (GA), Local Search (LS), and Variable Neighborhood Search (VNS). Each of these approaches have already had successful applications in solving various combinatorial optimization problems [7–9,21]. Figure 2 shows a flowchart of the proposed algorithm. On this flowchart, the components are colored in different colors in order to indicate to which approach they belong. Mainly, the red colored parts consist of the population initialization procedures (Initial Tour, Initial Packing Plan, Initial Two-Stage LS (2-opt, bitflip)). The parts that are colored in blue are the main genetic algorithm (parent selection, reproduction operators, population replacement). And finally, the parts that are colored in green are essentially hill-climbing algorithms (including VNS) that are not directly related to the genetic algorithm. In the following, we give a concise description of the different components of our algorithm.

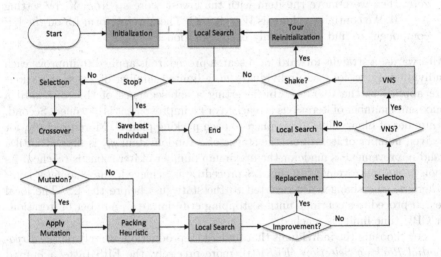

Fig. 2. A flowchart showing the basic components of the proposed algorithm. (Color figure online)

Assume that Pop is a collection of a given number of individuals. For each individual $ind \in Pop$, a tour π_{ind} is generated using the nearest-neighbor-heuristic (NNH) [7] and the packing plan is initialized to $Y_{ind} = \{\}$. The NNH simply starts at $V_1 \in G$, finds another random vertex $V_i \in G, V_i \neq V_1$ as the second vertex in π and then completes the tour by connecting each subsequent vertex using the nearest neighbor, that is not yet in π. Afterwards, for each individual ind, an initial packing plan Y_{ind} is generated using the packing heuristic (PH), that is outlined in Algorithm 1. The PH is based on the work in [12,13] (with slight modification) and starts by evaluating the current weight W according to Y_{ind} and then evaluates the remaining tour length d_i at the i-th vertex of π_{ind}. Then, the PH differentiates between a situation where the current weight is below or above the capacity C. For either case, we assign a score s_{ij} to each item. The heuristic value s_{ij} is an indication of the change in the objective value if the item is picked [12,13].

- If $0 \leq W < C$, we set the score of each item j at node i that we picked already (i.e., $y_{ij} = 1$) to $s_{ij} = 0$. If an item with a score $s_{ij} > 0$ still exists, we add the most profitable item to Y if it still fits and the worst-case gain is larger than zero. If the worst-case gain is not larger than zero, we calculate an average gain, where $v_{avg(i)} = v_{max} - (v_{max} - \frac{v_{min}-v_{max}}{C} \cdot W \cdot \frac{i}{n})$.
  We continue until the knapsack is either at full capacity or no more profitable item remains.
- If $W > C$, we set the score of each item j at vertex i, that we have not already picked up, to a sufficiently large number by using the value M (where M is a sufficiently large positive number). If the worst-case gain is smaller than zero, then we remove the item with the lowest score s_{ij} from Y, by setting $y_{ij} = 0$. We continue as long as $W > 0.9 \cdot C$. Then, we perform LS on the KP component to find a potentially better solution.

Afterwards, a straight forward local search procedure is applied to improve each individual. First, for each individual ind, a limited number of 2-opt operations are applied to the tour π_{ind} by inverting a sub-sequence of the tour until a maximum number of iterations is reached or an improvement is obtained. Second, a number of binary flips are applied to each packing plan Y. More precisely, for a given number of iterations, the status of a random item y_{ij} is flipped until a valid improvement is made or the maximum number of iterations is reached. As soon as an improvement in any LS procedure is achieved, the change is kept; otherwise, the changes are reverted to the state just before the LS. The local search procedures continue until a stopping criterion (e.g., number of iterations or CPU time limit) is reached.

For choosing the individuals that undergo reproduction, we rely on the *Exponential Ranking Selection (ERS)* [1]; more precisely, the ERS takes a limited number of r randomly chosen individuals from the population and orders them in a descending way, based on their fitness values (where, *fitness value* of an individual is its objective value, according to (2), for its corresponding TSP tour and packing plan). In order to select individuals, we start by assigning to each individual i (in the ordered list) a probability that is defined as follows [1]:

$$P(i,r) = \frac{(1-p)^{(r-i)}}{\sum_{j=1}^{r}(1-p)^{(r-j)}} \tag{4}$$

After selecting the parents, children are generated using an adapted Partially-Matched Crossover (PMX) algorithm [15]. The PMX works in the following way: Given two parents, the PMX selects randomly two crossover points which mark the crossover interval $\mathcal{I}$. The child is then constructed by starting with a

---

Algorithm 1. Packing Heuristic

1: **procedure** PACKING HEURISTIC(π, Y)
2:     $W \leftarrow currentWeight(Y)$;   ▷ Evaluate the current weight according to Y
3:     $d_i(\pi) \leftarrow remainingTourLength(\pi) \quad \forall i$;
4:     $s_{ij} = \frac{p_{ij} - R \cdot d_i}{w_{ij}} \quad : \forall i, j$;
5:     **if** $0 \leq W \leq C$ **then**   ▷ Add the most profitable items
6:         $s_{ij} = s_{ij} \cdot (1 - y_{ij}) \quad : \forall i, j$;
7:         Sort s_{ij} in a descending way and take the first element in the list
8:         **while** $0 \leq W < C$ **and** $s_{ij} > 0$ **do**
9:             **for** the current item **do**
10:                **if** $W + w_{ij} \leq C$ **then**
11:                    **if** $p_{ij} - \frac{R \cdot d_i(\pi)}{v_{min}} > 0$ **then**   ▷ Worst-case-gain > 0
12:                        $y_{ij} = 1$;
13:                        $W = W + w_{ij}$;
14:                    **else if** $p_{ij} - \frac{R \cdot d_i(\pi)}{v_{avg}(i)} > 0$ **then**
15:                        $y_{ij} = 1$;
16:                        $W = W + w_{ij}$;
17:                    **end if**
18:                **end if**
19:            **end for**
20:            Take the next item in the list
21:        **end while**
22:    **else**   ▷ Remove the least profitable items
23:        $s_{ij} = s_{ij} + (1 - y_{ij}) \cdot M \quad : \forall i, j$;
24:        sort the items in a list according to their value of s_{ij}
25:        **while** $W > 0.9 \cdot C$ **do**
26:            $W \leftarrow currentWeight(Y)$;
27:            **for** the current item **do**
28:                **if** $p_{ij} - \frac{R \cdot d_i(\pi)}{v_{min}} > 0$ **then**   ▷ Worst-case-gain > 0
29:                    Break   ▷ Break the for loop
30:                **else**
31:                    $y_{ij} = 0$;
32:                    $W = W - w_{ij}$;
33:                **end if**
34:            **end for**
35:            Take the next item in the list
36:        **end while**
37:    **end if**
38: **end procedure**

---

copy of parent 2. Within the crossover interval $\mathcal{I}$, the information of parent 2 is overwritten with that of parent 1. For the tour to remain valid, some positions from parent 2 are rearranged. We form the packing plan of the child Y_{child} based on inheritance [5]. The packing plan Y of the child is created by taking the y_{ij} according to the following equation:

$$y_{ij} = \begin{cases} y_{ij}(\text{Parent 1}) & \text{if node i was received from Parent 1,} \\ y_{ij}(\text{Parent 2}) & \text{otherwise.} \end{cases} \quad (5)$$

Then, based on a given probability λ, the mutation operator will simply invert the tour by changing the sequence in which vertices 2 to n are visited (through π) and empties the packing plan by setting $Y = \{\}$.

After applying the PMX, and, depending on λ, the mutation operator, it is necessary to apply the packing heuristic to refine the solution or possibly remove excessive weight from the knapsack to make the packing plan *valid* again. The adapted PMX always generates a valid *tour*; however, *the packing plan Y for the given tour may be far from optimal or even valid.*

The straight-forward local search procedures for the KP and TSP component are then applied to refine the solution. Afterwards, the parent with the least fitness values is replaced by the offspring, if the offspring has a higher fitness value than the parent.

In order to update the population, we use a *steady-state* approach, i.e., the population is not replaced entirely by a new generation but rather, incrementally some individuals (e.g., only 1 individual or a certain percentage of the total population) are replaced by newly produced offspring.

Whenever an offspring has been generated and possibly added into the population, we use LS and VNS as hill-climbing algorithms [7–9, 15, 21]. More precisely, the hill-climbing algorithms take solutions and try to improve them. The objective consists in giving the solutions a chance to possibly escape from a local maximum. For this purpose, first, we select a random individual from the population. Next, another random individual amongst the 5 best individuals of the population, found so far, is selected. Then, based on a given probability, we perform either LS or *Variable Neighborhood Search* (VNS) on both individuals. Algorithm 2 describes different steps of our VNS algorithm.

Generally, VNS consists of generating a sequence of neighborhoods close to the current solution and exploring them. Furthermore, in order to diversify the search, a random procedure (here, a randomly generated neighborhood), called *shaking* step, is used. This step is integrated, via random selection of neighborhood, through 2-opt operations (lines 5 & 6 of Algorithm 2). Furthermore, within neighborhoods, we apply LS to search for a better solution ((lines 7–20) of Algorithm 2) [7–9].

We should note that, for a given tour π, the packing plan Y might not be able to be improved. In a similar fashion, for a given packing plan Y, the corresponding tour π might be optimal. Therefore we need to consider the interdependence between both problems by changing the neighborhood (changing π) and then apply local search by investigating changes in Y. Since VNS is much more time consuming than LS, we rely on VNS only on a limited number of iterations. For this purpose, in order to generate a neighborhood $\pi' \in \mathcal{N}(\pi)$ (where, $\mathcal{N}$ stands for neighborhood of a given tour π) the 2-opt operator is used. The packing plan is then improved using LS on the KP component of the problem. If an improvement is made, VNS terminates; otherwise, we try a re-packing of items for the given tour by setting $Y = \{\}$ and applying the packing heuristic. If an improvement is made VNS terminates; otherwise, we continue until the stopping criterion is met. According to our observations, adjusting (dynamically) the number of iterations of LS or VNS can lead to an improvement compared to a fixed number of iterations. Hence, at each iteration of the EA, we adjust, dynamically, the limit on number of iterations of each LS and VNS procedure by using the following equation:

$$\text{num} = min[5, mod(\text{IterationCounter}, 60)], \qquad (6)$$

where, *IterationCounter* is the counter indicating current iteration of the EA and, throughout this iteration, *num* defines the limit on the number of LS or VNS iterations. Indeed, according to our observations, this dynamic adjustment creates higher diversifications in searching for high quality solutions.

Finally, the population is monitored based on:

- the number of iterations that have passed, since the last change in the best individual,
- and the standard deviation σ of the fitness of the population.

When both, the standard deviation σ becomes sufficiently small and the number of iterations since the last improvement to the best individual becomes sufficiently large, we execute a *population shaking* step in order to avoid premature convergence. More precisely, a part of the population is randomly re-initialized. This process creates a new neighborhood close to the current solution.

Algorithm 2. Variable Neighbourhood Search

1: **procedure** VARIABLENEIGHBOURHOODSEARCH(π, Y)
2:     $Y_{init} = Y$;
3:     $\pi_{init} = \pi$;
4:     **while** Stopping criterion is not met **do**
5:         $\pi \leftarrow \mathcal{N}(\pi)$ using 2-opt
6:         $Y' \leftarrow Y_{init}$
7:         **for** π **do**
8:             $Y \leftarrow LocalSearchKP(\pi, Y_{init})$
9:             **if** fitness$(\pi, Y) >$ fitness(π_{init}, Y_{init}) **then**
10:                 return π, Y;
11:                 break;
12:             **else**
13:                 $Y = \{\}$
14:                 $Y \leftarrow packingHeuristic(\pi)$
15:                 **if** fitness$(\pi, Y) >$ fitness(π_{init}, Y_{init}) **then**
16:                     return π, Y;
17:                     break;
18:                 **end if**
19:             **end if**
20:         **end for**
21:     **end while**
22:     return π_{init}, Y_{init};
23: **end procedure**

---

4   Numerical Experiments

4.1   Experimental Settings

Based on the well known TSPLIB[1] [19], Polyakovskiy et al. have generated test instances for the TTP [16]. In all instances, the values for the maximum speed v_{max} and minimum speed v_{min} are equal to 1.0 and 0.1, respectively. However, each instance has a given individual renting rate R.

Unfortunately, from the 9720 available TTP instances, there is not yet a representative subset of instances. However, Wagner et al. have done an extensive computational study on TTP instances [21]. Indeed, they have solved all available TTP instances [16] by all 21 different TTP published algorithms. Their work makes it easier to benchmark algorithms and compare the new results.[2] For the purpose of comparing the results, in a reasonable dimension, we selected a limited number of instances and compared our results with respect to the best recorded TTP values (as benchmark values) in [21]. More precisely, we used 40 TTP instances that are related to two classes of TSP instances (*eil51* and *eil76*) with item factors $m \in \{1, 3\}$ (*item factor* indicates the number of available items at each vertex) and *capacity factor* $c = \{1, \ldots, 10\}$, where the capacity factor is related to the capacity of knapsack as follows:

$$C = \frac{c}{11} \cdot \sum_{i=2}^{n} \sum_{j=1}^{m} w_{ij}. \qquad (7)$$

We implemented our EA in MATLAB® and carried out the experiments on a 2.5 GHz dual core machine. In a similar way to the previous works on the TTP, we considered a time-limit of 10 min [21]. Furthermore, due to the randomness in the EA, we ran the algorithm 10 times (each run for 10 min).

For the GA part of our EA, we set the population size to 100, the mutation rate $\lambda = 0.1$, the ERS probability $p = 0.3$, and the ERS size $r = 5$. In the VNS component of the EA, we set VNS probability equal to 5 % (the probability under which the VNS component is called). The following parameters were used to handle the monitoring and population shaking step:

– Monitoring:
   • If the standard deviation of the population's fitness $\sigma < 5$ % of the fitness of the best individual found so far,
   • and no improvement in the best individual's fitness was observed for the last 5000 iterations;
– then, apply a population shaking step by:
   • Randomly re-initializing the tour of 90 % of the worst individuals in the population.
   • Set $Y = \{\}$ and apply the PH for each of the 90 individuals.

---

[1] TSP Test Data, see comopt.ifi.uni-heidelberg.de/software/TSPLIB95/index.html.
[2] Wagner et al. have done their computational experiments on an Xeon 2.66 GHz quad core CPU and ran each algorithm for 10 min [21].

4.2   Numerical Results

We analyzed the results of our experiments from different aspects.

Analysis of the Fitness Values: Table 1 highlights the numerical results of our algorithm. The first row of the table indicates the class of the instances, i.e., eil51 and eil76 with 51 and 76 vertices, respectively. The itm_c columns show the number of items (itm) and capacity factor (c) for each corresponding instance. The column *bench.* gives the best known benchmark values (in the literature) for each instance [21]. The best values, obtained through 10 runs of our hybrid EA, are reported in the columns *best*. In the columns *change*, we present the quantity of improvements by the hybrid EA with respect to best known values. The columns *avg* are used to report the average fitness values provided by the hybrid EA through 10 runs. According to the results in Table 1, our algorithm performs significantly well on the eil51_n50 instances. In these instances, the average fitness *avg* of the best individuals from 10 runs is often better than the benchmark value. All obtained improved results are highlighted in Table 1.

Table 1. The results of our hybrid EA on TTP benchmark instances.

eil51					eil76				
itm_c	bench.	EA			itm_c	bench.	EA		
		best	change	avg			best	change	avg
50_01	4269	4409	3.3%	4335	75_01	4109	4332	5,4%	4045
50_02	5571	5903	6.0%	5671	75_02	7604	7610	0,1%	7316
50_03	6058	6106	0,8%	5682	75_03	11341	11558	1,9%	11067
50_04	6609	6773	2,5%	6574	75_04	10720	10193	-4,9%	9415
50_05	5236	5743	9,7%	5055	75_05	8074	7157	-11,4%	6113
50_06	7083	7962	12,4%	7027	75_06	12370	11369	-8,1%	10350
50_07	8427	9306	10,4%	8682	75_07	12530	12447	-0,7%	10766
50_08	7424	8848	19,2%	7911	75_08	15027	13543	-9,9%	12847
50_09	7205	8243	14,4%	7784	75_09	13766	12034	-12,6%	10639
50_10	11439	12861	12,4%	12231	75_10	14920	13125	-12,0%	11886
150_01	7532	7821	3,8%	7488	225_01	10108	10162	0,5%	9461
150_02	14359	14599	1,7%	13364	225_02	20335	18180	-10,6%	17451
150_03	18277	18027	-1,4%	16563	225_03	30237	29189	-3,5%	25525
150_04	24195	23921	-1,1%	22253	225_04	32095	29210	-9,0%	24963
150_05	25895	25176	-2,8%	24126	225_05	36913	33568	-9,1%	29931
150_06	24768	23782	-4,0%	21012	225_06	44041	40290	-8,5%	36943
150_07	25860	24672	-4,6%	22319	225_07	46535	44642	-4,1%	37813
150_08	25951	25177	-3,0%	22362	225_08	50394	46501	-7,7%	41973
150_09	29024	27805	-4,2%	26284	225_09	47192	42140	-10,7%	36548
150_10	30340	29836	-1,7%	27070	225_10	52156	47982	-8,0%	40800

Analysis of the TSP and KP Components: It is also interesting to study the structure of the underlying TSP and KP components of the TTP problem. For this purpose, we used the following measures:

- $t(\pi, Y = \{\})_{scaled}$: First, we compute the time to complete the tour π, for a given constant speed (i.e., no items are picked). The value is then scaled to the best known TSP solution (the best known TSP solutions for instances *eil51* and *eil76* are 426 and 538, respectively[3]).
- $t(\pi, Y)_{scaled}$: First, we compute the time to complete the tour π while picking up items according to Y, with an adjustment to the velocity v_i at each vertex of the tour. Therefore, this is the minuend in Eq. (2) divided by the renting ratio R. The value is then scaled to the best known TSP solution.
- The *used capacity*: indicates the total weight at the end of the tour i.e., W_{total}, divided by the capacity limit, i.e., C.

For different values of the capacity factor c, Fig. 3 gives an insight into the variation in the TSP tour length, TTP tour length, and the *used capacity* for all tested instances.

According to the results, depicted on Fig. 3, the time to complete the TTP tour $t(\pi, Y)_{scaled}$ (dashed curves) is always larger than the TSP tour. This is justified by the fact that the velocity, on at least one edge, will always be less than the maximum velocity. The reduction in velocity is due to picking up at least one item. It is interesting to note that the tour length at constant speed $t(\pi, Y = \{\})_{scaled}$ (bold curves) is reasonably close to the optimal solution for most instances and is almost independent from the capacity factor c. This helps promote the claim that for finding good TTP solutions it is generally beneficial to find good tours.

Another interesting fact (on Fig. 3) is in the overall time to complete the TTP tour: The time grows considerably with the capacity factor and has (almost) a logarithmic growth. Indeed, a low capacity factor c allows the thief to pick up only very few items during his tour. However, even lightweight items will slow the thief down considerably, as the change in speed does not just depend on the weight of the item, but also on the available capacity C, which in return is a function of the capacity factor c (see also [16]). However, due to the little available capacity (with a low capacity factor) and the high change in velocity, whenever an item is picked up, the TTP tours tend to be only slightly longer than the associated TSP tours. The reason behind this is the fact that only very few transitions tend to be affected by this change in speed, as items tend to be picked up very late during the tour (see also Fig. 4).

A higher capacity factor c allows the thief to pick up more items during his tour. Additionally, lightweight items will only slow down the thief slightly (due to the relation of the velocity to the current weight and capacity [16]). However, in these scenarios more transitions are affected by the change in velocity and, consequently, the change in the tour length becomes significant.

---

[3] Solutions available at comopt.ifi.uni-heidelberg.de/software/TSPLIB95/STSP.html.

On Fig. 3, the dependency of the used capacity (dotted curve) to the capacity factor can be explained with a similar argument and is, indeed, the reason for the logarithmic growth in the TTP tour length. Since only few transitions are affected by the change in velocity with a low capacity factor, the decline in the profit of the items based on changing speed are not significant and it seems beneficial to make use of the available capacity in most cases. With increments in the capacity factor c, more and more transitions are affected by the change in velocity and consequently, items tend to be less valuable (i.e., smaller positive impact on the value of the objective function) resulting in higher *unused* capacity (less than 100% of the total capacity is used). This claim can be supported by Fig. 4. For 3 different values of c, (indeed, for $c = 1$, $c =$ average for all c values, and $c = 10$), Fig. 4 shows the average remaining tour length and also *used capacity* at each vertex of the tour. In order to explain this figure, we note that, generally, one can have two feasible assumptions. First, we may assume that most items are picked during the end of the tour. This can be expected in order to travel at a velocity close to or at the maximum velocity for as long as possible. This helps minimize the penalty by minimizing the total travel time. Hence, the used capacity should show a convex growth at each vertex of the tour π. Second, it is also possible to assume that in any tour, comparatively long transitions are completed at the start of the tour. If long edges are passed while the velocity is still at or close to maximum velocity (previous assumption), this will also help minimize the penalty, by minimizing the total travel time. Consequently,

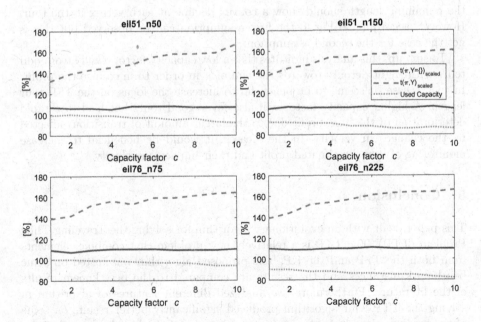

Fig. 3. Time to complete the TSP $(t(\pi, Y = \{\}))$ and TTP $(t(\pi, Y))$ tours scaled to the best known TSP solution as well as the used capacity versus the capacity factor c.

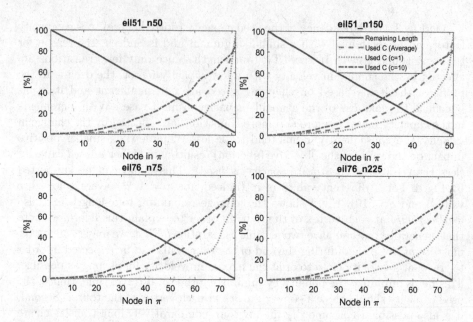

Fig. 4. The remaining distance in π and the used capacity ($c = 1$, $c =$ average of all c values, $c = 10$) per node on eil51 and eil76.

the remaining length should show a convex decline at each vertex in the tour. However, according to Fig. 4, the first assumption can be observed but this is not the case for the second assumption.

To sum up, this analysis indicates that a low capacity factor c will favor good tours with slight detours towards the end of π in order to maximize the profit. In such a case, it seems to be beneficial to increase the focus on the TSP tour length. A higher capacity factor will also favor good tours but lead to a rapid deterioration of the item profit due to the larger amount of transitions affected by the changes in velocity. In these cases, it would be beneficial to increase monitoring of items' estimated profit and their impact on velocity.

5   Conclusion

This paper dealt with an evolutionary algorithm for solving the Traveling Thief Problem (TTP). The TTP is a relatively new problem that combines elements from both the TSP and the KP. The proposed algorithm was tested on some benchmark instances and the results are compared to the best known results of the literature. Furthermore, we analyzed different features of algorithm in solving the TTP. Our algorithm produced significantly better results on some of the medium sized instances and improves some of the best known solutions; however, the algorithm must be improved for producing better results on large-scale instances. As a future work, we aim at improving the structure of the current algorithm in order to tackle large-scale instances. The research in this direction is in progress and the results will be reported in future.

References

1. Blickle, T., Thiele, L.: A comparison of selection schemes used in evolutionary algorithms. Evol. Comput. **4**(4), 361–394 (1996)
2. Bonyadi, M.R., Michalewicz, Z., Barone, L.: The travelling thief problem: the first step in the transition from theoretical problems to realistic problems. In: IEEE Congress on Evolutionary Computation (CEC), pp. 1037–1044 (2013)
3. Bonyadi, M.R., Michalewicz, Z., Przybyoek, M.R., Wierzbicki, A.: Socially inspired algorithms for the travelling thief problem. In: GECCO 2014, pp. 421–428 (2014)
4. Chand, S., Wagner, M.: Fast heuristics for the multiple traveling thieves problem In: GECCO 2016, pp. 293–300 (2016)
5. El Yafrani, M., Ahiod, B.: Population-based vs. single-solution heuristics for the travelling thief problem. In: GECCO 2016, pp. 317–324 (2016)
6. Faulkner, H., Polyakovskiy, S., Schultz, T., Wagner, M.: Approximate approaches to the traveling thief problem. In: GECCO 2015, pp. 385–392 (2015)
7. Floudas, C.A., Pardalos, P.M. (eds.): Encyclopedia of Optimization. Springer, New York (2009)
8. Gendreau, M., Potvin, J.-Y.: Handbook of Metaheuristics. Springer, New York (2010)
9. Goeke, D., Moeini, M., Poganiuch, D.: A variable neighborhood search heuristic for the maximum ratio clique problem. Comput. Oper. Res. 1–9 (2017, forthcoming). doi:10.1016/j.cor.2017.01.010
10. Gupta, B.C., Prakash, V.P.: Greedy heuristics for the travelling thief problem. In: 39th National Systems Conference (NSC), pp. 1–5 (2015)
11. Gueye, S., Michel, S., Moeini, M.: Adjacency variables formulation for the minimum linear arrangement problem. In: Pinson, E., Valente, F., Vitoriano, B. (eds.) ICORES 2014. CCIS, vol. 509, pp. 95–107. Springer, Cham (2015). doi:10.1007/978-3-319-17509-6_7
12. Mei, Y., Li, X., Yao, X.: Improving efficiency of heuristics for the large scale traveling thief problem. In: Dick, G., et al. (eds.) SEAL 2014. LNCS, vol. 8886, pp. 631–643. Springer, Cham (2014). doi:10.1007/978-3-319-13563-2_53
13. Mei, Y., Li, X., Salim, F., Yao, X.: Heuristic evolution with genetic programming for traveling thief problem. In: IEEE Congress on Evolutionary Computation (CEC), pp. 2753–2760 (2015)
14. Mei, Y., Li, X., Yao, X.: On investigation of interdependence between sub-problems of the travelling thief problem. Soft. Comput. **20**(1), 157–172 (2016)
15. Lourenço, N., Pereira, F.B., Costa, E.: An evolutionary approach to the full optimization of the traveling thief problem. In: EvoCOP, pp. 34–45 (2016)
16. Polyakovskiy, S., Bonyadi, M.R., Wagner, M., Michalewicz, Z., Neumann, F.: A comprehensive benchmark set and heuristics for the traveling thief problem. In: GECCO 2014, pp. 477–484 (2014)
17. Polyakovskiy, S., Neumann, F.: Packing while traveling: mixed integer programming for a class of nonlinear knapsack problems. In: CPAIOR, pp. 332–346 (2015)
18. Polyakovskiy, S., Neumann, F.: The packing while traveling problem. Eur. J. Oper. Res. **258**, 424–439 (2017)

19. Reinelt, G.: TSPLIB-a traveling salesman problem library. ORSA J. Comput. **4**(3), 376–384 (1991)
20. Wagner, M.: Stealing items more efficiently with ants: a swarm intelligence approach to the travelling thief problem. In: Dorigo, M., Birattari, M., Li, X., López-Ibáñez, M., Ohkura, K., Pinciroli, C., Stützle, T. (eds.) ANTS 2016. LNCS, vol. 9882, pp. 273–281. Springer, Cham (2016). doi:10.1007/978-3-319-44427-7_25
21. Wagner, M., Lindauer, M., Misir, M., Nallaperuma, S., Hutter, F.: A case study of algorithm selection for the traveling thief problem. arXiv:1609.00462v1, pp. 1–23 (2016)

A Feasibility Pump and a Local Branching Heuristics for the Weight-Constrained Minimum Spanning Tree Problem

Cristina Requejo[1(✉)] and Eulália Santos[2]

[1] University of Aveiro, 3810-193 Aveiro, Portugal
crequejo@ua.pt
[2] ISLA-Higher Institute of Santarém and Leiria, 2414-017 Leiria, Portugal

Abstract. The Weight-constrained Minimum Spanning Tree problem (WMST) is a combinatorial optimization problem aiming to find a spanning tree of minimum cost with total edge weight not exceeding a given specified limit. This problem has important applications in the telecommunications network design and communication networks.

In order to obtain optimal or near optimal solutions to the WMST problem we use heuristic methods based on formulations for finding feasible solutions. The feasibility pump heuristic starts with the LP solution, iteratively fixes the values of some variables and solves the corresponding LP problem until a feasible solution is achieved. In the local branching heuristic a feasible solution is improved by using a local search scheme in which the solution space is reduced to the neighborhood of a feasible solution that is explored for a better feasible solution. Extensive computational results show that these heuristics are quite effective in finding feasible solutions and present small gap values. Each heuristic can be used independently, however the best results were obtained when they are used together and the feasible solution obtained by the feasibility pump heuristic is improved by the local branching heuristic.

Keywords: Weighted MST · Minimum spanning tree · Feasibility Pump · Local Branching · Heuristics

1 Introduction

Given a graph with edge costs and edge weights, the aim of the Weight-constrained Minimum Spanning Tree (WMST) problem is to find a spanning tree with minimum cost, such that its total weight does not exceed a given specified integer positive limit W. The WMST problem is a NP-hard [2,21] combinatorial optimization problem with important applications in the telecommunications network design and communication networks.

Several algorithms have already been proposed to the problem, either with exact or approximation approaches for determining a feasible solution. Aggarwal et al. [2] and Shogan [18] propose exact algorithms that use a Lagrangian relaxation to approximate a solution combined with a Branch and Bound strategy.

© Springer International Publishing AG 2017
O. Gervasi et al. (Eds.): ICCSA 2017, Part II, LNCS 10405, pp. 669–683, 2017.
DOI: 10.1007/978-3-319-62395-5_46

Ravi and Goemans [15], Xue [20], Hong et al. [11] and Hassin and Levin [9] propose approximation schemes. A compilation of results and existing algorithms to solve the problem can be found in Henn [10].

Requejo et al. [16] describe several Integer Linear Programming (ILP) formulations, Agra et al. [4] describe new valid inequalities for the WMST problem, and Requejo and Santos [17] discuss algorithms based on Lagrangian relaxations and propose a Lagrangian relaxation combined with valid inequalities.

Another approach to the problem is to include the weight of the tree as a second objective instead of a hard constraint. The resulting problem is the bicriteria/biobjective spanning tree problem. Many references can be found for this approach, see [19] among many others.

In this paper we present heuristics for the WMST problem that use mixed integer models of the problem and allow interaction with a mixed integer programming (MIP) solver. We describe two heuristics and, additionally these heuristics are used together in one heuristic procedure. In the first one, the heuristic should provide a "good" feasible solution. In the second one, the heuristic should allow the improvement of an available feasible solution. For the first heuristic, a Feasibility Pump scheme is used. Such scheme was proposed by Fischetti, Glover and Lodi [6] with the goal of finding feasible solutions (if any exists) for generic Mixed Integer Linear Programming (MILP) problems and improved by Fischetti, Bertacco and Lodi [5] and by Achterberg and Berthold [1]. A mixed integer formulation for the WMST problem together with a MIP solver are used to obtain (fractional) linear relaxation solutions. These fractional solutions are rounded such that a feasible solution (if any exists) is found. For the second heuristic we consider a Local Branching method proposed by Fischetti and Lodi [7] to solve MIP problems. This enumerative scheme constructs a sequence of feasible solutions with improving (decreasing) value of costs which is considered a very effective improving method for large scale problems. Again a mixed integer formulation for the WMST problem together with a MIP solver are used with the objective to explore reduced feasible regions.

The structure of this paper is as follows. In Sect. 2 we describe the WMST problem and a general formulation that will be used in the heuristic schemes. In Sect. 3 we present a Feasibility Pump heuristic applied to the WMST problem. In Sect. 4 we present a Local Branching heuristic applied to the WMST problem. In Sect. 5, we present computational results for the Feasibility Pump and Local Branching schemes applied to the WMST problem. In the last section, Sect. 6, we conclude the paper.

2   The Weighted-Constrained Minimum Spanning Tree Problem

To define the WMST problem we consider an undirected complete graph $G = (V, E)$, with node set $V = \{0, 1, \ldots, n-1\}$ and edge set $E = \{\{i, j\}, i, j \in V, i \neq j\}$. Associated with each edge $e = \{i, j\} \in E$ consider positive integer costs c_e and weights w_e. The WMST problem is to find a spanning tree $T = (V, E_T)$

in G, $E_T \subset E$, of minimum cost $C(T) = \sum_{e \in E_T} c_e$ and with total weight $W(T) = \sum_{e \in E_T} w_e$ not exceeding a given limit W.

To obtain formulations for the WMST problem one can easily adapt a Minimum Spanning Tree (MST) formulation. For the MST several formulations are well known (see Magnanti and Wolsey [12]) and in [16] natural and extended formulations for the WMST problem are discussed.

It is well known (see Magnanti and Wolsey [12]) that oriented formulations (based on the underlying directed graph) leads, in general, to tighter formulations (formulations whose lower bounds provided by the linear relaxations are closer to the optimum values). Thus, henceforward consider the corresponding directed graph, with root node 0, where each edge $e = \{0,j\} \in E$ is replaced with arc $(0,j)$ and each edge $e = \{i,j\} \in E, i \neq 0$, is replaced with two arcs, arc (i,j) and arc (j,i). Thus we obtain the arc set $A = \{(i,j),\ i \in V, j \in V \setminus \{0\}, i \neq j\}$. Each arc $(i,j) \in A$ inherits the cost and weight of the corresponding ancestor edge $\{i,j\}$.

Consider the binary variables x_{ij} (for all $(i,j) \in A$) indicating whether arc (i,j) is in the MST solution. Two classical formulations for the MST on the space of the binary variables x_{ij} can be considered [12]. To prevent the existence of circuits in the feasible solutions, and thus ensuring the connectivity of the feasible solutions, one formulation uses the cut-set inequalities and the other formulation uses the circuit elimination inequalities. The linear relaxation of both models provide the same bound value [12]. However the number of inequalities in both sets of constraints increase exponentially with the size of the model. In order to ensure connectivity/prevent circuits, instead of using one of those families with an exponential number of inequalities, one can use compact extended formulations. The well-known multicommodity flow formulation using additional flow variables can be considered. In this formulation the connectivity of the solution is ensured through the flow conservation constraints together with the connecting constraints, see [12]. These three formulations for the MST are easily adapted for the WMST problem through the inclusion of a weight constraint.

A generic formulation for the WMST problem is as follows.

$$\text{(WMST):} \quad \min \quad \sum_{(i,j) \in A} c_{ij} x_{ij} \tag{1}$$

$$\text{s.t.} \quad x \in (MST) \tag{2}$$

$$\sum_{(i,j) \in A} w_{ij} x_{ij} \leq W \tag{3}$$

Where $x = (x_{ij}) \in \mathbb{R}^{|A|}$ is the solution vector and (MST) represents a set of inequalities describing the convex hull of the integer solutions of the MST. As referred, several sets of inequalities can be used. We use the multicommodity flow formulation. Consider the additional set of flow variables $f_{ij}^k \geq 0$, for all $(i,j) \in A$ and $k \in V \setminus \{0,i\}$, indicating weather arc $(i,j) \in A$ is used in the path from the root node to node k. The following flow conservation constraints

$$\sum_{i \in V \setminus \{k\}} f_{ij}^k - \sum_{i \in V \setminus \{0\}} f_{ji}^k = \begin{cases} -1 \ j = 0 \\ \ 0 \ j \neq 0, k \ , \\ \ 1 \ j = k \end{cases} \qquad j \in V, \ k \in V \setminus \{0\},$$

and the connecting constraints $f_{ij}^k \leq x_{ij}$, for all $(i,j) \in A, k \in V \setminus \{0,i\}$ are used to ensure the connectivity of the solution and to represent the set (2) of constraints, together with the set of constraints $\sum_{i \in V} x_{ij} = 1$, for all $j \in V$, guarantying each non-root node receives an arc. Additionally, include the variables integrality constraints $x_{ij} \in \{0,1\}$, for all $(i,j) \in A$ and the variables non-negativity constraints $f_{ij}^k \geq 0$, for all $(i,j) \in A$ and $k \in V \setminus \{0,i\}$. Constraint (3) is the weight constraint.

3   The Feasibility Pump Heuristic for the WMST Problem

In this heuristic procedure the objective is to obtain a "good" feasible solution to the WMST problem. This objective is achieved through a relax-and-fix approach. We use the Feasibility Pump scheme, proposed by Fischetti, Glover and Lodi [6].

The first step of this procedure, the relax step, is to obtain the optimal solution of a linear relaxation formulation for the WMST problem, the LP solution. Let $\bar{x}$ be such solution. If $\bar{x}$ is an integer solution, then it is the optimal solution of the problem.

The second step is the rounding and fixing variables value step. Let the rounding $\tilde{x}$ of vector x be obtained by setting $\tilde{x}_{ij} = round(x_{ij})$, the usual scalar rounding, for all $(i,j) \in A$. The obtained vector $\tilde{x}$ is integer and, in general, $\tilde{x}$ is not a feasible solution.

The third step is to obtain the closest feasible solution to vector $\tilde{x}$. For that define a distance function $\Delta(x, \tilde{x})$ between a generic vector x and the given integer solution $\tilde{x}$. To define the distance function use the following L_1-norm function $\Delta(x, \tilde{x}) := \sum_{(i,j) \in A} |x_{ij} - \tilde{x}_{ij}|$. Notice that this function is equivalent to the linear function defined by $\sum_{(i,j) \in S}(1 - x_{ij}) + \sum_{(i,j) \in \overline{S}} x_{ij}$, with set $S = \{(i,j) \in A : \tilde{x}_{ij} = 1\}$ and its complement set $\overline{S} = \{(i,j) \in A : \tilde{x}_{ij} = 0\}$. Thus consider the distance function

$$\Delta(x, \tilde{x}) := \sum_{(i,j) \in S}(1 - x_{ij}) + \sum_{(i,j) \in \overline{S}} x_{ij}. \qquad (4)$$

Hence, given an integer $\tilde{x}$, the closest vector x satisfying constraints (2) and (3) can be determined by minimizing the value of the distance function $\Delta(x, \tilde{x})$ as follows

$$\text{(D-WMST):} \qquad \min \ \Delta(x, \tilde{x}) \qquad (5)$$
$$\text{s.t.} \quad (2), (3).$$

Let $\hat{x}$ be the optimal solution of the LP relaxation of problem (5), the D-WMST problem, and the value $\Delta(\hat{x}, \tilde{x})$ is its optimal value. The vector $\hat{x}$ is the closest solution to the integer $\tilde{x}$ and two cases may occur. If $\Delta(\hat{x}, \tilde{x}) = 0$, then $\hat{x} = \tilde{x}$ is an integer feasible solution for the WMST problem. If $\Delta(\hat{x}, \tilde{x}) > 0$, then a new integer $\tilde{\tilde{x}}$ is going to be obtained by rounding $\hat{x}$. By rounding $\hat{x}$ two cases may occur. Either $\tilde{\tilde{x}} \neq \tilde{x}$, a different vector was obtained, or $\tilde{\tilde{x}} = \tilde{x}$. When $\tilde{\tilde{x}} \neq \tilde{x}$ occurs the iterative process continues by solving the D-WMST problem again and a new solution closest to the new integer vector $\tilde{\tilde{x}}$ is going to be obtained. When $\tilde{\tilde{x}} = \tilde{x}$, the same solution was obtained and a cycle occurs. To avoid cycling the following perturbation mechanism [6] is applied to an integer solution $\tilde{x}$. For a given parameter $\delta > 0$, modify $\tilde{x}$ such that $|x_{ij} - \tilde{x}_{ij}| + max\{\rho_{ij}, 0\} > \delta$, for all $(i, j) \in A$ and ρ_{ij} randomly selected in $[-0.3, 0.7]$. Notice that this perturbation mechanism gives the possibility to modify the variables such that $|x_{ij} - \tilde{x}_{ij}| = 0$.

Iteratively another step is performed in order to further reduce the distance $\Delta(\hat{x}, \tilde{x})$ between the vector $\tilde{x}$, the integer rounded vector, and the closer solution $\hat{x}$ of the LP problem (5). Therefore the pair $(\hat{x}, \tilde{x})$ is iteratively updated by performing this relax-and-fix procedure until a feasible solution to the WMST problem is found.

In practice it may be much time consuming to achieve such feasible solution. Therefore a *time limit*, denoted maxtime, and a *maximum number of iterations*, denoted maxiter, are imposed. Hence we obtain a heuristic procedure, the feasibility pump heuristic which is briefly described in Algorithm 1.

Algorithm 1. Feasibility Pump heuristic for the WMST problem (FP)

Require: problem data (graph $G = (V, E)$; costs c_{ij}; weights w_{ij}; weight limit W);
    parameters (maxtime; maxiter).
1: solve the LP relaxation of a formulation to the WMST problem
2: let $\hat{x}$ be its optimal solution
3: **if** $\hat{x}$ is integer **then**
4:    **stop**, $\hat{x}$ is the optimal solution of the WMST problem
5: **end if**
6: $\tilde{x}^0 \leftarrow round(\hat{x})$
7: $t \leftarrow 0$
8: **while** ((time $<$ maxtime) and ($t <$ maxiter)) **do**
9:    obtain $\hat{x}$, the optimal solution of the LP relaxation of D-WMST problem with
    $\Delta(x, \tilde{x}^t)$ as objective function
10:    let $\Delta(\hat{x}, \tilde{x}^t)$ be its optimal value
11:    **if** $\Delta(\hat{x}, \tilde{x}^t) = 0$ **then**
12:        **stop**, $\hat{x}$ is an integer feasible solution for the WMST problem
13:    **else**
14:        $t \leftarrow t + 1$
15:        $\tilde{x}^t \leftarrow round(\hat{x})$
16:        **if** $\tilde{x}^t = \tilde{x}^{t-1}$ **then**
17:            apply perturbation mechanisms
18:        **end if**
19:    **end if**
20: **end while**
    return feasible solution $\hat{x}$ and integer solution $\tilde{x}^t$

In the first line of Algorithm 1 a solver is used to obtain the LP solution $\hat{x}$. A solver is again used in line 9 to obtain a solution to the D-WMST problem.

In Algorithm 1 all the solutions, except the first LP solution that uses the objective function of the WMST problem, are obtained by using as objective function the distance function (4). This objective function does not take into account the objective function of the WMST problem. As a consequence the solution obtained at the end of the procedure may have a value far from the best objective value of the WMST problem. To overcome this disadvantage, Achterberg and Berthold [1] propose the use of a different objective function that is a convex linear combination of the objective function (1) of the WMST problem and the distance function (4) of the D-WMST problem. The proposed objective function is

$$\Delta_\alpha(x, \tilde{x}) := (1 - \alpha)\Delta(x, \tilde{x}) + \alpha \frac{\sqrt{|A|}}{||c||} \sum_{(i,j)\in A} c_{ij}x_{ij}, \tag{6}$$

with $\alpha \in [0, 1]$ and where $||c||$ is the Euclidean norm of the cost vector c and $|A|$ is the cardinality of set A. Thus $\sqrt{|A|}$ is the Euclidean norm of the objective function vector in (4). In the objective function (6), the influence of the objective function (1) of the WMST problem is controlled by the α parameter. For values of α near to 1 the influence of the objective function is high.

In order to obtain an Improved Feasibility Pump heuristic for the WMST problem (IFP) one has to obtain, in line 9 of the Algorithm 1, the optimal solution $\hat{x}$ of (5) with (6) as the objective function.

Both the FP and the IFP return an integer solution, which may not satisfy constraints (2) and (3), and a solution satisfying constraints (2) and (3), but not necessarily integer. Two trajectories of solutions, hopefully convergent, are constructed. One is formed by a sequence of solutions satisfying constraints (2) and (3), solutions $\hat{x}$ that may not be integer. The other sequence is formed by integer solutions $\tilde{x}^t$ that may not satisfy constraints (2) and (3).

4   The Local Branching for the WMST Problem

The improvement of a previously obtained feasible solution to the WMST problem is the objective of the second heuristic we describe. The improvement is done through a local search scheme that uses a local branching method based on the method proposed by Fischetti and Lodi [7].

Consider, as a reference solution, a previously obtained feasible solution $\tilde{x}$ to the WMST problem. This integer solution corresponds to a spanning tree $T_{\tilde{x}}$ with cost $C(\tilde{x}) = C(T_{\tilde{x}})$ and weight $W(\tilde{x}) = W(T_{\tilde{x}})$. Define two sets, set $S = \{(i, j) \in A : \tilde{x}_{ij} = 1\}$ and its complement set $\overline{S} = \{(i, j) \in A : \tilde{x}_{ij} = 0\}$. For a given positive integer parameter k', define the *neighborhood* of $\tilde{x}$ as the set of feasible solutions of the WMST problem satisfying the additional *local branching constraint*:

$$\Delta(x, \tilde{x}) = \sum_{(i,j)\in S} (1 - x_{ij}) + \sum_{(i,j)\in\overline{S}} x_{ij} \leq k'. \tag{7}$$

The linear constraint (7) limits to k' the total number of binary variables flipping their value with respect to the solution $\tilde{x}$, either from 1 to 0 or from 0 to 1, respectively. Notice that the local branching constraint uses the same distance function $\Delta(x, \tilde{x})$ that is used in the FP heuristic as the objective function (4).

For every feasible solution to the WMST problem the cardinality of the set S is constant and equal to the number of edges of the corresponding feasible tree $T_{\tilde{x}}$. Further the number of variables exchanging from 1 to 0 must be equal to the number of variables exchanging from 0 to 1. Thus the *local branching constraint* may assume the asymmetric form:

$$\Lambda(x, \tilde{x}) = \sum_{(i,j) \in S} (1 - x_{ij}) \leq k \qquad (8)$$

with $k = \frac{k'}{2}$. Constraint (8) can be used as a branching criterion within an enumerative scheme and the solution space associated with the current branching node can be partitioned by means of the disjunction

(i) $\Lambda(x, \tilde{x}) \leq k$ (left branch)   or   (ii) $\Lambda(x, \tilde{x}) \geq k + 1$ (right branch).

With each one of those constraints the solution space is reduced. Define the *neighborhood* $\mathcal{N}(\tilde{x}, k)$ of $\tilde{x}$ as the set of feasible solutions of the WMST problem satisfying the additional local branching constraint $\Lambda(x, \tilde{x}) \leq k$, and the *neighborhood* $\mathcal{N}_+(\tilde{x}, k)$ of $\tilde{x}$ as the set of feasible solutions of the WMST problem satisfying the additional local branching constraint $\Lambda(x, \tilde{x}) \geq k + 1$. The choice of the size of the neighborhoods given by the parameter k is a problem which may depend on the size and structure of the instances used. On one hand, the k must be large enough so that the neighborhood $\mathcal{N}(\tilde{x}, k)$ contains better valued solutions than $\tilde{x}$. On the other hand, the k should be small enough to ensure that the neighborhood $\mathcal{N}(\tilde{x}, k)$ is quickly explored. Note that neighborhood $\mathcal{N}(\tilde{x}, k)$ of $\tilde{x}$ has solutions similar to $\tilde{x}$ and neighborhood $\mathcal{N}_+(\tilde{x}, k)$ contains solutions that differ from $\tilde{x}$ in more than $2 \times (k + 1)$ variables. The method explores both neighborhoods, but the neighborhood $\mathcal{N}_+(\tilde{x}, k)$ is only explored when a feasible solution better valued than $\tilde{x}$ is not found in the neighborhood $\mathcal{N}(\tilde{x}, k)$. Algorithm 2 displays a brief description of the local branching scheme applied to the WMST problem.

In the first line of Algorithm 2 a solver is used to obtain the first feasible integer solution $\tilde{x}^1$ which is taken as a reference solution. A solver is again used in lines 7 and 13 to obtain the solutions in the reduced solution space. The sequence of solutions $\tilde{x}^t$ generated by the LB, corresponds to a decreasing sequence of costs.

The Local Branching is a MIP technique planned to be an exact method, but that acts as a heuristic method when a time limit is set and reached before the optimal solution is found [8]. In case the time limit is exceeded, the obtained solution $\tilde{x}$ is not the optimal solution and the exploration of the neighborhood is not complete. In that case the size of the neighborhood that is to be explored has to be modified in order to either reduce or enlarge the region where the

---

Algorithm 2. Local branching scheme for the WMST problem (LB)

Require: problem data (graph $G = (V, E)$; costs c_{ij}; weights w_{ij}; weight limit W);
    parameters (k; maxtime; maxiter).
1: obtain a feasible solution $\tilde{x}$, the reference solution
2: $t \leftarrow 1$
3: $\tilde{x}^t \leftarrow \tilde{x}$
4: **while** ((time < maxtime) and (t < maxiter)) **do**
5:     define k
6:     introduce the local branching constraint $\Lambda(x, \tilde{x}^t) \leq k$ in the WMST problem
7:     solve the problem and let $\tilde{x}$ be its optimal solution within the neighborhood
    $\mathcal{N}(\tilde{x}^t, k)$
8:     **if** $C(\tilde{x}) < C(\tilde{x}^t)$ **then**
9:         $t \leftarrow t + 1$
10:        $\tilde{x}^t \leftarrow \tilde{x}$
11:     **else**
12:        introduce the local branching constraint $\Lambda(x, \tilde{x}^t) \geq k + 1$ in the WMST
    problem
13:        solve the problem and let $\tilde{x}$ be its optimal solution within the neighborhood
    $\mathcal{N}_+(\tilde{x}^t, k)$
14:        **if** $C(\tilde{x}) < C(\tilde{x}^t)$ **then**
15:           $t \leftarrow t + 1$
16:           $\tilde{x}^t \leftarrow \tilde{x}$
17:        **end if**
18:     **end if**
19: **end while**
    return integer solution $\tilde{x}^t$

---

solution is sought. The following mechanisms, see [13], modify the size of the neighborhood.

Intensification mechanism. The intensification mechanism aims to reduce the size of the neighborhood in an attempt to speed-up its exploration. The right hand side of the constraint $\Lambda(x, \tilde{x}) \leq k$ is reduced to $\lfloor \frac{k}{2} \rfloor$.

Diversification mechanism. The diversification mechanism aims to enlarge the size of the neighborhood. However, and consequently the exploration time is also increased. First apply a "weak" diversification mechanism, in which the right hand side of the constraint $\Lambda(x, \tilde{x}) \leq k$ is increased by $\lceil \frac{k}{2} \rceil$, i.e., it is introduced the constraint $\Lambda(x, \tilde{x}) \leq k + \lceil \frac{k}{2} \rceil$. In case an improved solution is not found, apply a "strong" diversification mechanism, in which the right hand side of the constraint $\Lambda(x, \tilde{x}) \leq k$ is increased with $2 \times \lceil \frac{k}{2} \rceil$, i.e., it is introduced the constraint $\Lambda(x, \tilde{x}) \leq k + 2\lceil \frac{k}{2} \rceil$.

When a time limit is exceeded and the obtained solution $\tilde{x}$ is not the optimal solution the following cases may occur. (i) If the solution $\tilde{x}$ has an improved value, then the reference solution is updated, but the value of parameter k is not modified. (ii) If the solution $\tilde{x}$ does not have an improved value, $C(\tilde{x}) > C(\tilde{x}^t)$, then apply the intensification mechanism in order to reduce the neighborhood.

If again an improved solution is not found, apply a "weak" diversification mechanism. (iii) If the solution $\tilde{x}$ is infeasible, then apply the "strong" diversification mechanism in order to enlarge the neighborhood.

5 Computational Experience

In this section we report computational tests of the FP and of the LB heuristics applied to the WMST problem. We compare the performance of the heuristics against an upper bound obtained by a branch-and-cut algorithm based on the strengthened weighted Miller-Tucker-Zemlin formulation (Agra et al. [3] and Requejo et al. [16]).

All the tests were performed using a computer with an Intel(R) Core(TM)2 Duo CPU (T7100) 2.00 GHz processor and 4 GB of RAM, and were conducted using the Xpress-Optimizer 23.01.03 solver with the default options.

We present computational results for instances to the WMST problem defined on complete graphs with a number of nodes varying between 10 and 1000, in a total of 215 instances.

5.1 Instances Generation Description

To generate an instance of the WMST problem, the cost c_e and the weight w_e of each edge e have to be defined. Afterwards, a (feasible) value to the weight limit W must also be defined. We built three sets of instances, constituting three different ways of generating costs and weights.

In a first set of instances, costs c_e and weights w_e are generated similarly to a set of instances described in Pisinger [14] and named therein as Spanner instances. A value for W is selected between 1000 and 3500 proportional to the number n of nodes of the instance. The costs and the weights are multiples of a small set (the spanner set in [14]) of costs and weights following one particular distribution, we use the Uncorrelated distribution (which is in the Pisinger's [14] proposed list of distributions), and the following two parameters $s = 2$ and $m = 10$. At the end, the weights of some edges are manipulated in such a way that the optimal solution has a desired predefined structure. After testing a few structures we obtained some challenging instances when the optimal structure of the WMST problem instance solution has large diameter values, almost $n-1$, but not equal to $n-1$, in such way that the tree is almost a path. Thus we name this instance's set as "Almost Path" (AP).

For the second set of instances, named Random (R) instances, the costs c_e and the weights w_e are uniformly generated in the interval $[1, 1000]$.

For the third set of instances, named Euclidean (E) instances, costs c_e and weights w_e are obtained using Euclidean distances. After randomly generating the coordinates of n points/nodes in a 100×100 grid, the cost c_e of each edge $e = \{i, j\}$ is the integer part of the Euclidean distance between points/nodes i and j. We proceed independently and similarly to obtain the weights.

To define a (feasible) value to the weight limit W for each instance of these two sets of instances (sets R and E), we start by obtaining the weight of the minimum spanning tree $W(T_c)$ and the weight of the minimum weight spanning tree $W(T_w)$ and we select W to be one of the values $W_i = \frac{W(T_c) + W(T_w)}{2^i}$, for $i \in \{1, \ldots, 10\}$.

A total of 215 instances were generated, 95 of the set AP and 60 of each set R and E. For each set AP and each instance size between 10 and 150 we have 10 instances and for each instance size between 200 and 1000 we have 5 instances. For each set R and E and each instance size we have 5 instances.

5.2   Computational Tests for the FP and IFP Heuristics

Computational tests performed in all groups of instances allow us to conclude that the heuristics FP and IFP, described in Sect. 3, obtain similar quality solutions. However the heuristic FP uses less computational time in all groups of instances, see Fig. 1. Also the number of iterations used by the IFP heuristic is approximately three times higher than the number of iterations used by the FP heuristic. On average the number of iterations of the FP is 4 and the number of iterations of the IFP is 14. Therefore, the remaining computational results will be presented only for the FP heuristic.

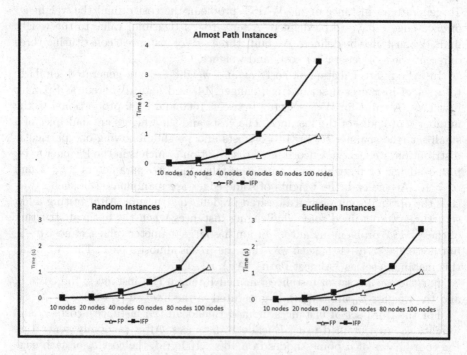

Fig. 1. Comparing the mean execution times (in seconds) for FP and IFP heuristics.

5.3   Computational Tests for the LB Heuristic

Computational experiments performed with the LB heuristic, described in Sect. 4, showed that when the value k of the local branching constraints (8) increases, the mean execution times also increases. To obtain the computational results of the LB heuristic we used the value $k = 5$ and several strategies were tested. First strategy (LB1): a time limit of 10000 s is imposed. Second strategy (LB2): in addition to the time limit, the solver is stopped when the first integer solution is obtained, and intensification and diversification mechanisms are used. Third strategy (LB3): in addition to the second strategy conditions, the reference solution is the first integer solution found by the solver and a time limit is imposed for the exploration of the neighborhoods. Fourth strategy (LB4): differs from the third strategy because the reference solution is obtained with the FP heuristic. To evaluate these four strategies some preliminary computational tests were performed to instances with up to $n = 100$ nodes in a total of 120 instances. Table 1 displays mean computational times and mean number of iterations for the four LB strategies. The LB1 strategy obtained good valued solutions, but it uses high computational times and 18.3% of the instances (13 AP instances and 9 E instances) exceed a computational time limit imposed.

Table 1. Mean execution time (in seconds) and mean number of iterations of the four LB strategies.

	n	LB1		LB2		LB3		LB4	
		Time	Iter	Time	Iter	Time	Iter	Time	Iter
Almost Path	10	0.42	3	0.58	5	0.40	3	0.40	3
	20	3.75	3	4.66	6	3.49	3	2.37	3
	40	64.06	4	29.27	8	24.32	5	20.00	4
	60	2273.08	4	17.85	7	34.10	4	24.00	3
	80	3443.43	3	41.93	7	61.95	3	51.05	3
	100	9999.25	3	67.04	7	123.66	4	98.63	3
Random	10	0.41	3	0.55	4	0.61	3	0.37	3
	20	2.68	3	4.73	6	2.82	3	2.35	3
	40	17.26	4	26.27	7	13.67	3	10.06	3
	60	18.89	4	20.32	8	20.04	4	13.47	3
	80	64.02	5	48.81	9	49.26	4	30.50	3
	100	1942.81	11	90.72	10	238.08	8	75.49	3
Euclidean	10	0.89	3	2.28	4	2.69	4	2.69	4
	20	6.52	4	14.92	9	13.41	7	6.85	5
	40	88.60	7	63.84	11	75.01	9	29.52	4
	60	3802.02	11	62.20	11	241.64	12	28.98	4
	80	8084.06	12	476.79	12	1927.29	14	44.58	4
	100	9102.89	18	319.51	14	3471.08	19	132.12	5

Table 2. Mean gap values (in percentage) of the HP procedure, the FP and the LB heuristics.

	n	AP			R			E		
		HP	FP	LB	HP	FP	LB	HP	FP	LB
Mean gap	10	0.00	7.14	0.00	0.00	3.84	0.50	0.00	5.27	0.00
	20	0.00	0.57	0.00	0.00	5.28	0.00	0.00	2.25	0.10
	40	0.00	0.58	0.00	0.00	2.74	0.00	0.00	1.96	0.12
	60	0.00	0.02	0.00	0.00	1.31	0.00	0.00	0.75	0.08
	80	0.00	0.06	0.09	0.00	1.50	0.00	0.00	0.58	0.08
	100	0.00	0.67	0.28	0.00	0.55	0.01	0.00	0.56	0.06
	150	0.00	0.77	0.00	0.00	0.79	0.39	0.00	0.28	0.02
	200	0.00	1.18	0.00	0.00	0.53	0.02	0.00	0.40	0.03
	300	0.00	1.20	0.00	0.00	0.24	0.16	0.00	0.22	0.02
	400	0.00	0.89	0.00	0.00	0.14	0.14	13.58	0.00	0.03
	500	0.00	1.06	0.00	0.00	0.22	0.14	*	0.00	0.00
	1000	0.00	1.41	0.00	0.04	0.07	0.07	*	0.00	0.00
σ_{gap}	10	0.00	8.08	0.00	0.00	3.43	1.12	0.00	3.49	0.00
	20	0.00	0.61	0.00	0.00	5.01	0.00	0.00	2.68	0.23
	40	0.00	0.61	0.00	0.00	1.68	0.00	0.00	0.39	0.00
	60	0.00	0.03	0.00	0.00	1.07	0.00	0.00	0.71	0.13
	80	0.00	0.05	0.11	0.00	1.32	0.00	0.00	0.43	0.15
	100	0.00	1.40	0.63	0.00	0.26	0.02	0.00	0.66	0.07
	150	0.00	1.27	0.00	0.00	0.68	0.86	0.00	0.19	0.02
	200	0.00	1.42	0.00	0.00	0.38	0.03	0.00	0.20	0.03
	300	0.00	0.88	0.00	0.00	0.62	0.21	0.00	0.16	0.02
	400	0.00	0.73	0.00	0.00	0.10	0.10	4.50	0.00	0.06
	500	0.00	1.32	0.00	0.00	0.14	0.19	*	0.00	0.00
	1000	0.00	0.43	0.00	0.08	0.08	0.08	*	0.00	0.00

The symbol * means we could not obtain an upper bound within the time limit imposed.

The LB2 and LB3 strategies use less computational time than strategy LB1, however the LB4 strategy uses less computational time, obtains the best valued solutions and the optimal solution was obtained in 83.3% of the instances (in 100 of the 120 instances). Table 1 shows that the LB4 strategy presents the lowest average execution times and also uses the smallest average number of iterations, for all group of instances. Therefore we report results obtained with this LB4 strategy, denoted as LB in what follows. Notice that this strategy corresponds to the FP heuristic followed by a LB heuristic.

Table 3. Mean execution time (in seconds) for the HP procedure, the FP and the LB heuristics.

	n	AP			R			E		
		HP	FP	LB	HP	FP	LB	HP	FP	LB
Mean time	10	0.11	0.01	0.40	0.19	0.01	0.37	0.11	0.01	2.69
	20	1.32	0.03	2.37	1.13	0.03	2.35	0.89	0.03	6.85
	40	41.67	0.12	20.00	4.13	0.12	10.06	5.60	0.12	29.52
	60	517.22	0.29	24.00	5.96	0.27	13.47	12.36	0.29	28.98
	80	2535.07	0.55	51.05	14.25	0.54	30.50	41.12	0.54	44.58
	100	10596.60	0.97	98.63	20.68	1.21	75.49	116.40	1.10	132.12
	150	10804.61	3.01	176.11	95.07	4.98	175.32	585.16	2.30	181.10
	200	10813.50	7.89	355.65	310.59	12.52	299.32	1905.27	4.94	314.05
	300	10866.94	40.86	688.63	1566.21	39.65	727.49	12368.56	18.87	1205.03
	400	10929.46	90.46	1525.50	2628.47	92.53	1406.42	31103.45	39.50	2310.27
	500	11057.40	268.78	3624.23	9225.34	311.44	2357.79	27611.01	98.89	4530.66
	1000	11988.70	1014.11	9621.59	17636.70	1724.17	10808.70	31071.30	1171.19	10833.06
σ_{time}	10	0.04	0.01	0.14	0.14	0.01	0.17	0.06	0.01	1.78
	20	0.87	0.00	0.69	0.62	0.00	0.27	0.56	0.01	3.26
	40	29.75	0.02	5.84	1.15	0.01	2.60	1.21	0.01	7.64
	60	338.83	0.01	5.32	1.19	0.01	3.96	4.65	0.01	6.31
	80	2095.42	0.02	10.49	2.32	0.04	12.24	13.01	0.05	6.02
	100	645.98	0.06	19.44	6.10	0.35	35.28	34.54	0.26	48.59
	150	1.82	0.15	37.14	37.46	1.18	84.89	181.11	0.11	26.14
	200	4.73	1.41	176.84	222.72	2.55	64.89	737.56	0.41	83.14
	300	34.44	7.36	177.79	866.04	15.76	62.21	3070.60	3.47	234.47
	400	79.30	4.31	173.93	519.69	45.28	92.43	21167.74	1.55	482.86
	500	132.57	76.00	327.01	2740.35	31.91	617.46	7483.64	20.97	381.97
	1000	488.97	88.88	1048.95	2945.74	264.51	421.43	86.23	136.92	13.18

5.4   Results Description and Analysis

We compare the performance of the FP and of the LB heuristics with the performance of the HP procedure.

In [3,16] the best results to obtain the optimal value using the software Xpress 7.3 (Xpress Release 2012 with Xpress-Optimizer 23.01.03 and Xpress-Mosel 3.4.0), were obtained with the Branch and Cut algorithm based on a weighted MTZ (Miller-Tucker-Zemlin) formulation with the inclusion of cuts preventing cycles at the root node. This procedure will be denoted by HP (Hybrid Procedure) and is used to access the quality of the solutions obtained with the FP and LB heuristics.

Having an upper bound on the value of the cost, the upper bound gap is $gap = \frac{UB - OPT}{OPT} \times 100$, where UB is the upper bound obtained through the considered method (HP procedure, FP heuristic, LB heuristic) and OPT is the optimal value obtained with the HP procedure or the best obtained value with this procedure within a time limit of 10000 s.

For each instance set AP, R, and E and each instance size set we display in the upper part of the Table 2 the average upper bound gap (in percentage) and the corresponding standard deviation values, in the lower part of the Table 2. In the upper part of the Table 3 are displayed the average execution times (in seconds) and corresponding standard deviation values, in the lower part of the Table 3.

The LB heuristic improves the solutions obtained by the FP heuristic in 68.37% of the instances (147 from 215 instances). In 75.34% of the instances (162 from 215 instances) the gap is null, i.e. the optimal solution was obtained, when using the LB heuristic. The set of instances with the highest gap is the E instance set, the Euclidean instances.

For every procedure the computational time increases as the size of the instance (number of nodes) increases. In 61.86% of the instances (133 from 215 instances), the computational time of the LB heuristic is smaller than the computational time of the HP procedure. The set of instances with the highest computational time is, again, the instance set E. Using the HP procedure it was not possible to obtain an upper bound within the time limit imposed for two E instances.

6   Conclusions

We describe two heuristic procedures to the WMST problem, the FP (feasibility pump) and the LB (local branching) heuristics. The FP heuristic is a constructive heuristic and uses a relax-and-fix strategy to obtain a good feasible solution. The LB heuristic uses a local branch strategy to improve a feasible solution. Our computational results show that the FP heuristic is fast in obtaining feasible solutions for the WMST problem, and the LB heuristic can be used to improve the obtained feasible solution. Both heuristics can be used independently, however the best strategy is to use the heuristics together, the FP followed by the LB. This process is faster than the HP procedure, a branch-and-cut procedure used to access the quality of the described heuristics, and obtains good valued solutions.

The FP and the LB heuristics are a good choice in obtaining feasible solutions for the WMST problem and can be used together for better quality solutions and with very competitive computational times.

Acknowledgements. The research of the authors has been partially supported by Portuguese funds through the *CIDMA (Center for Research and Development in Mathematics and Applications)* and the FCT, the Portuguese Foundation for Science and Technology, within project UID/MAT/04106/2013.

References

1. Achterberg, T., Berthold, T.: Improving the feasibility pump. Discrete Optim. **4**(1), 77–86 (2007)
2. Aggarwal, V., Aneja, Y.P., Nair, K.P.K.: Minimal spanning tree subject to a side constraint. Comput. Oper. Res. **9**, 287–296 (1982)

3. Agra, A., Cerveira, A., Requejo, C., Santos, E.: On the weight-constrained minimum spanning tree problem. In: Pahl, J., Reiners, T., Voß, S. (eds.) INOC 2011. LNCS, vol. 6701, pp. 156–161. Springer, Heidelberg (2011). doi:10.1007/978-3-642-21527-8_20

4. Agra, A., Requejo, C., Santos, E.: Implicit cover inequalities. J. Combin. Optim. **31**(3), 1111–1129 (2016)

5. Bertacco, L., Fischetti, M., Lodi, A.: A feasibility pump heuristic for general mixed-integer problems. Discrete Optim. **4**(1), 63–76 (2007)

6. Fischetti, M., Glover, F., Lodi, A.: The feasibility pump. Math. Program. **104**(1), 91–104 (2005)

7. Fischetti, M., Lodi, A.: Local branching. Math. Program. **98**(1–3), 23–47 (2003)

8. Hansen, P., Mladenović, N., Urošević, D.: Variable neighborhood search and local branching. Comput. Oper. Res. **33**(10), 3034–3045 (2006)

9. Hassin, R., Levin, A.: An efficient polynomial time approximation scheme for the constrained minimum spanning tree problem using matroid intersection. SIAM J. Comput. **33**(2), 261–268 (2004)

10. Henn, S.: Weight-constrained minimum spanning tree problem. Master's thesis, University of Kaiserslautern, Kaiserslautern, Germany (2007)

11. Hong, S., Chung, S., Park, B.H.: A fully polynomial bicriteria approximation scheme for the constrained spanning tree problem. Oper. Res. Lett. **32**, 233–239 (2004)

12. Magnanti, T.L., Wolsey, L.A.: Optimal trees. In: Ball, M., Magnanti, T.L., Monma, C., Nemhauser, G.L. (eds.) Network Models, Handbooks in Operations Research and Management Science, vol. 7, pp. 503–615. Elsevier Science Publishers, North-Holland (1995)

13. Mladenović, N., Hansen, P.: Variable neighborhood search. Comput. Oper. Res. **24**(11), 1097–1100 (1997)

14. Pisinger, D.: Where are the hard knapsack problems? Comput. Oper. Res. **32**(9), 2271–2284 (2005)

15. Ravi, R., Goemans, M.X.: The constrained minimum spanning tree problem. In: Karlsson, R., Lingas, A. (eds.) SWAT 1996. LNCS, vol. 1097, pp. 66–75. Springer, Heidelberg (1996). doi:10.1007/3-540-61422-2_121

16. Requejo, C., Agra, A., Cerveira, A., Santos, E.: Formulations for the weight-constrained minimum spanning tree problem. AIP Conf. Proc. **1281**, 2166–2169 (2010)

17. Requejo, C., Santos, E.: Lagrangian based algorithms for the weight-constrained minimum spanning tree problem. In: Proceedings of the VII ALIO/EURO Workshop on Applied Combinatorial Optimization, pp. 38–41 (2011)

18. Shogan, A.: Constructing a minimal-cost spanning tree subject to resource constraints and flow requirements. Networks **13**, 169–190 (1983)

19. Sourd, F., Spanjaard, O.: A multiobjective branch-and-bound framework: application to the biobjective spanning tree problem. INFORMS J. Comput. **20**(3), 472–484 (2008)

20. Xue, G.: Primal-dual algorithms for computing weight-constrained shortest paths and weight-constrained minimum spanning trees. In: Proceedings of the IEEE International Conference on Performance, Computing, and Communications, pp. 271–277 (2000)

21. Yamada, T., Watanabe, K., Kataoka, S.: Algorithms to solve the knapsack constrained maximum spanning tree problem. Int. J. Comput. Math. **82**, 23–34 (2005)

A Decomposition Algorithm for Robust Lot Sizing Problem with Remanufacturing Option

Öykü Naz Attila[1]([✉]) [iD], Agostinho Agra[2][iD], Kerem Akartunalı[1][iD],
and Ashwin Arulselvan[1][iD]

[1] Department of Management Science, University of Strathclyde, Glasgow, UK
{oyku.attila,kerem.akartunali,ashwin.arulselvan}@strath.ac.uk
[2] Department of Mathematics and CIDMA,
University of Aveiro, Aveiro, Portugal
aagra@ua.pt

Abstract. In this paper, we propose a decomposition procedure for constructing robust optimal production plans for reverse inventory systems. Our method is motivated by the need of overcoming the excessive computational time requirements, as well as the inaccuracies caused by imprecise representations of problem parameters. The method is based on a min-max formulation that avoids the excessive conservatism of the dualization technique employed by Wei et al. (2011). We perform a computational study using our decomposition framework on several classes of computer generated test instances and we report our experience. Bienstock and Özbay (2008) computed optimal base stock levels for the traditional lot sizing problem when the production cost is linear and we extend this work here by considering return inventories and setup costs for production. We use the approach of Bertsimas and Sim (2004) to model the uncertainties in the input.

Keywords: Robust lot sizing · Remanufacturing · Decomposition

1 Introduction

Traditional lot sizing problems mainly aim to construct production plans that minimize the total operational cost for a specific production system, while ensuring that demand in each time period is satisfied. For such models to remain applicable for present production systems, recent shifts in manufacturing practices have to be taken into consideration through revising model structure and assumptions. A practice that has been increasingly applied is the reuse of deformed items to manufacture as-good-as-new products, motivated by the increasing interest of implementing recycling activities. More specifically, such production systems with item recoveries are expected to have reduced overall production costs and waste through restoring deformed products to their usable state. Recovery of these items can be undertaken in several ways (see Thierry et al. 1995). In our work, we are interested in investigating the option of recovering these items through remanufacturing. Applications of remanufacturing are

© Springer International Publishing AG 2017
O. Gervasi et al. (Eds.): ICCSA 2017, Part II, LNCS 10405, pp. 684–695, 2017.
DOI: 10.1007/978-3-319-62395-5_47

observed in the production of a wide range of products, such as electronic goods and industrial items (see, Thierry et al. 1995; Guide and Van Wassenhove 2009; Agrawal et al. 2015). Our main focus is to consider the additional decisions regarding remanufacturing while constructing an optimal production plan for a discrete and finite time horizon, where the exact values for demands and returned items are known to be uncertain.

Despite the wide range of research on lot sizing problems (see, e.g., Akartunalı et al. 2016), very few studies have focused on lot sizing problems with remanufacturing (LSR). Preliminary research on LSR problems includes the implementation of the Wagner-Whitin algorithm by Richter and Sombrutzki (2000), which was later extended to one with manufacturing and remanufacturing costs by Richter and Weber (2001). An economic LSR formulation (ELSR) with disposal costs was introduced by Golany et al. (2001), where the ELSR problem was shown to be NP-complete. A dynamic programming algorithm was presented by Teunter et al. (2006), which solves the ELSR problem in $O(T^4)$ time for a special case of the problem. In more recent work, Helmrich et al. (2014) have introduced alternative formulations for the ELSR problem, and have shown that the problem with joint or separate setups is NP-hard. The work of Akartunalı and Arulselvan (2016) has shown the tractability of a polynomial time special case and have introduced two classes of valid inequalities for the capacitated version of the problem. However, there is a lack of literature on the impact of uncertainty on these formulations, with the exception of Wei et al. (2011). The present study aims to contribute to the growing research on ELSR problems by studying the implications of parameter uncertainties within the framework of robust optimization.

Robust optimization was first introduced by Soyster (1973), where uncertain parameters are defined through uncertainty sets and a robust optimal solution is defined as one that remains optimal for every parameter representation in an uncertainty set. More recent studies relaxed this conservative assumption, with the seminal work of Ben-Tal and Nemirovski (1998, 1999) constructing uncertainty sets as ellipsoids. Later, Bertsimas and Sim (2004) defined the uncertainty sets as budgeted polytopes, where robust parameter representations are constrained by a specified value. Their approach was applied to traditional lot sizing problems by Bertsimas and Thiele (2006) and adapted to the robust ELSR in Wei et al. (2011). Bienstock and Özbay (2008) propose a decomposition approach for solving a min-max formulation of the special lot sizing problem consisting in the computation of basestock levels. Robust lot sizing problems have also been considered as particular cases of the general problems addressed in Agra et al. (2016), and Atamtürk and Zhang (2007). For comprehensive books on robust optimization we refer the reader to Bertsimas and Sim (2004) and Ben-Tal et al. (2009). For concise overviews on robust optimization methods see Bertsimas and Thiele (2011); Gabrel et al. (2014); Gorissen et al. (2015).

Here we consider a min-max formulation for the robust ELSR problem, since the approach followed in Wei et al. (2011) is known to be too conservative and uses many dual variables which restricts its applicability. For a detailed explanation of this conservativeness, see Bienstock and Özbay (2008). A common approach to handle min-max robust optimization problems is to use a variant of the Benders' decomposition, (see Thiele et al. 2010). Frequently, this decomposition results in the iterative inclusion of rows and columns (Agra et al. 2013; Zeng and Zhao 2013). Such approach is also known as the *Adversarial* approach (Gorissen et al. 2015). For solving robust inventory problems, the decomposition framework was introduced by Bienstock and Özbay (2008) and revisited later by Agra et al. (2016), where demand in each time period is assumed to be uncertain. Our approach for the robust ELSR problem is also motivated by these studies, where our objective is to generate optimal production plans when demands and returns are uncertain. We use the approach of Bertsimas and Sim (2004) to model the uncertainty set as budgeted polytopes, where the variation of the demands and returns in relation to their nominal values is constrained by a specified value. A robust model with recourse is considered where the inventory levels are allowed to adjust to the realization of the uncertain parameters. Our contribution is two-fold. Firstly, we model an extended version of the lot sizing problem, wherein we consider uncertainty in both the return and demand sets with set up costs for production. Our second contribution lies in reporting our computational experience with several input classes of costs and inventory levels.

The remainder of this paper is organized as follows: In Sect. 2, we introduce the deterministic and robust formulations for the ELSR problem. We introduce the robust decomposition algorithm in Sect. 3, and finally we conclude by presenting preliminary performance results in Sect. 4.

2   Problem Definition

The main problem addressed in this study is the economic lot sizing problem with remanufacturing and joint setups in a robust setting. Throughout the paper, the term "robust" refers to probability-free uncertainty. Problem assumptions and notations, the deterministic ELSR formulation, and a detailed description of parameter uncertainties are presented prior to the robust formulation. The decomposition algorithm introduced in Sect. 3 is based on the robust min-max formulation given in Sect. 2.2. Thus, the assumptions and notations given under this section remain valid for the decomposition algorithm. The objective of our problem is to produce a production plan detailing the amounts to be manufactured, remanufactured, kept in inventory, backlogged and disposed of, where total operational costs are minimized for the maximum value of return and demand deviations. Our assumptions are: (a) remanufacturing is a single operation (has no accompanying inspection/disassembly) (b) remanufactured items are as good as manufactured ones (c) serviceable inventory (ready to serve demand) can either be positive (incurs holding cost) or negative (incurs a backlogging cost) (d) return items can be disposed at a cost (e) manufacturing and remanufacturing are not capacitated and incur a joint setup cost of K (f) The production plan is generated for a finite and discrete time horizon, T.

In addition, we assume that all problem parameters are known. The values of demands and returns are inexact, however, the inputs used to construct the relevant uncertainty sets are known. All cost parameters are time-invariant, and serviceables have a greater holding cost than returned items. Similarly, manufacturing an item is more expensive than remanufacturing a returned item.

Manufacturing, remanufacturing, disposal and backlogging costs of a single item are represented as m, r, f and b, respectively. The unit holding cost of serviceable (returned) goods are shown as h^s (h^r). Let the demands (returns) for periods $t = 1, \ldots, T$ be D_t (R_t). For modelling the set up decision, we introduce a binary variable y_t and a sufficiently big M_t for all t. Variable x_t^m (x_t^r) indicates the number of items manufactured (remanufactured) in time t. Let the number of items disposed at the end of period t be d_t. Finally, Z_t^D (Z_t^R) models the scaled deviation of demands (returns) from the nominal value in period t. We might drop the time index t, to denote the corresponding vector. For instance, Γ^R will denote the vector in T dimensions with the tth component being Γ_t^R.

2.1 Classical Deterministic Model

The ELSR problem can be written as:

$$\min_{(\mathbf{x}, y) \in \mathcal{P}} \theta^{D,R}(\mathbf{x}, y) \tag{1}$$

where

$$\theta^{D,R}(\mathbf{x}, y) = \sum_{t=1}^{T} (K y_t + m x_t^m + r x_t^r + f d_t + H_t^s + H_t^r), \tag{2}$$

and $\mathbf{x} = (x^m, x^r)$ and y are vectors specifying a feasible production plan that belongs to the set

$$\mathcal{P} := \{(x_t^m, x_t^r, y_t) \in \mathbb{R}_+^{2T} \times \mathbb{Z}_+^T : I_0^r + \sum_{i=1}^{t} (R_i - x_i^r - d_i) \geq 0, \quad \forall t = 1, \ldots, T \tag{3}$$

$$M_t y_t \geq x_t^m + x_t^r, \quad \forall t = 1, \ldots, T \}$$

As reverse flows do not exist for returns, their inventory levels are restricted to be nonnegative and we ensure that the setup cost K is incurred for the time period t when $y_t = 1$, where $M_t = \sum_{i=t}^{T} D_i$. Variables H_t^s (H_t^r) model the total cost of serviceable (return) inventory held in period t and is given by

$$H_t^s = \max\{h^s[I_0^s + \sum_{i=1}^{t} (x_i^m + x_i^r - D_i)], -b[I_0^s + \sum_{i=1}^{t} (x_i^m + x_i^r - D_i)]\} \tag{4}$$

$$H_t^r = h^r[I_0^r + \sum_{i=1}^{t} (R_i - x_i^r - d_i)] \tag{5}$$

2.2   Uncertainty

In practical cases some of the parameters may not be known in advance. Here we assume the demands D_t and the returns R_t are uncertain, and consider a twostage robust model. The number of items manufactured, remanufactured and disposals (and consequently the set-up decisions) are assumed to be first stage or "here-and-now" decisions. Thus, such decisions are taken before the value of the uncertain parameters is revealed. While the serviceable and return inventory levels are second-stage variables since they are allowed to adjust to the value of the parameters.

We apply the robust optimization approach of Bertsimas and Sim (2004) defining uncertainty sets as budgeted polytopes. The uncertainty on demand and return parameters is considered to be independent from each other. Therefore, an independent uncertainty set for demands (U^D), and returns (U^R) exist. For each time period $t = 1, \ldots, T$, parameters $\Gamma_t^D, \bar{D}_t, \hat{D}_t$ $(\Gamma_t^R, \bar{R}_t, \hat{R}_t)$ are the budget of uncertainty for demands (returns), nominal demands (returns) and maximum deviation in demands (returns) respectively. The robust parameter D_t takes its value in the interval $[\bar{D}_t, \bar{D}_t + \hat{D}_t]$. Similarly, R_t takes its value in the interval $[\bar{R}_t, \bar{R}_t + \hat{R}_t]$. Hence, our uncertainty sets are defined as:

$$U^D(\Gamma^D) := \{D \in \mathbb{R}_+^T : \ D_t = \bar{D}_t + \hat{D}_t z_t^D, \quad \forall t = 1, \ldots, T, \ z_t^D \in Z_t^D(\Gamma_t^D)\} \quad (6)$$

$$U^R(\Gamma^R) := \{R \in \mathbb{R}_+^T : \ R_t = \bar{R}_t + \hat{R}_t z_t^R, \quad \forall t = 1, \ldots, T, \ z_t^R \in Z_t^R(\Gamma_t^R)\} \quad (7)$$

The variables z_t^D and z_t^R in (6) and (7) take their values in the interval $[0, 1]$ and are used to indicate a given proportion of the maximum deviations $\hat{D}_t$ and $\hat{R}_t$. In order to avoid overconservative parameter representations, the parameters Γ_t^D and Γ_t^R are introduced to constrain z_t^D and z_t^R. More specifically, the cumulative values of scaled deviation variables for demands and returns are required to be strictly less than or equal to Γ_t^D and Γ_t^R, hence we obtain:

$$Z_t^D(\Gamma_t^D) := \{z_t^D \in [0, 1]^t : \sum_{i=1}^{t} z_t^D \leq \Gamma_t^D, \forall t = 1, \ldots, T\} \quad (8)$$

$$Z_t^R(\Gamma_t^R) := \{z_t^R \in [0, 1]^t : \sum_{i=1}^{t} z_t^R \leq \Gamma_t^R, \forall t = 1, \ldots, T\} \quad (9)$$

As the inventory levels are allowed to adjust to the uncertain parameters, the variables H_t^s and H_t^r will depend on the demands and returns. So, for each $t = 1, \ldots, T$ and $D \in U^D$, we have $H_t^s(D)$ given from (4), and for each $t = 1, \ldots, T$ and $R \in U^R$, we have $H_t^r(R)$ given from (5).

We can now extend the deterministic ELSR problem to this uncertain case as a robust min-max formulation:

$$\min_{(\mathbf{x},\mathbf{y}) \in \mathcal{P}} \max_{\substack{D \in U^D(\Gamma^D) \\ R \in U^R(\Gamma^R)}} \theta^{D,R}(\mathbf{x},\mathbf{y}) \quad (10)$$

where $\theta^{D,R}(\mathbf{x}, y)$ is extended as follows:

$$\theta^{D,R}(\mathbf{x}, y) = \sum_{t=1}^{T}(Ky_t + mx_t^m + rx_t^r + fd_t + H_t^s(D) + H_t^r(R)) \qquad (11)$$

3   Decomposition Approach

As the number of variables $H_t^s(D)$ and $H_t^r(R)$ is not finite, the inner maximization problem is not finite. However, practical experience based on decomposition algorithms for related problems (see, for instance, Agra et al. (2013) for the case of the robust vehicle routing problem with time windows, Agra et al. (2016) for a general class of problems including the robut lot-sizing problem, and Bienstock and Özbay (2008) for the problem of computing robust basestock levels) has shown that only a few of the values of the uncertainty sets $U^D(\Gamma^D)$ and $U^R(\Gamma^R)$ are necessary to solve the problem.

Here we present a decomposition algorithm that iteratively solves a restricted version of the robust min-max problem (10) with respect to a subset of points of $U^D(\Gamma^D)$ and of $U^R(\Gamma^R)$ which will be denoted by $\tilde{U}^D$ and $\tilde{U}^R$, respectively. We call this restricted version of (10) as "Decision Maker's" problem (DMP). Given an optimal solution $(\mathbf{x}^*, y^*) \in \mathcal{P}$ to the DMP, we solve a certain maximization problem, which seeks a demand $D \in U^D(\Gamma^D)$ and return $R \in U^R(\Gamma^R)$ that maximises the total inventory and backlogging costs for the production plan $(\mathbf{x}^*, y^*) \in \mathcal{P}$. We refer to this subproblem as the "Adversarial Problem"(AP). The extreme point D^*, R^* generated by AP is used to update $\tilde{U}^D$ and $\tilde{U}^R$ and the process is repeated. Convergence is guaranteed through the finiteness of the number of extreme points of the uncertainty sets $U^D(\Gamma^D)$ and $U^R(\Gamma^D)$. The formal description of this idea is given in Algorithm 1.

Initialize $UB = +\infty$, $LB = 0$, $\tilde{U}^D = \{\bar{D}\}$, $\tilde{U}^D = \{\bar{R}\}$
while $(UB - LB)/LB \geq \epsilon$ **do**
   1. Solve DMP
    a. $(\mathbf{x}^*, y^*)$ be the solution of $\min_{(\mathbf{x},y)\in\mathcal{P}} \max_{D,R\in\tilde{U}^D\times\tilde{U}^R} \theta^{D,R}(\mathbf{x}, y)$
    b. Set $LB = \max_{D,R\in\tilde{U}^D\times\tilde{U}^R} \theta^{D,R}(\mathbf{x}^*, y^*)$
   2. Solve AP
    a. $(D^*, R^*) = \arg\max_{D,R\in U^D\times U^R} \theta^{D,R}(\mathbf{x}^*, y^*)$
    b. $\tilde{U}^D = \tilde{U}^D \cup \{D^*\}$, $\tilde{U}^R = \tilde{U}^R \cup \{R^*\}$
    c. $UB = \min\{UB, \theta^{D^*,R^*}(\mathbf{x}^*, y^*)\}$
end

Algorithm 1: Robust decomposition algorithm

For the sake of completeness, we give the DMP and the AP. In order to model the DMP, notice that the inner maximization problem in (10) defined for the restricted set $\tilde{U}^D \times \tilde{U}^R$, $\max_{D,R\in\tilde{U}^D\times\tilde{U}^R} \theta^{D,R}(\mathbf{x}, \mathbf{y})$, can be written as:

$$\sum_{t=1}^{T}(Ky_t + mx_t^m + rx_t^r + fd_t) + \max_{D,R \in \tilde{U}^D \times \tilde{U}^R} \sum_{t=1}^{T}(H_t^s(D) + H_t^r(R))). \qquad (12)$$

Introducing variable π to indicate the maximum value of the total inventory and backlogging costs over all possible realizations of demands and returns, the DMP can be written as follows:

$$\min \sum_{t=1}^{T}(Ky_t + mx_t^m + rx_t^r + fd_t) + \pi \qquad (13)$$

$$\text{s.t. } \pi \geq \sum_{t=1}^{T}(H_t^s(D) + H_t^r(R)) \qquad \begin{matrix} \forall D \in \tilde{U}^D \\ \forall R \in \tilde{U}^R \end{matrix} \qquad (14)$$

$$H_t^s(D) \geq h^s\left(I_0^s + \sum_{i=1}^{t}(x_i^m + x_i^r - D_i)\right) \qquad \begin{matrix} \forall t = 1,...,T \\ \forall D \in \tilde{U}^D \end{matrix} \qquad (15)$$

$$H_t^s(D) \geq -b\left(I_0^s + \sum_{i=1}^{t}(x_i^m + x_i^r - D_i))\right) \qquad \begin{matrix} \forall t = 1,...,T \\ \forall D \in \tilde{U}^D \end{matrix} \qquad (16)$$

$$H_t^r(R) = h^r\left(I_0^r + \sum_{i=1}^{t}(R_i - x_i^r - d_i)\right), \qquad \begin{matrix} \forall t = 1,...,T \\ \forall R \in \tilde{U}^R \end{matrix} \qquad (17)$$

$$\sum_{i=1}^{t}(R_i - d_i - x_i^r) \geq 0 \qquad \begin{matrix} \forall t = 1,...,T \\ \forall R \in \tilde{U}^R \end{matrix} \qquad (18)$$

$$M_t y_t \geq x_t^m + x_t^r \qquad \forall t = 1,...,T \qquad (19)$$

$$(x,y) \in P$$

Note that variables $H_t^r(R)$ can be eliminated using Eq. (17). Given a solution for variables x_i^m, x_i^r, d_i, the AP is formulated as follows:

$$\max \pi \qquad (20)$$

$$\text{s.t. } \pi \leq \sum_{t=1}^{T}(H_t^s + h^r\sum_{i=1}^{t}(\bar{R}_i + \hat{R}_i z_i^R - d_i - x_i^r)) \qquad (21)$$

$$H_t^s = \max\left\{h^s\left(I_0^s + \sum_{i=1}^{t}(x_i^m + x_i^r - (\bar{D}_i + \hat{D}_i z_i^D))\right),\right.$$

$$\left. -b\left(I_0^s + \sum_{i=1}^{t}(x_i^m + x_i^r - (\bar{D}_i + \hat{D}_i z_i^D))\right)\right\} \qquad \forall t = 1,...,T \qquad (22)$$

$$I_0^r + \sum_{i=1}^{t}(\bar{R}_i + \hat{R}_i z_i^R - x_i^r - d_i) \geq 0 \qquad \forall t = 1,...,T \qquad (23)$$

$$\sum_{i=1}^{t}z_i^D \leq \Gamma_t^D, \quad \sum_{i=1}^{t}z_i^R \leq \Gamma_t^R \qquad \forall t = 1,...,T \qquad (24)$$

$$0 \leq z_t^{Dj} \leq 1, \quad 0 \leq z_t^{Rj} \leq 1 \qquad \forall t = 1,...,T \qquad (25)$$

In order to linearize (22), we introduce binary variable s_t indicating whether inventory is kept or demand is backlogged, and rewrite it as:

$$H_t^s \leq h^s \left(I_0^s + \sum_{i=1}^{t} (x_i^m + x_i^r - (\bar{D}_i + \hat{D}_i z_i^D)) \right) + M_{1t}(1 - s_t) \quad \forall t = 1, ..., T \quad (26)$$

$$H_t^s \leq -b \left(I_0^s + \sum_{i=1}^{t} (x_i^m + x_i^r - (\bar{D}_i + \hat{D}_i z_i^D)) \right) + M_{2t} s_t \quad \forall t = 1, ..., T \quad (27)$$

4 Experiments

The proposed decomposition algorithm was implemented in Java using Eclipse Mars. Our formulations were implemented and solved as MIPs using Java API for CPLEX 12.6 on an Intel Core i5, 3.30 GHz CPU, 3.29 GHz, 8 GB RAM machine.

Additionally, each run has been restricted to a total running time of 10,000 s. The terminating condition for instances with a smaller running time is set as $\epsilon = 0.01$, where $\epsilon = \frac{UB - LB}{LB}$.

4.1 Data Generation

Data sets have been generated for different levels of four parameter types: number of returns, probability of constraint violation caused by Γ_t^D and Γ_t^R, the setup cost and the disposal cost. We consider three different levels for each group, except for disposal costs: low, medium and high. For disposal costs, we are interested in observing two different cases, namely when the disposal cost is greater or less than the remanufacturing cost. Throughout this section, the data sets are abbreviated as "ABCD_T", where each letter indicates the levels of the aforementioned parameters in their given order, with T time periods.

For all data sets, nominal demand is generated randomly in the interval $[50, 100]$. Likewise, returns are generated randomly in intervals $[15, 30]$, $[25, 50]$ and $[35, 70]$, for low, medium and high levels, respectively. Maximum demand and return deviations are calculated as $\hat{D}_t = 0.1\bar{D}_t$ and $\hat{R}_t = 0.1\bar{R}_t$. In order to determine Γ_t^D and Γ_t^R, we use the probabilistic bounds given by Bertsimas and Sim (2004). We set the probability of constraint violation as 0.01, 0.05 and 0.10, for low, medium and high levels, respectively. To determine the setup cost, we use the following equations: $K = 0.1\bar{D}_{min}h^s$, $K = 2\bar{D}_{med}h^s$ and $K = 5\bar{D}_{max}h^s$, where $\bar{D}_{min} = 50, \bar{D}_{med} = 75$ and $\bar{D}_{max} = 100$, for low, medium and high levels. Finally, the disposal cost is set as $d = 0.5r$ when it is less than the remanufacturing cost, and as $d = 2r$ otherwise.

In addition, the holding cost of serviceables is generated in the interval $[5, 10]$, through which the remaining cost parameters are defined. We set the holding cost for returns as $h^r = 0.1h^s$, the backlogging cost as $b = 4h^s$, the manufacturing cost as $m = 2h^r$, and the remanufacturing cost as $r = 2h^r$.

4.2 Preliminary Results

To observe the performance of our decomposition algorithm, we analyse the total time requirements for obtaining the smallest possible $UB - LB$ gap. The following performance measures are preliminary results that are obtained from 16 different data sets for $T = 10$, and 5 different data sets for $T = 50$. A total of 10 instances were solved for each data set.

When $T = 10$, all instances can be solved to $\epsilon = 0.07$ or better. It is possible to achieve $\epsilon = 0.001$ very quickly for the vast majority of the data sets. However, a few instances were aborted due to the time limit, for which the gap remains relatively large. For each data set, the total running time, final gap and total iterations for $T = 10$ are given in Table 1. The overall performance of cases when the gap could not be reduced down to $\epsilon = 0.001$ under 10,000 s are given in Figs. 1 and 2.

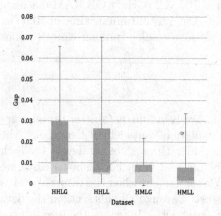

Fig. 1. Total $UB - LB$ gap for $T = 10$ (when gap is greater than 0.01).

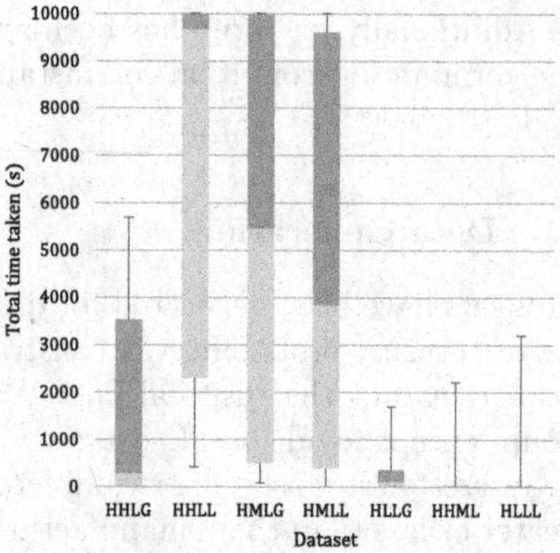

Fig. 2. Total running time for $T = 10$ (when $\epsilon > 0.01$ could not be achieved under 10,000 s).

As the detailed results in Table 1 indicate, the gaps achieved and number of iterations needed are in general consistent across different data sets, in addition to being in general very small (e.g., the highest maximum gap is still under 0.1, and no more than 8 iterations were necessary for any instance). On the other hand, as it can be also observed from the Figs. 1 and 2, the time performance can vary significantly not only among different datasets but also among different instances of most datasets.

When we look into the datasets with $T = 50$, a greater number of instances naturally run until the maximum time limit is reached as a consequence of the increased number of periods. However, the algorithm is still able to close the gap up to $\epsilon = 0.003$ for some instances.

In comparison to $T = 10$, the total number of iterations reduce as the time limit is reached in earlier iterations. We also observe a greater variety in terms

Table 1. Gap, time, iteration performance for T = 10.

Dataset	Gap (ϵ)		Time performance (s)		Number of iterations	
	Avg.	Std. Dev.	Avg.	Std. Dev.	Avg.	Std. Dev.
HHHG	0.004	0.002	0.4	0.1	4.2	0.9
HHHL	0.003	0.003	0.7	0.3	4.9	1.1
HHMG	0.005	0.003	27.8	40.5	5.9	1.2
HHML	0.003	0.002	236.2	692.5	5.3	0.9
HHLG	0.020	0.021	1706.0	2314.5	4.2	1.5
HHLL	0.018	0.022	6783.9	4337.9	4.2	1.2
HMHG	0.004	0.004	0.4	0.2	3.7	0.9
HMHL	0.005	0.004	0.3	0.2	3.9	0.3
HMLG	0.007	0.008	5247.2	4542.0	4.6	1.0
HMLL	0.008	0.013	4713.2	4512.3	4.5	1.2
HLHG	0.007	0.004	0.3	0.1	2.8	0.4
HLHL	0.005	0.004	0.4	0.2	3.0	0.5
HLMG	0.005	0.003	32.7	54.7	3.8	0.4
HLML	0.004	0.005	19.8	34.1	3.9	0.6
HLLG	0.002	0.004	401.4	617.4	4.2	0.6
HLLL	0.002	0.003	433.5	1024.5	4.2	0.8

of the total solution time for several data sets (see Fig. 4). For groups where the total running time is invariant, the final gap remains larger compared to others, in which case ϵ is less than 0.05. The gap, time and iteration performances for $T = 50$ are presented in Table 2. Although many instances exhausted the time limit, it is encouraging to see that the maximum gaps still remain very small (Fig. 3).

Table 2. Gap, time, iteration performance for T = 50.

Dataset	Gap (ϵ)		Time performance (s)		Number of iterations	
	Avg.	Std. Dev.	Avg.	Std. Dev.	Avg.	Std. Dev.
HHHG	0.010	0.006	8203.7	2092.2	3.5	0.7
HHHL	0.012	0.006	7595.5	3305.4	2.8	0.8
HHLG	0.159	0.043	10002.9	1.1	2.1	0.3
HHLL	0.141	0.028	10004.3	1.1	2.0	0.0
HHMG	0.029	0.008	9044.4	3033.3	2.7	0.5

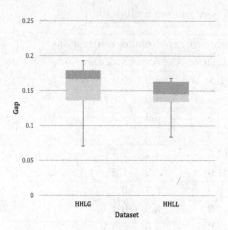

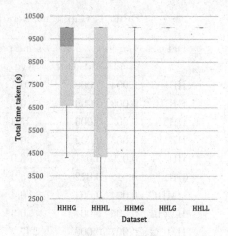

Fig. 3. Total UB − LB gap for T = 50 (when gap is greater than 0.05).

Fig. 4. Total running time for T = 50.

5 Conclusions

In this paper, we studied the robust lot sizing problem with remanufacturing option, where, to the best of our knowledge, the literature is at best scarce. We proposed a simple but effective decomposition procedure for constructing robust optimal production plans, where the approach of Bertsimas and Sim (2004) was used to model the uncertainties in input. Preliminary computational results on various datasets indicate that this procedure can work effectively, in particular to address current issues caused by imprecise representations of problem parameters.

Future work includes further development and improvement of the current computational framework in order to achieve more effective results, in particular of the computational times. It is also important to perform extensive computational testing to allow a thorough statistical analysis of the performance.

References

Agra, A., Santos, M.C., Nace, D., Poss, M.: A dynamic programming approach for a class of robust optimization problems. SIAM J. Optim. **26**(3), 1799–1823 (2016)

Agra, A., Christiansen, M., Figueiredo, R., Hvattum, L.M., Poss, M., Requejo, C.: The robust vehicle routing problem with time windows. Comput. Oper. Res. **40**(3), 856–866 (2013)

Agrawal, V.V., Atasu, A., Van Ittersum, K.: Remanufacturing, third-party competition, and consumers' perceived value of new products. Manag. Sci. **61**(1), 60–72 (2015)

Akartunalı, K., Fragkos, I., Miller, A., Wu, T.: Local cuts and two-period convex hull closures for big-bucket lot-sizing problems. INFORMS J. Comput. **28**(4), 766–780 (2016)

Akartunalı, K., Arulselvan, A.: Economic lot-sizing problem with remanufacturing option: complexity and algorithms. In: Pardalos, P.M., Conca, P., Giuffrida, G., Nicosia, G. (eds.) MOD 2016. LNCS, vol. 10122, pp. 132–143. Springer, Cham (2016). doi:10.1007/978-3-319-51469-7_11

Atamtürk, A., Zhang, M.: Two-stage robust network flow and design under demand uncertainty. Oper. Res. **55**(4), 662–673 (2007)

Ben-Tal, A., El Ghaoui, L., Nemirovski, A.: Robust Optimization. Princeton Series in Applied Mathematics. Princeton University Press, Princeton (2009)

Ben-Tal, A., Nemirovski, A.: Robust convex optimization. Math. Oper. Res. **23**(4), 769–805 (1998)

Ben-Tal, A., Nemirovski, A.: Robust solutions of uncertain linear programs. Oper. Res. Lett. **25**(1), 1–13 (1999)

Bertsimas, D., Brown, D.B., Caramanis, C.: Theory and applications of robust optimization. SIAM Rev. **53**, 464–501 (2011)

Bertsimas, D., Sim, M.: The price of robustness. Oper. Res. **52**(1), 35–53 (2004)

Bertsimas, D., Thiele, A.: A robust optimization approach to inventory theory. Oper. Res. **54**(1), 150–168 (2006)

Bienstock, D., Özbay, N.: Computing robust basestock levels. Discrete Optim. **5**(2), 389–414 (2008)

Gabrel, V., Murat, C., Thiele, A.: Recent advances in robust optimization: an overview. Eur. J. Oper. Res. **235**(3), 471–483 (2014)

Golany, B., Yang, J., Yu, G.: Economic lot-sizing with remanufacturing options. IIE Trans. **33**(11), 995–1003 (2001)

Gorissen, B.L., Yanikoğlu, I., den Hertog, D.: A practical guide to robust optimization. Omega **53**, 124–137 (2015)

Guide Jr., V.D.R., Van Wassenhove, L.N.: The evolution of closed-loop supply chain research. Oper. Res. **57**(1), 10–18 (2009)

Retel Helmrich, M.J., Jans, R., van den Heuvel, W., Wagelmans, A.P.: Economic lot-sizing with remanufacturing: complexity and efficient formulations. IIE Trans. **46**(1), 67–86 (2014)

Richter, K., Sombrutzki, M.: Remanufacturing planning for the reverse Wagner/Within models. Eur. J. Oper. Res. **121**(2), 304–315 (2000)

Richter, K., Weber, J.: The reverse Wagner/Within model with variable manufacturing and remanufacturing cost. Int. J. Prod. Econ. **71**(1), 447–456 (2001)

Soyster, A.L.: Technical note–convex programming with set-inclusive constraints and applications to inexact linear programming. Oper. Res. **21**(5), 1154–1157 (1973)

Teunter, R.H., Bayindir, Z.P., Den Heuvel, W.V.: Dynamic lot sizing with product returns and remanufacturing. Int. J. Prod. Res. **44**(20), 4377–4400 (2006)

Thiele, A., Terry, T., Epelman, M.: Robust linear optimization with recourse. Technical report. Available in Optimization-Online (2010)

Thierry, M., Salomon, M., Van Nunen, J., Van Wassenhove, L.: Strategic issues in product recovery management. Calif. Manag. Rev. **37**(2), 114–135 (1995)

Wei, C., Li, Y., Cai, X.: Robust optimal policies of production and inventory with uncertain returns and demand. Int. J. Prod. Econ. **134**(2), 357–367 (2011)

Zeng, B., Zhao, L.: Solving two-stage robust optimization problems using a column-and-constraint generation method. Oper. Res. Lett. **41**(5), 457–461 (2013)

Dual-RAMP for the Capacitated Single Allocation Hub Location Problem

Telmo Matos[✉][ID] and Dorabela Gamboa[✉][ID]

CIICESI, Escola Superior de Tecnologia e Gestão,
Instituto Politécnico do Porto, Felgueiras, Portugal
{tsm, dgamboa}@estg.ipp.pt

Abstract. We consider the Capacitated Single Allocation Hub Location Problem (CSAHLP) in which the objective is to choose the set of hubs from all nodes in a given network in such way that the allocation of all the nodes to the chosen hubs is optimal. We propose a Relaxation Adaptive Memory Programming (RAMP) approach for the CSAHLP. Our method combines Lagrangean Subgradient search with an improvement method to explore primal-dual relationships and create advanced memory structures that integrate information from both primal and dual solutions spaces. The algorithm was tested on the standard dataset and produced extremely competitive results that include new best-known solutions. Comparisons with the current best performing algorithms for the CSAHLP show that our RAMP algorithm exhibits excellent results.

Keywords: RAMP · Adaptive Memory Programming · CSAHLP · Hub Location Problem · Lagrangean Relaxation

1 Introduction

The Capacitated Single Allocation Hub Location Problem (CSAHLP) is a well-known combinatorial optimization problem belonging to the class of the NP-Hard problems [1]. The model used in this work was given by Contreras et al. [2] and is described as follows. Consider the complete graph $G = (N, A)$, where N is the set of nodes $N = \{1, 2, \ldots, n\}$, that correspond to origins/destinations as well as potential hub locations. Let w_{ij} be the flow between i and j, $O_i = \sum_{j \in N} w_{ij}$ be the outgoing flow from node $i \in N$, and $D = \sum_{i \in N} O_i$ the total flow generated in the graph. For each node $i \in N$, let b_i denote the capacity and f_i the fixed set-up cost of hub i. The capacity of a hub represents an upper bound on the total incoming flow that can be processed in the hub. Thus, it refers to the sum of the flow generated at the nodes that are assigned to the hub. The distance between nodes i and j is assumed to satisfy the triangle inequality, and is denoted by d_{ij}. We will use these distances as a measure of the per unit flow transportation costs along the links of the graph. These distances are weighted by some discount factors, denoted χ, α and δ, to represent the collection, transfer and distribution costs per unit of flow, respectively. Being single allocation implies that each node can only be assigned to one hub. The objective consists in choosing the set of nodes to be established as hubs, while minimize the total cost of assigning all the non-hubs to the

© Springer International Publishing AG 2017
O. Gervasi et al. (Eds.): ICCSA 2017, Part II, LNCS 10405, pp. 696–708, 2017.
DOI: 10.1007/978-3-319-62395-5_48

chosen hubs, and without violating the capacity constraint of the hubs. The total cost of routing the flow along the path $i - j - k - m$ (these are the paths between origin destination pairs, where i and j represents the origin and destination, respectively, and k and m are the hubs to which i and j are allocated, respectively) is given by:

$$F_{ijkm} = w_{ij}(\chi d_{ik} + \alpha d_{km} + \delta d_{mj}) \tag{1}$$

and for each pair $i, k \in N$ the following sets of binary decision variables are defined by:

$$Z_{ik} = \begin{cases} 1 \text{ if node } i \text{ is assigned to hub } k; \\ 0 \text{ otherwise.} \end{cases} \tag{2}$$

Variable Z_{kk} denotes the establishment or not of a hub at node k, when $i = k$. An additional set of binary variables will be defined. These variables indicate if there is flow through each link of the graph. For each $i, j, k, m \in N$ the existence of flow is defined by:

$$X_{ijkm} = \begin{cases} 1 \text{ if flow from } i \text{ to } j \text{ goes via hubs } k \text{ and } m; \\ 0 \text{ otherwise.} \end{cases} \tag{3}$$

The mathematical formulation for CSAHLP is:

$$min \sum_{k \in N} f_k Z_{kk} + \sum_{i \in N} \sum_{j \in N} \sum_{k \in N} \sum_{m \in N} F_{ijkm} X_{ijkm} \tag{4}$$

$$s.t. \ \sum_{k \in N} \sum_{m \in N} X_{ijkm} = 1 \ \forall \, i, j \in N \tag{5}$$

$$Z_{ik} \le Z_{kk} \ \forall \, i, k \in N \tag{6}$$

$$\sum_{m \in N} X_{ijkm} = Z_{ik} \ \forall \, i, j, k \in N \tag{7}$$

$$\sum_{k \in N} X_{ijkm} = Z_{jm} \ \forall \, i, j, m \in N \tag{8}$$

$$\sum_{i \in N} O_i Z_{ik} \le b_k Z_{kk} \ \forall \, k \in N \tag{9}$$

$$\sum_{k \in N} b_k Z_{kk} \ge D \tag{10}$$

$$Z_{ik} \in \{0, 1\} \ \forall \, i, k \in N \tag{11}$$

$$X_{ijkm} \in \{0, 1\} \ \forall \, i, j, k, m \in N \tag{12}$$

Constraints (5) assure that every single node is assigned to one hub, whereas constraints (6) guarantee that a non-hub node is assigned to a hub node. Constraints (7) insure that if node i is assigned to hub k then all the flow from this node to any other (fixed) node j must go through some other hub m. Constraints (8) state a similar interpretation regarding to the flow sent to a node j, assigned to hub m, from some node i.

Constraints (9) guarantee that the total incoming flow of nodes assigned to a hub does not overdo its capacity. Constraint (10) concerns the demand constraint and satisfying this aggregated (demand) constraint is necessary to assure the feasibility of the Z variables. This can also be achieved by adding up all constraints (9), and taking into consideration equalities (5) and (7). Constraint (10) is redundant for the formulation but we include it in order to apply the Lagrangean Relaxation, as referred in [2].

2 Related Work

The Hub Location Problem (HLP) has been extensively studied over the past years (as described in several surveys [3–6]), resulting in a great number of different algorithmic approaches. Numerous exact and heuristic techniques were proposed for the CSAHLP with competitive results. With regard to the exact approaches, a Branch & Bound [7] technique is found in the work of Ernst and Krishnamoorthy [8], in which the authors presented a modified problem formulation of the CSApHLP (where p is the number of necessary hubs to be opened) to solve the CSAHLP. Correia et al. [9] explored the problem formulation leading to the inclusion of additional sets of constraints that helped to decrease the computational time needed to solve the problem to optimality for small instances. Labbé et al. [10] examined the polyhedral properties and produced a Branch & Cut algorithm for this problem. Costa et al. [11] presented a second objective function to the model with an interactive procedure to generate non-dominated solutions. In this work, some computational results are shown with a comparison between single and double objective functions. Almeida et al. [12] proposed a Local Branching to solve the CSAHLP. This method is based on Branch & Cut with some heuristic and local search techniques to explore neighborhood solutions. Stanojević and Marić [13] produced an algorithm to solve the CSAHLP and the USAHLP with exact and heuristic methods. The main idea is to search the entire hub configuration and, for each configuration, an exact approach (the authors used the commercial solver CPLEX) is used to solve the allocation subproblem. They have tested the algorithm on small and large instances obtaining very good results but with expensive computational times. Being NP-Hard, the time needed to solve this problem is very costly. Due to this fact, several heuristic procedures are presented in the literature. Ernst and Krishnamoorthy [8] presented a random descent and a simulated annealing [14] algorithm for this problem. Population based heuristics can also be found in the literature, in the works of Stanimirović [15], Almeida et al. [16] and Baker [17] using Genetic Algorithms, and using Ant Colony optimization in Stutzle and Dorigo [18] and in Randall [19]. Contreras et al. [2] introduced a Lagrangean Relaxation (LR) and reduction tests based on LR bounds that reduced the size of the problem and consequently the computational time. More recently, Marić [20] presented a Variable Neighborhood Search [21] for choosing the set of hubs and its locations. The author also used a commercial solver to solve the allocation part. The reported computational results refer only to small instances. Finally, Stenojević et al. [22] proposed a hybrid procedure combining a parallel Branch & Bound technique with an evolutionary approach to solve this problem. This method proved to be very effective for small and large instances.

3   RAMP Algorithm for the CSAHLP

Relaxation Adaptive Memory Programming (RAMP) is a relatively recent meta-heuristic framework proposed by Rego [23]. The method has proved to be extremely effective in finding optimal and near-optimal solutions for a variety of combinatorial optimization problems such as the capacitated minimum spanning tree [24], the linear ordering problem [25] and the resource constrained project scheduling problem [26], among others. The underlying feature of the RAMP method is the creation of advanced memory structures derived from primal-dual relationships and their use in guiding the algorithm search process. This is achieved by exploring the dual solutions space of an associated relaxation problem and its primal counterpart through adaptive memory local search and evolutionary methods. The RAMP method may be implemented using two stages representing different levels of sophistication. At a first level of sophistication a Dual-RAMP variant may be implemented by combining dual search with a local search based adaptive memory approach such as tabu search or path-relinking to explore the primal search space. Primal and dual procedures are interconnected by memory structures used to guide the search at a higher level. At a second level of sophistication, the method may be extended with an evolutionary approach to implement a Primal-Dual RAMP designed for exploring primal-dual relationships more extensively.

We propose a Dual-RAMP algorithm for the CSAHLP that uses a Lagrangean Relaxation on the dual side and an improvement method on the primal side. With this simple level of sophistication, the algorithm explores and combines dual and primal solutions spaces achieving very good results for small and large instances.

3.1   Dual Phase

The Dual Phase proposed in this algorithm relies on the exploration of the dual solutions space with subgradient optimization to solve the dual problem obtained by the Lagrangean Relaxation. At each iteration of the subgradient optimization, a solution for the relaxed problem is obtained, then a Projection Method projects it to the primal solutions space and an Improvement Method tries to improve it.

Specifically, if we relax constraints (7) and (8), we obtain the following optimization problem:

$$L(u,v) = min \sum_{k \in N} f_k Z_{kk}$$

$$+ \sum_{i \in N} \sum_{j \in N} \sum_{k \in N} \sum_{m \in N} F_{ijkm} X_{ijkm} + \sum_{i \in N} \sum_{j \in N} \sum_{k \in N} u_{ijk} \left(\sum_{m \in N} X_{ijkm} \right.$$

$$\left. - Z_{jm} \right) + \sum_{i \in N} \sum_{j \in N} \sum_{m \in N} v_{ijm} \left(\sum_{k \in N} X_{ijkm} - Z_{jm} \right)$$

$$(13)$$

s.t. (5), (6), (9), (10), (11) *and* (12).

The remaining problem can be separated in two subproblems, denoted as $L_Z(u,v)$ and $L_X(u,v)$, for Z space variables and for X space variables, respectively.

For Z space variables, decomposing the main problem $L(u, v)$ in respect to these variables, we get the following subproblem:

$$L_Z(u, v) = min \sum_{k \in N} f_k Z_{kk} - \sum_{i \in N} \sum_{k \in N} \left(\sum_{j \in N} \left(u_{ijk} + v_{jik} \right) \right) Z_{ik} \qquad (14)$$

$$s.t. \ \ Z_{ik} \le Z_{kk} \ \forall i, k \in N \qquad (15)$$

$$\sum_{i \in N} O_i Z_{ik} \le b_k Z_{kk} \ \forall k \in N \qquad (16)$$

$$\sum_{k \in N} b_k Z_{kk} \ge D \qquad (17)$$

$$Z_{ik} \in \{0, 1\} \ \forall i, k \in N \qquad (18)$$

The solution $L_Z(u, v)$ is given by a set of hubs to be open ($Z_{kk} = 1$) and by the allocation of nodes to these hubs ($Z_{ik} = 1, i \ne k$). In this subproblem, and as mentioned in [2], a feasible solution does not require nodes to be assigned to just one open hub. So, it can happen that a node is not assigned to any hub, or that it is assigned to more than one open hub. If a hub is in $k \in N$, so that $Z_{kk} = 1$, the remaining nodes that will be assigned to hub k can be identified by solving the following binary knapsack problem:

$$KP_k = min \sum_{i \ne k} \left(\sum_j \left(u_{ijk} + v_{jik} \right) \right) Z_{ik} \qquad (19)$$

$$s.t. \ \sum_{i \ne k} O_i z_{ik} \le (b_k - O_k) \ \forall k \in N \qquad (20)$$

$$Z_{ik} \in \{0, 1\} \ \forall i, k \in N, i \ne k \qquad (21)$$

Each KP_k is a binary knapsack problem that assesses the maximum gain of opening a hub at node k with respect to the coefficients of gain $\sum_{j \in N} - \left(u_{ijk} + v_{jik} \right)$. Constraints (20) represent the capacity of hub k (when it is open) that is accessible for the rest of the nodes. The optimal set of hubs in $L_Z(u, v)$ can be obtained by solving the problem:

$$L_Z(u, v) = min \sum_{k \in N} \left(f_k - \sum_j \left(u_{kjk} + v_{jkk} \right) - KP_k \right) Z_{kk} \qquad (22)$$

$$s.t. \ \sum_{k \in N} b_k Z_{kk} \ge D \qquad (23)$$

$$Z_{kk} \in \{0, 1\} \forall k \in N \qquad (24)$$

Consequently, we can obtain the solution to $L_Z(u, v)$ by solving a series of $|N| + 1$ knapsack problems. For this we follow the same strategy as Contreras et al. [2] and use the algorithm of Martello, Pisinger and Toth [27].

For X space variables, decomposing the main problem $L(u, v)$ in respect to the space of X variables, we get the following subproblem:

$$L_X(u, v) = min \sum_{i \in N} \sum_{j \in N} \sum_{k \in N} \sum_{m \in N} (F_{ijkm} + u_{ijk} + v_{ijm}) X_{ijkm} \qquad (25)$$

$$s.t. \sum_{k \in N} \sum_{m \in N} X_{ijkm} = 1 \, \forall \, i, j \in N \qquad (26)$$

$$X_{ijkm} \in \{0, 1\} \, \forall \, i, j, k, m \in N \qquad (27)$$

It is trivial to note that $L_X(u, v)$ is a semi-assignment problem. Thus, for each pair (i, j), it can be decomposed into $(N - 1)^2$ independent semi-assignment problems of the form:

$$SAP_{ij} = min \sum_{k \in N} \sum_{m \in N} (F_{ijkm} + u_{ijk} + v_{ijm}) X_{ijkm} \qquad (28)$$

$$s.t. \sum_{k \in N} \sum_{m \in N} X_{ijkm} = 1 \qquad (29)$$

$$X_{ijkm} \in \{0, 1\} \, \forall \, k, m \in N \qquad (30)$$

For a given pair (i, j), SAP_{ij} can be solved by the following rule:

$$X_{ijkm} = 1 \text{ for } F_{ijkm} = min\{F_{ijkm} + u_{ijk} + v_{ijm} | k, m \in N\} \qquad (31)$$

$$X_{ijkm} = 0, \text{Otherwise} \qquad (32)$$

The relaxed problem $L(u, v)$ can be obtained by the sum of the two subproblems, one for each space variables as shown above ($L_Z(u, v) + L_X(u, v)$). The lower bound computation, Z_D will be $max_{u,v} L(u, v)$. After solving the dual problem, a classical subgradient optimization method will be used. At each iteration, the solution for the relaxed problem $L(u, v)$ is obtained, which is projected to the primal solutions space by a projection method and then an improvement method tries to improve it. We start with the lower bound $Z_D = 0$, an initial value for the initial upper bound Z_{UB} (that we will see in the Primal Phase) and the lagrangean multipliers $\lambda_{u,v} = 0$. The agility parameter π is initialized with the value 2, it is divided by 2 every 25 consecutive iterations without improving Z_D and it is reset to the original value every 100 iterations in order to decrease the step size (Δ). The step size (Δ) is calculated as $\Delta = \frac{\pi(Z_{UB} - L(u,v))}{||\delta||}$. For a given vector (u, v), let $z(u, v)$ and $x(u, v)$ represent the optimal solution to $L(u, v)$. Then, a subgradient $\delta(u, v)$ of $L(u, v)$, is given by the following expression:

$$\delta(u, v) = \left(\left(\sum_m X_{ijkm}(u) - Z_{ik} \right)_{i,j,k}, \left(\sum_k X_{ijkm}(u) - Z_{jm} \right)_{i,j,m} \right) \qquad (33)$$

Finally, to determine the set of multipliers $\lambda_{u,v}$ that maximizes the Lagrangean function $L(u, v)$, the subgradient method requires generating a new sequence of

multipliers, one for each iteration of the Lagrangean Relaxation, given by
$\lambda_{u,v}^{iter+1} = \lambda_{u,v} + \Delta\delta(u,v)$.

3.2   Primal Phase

The primal phase starts after the projection method projects the dual solution onto the primal solutions space. For X solutions space we have to solve the $|N| + 1$ knapsack problems (Eqs. 19–24) and for Z solutions space we solve the semi-assignment problems associated with each pair (i,j) (Eqs. 28–32). At each iteration, the RAMP algorithm starts the primal exploration with an improvement method that uses the primal feasible solution that we get from the dual phase as the initial solution. Often a feasible solution is not produced by the dual phase iteration, so we use a simple method to transform an infeasible solution into a feasible one. This method is divided into two stages. In the first stage, the projection method tries to open hubs. If it fails to open at least one hub, then it chooses to open the hub with the highest supply (and all the nodes are assigned to it). The second stage assumes that the projection method opened at least one hub, and verifies if the opened hubs can supply all nodes assigned to them. If not, it opens hubs until the nodes demand is met without violating the hubs total supply. The RAMP algorithm proceeds with the primal phase by trying to improve the feasible primal solution with a simple improvement method. This method has three neighborhood structures: shift, swap and close, that we can describe as follows:

- Shift (nodes): the shift procedure shifts nodes from one open hub to another, until no more improvement can be made to the objective function. Basically (for every node), the procedure tries to improve the objective function by shifting a node from one open hub to another;
- Swap (nodes): the swap procedure swaps nodes from open hubs until no more improvement can be made to the objective function. For every node, this procedure analysis the benefit of swapping two nodes assigned to two different hubs and does the swapping if it improves the objective function;
- Swap (hubs): the swap procedure swaps open hubs until no more improvement can be made to the objective function. For every hub, this procedure analysis the gain of swapping two open hubs and does the swapping (by reassigning all the nodes from one open hub to the other) if it improves the objective function;
- Close (hubs): the close procedure looks for profit in closing an open hub and reassigning its nodes to the remaining open hubs based on minimum distances.

The improvement method performs the procedures (Shift-nodes, Swap-nodes, Swap-hubs and Close-hubs), one at a time, until no improvement to the objective function is possible. The algorithm continues alternating between the dual and the primal solutions spaces until one of the four stopping criteria is achieved:

- The agility parameter is less than 0.005;
- The norm reaches the value 0;
- The maximum number of iterations is reached;
- The difference between the upper bound (Z_{UB}) and lower bound (Z_D) is less than 1.

4 Computational Results

The performance of the RAMP algorithm was evaluated on a standard AP (Australia Post) data benchmark, composed by 56 instances, introduced by Ernst and Krishnamoorthy [8] and obtained in Beasley's OR-Library [28]. These instances result from the mail flows in an Australian city and contain up to 200 nodes. The flow is not symmetric, that is, $W_{ij} \neq W_{ji}$ and the collection, transfer and distribution ratios are $\chi = 3, \alpha = 0.75$ and $\delta = 2$. These instances are divided in small, medium and large size and have two different types of setup costs and hubs capacities. We have tight (T) setup costs when these costs increase with the amount of flow generated in the node and we have loose (L) setup costs when the instances do not have this characteristic. The tight setup costs are supposed to be more difficult to solve. For the hubs capacities we have the same representation, but in this case related to how tightly (T) or loosely (L) constrained the instances are. For more information about the instances, please refer to Ernst and Krishnamoorthy [8] or Contreras et al. [2]. Each group has four instances corresponding to the four possible combinations of setup costs and capacities (LL, LT, TL or TT). For instances up to 50 nodes the optimal solution is known. For all the other instances, we used the best-known solutions reported in Contreras et al. [2]. The algorithm was coded in C programming language and run on an Intel Pentium I7 2.40 GHz (only one processor was used) with 8 GB RAM under Ubuntu operating system. The Dual-RAMP algorithm was compared with the state of the art algorithms for the solution of the CSAHLP. All results tables show the average percent deviation from the optimal/best-known solution (gap) and the associated computational time (cpu) in seconds. The best-known approaches for the CSAHLO are a Genetic Algorithm (GA) [15], an exact approach (Local Branching - LB) [12], a Variable Neighborhood Search (VNS) [20] and an Evolutionary Algorithm and parallel Branch-and-Bound approach (EA) [22]. We cannot compare our computational times with the EA approach since our algorithm is not parallel and we only use one processor. Nevertheless, we include the solutions obtained with this approach in the results tables. For the GA and the EA algorithms, the average results were achieved in a specific number of runs with different parameters. Our algorithm obtained the reported solutions with a single run. The first table (Table 1) summarizes the results for small, medium and large AP data instances grouped by dimension. The algorithms that we are comparing do not provide results for all AP datasets. In all results tables the value of the "gap" column was computed as $(UB - Z^*)/UB * 100$ (Z^* is the optimal/best-known solution and UB is the obtained upper bound) and the value of the "cpu" column is the computational time (in seconds) needed to achieve Z^*. Table 1 clearly shows that the RAMP approach achieved excellent quality results for all small instances. Only the LB algorithm reaches a better average gap in one specific instance, as we will see in the detailed results. However, the LB approach needs a higher computational effort to achieve such solutions. For medium and large instances (100 and 200 nodes, respectively), our algorithm outperforms all other algorithms (GA and EA) in solution quality (managing to find a new best-known solution, as we will see in the following tables) with reasonable computation times.

Table 1. Aggregated results for the AP data instances

AP Data	GA		LB		VNS		EA	RAMP	
	gap	cpu	gap	cpu	gap	cpu	gap	gap	cpu
10	0.07	1.02	0.00	1.56	0.05	0.17	0.00	**0.00**	**0.03**
20	0.09	2.33	0.00	4.45	0.07	1.67	0.00	**0.00**	**0.93**
25	1.21	3.03	0.00	9.11	0.10	7.71	0.12	**0.00**	**2.52**
40	1.97	5.77	0.00	1054.88	1.01	58.68	0.00	**0.00**	**19.22**
50	1.07	8.96	0.04	15694.55	2.08	245.37	0.21	**0.12**	**54.30**
Average	0.88	4.22	0.01	3352.91	0.66	62.72	0.07	**0.02**	**15.40**
100	2.36	70.07	–	–	–	–	0.48	**0.01**	**635.68**
200	1.75	397.09	–	–	–	–	0.91	**0.01**	**8815.52**

Tables 2, 3, 4 and 5 show, in detail, the computational results for small, medium and large instances. For the small instances (whose optimal solution is known), we can see from Table 2 that RAMP achieved the optimal solution in all instances except for the 50TT instance, for which RAMP obtained a 0.48% deviation from the optimal solution. This is the most difficult instance of the small ones, as demonstrated by the results produced by all algorithms, since none attained the optimal solution. Nevertheless, the proposed algorithm achieved almost every optimal solution in extremely reduced computational times, under 16 s in average (note that a direct comparison is not possible because the algorithms run on different machines).

Table 3 shows the results for the medium instances (100 nodes) whose optimal solution is not known. For this group of instances we can see that the computational time increases. Our algorithm is compared with EA and GA and RAMP outperforms these algorithms in terms of solutions quality, obtaining a 0.01% average deviation from the best-known solutions. RAMP also manages to improve the best-known solution for the 100TT instance.

Table 4 displays the results for the large instances (200 nodes) whose optimal solution is not known. For these instances, considered the most difficult ones, we can see that RAMP achieves the same quality level as with smaller instances, demonstrating that our algorithm is the best approach for this problem. For the 200LT instance, we improve the best-known solution about 0.42%. Comparing with the best algorithms in the literature, the RAMP approach outperformed all in terms of solutions quality, achieving a 0.01% average deviation from the best-known solutions. Comparing with LR, the proposed algorithm achieved equal or better solutions quality, except for the 200TT instance.

Table 5 shows the improvement that RAMP and EA algorithms obtained on the best-known solutions for instances 100TT, 200LT and 200TT. For the EA approach, these improvements cannot be seen in the previous tables because, as explained previously, the EA results are averages of a specific number (20) of runs. In Table 5, we report the results obtained in individual runs that allowed EA to improve some best-known solutions. The "improved" column displays the upper bound obtained by the RAMP algorithm. EA improved the best-known solutions for instances 100TT, 200LT and 200TT but RAMP managed to find an even better solution for the instance

Table 2. AP data results – small instances

AP Data (small)	LR		GA		LB		VNS		EA	RAMP	
	optimal	cpu	gap	cpu	gap	cpu	gap	cpu	gap	gap	cpu
10LL	224250.05	0.03	0.13	0.94	0	2.04	0	0.16	0	**0**	**0.02**
10LT	250992.26	0.03	0.00	1.11	0	1.63	0	0.42	0	**0**	**0.03**
10TL	263399.94	0.06	0.12	1.07	0	1.03	0	0.07	0	**0**	**0.04**
10TT	263399.94	0.05	0.05	0.97	0	1.52	0.18	0.03	0	**0**	**0.03**
20LL	234690.96	0.38	0	2.17	0	3.33	0	0.91	0	**0**	**0.13**
20LT	253517.4	1.64	0.07	2.41	0	5.39	0	0.79	0	**0**	**1.78**
20TL	271128.18	0.19	0	2.57	0	2.85	0.26	1.82	0	**0**	**0.11**
20TT	296035.4	0.16	0.30	2.16	0	6.22	0	3.16	0	**0**	**1.71**
25LL	238977.95	1.97	0.78	3.07	0	7.56	0	6.06	0	**0**	**1.44**
25LT	276372.5	2.55	0.98	2.97	0	10.48	0.2	6.31	0.20	**0**	**3.91**
25TL	310317.64	1.05	2.25	3.08	0	6.46	0	7.62	0	**0**	**0.79**
25TT	348369.15	2.47	0.85	3.00	0	11.93	0.21	10.84	0.28	**0**	**3.95**
40LL	241955.71	7.24	0.08	5.27	0	29.63	0	32.85	0	**0**	**21.51**
40LT	272218.32	6.81	3.54	6.46	0	70.12	1.67	68.13	0	**0**	**20.58**
40TL	298919.01	1.59	0.00	5.62	0	22.93	0	38.86	0	**0**	**11.16**
40TT	354874.1	11.39	4.25	5.75	0	4096.85	2.37	94.86	0	**0**	**23.63**
50LL	238520.59	13.00	0.27	7.45	0	55.19	0.36	89.79	0	**0**	**45.25**
50LT	272897.49	63.98	0.52	8.53	0	512	1.83	171.54	0.53	**0**	**57.96**
50TL	319015.77	20.02	0.72	8.87	0	92.32	0.48	82.4	0	**0**	**52.02**
50TT	417440.99	77.16	2.79	10.98	0.16	62118.7	5.63	637.74	0.33	**0.48**	**61.95**
Average		10.59	0.88	4.22	0.01	3352.91	0.66	62.72	0.07	**0.02**	**15.40**

Table 3. AP data results – medium instances

APData (medium)	LR		GA		EA	RAMP	
	Best-known	cpu	gap	cpu	gap	gap	cpu
100LL	246713.97	873.80	1.07	56.05	0.96	**0.00**	**612.47**
100LT	256155.33	1039.60	3.19	81.97	0.65	**0.00**	**694.01**
100TL	362950.09	397.02	0.77	66.29	0.10	**0.05**	**558.54**
100TT	474186.73	347.87	4.39	75.96	0.20	**−0.03**	**677.68**
Average		664.57	2.36	70.07	0.48	**0.01**	**635.68**

Table 4. AP data results – large instances

APData (large)	LR		GA		EA	RAMP	
	Best-known	cpu	gap	cpu	gap	gap	cpu
200LL	231069.5	19590.75	0.70	424.52	0.77	**0**	**8123.18**
200LT	268820.57	23212.58	1.86	410.41	1.42	**−0.42**	**10344.32**
200TL	273443.81	2801.48	3.63	325.88	0.45	**0**	**7362.94**
200TT	290841.84	3111.90	0.81	427.57	1.02	**0.47**	**9431.64**
Average		12179.18	1.75	397.09	0.91	**0.01**	**8815.52**

Table 5. Improved solutions over the best-known

APData (improved instances)	LR	EA	Dual RAMP		
	Best-known	gap	Improved	gap	cpu
100TT	474186.73	−0.02	**474068.91**	**−0.03**	**677.68**
200LT	268820.57	−0.59	267702.59	**−0.42**	**7362.94**
200TT	290841.84	−0.03	–	**0.47**	**10344.32**

100TT. Our algorithm also improved the best-known solution for the 200LT instance in 0.42% but, in this case, EA achieved a higher improvement of 0.59%. For the 200TT, we obtained a 0.47% deviation from the best-known and EA managed to improve it in 0.03%. Nevertheless, and taking in consideration that the RAMP algorithm is sequential, we believe that combining parallel computing with the RAMP framework will produce even better results than the ones presented in this paper.

5   Conclusions

This paper describes a simple Dual-RAMP algorithm for the CSAHLP that combines a Lagrangean Relaxation technique with an Improvement Method to effectively explore the dual and primal solutions spaces. This approach competes with the best-known algorithms for the solution of this problem. Despite the fact that only the first level of sophistication of the RAMP method was implemented, the proposed algorithm managed to produce excellent results for the AP dataset in reasonable computational times, even improving some of the best-known reported solutions. We conjecture that taking advantage of the exploration of primal-dual relationships provides an effective interaction between intensification and diversification processes that is absent in search methods confined to the primal solutions space. The results obtained with the RAMP application to hard combinatorial problems certainly invite further studies in the application of the method to other hub location problems and even to other location problems.

References

1. O'Kelly, M.E.: A quadratic integer program for the location of interacting hub facilities. Eur. J. Oper. Res. **32**, 393–404 (1987). doi:10.1016/S0377-2217(87)80007-3
2. Contreras, I., Díaz, J.A., Fernández, E.: Lagrangean relaxation for the capacitated hub location problem with single assignment. OR Spectr. **31**, 483–505 (2009). doi:10.1007/s00291-008-0159-y
3. Farahani, R.Z., Hekmatfar, M., Arabani, A.B., Nikbakhsh, E.: Hub location problems: a review of models, classification, solution techniques, and applications. Comput. Ind. Eng. **64**, 1096–1109 (2013). doi:10.1016/j.cie.2013.01.012
4. Alumur, S., Kara, B.Y.: Network hub location problems: the state of the art. Eur. J. Oper. Res. **190**, 1–21 (2008). doi:10.1016/j.ejor.2007.06.008

5. O'Kelly, M.E., Bryan, D., Skorin-Kapov, D., Skorin-Kapov, J.: Hub network design with single and multiple allocation: a computational study. Locat. Sci. **4**, 125–138 (1996). doi:10.1016/S0966-8349(96)00015-0
6. Campbell, J.F.: A survey of network hub location. Stud. Locat. Anal. **6**, 31–49 (1994)
7. Lawler, E., Wood, D.: Branch-and-bound methods: a survey. Oper. Res. **14**, 699–719 (1966). doi:http://dx.doi.org/10.1287/opre.14.4.699
8. Ernst, A.T., Krishnamoorthy, M.: Solution algorithms for the capacitated single allocation hub location problem. Ann. Oper. Res. **86**, 141–159 (1999). doi:10.1023/A:1018994432663
9. Correia, I., Nickel, S., Saldanha-da-Gama, F.: The capacitated single-allocation hub location problem revisited: a note on a classical formulation. Eur. J. Oper. Res. **207**, 92–96 (2010). doi:10.1016/j.ejor.2010.04.015
10. Labbé, M., Yaman, H., Gourdin, E.: A branch and cut algorithm for hub location problems with single assignment. Math. Program. **102**, 371–405 (2005). doi:10.1007/s10107-004-0531-x
11. da Graça Costa, M., Captivo, M.E., Clímaco, J.: Capacitated single allocation hub location problem-A bi-criteria approach. Comput. Oper. Res. **35**, 3671–3695 (2008). doi:10.1016/j.cor.2007.04.005
12. Almeida, W.G., Yanasse, H.H., Senne, E.L.F.: Uma abordagem exata para problema de localização de concentradores capacitado. In: 43 Simpósio Bras. Pesqui. Operacional, pp. 2192–2203 (2011)
13. Stanojević, P., Marić, M.: Solving large scale instances of hub location problems with a sub-problem using an exact method. IPSI BgD Trans. Internet Res. **1**, 6 (2014). doi:10.1142/9789812709691_0055
14. Koulamas, C., Antony, S., Jaen, R.: A survey of simulated annealing applications to operations research problems. Omega **22**, 41–56 (1994). doi:10.1016/0305-0483(94)90006-X
15. Stanimirović, Z.: Solving the capacitated single allocation hub location problem using genetic algorithm. In: Recent Advances in Stochastic Modelling and Data Analysis, pp. 464–471. World Scientific (2007)
16. Almeida, W., Senne, E.L.F., Yanasse, H.: Abordagens meta-heurísticas para o problema de localização de concentradores com restrições de capacidade. In: X Worcap, p. 12 (2010)
17. Baker, J.: Adaptive selection methods for genetic algorithms. In: Proceedings 1st International Conference Genetic Algorithms, pp. 101–111 (1985)
18. Stützle, T., Dorigo, M.: The ant colony optimization metaheuristic: algorithms, applications and advances. In: Glover, F., Kochenberger, G.A. (eds.) Handbook Metaheuristics, vol. 57, pp. 250–285. Springer, New York (2003)
19. Randall, M.: Solution approaches for the capacitated single allocation hub location problem using ant colony optimisation. Comput. Optim. Appl. **39**, 239–261 (2008). doi:10.1007/s10589-007-9069-1
20. Marić, M.: Variable neighborhood search for solving the capacitated single allocation hub location problem. Serdica J. Comput. **7**, 343–354 (2013)
21. Mladenović, N., Hansen, P.: Variable neighborhood search. Comput. Oper. Res. **24**, 1097–1100 (1997). doi:10.1016/S0305-0548(97)00031-2
22. Stanojević, P., Marić, M., Stanimirović, Z.: A hybridization of an evolutionary algorithm and a parallel branch and bound for solving the capacitated single allocation hub location problem. Appl. Soft Comput. **33**, 24–36 (2015). doi:10.1016/j.asoc.2015.04.018
23. Rego, C.: RAMP: a new metaheuristic framework for combinatorial optimization. In: Rego, C., Alidaee, B. (eds.) Metaheuristic Optimization via Memory and Evolution. Tabu Search and Scatter Search. pp. 441–460. Kluwer Academic Publishers, New York (2005)
24. Rego, C., Mathew, F., Glover, F.: RAMP for the capacitated minimum spanning tree problem. Ann. Oper. Res. **181**, 661–681 (2010). doi:10.1007/s10479-010-0800-4

25. Gamboa, D.: Adaptive memory algorithms for the solution of large scale combinatorial optimization problems. Ph.D. thesis (in portuguese). Instituto Superior Técnico, Universidade Técnica de Lisboa (2008)
26. Riley, C., Rego, C., Li, H.: A simple dual-RAMP algorithm for resource constraint project scheduling. In: ACM Southeast Region Conference, p. 67 (2010)
27. Martello, S., Pisinger, D., Toth, P., et al.: New trends in exact algorithms for the 0–1 knapsack problem. Eur. J. Oper. Res. **123**, 325–332 (2000). doi:10.1016/S0377-2217(99)00260-X
28. Beasley, J.: OR-Library: distributing test problems by electronic mail. J. Oper. Res. Soc. **65**, 1069–1072 (1990)

Variable Neighborhood Search for Integrated Planning and Scheduling

Mário Leite[(✉)], Cláudio Alves, and Telmo Pinto

Escola de Engenharia, Universidade do Minho, 4710-057 Braga, Portugal
maleite.cbc@gmail.com, {claudio,telmo}@dps.uminho.pt

Abstract. In this paper, we consider the integrated planning and scheduling problem on parallel and identical machines. The problem is composed by two parts which are simultaneously solved in an integrated form. The first is the planning part, which consists in determining the jobs that should be processed in each period of time. The second is the scheduling part, which consists in assigning the jobs to the machines according to their release dates. We present new optimization approaches based on local search heuristics and metaheuristic methods based on variable neighborhood search using two neighborhood structures. Two different algorithms were implemented in the construction of initial solutions and combined with fifteen variants of the initial sequence of jobs. Computational experiments were performed with benchmark instances from the literature in order to assess the proposed methods.

Keywords: Planning · Scheduling · Integrated optimization problems · Heuristics · Metaheuristics

1 Introduction

The current market is becoming more competitive, requiring that companies seek for effective alternatives in order to face competition. This concern is not new since several case studies reported that the adaptation of companies to new circumstances was decisive for achieving success. Reducing costs by suppressing waste while preserving or even improving the quality of products is a daily challenge, due to the rigorous requirements of customers.

The importance and the potential benefit of integrated optimization in the supply chain is well known. Optimizing operations throughout different functional areas of the companies contributes both to the cost reduction and to the performance improvement. In this sense, many authors suggested a stronger interaction between planning and scheduling, from the acquisition of raw materials to the delivery plan of final products [6]. Particularly in the last decades, the research in integrated models has been increasing, helping to achieve the best possible global solutions [1,2,5].

Currently, companies aim to produce only what is strictly necessary while using fewer resources and keeping only which is essential to produce. Hence, it

© Springer International Publishing AG 2017
O. Gervasi et al. (Eds.): ICCSA 2017, Part II, LNCS 10405, pp. 709–724, 2017.
DOI: 10.1007/978-3-319-62395-5_49

can be possible to reduce the surplus and then minimize the production costs. As a result, production is only performed if required, at the moment it is needed while minimizing inventory costs [14]. Some reasons can explain the motivation of companies to adopt a "zero inventory" policy: consumers with a very diversified demand, products with a short lifespan and finally, the objective of reducing costs. However, it is necessary to have products in stock in order to quickly satisfy customers' demand. In many cases, waiting time can induce loss of customers [7].

Recurrently, managers of the companies have several questions to which they have to answer quickly, such as: What to buy?, When to buy?, Where to save or to place?, What to sell?, What to produce?, When to produce?, Where to produce?, among others. There are no fixed answers since they constantly changing according to the customer demand, to the amount variation of raw materials and, implicitly, to their price [13].

In this paper, we consider the integrated planning and scheduling problem on parallel and identical machines, analyzed in [8] and more recently in [10,11]. This problem is composed by two parts which are simultaneously solved in an integrated way: the output of the first part is the input of the second one. The first part, which is relative to the planning, has to be solved for a given time horizon that is divided into periods of time. The objective is to determine the jobs that will be processed in each period. The second part is related to the scheduling and it consists in assigning jobs to the available machines in each period, taking into account the release dates on which jobs are available to be processed. All jobs should be processed until the end of the planning horizon, without exceeding the number of machines. Furthermore, a job that is done before or after its due date will incur in a penalty. These two parts correspond to two separate activities with different functions, but they are deeply interdependent since both have the same purpose and, therefore, they must be solved in an integrated form [12]. The objective of this integrated problem is to determine the planning and scheduling that minimize the sum of the penalties.

Since the jobs cannot be split into periods or into machines, the problem can be seen as a cutting and packing problem [3], where the objective is to distribute the jobs (small objects) by the existing resources (large objects). According to the typology of [15], it corresponds to a 1-dimension problem, and the kind of assignment is the input minimization in which jobs should be assigned to a subset of periods of time. The jobs diversity in this context is defined by the processing time of jobs which can be strongly heterogeneous. Therefore, the basic problem type can be defined as a bin-packing problem. This problem is NP-hard [4], and thus the computational time to find an optimal solution increases exponentially with the size of the problem. In this context, it can be profitable to obtain solutions through heuristics and metaheuristic methods.

The approaches presented in this paper are based on heuristic methods of local search, on the Variable Neighborhood Search (VNS) metaheuristic [9] using two neighborhood structures, and on a variant of the multistart. The initial solutions are obtained through two different algorithms that make use of fifteen ordering criteria for the job allocation to the machines, leading to a wide space of solutions.

This paper is organized as follows. In Sect. 2, we present both the problem definition and the used notation. In Sect. 3 we describe the local search heuristic methods, while in Sect. 4 we present the VNS approach that aims to achieve better solutions. The computational experiments and the analysis of the obtained results are presented in Sect. 5. Finally, in Sect. 6 we present the main conclusions.

2 Planning and Scheduling Problem

The problem addressed in this paper can be formulated as follows. The entire time horizon T is divided into τ time periods of length equal to $P \in \mathbb{N}$. Then, $T = \{1, \ldots, \tau\}$. There are M parallel and identical machines, where the set of jobs N should be processed. Each job $j \in N$ has five attributes:

(1) processing time ($p_j \in \mathbb{N} \backslash \{0\}$);
(2) release date ($r_j \in T$);
(3) due date ($d_j \in T$);
(4) penalty factor incurred when a job is completed before its due date ($e_j \in \mathbb{N}$);
(5) penalty factor incurred when a job is completed after its due date ($l_j \in \mathbb{N}$).

The processing time p_j corresponds to the units of time necessary to complete a given job $j \in N$. The release date r_j represents the minimum period of time in T, in which a job is available to be processed, $i.e.$ a job j can only be processed in a period of time $t \in T$ if $r_j \leq t$. The due date d_j is the established period of time to process the job without associating penalties. Usually, this date is agreed with the client, providing that $d_j \geq r_j$. Penalty factors are applied whenever a job $j \in N$ is processed in a period of time before or after its due date, according to a earliness or tardiness function, respectively. If a job $j \in N$ is performed in the period of time $t \in T$, then it will incur in a penalty w_t^j according to the following formula (1):

$$w_t^j = e_j \times \underbrace{max\{0, d_j - t\}}_{earliness} + l_j \times \underbrace{max\{0, t - d_j\}}_{tardiness}. \tag{1}$$

Each job is only processed in one machine, and one machine can only process one job at a time. A job interruption is not allowed and, in addition, it cannot be processed in two different periods, even if they are consecutive. Finally, the sum of the processing time of all jobs processed in a given period of time $t \in T$ cannot exceed the length P. As a consequence, $P \geq max\{p_j : j \in N\}$. The main objective is to minimize the sum of the penalties associated to the advance or to the delay of the jobs processing while satisfying all the constraints referred to above.

Example 1. It is considered as a time horizon one working day (24 h), and 10 jobs to be processed in 2 identical parallel machines, in 3 shifts of 8 h each, then $N = \{1, 2, 3, \dots, 10\}$, $M = 2$, $P = 8$ and $\tau = 3$. So we have the set $T = \{1, 2, 3\}$. In this example, only one day is considered as the time horizon. However, according to each situation, the example can be easily adapted. For example, the time horizon can be based in weeks which is commonly used in planning and scheduling problems.

3   Local Search

Local search methods are commonly used due to their simplicity of implementation and to the satisfactory results for many situations in which they are applied. Local search consists in an iterative process, reaching a local optimum, with three main steps: to obtain an initial solution, to generate neighborhood solutions and to define the evaluation measure between solutions (in our case, the value of the penalties). However, one negative issue of these methods is that they can find a solution which is a local optimum, but still can be distant from the global optimal solution. It is easy to understand the importance that the quality of the initial solution has in the analyzed solution space. Therefore, in this section we will describe the main elements of this approach: the construction of the initial solutions and the different procedures to generate neighborhood solutions.

All solutions generated are feasible or viable solutions. In the construction of the initial solution or when reaching a neighborhood solution, all the constraints of the problem are satisfied, and infeasible solutions are not generated since non-viable solutions are excluded.

3.1   Initial Solutions

Initial solutions have an important role throughout the subsequent local search process. Therefore, different problem approaches have been considered. Computational experiments have been performed to find initial solutions through four methods obtained by combining the type of construction and the order of the jobs and another method in which the best initial solution among the other approaches is selected. These combinations are summarized in Table 1. There are two different algorithms to build the initial solution: (1) the construction of the solution is done from the release date of each job; (2) the construction is done based on the due date of each job. In other words, the first algorithm tries to allocate the jobs as soon as they are available and if it is not possible, it tries to allocate the job as much as closer as possible to this date. The second algorithm tries, if possible, to allocate each job to its due date, or in the period that results in the lowest penalty for the job. The jobs can be ordered according to the priority rules (to be latter presented) or without priority rules, and in this case the jobs have the order in which they arise in the problem.

The construction type from the release date allocates, if possible, the jobs in the period corresponding to their release date. If the machines do not have capacity in the period corresponding to that date, $i.e.$ they do not have enough time to process the job in that period, then the job is allocated to the following

Table 1. Initial solutions construction

CSI	Type of construction	Order of jobs
CSI1	From the release date	No priority rules
CSI2	From the release date	With priority rules
CSI3	Based on the due date	No priority rules
CSI4	Based on the due date	With priority rules
CSI5	The best initial solution between CSI1, CSI2, CSI3 and CSI4	

period, while ensuring time to process the job in the next period. As soon as the jobs are available (from r_j), they are assigned to the first available period t with $t \geq r_j$. This type of construction does have any concern about the possible penalties.

The initial solutions obtained through the type of construction based on the due date (CSI3 and CSI4) aims to allocate each job to the available periods and machines that result in the lowest penalty associated to each job. When the allocation is done, previous allocations are not modified. Thus, jobs are allocated, if possible, on their due dates aiming to meet their deadlines. If there are no available periods corresponding to the due date of the job, the allocation must be advanced or delayed. However, in a first phase, all the jobs that can be allocated to their due dates are assigned, and only then other jobs are advanced or delayed according to the following principles. If $r_j = d_j$, the job j must be postponed and therefore assigned to one of the following periods that result in the lowest penalty for the job j, only if the period is available to process it. If $r_j \neq d_j$, the job j is placed in a period before or after its due date. In both cases, the process of assigning job j seeks for the period that reaches the lowest penalty for the job. This type of construction of the initial solution can be seen as more greedy than the previous one, since it is always concerned with the penalty taken by the jobs in each allocation.

For the initial solutions built without associating priority rules (CSI1 and CSI3), the order of job allocation is determined by the order of the input of the problem.

The priority rules consider criteria to reorder the job list and they determine the order in which each job is to be allocated to the periods and machines. This means that the highest priority job (the first job in the list) will be the first one to be placed, and so on until all the jobs of set N are assigned.

In Table 2, we present the 14 sorting variants, from *heu*1 to *heu*14. In each variant, sequences are defined according to *Criterion 1* by the *Order 1*, breaking ties by *Criterion 2* by the *Order 2*.

In the variants *heu*1, *heu*2, *heu*3 and *heu*4, *Criterion 1* considers the quotient of the integer division between the penalty factors and the processing time, and it can be seen as penalty per unit of processing time of the job.

All the variants consider the processing time as a criterion. This analysis becomes important because both ascending and descending order have

Table 2. Priority rules

Variant	Criterion 1	Order 1	Criterion 2	Order 2
heu1	$(e_j + l_j)/p_j$	Descending	p_j	Descending
heu2	$(e_j + l_j)/p_j$	Descending	p_j	Ascending
heu3	l_j/p_j	Descending	p_j	Descending
heu4	l_j/p_j	Descending	p_j	Ascending
heu5	p_j	Descending		
heu6	p_j	Ascending		
heu7	$e_j + l_j$	Descending	p_j	Descending
heu8	$e_j + l_j$	Descending	p_j	Ascending
heu9	l_j	Descending	p_j	Descending
heu10	l_j	Descending	p_j	Ascending
heu11	$p_j/(e_j + l_j)$	Ascending	p_j	Descending
heu12	$p_j/(e_j + l_j)$	Ascending	p_j	Ascending
heu13	p_j/l_j	Ascending	p_j	Descending
heu14	p_j/l_j	Ascending	p_j	Ascending

advantages and disadvantages relying in each input. Regarding the increasing order (jobs of shorter duration first) can lead to a smaller number of jobs that have penalties. This can be suitable for problems in which the objective is to reduce the number of jobs delayed, but delaying larger jobs can postpone them to periods long after their due dates, increasing the value of the objective function. This happens because their duration can lead to the existence of large gaps in remote periods. On the other hand, if decreasing order of processing time is considered (*i.e.* the jobs of longer duration are allocated first), there may be less jobs assigned to their due dates, but the remaining ones may have lower penalties as these may be closer to their due dates.

The variants *heu7* to *heu10* consider only the penalty factors without existing a division by the processing time. In cases with many jobs having a processing time higher than the penalty factors, the value of this integer division would always be zero. This means that the variants *heu1*, *heu3* and *heu5* would present the job list in exactly the same order based only on the decreasing order of the processing time. Similarly, *heu2*, *heu4* and *heu6*, also present the same order according to the ascending order of processing time. Hence, these variants aim to give priority to jobs with higher penalty factors, even if they also have large processing times, which would not happen in the other variants. Therefore, and in order to consider the relation between the processing time and the penalty factors for instances in which the jobs have longer processing time, the variants *heu11*, *heu12*, *heu13* and *heu14* are also considered, in ascending order.

In order to achieve the best initial solution, *i.e.* the one that presents the best value for the objective function in the construction of the initial solution, we also suggest other approach (CSI5) where the best solution among (CSI1, CSI2, CSI3 and CSI4) is selected. This approach is similar to the multistart metaheuristic which search among the initial solutions in order to escape from local optimum and then achieve the best possible final value. However, in our approach, CSI5 aims to find the solution that is the best alternative to the initial solution and, from this solution proceed to local search and VNS methods. It should be noted that the initial solution in all approaches is obtained in few milliseconds.

3.2 Neighborhood Solutions

In this subsection, we describe the process for obtaining neighborhood solutions $N(s)$ from a solution s in the local search method and in the VNS algorithm.

In a first step and for a given a solution s, the list of jobs is sorted according to the decreasing order of penalties. Let job j be the first job of this list. This job is selected to change its position if possible. In other words, the most penalizing job is selected and it is verified if it can be assigned to a period of time within it will incur in a lower penalty. Firstly, this process tries to exchange job j with some other job k which is allocated to a period of time d_j if this exchange is profitable. In order to have a profitable exchange, the sum of the penalties of both job j and job k in their new periods should be less than to the sum of their initial penalties. Furthermore, job j can only be exchanged with job k if the time limit of the instance is not exceeded. Thus, it must be ensured that:

$$t_{rest\_k} - p_k + p_j \geq 0, \tag{2}$$

where t_{rest_k} corresponds to the remaining time of the period in which job k is allocated.

One exchange is not necessarily a direct exchange between two jobs. For instance, consider that job j was in a period X and it is exchanged to a period Y. If a job k was assigned to Y, it can be exchanged to other period of time Z, incurring in a lower penalty. In the worst case, the penalty is less than or equal to the penalty of assigning job k to the period X, since after this period the penalties will not be better. On the contrary, in the best case, the job k is assigned to the period of its due date. Only if the best possible place for the job k is the period X (first location of the job j) that a direct exchange occurs.

If in the period d_j there is any job that can be feasibly exchanged with job j or if this exchanged is not profitable, then the following or the previous period is tested according to the penalties e_j and l_j. If the penalties of advancing the job j are less than the ones of postponing this job ($e_j < l_j$) then the previous period is tested ($d_j - 1$). Otherwise it is desirable to postpone it to the next period ($d_j + 1$).

If the exchanges using the selected period are unfavorable, it is analyzed whether it is better to advance or to postpone the job, taking into account the values of advancing or delaying a job are no longer the same. If at the beginning the penalty to advance and delay were, respectively, e_j and l_j, now the penalty values are different.

Considering that in the first situation it compensates to advance, and in the period $d_j - 1$ there were no favorable exchanges, then it is necessary to define a new period for the search. Thus, the new values of the penalties associated to the job j to advance (two periods, since advancing one period has already been tested) or delay (one period) are $e_j \times 2$ and l_j. If $e_j \times 2 < l_j$, it will continue to compensate to advance. On the contrary, if in the first situation l_j is less than e_j, then period $d_j + 1$ is tested, *i.e.* it compensates to postpone. If any favorable exchange is found in this period, and if e_j is less than $l_j \times 2$, it is profitable to analyze the period that it before $(d_j - 1)$, otherwise it compensates to postpone and verify the period $d_j + 2$.

After analyzing all the possible jobs that can be exchanged with j, if there are no profitable modifications, the possibility of assigning job j to other period is verified providing that a lower penalty is achieved. There is only the change of period to the job j, without exchanging it with another job. The methods for obtaining initial solutions have similarly followed this last principle.

In this iterative local search process, the selection of the neighbor is made as soon as there is an improvement in the value of the solution (first improvement policy). Whenever there is an exchange for at least one job (*i.e.* when there is an improvement), the list of jobs is reordered and the first job of the list is selected one more time. If there are no profitable exchanges, the second job of the list is selected by doing the same process and so on until there are no further improvements.

In the VNS algorithm, we consider two neighborhood structures.

- $k = 1$ – Exchange between two jobs not necessarily from adjacent periods;
- $k = 2$ – Exchange between two jobs from adjacent periods.

The first neighborhood structure ($k = 1$) allows the exchange between any two jobs except if they are in the same period. This exchange is not considered because it is known in advance that penalty value would remain the same. In the second structure ($k = 2$), it is only possible to exchange two jobs that are in adjacent periods, *i.e.*, the jobs must be in two consecutive periods. In both structures, only feasible solutions are explored, taking into account all the constraints.

4   Variable Neighborhood Search (VNS)

The VNS is a metaheuristic that consists in a systematic neighborhood exchange combined with local search. There are several VNS methods, and the one addressed in this paper is the Basic VNS (BVNS) [9] that combines deterministic changes of neighborhoods and stochastic changes. The latter provide randomness to the method in order to avoid possible cycles and to increase the covered space of solutions.

The proposed BVNS is summarized in *Algorithm* 1. It has three input parameters: an initial solution (s), the maximum number of neighborhood structures (k_{max}) and the maximum execution time of the algorithm t_{max}.

Algorithm 1. BVNS steps

1: **function** BASICVNS(s, k_{max}, t_{max})
2:     $t := 0$;
3:     **while** $(t \leq t_{max})$ **do**
4:         $k := 1$;
5:         **while** $(k \leq k_{max})$ **do**
6:             $s' := shake(s,k)$;
7:             $s'' := firstImprovement(s')$;
8:             **if** $(f(s'') < f(s))$ **then**
9:                 $s := s''$;
10:                $k := 1$;
11:            **else**
12:                $k := k + 1$;
13:            $t := CPUTIME()$
14:        $s := localSearch(s)$;

In line 6, the *shake* function is applied in order to stir the solution s according to the neighborhood structure defined for a given value of k. In line 7, a neighbor of s' is selected through the iterative local search process with using a first improvement strategy: there is only an improvement of the solution in the previous step, resulting in a solution s''. If the new solution s'' is better than the initial solution s, then s is set to s'' and, in the next iteration, the neighborhood structure to be considered is the first one ($k = 1$). Otherwise, the following neighborhood structure is explored ($k = k + 1$). If there is any neighborhood solution achieving a better result (reaching k_{max}) the process is repeated only if the execution time of the algorithm t_{max} was not exceeded. The *shake* function consists in a random process that makes it possible to achieve different results in the same process. In line 14, successive local search iterations are performed until the local minimum has been reached, since some exchange may have occurred, providing space for some jobs to be assigned to better periods. In opposition to the first improvement strategy in line 7, in this case a full exploration is performed until there are no further improvements.

The successive neighborhood exchange that characterizes this approach is an important issue. The shake function is summarized in *Algorithm* 2 that presents also the use of the neighborhood structures and how the jobs are selected to be exchanged.

In both structures, randomness is present in the choice of the first (line 4) and second (line 10) jobs. However, the choice of the second job relies on the first selected job in order to satisfy all the constraints of the problem. The second job is randomly selected among all the possible (eligible) jobs that can be exchanged with the first one. That is, the release date of the second job must be equal to or less than the current period of the first job and, similarly, the release date of the first job must be equal to or less than the current period of the second job.

In line 6, all the jobs that can be exchanged with *job*1 are selected, according to the structure $k = 1$. In line 8, all the adjacent jobs that can be exchanged with

Algorithm 2. Steps of the *shake* function used in *basicVNS*

```
1: function SHAKE(s, k)
2:     foundExchange := 0;
3:     while (foundExchange = 0) do
4:         job1 := random(jobs);
5:         if (k=1) then
6:             eligibleJobs := getEligibleJobs(job1);
7:         else
8:             eligibleJobs := getAdjacentEligibleJobs(job1);
9:         if (eligibleJobs.size > 0) then
10:            job2 := random(eligibleJobs);
11:            s := swapJobs(s, job1, job2);
12:            foundEnchange := 1;
```

*job*1 are selected, as defined by the structure $k = 2$. The second job (*job*2) is randomly chosen among the possible jobs that can exchange with *job*1. Finally, in the solution *s*, direct exchange is made between the *job*1 and *job*2.

5   Computational Experiments

For the different approaches (CSI1, CSI2, CSI3, CSI4 and CSI5, and their possible variants), computational experiments were performed in order to evaluate and compare their performance. All structures and algorithms were implemented in the Java programming language and run on a computer with an Intel Core i7 2.70 GHz processor and 8 GB of RAM.

As the tested algorithms have a component with some randomness in the *shake* function, the results obtained in a single execution may not be a good example of the performance and effectiveness of the algorithm. Therefore, in order to overcome this issue, experiments were performed 5 times, thus obtaining a larger sample and confirming or not, the consistency of the results. Generally, an algorithm based on the BVNS method has a time limit or maximum number of iterations to perform the search. In our approach, the time limit of 5 s ($t_{max} = 5$) was imposed.

The computational experiments were performed on benchmark instances proposed in [8], and the results are compared with other works addressing the same problem using exact methods (executing each instance during 1200 s) [11] and heuristics [10]. Thus, these works defined our lower and upper bound, respectively. The different issues of the considered instances allowed us to determine the situations in which our algorithms have a more efficient procedure.

5.1   Description of the Instances

The instances consist in problems that contain small jobs, problems with small and large jobs, and others with only large jobs. According to these specific issues, and regarding the processing time of the jobs, the instances were divided in three large sets A, B and C, respectively. In all instances the size set for each time period was 100, so $P = 100$. The small jobs (set A) are those that have a processing time less than one third of P ($p_j \leq 33$) while large jobs (set C) are those having a processing time greater than 33 up to the value of P ($34 \leq p_j \leq 100$). Half of set B is composed by small jobs and the other half by large ones.

Each set (A, B, and C) has 5 different instances for the combination of the number of jobs (N), the number of machines (M) and the number of periods (τ_0) in which jobs have their release and due dates (ranging from 1 to τ_0). The number of machines and the number of periods τ_0 considered for each set were equal: $M \in \{2, 6, 10\}$ and $\tau_0 \in \{2, 6, 10\}$. The number of jobs is different for each set due to the different complexity. Thus, for set A, the number of jobs was $N \in \{100, 150, 200, 250, 300\}$. In set B, $N \in \{50, 100, 150, 200, 250, 300\}$. And, finally, in set C $N \in \{40, 60, 80, 100\}$. To ensure that all problems were feasible for their different issues, in [8] the number of periods (τ) is defined on each machine, such as $\tau = \tau_0 + \lceil 2 \times \sum_{j=1}^{N} \frac{p_j}{(P \times M)} \rceil$.

Considering these elements (5 different instances and N, M and τ_0) 225, 270 and 180 instances are obtained for set A, B and C, respectively, with a total of 675 instances.

5.2   Computational Results and Critical Analysis

The algorithms run 5 times for each instance for the different approaches. For CSI1 and CSI3 approaches, which do not consider the priority rules, 6750 results are obtained ($5 \times 675 + 5 \times 675$). For CSI2 and CSI4 approaches, which consider the 14 priority rules, we have 94500 results ($5 \times 675 \times 14 + 5 \times 675 \times 14$). The CSI5 approach does not have variants since it selects the best one, so we have $5 \times 675 = 3375$ results. In total, 104625 results are obtained for the considered set of instances, with 155 results analyzed per instance.

For each of the three sets, A, B, and C, we collected the average results of the 15 existing instances for problems with the same number of jobs N and the same number of machines M. In Tables 3, 4 and 5 the average computational results are presented for the set A, B and C, respectively. The meaning of the columns in these tables is the following:

- *bestlb*: average of the best lower bound values from the exact method in [11];
- *bestub*: average of the best upper bound values obtained in [10];
- *gap*: average optimality gap (in percentage) between *bestlb* and *bestub* and it is computed as follows: $((bestlb-bestub)/(bestlb))$. If there is no difference between *bestlb* and *bestub*, the symbol "-" appears instead;
- CSI1 to CSI5: average of the best solutions obtained through the BVNS algorithm in each of the approaches from CSI1 to CSI5;
- *vbest*: average of the best solutions obtained in all considered approaches;
- *gap'*: difference between the gap of *bestlb* and *vbest*, and the column *gap* referred to above (also in percentage);
- CSI1 *best* to CSI5 *best*: number of times the solution found through the BVNS algorithm is equal to the lower limit in each approach from CSI1 to CSI5.

Table 3. Computational results for the instances of the set A

|N|	|M|	best lb	best ub	gap	CSI1	CSI2	CSI3	CSI4	CSI5	vbest	gap'	CSI1 best	CSI2 best	CSI3 best	CSI4 best	CSI5 best
100	2	449	458	3.1	506	486	499	466	474	465.3	2.5	0	0	0	0	0
100	6	36	36	0.1	39	38	38	37	37	37.0	1.0	8	10	8	10	9
100	10	7	7	0.0	8	8	8	7	8	7.3	1.0	11	11	11	12	11
150	2	1432	1618	26.7	1618	1552	1602	1492	1508	1489.8	−20.7	0	0	0	0	0
150	6	128	129	0.3	142	140	143	134	135	134.1	3.1	5	5	4	6	5
150	10	38	38	0.0	42	40	41	40	41	39.8	1.6	9	10	9	9	9
200	2	2976	3319	20.2	3402	3246	3353	3089	3105	3086.2	−14.4	0	0	0	0	0
200	6	345	366	11.0	387	390	391	360	361	359.0	−3.6	0	2	2	3	1
200	10	97	98	0.5	110	105	109	102	102	102.0	1.4	10	10	10	10	10
250	2	4439	5606	7.2	5978	5653	5915	5347	5399	5346.3	−4.3	0	0	0	0	0
250	6	679	758	21.7	776	755	763	705	712	705.4	−13.7	0	0	0	0	0
250	10	207	218	1.8	233	232	233	219	220	218.6	4.6	5	5	5	5	5
300	2	0	9079	-	9680	9175	9508	8733	8826	8731.3	-	0	0	0	0	0
300	6	1185	1360	48.1	1369	1311	1347	1235	1244	1234.5	−39.0	0	0	0	0	0
300	10	416	448	17.3	469	467	471	437	438	436.9	−11.2	4	4	3	4	4

Table 4. Computational results for the instances of the set B

|N|	|M|	best lb	best ub	gap	CSI1	CSI2	CSI3	CSI4	CSI5	vbest	gap'	CSI1 best	CSI2 best	CSI3 best	CSI4 best	CSI5 best
50	2	346	346	0.0	392	369	385	364	383	362.3	4.9	0	1	0	1	0
50	6	31	31	0.0	39	34	40	34	36	33.3	4.3	5	7	3	9	7
50	10	9	9	0.0	10	10	12	9	11	9.5	2.3	10	10	10	10	10
100	2	1957	1975	1.2	2210	2132	2138	2093	2122	2088.1	6.6	0	0	0	0	0
100	6	285	285	0.2	337	321	333	310	317	309.1	13.7	0	0	0	0	0
100	10	90	90	0.0	106	101	111	99	104	98.5	8.1	2	4	1	5	5
150	2	5499	6235	14.8	6028	5903	6001	5824	5846	5808.7	−8.6	0	0	0	0	0
150	6	875	879	0.7	1009	973	984	950	960	948.7	11.9	0	0	0	0	0
150	10	330	330	0.2	386	374	384	359	364	359.1	17.9	0	0	0	0	0
200	2	10809	12011	12.1	11761	11491	11640	11340	11375	11328.5	−6.9	0	0	0	0	0
200	6	1998	2016	1.5	2300	2211	2255	2174	2190	2168.7	10.6	0	0	0	0	0
200	10	711	713	0.7	821	801	812	777	786	774.5	17.7	0	0	0	0	0
250	2	2756	19766	5.7	19398	18971	19396	18844	18852	18815.2	−2.9	0	0	0	0	0
250	6	3571	3913	13.7	4094	3946	3975	3848	3866	3841.6	−3.2	0	0	0	0	0
250	10	1401	1414	1.6	1609	1565	1596	1535	1546	1528.8	13.4	0	0	0	0	0
300	2	0	28787	-	28507	27959	28515	27718	27749	27680.3	-	0	0	0	0	0
300	6	5761	6577	18.6	6494	6298	6378	6186	6198	6176.7	−9.4	0	0	0	0	0
300	10	2446	2669	10.8	2810	2718	2752	2662	2688	2654.7	2.4	0	0	0	0	0

For set A, 27.11% of our solutions have exactly the same values as *bestlb*, while in set B and C this happens, respectively, in 9.26% and 35.56% of the obtained solutions. In order to compare the obtained results with the upper limit given in [10], we present in Fig. 1 the percentage of instances in which our results were better, the same or worse, for each set.

Figure 2 presents the average gap value (in percentage) between the lower limit (*bestlb*) and both the best solutions (vBest) and the *bestub* solutions (UB), in the different sets.

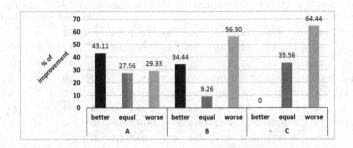

Fig. 1. Comparison of the best computational results obtained with *bestub*

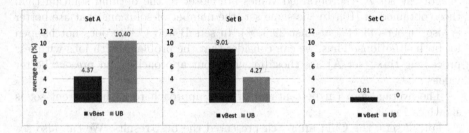

Fig. 2. Comparison of average gap (%) for the sets A, B and C

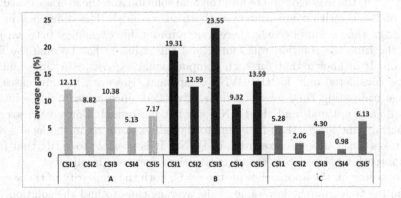

Fig. 3. Average gap (%) of each approach in the three sets

Table 5. Computational results for the instances of the set C

|N|	|M|	best lb	best ub	gap	CSI1	CSI2	CSI3	CSI4	CSI5	vbest	gap'	CSI1 best	CSI2 best	CSI3 best	CSI4 best	CSI5 best
40	2	741	741	0.0	773	756	768	744	765	743.7	0.4	0	2	1	3	1
40	6	86	86	0.0	89	87	88	86	88	86.3	0.1	8	12	8	14	9
40	10	21	21	0.0	21	21	21	21	22	20.7	0.0	10	15	11	15	10
60	2	1868	1868	0.0	1942	1909	1920	1889	1939	1885.5	1.1	0	0	0	0	0
60	6	291	291	0.0	302	296	301	294	304	293.3	0.4	3	7	3	6	1
60	10	92	92	0.0	95	93	95	92	96	92.1	0.2	5	10	7	10	4
80	2	3779	3779	0.0	3974	3848	3942	3813	3932	3810.3	0.9	0	0	0	0	0
80	6	653	653	0.0	677	668	677	661	678	660.5	1.5	0	0	0	1	0
80	10	219	219	0.0	227	223	230	221	225	220.4	0.6	1	5	2	7	2
100	2	6474	6474	0.0	6731	6607	6728	6530	6628	6530.0	0.9	0	0	0	0	0
100	6	1141	1141	0.0	1201	1177	1195	1157	1198	1155.5	1.7	0	0	0	0	0
100	10	421	421	0.0	439	435	440	428	443	426.9	1.7	1	2	1	3	1

For the set A, the obtained values are closer to the optimal solution than those obtained in [10]. In the same set, the number of solutions that are better is also higher (43.11% against 29.33%). In sets B and C this does not happen for both situations. Thus, we can consider that the problems with jobs with less processing time (set A) are those in which our approaches can present better values.

The average gap (in percentage) of each approach to each different set is shown in Fig. 3.

For all sets, the CSI4 approach presented the best results. We can also see that the approaches that consider priority rules for the order of jobs are those that achieve the most profitable solutions. As expected, the CSI5 approach is the one that presents the lowest average values of the initial solutions. However, as we can see in Fig. 3, this is not the approach that reaches the best final solution. In the case of set C, the CSI5 approach has worse results.

In Fig. 4, the average time to find the final solution and the average execution time of the algorithm are presented (both in seconds). The difference in the amount of time occurs because there were unprofitable exchanges between jobs after the time the solution was found, and this solution has become the final solution. It is known that for each computational experience a time limit of 5 s was considered only for the BVNS component. However, in some cases, an average computing time lower than 5 s is verified since some initial solutions were already optimal, and the BVNS was not performed. Conversely, average computing time longer than 5 s is due to the full exploration in the local search that takes place after the execution of BVNS. It is possible to verify that from the approach CSI1 to CSI5, the execution time decreases in all sets.

Regarding all the approaches for the set C, both the similarity of the average computing time and the low value of the average time to find the solution, can denote that the considered neighborhood structures may not be able to produce the necessary modifications for this particular type of problems.

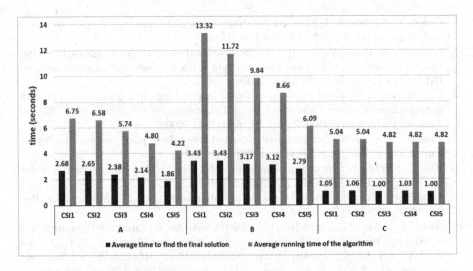

Fig. 4. Average time to find the solution and average running time of the algorithm

6   Conclusions

This paper focused on new optimization approaches to the integrated of planning and scheduling problem. The new approaches are based on heuristic methods of local search and in the metaheuristic of variable neighborhood search.

After the analysis of the computational experiments, the different approaches could be compared to each other and to the other approaches presented in literature. The use of benchmark instances allows to verify if the obtained results are competitive with those previous contributions from literature.

Our approaches have been more effective for problems with jobs whose processing time is relatively short when compared to the total time of the period (instances of set A). Our best approach was CSI4, that is, the initial solution is built based on the due date using priority rules. The variants *heu*13 and *heu*14 are the ones that have been able to often achieve better results.

Acknowledgements. This work has been supported by COMPETE: POCI-01-0145-FEDER-007043 and FCT – Fundação para a Ciência e Tecnologia within the Project Scope: UID/CEC/00319/2013.

References

1. Baldea, M., Harjunkoski, I.: Integrated production scheduling and process control: a systematic review. Comput. Chem. Eng. **71**, 377–390 (2014)
2. Chang, Y., Lee, C.: Machine scheduling with job delivery coordination. Eur. J. Oper. Res. **158**, 470–487 (2004)
3. Coffman, E., Garey, M., Johnson, D.: An application of bin-packing to multiprocessor scheduling. SIAM J. Comput. **7**, 1–17 (1978)

4. Garey, M., Johnson, D.: Strong NP-completeness results: motivation, examples, and implications. J. ACM **25**, 499–508 (1978)
5. Geismar, H., Laporte, G., Lei, L., Sriskandarajah, C.: The integrated production and transportation scheduling problem for a product with a short lifespan. INFORMS J. Comput. **20**, 21–33 (2008)
6. Grossmann, I.: Research challenges in planning and scheduling for enterprise-wide optimization of process industries. In: 10th International Symposium on Process Systems Engineering: Part A, vol. 27, pp. 15–21 (2009)
7. Jain, A., Meeran, S.: Deterministic job-shop scheduling: past, present and future. Eur. J. Oper. Res. **113**, 390–434 (1999)
8. Kis, T., Kovács, A.: A cutting plane approach for integrated planning and scheduling. Comput. Oper. Res. **39**, 320–327 (2012)
9. Mladenović, N., Hansen, P.: Variable neighborhood search. Comput. Oper. Res. **24**, 1097–1100 (1997)
10. Rietz, J., Alves, C., Carvalho, J.V.: Fast heuristics for integrated planning and scheduling. In: Gervasi, O., Murgante, B., Misra, S., Gavrilova, M.L., Rocha, A.M.A.C., Torre, C., Taniar, D., Apduhan, B.O. (eds.) ICCSA 2015. LNCS, vol. 9156, pp. 413–428. Springer, Cham (2015). doi:10.1007/978-3-319-21407-8_30
11. Rietz, J., Alves, C., Braga, N., Carvalho, J.: An exact approach based on a new pseudo-polynomial network flow model for integrated planning and scheduling. Comput. Oper. Res. **76**, 183–194 (2016)
12. Shen, W., Wang, L., Hao, Q.: Agent-based distributed manufacturing process planning and scheduling: a state-of-the-art survey. IEEE Trans. Syst. Man Cybern. Part C (Appl. Rev.) **36**, 563–577 (2006)
13. Shobrys, D., White, D.: Planning, scheduling and control systems: why can they not work together. Comput. Chem. Eng. **24**, 163–173 (2000)
14. Sugimori, Y., Kusunoki, K., Cho, F., Uchikawa, S.: Toyota production system and kanban system materialization of just-in-time and respect-for-human system. Int. J. Prod. Res. **15**, 553–564 (1977)
15. Wäscher, G., Haußner, H., Schumann, H.: An improved typology of cutting and packing problems. Eur. J. Oper. Res. **183**, 1109–1130 (2007)

Author Index

Printed in the United States
By Bookmasters